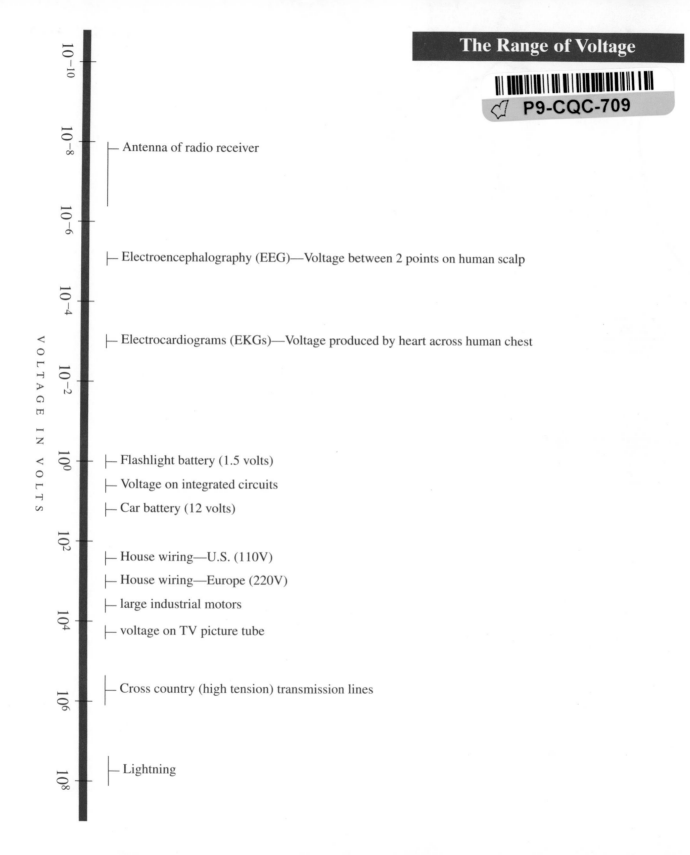

The Range of Voltage

VOLTAGE IN VOLTS

10^{-10}

10^{-8} — Antenna of radio receiver

10^{-6}

— Electroencephalography (EEG)—Voltage between 2 points on human scalp

10^{-4}

— Electrocardiograms (EKGs)—Voltage produced by heart across human chest

10^{-2}

10^{0} — Flashlight battery (1.5 volts)
— Voltage on integrated circuits
— Car battery (12 volts)

10^{2} — House wiring—U.S. (110V)
— House wiring—Europe (220V)
— large industrial motors
10^{4} — voltage on TV picture tube

10^{6} — Cross country (high tension) transmission lines

10^{8} — Lightning

Basic Engineering Circuit Analysis

Basic Engineering Circuit Analysis

Sixth Edition

J. David Irwin
Auburn University

Chwan-Hwa Wu
Auburn University

PRENTICE-HALL, Upper Saddle River, New Jersey 07458

Irwin, J. David
    Basic engineering circuit analysis / J. David Irwin, Chwan-Hwa Wu.
    —6th ed.
      p.   cm.
    Includes bibliographical references and index.
    ISBN 0-13-792714-2
    1. Electric circuit analysis.   I. Wu, Chwan-Hwa.   II. Title.
  TK454.I78   1998
  621.319'2—dc21   98–48510
                                                      CIP

Publisher: *Tom Robbins*
Editorial Director: *Tim Bozik*
Editor-in-Chief: *Marcia Horton*
Assistant Vice President of Production and Manufacturing: *David W. Riccardi*
Editorial/Production Supervision: *Rose Kernan*
Managing Editor: *Eileen Clark*
Marketing Manager: *Danny Hoyt*
Design Director: *Paula Maylahn*
Art Director: *Ann France*
Cover and Interior Design: *Ann France*
Manufacturing Buyer: *Pat Brown*
Editorial Assistant: *Dan De Pasquale*

© 1999, 1996 by Prentice-Hall, Inc.
Upper Saddle River, New Jersey 07458

The author and publisher of this book have used their best efforts in preparing this book. These efforts include the development, research, and testing of the theories and programs to determine their effectiveness. The author and publisher shall not be liable in any event for incidental or consequential damages in connection with, or arising out of, the furnishing, performance, or use of these programs.

Printed in the United States of America
10 9 8 7 6 5 4 3 2

Previous editions published by Macmillian Publishing Company

ISBN 0-13-792714-2

Prentice-Hall International (UK) Limited, *London*
Prentice-Hall of Australia Pty. Limited, *Sydney*
Prentice-Hall Canada Inc., *Toronto*
Prentice-Hall Hispanoamericana, S. A., *Mexico*
Prentice-Hall of India Private Limited, *New Delhi*
Prentice-Hall of Japan, Inc., *Tokyo*
Pearson Education Asia Pte. Ltd., *Singapore*
Editora Prentice-Hall do Brasil, Ltda., *Rio de Janeiro*

Brief Contents

Contents

Preface

Five successful previous editions of this text have provided readers with a thorough understanding of the fundamental concepts of circuit analysis and their application to real-world problems. The basic organization of the previous editions has continued to evolve in upgrading the presentation. This sixth edition makes some important changes which have resulted from the way introductory circuits is currently taught, the experience of teaching from earlier editions, and the comments of colleagues who have taught from the book. I am particularly pleased that Chwan-Hwa Wu has joined as co-author of the sixth edition. He has made significant contributions to the preparation of this new edition. We have greatly enjoyed working together and he is a welcome addition to the project.

This book is designed as a core text for students majoring in electrical and computer engineering, as well as a basic introduction to electric circuits for students in other engineering disciplines. The presentation begins with a discussion of circuit elements, followed by resistive circuits including operational amplifiers, then transients, sinusoidal steady state, complex frequency, ending with Laplace and Fourier methods. The text is written at a level suitable for students who have completed beginning college physics and calculus courses. Our goal for this edition is to provide an effective and efficient environment for students to obtain a thorough understanding of the analysis as well as an introduction to the design of linear electric circuits. We have taken great pains to provide complete and careful discussions, as well as an abundant supply of new learning aids. Apart from the authors and reviewers, the text and the problems have been checked for clarity and accuracy of the solutions by Kurt Norlin of Laurel Technical Services.

The organization and presentation of the material is designed to enable students to understand and apply the fundamentals as quickly as possible. The organization allows for its use in a variety of different course settings. Some sections or chapters can be skipped completely in a natural progression from start to finish in a coherent presentation. For example, those instructors who do not wish to cover PSPICE can skip it with no loss of continuity.

The overall tutorial nature of the book attempts to follow the outstanding example set by the late Mac Vac Valkenburg who has influenced a whole generation of circuit theory instructors and authors.

CHAPTER ORGANIZATION AND LEARNING ENHANCEMENTS

Each chapter is organized in the following format.

Chapter Openers provide a topic preview for the topics within the chapter.

A Side Column containing comments has been added to this edition to help shorten the learning curve. These comments in the margin provide guidance for understanding different facets of the presentation including text, examples, figures, and problems of all types. The strategic use of the side column for explanation, coupled with myriad examples, provides readers with a companion tutor. Additionally, it aids the instructor and the student, conveying some of the subtleties which are typically implicit in lecture or traditional presentation.

Examples and Applications A large number of worked-out examples have been the hallmark feature of this text. Examples, more often than any other component, provide students with the means for acquiring new knowledge and evaluating it. **Examples**, many of which are application oriented, appear throughout the text. Practical examples, labeled **Applications** in many sections of the text and at the end of almost every chapter frequently deal with design issues. Applications range from simple ac power circuit safety issues to a modeling of the collapse of the Tacoma Narrows Bridge.

Drill Problems and Extension Exercises are assessment tools, which are coordinated with the text, and guide the students in the proper problem-solving techniques needed to solve the end-of-chapter problems. Extension exercises are a new element to the presentation. They provide practice in applying basic concepts and guidance for the reader to probe new issues or make extensions. Answers for both appear in the text.

Problem-solving Strategies are strategically placed to assist the student in selecting the proper solution technique, or combination of techniques applicable in a particular situation. This assistance not only helps in understanding the subtle differences among various techniques in their application to a particular problem, but tends to eliminate the psychological barrier that often exists in determining a suitable method of attack.

Computer-aided Design Tools are applied to the analysis and design of electric circuits. PSPICE® from OrCAD is employed in the text in Chapters 4, 7, and 8 to illustrate the use of this CAD tool in dc, transient, and ac circuits. PSPICE®, MATLAB®, Microsoft EXCEL® and Electronic Workbench® are included in the accompanying study guide. A CD bound in the study guide includes circuit simulations and five easy-to-use video segments demonstrating PSPICE solutions.

Circuit Design sections appear at the end of each applicable chapter. This features provides the readers with an understanding of how to apply what they have learned to the design of circuits. The use of engineering design in a curriculum is a major component of the ABET criteria.

Chapter Summaries appear at the end of every chapter. The important topics are summarized in a concise manner as as quick reminder for readers.

End-of-chapter Problems are divided into basic and advanced problems. The basic problems are further subdivided and referenced to each section within the chapter. Within each subdivision, the problems are graduated in difficulty and assistance in solving the problems is provided in the side column. In order to stimulate independent thinking, no assistance is provided for advanced problems. Answers to selected problems appear in the back of the book.

SUPPLEMENTS

A Study Guide is available to serve as an adjunct to *Basic Engineering Circuit Analysis, 6/E.* The contents of the study guide along with the CD have been carefully selected in order to address four areas of student development:

1. competence in problem solving
2. similarities and differences between commercial products with model elements
3. emphasis of principle issues in circuit analysis
4. simulation techniques

The study guide contains over 100 representative problems with complete solutions. Computer simulation is a major topic in the study guide. Simulations are conducted utilizing four widely available software packages: PSPICE®, MATLAB®, Microsoft EXCEL® and Electronic Workbench®. The five PSPICE® videos demostrating PSPICE® 8.0 are executable files requiring no additional software to run. MATLAB® simulations demonstrate how to simulate a particular circuit by writing code. EXCEL® files provide additional visualization opportunities. These include a Bode plotter and a notch filter design spreadsheet. True circuit simulations are performed with Electronic Workbench and PSPICE 8.0. Topics such as advanced use of the PROBE® plotting program and defining/sweeping global variables are treated. A custom library containing simple op amps, ideal transformers, and balanced three-phase loads and sources is included in the CD. Fourier analysis and the Fast-Fourier Transform are also discussed.

A Companion Web Site for the book contains an array of support material for both students and instructors. This site (http://www.prenhall.com/irwin) includes a syllabus builder, sample tests, down loadable figures appropriate for use as overheads, and much more.

A Solutions Manual containing fully worked solutions to the exercises in the book is available from Prentice

Hall only for instructors who have adopted the text for classroom use. Answers to drill problems and extension exercises appear in the text, and answers to selected end-of-chapter problems are given at the end of the book.

ACKNOWLEDGEMENTS

We wish to acknowledge with our grateful thanks the people who have contributed to the development of this edition. In particular, Mr. William C. Dillard prepared the study guide and the solutions manual. He has been extremely helpful in the preparation of this text. Mr. Kurt Norlin reviewed portions of the manuscript and checked a variety of the problems. Tom Robbins, our Editor, made numerous helpful suggestions that improved the manuscript, and Rose Kernan checked the entire book and kept the development on schedule.

It is also a pleasure to gratefully acknowledge the support of friends and colleagues at Auburn who have helped in the preparation of earlier editions. They are Paula R. Marino, Thomas A. Baginski, Charles A. Gross, James L. Lowry, Zhi Ding, David C. Hill, Henry Cobb, Les Simonton, Betty Kelley, E. R. Graf, L. L. Grigsby, M. A. Honnell, R. G. Jaeger, M. S. Morse, C. L. Rogers, Travis Blalock, Kevin Driscoll, Keith Jones, George Lindsey, David Mack, John Parr, Monty Rickels, James Trivltayakhun, Susan Williamson, and Jacinda Woodward.

In the past, the following professors have made numerous suggestions for improving this book. M. E. Shafeei, Pennsylvania State University at Harrisburg; Leonard J. Tung, Florida A&M University/Florida State University; Darrell L. Vines, Texas Tech University; David Anderson, University of Iowa; Richard L. Baker, University of California, Los Angeles; James L. Dodd, Professor Emeritus, Mississippi State University; Earl D. Eyman, University of Iowa; Arvin Grabel, Northeastern University; Paul Gray, University of Wisconsin, Platteville; Mohammad Habli, University of New Orleans; John Hadjilogiou, Florida Institute of Technology; Ralph Kinney, Louisiana State University; K.S.P. Kumar, University of Minnesota; James Luster, Snow College; Ian McCausland, University of Toronto; Arthur C. Moeller, Marquette University; M. Paul Murray, Mississippi State University; Burks Oakley, II University of Illinois at Champaign-Urbana; John O'Malley, University of Florida; William R. Parkhurst, The Wichita State University; George Prans, Manhattan College; James Rowland, University of Kansas; Robert N. Sackett, Normandale Community College; Richard Sanford, Clarkson University; Ronald Schultz, Cleveland State University; Karen M. St. Germaine, University of Nebraska; Janusz Strazyk, Ohio University; Saad Tabet, Florida State University; and Seth Wolpert, University of Maine.

Finally, we wish to express our deepest appreciation to our wives, Edie and Shau-Hwa, without whose help and support this book would not have been possible.

J. David Irwin
Chwan-Hwa (John) Wu

Basic Concepts

1

*T*oday we live in a predominantly electrical world. While this statement may sound strange at first, a moment's reflection will indicate its inherent truth. The two primary areas of electrotechnology that permeate essentially every aspect of our lives are power and information. Without them, life as we know it would undergo stupendous changes. And yet, we have learned to generate, convert, transmit, and utilize these technologies for the enhancement of the whole human race.

Electrotechnology is a driving force in the changes that are occurring in every engineering discipline. For example, surveying is now done with lasers and electronic range finders, and automobiles employ electronic dashboards and electronic ignition systems. Industrial processes that range from chemical refineries and metal foundries to waste water treatment plants use (1) electronic sensors to obtain information about the process, (2) instrumentation systems to gather the information, and (3) computer control systems to process the information and generate electronic commands to actuators, which correct and control the process.

Fundamental to electrotechnology is the area of circuit analysis. A thorough knowledge of this subject provides an understanding of such things as cause and effect, amplification and attenuation, feedback and control, and stability and oscillation. Of critical importance is the fact that these principles can be applied, not only to engineering systems, but also to economic and social systems. Thus, the ramifications of circuit analysis are immense, and a solid understanding of this subject is well worth the effort expended to obtain it.

1.1 SYSTEM OF UNITS

The system of units we employ is the international system of units, the Système International des Unités, which is normally referred to as the SI standard system. This system, which is composed of the basic units meter (m), kilogram (kg), second (s), ampere (A), degree kelvin (°K), and candela (cd), is defined in all modern physics texts and therefore will not be defined here. However, we will discuss the units in some detail as we encounter them in our subsequent analyses.

The standard prefixes that are employed in SI are shown in Fig. 1.1. Note the decimal relationship between these prefixes. These standard prefixes are employed throughout our study of electric circuits.

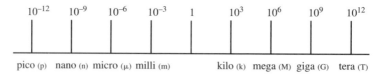

Figure 1.1 Standard SI prefixes.

Circuit technology has changed drastically over the years. For example, in the early 1960s the space on a circuit board occupied by the base of a single vacuum tube was about the size of a quarter (25-cent coin). Today that same space could be occupied by an Intel Pentium integrated circuit chip containing 3.1 million transistors. These chips are the engine for a host of electronic equipment.

1.2 BASIC QUANTITIES

Before we begin our analysis of electronic circuits, we must define terms that we will employ. However, in this chapter and throughout the book our definitions and explanations will be as simple as possible in order to foster an understanding of the use of the material. No attempt will be made to give complete definitions of many of the quantities because such definitions are not only unnecessary at this level but are often confusing. Although most of us have an intuitive concept of what is meant by a circuit, we will simply refer to an *electric circuit* as an interconnection of electrical components.

It is very important at the outset that the reader understand the basic strategy that we will employ in our analysis of electric circuits. This strategy is outlined in Fig. 1.2, and we will be concerned only with the portion of the diagram to the right of the dashed line. In subsequent courses or further study the reader will learn how to model physical devices such as electronic components. Our procedure here will be to employ linear models for our circuit components, then define the variables that are used in the appropriate circuit equations to yield solution values, and finally, interpret the solution values for the variables in order to determine what is actually happening in the physical circuit. The variables may be time varying or constant depending on the nature of the physical parameters they represent.

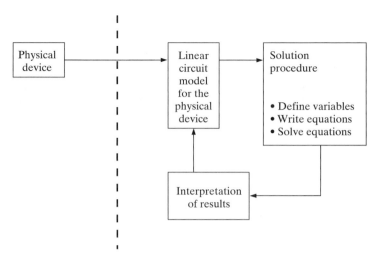

Figure 1.2 Basic strategy employed in circuit analysis.

The most elementary quantity in an analysis of electric circuits is the electric *charge*. We know from basic physics that the nature of charge is based on concepts of atomic theory. We view the atom as a fundamental building block of matter that is composed of a positively charged nucleus surrounded by negatively charged electrons. In the metric system, charge is measured in coulombs (C). The charge on an electron is negative and equal in magnitude to 1.602×10^{-19} C. However, our interest in electric charge is centered around its motion, since charge in motion results in an energy transfer. Of particular interest to us are those situations in which the motion is confined to a definite closed path.

An electric circuit is essentially a pipeline that facilitates the transfer of charge from one point to another. The time rate of change of charge constitutes an electric *current*. Mathematically, the relationship is expressed as

$$i(t) = \frac{dq(t)}{dt} \quad \text{or} \quad q(t) = \int_{-\infty}^{t} i(x)\,dx \qquad \textbf{1.1}$$

where i and q represent current and charge, respectively (lowercase letters represent time dependency and capital letters are reserved for constant quantities). The basic unit of current is the ampere (A) and 1 ampere is 1 coulomb per second.

Although we know that current flow in metallic conductors results from electron motion, the conventional current flow, which is universally adopted, represents the movement of positive charges. It is important that the reader think of current flow as the movement of positive charge regardless of the physical phenomena that take place. The symbolism that will be used to represent current flow is shown in Fig. 1.3. The variable representing the current in the wire in Fig. 1.3a was defined as I_1 flowing in the wire from left to right in the figure. A set of equations was written for the circuit and a solution value of 2 A was obtained; that is, $I_1 = 2$ A. This means that the physical current in the wire is flowing from left to right, in the direction of our variable, and is 2 A. $I_1 = 2$ A in Fig. 1.3a indicates that at any point in the wire shown, 2 C of charge pass from left to right each second. The same procedure was followed for the wire shown in Fig. 1.3b and the variable

I_2 was defined in the same manner. A set of equations was written and a solution value of -3 A was obtained. An interpretation of this result is that the physical current in the wire is from right to left, opposite to our reference direction for I_1; its value is 3 A. $I_2 = -3$ A in Fig. 1.3b indicates that at any point in the wire shown, 3 C of charge pass from right to left each second. Therefore, it is important to specify not only the magnitude of the variable representing the current, but also its direction. After a solution is obtained for the current variable, the actual physical current is known.

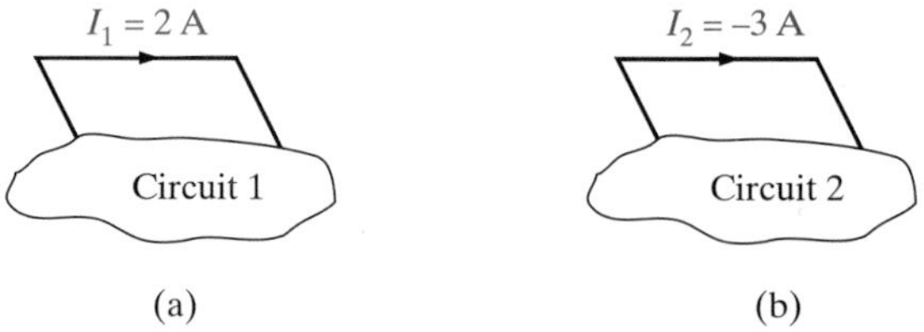

$I_1 = 2$ A   Circuit 1   (a) $I_2 = -3$ A   Circuit 2   (b)

Figure 1.3 Conventional current flow: (a) positive current flow; (b) negative current flow.

DRILL

D1.1   Determine the amount of time required for 100 C of charge to pass through the circuit in Fig. 1.3a.

ANSWER: 50 s

There are two types of current that we encounter often in our daily lives, alternating current (ac) and direct current (dc), which are shown as a function of time in Fig. 1.4. *Alternating current* is the common current found in every household, used to run the refrigerator, stove, washing machine, and so on. Batteries, which are used in automobiles or flashlights, are one source of *direct current*. In addition to these two types of currents, which have a wide variety of uses, we can generate many other types of currents. We will examine some of these other types later in the book. In the meantime, it is interesting to note that the magnitude of currents in elements familiar to us ranges from soup to nuts, as shown in Fig. 1.5.

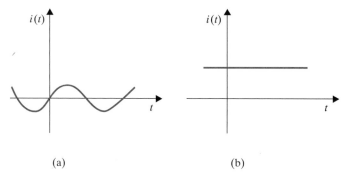

(a)   (b)

Figure 1.4 Two common types of current: (a) alternating current (ac); (b) direct current (dc).

We have indicated that charges in motion yield an energy transfer. Now we define the *voltage* (also called the *electromotive force* or *potential*) between two points in a circuit as the difference in energy level of a unit charge located at each of the two points. Work or energy, $w(t)$ or W, is measured in joules (J); 1 joule is 1 newton meter (N·m). Hence, voltage [$v(t)$ or V] is measured in volts (V) and 1 volt is 1 joule per coulomb; that is, 1 volt = 1 joule per coulomb = 1 newton meter per coulomb.

$$V = \frac{W}{q}$$

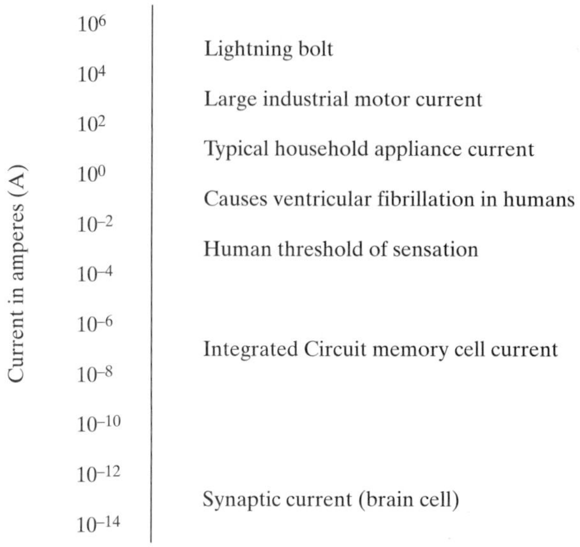

Current in amperes (A)

10^6

10^4 — Lightning bolt

10^2 — Large industrial motor current

10^0 — Typical household appliance current

10^{-2} — Causes ventricular fibrillation in humans

10^{-4} — Human threshold of sensation

10^{-6}

10^{-8} — Integrated Circuit memory cell current

10^{-10}

10^{-12}

10^{-14} — Synaptic current (brain cell)

Figure 1.5 Typical current magnitudes.

If a unit positive charge is moved between two points, the energy required to move it is the difference in energy level between the two points and is the defined voltage. It is extremely important that the variables that are used to represent voltage between two points be defined in such a way that the solution will let us interpret which point is at the higher potential with respect to the other.

In Fig. 1.6a the variable that represents the voltage between points A and B has been defined as V_1, and it is assumed that point A is at a higher potential than point B, as indicated by the $+$ and $-$ signs associated with the variable and defined in the figure. The $+$ and $-$ signs define a reference direction for V_1. The equations for the circuit were written and a solution value of 2 V was obtained; that is $V_1 = 2$ V, as shown in the figure. The physical interpretation of this is that the difference in potential of points A and B is 2 V and point A is at the higher potential. If a unit positive charge is moved from point A through the circuit to point B, it will give up energy to the circuit and have 2 J less energy when it reaches point B. If a unit positive charge is moved from point B to point A, extra energy must be added to the charge by the circuit, and hence the charge will end up with 2 J more energy at point A than it started with at point B.

DRILL

D1.2 Determine the energy required to move 120 C of charge from point B to point A in the network in Fig. 1.6a.

ANSWER: 240 J

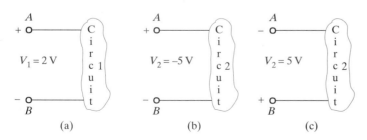

(a)   (b)   (c)

Figure 1.6 Voltage representations.

The same procedure was followed for the circuit in Fig. 1.6b, and the variable V_2 was defined in the same manner. A set of equations was written and a solution value of -5 V was obtained; that is, $V_2 = -5$ V. The physical interpretation of $V_2 = -5$ V is that the potential between points A and B is 5 V and point B is at the higher potential. If the circuit in Fig. 1.6b was reworked with the variable for the voltage defined as shown in Fig. 1.6c, the solution would have been $V_2 = 5$ V. The physical interpretation of this is that the difference in potential of points A and B is 5 V, with point B at the higher potential.

Note that it is important to define a variable with a reference direction so that the answer can be interpreted to give the physical condition in the circuit. We will find that it is not possible in many cases to define the variable so that the answer is positive, and we will also find that it is not necessary to do so. A negative number for a given variable gives exactly the same information as a positive number for a new variable that is the same as the old variable, except that it has an opposite reference direction. Hence, when we define either current or voltage, it is absolutely necessary that we specify both magnitude and direction. Therefore, it is incomplete to say that the voltage between two points is 10 V or the current in a line is 2 A, since only the magnitude and not the direction for the variables has been defined.

The range of magnitudes for voltage, equivalent to that for currents in Fig. 1.5, is shown in Fig. 1.7. Once again, note that this range spans many orders of magnitude.

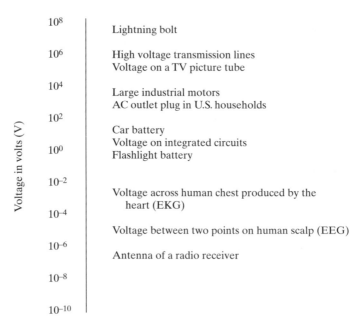

Figure 1.7 Typical voltage magnitudes.

At this point we have presented the conventions that we employ in our discussions of current and voltage. *Energy* is yet another important term of basic significance. Figure 1.8 illustrates the voltage–current relationships for energy transfer. In this figure, the block representing a circuit element has been extracted from a larger circuit for examination.

In Fig. 1.8a, energy is being supplied *to* the element by whatever is attached to the terminals. Note that 2 A, that is, 2 C, of charge are moving from point *A* to point *B* through the element each second. Each coulomb loses 3 J of energy as it passes through the element from point *A* to point *B*. Therefore, the element is absorbing 6 J of energy per second. Note that when the element is *absorbing* energy, a positive current enters the positive terminal. In Fig. 1.8b energy is being supplied *by* the element to whatever is connected to terminals *A-B*. In this case, note that when the element is *supplying* energy, a positive current enters the negative terminal and leaves via the positive terminal. In this convention a negative current in one direction is equivalent to a positive current in the opposite direction, and vice versa. Similarly, a negative voltage in one direction is equivalent to a positive voltage in the opposite direction.

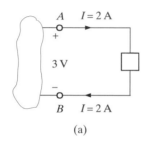 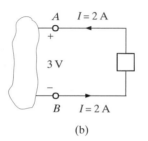

Figure 1.8 Voltage-current relationships for energy absorbed (a) and energy supplied (b).

(a)   (b)

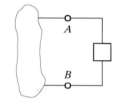

EXAMPLE 1.1

Suppose that your car will not start. To determine if the battery is faulty, you turn on the light switch and find that the lights are very dim, indicating a weak battery. You borrow a friend's car and a set of jumper cables. However, how do you connect his car's battery to yours? What do you want his battery to do?

SOLUTION   Essentially, his car's battery must supply energy to yours, and therefore it should be connected in the manner shown in Fig. 1.9. Note that the positive current leaves the positive terminal of the good battery (supplying energy) and enters the positive terminal of the weak battery (absorbing energy). Note that the same connections are used when charging a battery.

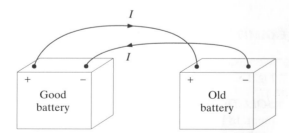

Figure 1.9 Diagram for Example 1.1.

In practical applications there are often considerations other than simply the electrical relations (e.g., safety). Such is the case with jump starting an automobile.

Automobile batteries produce explosive gases that can be ignited accidentally, causing severe physical injury. Be safe—follow the procedure described in your auto owner's manual. ❑

We have defined voltage in joules per coulomb as the energy required to move a positive charge of 1 C through an element. If we assume that we are dealing with a differential amount of charge and energy, then

$$v = \frac{dw}{dq} \qquad\qquad \textbf{1.2}$$

Multiplying this quantity by the current in the element yields

$$vi = \frac{dw}{dq}\left(\frac{dq}{dt}\right) = \frac{dw}{dt} = p \qquad\qquad \textbf{1.3}$$

which is the time rate of change of energy or power measured in joules per second, or watts (W). Since, in general, both v and i are functions of time, p is also a time-varying quantity. Therefore, the change in energy from time t_1 to time t_2 can be found by integrating Eq. (1.3); that is,

$$w = \int_{t_1}^{t_2} p\, dt = \int_{t_1}^{t_2} vi\, dt \qquad\qquad \textbf{1.4}$$

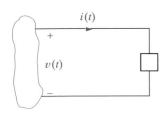

Figure 1.10 Sign convention for power.

The passive sign convention is used to determine whether power is being absorbed or supplied.

At this point, let us summarize our sign convention for power. To determine the sign of any of the quantities involved, the variables for the current and voltage should be arranged as shown in Fig. 1.10. The variable for the voltage $v(t)$ is defined as the voltage across the element with the positive reference at the same terminal that the current variable $i(t)$ is entering. This convention is called the *passive sign convention* and will be so noted in the remainder of this book. The product of v and i, with their attendant signs, will determine the magnitude and sign of the power. If the sign of the power is positive, power is being absorbed by the element; if the sign is negative, power is being supplied by the element.

EXAMPLE 1.2

We wish to determine the power absorbed by, or supplied by, the elements in Fig. 1.8.

SOLUTION   In Fig. 1.8a, $P = VI = (3\text{ V})(2\text{ A}) = 6\text{ W}$ is absorbed by the element. In Fig. 1.8b, $P = VI = (3\text{ V})(-2\text{ A}) = -6\text{ W}$ is absorbed by the element, or $+6\text{ W}$ is supplied by the element. ❑

EXAMPLE 1.3

Given the two diagrams shown in Fig. 1.11, determine whether the element is absorbing or supplying power and how much.

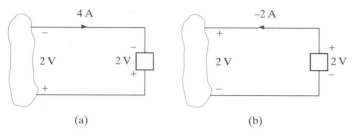

(a)   (b)

Figure 1.11 Elements for Example 1.3.

SOLUTION   In Fig. 1.11a the power is $P = (2 \text{ V})(-4 \text{ A}) = -8$ W. Therefore, the element is supplying power. In Fig. 1.11b, the power is $P = (2 \text{ V})(2 \text{ A}) = 4$ W. Therefore, the element is absorbing power.   ❏

EXTENSION EXERCISE

E1.1   Determine the amount of power absorbed or supplied by the elements in Fig. E1.1.

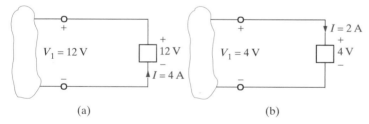

(a)   (b)

Figure E1.1

ANSWER: (a) $P = -48$ W; (b) $P = 8$ W.

EXAMPLE 1.4

We wish to determine the unknown voltage or current in Fig. 1.12.

SOLUTION   In Fig. 1.12a, a power of -20 W indicates that the element is delivering power. Therefore, the current enters the negative terminal (terminal A), and from Eq. (1.3) the voltage is 4 V.

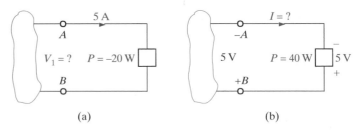

Figure 1.12 Elements for Example 1.4.

In Fig 1.12b, a power of +40 W indicates that the element is absorbing power and, therefore, the current should enter the positive terminal B. The current thus has a value of −8 A, as shown in the figure.   ❏

EXTENSION EXERCISE

E1.2   Determine the unknown variables in Fig. E1.2.

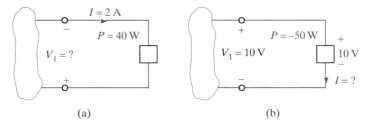

Figure E1.2

ANSWER: (a) $V_1 = -20$ V; (b) $I = -5$ A.

Finally, it is important to note that these electrical networks satisfy the principle of conservation of energy. For our present purposes this means that the power that is supplied in a network is exactly equal to the power absorbed.

1.3 CIRCUIT ELEMENTS

It is important to realize at the outset that in our discussions we will be concerned with the behavior of a circuit element, and we will describe that behavior by a mathematical model. Thus, when we refer to a particular circuit element, we mean the *mathematical model* that describes its behavior.

Thus far we have defined voltage, current, and power. In the remainder of this chapter we will define both independent and dependent current and voltage sources. Although we will assume ideal elements, we will try to indicate the shortcomings of these assumptions as we proceed with the discussion.

In general, the elements we will define are terminal devices that are completely characterized by the current through the element and/or the voltage across it. These elements, which we will employ in constructing electric circuits, will be broadly classified as being either active or passive. The distinction between these two classifications depends essentially upon one thing—whether they supply or absorb energy. As the words themselves imply, an *active* element is capable of generating energy and a *passive* element cannot generate energy.

However, we will show later that some passive elements are capable of storing energy. Typical active elements are batteries, generators, and transistor models. The three common passive elements are resistors, capacitors, and inductors.

In Chapter 2 we will launch an examination of passive elements by discussing the resistor in detail. However, before proceeding with that element, we first present some very important active elements.

1. Independent voltage source

2. Independent current source

3. Two dependent voltage sources

4. Two dependent current sources

Independent Sources

An *independent voltage source* is a two-terminal element that maintains a specified voltage between its terminals *regardless of the current through it*. The general symbol for an independent source, a circle, is shown in Fig. 1.13a. As the figure indicates, terminal A is

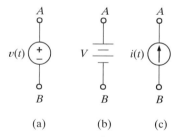

Figure 1.13 Symbols for (a) independent voltage source; (b) constant voltage source; (c) independent current source.

$v(t)$ volts positive with respect to terminal B. The word *positive* may be somewhat misleading. What is meant in this case is that $v(t)$ is referenced positive at terminal A and that the physical voltage across the device must be interpreted from the numerical value of $v(t)$. That is, if $v(t)$ at $t = 2$ s is -10 V, point B is at a higher potential than point A at $t = 2$ s. The symbol $v(t)$ is normally employed for time-varying voltages. However, if the voltage is time invariant (i.e., constant), the symbol shown in Fig. 1.13b is sometimes used. This symbol, which is used to represent a battery, illustrates that terminal A is V volts positive with respect to terminal B, where the long line on the top and the short line on the bottom indicate the positive and negative terminals, respectively, and thus the polarity of the element.

In contrast to the independent voltage source, the *independent current source* is a two-terminal element that maintains a specified current *regardless of the voltage across its terminals*. The general symbol for an independent current source is shown in Fig. 1.13c, where $i(t)$ is the specified current and the arrow indicates the positive direction of current flow.

Although sources come in a variety of forms, it is informative to note that there are modern high-precision instruments on the market that are capable of functioning as sources. The unit shown in Fig. 1.14 is such an instrument. This very versatile device can function as either a voltage source or a current source. It has a number of other important features, and they will be specifically addressed in the following chapter.

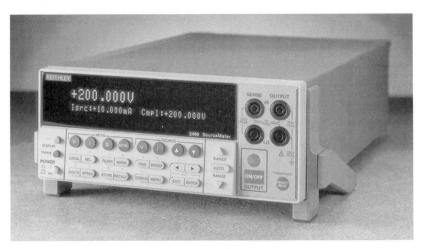

Figure 1.14 Keithley Model 2400 Source Meter. (Courtesy of Keithley Instruments, Inc.)

In their normal mode of operation, independent sources supply power to the remainder of the circuit. However, they may also be connected into a circuit in such a way that they absorb power. A simple example of this latter case is a battery charging circuit such as that shown in Example 1.1.

It is important that we pause here to interject a comment concerning a shortcoming of the models. In general, mathematical models approximate actual physical systems only under a certain range of conditions. Rarely does a model accurately represent a physical system under every set of conditions. To illustrate this point, consider the model for the voltage source in Fig. 1.13a. We assume that the voltage source delivers v volts regardless of what is connected to its terminals. Theoretically, we could adjust the external circuit so that an infinite amount of current would flow, and therefore the voltage source would deliver an infinite amount of power. This is, of course, physically impossible. A similar argument could be made for the independent current source. Hence, the reader is cautioned to keep in mind that models have limitations and are thus valid representations of physical systems only under certain conditions.

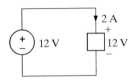

EXAMPLE 1.5

Determine the power absorbed or supplied by the elements in the network in Fig. 1.15.

Technique

Elements that are connected in series have the same current.

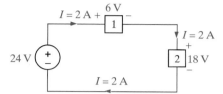

Figure 1.15 Network for Example 1.5.

SOLUTION   The current flow is out of the positive terminal of the 24-V source, and therefore this element is supplying (2)(24) = 48 W of power. The current is into the positive terminals of elements 1 and 2, and therefore elements 1 and 2 are absorbing (2)(6) = 12 W and (2)(18) = 36 W, respectively. Note that the power supplied is equal to the power absorbed.   ❏

EXTENSION EXERCISE

E1.3   Find the power that is absorbed or supplied by the elements in Fig. E1.3.

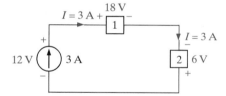

Figure E1.3

ANSWER: Current source supplies 36 W, element 1 absorbs 54 W, and elements 2 supplies 18 W.

Dependent Sources

In contrast to the independent sources, which produce a particular voltage or current completely unaffected by what is happening in the remainder of the circuit, dependent sources generate a voltage or current that is determined by a voltage or current at a specified location in the circuit. These sources are very important because they are an integral part of the mathematical models used to describe the behavior of many electronic circuit elements.

For example, both metal-oxide-semiconductor field-effect transistors (MOSFETs) and bipolar transistors, both of which are commonly found in a host of electronic equipment, are modeled with dependent sources, and therefore the analysis of electronic circuits involves the use of these controlled elements.

In contrast to the circle used to represent independent sources, a diamond is used to represent a dependent or controlled source. Figure 1.16 illustrates the four types of dependent sources. The input terminals on the left represent the voltage or current that controls the dependent source, and the output terminals on the right represent the output current or voltage of the controlled source. Note that in Figs. 1.16a and d the quantities μ and β are dimensionless constants because we are transforming voltage to voltage and current to current. This is not the case in Figs. 1.16b and c; hence when we employ these elements a short time later, we must describe the units of the factors r and g.

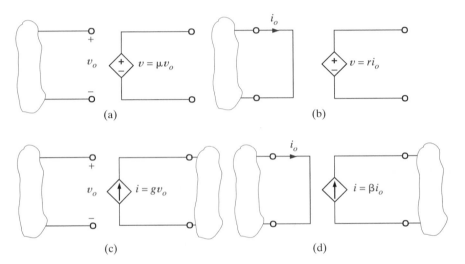

Figure 1.16 Four different types of dependent sources.

EXAMPLE 1.6

Given the two networks shown in Fig. 1.17, we wish to determine the outputs.

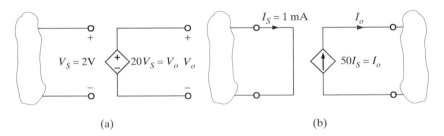

Figure 1.17 Circuits for Example 1.6.

SOLUTION In Fig. 1.17a the output voltage is $V_o = \mu V_S$ or $V_o = 20\,V_S = (20)(2\text{ V}) = 40\text{ V}$. Note that the output voltage has been amplified from 2 V at the input terminals to 40 V at the output terminals; that is, the circuit is an amplifier with an amplification factor of 20.

In Fig. 1.17b, the output current is $I_o = \beta I_S = (50)(1\text{ mA}) = 50\text{ mA}$; that is, the circuit has a current gain of 50, meaning that the output current is 50 times greater than the input current. ❏

EXTENSION EXERCISE

E1.4   Determine the power supplied by the dependent sources in Fig. E1.4.

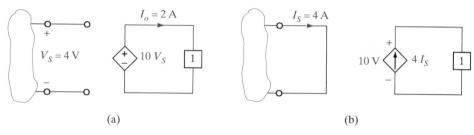

(a)   (b)

Figure E1.4

ANSWER: (a) Power supplied $=$ 80 W; (b) power supplied $=$ 160 W.

EXAMPLE 1.7

Let us compute the power that is absorbed or supplied by the elements in the network in Fig. 1.18.

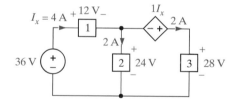

Figure 1.18 Circuit for Example 1.7.

SOLUTION   Using the sign convention for power illustrated in Fig. 1.10, we find the following:

$$P_{36\text{ V}} = (36)(-4) = -144\text{ W}$$
$$P_1 = (12)(4) = 48\text{ W}$$
$$P_2 = (24)(2) = 48\text{ W}$$
$$P_{DS} = (1I_x)(-2) = (4)(-2) = -8\text{ W}$$
$$P_3 = (28)(2) = 56\text{ W}$$

Note that the 36-V source and the dependent source are supplying power to the network and the remaining elements are absorbing power. Furthermore, energy is conserved, since the power supplied to the network is identically equal to the power absorbed by the network. ❑

EXAMPLE 1.8

Let us find the current I_o in the network in Fig. 1.19.

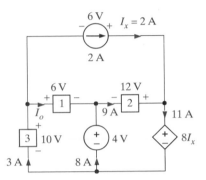

Figure 1.19 Circuit used in Example 1.8.

SOLUTION First, we must determine the power absorbed or supplied by each element in the network. Using the sign convention for power, we find

$$P_{2\,A} = (6)(-2) = -12 \text{ W}$$
$$P_1 = (6)(I_o) = 6I_o \text{ W}$$
$$P_2 = (12)(-9) = -108 \text{ W}$$
$$P_3 = (10)(-3) = -30 \text{ W}$$
$$P_{4\,V} = (4)(-8) = -32 \text{ W}$$
$$P_{DS} = (8I_x)(11) = (16)(11) = 176 \text{ W}$$

Since energy must be conserved,

$$-12 + 6I_o - 108 - 30 - 32 + 176 = 0$$

or

$$6I_o + 176 = 12 + 108 + 30 + 32$$

Hence,

$$I_o = 1 \text{ A}$$ ❑

EXTENSION EXERCISE

E1.5 Find the power that is absorbed or supplied by the circuit elements in the network in Fig. E1.5.

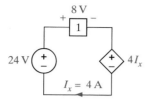

Figure E1.5

ANSWER: $P_{24\text{V}} = 96$ W supplied, $P_1 = 32$ W absorbed, $P_{4I_x} = 64$ W absorbed.

As we conclude this chapter, a number of important comments are in order. The reader must thoroughly understand our approach as outlined in Fig. 1.2 because in the following chapters we will simply define and refer to voltages and currents at specific locations within a circuit. Before any equations are written, readers should think carefully about what kinds of answers they expect. For example, if all the sources are constant, all the voltages and currents in the circuit will be constant, and therefore capital letters should be used for all the variables representing the voltages and currents. It is also useful for readers to think ahead about how variables are related to one another and define the variables so that a minimum number of minus signs are used. Although this is not important in the final solution, as indicated earlier, it does force readers to think about the problem from beginning to end before attacking it so that the solution will not come as a surprise.

1.4 SUMMARY

The standard prefixes employed

$$p = 10^{-12} \qquad k = 10^3$$

$$n = 10^{-9} \qquad M = 10^6$$

$$\mu = 10^{-6} \qquad G = 10^9$$

$$m = 10^{-3} \qquad T = 10^{12}$$

The relationships between current and charge

$$i(t) = \frac{dq(t)}{dt} \qquad \text{or} \qquad q(t) = \int_{-\infty}^{t} i(x)\, dx$$

The relationships among power, energy, current, and voltage are

$$p = \frac{dw}{dt} = vi$$

$$W = \int_{t_1}^{t_2} p \, dt = \int_{t_1}^{t_2} vi \, dt$$

The passive sign convention The passive sign convention states that if the voltage and current associated with an element are as shown in Fig. 1.10, the product of v and i, with their attendant signs, determines the magnitude and sign of the power. If the sign is positive, power is being absorbed by the element, and if the sign is negative, the element is supplying power.

Independent and dependent sources An ideal independent voltage (current) source is a two-terminal element that maintains a specified voltage (current) between its terminals regardless of the current (voltage) through (across) the element. Dependent or controlled sources generate a voltage or current that is determined by a voltage or current at a specified location in the circuit.

Conservation of energy The electric circuits under investigation satisfy the conservation of energy.

PROBLEMS

Section 1.2

Use the relationship between current and charge.

1.1 In an electric conductor, a charge of 300 C passes any point in a 5-s interval. Determine the current in the conductor.

Similar to P1.1

1.2 The current in a conductor is 1.5 A. How many coulombs of charge pass any point in a time interval of 1.5 min?

Similar to P1.1

1.3 Determine the number of coulombs of charge produced by a 12-A battery charger in an hour.

Similar to P1.1

1.4 Determine the time required for a 12-A battery charger to deliver a charge of 10,000 C.

Similar to P1.1

1.5 A lightning bolt carrying 20,000 A lasts for 70 μs. If the lightning strikes a tractor, determine the charge deposited on the tractor if the tires are assumed to be perfect insulators

$V = W/q$

1.6 A 10^8-V lightning bolt carrying 20,000 A lasts for 10 μs. If the lightning strikes a tree and burns it, determine the energy deposited on the tree.

Similar to P1.6

1.7 Determine the energy required to move 1000 C of charge through 12 V.

Similar to P1.6

1.8 If a 12-V battery supplies 10 A, find the amount of energy delivered.

1.9   If a 12-V battery delivers 100 J in 5 s, find (a) the amount of charge delivered and (b) the current produced.

Similar to P1.6 and P1.1

1.10 Five coulombs of charge pass through the element in Fig. P1.10 from point A, to point B. If the energy absorbed by the element is 120 J, determine the voltage across the element.

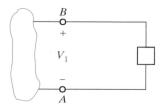

Similar to P1.6

Figure P1.10

1.11 A charge of 50 C passes through the element in Fig. P1.10 from point A to B. If $V_1 = 90$ V, determine the power required by this element.

Use the relationship between P, W, I, and V

1.12 Determine the amount of power absorbed or supplied by the element in Fig. P1.12. if

(a) $V_1 = 9$ V and $I = 2$ A.

(b) $V_1 = 9$ V and $I = -3$ A.

(c) $V_1 = -12$ V and $I = 2$ A.

(d) $V_1 = -12$ V and $I = -3$ A.

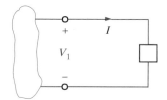

Similar to P1.11

Figure P1.12

1.13 Determine the magnitude and direction of the voltage across the elements in Fig. P1.13.

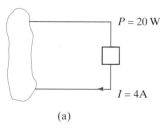

(a)

$P = 20$ W

$I = 4$A

$P = 20$ W

$I = 4$A

(b)

$P = -20$ W

$I = 4$A

(c)

$P = 20$ W

$I = 4$A

(d)

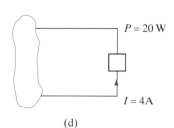

Similar to P1.11

Figure P1.13

1.14 Determine the power supplied to the elements in Fig. P1.14.

Similar to P1.11. Sum the power for each element.

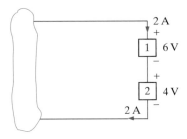

Figure P1.14

1.15 Determine the power supplied to the elements in Fig. P1.15.

Similar to P1.14

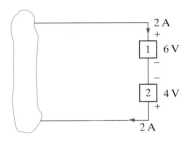

Figure P1.15

1.16 Two elements are connected in series, as shown in Fig. P1.16. Element 1 supplies 24 W of power. Is element 2 absorbing or supplying power, and how much?

1. Find the current.
2. Find the power.

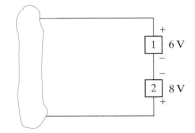

Figure P1.16

Section 1.3

Important note: The values used in the remaining problems of Section 1.3 are not arbitary. They have been selected to satisfy the basic laws of circuit analysis that will be studied in the following chapters.

1.17 Determine the power that is absorbed or supplied by the circuit elements in Fig. P1.17.

Similar to P1.14

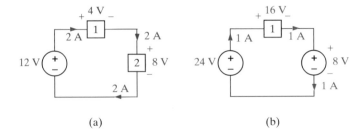

(a)   (b)

Figure P1.17

1.18 Find the power that is absorbed or supplied by the circuit elements in Fig. P1.18.

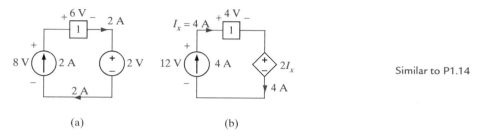

(a)   (b)

Similar to P1.14

Figure P1.18

1.19 Find the power that is absorbed or supplied by the network elements in Fig. P1.19.

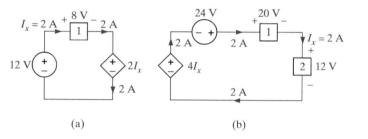

(a)   (b)

Similar to P1.14

Figure P1.19

Advanced Problems

1.20 The charge entering an element is shown in Fig. P1.20. Find the current in the element in the time interval $0 \leq t \leq 0.5$ s. [*Hint:* The equation for $q(t)$ is $q(t) = 1 + (1/0.5)t, t \geq 0$.]

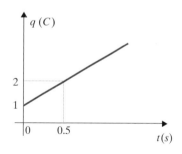

Figure P1.20

1.21 The current that enters an element is shown in Fig. 1.21. Find the charge that enters the element in the time interval $0 < t < 20$ s.

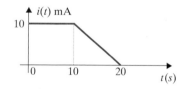

Figure P1.21

1.22 Is the source V_S in the network in Fig. P1.22 absorbing or supplying power, and how much?

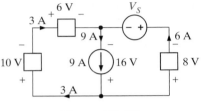

Figure P1.22

1.23 Find V_x in the network in Fig. P1.23.

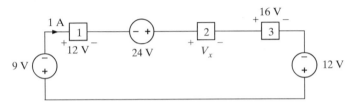

Figure P1.23

1.24 Find I_x in the network in Fig. P1.24

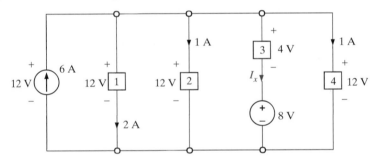

Figure P1.24

1.25 Find I_s such that the power absorbed by the two elements in Fig. P1.25 is 24 W.

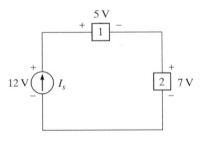

Figure P1.25

1.26 Find V_x in the network Fig. P1.26.

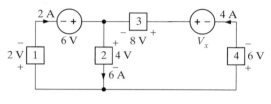

Figure P1.26

1.27 Find I_o in the network in Fig. P1.27.

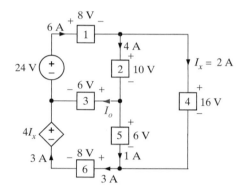

Figure P1.27

1.28 The charge entering the positive terminal of an element is $q(t) = -30e^{-4t}$ mC. If the voltage across the element is $120e^{-2t}$ V, determine the energy delivered to the element in the time interval $0 < t < 50$ ms.

1.29 The charge entering the positive terminal of an element is given by the expression $q(t) = -12e^{-2t}$ mC. The power delivered to the element is $p(t) = 2.4e^{-3t}$ W. Compute the current in the element, the voltage across the element, and the energy delivered to the element in the time interval $0 < t < 100$ ms.

1.30 The voltage across an element is $12e^{-2t}$ V. The current entering the positive terminal of the element is $2e^{-2t}$ A. Find the energy absorbed by the element in 1.5 s starting from $t = 0$.

Resistive Circuits

*I*n this chapter we introduce the basic concepts and laws that are fundamental to circuit analysis. These laws are Ohm's law, Kirchhoff's current law (KCL), and Kirchhoff's voltage law (KVL). We cannot overemphasize the importance of these three laws because they will be used extensively throughout our entire study of circuit analysis. The reader who masters their use quickly will not only find the material in this text easy to learn, but will be well positioned to grasp subsequent topics in the field of electrical engineering.

As a general rule, most of our activities will be confined to analysis; that is, to the determination of a specific voltage, current, or power somewhere in a network. The techniques we introduce have wide application in circuit analysis, even though we will discuss them within the framework of simple networks.

Our approach here is to begin with the simplest passive element, the resistor, and the mathematical relationship that exists between the voltage across it and the current through it, as specified by Ohm's law. As we build our confidence and proficiency by successfully analyzing some elementary circuits, we will introduce other techniques, such as voltage division and current division, that will accelerate our work.

We will introduce circuits containing dependent sources, which are used to model active devices such as transistors. Thus, our study of circuit analysis provides a natural introduction to many topics in the area of electronics.

The devices and circuits used in dc measurements are introduced, and thus we learn the fundamentals of measuring such quantities as voltage, current, and resistance.

Finally, we present some real-world applications that begin to indicate the usefulness of circuit analysis, and then we briefly introduce the topic of circuit design in an elementary fashion. In future chapters these topics will be revisited often to present some fascinating examples that describe problems we encounter in our everyday lives.

2.1 OHM'S LAW

Ohm's law is named for the German physicist Georg Simon Ohm, who is credited with establishing the voltage–current relationship for resistance. As a result of his pioneering work, the unit of resistance bears his name.

Ohm's law states that the voltage across a resistance is directly proportional to the current flowing through it. The resistance, measured in ohms, is the constant of proportionality between the voltage and current.

The passive sign convention will be employed in conjunction with Ohm's law.

A circuit element whose electrical characteristic is primarily resistive is called a resistor and is represented by the symbol shown in Fig. 2.1a. A resistor is a physical device that can be purchased in certain standard values in an electronic parts store. These resistors, which find use in a variety of electrical applications, are normally carbon composition or wire-wound. In addition, resistors can be fabricated using thick oxide or thin metal films for use in hybrid circuits, or they can be diffused in semiconductor integrated circuits. Some typical resistors are shown in Fig. 2.1b.

The mathematical relationship of Ohm's law is illustrated by the equation

$$v(t) = R \times i(t) \quad \text{where} \quad R \geq 0 \qquad \textbf{2.1}$$

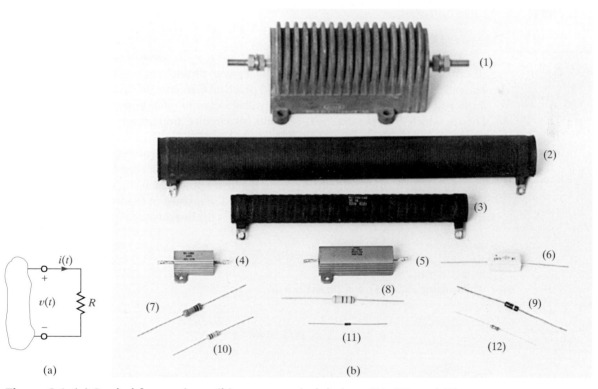

Figure 2.1 (a) Symbol for a resistor; (b) some practical devices. (1), (2), and (3) are high-power resistors. (4) and (5) are variable resistors. (6) is a high-precision resistor. (7)–(12) are fixed resistors with different power ratings.

D2.1 Determine $v(t)$ in the two circuits

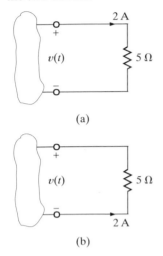

(a)

(b)

ANSWER: (a) 10 V

(b) −10 V

or equivalently, by the voltage–current characteristic shown in Fig. 2.2a. Note carefully the relationship between the polarity of the voltage and the direction of the current. In addition, note that we have tacitly assumed that the resistor has a constant value and therefore that the voltage–current characteristic is linear.

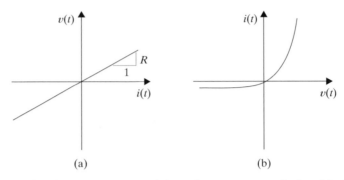

Figure 2.2 Graphical representation of the voltage–current relationship for (a) a linear resistor and (b) a diode.

The symbol Ω is used to represent ohms, and therefore,

$$1\ \Omega = 1\ \text{V/A}$$

Although in our analysis we will always assume that the resistors are *linear* and are thus described by a straight-line characteristic that passes through the origin, it is important that readers realize that some very useful and practical elements do exist that exhibit a *nonlinear* resistance characteristic; that is, the voltage–current relationship is not a straight line. Diodes, which are used extensively in electric circuits, are examples of nonlinear resistors. A typical diode characteristic is shown in Fig. 2.2b.

Since a resistor is a passive element, the proper current–voltage relationship is illustrated in Fig. 2.1a. The power supplied to the terminals is absorbed by the resistor. Note that the charge moves from the higher to the lower potential as it passes through the resistor and the energy absorbed is dissipated by the resistor in the form of heat. As indicated in Chapter 1, the rate of energy dissipation is the instantaneous power, and therefore

$$p(t) = v(t)i(t) \tag{2.2}$$

which, using Eq. (2.1), can be written as

$$p(t) = Ri^2(t) = \frac{v^2(t)}{R} \tag{2.3}$$

This equation illustrates that the power is a nonlinear function of either current or voltage and that it is always a positive quantity.

Conductance, represented by the symbol G, is another quantity with wide application in circuit analysis. By definition, conductance is the inverse of resistance; that is,

$$G = \frac{1}{R}$$ **2.4**

The unit of conductance is the siemens, and the relationship between units is

$$1\,S = 1\,A/V$$

Using Eq. (2.4), we can write two additional expressions,

$$i(t) = Gv(t)$$ **2.5**

and

$$p(t) = \frac{i^2(t)}{G} = Gv^2(t)$$ **2.6**

Equation (2.5) is another expression of Ohm's law.

Two specific values of resistance, and therefore conductance, are very important: $R = 0$ and $R = \infty$.

In examining the two cases, consider the network in Fig. 2.3a. The variable resistance symbol is used to describe a resistor such as the volume control on a radio or television set. As the resistance is decreased and becomes smaller and smaller, we finally reach a point where the resistance is zero and the circuit is reduced to that shown in Fig. 2.3b; that is, the resistance can be replaced by a short circuit. On the other hand, if the resistance is increased and becomes larger and larger, we finally reach a point where it is essentially infinite and the resistance can be replaced by an open circuit, as shown in Fig. 2.3c. Note that in the case of a short circuit where $R = 0$,

$$v(t) = Ri(t)$$
$$= 0$$

DRILL

D2.2   Determine $i(t)$ in the following two circuits:

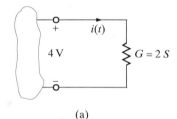

(a)

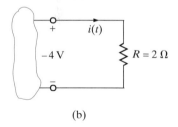

(b)

ANSWER: (a) 8 A

(b) −2 A

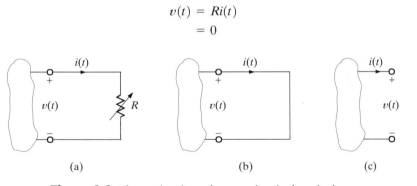

(a)   (b)   (c)

Figure 2.3 Short-circuit and open-circuit descriptions.

Therefore, $v(t) = 0$, although the current could theoretically be any value. In the open-circuit case where $R = \infty$,

$$i(t) = v(t)/R$$
$$= 0$$

Therefore, the current is zero regardless of the value of the voltage across the open terminals.

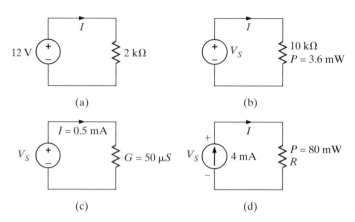

(a)   (b)

(c)   (d)

Figure 2.4 Circuits for Examples 2.1 to 2.4.

EXAMPLE 2.1

In the circuit in Fig. 2.4a, determine the current and the power absorbed by the resistor.

SOLUTION Using Eq. (2.1), we find the current to be

$$I = V/R = 12/2k = 6 \text{ mA}$$

Note that because many of the resistors employed in our analysis are in kΩ, we will use k in the equations in place of 1000. The power absorbed by the resistor is given by Eq. (2.2) or (2.3) as

$$P = VI = (12)(6 \times 10^{-3}) = 0.072 \text{ W}$$
$$= I^2R = (6 \times 10^{-3})^2(2k) = 0.072 \text{ W}$$
$$= V^2/R = (12)^2/2k = 0.072 \text{ W} \qquad \square$$

EXAMPLE 2.2

The power absorbed by the 10 kΩ resistor in Fig. 2.4b is 3.6 mW. Determine the voltage and the current in the circuit.

SOLUTION Using the power relationship, we can determine either of the unknowns.

$$V_S^2/R = P$$
$$V_S^2 = (3.6 \times 10^{-3})(10k)$$
$$V_S = 6 \text{ V}$$

and

$$I^2R = P$$
$$I^2 = (3.6 \times 10^{-3})/10k$$
$$I = 0.6 \text{ mA}$$

Furthermore, once V_S is determined, I could be obtained by Ohm's law and likewise once I is known, then Ohm's law could be used to derive the value of V_S. Note carefully that the equations for power involve the terms I^2 and V_S^2. Therefore, $I = -0.6$ mA and $V_S = -6$ V also satisfy the mathematical equations and, in this case, the direction of *both* the voltage and current is reversed. ❏

EXAMPLE 2.3

Given the circuit in Fig. 2.4c, we wish to find the value of the voltage source and the power absorbed by the resistance.

SOLUTION The voltage is

$$V_S = I/G = (0.5 \times 10^{-3})/(50 \times 10^{-6}) = 10 \text{ V}$$

The power absorbed is then

$$P = I^2/G = (0.5 \times 10^{-3})^2/(50 \times 10^{-6}) = 5 \text{ mW}$$

Or we could simply note that

$$R = 1/G = 20 \text{ k}\Omega$$

and therefore

$$V_S = IR = (0.5 \times 10^{-3})(20\text{k}) = 10 \text{ V}$$

and the power could be determined using $P = I^2R = V_S^2/R = V_S I$. ❏

EXAMPLE 2.4

Given the network in Fig. 2.4d, we wish to find R and V_S.

SOLUTION Using the power relationship, we find that

$$R = P/I^2 = (80 \times 10^{-3})/(4 \times 10^{-3})^2 = 5 \text{ k}\Omega$$

The voltage can now be derived using Ohm's law as

$$V_S = IR = (4 \times 10^{-3})(5\text{k}) = 20 \text{ V}$$

The voltage could also be obtained from the remaining power relationships in Eqs. (2.2) and (2.3). ❏

Before leaving this initial discussion of circuits containing sources and a single resistor, it is important to note a phenomenon that we will find to be true in circuits containing many sources and resistors. The presence of a voltage source between a pair of terminals tells us precisely what the voltage is between the two terminals regardless of what is happening in the balance of the network. What we do not know is the current in the voltage source. We must apply circuit analysis to the entire network to determine this current. Likewise, the presence of a current source connected between two terminals specifies the exact value of the current through the source between the terminals. What we do not know is the value of the voltage across the current source. This value must be calculated by applying circuit analysis to the entire network. Furthermore, it is worth emphasizing that when applying Ohm's law, the relationship $V = IR$ specifies a relationship between the voltage *directly across* a resistor R and that current that is *present in* this resistor. Ohm's law does not apply when the voltage is present in one part of the network and the current exists in another. This is a common mistake made by students who try to apply $V = IR$ to a resistor R in the middle of the network while using a V at some other location in the network.

EXTENSION EXERCISES

E2.1 Given the circuits in Fig. E2.1, find (a) the current I and the power absorbed by the resistor in Fig. E2.1a, and (b) the voltage across the current source and the power supplied by the source in Fig. E2.1b.

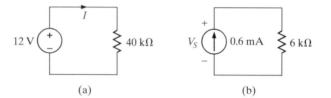

Figure E2.1   (a)   (b)

ANSWER: (a) $I = 0.3$ mA, $P = 3.6$ mW, (b) $V_S = 3.6$ V, $P = 2.16$ mW.

E2.2 Given the circuits in Fig. E2.2, find (a) R and V_S in the circuit in Fig. E2.2a, and (b) find I and R in the circuit in Fig. E2.2b.

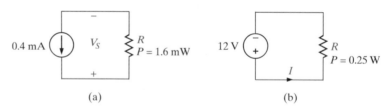

(a)   (b)

Figure E2.2

ANSWER: (a) $R = 10$ kΩ, $V_S = 4$ V, (b) $I = 20.8$ mA, $R = 576$ Ω.

2.2 KIRCHHOFF'S LAWS

The previous circuits that we have considered have all contained a single resistor and were analyzed using Ohm's law. At this point we begin to expand our capabilities to handle more complicated networks that result from an interconnection of two or more of these simple elements. We will assume that the interconnection is performed by electrical conductors (wires) that have zero resistance; that is, perfect conductors. Because the wires have zero resistance, the energy in the circuit is in essence lumped in each element, and we employ the term *lumped-parameter circuit* to describe the network.

To aid us in our discussion, we will define a number of terms that will be employed throughout our analysis. As will be our approach throughout this text, we will use examples to illustrate the concepts and define the appropriate terms. For example, the circuit shown in Fig. 2.5 will be used to describe the terms *node*, *loop*, and *branch*. A *node* is simply a point of connection of two or more circuit elements. The reader is cautioned to note that although one node can be spread out with perfect conductors, it is still only one node. For example, node 5 consists of the entire bottom connector of the circuit. In other words, if we start at some point in the circuit and move along perfect conductors in any direction until we encounter a circuit element, the total path we cover represents a single node. Therefore, we can assume that a node is one end of a circuit element together with all the perfect conductors that are attached to it. Examining the circuit, we note that there are numerous paths through it. A *loop* is simply any *closed path* through the circuit in which no node is encountered more than once. For example, starting from node 1, one loop would contain the elements R_1, v_2, R_4, and i_1; another loop would contain R_2, v_1, v_2, R_4, and i_1; and so on. However, the path R_1, v_1, R_5, v_2, R_3, and i_1 is not a loop because we have encountered node 3 twice. Finally, a *branch* is a portion of a circuit containing only a single element and the nodes at each end of the element. The circuit in Fig. 2.5 contains eight branches.

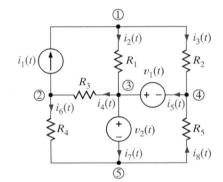

Figure 2.5 Circuit used to illustrate KCL.

Given the previous definitions, we are now in a position to consider Kirchhoff's laws, named after German scientist Gustav Robert Kirchhoff. These two laws are quite simple but extremely important. We will not attempt to prove them because the proofs are beyond our current level of understanding. However, we will demonstrate their usefulness and attempt to make the reader proficient in their use. The first law is *Kirchhoff's current law*, which states that *the algebraic sum of the currents entering any node is zero*. In mathematical form the law appears as

KCL is an extremely important and useful law.

$$\sum_{j=1}^{N} i_j(t) = 0 \qquad\qquad \textbf{2.7}$$

where $i_j(t)$ is the jth current entering the node through branch j and N is the number of branches connected to the node. To understand the use of this law, consider node 3 shown in Fig. 2.5. Applying Kirchhoff's current law to this node yields

$$i_2(t) - i_4(t) + i_5(t) - i_7(t) = 0$$

We have assumed that the algebraic signs of the currents entering the node are positive and, therefore, that the signs of the currents leaving the node are negative.

If we multiply the foregoing equation by -1, we obtain the expression

$$-i_2(t) + i_4(t) - i_5(t) + i_7(t) = 0$$

which simply states that *the algebraic sum of the currents leaving a node is zero*. Alternatively, we can write the equation as

$$i_2(t) + i_5(t) = i_4(t) + i_7(t)$$

which states that *the sum of the currents entering a node is equal to the sum of the currents leaving the node*. Both of these italicized expressions are alternative forms of Kirchhoff's current law.

Once again it must be emphasized that the latter statement means that the sum of the *variables* that have been defined entering the node is equal to the sum of the *variables* that have been defined leaving the node, not the actual currents. For example, $i_j(t)$ may be defined entering the node, but if its actual value is negative, there will be positive charge leaving the node.

Note carefully that Kirchhoff's current law states that the *algebraic* sum of the currents either entering or leaving a node must be zero. We now begin to see why we stated in Chapter 1 that it is critically important to specify both the magnitude and the direction of a current.

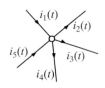

EXAMPLE 2.5

Let us write KCL for every node in the network in Fig. 2.5 assuming that the currents leaving the node are positive.

SOLUTION The KCL equations for nodes 1 through 5 are

$$-i_1(t) + i_2(t) + i_3(t) = 0$$
$$i_1(t) - i_4(t) + i_6(t) = 0$$
$$-i_2(t) + i_4(t) - i_5(t) + i_7(t) = 0$$
$$-i_3(t) + i_5(t) - i_8(t) = 0$$
$$-i_6(t) - i_7(t) + i_8(t) = 0$$

Note carefully that if we add the first four equations we obtain the fifth equation. What does this tell us? Recall that this means that this set of equations is not linearly independent. We can show that the first four equations are, however, linearly independent. Store this idea in memory because it will become very important when we learn how to write the equations necessary to solve for all the currents and voltages in a network in the following chapter. ❏

EXAMPLE 2.6

The network in Fig. 2.5 is represented by the topological diagram shown in Fig. 2.6. We wish to find the unknown currents in the network.

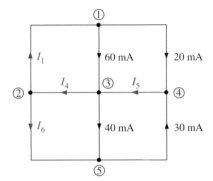

Figure 2.6 Topological diagram for the circuit in Fig. 2.5.

SOLUTION   Assuming the currents leaving the node are positive, the KCL equations for nodes 1 through 4 are

$$-I_1 + 0.06 + 0.02 = 0$$
$$I_1 - I_4 + I_6 = 0$$
$$-0.06 + I_4 - I_5 + 0.04 = 0$$
$$-0.02 + I_5 - 0.03 = 0$$

The first equation yields I_1 and the last equation yields I_5. Knowing I_5 we can immediately obtain I_4 from the third equation. Then the values of I_1 and I_4 yield the value of I_6 from the second equation. The results are $I_1 = 80$ mA, $I_4 = 70$ mA, $I_5 = 50$ mA, and $I_6 = -10$ mA.

   As indicated earlier, dependent or controlled sources are very important because we encounter them when analyzing circuits containing active elements such as transistors. The following example presents a circuit containing a current-controlled current source.   ❑

EXAMPLE 2.7

Let us write the KCL equations for the circuit shown in Fig. 2.7.

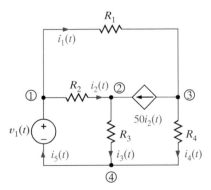

Figure 2.7 Circuit containing a dependent current source.

SOLUTION The KCL equations for nodes 1 through 4 follow.

$$i_1(t) + i_2(t) - i_5(t) = 0$$
$$-i_2(t) + i_3(t) - 50i_2(t) = 0$$
$$-i_1(t) + 50i_2(t) + i_4(t) = 0$$
$$i_5(t) - i_3(t) - i_4(t) = 0$$

If we added the first three equations, we would obtain the negative of the fourth. What does this tell us about the set of equations? ❏

Finally, it is possible to generalize Kirchhoff's current law to include a closed surface. By a closed surface we mean some set of elements completely contained within the surface that are interconnected. Since the current entering each element within the surface is equal to that leaving the element (i.e., the element stores no net charge), it follows that the current entering an interconnection of elements is equal to that leaving the interconnection. Therefore, Kirchhoff's current law can also be stated as follows: *The algebraic sum of the currents entering any closed surface is zero.*

EXAMPLE 2.8

Let us find I_4 and I_1 in the network represented by the topological diagram in Fig. 2.6.

SOLUTION This diagram is redrawn in Fig. 2.8, and node 1 is enclosed in surface 1 and nodes 3 and 4 are enclosed in surface 2. A quick review of the previous example indicates that we derived a value for I_4 from the value of I_5. However, I_5 is now completely enclosed in surface 2. If we apply KCL to surface 2, assuming the currents out of the surface are positive, we obtain

$$I_4 - 0.06 - 0.02 - 0.03 + 0.04 = 0$$

or

$$I_4 = 70 \text{ mA}$$

which we obtained without any knowledge of I_5. Likewise for surface 1, what goes in must come out and, therefore, $I_1 = 80$ mA. The reader is encouraged to cut the

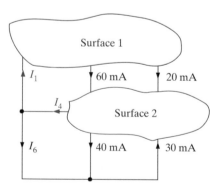

Figure 2.8 Diagram used to demonstrate KCL for a surface.

network in Fig. 2.6 into two pieces in any fashion and show that KCL is always satisfied at the boundaries.   ❑

EXTENSION EXERCISES

E2.3   Given the networks in Fig. E2.3, find (a) I_1 in Fig. E2.3a and (b) I_T in
Fig. E2.3b.

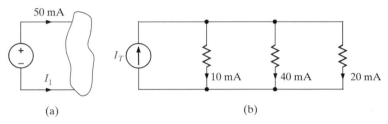

(a)   (b)

Figure E2.3

ANSWER: (a) $I = -50$ mA, (b) $I = 70$ mA.

E2.4   Find (a) I_1 in the network in Fig. E2.4a and (b) I_1 and I_2 in the circuit in
Fig. E2.4b.

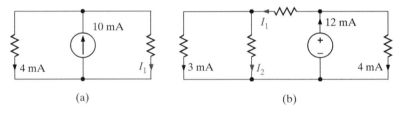

(a)   (b)

Figure E2.4

ANSWER: (a) $I_1 = 6$ mA, (b) $I_1 = 8$ mA and $I_2 = 5$ mA.

E2.5   Find the current i_x in the circuits in Fig. E2.5.

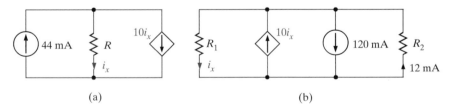

(a)   (b)

Figure E2.5

ANSWER: (a) $i_x = 4$ mA, (b) $i_x = 12$ mA.

DRILL

D2.4 Write the KVL equation for the following loop, traveling clockwise:

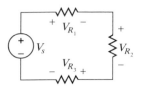

ANSWER: $-V_S + V_{R_1}$
$+ V_{R_2} + V_{R_3} = 0$

Kirchhoff's second law, called *Kirchhoff's voltage law*, states that *the algebraic sum of the voltages around any loop is zero*. As was the case with Kirchhoff's current law, we will defer the proof of this law and concentrate on understanding how to apply it. Once again the reader is cautioned to remember that we are dealing only with lumped-parameter circuits. These circuits are conservative, meaning that the work required to move a unit charge around any loop is zero.

Recall that in Kirchhoff's current law, the algebraic sign was required to keep track of whether the currents were entering or leaving a node. In Kirchhoff's voltage law the algebraic sign is used to keep track of the voltage polarity. In other words, as we traverse the circuit, it is necessary to sum to zero the increases and decreases in energy level. Therefore, it is important we keep track of whether the energy level is increasing or decreasing as we go through each element.

EXAMPLE 2.9

Consider the circuit shown in Fig. 2.9. If V_{R_1} and V_{R_2} are known quantities, let us find V_{R_3}.

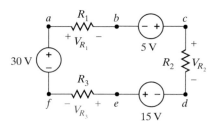

Figure 2.9 Circuit used to illustrate KVL.

SOLUTION In applying KVL, we must traverse the circuit and sum to zero the increases and decreases in energy level. Since the network is a single loop, we have only one closed path. We will adopt a policy of considering an increase in energy level as negative and a decrease in energy level as positive. Using this policy and starting at point a in the network and traversing it in a clockwise direction, we obtain the equation

$$+V_{R_1} - 5 + V_{R_2} - 15 + V_{R_3} - 30 = 0$$

which can be written as

$$+V_{R_1} + V_{R_2} + V_{R_3} = 5 + 15 + 30$$
$$= 50$$

Now suppose that V_{R_1} and V_{R_2} are known to be 18 V and 12 V, respectively. Then $V_{R_3} = 20$ V. ❏

EXAMPLE 2.10

Consider the network in Fig. 2.10.

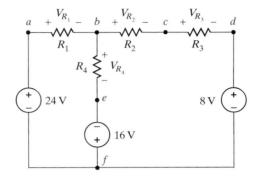

Figure 2.10 Circuit used to explain KVL.

Let us demonstrate that only two of the three possible loop equations are linearly independent.

SOLUTION   Note that this network has three closed paths: the left loop, right loop, and outer loop. Applying our policy for writing KVL equations and traversing the left loop starting at point a, we obtain

$$V_{R_1} + V_{R_4} - 16 - 24 = 0$$

The corresponding equation for the right loop starting at point b is

$$V_{R_2} + V_{R_3} + 8 + 16 - V_{R_4} = 0$$

The equation for the outer loop starting at point a is

$$V_{R_1} + V_{R_2} + V_{R_3} + 8 - 24 = 0$$

Note that if we add the first two equations, we obtain the third equation. Therefore, as we indicated in Example 2.5, the three equations are not linearly independent. Once again, we will address this issue in the next chapter and demonstrate that we need only the first two equations to solve for the voltages in the circuit.   ❏

Finally, we employ the convention V_{ab} to indicate the voltage of point a with respect to point b: that is, the variable for the voltage between point a and point b, with point a considered positive relative to point b. Since the potential is measured between two points, it is convenient to use an arrow between the two points with the head of the arrow located at the positive node. Note that the double-subscript notation, the + and − notation, and the single-headed arrow notation are all the same if the head of the arrow is pointing toward the positive terminal and the first subscript in the double-subscript notation. All of these equivalent forms for labeling voltages are shown in Fig. 2.11.

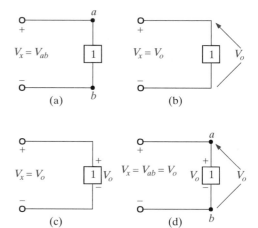

Figure 2.11 Equivalent forms for labeling voltage.

EXAMPLE 2.11

Consider the network in Fig. 2.12a. Let us apply KVL to determine the voltage between two points. Specifically, in terms of the double-subscript notation, let us find V_{ae} and V_{ec}.

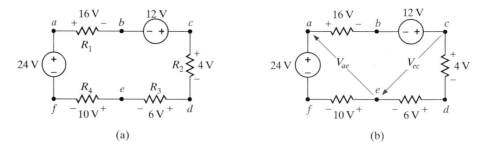

(a)   (b)

Figure 2.12 Network used in Example 2.11.

SOLUTION The circuit is redrawn in Fig. 2.12b. Since points a and e as well as e and c are not physically close, the arrow notation is very useful. Our approach to determining the unknown voltage is to apply KVL with the unknown voltage in the closed path. Therefore, to determine V_{ae} we can use the path $aefa$ or $abcdea$. The equations for the two paths in which V_{ae} is the only unknown are

$$V_{ae} + 10 - 24 = 0$$

and

$$16 - 12 + 4 + 6 - V_{ae} = 0$$

Note that both equations yield $V_{ae} = 14$ V. Even before calculating V_{ae}, we could calculate V_{ec} using the path *cdec* or *cefabc*. However, since V_{ae} is now known, we can also use the path *ceabc*. KVL for each of these paths is

$$4 + 6 + V_{ec} = 0$$
$$-V_{ec} - V_{ae} + 16 - 12 = 0$$

and

$$-V_{ec} + 10 - 24 + 16 - 12 = 0$$

Each of these equations yields $V_{ec} = -10$ V.   ❑

In general, the mathematical representation of Kirchhoff's voltage law is

> KVL is an extremely important and useful law.

$$\sum_{j=1}^{N} v_j(t) = 0 \qquad\qquad \textbf{2.8}$$

where $v_j(t)$ is the voltage across the *j*th branch (with the proper reference direction) in a loop containing N voltages. This expression is analogous to Eq. (2.7) for Kirchhoff's current law.

EXAMPLE 2.12

Given the network in Fig. 2.13 containing a dependent source, let us write the KVL equations for the two closed paths *abda* and *bcdb*.

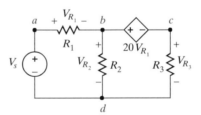

Figure 2.13 Network containing a dependent source.

SOLUTION   The two KVL equations are

$$V_{R_1} + V_{R_2} - V_S = 0$$
$$20V_{R_1} + V_{R_3} - V_{R_2} = 0 \qquad\qquad ❑$$

EXTENSION EXERCISES

E2.6   In the circuit in Fig. E2.6, find V_{ad}, V_{ac} and V_{bd}.

ANSWER: $V_{ad} = 12$ V, $V_{ac} = 10$ V, and $V_{bd} = 6$ V.

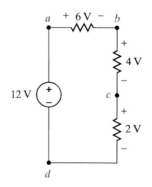

Figure E2.6

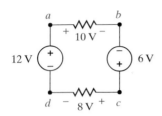

Figure E2.7

E2.7   In the network in Fig. E2.7, find V_{ac} and V_{db}.

ANSWER: $V_{ac} = 4$ V, and $V_{db} = -2$ V.

E2.8   Find V_{ad} and V_{eb} in the network in Fig. E2.8.

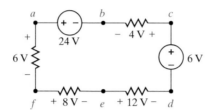

Figure E2.8

ANSWER: $V_{ad} = 26$ V, $V_{eb} = 10$ V.

E2.9   Find V_{bd} in the circuit in Fig. E2.9.

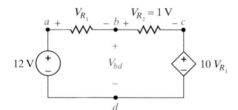

Figure E2.9

ANSWER: $V_{bd} = 11$ V.

The subtleties associated with Ohm's law, as described here, are important and must be adhered to in order to ensure that the variables have the proper sign.

Before proceeding with the analysis of simple circuits, it is extremely important that we emphasize a subtle but very critical point. Ohm's law as defined by the equation $V = IR$ refers to the relationship between the voltage and current as defined in Fig. 2.14a. If the direction of either the current or the voltage, but not both, is reversed, the relationship between the current and the voltage would be $V = -IR$. In a similar manner, given the circuit in Fig. 2.14b, if the polarity of the voltage between the terminals A and B is specified as shown, then the direction of the current I is from point B through R to

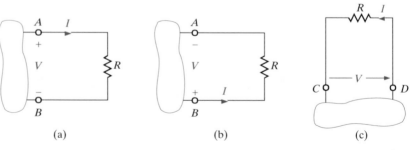

Figure 2.14 Circuits used to explain Ohm's law.

point A. Likewise, in Fig. 2.14c, if the direction of the current is specified as shown, then the polarity of the voltage must be such that point D is at a higher potential than point C and, therefore, the arrow representing the voltage V is from point C to point D.

2.3 SINGLE-LOOP CIRCUITS

Voltage Division

At this point we can begin to apply the laws we have presented earlier to the analysis of simple circuits. To begin, we examine what is perhaps the simplest circuit—a single closed path, or loop, of elements. The elements of a single loop carry the same current and, therefore, are said to be in *series*. However, we will apply Kirchhoff's voltage law and Ohm's law to the circuit to determine various quantities in the circuit.

Our approach will be to begin with a simple circuit and then generalize the analysis to more complicated ones. The circuit shown in Fig. 2.15 will serve as a basis for discussion. This circuit consists of an independent voltage source that is in series with two resistors. We have assumed that the current flows in a clockwise direction. If this assumption is correct, the solution of the equations that yields the current will produce a positive value. If the current is actually flowing in the opposite direction, the value of the current variable will simply be negative, indicating that the current is flowing in a direction opposite to that assumed. We have also made voltage polarity assignments for v_{R_1} and v_{R_2}. These assignments have been made using the convention employed in our discussion of Ohm's law and our choice for the direction of $i(t)$; that is, the convention shown in Fig. 2.14a.

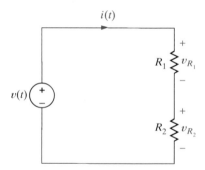

Figure 2.15 Single-loop circuit.

Applying Kirchhoff's voltage law to this circuit yields

$$-v(t) + v_{R_1} + v_{R_2} = 0$$

or

$$v(t) = v_{R_1} + v_{R_2}$$

However, from Ohm's law we know that

$$v_{R_1} = R_1 i(t)$$
$$v_{R_2} = R_2 i(t)$$

Therefore,

$$v(t) = R_1 i(t) + R_2 i(t)$$

Solving the equation for $i(t)$ yields

$$i(t) = \frac{v(t)}{R_1 + R_2} \qquad \textbf{2.9}$$

Knowing the current, we can now apply Ohm's law to determine the voltage across each resistor:

$$
\begin{aligned}
v_{R_1} &= R_1 i(t) \\
&= R_1 \left[\frac{v(t)}{R_1 + R_2} \right] \\
&= \frac{R_1}{R_1 + R_2} v(t) \qquad \textbf{2.10}
\end{aligned}
$$

Similarly,

$$v_{R_2} = \frac{R_2}{R_1 + R_2} v(t) \qquad \textbf{2.11}$$

The manner in which voltage divides between two series resistors

Although simple, Eqs. (2.10) and (2.11) are very important because they describe the operation of what is called a *voltage divider*. In other words, the source voltage $v(t)$ *is divided between the resistors R_1 and R_2 in direct proportion to their resistances.*

In essence, if we are interested in the voltage across the resistor R_1, we bypass the calculation of the current $i(t)$ and simply multiply the input voltage $v(t)$ by the ratio $R_1(R_1 + R_2)$. As illustrated in Eq. (2.10), we are using the current in the calculation, but not explicitly.

Note that the equations satisfy Kirchhoff's voltage law, since

$$-v(t) + \frac{R_1}{R_1 + R_2} v(t) + \frac{R_2}{R_1 + R_2} v(t) = 0$$

EXAMPLE 2.13

Consider the circuit shown in Fig. 2.16. The circuit is identical to Fig. 2.15 except that R_1 is a variable resistor such as the volume control for a radio or television set. Suppose that $V_S = 9$ V, $R_1 = 90$ kΩ and $R_2 = 30$ kΩ.

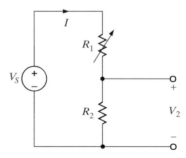

Figure 2.16 Voltage-divider circuit.

Let us examine the change in both the voltage across R_2 and the power absorbed in this resistor as R_1 is changed from 90 kΩ to 15 kΩ.

SOLUTION   Since this is a voltage-divider circuit, the voltage V_2 can be obtained directly as

$$V_2 = \left[\frac{R_2}{R_1 + R_2}\right]V_S$$

$$= \left[\frac{30k}{90k + 30k}\right](9)$$

$$= 2.25 \text{ V}$$

Now suppose that the variable resistor is changed from 90 kΩ to 15 kΩ. Then

$$V_2 = \left[\frac{30k}{30k + 15k}\right]9$$

$$= 6 \text{ V}$$

The direct voltage-divider calculation is equivalent to determining the current I and then using Ohm's law to find V_2. Note that the larger voltage is across the larger resistance. This voltage-divider concept and the simple circuit we have employed to describe it are very useful because, as will be shown later, more complicated circuits can be reduced to this form.

Finally, let us determine the instantaneous power absorbed by the resistor R_2 under the two conditions $R_1 = 90$ kΩ and $R_1 = 15$ kΩ. For the case $R_1 = 90$ kΩ, the power absorbed by R_2 is

$$P_2 = I^2 R_2 = \left(\frac{9}{120k}\right)^2 (30k)$$

$$= 0.169 \text{ mW}$$

In the second case

$$P_2 = \left(\frac{9}{45k}\right)^2 (30k)$$

$$= 1.2 \text{ mW}$$

The current in the first case is 75 μA, and in the second case it is 200 μA. Since the power absorbed is a function of the square of the current, the power absorbed in the two cases is quite different.   ❏

Let us now demonstrate the practical utility of this simple voltage-divider network.

EXAMPLE 2.14

Consider the circuit in Fig. 2.17a, which is an approximation of a high-voltage dc transmission facility. We have assumed that the bottom portion of the transmission line is a perfect conductor and will justify this assumption in the next chapter. The load can be represented by a resistor of value 183.5 Ω. Therefore, the equivalent circuit of this network is shown in Fig. 2.17b.

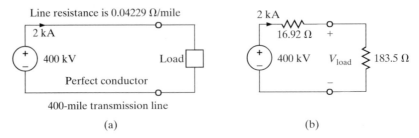

Line resistance is 0.04229 Ω/mile

400-mile transmission line

(a)   (b)

Figure 2.17 A high-voltage dc transmission facility.

Let us determine both the power delivered to the load and the power losses in the line.

SOLUTION Using voltage division, the load voltage is

$$V_{load} = \left[\frac{183.5}{183.5 + 16.92} \right] 400k$$

$$= 366.24 \text{ kV}$$

The input power is 800 MW and the power transmitted to the load is

$$P_{load} = I^2 R_{load}$$

$$= 734 \text{ MW}$$

Therefore, the power loss in the transmission line is

$$P_{line} = P_{in} - P_{load} = I^2 R_{line}$$

$$= 66 \text{ MW}$$

Since $P = VI$, suppose now that the utility company supplied power at 200 kV and 4 kA. What effect would this have on our transmission network? Without making a single calculation, we know that because power is proportional to the square of the current, there would be a large increase in the power loss in the line and, therefore, the efficiency of the facility would decrease substantially. That is why, in general, we transmit power at high voltage and low current.   ❏

Multiple Source/Resistor Networks

At this point we wish to extend our analysis to include a multiplicity of voltage sources and resistors. For example, consider the circuit shown in Fig. 2.18a. Here we have assumed that the current flows in a clockwise direction, and we have defined the variable $i(t)$ accordingly. This may or may not be the case, depending on the value of the various voltage sources. Kirchhoff's voltage law for this circuit is

$$+v_{R_1} + v_2(t) - v_3(t) + v_{R_2} + v_4(t) + v_5(t) - v_1(t) = 0$$

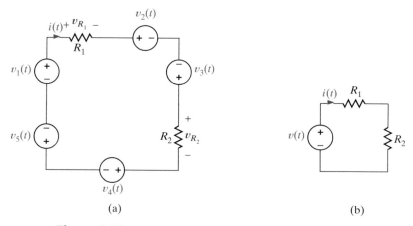

Figure 2.18 Equivalent circuits with multiple sources.

or, using Ohm's law,

$$(R_1 + R_2)i(t) = v_1(t) - v_2(t) + v_3(t) - v_4(t) - v_5(t)$$

which can be written as

$$(R_1 + R_2)i(t) = v(t)$$

where

$$v(t) = v_1(t) + v_3(t) - \left[v_2(t) + v_4(t) + v_5(t)\right]$$

so that under the preceding definitions, Fig. 2.18a is equivalent to Fig. 2.18b. In other words, the sum of several voltage sources in series can be replaced by one source whose value is the algebraic sum of the individual sources. This analysis can, of course, be generalized to a circuit with N series sources.

Now consider the circuit with N resistors in series, as shown in Fig. 2.19a. Applying Kirchhoff's voltage law to this circuit yields

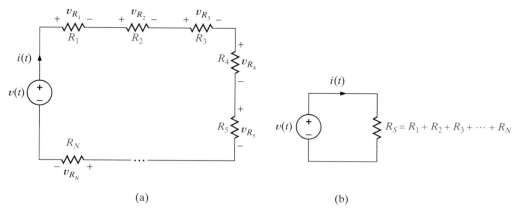

Figure 2.19 Equivalent circuits.

$$v(t) = v_{R_1} + v_{R_2} + \cdots + v_{R_N}$$
$$= R_1 i(t) + R_2 i(t) + \cdots + R_N i(t)$$

and therefore,

$$v(t) = R_S i(t) \qquad\qquad \textbf{2.12}$$

where

$$R_S = R_1 + R_2 + \cdots + R_N \qquad\qquad \textbf{2.13}$$

and hence,

$$i(t) = \frac{v(t)}{R_S} \qquad\qquad \textbf{2.14}$$

Note also that for any resistor R_i in the circuit, the voltage across R_i is given by the expression

$$v_{R_i} = \frac{R_i}{R_S} v(t) \qquad\qquad \textbf{2.15}$$

which is the voltage-division property for multiple resistors in series.

Equation (2.13) illustrates that *the equivalent resistance of N resistors in series is simply the sum of the individual resistances.* Thus, using Eq. (2.13), we can draw the circuit in Fig. 2.19b as an equivalent circuit for the one in Fig. 2.19a.

EXAMPLE 2.15

Given the circuit in Fig. 2.20, let us find I, V_{bd}, and the power absorbed by the 30-kΩ resistor.

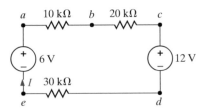

Figure 2.20 Circuit used in Example 2.15.

SOLUTION   KVL for the network yields the equation

$$10kI + 20kI + 12 + 30kI - 6 = 0$$
$$60kI = -6$$
$$I = -0.1 \text{ mA}$$

Therefore, the magnitude of the current is 0.1 mA, but its direction is opposite to that assumed.

The voltage V_{bd} can be calculated using either of the closed paths *abdea* or *bcdb*. The equations for both cases are

$$10kI + V_{bd} + 30kI - 6 = 0$$

and

$$20kI + 12 - V_{bd} = 0$$

Using $I = -0.1$ mA in either equation yields $V_{bd} = 10$ V. Finally, the power absorbed by the 30 kΩ resistor is

$$P = I^2 R = 0.3 \text{ mW}$$

❏

EXAMPLE 2.16

A dc transmission facility is modeled by the approximate circuit shown in Fig. 2.21. If the load voltage is known to be $V_{\text{load}} = 458.3$ kV, we wish to find the voltage at the sending end of the line and the power loss in the line.

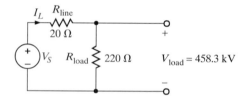

Figure 2.21 Circuit used in Example 2.16.

SOLUTION   Knowing the load voltage and load resistance, we can obtain the line current using Ohm's law:

$$I_{\text{L}} = 458.3\text{k}/220$$
$$= 2.083 \text{ kA}$$

The voltage drop across the line is

$$V_{\text{line}} = (I_L)(R_{\text{line}})$$
$$= 41.66 \text{ kV}$$

Now, using KVL,

$$V_S = V_{\text{line}} + V_{\text{load}}$$
$$= 500 \text{ kV}$$

Note that since the network is simply a voltage-divider circuit, we could obtain V_S immediately from our knowledge of R_{line}, R_{load}, and V_{load}. That is,

$$V_{\text{load}} = \left[\frac{R_{\text{load}}}{R_{\text{load}} + R_{\text{line}}} \right] V_S$$

and V_S is the only unknown in this equation.

The power absorbed by the line is

$$P_{\text{line}} = I_L^2 R_{\text{line}}$$
$$= 86.79 \text{ MW}$$

❑

EXTENSION EXERCISES

E2.10   Find I and V_{bd} in the circuit in Fig. E2.10.

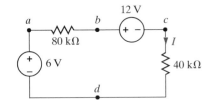

Figure E2.10

ANSWER: $I = -0.05$ mA and $V_{bd} = 10$ V.

E2.11   In the network in Fig. E2.11, if V_{ad} is 3 V, find V_S.

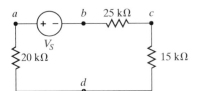

Figure E2.11

ANSWER: $V_S = 9$ V.

2.4 SINGLE-NODE-PAIR CIRCUITS

Current Division

An important circuit is the single-node-pair circuit. In this case the elements have the same voltage across them and, therefore, are in *parallel*. We will, however, apply Kirchhoff's current law and Ohm's law to determine various unknown quantities in the circuit.

Following our approach with the single-loop circuit, we will begin with the simplest case and then generalize our analysis. Consider the circuit shown in Fig. 2.22. Here we have an independent current source in parallel with two resistors.

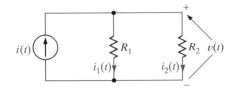

Figure 2.22 Simple parallel circuit.

Since all of the circuit elements are in parallel, the voltage $v(t)$ appears across each of them.

Furthermore, an examination of the circuit indicates that the current $i(t)$ is into the upper node of the circuit and the currents $i_1(t)$ and $i_2(t)$ are out of the node. Since KCL essentially states that what goes in must come out, the question we must answer is how $i_1(t)$ and $i_2(t)$ divide the input current $i(t)$.

Applying Kirchhoff's current law to the upper node, we obtain

$$i(t) = i_1(t) + i_2(t)$$

And, employing Ohm's law, we have

$$
\begin{aligned}
i(t) &= \frac{v(t)}{R_1} + \frac{v(t)}{R_2} \\
&= \left(\frac{1}{R_1} + \frac{1}{R_2} \right) v(t) \\
&= \frac{v(t)}{R_p}
\end{aligned}
$$

where

$$\frac{1}{R_p} = \frac{1}{R_1} + \frac{1}{R_2}$$

2.16   The parallel resistance equation

$$R_p = \frac{R_1 R_2}{R_1 + R_2}$$

2.17

Therefore, the equivalent resistance of two resistors connected in parallel is equal to the product of their resistances divided by their sum. Note also that this equivalent resistance R_p is always less than either R_1 or R_2. Hence, by connecting resistors in parallel we reduce the overall resistance. In the special case when $R_1 = R_2$, the equivalent resistance is equal to half of the value of the individual resistors.

The manner in which the current $i(t)$ from the source divides between the two branches is called *current division* and can be found from the preceding expressions. For example,

$$
\begin{aligned}
v(t) &= R_p i(t) \\
&= \frac{R_1 R_2}{R_1 + R_2} i(t)
\end{aligned}
$$

2.18

and

$$i_1(t) = \frac{v(t)}{R_1}$$

The manner in which current divides between two parallel resistors

$$i_1(t) = \frac{R_2}{R_1 + R_2} i(t)$$

2.19

and

DRILL

D2.5   Find I_1 and I_2 in the
following circuit:

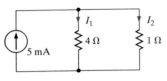

ANSWER: $I_1 = 1\,\text{mA}$

$I_2 \overset{\bullet}{=} 4\,\text{mA}$

$$i_2(t) = \frac{v(t)}{R_2}$$

$$= \frac{R_1}{R_1 + R_2}\,i(t) \qquad\qquad \textbf{2.20}$$

Equations (2.19) and (2.20) are mathematical statements of the current-division rule.

EXAMPLE 2.17

Given the network in Fig. 2.23a, let us find I_1, I_2 and V_o.

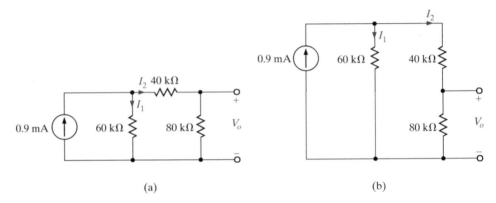

(a)   (b)

Figure 2.23 Circuits used in Example 2.17.

SOLUTION   First, it is important to recognize that the current source feeds two parallel paths. To emphasize this point, the circuit is redrawn as shown in Fig. 2.23b. Applying current division, we obtain

$$I_1 = \left[\frac{40\text{k} + 80\text{k}}{60\text{k} + (40\text{k} + 80\text{k})}\right](0.9 \times 10^{-3})$$

$$= 0.6\,\text{mA}$$

and

$$I_2 = \left[\frac{60\text{k}}{60\text{k} + (40\text{k} + 80\text{k})}\right](0.9 \times 10^{-3})$$

$$= 0.3\,\text{mA}$$

Note that the larger current flows through the smaller resistor, and vice versa. In addition, one should note that if the resistances of the two paths are equal, the current will divide equally between them. KCL is satisfied since $I_1 + I_2 = 0.9\,\text{mA}$.

The voltage V_o can be derived using Ohm's law as

$$V_o = 80kI$$
$$= 24 \text{ V} \qquad \square$$

EXAMPLE 2.18

A typical car stereo consists of a 2-W audio amplifier and two speakers represented by the diagram shown in Fig. 2.24a. The output circuit of the audio amplifier is in essence a 430-mA current source and the speakers each have a resistance of 4 Ω. Let us determine the power absorbed by the speakers.

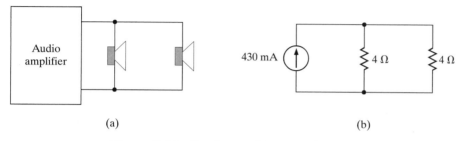

(a)   (b)

Figure 2.24 Circuits used in Example 2.18.

SOLUTION   The audio system can be modeled as shown in Fig. 2.24b. Since the speakers are both 4-Ω devices, the current will split evenly between them and the power absorbed by each speaker is

$$P = I^2R$$
$$= (215 \times 10^{-3})^2(4)$$
$$= 184.9 \text{ mW} \qquad \square$$

EXTENSION EXERCISE

E2.12   Find the currents I_1 and I_2 and the power absorbed by the 40-kΩ resistor in the network in Fig. E2.12.

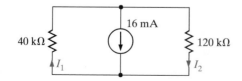

Figure E2.12

ANSWER: $I_1 = 12$ mA, $I_2 = -4$ mA, and $P_{40\text{k}\Omega} = 5.76$ W.

Multiple Source/Resistor Networks

Let us now extend our analysis to include a multiplicity of current sources and resistors in parallel. For example, consider the circuit shown in Fig. 2.25a. We have assumed that the upper node is $v(t)$ volts positive with respect to the lower node. Applying Kirchhoff's current law to the upper node yields

$$i_1(t) - i_2(t) - i_3(t) + i_4(t) - i_5(t) - i_6(t) = 0$$

or

$$i_1(t) - i_3(t) + i_4(t) - i_6(t) = i_2(t) + i_5(t)$$

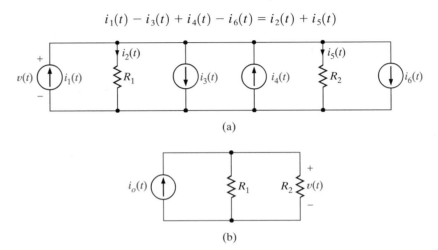

(a)

(b)

Figure 2.25 Equivalent circuits.

The terms on the left side of the equation all represent sources that can be combined algebraically into a single source; that is,

$$i_o(t) = i_1(t) - i_3(t) + i_4(t) - i_6(t)$$

which effectively reduces the circuit in Fig. 2.25a to that in Fig. 2.25b. We could, of course, generalize this analysis to a circuit with N current sources. Using Ohm's law, the currents on the right side of the equation can be expressed in terms of the voltage and individual resistances so that the KCL equation reduces to

$$i_o(t) = \left(\frac{1}{R_1} + \frac{1}{R_2}\right)v(t)$$

Now consider the circuit with N resistors in parallel, as shown in Fig. 2.26a. Applying Kirchhoff's current law to the upper node yields

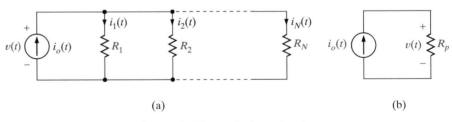

(a)

(b)

Figure 2.26 Equivalent circuits.

$$i_o(t) = i_1(t) + i_2(t) + \cdots + i_N(t)$$

$$= \left(\frac{1}{R_1} + \frac{1}{R_2} + \cdots + \frac{1}{R_N} \right) v(t) \qquad \textbf{2.21}$$

or

$$i_o(t) = \frac{v(t)}{R_p} \qquad \textbf{2.22}$$

where

$$\frac{1}{R_p} = \sum_{i=1}^{N} \frac{1}{R_i} \qquad \textbf{2.23}$$

so that as far as the source is concerned, Fig. 2.26a can be reduced to an equivalent circuit, as shown in Fig. 2.26b.

The current division for any branch can be calculated using Ohm's law and the preceding equations. For example, for the jth branch in the network of Fig. 2.26a,

$$i_j(t) = \frac{v(t)}{R_j}$$

Using Eq. (2.22), we obtain

$$i_j(t) = \frac{R_p}{R_j} i_o(t) \qquad \textbf{2.24}$$

which defines the current-division rule for the general case.

EXAMPLE 2.19

Given the circuit in Fig. 2.27a, we wish to find the current in the 12-kΩ load resistor.

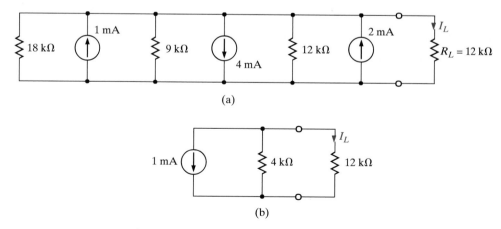

Figure 2.27 Circuits used in Example 2.19.

SOLUTION   In order to simplify the network in Fig. 2.27a, we add the current sources algebraically and combine the parallel resistors in the following manner:

$$\frac{1}{R_p} = \frac{1}{18k} + \frac{1}{9k} + \frac{1}{12k}$$

$$R_p = 4\text{ k}\Omega$$

Using these values we can reduce the circuit in Fig. 2.27a to that in Fig. 2.27b. Now, applying current division, we obtain

$$I_L = -\left[\frac{4k}{4k + 12k}\right](1 \times 10^{-3})$$

$$= -0.25\text{ mA} \qquad \qquad ❏$$

EXTENSION EXERCISE

E2.13   Find the power absorbed by the 6-kΩ resistor in the network in Fig. E2.13.

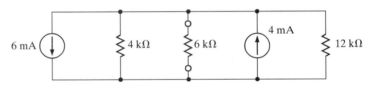

Figure E2.13

ANSWER: $P = 2.67\text{ mW}$.

2.5 SERIES AND PARALLEL RESISTOR COMBINATIONS

We have shown in our earlier developments that the equivalent resistance of N resistors in series is

$$R_S = R_1 + R_2 + \cdots + R_N \qquad\qquad \textbf{2.25}$$

and the equivalent resistance of N resistors in parallel is found from

$$\frac{1}{R_p} = \frac{1}{R_1} + \frac{1}{R_2} + \cdots + \frac{1}{R_N} \qquad\qquad \textbf{2.26}$$

Let us now examine some combinations of these two cases.

EXAMPLE 2.20

We wish to determine the resistance at terminals A-B in the network in Fig. 2.28a.

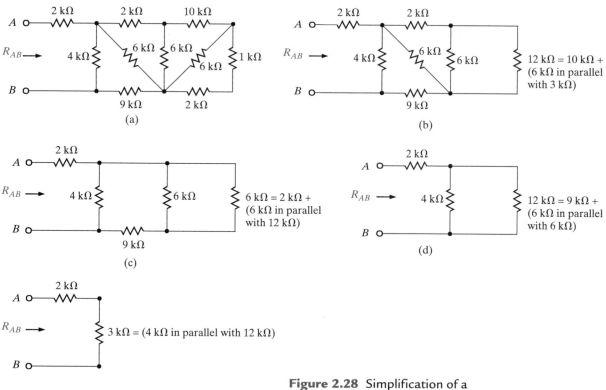

Figure 2.28 Simplification of a resistance network.

SOLUTION Starting at the opposite end of the network from the terminals and combining resistors as shown in the sequence of circuits in Fig. 2.28, we find that the equivalent resistance at the terminals is 5 kΩ. ❏

EXTENSION EXERCISES

E2.14 Find the equivalent resistance at the terminals A-B in the network in Fig. E2.14.

ANSWER: $R_{AB} = 22$ kΩ.

E2.15 Find the equivalent resistance at the terminals A-B in the circuit in Fig. E2.15.

ANSWER: $R_{AB} = 3$ kΩ

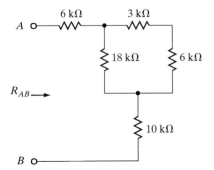

Figure E2.14

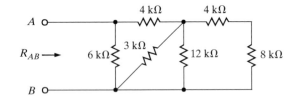

Figure E2.15

A problem-solving strategy for simplifying resistor combinations

When trying to determine the equivalent resistance at a pair of terminals of a network composed of an interconnection of numerous resistors, it is recommended that the analysis begin at the end of the network opposite the terminals. Two or more resistors are combined to form a single resistor, thus simplifying the network by reducing the number of components as the analysis continues in a steady progression toward the terminals. The simplification involves the following:

1. *Resistors in series.* Resistors R_1 and R_2 are in series if they are connected in tandem and carry exactly the same current. They can then be combined into a single resistor R_S, where $R_S = R_1 + R_2$.

2. *Resistors in parallel.* Resistors R_1 and R_2 are in parallel if they are connected to the same two nodes and have exactly the same voltage across their terminals. They can then be combined into a single resistor R_P, where $R_P = R_1 R_2/(R_1 + R_2)$.

These two combinations are used repeatedly, as needed, to reduce the network to a single resistor at the pair of terminals.

As a practical example of the use of series and parallel resistors, consider the following example.

EXAMPLE 2.21

A standard dc current-limiting power supply shown in Fig. 2.29a provides 0–18 V at 3 A to a load. The voltage drop, V_R, across a resistor, R, is used as a current-sensing device, fed back to the power supply and used to limit the current I. That is, if

the load is adjusted so that the current tries to exceed 3 A, the power supply will act
to limit the current to that value. The feedback voltage, V_R, should typically not exceed 600 mV.

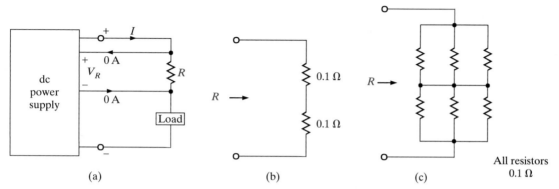

Figure 2.29 Circuits used in Example 2.21.

If we have a box of standard 0.1-Ω, 5-W resistors, let us determine the configuration of these resistors that will provide a $V_R = 600$ mV when the current is 3 A.

SOLUTION   Using Ohm's law, the value of R should be

$$R = \frac{V_R}{I}$$

$$= \frac{0.6}{3}$$

$$= 0.2 \ \Omega$$

Therefore, two 0.1-Ω resistors connected in series, as shown in Fig. 2.29b, will
provide the proper feedback voltage. Suppose, however, that the power supply current is to be limited to 9 A. The resistance required in this case to produce
$V_R = 600$ mV is

$$R = \frac{0.6}{9}$$

$$= 0.0667 \ \Omega$$

We must now determine how to interconnect the 0.1-Ω resistors to obtain
$R = 0.0667 \ \Omega$. Since the desired resistance is less than the components available
(i.e., 0.1 Ω), we must connect the resistors in some type of parallel configuration.
Since all the resistors are of equal value, note that three of them connected in parallel would provide a resistance of one-third their value, or 0.0333 Ω. Then two such
combinations connected in series, as shown in Fig. 2.29c, would produce the proper resistance.

Finally, we must check to ensure that the configurations in Figs. 2.29b and c have
not exceeded the power rating of the resistors. In the first case, the current $I = 3$ A

is present in each of the two series resistors. Therefore, the power absorbed in each resistor is

$$P = I^2 R$$
$$= (3)^2(0.1)$$
$$= 0.9 \text{ W}$$

which is well within the 5-W rating of the resistors.

In the second case, the current $I = 9$ A. The resistor configuration for R in this case is a series combination of two sets of three parallel resistors of equal value. Using current division, we know that the current I will split equally among the three parallel paths and, hence, the current in each resistor will be 3 A. Therefore, once again, the power absorbed by each resistor is within its power rating. ❏

2.6 CIRCUITS WITH SERIES-PARALLEL COMBINATIONS OF RESISTORS

At this point we have learned many techniques that are fundamental to circuit analysis. Now we wish to apply them and show how they can be used in concert to analyze circuits. We will illustrate their application through a number of examples that will be treated in some detail.

EXAMPLE 2.22

We wish to find all the currents and voltages labeled in the ladder network shown in Fig. 2.30a.

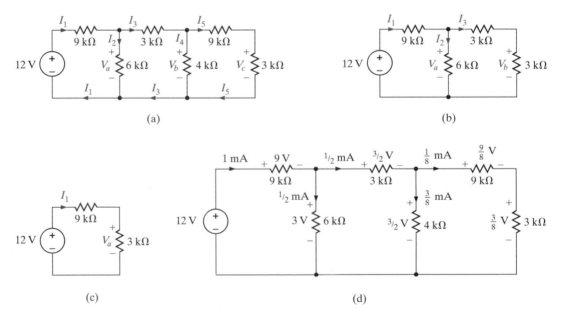

Figure 2.30 Analysis of a ladder network.

SOLUTION   To begin our analysis of the network, we start at the right end of the circuit and combine the resistors to determine the total resistance seen by the 12-V source. This will allow us to calculate the current I_1. Then employing KVL, KCL, Ohm's law, and/or voltage and current division, we will be able to calculate all currents and voltages in the network.

At the right end of the circuit, the 9-kΩ and 3-kΩ resistors are in series and, thus, can be combined into one equivalent 12-kΩ resistor. This resistor is in parallel with the 4-kΩ resistor, and their combination yields an equivalent 3-kΩ resistor, shown at the right edge of the circuit in Fig. 2.30b. In Fig. 2.30b the two 3-kΩ resistors are in series and their combination is in parallel with the 6-kΩ resistor. Combining all three resistances yields the circuit shown in Fig. 2.30c.

Applying Kirchhoff's voltage law to the circuit in Fig. 2.30c yields

$$I_1(9k + 3k) = 12$$
$$I_1 = 1 \text{ mA}$$

V_a can be calculated from Ohm's law as

$$V_a = I_1(3k)$$
$$= 3 \text{ V}$$

or, using Kirchhoff's voltage law,

$$V_a = 12 - 9kI_1$$
$$= 12 - 9$$
$$= 3 \text{ V}$$

Knowing I_1 and V_a, we can now determine all currents and voltages in Fig. 2.30b. Since $V_a = 3$ V, the current I_2 can be found using Ohm's law as

$$I_2 = \frac{3}{6k}$$
$$= \frac{1}{2} \text{ mA}$$

Then, using Kirchhoff's current law, we have

$$I_1 = I_2 + I_3$$
$$1 \times 10^{-3} = \frac{1}{2} \times 10^{-3} + I_3$$
$$I_3 = \frac{1}{2} \text{ mA}$$

Note that the I_3 could also be calculated using Ohm's law:

$$V_a = (3k + 3k)I_3$$
$$I_3 = \frac{3}{6k}$$
$$= \frac{1}{2} \text{ mA}$$

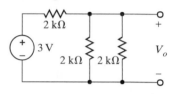
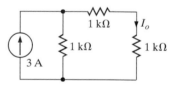

Applying Kirchhoff's voltage law to the right-hand loop in Fig. 2.30b yields

$$V_a - V_b = 3kI_3$$

$$3 - V_b = \frac{3}{2}$$

$$V_b = \frac{3}{2} \text{ V}$$

or, since V_b is equal to the voltage drop across the 3-kΩ resistor, we could use Ohm's law as

$$V_b = 3kI_3$$

$$= \frac{3}{2} \text{ V}$$

We are now in a position to calculate the final unknown currents and voltages in Fig. 2.30a. Knowing V_b, we can calculate I_4 using Ohm's law as

$$V_b = 4kI_4$$

$$I_4 = \frac{\dfrac{3}{2}}{4k}$$

$$= \frac{3}{8} \text{ mA}$$

Then, from Kirchhoff's current law, we have

$$I_3 = I_4 + I_5$$

$$\frac{1}{2} \times 10^{-3} = \frac{3}{8} \times 10^{-3} + I_5$$

$$I_5 = \frac{1}{8} \text{ mA}$$

We could also have calculated I_5 using the current-division rule. For example,

$$I_5 = \frac{4k}{4k + (9k + 3k)} I_3$$

$$= \frac{1}{8} \text{ mA}$$

Finally, V_c can be computed as

$$V_c = I_5(3k)$$

$$= \frac{3}{8} \text{ V}$$

V_c can also be found using voltage division (i.e., the voltage V_b will be divided between the 9-kΩ and 3-kΩ resistors). Therefore,

$$V_c = \left[\frac{3k}{3k + 9k}\right]V_b$$

$$= \frac{3}{8}\,V$$

Note that Kirchhoff's current law is satisfied at every node and Kirchhoff's voltage law is satisfied around every loop, as shown in Fig. 2.30d.   ❏

The following example is, in essence, the reverse of the previous example in that we are given the current in some branch in the network and asked to find the value of the input source.

EXAMPLE 2.23

Given the circuit in Fig. 2.31 and $I_4 = \frac{1}{2}$ mA, let us find the source voltage V_o.

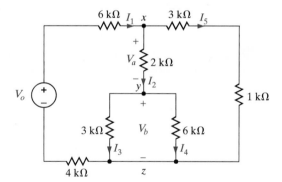

Figure 2.31 Example circuit for analysis.

SOLUTION   If $I_4 = \frac{1}{2}$ mA, then from Ohm's law, $V_b = 3$ V. V_b can now be used to calculate $I_3 = 1$ mA. Kirchhoff's current law applied at node y yields

$$I_2 = I_3 + I_4$$
$$= 1.5\ \text{mA}$$

Then, from Ohm's law, we have

$$V_a = (1.5 \times 10^{-3})(2k)$$
$$= 3\ \text{V}$$

Since $V_a + V_b$ is now known, I_5 can be obtained:

$$I_5 = \frac{V_a + V_b}{3k + 1k}$$
$$= 1.5\ \text{mA}$$

Applying Kirchhoff's current law at node x yields

$$I_1 = I_2 + I_5$$
$$= 3 \text{ mA}$$

Now KVL applied to any closed path containing V_o will yield the value of this input source. For example, if the path is the outer loop, KVL yields

$$-V_o + 6kI_1 + 3kI_5 + 1kI_5 + 4kI_1 = 0$$

Since $I_1 = 3 \text{ mA}$ and $I_5 = 1.5 \text{ mA}$,

$$V_o = 36 \text{ V}$$

If we had selected the path containing the source and the points x, y, and z, we would obtain

$$-V_o + 6kI_1 + V_a + V_b + 4kI_1 = 0$$

Once again, this equation yields

$$V_o = 36 \text{ V} \qquad \blacksquare$$

EXTENSION EXERCISES

E2.16   Find V_o in the network in Fig. E2.16.

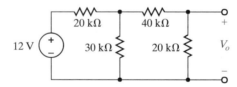

Figure E2.16

ANSWER: $V_o = 2 \text{ V}$.

E2.17   Find V_S in the circuit in Fig. E2.17.

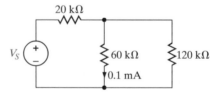

Figure E2.17

ANSWER: $V_S = 9 \text{ V}$.

E2.18   Find I_S in the circuit in Fig. E2.18.

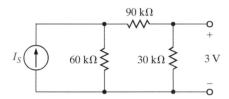

Figure E2.18

ANSWER: $I_S = 0.3$ mA.

Problem-solving strategy for analyzing circuits containing a single source and a series-parallel interconnection of resistors

Step 1.   Systematically reduce the resistive network so that the resistance seen by the source is represented by a single resistor.

Step 2.   Determine the source current for a voltage source or the source voltage if a current source is present.

Step 3.   Expand the network, retracing the simplification steps, and apply Ohm's law, KVL, KCL, voltage division, and current division to determine all currents and voltages in the network.

2.7 CIRCUITS WITH DEPENDENT SOURCES

In Chapter 1 we outlined the different kinds of dependent sources. These controlled sources are extremely important because they are used to model physical devices such as *npn* and *pnp* bipolar junction transistors (BJTs) and field-effect transistors (FETs) that are either metal-oxide-semiconductor field-effect transistors (MOSFETs) or insulated-gate field-effect transistors (IGFETs). These basic structures are, in turn, used to make analog and digital devices. A typical analog device is an operational amplifier (op-amp). Typical digital devices are random access memories (RAMs), read-only memories (ROMs), and microprocessors. We will now show how to solve simple one-loop and one-node circuits that contain these dependent sources. Although the following examples are fairly simple, they will serve to illustrate the basic concepts.

Problem-solving strategy for circuits with dependent sources

Step 1.   When writing the KVL and/or KCL equations for the network, treat the dependent source as though it were an independent source.

Step 2.   Write the equation that specifies the relationship of the dependent source to the controlling parameter.

Step 3.   Solve the equations for the unknowns. Be sure that the number of linearly independent equations matches the number of unknowns.

The following example employs a current-controlled voltage source.

EXAMPLE 2.24

Let us determine the voltage V_o in the circuit in Fig. 2.32.

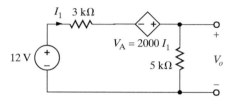

Figure 2.32 Circuit used in
Example 2.24.

SOLUTION   Applying KVL, we obtain

$$-12 + 3kI_1 - V_A + 5kI_1 = 0$$

where

$$V_A = 2000I_1$$

and the units of the multiplier, 2000, are ohms. Solving these equations yields

$$I_1 = 2 \text{ mA}$$

Then

$$V_o = (5k)I_1$$

$$= 10 \text{ V}$$ ❏

EXAMPLE 2.25

Given the circuit in Fig. 2.33 containing a current-controlled current source, let us
find the voltage V_o.

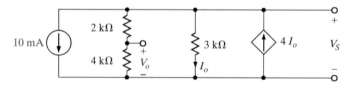

Figure 2.33 Circuit used in Example 2.25.

SOLUTION   Applying KCL at the top node, we obtain

$$10 \times 10^{-3} + \frac{V_S}{2k + 4k} + \frac{V_S}{3k} - 4I_o = 0$$

where

$$I_o = \frac{V_S}{3k}$$

Substituting this expression for the controlled source into the KCL equation yields

$$\frac{10}{1k} + \frac{V_S}{6k} + \frac{V_S}{3k} - \frac{4V_S}{3k} = 0$$

Solving this equation for V_S, we obtain

$$V_S = 12 \text{ V}$$

The voltage V_o can now be obtained using a simple voltage divider; that is,

$$V_o = \left[\frac{4k}{2k + 4k}\right]V_S$$

$$= 8 \text{ V} \qquad \qquad ❑$$

EXAMPLE 2.26

The network in Fig. 2.34 contains a voltage-controlled voltage source. We wish to find V_o in this circuit.

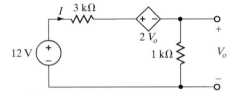

Figure 2.34 Circuit used in Example 2.26.

SOLUTION   Applying KVL to this network yields

$$-12 + 3kI + 2V_o + 1kI = 0$$

where

$$V_o = 1kI$$

Hence, the KVL equation can be written as

$$-12 + 3kI + 2kI + 1kI = 0$$

or

$$I = 2 \text{ mA}$$

Therefore,

$$V_o = 1kI$$

$$= 2 \text{ V} \qquad \qquad ❑$$

EXAMPLE 2.27

An equivalent circuit for a FET common-source amplifier or BJT common-emitter amplifier can be modeled by the circuit shown in Fig. 2.35a. We wish to determine an expression for the gain of the amplifier, which is the ratio of the output voltage to the input voltage.

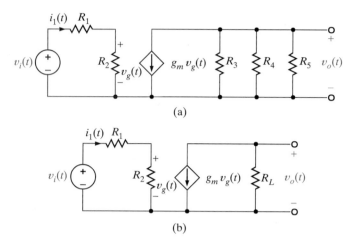

(a)

(b)

Figure 2.35 Example circuit containing a voltage-controlled current source.

SOLUTION Note that although this circuit, which contains a voltage-controlled current source, appears to be somewhat complicated, we are actually in a position now to solve it with techniques we have studied up to this point. The loop on the left, or input to the amplifier, is essentially detached from the output portion of the amplifier on the right. The voltage across R_2 is $v_g(t)$, which controls the dependent current source.

To simplify the analysis, let us replace the resistors R_3, R_4, and R_5 with R_L such that

$$\frac{1}{R_L} = \frac{1}{R_3} + \frac{1}{R_4} + \frac{1}{R_5}$$

Then the circuit reduces to that shown in Fig. 2.35b. Applying Kirchhoff's voltage law to the input portion of the amplifier yields

$$v_i(t) = i_1(t)(R_1 + R_2)$$

and

$$v_g(t) = i_1(t)R_2$$

Solving these equations for $v_g(t)$ yields

$$v_g(t) = \frac{R_2}{R_1 + R_2} v_i(t)$$

From the output circuit, note that the voltage $v_o(t)$ is given by the expression

$$v_o(t) = -g_m v_g(t)R_L$$

Combining this equation with the preceding one yields

$$v_o(t) = \frac{-g_m R_L R_2}{R_1 + R_2} v_i(t)$$

Therefore, the amplifier gain, which is the ratio of the output voltage to the input voltage, is given by

$$\frac{v_o(t)}{v_i(t)} = -\frac{g_m R_L R_2}{R_1 + R_2}$$

Reasonable values for the circuit parameters in Fig. 2.35a are $R_1 = 100\ \Omega$, $R_2 = 1\ k\Omega, g_m = 0.04\ S, R_3 = 50\ k\Omega,$ and $R_4 = R_5 = 10\ k\Omega$. Hence, the gain of the amplifier under these conditions is

$$\frac{v_o(t)}{v_i(t)} = \frac{-(0.04)(4.545)(10^3)(1)(10^3)}{(1.1)(10^3)}$$

$$= -165.29$$

Thus, the magnitude of the gain is 165.29.   ❏

At this point it is perhaps helpful to point out again that when analyzing circuits with dependent sources, we first treat the dependent source as though it were an independent source when we write a Kirchhoff's current or voltage law equation. Once the equation is written, we then write the controlling equation that specifies the relationship of the dependent source to the unknown variable. For instance, the first equation in Example 2.25 treats the dependent source like an independent source. The second equation in the example specifies the relationship of the dependent source to the voltage, which is the unknown in the first equation.

EXTENSION EXERCISES

E2.19   Find V_o in the circuit in Fig. E2.19.

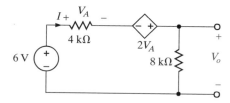

Figure E2.19

ANSWER: $V_o = 12\ V.$

E2.20   Find V_o in the network in Fig. E2.20.

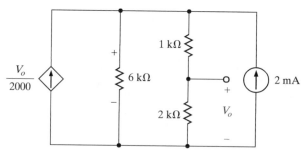

Figure E2.20

ANSWER: $V_o = 8\ V.$

2.8 DC MEASUREMENTS

We now describe the instruments that are employed to physically measure current, voltage, resistance, and power. However, before proceeding with a discussion of these devices, it is instructive to describe the meter movement used in many instruments for electrical measurements.

The D'Arsonval Meter Movement

The D'Arsonval meter movement, which is a fundamental component in many electric instruments, is shown in Fig. 2.36. This is a permanent-magnet-moving-coil (PMMC) mechanism in which a coil of fine wire is allowed to rotate between the poles of a permanent magnet. If a current is passed through the moving coil, the resulting magnetic field reacts with the magnetic field of the permanent magnet, producing a torque, which is counterbalanced by a restoring spring. The deflection of the pointer attached to the coil is proportional to the coil current, produced by the quantity being measured. A meter rated at 100 mV, 30 mA is an example of a typical standard meter.

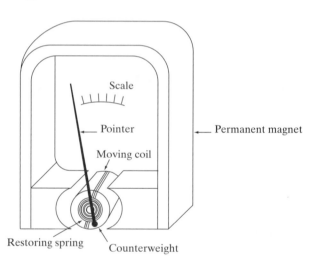

Figure 2.36 D'Arsonval meter movement.

Current Measurement. The ammeter is a device employed to measure current, and a variety of meter movements is commercially available. When connected in a network, the ammeter is inserted directly into the path of the current to be measured. DC ammeters should be connected such that the current enters the positive terminal for upscale deflection.

In the ideal case the internal resistance of an ammeter would be zero. However, practical ammeters do have some small internal resistance, R_m, which can produce a loading effect in low-voltage, low-resistance circuits; that is, the current measured by the meter is changed by inserting the meter. Therefore, it is convenient to represent the ammeter as shown in Fig. 2.37, in which all the resistance of the meter is lumped into R_m.

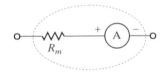

Figure 2.37 Ammeter circuit representation.

EXAMPLE 2.28

Let us determine the internal resistance of a standard ammeter that has a full-scale rating of 30 mA at 100 mV.

SOLUTION   The internal resistance for this meter is

$$R_m = \frac{100 \times 10^{-3}}{30 \times 10^{-3}}$$

$$= 3.33 \ \Omega \qquad \square$$

Suppose now that we wish to use the meter mechanism in Example 2.28 to measure a current, as shown in Fig. 2.38a, which is larger than the meter's current rating. We can do this by shunting a large part of the current to be measured around the meter movement, as shown in Fig. 2.38b. Note that the parallel combination of R_m and R_{SH} form a current divider. By selecting the proper value of the shunt resistor, we can use the meter movement to measure a current larger than the rating of the meter mechanism. For example, the configuration shown in Fig. 2.38c is typical of a multirange ammeter. The following example illustrates the selection of a shunt resistor to increase the current range.

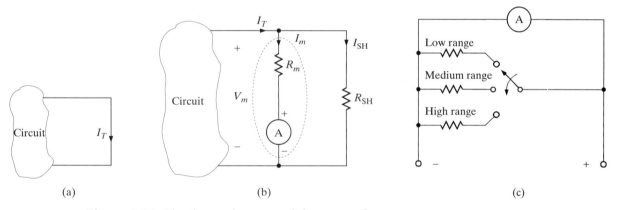

(a)   (b)   (c)

Figure 2.38   Circuits used to expand the range of a meter.

EXAMPLE 2.29

We wish to employ a 5-V, 500-mA dc adjustable current power supply to provide the power for the ringing circuit for a doorbell, as shown in Fig. 2.39a. The coil for the ringing circuit has a current limit of 250 mA. The only meter movement available to us is the 30-mA, 100-mV meter described previously. Therefore, we must select a shunt resistance, which when placed in parallel with the meter will permit us to measure 250 mA full scale.

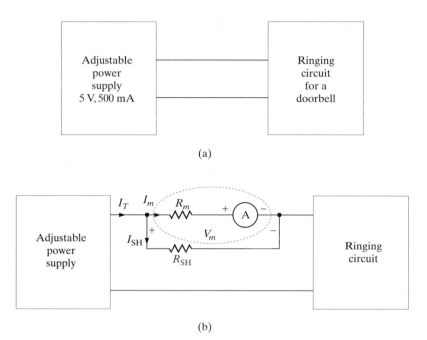

(a)

(b)

Figure 2.39 Circuits used in Example 2.29.

SOLUTION With reference to Fig. 2.39b, the total current $I_T = 250$ mA. The current in the meter movement $I_m = 30$ mA. Therefore, using KCL

$$I_{SH} = I_T - I_m$$
$$= 220 \text{ mA}$$

Since the voltage $V_m = 100$ mV is across both the meter resistance and the shunt resistance, the required value of R_{SH} is

$$R_{SH} = \frac{V_m}{I_{SH}}$$

$$= \frac{0.1}{0.22}$$

$$= 0.4545 \ \Omega$$

Note that if the meter is removed the power supply must be dialed back by 0.1 V in order to limit the current to 250 mA. This illustrates the point made earlier about the loading effect in low voltage circuits. ❏

EXTENSION EXERCISES

E2.21   A dc ammeter has a full-scale rating of 200 mV, 20 mA. Find the meter resistance.

ANSWER: $R_m = 10\ \Omega$

E2.22   Determine the amount of power that the meter in drill problem E2.21 dissipates at full-scale deflection.

ANSWER: $P = 4$ mW.

Voltage Measurement

A voltmeter is the basic device used to measure voltage and is connected in parallel with the element whose voltage is desired. In order to obtain an upscale meter deflection, the + terminal of the voltmeter should be connected to the point of higher potential.

Once again the resistance of the meter movement is R_m and is represented as shown in Fig. 2.37. A resistance R_{se}, called a multiplier, is connected in series with the meter movement. This resistance R_{se} and the meter movement form a voltage divider and permit the meter to be used in a manner that extends the voltage range of the meter. The following example illustrates this technique.

EXAMPLE 2.30

The remote control for our television set does not work, and we are suspicious that the 9-V battery in the remote is bad. We wish to measure this voltage, but all we have available is a 100-mV, 30-mA D'Arsonval meter movement. Let us determine the value of the resistor, R_{se}, which must be placed in series with the meter movement to measure this voltage.

SOLUTION   The circuit for this application is shown in Fig. 2.40 and should produce full-scale deflection of 9 V using a 100-mV, 30-mA meter movement. For full-scale deflection, $V_m = 100$ mV and, hence, using KVL,

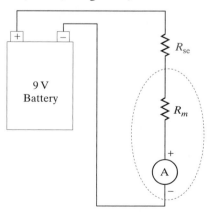

Figure 2.40 Circuit used in Example 2.30.

$$9 = I_m R_{se} + V_m$$

and since $V_m = 100$ mV and $I_m = 30$ mA,

$$R_{se} = \frac{9 - 0.100}{0.03}$$

$$= 296.7 \ \Omega$$

Recall from Example 2.28 that the meter resistance is 3.33 Ω. Therefore, if the battery voltage has dropped to 6 V, the current read by the meter would be

$$I_m = \frac{6}{296.7 + 3.33}$$

$$= 20 \ \text{mA}$$

which is 6/9 or 2/3 of full scale.   ❏

EXTENSION EXERCISE

E2.23   A meter has a resistance of 50 Ω and a full-scale current of 1 mA. Select the multiplier resistance needed to measure 12 V full scale.
ANSWER: $R_{se} = 11.95$ kΩ.

Finally, the voltmeter, like the ammeter, has a loading effect on the circuit. The magnitude of the effect is dependent upon the resistance of the voltmeter and the resistance of the circuit elements the voltmeter is connected across. When a voltmeter is connected across an element, some current will flow in the meter. This current will be negligible if the resistance of the meter is much higher than that of the elements across which it is connected.

Voltmeter sensitivity is a quantitative measure of the loading effect. It is defined as the reciprocal of the current required for full-scale deflection; that is,

$$\text{Sensitivity} = \frac{1}{I_m} \ \Omega/\text{V}$$

Therefore, a voltmeter with a 50-V range, which has a 100-μA meter movement, has a sensitivity of

$$\text{Sensitivity} = \frac{1}{100 \times 10^{-6}}$$

$$= 10,000 \ \Omega/\text{V}$$

and, hence, the meter resistance is

$$R_m = (\text{sensitivity})(\text{voltage rating})$$

$$= (10,000)(50)$$

$$= 0.5 \ \text{M}\Omega$$

The sensitivity of a meter in ohms/volt is normally printed on the meter scale, which aids us in selecting a meter that will not load a circuit and cause measurement errors.

Resistance Measurement

An ohmmeter is an instrument used to measure resistance. An ohmmeter circuit, shown in Fig. 2.41a, typically consists of a meter movement in series with a battery and regulating resistor. The meter scale typically reads in ohms from right to left, as shown in Fig. 2.41b. If $R_x = \infty$ (that is, the ohmmeter leads or terminals are open), there is no meter deflection. If, however, $R_x = 0$ (that is, the meter leads are shorted), maximum current flows in the circuit and full-scale deflection occurs. The regulating resistor R_{reg} is a calibration device, which allows us to zero-adjust the meter movement when the leads are shorted and, thus, compensate for such things as meter resistance.

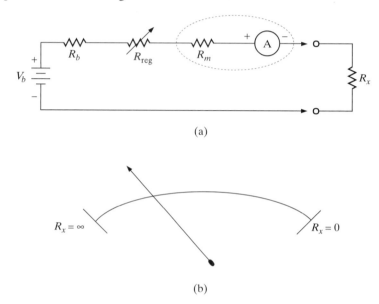

(a)

(b)

Figure 2.41 Ohmmeter circuit and scale.

The ohmmeter differs from the ammeter and voltmeter in one important respect: It must never be connected to an energized circuit. That is, the element to be measured must be disconnected from the original circuit when the measurement is made.

Although this type of ohmmeter is very useful, it is not a precision instrument. A much more accurate device for measuring resistance values over a wide range is the *Wheatstone bridge*. The Wheatstone bridge circuit is shown in Fig. 2.42. In this network, the resistors R_1, R_2, and R_3 are known. R_x is the unknown resistor. The device in the center leg of the bridge is a sensitive D'Arsonval mechanism in the microamp range called a galvanometer.

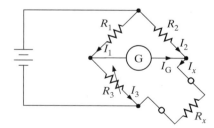

Figure 2.42 The Wheatstone bridge circuit.

The bridge is used in the following manner. The unknown resistance R_x is connected as shown in Fig. 2.42, and then R_3 is adjusted until there is no current in the galvanometer. At this point the bridge is said to be balanced. Under this balanced condition, $I_G = 0$ and, hence, KCL applied to the center nodes of the bridge yields

$$I_1 = I_3$$

and

$$I_2 = I_x$$

Furthermore, since $I_G = 0$, there is no voltage drop across the galvanometer and, therefore, KVL requires that

$$I_1 R_1 = I_2 R_2$$

and

$$I_3 R_3 = I_x R_x$$

Dividing one equation by the other and using the fact that $I_1 = I_3$ and $I_2 = I_x$ yields the relationship

$$\frac{R_1}{R_3} = \frac{R_2}{R_x}$$

so that

$$R_x = \left(\frac{R_2}{R_1} \right) R_3$$

It is interesting to note that R_x is independent of the voltage source and dependent only upon the ratio of R_2 to R_1 and not their specific values. In commercial units the ratio is made equal to a power of 10 by switching into the bridge circuit resistors with values of 1, 10, 100, and so on.

In addition to using the Wheatstone bridge to measure an unknown resistance, it is employed by engineers to measure strain in a solid material, as illustrated in the following example.

EXAMPLE 2.31

A system used to determine the weight of a truck is shown in Fig. 2.43a. The platform is supported by cylinders upon which are mounted strain gauges. The strain gauges, which measure strain when the cylinder deflects under load, are connected to a Wheatstone bridge as shown in Fig. 2.43b. The strain gauge has a resistance of 120 Ω under no-load conditions and changes value under load. The variable resistor in the bridge is a calibrated precision device.

Weight is determined in the following manner. The ΔR_3 required to balance the bridge represents the Δ strain, which when multiplied by the modulus of elasticity yields the Δ stress. The Δ stress multiplied by the cross-sectional area of the cylinder produces the Δ load, which is used to determine weight.

Let us determine the value of R_3 under no load when the bridge is balanced and its value when the resistance of the strain gauge changes to 120.24 Ω under load.

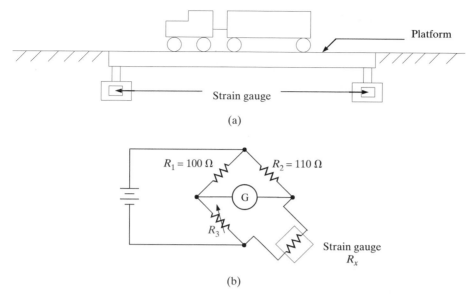

Figure 2.43 Diagrams used in Example 2.31.

SOLUTION   Using the balance equation for the bridge, the value of R_3 at no load is

$$R_3 = \left(\frac{R_1}{R_2}\right)R_x$$

$$= \left(\frac{100}{110}\right)(120)$$

$$= 109.0909 \ \Omega$$

Under load, the value of R_3 is

$$R_3 = \left(\frac{100}{110}\right)(120.24)$$

$$= 109.3091 \ \Omega$$

Therefore, the ΔR_3 is

$$\Delta R_3 = 109.3091 - 109.0909$$

$$= 0.2182 \ \Omega$$ ❑

EXTENSION EXERCISE

E2.24   Given the bridge circuit in Fig. 2.42, find the value of R_x if the bridge is balanced at $R_1 = 200 \ \Omega$, $R_2 = 240 \ \Omega$, and $R_3 = 220 \ \Omega$.

ANSWER: $R_x = 264 \ \Omega$.

Power Measurements

In a dc circuit, power is the product of voltage and current. Therefore, power can be derived from current and voltage measurements. DC power can be measured directly using a wattmeter. However, since very accurate measurements can be obtained using a voltmeter in conjunction with an ammeter, the wattmeter is normally not used in dc measurements. The wattmeter is used extensively to measure power in ac circuits, and we will discuss this device in some detail in ac circuit analysis later in this text.

There are many instruments on the market that are capable of measuring voltage, current, and resistance. Two such devices are the digital multimeters shown in Fig. 2.44a. The source meter shown in Fig. 1.14 is also capable of these measurements. Instruments are also available for measurements in a specific range. For example, the high-precision instrument shown in Fig. 2.44b makes micro-ohm measurements. Thus, the modern instruments available today provide us with enormous capability in our analysis of electric circuits.

2.9 APPLICATIONS

Throughout this book we endeavor to present a wide variety of examples that demonstrate the usefulness of the material under discussion in a practical environment. In order to enhance our presentation of the practical aspects of circuit analysis and design, we have dedicated sections, such as this one, in most chapters for the specific purpose of presenting additional application-oriented examples.

EXAMPLE 2.32

The eyes (heating elements) of an electric range are frequently made of resistive nichrome strips. Operation of the eye is quite simple. A current is passed through the heating element causing it to dissipate power in the form of heat. Also, a four-position selector switch, shown in Fig. 2.45, controls the power (heat) output. In this case the eye consists of two nichrome strips modeled by the resistors R_1 and R_2, where $R_1 < R_2$.

1. How should positions A, B, C, and D be labeled with regard to high, medium, low, and off settings?

2. If we desire that high and medium correspond to 2000 W and 1200 W power dissipation, respectively, what are the values of R_1 and R_2?

3. What is the power dissipation at the low setting?

SOLUTION   Position A is the off setting since no current flows to the heater elements. In position B, current flows through R_2 only, while in position C current flows through R_1 only. Since $R_1 < R_2$, more power will be dissipated when the switch is at position C. Thus, position C is the medium setting, B is the low setting, and, by elimination, position D is the high setting.

When the switch is at the medium setting, only R_1 dissipates power, and we can write R_1 as

$$R_1 = \frac{V_s^2}{P_1} = \frac{230^2}{1200}$$

(a)

(b)

Figure 2.44 (a) Two modern digital multimeters. (Courtesy of Fluke Corporation and Keithley Instruments, Inc.) (b) A micro-ohm meter. (Courtesy of Keithley Instruments, Inc.)

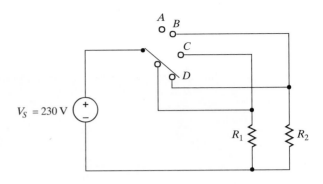

Figure 2.45 Simple resistive heater selector circuit.

or

$$R_1 = 44.08 \ \Omega$$

On the high setting, 2000 W of total power is delivered to R_1 and R_2. Since R_1 dissipates 1200 W, R_2 must dissipate the remaining 800 W. Therefore, R_2 is

$$R_2 = \frac{V_S^2}{P_2} = \frac{230^2}{800}$$

or

$$R_2 = 66.13 \ \Omega$$

Finally, at the low setting, only R_2 is connected to the voltage source; thus, the power dissipation at this setting is 800 W. ❏

EXAMPLE 2.33

An equivalent circuit for a transistor amplifier used in a portable tape player is shown in Fig. 2.46. The ideal independent source, V_S, and R_S represent the magnetic head playback circuitry. The dependent source, R_o and R_{in}, model the transistor. Finally, the resistor R_L models the load, which in this case is another transistor circuit. Let us find the voltage gain of the network.

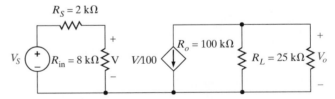

Figure 2.46 Transistor amplifier circuit model.

SOLUTION The gain of the transistor amplifier can be derived as follows. The output voltage can be expressed as

$$V_o = -\frac{V}{100}(R_o//R_L)$$

where $(R_o//R_L)$ represents R_o in parallel with R_L.

From the input voltage divider, we can express V as a function of V_S.

$$V = V_S\left(\frac{R_{in}}{R_{in} + R_S}\right)$$

Therefore,

$$V_o = -\frac{V_S}{100}\left(\frac{R_{in}}{R_{in} + R_S}\right)(R_o//R_L)$$

Given the component values in Fig. 2.46, the voltage gain is

$$A_V = \frac{V_o}{V_S} = -160 \qquad \square$$

2.10 CIRCUIT DESIGN

The vast majority of this text is concerned with circuit analysis; that is, given a circuit in which all the components are specified, analysis involves finding such things as the voltage across some element or the current through another. Furthermore, the solution of an analysis problem is generally unique. In contrast, design involves determining the circuit configuration that will meet certain specifications. In addition, the solution is generally not unique in that there may be many ways to satisfy the circuit/performance specifications. It is also possible that there is no solution that will meet the design criteria.

In addition to meeting certain technical specifications, designs normally must also meet other criteria, such as economic, environmental, and safety constraints. For example, if a circuit design that meets the technical specifications is either too expensive or unsafe, it is not viable regardless of its technical merit.

At this point, the number of elements that we can employ in circuit design is limited primarily to the linear resistor and the active elements we have presented. However, as we progress through the text we will introduce a number of other elements (for example, the op-amp, capacitor, and inductor), which will significantly enhance our design capability.

We begin our discussion of circuit design by considering a couple of simple examples that demonstrate the selection of specific components to meet certain circuit specifications.

EXAMPLE 2.34

An electronics hobbyist who has built his own stereo amplifier wants to add a back-lit display panel to his creation for that professional look. His panel design requires seven light bulbs—two operate at 12 V/15 mA and five at 9 V/5 mA. Luckily, his stereo design already has a quality 12-V dc supply; however, there is no 9-V supply. Rather than building a new dc power supply, let us use the inexpensive circuit shown in Fig. 2.47a to design a 12-V to 9-V converter with the restriction that the

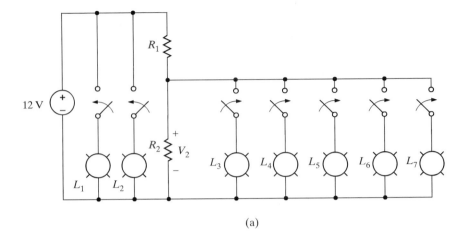

(a)

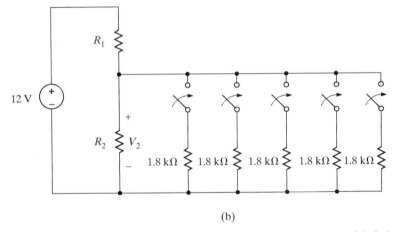

(b)

Figure 2.47 12-V to 9-V converter circuit for powering panel lighting.

variation in V_2 be no more than ±5%. In particular, we must determine the necessary values of R_1 and R_2.

SOLUTION First, lamps L_1 and L_2 have no effect on V_2. Second, when lamps L_3–L_7 are on, they each have an equivalent resistance of

$$R_{eq} = \frac{V_2}{I} = \frac{9}{0.005} = 1.8 \text{ k}\Omega$$

As long as V_2 remains fairly constant, the lamp resistance will also be fairly constant. Thus, the requisite model circuit for our design is shown in Fig. 2.47b. The voltage V_2 will be at its maximum value of $9 + 5\% = 9.45$ V when L_3–L_7 are all off. In this case R_1 and R_2 are in series and V_2 can be expressed by simple voltage division as

$$V_2 = 9.45 = 12\left[\frac{R_2}{R_1 + R_2}\right]$$

Rearranging the equation yields

$$\frac{R_1}{R_2} = 0.27$$

A second expression involving R_1 and R_2 can be developed by considering the case when L_3–L_7 are all on, which causes V_2 to reach its minimum value of $9 - 5\%$, or 8.55 V. Now, the effective resistance of the lamps is five 1.8-kΩ resistors in parallel, or 360 Ω. The corresponding expression for V_2 is

$$V_2 = 8.55 = 12 \left[\frac{R_2 // 360}{R_1 + (R_2 // 360)} \right]$$

which can be rewritten in the form

$$\frac{\dfrac{360 R_1}{R_2} + 360 + R_1}{360} = \frac{12}{8.55} = 1.4$$

Substituting the value determined for R_1/R_2 into the preceding equation yields

$$R_1 = 360[1.4 - 1 - 0.27]$$

or

$$R_1 = 48.1\ \Omega$$

and so for R_2

$$R_2 = 178.3\ \Omega \qquad \qquad \square$$

EXAMPLE 2.35

A circuit board within a stereo amplifier requires the use of three different voltages at the circuit nodes. The voltages needed are 4 V, 16 V, and 24 V. Criteria such as weight, cost, and size dictate that these voltages be supplied with a single voltage source. Furthermore, heat constraints dictate that the supply must dissipate less than 5 W. Assuming no loading effect from the balance of the network, design a voltage string (multiple-resistor voltage divider) that will satisfy the requirements.

SOLUTION   A 24-V source can directly provide one of the voltages, and the voltage divider in Fig. 2.48 can be used to generate the other two. The power requirement limits our choices to current less than 208.33 mA. The exact value we choose is somewhat arbitrary and, therefore, we select $I = 200$ mA. The total resistance is then

$$R_1 + R_2 + R_3 = \frac{24}{0.2} = 120\ \Omega$$

Using voltage division, we find that

$$\frac{R_3}{R_1 + R_2 + R_3} = \frac{R_3}{120} = \frac{4}{24}$$

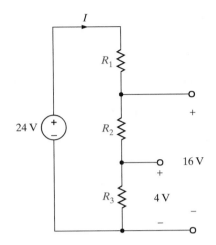

Figure 2.48 Voltage string for generating intermediate voltages.

or

$$R_3 = 20 \ \Omega$$

Similarly,

$$\frac{R_2}{120} = \frac{16 - 4}{24}$$

or

$$R_2 = 60 \ \Omega$$

Finally, since the total resistance is 120 Ω, $R_1 = 40 \ \Omega$. ❏

EXAMPLE 2.36

The network in Fig. 2.49 is an equivalent circuit for a transistor amplifier used in a stereo preamplifier. The input circuitry, consisting of a 2-mV source in series with a 500-Ω resistor, models the output of a compact disk player. The dependent source, R_{in}, and R_o model the transistor, which amplifies the signal and then sends it to the power amplifier. The 10-kΩ load resistor models the input to the power amplifier that actually drives the speakers. We must design a transistor amplifier as shown in Fig. 2.49 that will provide an overall gain of −200. In practice we do not actually vary the device parameters to achieve the desired gain; rather we select a transistor from the manufacturer's data books that will satisfy the required specification. The model parameters for three different transistors are listed as follows:

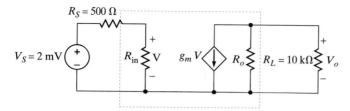

Figure 2.49 Transistor amplifier circuit model.

Manufacturer's transistor parameter values

Part Number	R_{in} (kΩ)	R_o (kΩ)	g_m (mA/V)
1	1.0	50	50
2	2.0	75	30
3	8.0	80	20

Design the amplifier by choosing the transistor that produces the most accurate gain. What is the percent error of your choice?

SOLUTION   The output voltage can be written

$$V_o = -g_m V (R_o // R_L)$$

Using voltage division at the input to find V,

$$V = V_S \left(\frac{R_{in}}{R_{in} + R_S} \right)$$

Combining these two expressions, we can solve for the gain:

$$A_V = \frac{V_o}{V_S} = -g_m \left(\frac{R_{in}}{R_{in} + R_S} \right) (R_o // R_L)$$

Using the parameter values for the three transistors, we find that

Part Number	Gain
1	−277.8
2	−211.8
3	−167.3

Obviously, the best alternative is transistor number 2, which has a gain error of

$$\text{Percent error} = \left(\frac{211.8 - 200}{200} \right) \times 100\% = 5.9\%$$

2.11 SUMMARY

Ohm's law $V = IR$

The passive sign convention with Ohm's law   The current enters the resistor terminal with the positive voltage reference.

Kirchhoff's current law (KCL)   The algebraic sum of the currents leaving (entering) a node is zero.

Kirchhoff's voltage law (KVL)   The algebraic sum of the voltages around any closed path is zero.

Solving a single-loop circuit   Determine the loop current by applying KVL and Ohm's law.

Solving a single-node-pair circuit   Determine the voltage between the pair of nodes by applying KCL and Ohm's law.

The voltage division rule   The voltage is divided between two series resistors in direct proportion to their resistance.

The current division rule   The current is divided between two parallel resistors in reverse proportion to their resistance.

The equivalent resistance of a network of resistors   Combine resistors in series by adding their resistances. Combine resistors in parallel by adding their conductances.

Short circuit   Zero resistance, zero voltage; the current in the short is determined by the rest of the circuit.

Open circuit   Zero conductance, zero current; the voltage across the open terminals is determined by the rest of the circuit.

PROBLEMS

Section 2.1

2.1   Determine the current and power dissipated in the resistor in Fig. P2.1.

Ohm's law is applicable.

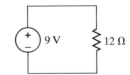

Figure P2.1

2.2   Determine the current and power dissipated in the resistors in Fig. P2.2.

Once the total resistance is known the problem is similar to P2.1.

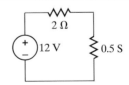

Figure P2.2

2.3 Determine the voltage across the resistor in Fig. P2.3 and the power dissipated.

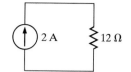

Figure P2.3

Similar to P2.1.

2.4 Given the circuit in Fig. P2.4, find the voltage across each resistor and the power dissipated in each.

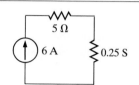

Figure P2.4

Similar to P2.2.

2.5 A model for a standard two D-cell flashlight is shown in Fig. P2.5. Find the power dissipated in the lamp.

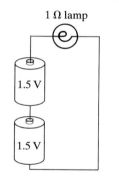

Figure P2.5

Find the total voltage across the lamp and use this voltage and the resistance to determine the power.

2.6 Determine the power dissipated in the resistor in Fig. P2.6.

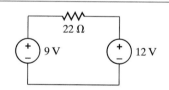

Figure P2.6

Similar to P2.5.

2.7 An automobile uses two halogen headlights connected as shown in Fig. P2.7. Determine the power supplied by the battery if each headlight draws 3 A of current.

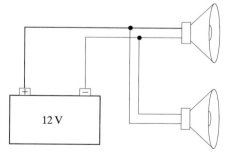

Figure P2.7

The current and voltage will yield the power.

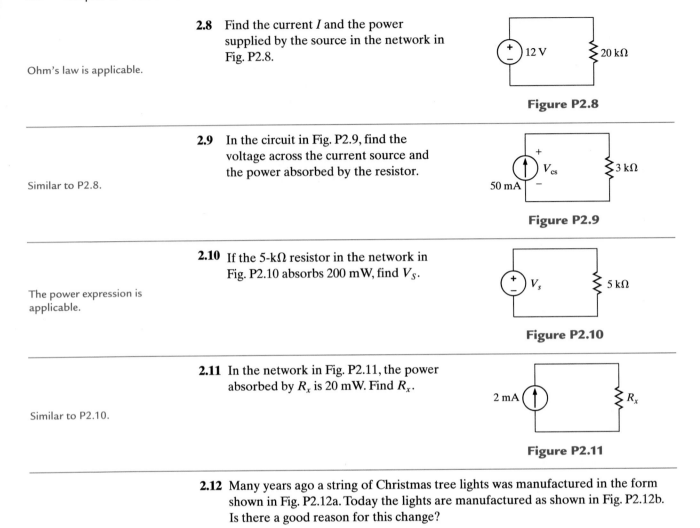

2.8 Find the current I and the power supplied by the source in the network in Fig. P2.8.

Ohm's law is applicable.

Figure P2.8

2.9 In the circuit in Fig. P2.9, find the voltage across the current source and the power absorbed by the resistor.

Similar to P2.8.

Figure P2.9

2.10 If the 5-kΩ resistor in the network in Fig. P2.10 absorbs 200 mW, find V_S.

The power expression is applicable.

Figure P2.10

2.11 In the network in Fig. P2.11, the power absorbed by R_x is 20 mW. Find R_x.

Similar to P2.10.

Figure P2.11

2.12 Many years ago a string of Christmas tree lights was manufactured in the form shown in Fig. P2.12a. Today the lights are manufactured as shown in Fig. P2.12b. Is there a good reason for this change?

What happens if a bulb malfunctions?

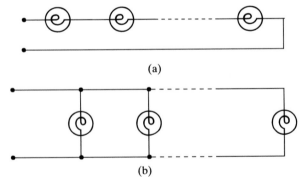

(a)

(b)

Figure P2.12

Section 2.2

2.13 Find I_1 in the network in Fig. P2.13.

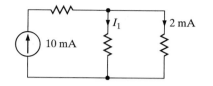

Figure P2.13

KCL is applicable.

2.14 Find I_1 in the network in Fig. P2.14.

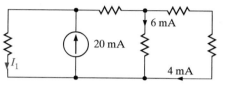

Figure P2.14

Similar to P2.13.

2.15 Find I_1 in the network in Fig. P2.15.

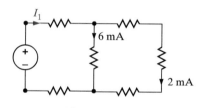

Figure P2.15

Similar to P2.13.

2.16 Find I_1 and I_2 in the circuit in Fig. P2.16.

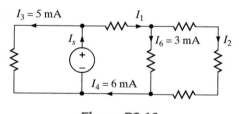

Figure P2.16

KCL is applicable.

2.17 Find I_1 and I_2 in the circuit in Fig. P2.17.

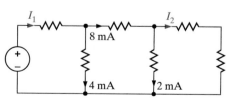

Figure P2.17

Similar to P2.16.

2.18 Find I_1 in the circuit in Fig. P2.18.

Similar to P2.16.

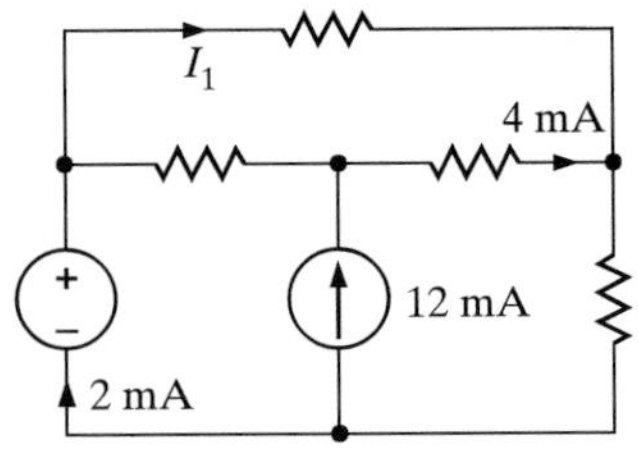

Figure P2.18

2.19 Find I_o in the circuit in Fig. P2.19.

Similar to P2.16.

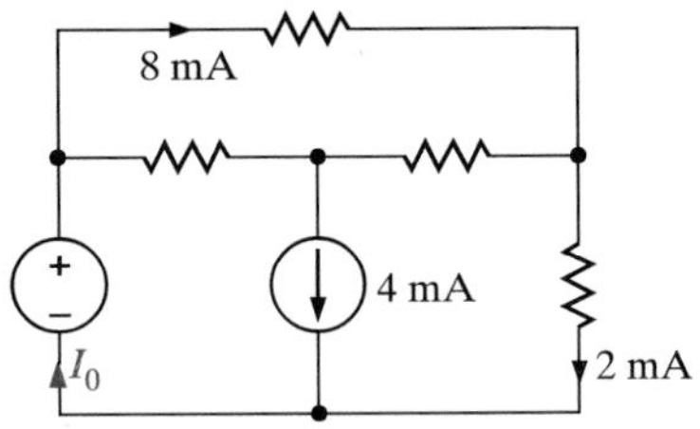

Figure P2.19

2.20 Find I_o and I_1 in the circuit in Fig. P2.20.

Similar to P2.16.

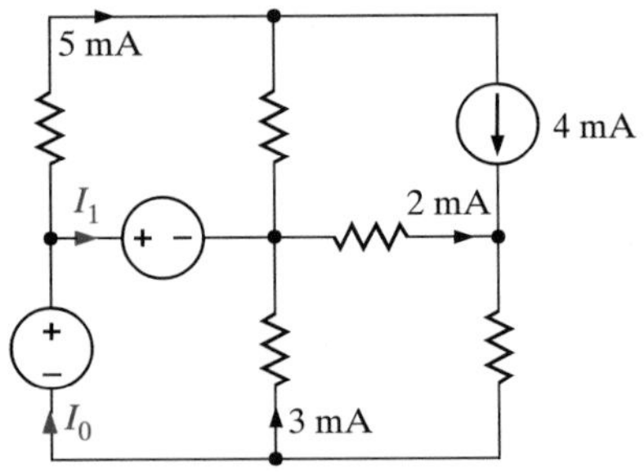

Figure P2.20

2.21 Find I_x in the circuit in Fig. P2.21.

KCL yields a solution.

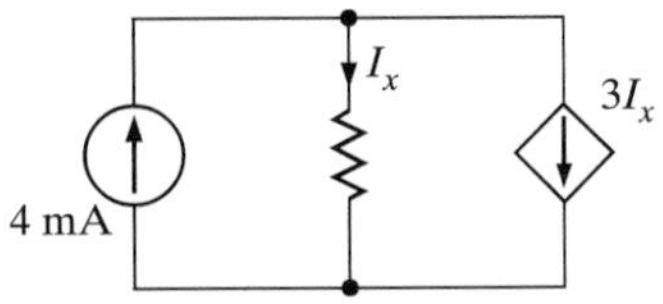

Figure P2.21

2.22 Find I_x in the circuit in Fig. P2.22.

Similar to P2.21.

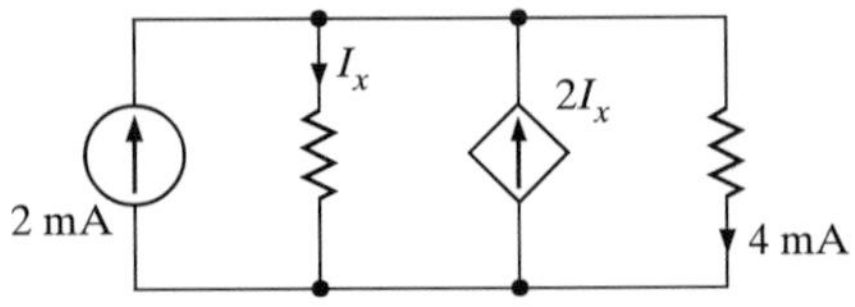

Figure P2.22

2.23 Find I_x in the circuit in Fig. P2.23.

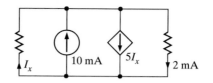

Figure P2.23

Similar to P2.21.

2.24 Find I_x in the circuit in Fig. P2.24.

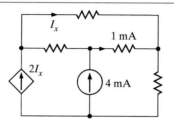

Figure P2.24

Similar to P2.21.

2.25 Find V_{bd} in the circuit in Fig. P2.25.

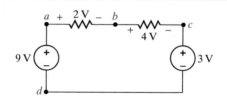

Figure P2.25

KVL is applicable in this case.

2.26 Find V_{ae} and V_{cf} in the circuit in Fig. P2.26.

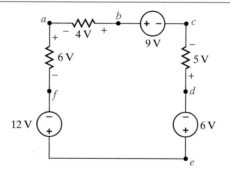

Figure P2.26

Similar to P2.25.

2.27 Find V_x in the circuit in Fig. P2.27.

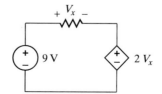

Figure P2.27

Similar to P2.25.

2.28 Find V_{ac} in the circuit in Fig. P2.28.

Similar to P2.25.

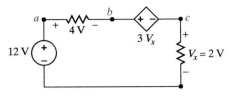

Figure P2.28

2.29 Find V_{ad} and V_{ce} in the circuit in Fig. P2.29.

Similar to P2.25.

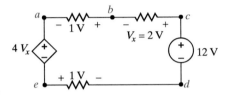

Figure P2.29

Section 2.3

2.30 Find V_{ab} in the network in Fig. P2.30.

The voltage division rule yields an immediate solution.

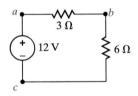

Figure P2.30

2.31 Find V_{ac} in the network in Fig. P2.31.

Similar to P2.30.

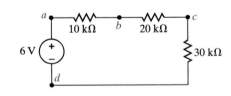

Figure P2.31

2.32 Find V_{ab} in the network in Fig. P2.32.

Combine sources and use the voltage division rule.

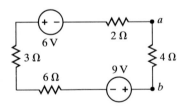

Figure P2.32

2.33 Find both I and V_{bd} in the circuit in Fig. P2.33.

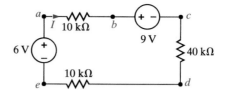

Figure P2.33

Similar to P2.32.

2.34 Find the power supplied by the 12-V source and the power absorbed by the 40-kΩ resistor in the network in Fig. P2.34

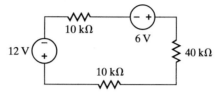

Figure P2.34

Similar to P2.32.

2.35 Find V_x in the network in Fig. P2.35.

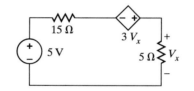

Figure P2.35

Use KVL to find the current in the loop that will, in turn, yield V_x.

2.36 Find V_1 in the network in Fig. P2.36.

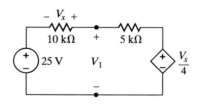

Figure P2.36

Similar to P2.25.

2.37 Find the power absorbed by the 30-kΩ resistor in the circuit in Fig. P2.37.

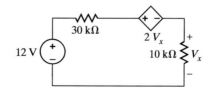

Figure P2.37

Similar to P2.35.

Section 2.4

2.38 Find I_o in the circuit in Fig. P2.38.

The current-division rule is immediately applicable.

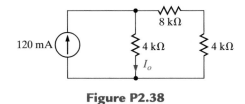

Figure P2.38

2.39 Determined I_L in the circuit in Fig. P2.39.

Combine sources and use the current-division rule.

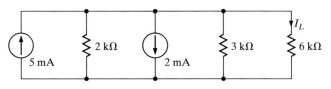

Figure P2.39

2.40 Determined I_L in the circuit in Fig. P2.40.

Assume the voltage across the parallel elements is V_L and use KCL.

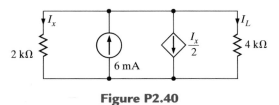

Figure P2.40

2.41 Determine I_L in the circuit in Fig. P2.41.

Similar to P2.40.

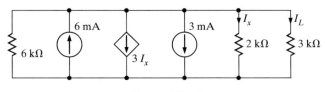

Figure P2.41

Section 2.5

2.42 If the conductance of each resistor in the network in Fig. P2.42 is 250 μS, find the resistance at the terminals A-B.

Use the problem-solving strategy for resistor combinations.

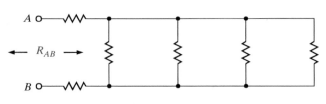

Figure P2.42

2.43 Find R_{AB} in the circuit in Fig. P2.43.

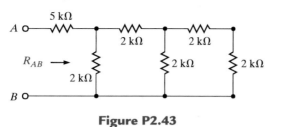

Figure P2.43

Similar to P2.42.

2.44 Find R_{AB} in the circuit in Fig. P2.44.

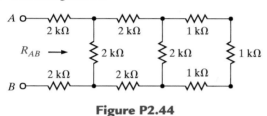

Figure P2.44

Similar to P2.42.

2.45 Find R_{AB} in the network in Fig. P2.45.

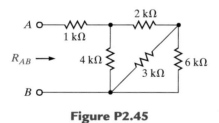

Figure P2.45

Similar to P2.42.

2.46 Determine R_{AB} in the circuit in Fig. P2.46.

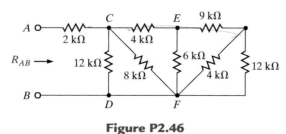

Figure P2.46

Similar to P2.42.

2.47 Find R_{AB} in the network in Fig. P2.47.

Similar to P2.42.

Figure P2.47

Section 2.6

2.48 Find I_1 and V_o in the circuit in Fig. P2.48.

Use the problem-solving strategy for circuits with series-parallel combinations of resistors.

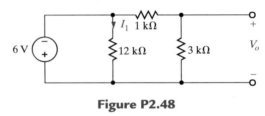

Figure P2.48

2.49 Determine V_o in the network in Fig. P2.49.

Similar to P2.48.

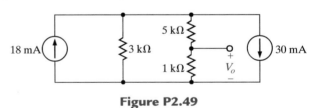

Figure P2.49

2.50 Find I_1 and V_o in the circuit in Fig. P2.50.

Similar to P2.48.

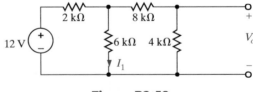

Figure P2.50

2.51 Find I_1 in the circuit in the Fig. P2.51.

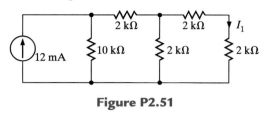

Figure P2.51

Similar to P2.48.

2.52 Determine I_o in the circuit in Fig. P2.52.

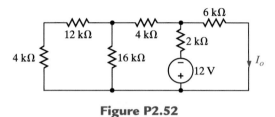

Figure P2.52

Similar to P2.48.

2.53 Determine V_o in the network in Fig. P2.53.

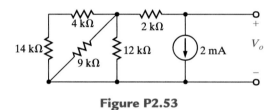

Figure P2.53

Similar to P2.48.

2.54 Determine V_o in the circuit in Fig. P2.54.

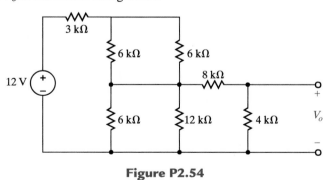

Figure P2.54

Similar to P2.48.

2.55 If $V_o = 3$ V in the circuit in Fig. P2.55, find V_S.

Reverse the voltage division rule or use Ohm's law and KVL to determine V_s.

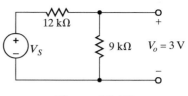

Figure P2.55

2.56 If $I_o = 5$ mA in the circuit in Fig. P2.56, find I_S.

Reverse the current-division rule or use Ohm's law and KCL to find I_S.

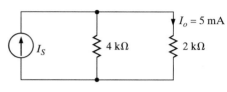

Figure P2.56

2.57 If $V_o = 2$ V in the circuit in Fig. P2.57, find V_S.

Similar to P2.55.

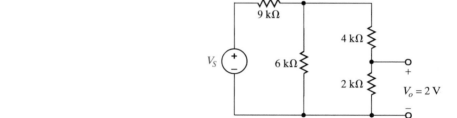

Figure P2.57

2.58 If the power absorbed by the 4-kΩ resistor in the circuit in Fig. P2.58 is 36 mW find I_o.

The current in the 4-kΩ resistor and Ohm's law yield a solution.

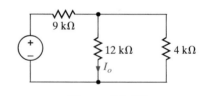

Figure P2.58

2.59 If $V_o = 6$ V in the circuit in Fig. P2.59, find I_S.

The current in the 3-kΩ resistor, Ohm's law, and KCL yield a solution.

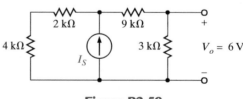

Figure P2.59

2.60 If $I_o = 2$ mA in the circuit in Fig. P2.60, find V_S.

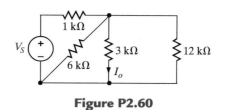

Figure P2.60

Similar to P2.59.

Section 2.7

2.61 Find V_o in the circuit in Fig. P2.61.

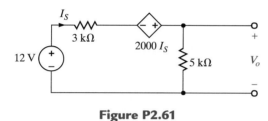

Figure P2.61

Use the problem-solving strategy for circuits with dependent sources.

2.62 Find V_o in the network in Fig. P2.62.

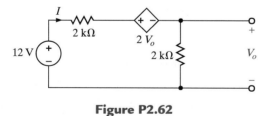

Figure P2.62

Similar to P2.61.

2.63 Find V_o in the network in Fig. P2.63.

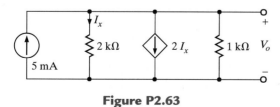

Figure P2.63

Similar to P2.61.

2.64 Find V_o in the circuit in Fig. P2.64.

Similar to P2.61.

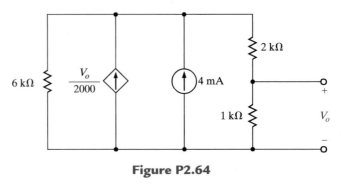

Figure P2.64

2.65 A single-stage transistor amplifier is modeled as shown in Fig. P2.65. Find the current in the load R_L.

Similar to P2.61.

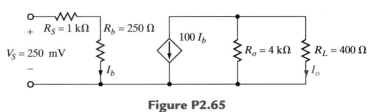

Figure P2.65

Section 2.8

Ohm's law yields the resistance.

2.66 Given a meter movement rated at 50 mV, 20 mA full-scale, find the meter resistance and power dissipated at full-scale deflection.

Similar to P2.66.

2.67 A 50-mV, 20-mA meter movement is used to measure currents between 0 and 150 mA. Find the shunt resistance required for the measurements.

Similar to P2.66.

2.68 A 30-mA, 100-mV meter movement is used in conjunction with a shunt resistance of 0.4 Ω. Find the magnitude of the current that can be measured by the meter at full scale.

Similar to P2.66.

2.69 A 100-mV, 30-mA meter movement is to be used to measure 100 mA full scale. Find the shunt resistance required for this application.

2.70 A 50-mV, 20-mA meter movement is to be used to check the voltage on 1.5-V flashlight batteries, as shown in Fig P2.70. Find the value of R_{se} required for this meter to read 1.5 V full scale.

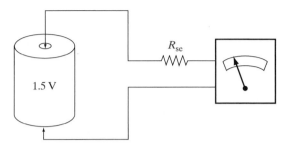

Similar to P2.66.

Figure P2.70

2.71 A resistance of 50 Ω is connected in series with a 50-mV, 20-mA meter movement. What full-scale voltage will this combination measure?

Similar to P2.66.

2.72 The ideal ammeter in Fig. P2.72 has no internal resistance and a maximum range of 10 mA. In conjunction with the series resistance R_{se}, it functions as a voltmeter to measure the voltage V_x. Choose R_{se} such that the ammeter reads its maximum value when V_x is 20 V. If the ammeter reads 2.7 mA, what is the voltage at V_x?

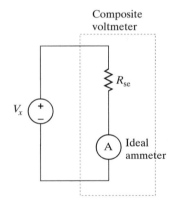

Similar to P2.66.

Figure P2.72

Advanced Problems

2.73 Find R_{AB} in the circuit in Fig. P2.73.

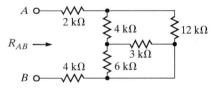

Figure P2.73

2.74 Find R_{AB} in the circuit in Fig. P2.74.

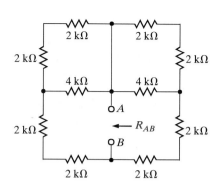

Figure P2.74

2.75 Find I_o in the circuit in Fig. P2.75.

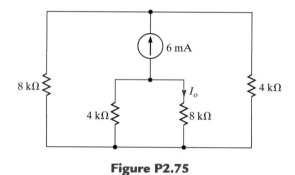

Figure P2.75

2.76 Find V_o In the circuit in Fig. P2.76.

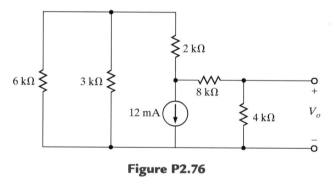

Figure P2.76

2.77 Find V_o in the circuit in Fig. P2.77.

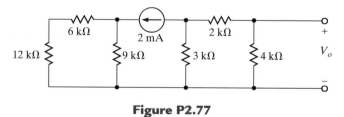

Figure P2.77

2.78 Find V_o in the circuit in Fig. P2.78.

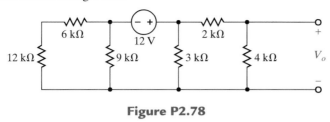

Figure P2.78

2.79 Find V_o in the circuit in Fig. P2.79.

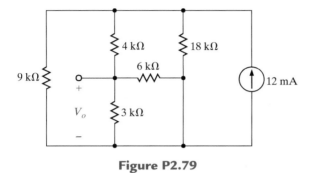

Figure P2.79

2.80 Find I_o in the circuit in Fig. P2.80.

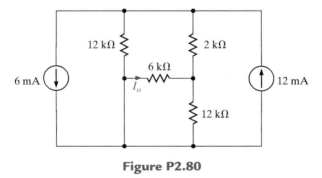

Figure P2.80

2.81 Find I_o in the circuit in Fig. P2.81.

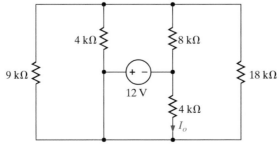

Figure P2.81

2.82 Find I_o in the circuit in Fig. P2.82.

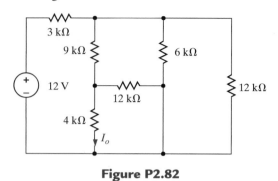

Figure P2.82

2.83 Find V_o in the circuit in Fig. P2.83.

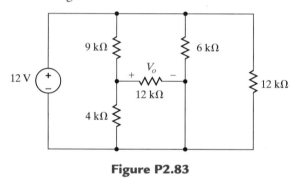

Figure P2.83

2.84 Find I_o in the circuit in Fig. P2.84.

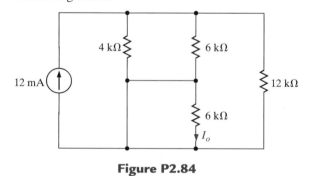

Figure P2.84

2.85 Determine the value of V_o in the circuit in Fig. P2.85

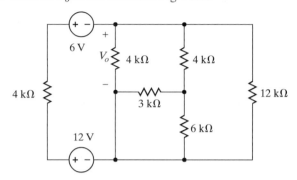

Figure P2.85

2.86 Find V_o in the circuit in Fig. P2.86.

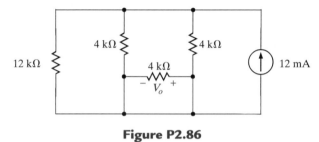

Figure P2.86

2.87 If $V_1 = 5$ V in the circuit in Fig. P2.87, find I_S.

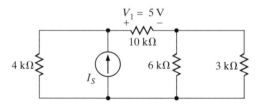

Figure P2.87

2.88 If $I_o = 2$ mA in the circuit in Fig. P2.88,
find V_S.

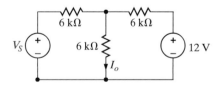

Figure P2.88

2.89 If $I_o = 4$ mA in the circuit in Fig. P2.89, find I_S.

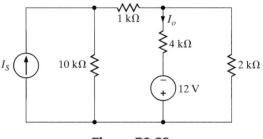

Figure P2.89

2.90 If $V_o = 6$ V in the circuit in Fig. P2.90, find V_S.

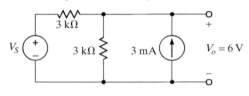

Figure P2.90

2.91 If $V_o = 2$ V in the network in Fig. P2.91, find V_S.

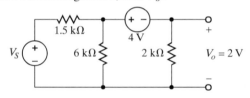

Figure P2.91

2.92 If $V_o = 6$ V in the circuit in Fig. P2.92, find I_S.

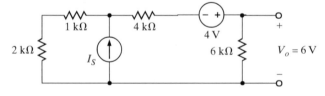

Figure P2.92

2.93 Given that $V_o = 4$ V in the network in Fig. P2.93, find V_S.

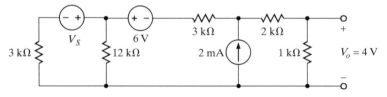

Figure P2.93

2.94 Find I_o in the circuit in Fig. P2.94.

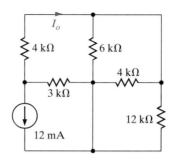

Figure P2.94

2.95 Find I_o in the circuit in Fig. P2.95.

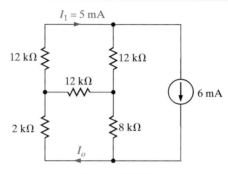

Figure P2.95

2.96 Find V_o in the circuit in Fig. P2.96.

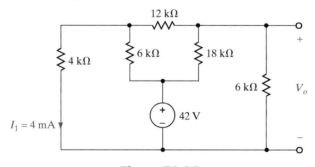

Figure P2.96

2.97 Find the power supplied by the 24-V source in the circuit in Fig. P2.97.

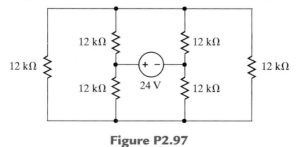

Figure P2.97

2.98 If all the resistors in the network in Fig. P2.98 are 12 kΩ, find V_o.

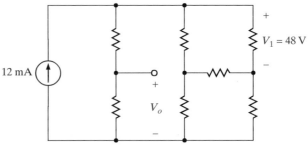

Figure P2.98

2.99 A typical transistor amplifier is shown in Fig. P2.99. Find the amplifier gain G (i.e., the ratio of the output voltage to the input voltage).

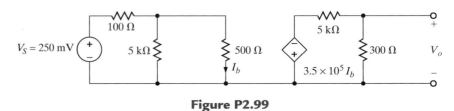

Figure P2.99

2.100 In many amplifier applications we are concerned not only with voltage gain, but also with power gain.

Power gain = A_p = (power delivered to the load)/(power delivered by the input)

Find the power gain for the circuit in Fig. P2.100, where $R_L = 50$ kΩ.

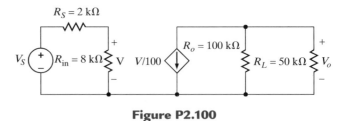

Figure P2.100

2.101 For the network in Fig. P2.101, choose the values of R_{in} and R_o such that V_o is maximized. What is the resulting ratio, V_o/V_S?

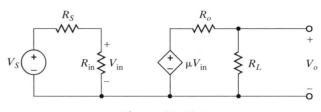

Figure P2.101

2.102 Find the power absorbed by the 12-kΩ resistor in the circuit in Fig. P2.102.

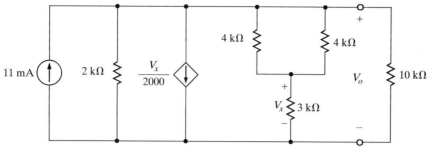

Figure P2.102

2.103 Find the power absorbed by the 12-kΩ resistor in the circuit in Fig. P2.103.

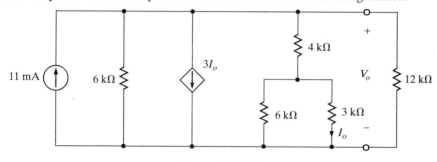

Figure P2.103

2.104 Find the power supplied by the 21-V source in the circuit in Fig. 2.104.

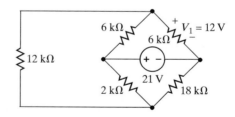

Figure P2.104

2.105 The Wheatstone bridge in Fig. 2.42 is used to measure an unknown resistor R_x. The bridge is balanced with $R_1 = 1\ \text{k}\Omega$, $R_2 = 1.2\ \text{k}\Omega$, and $R_3 = 1.8\ \text{k}\Omega$. Find the value of R_x.

2.106 If the Wheatstone bridge in Fig. 2.42 balanced with $R_x = 140\ \Omega$, $R_1 = 100\ \Omega$, and $R_2 = 200\ \Omega$, what is the value of the variable resistor R_3?

2.107 Redesign the voltage divider of Example 2.35 if the node voltage values are 18 V, 15 V, and 4 V with 60 mW of power dissipation.

Nodal and Loop Analysis Techniques

3

*I*n Chapter 2 we analyzed the simplest possible circuits, those containing only a single-node pair or a single loop. We found that these circuits can be completely analyzed via a single algebraic equation. In the case of the single-node-pair circuit (i.e., one containing two nodes, one of which is a reference node), once the node voltage is known, we can calculate all the currents. In a single-loop circuit, once the loop current is known, we can calculate all the voltages.

In this chapter we extend our capabilities in a systematic manner so that we can calculate all currents and voltages in circuits that contain multiple nodes and loops. Our analyses are based primarily on two laws with which we are already familiar: Kirchhoff's current law (KCL) and Kirchhoff's voltage law (KVL). In a nodal analysis we employ KCL to determine the node voltages, and in a loop analysis we use KVL to determine the loop currents.

A very important commercially available circuit known as the operational amplifier, or op-amp, is presented and discussed. Op-amps are used in literally thousands of applications, including such things as compact disk (CD) players, random access memories (RAMs), analog-to-digital (A/D) and digital-to-analog (D/A) converters, headphone amplifiers, and electronic instrumentation of all types. Finally, we will discuss the terminal characteristics of this circuit and demonstrate its use in practical applications as well as circuit design.

3.1 NODAL ANALYSIS

In a nodal analysis the variables in the circuit are selected to be the node voltages. The node voltages are defined with respect to a common point in the circuit. One node is selected as the reference node, and all other node voltages are defined with respect to that node. Quite often this node is the one to which the largest number of branches are connected. It is commonly called *ground* because it is said to be at ground-zero potential, and it sometimes represents the chassis or ground line in a practical circuit.

We will select our variables as being positive with respect to the reference node. If one or more of the node voltages are actually negative with respect to the reference node, the analysis will indicate it.

In order to understand the value of knowing all the node voltages in a network, we consider once again the network in Fig. 2.30, which is redrawn in Fig. 3.1. The voltages, V_S, V_a, V_b, and V_c, are all measured with respect to the bottom node, which is selected as the reference and labeled with the ground symbol $\perp$. Therefore, the voltage at node 1 is $V_S = 12$ V with respect to the reference node 5; the voltage at node 2 is $V_a = 3$ V with respect to the reference node 5, and so on. Now note carefully that once these node voltages are known, we can immediately calculate any branch current or the power supplied or absorbed by any element, since we know the voltage across every element in the network. For example, the voltage V_1 across the left-most 9-kΩ resistor is the difference in potential between the two ends of the resistor; that is,

$$V_1 = V_S - V_a$$
$$= 12 - 3$$
$$= 9 \text{ V}$$

This equation is really nothing more than an application of KVL around the left-most loop; that is,

$$- V_S + V_1 + V_a = 0$$

In a similar manner, we find that

$$V_3 = V_a - V_b$$

and

$$V_5 = V_b - V_c$$

Then the currents in the resistors are

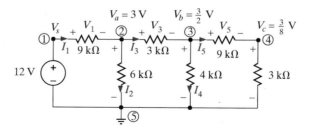

Figure 3.1 Circuit with known node voltages.

$$I_1 = \frac{V_1}{9k} = \frac{V_S - V_a}{9k}$$

$$I_3 = \frac{V_3}{3k} = \frac{V_a - V_b}{3k}$$

$$I_5 = \frac{V_5}{9k} = \frac{V_b - V_c}{9k}$$

In addition,

$$I_2 = \frac{V_a - 0}{6k}$$

$$I_4 = \frac{V_b - 0}{4k}$$

since the reference node 5 is at zero potential.

Thus, as a general rule, if we know the node voltages in a circuit, we can calculate the current through any resistive element using Ohm's law; that is,

$$i = \frac{v_m - v_N}{R} \qquad \textbf{3.1}$$

as illustrated in Fig. 3.2.

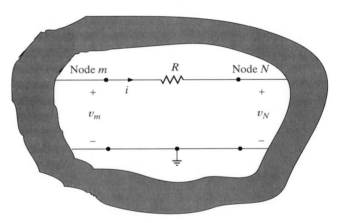

Figure 3.2 Circuit used to illustrate Ohm's law in a multiple-node network.

In Example 2.5 we illustrated that the number of linearly independent KCL equations for an N-node network was $N - 1$. Furthermore, we found that in a two-node circuit, in which one node was the reference node, only one equation was required to solve for the unknown node voltage. What is illustrated in this simple case is true in general; that is, in an N-node circuit one linearly independent KCL equation is written for each of the $N - 1$ nonreference nodes, and this set of $N - 1$ linearly independent simultaneous equations, when solved, will yield the $N - 1$ unknown node voltages.

It is instructive to treat nodal analysis by examining several different types of circuits and illustrating the salient features of each. We begin with the simplest case. However, as a prelude to our discussion of the details of nodal analysis, experience indicates that it is worthwhile to diverge for a moment to ensure that the concept of node voltage is clearly understood.

At the outset it is important to specify a reference. For example, to state that the voltage at node A is 12 V means nothing unless we provide the reference point; that is, the voltage at node A is 12 V with respect to what. The circuit in Fig. 3.3 illustrates a portion of a network containing three nodes, one of which is the reference node.

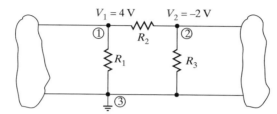

Figure 3.3 An illustration of node voltages.

The voltage $V_1 = 4$ V is the voltage at node 1 with respect to the reference node 3. Similarly, the voltage $V_2 = -2$ V is the voltage at node 2 with respect to node 3. In addition, however, the voltage at node 1 with respect to node 2 is +6 V and the voltage at node 2 with respect to node 1 is −6 V. Furthermore, since the current will flow from the node of higher potential to the node of lower potential, the current in R_1 is from top to bottom, the current in R_2 is from left to right, and the current in R_3 is from bottom to top.

These concepts have important ramifications in our daily lives. If a man were hanging in midair with one hand on one line and one hand on another and the dc line voltage of each line was exactly the same, the voltage across his heart would be zero and he would be safe. If, however, he let go of one line and let his feet touch the ground, the dc line voltage would then exist from his hand to his foot with his heart in the middle. He would probably be dead the instant his foot hit the ground.

In the town where we live, a young man tried to retrieve his parakeet that had escaped its cage and was outside sitting on a power line. He stood on a metal ladder and with a metal pole reached for the parakeet; when the metal pole touched the power line, the man was killed instantly. Electric power is vital to our standard of living, but it is also very dangerous. The material in this book *does not* qualify you to handle it safely. Therefore, always be extremely careful around electric circuits.

Now as we begin our discussion of nodal analysis, our approach will be to begin with simple cases and proceed in a systematic manner to those that are more challenging. Numerous examples will be the vehicle used to demonstrate each facet of this approach. Finally, at the end of this section, we will outline a strategy for attacking any circuit using nodal analysis.

Circuits Containing Only Independent Current Sources

Consider the network shown in Fig. 3.4. There are three nodes, and the bottom node is selected as the reference node. The branch currents are assumed to flow in the directions indicated in the figures. If one or more of the branch currents are actually flowing in a direction opposite to that assumed, the analysis will simply produce a branch current that is negative.

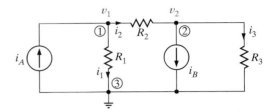

Figure 3.4 A three-node circuit.

Applying KCL at node 1 yields

$$-i_A + i_1 + i_2 = 0$$

Employing the passive sign convention.

Using Ohm's law ($i = Gv$) and noting that the reference node is at zero potential, we obtain

$$-i_A + G_1(v_1 - 0) + G_2(v_1 - v_2) = 0$$

or

$$(G_1 + G_2)v_1 - G_2v_2 = i_A$$

KCL at node 2 yields

$$-i_2 + i_B + i_3 = 0$$

or

$$-G_2(v_1 - v_2) + i_B + G_3(v_2 - 0) = 0$$

which can be expressed as

$$-G_2 v_1 + (G_2 + G_3)v_2 = -i_B$$

Therefore, the two equations for the two unknown node voltages v_1 and v_2 are

$$(G_1 + G_2)v_1 - G_2v_2 = i_A$$
$$-G_2 v_1 + (G_2 + G_3)v_2 = -i_B$$

3.2

Note that the analysis has produced two simultaneous equations in the unknowns v_1 and v_2. They can be solved using any convenient technique, and modern calculators and personal computers are very efficient tools for their application. Two such techniques, Gaussian elimination and matrix analysis, will be demonstrated here. For example, Eq. (3.2) may be rewritten in matrix form as

$$\mathbf{Av} = \mathbf{i}$$

3.3

where

$$\mathbf{A} = \begin{bmatrix} G_1 + G_2 & -G_2 \\ -G_2 & G_2 + G_3 \end{bmatrix}, \quad \mathbf{v} = \begin{bmatrix} v_1 \\ v_2 \end{bmatrix}, \quad \mathbf{i} = \begin{bmatrix} i_A \\ -i_B \end{bmatrix}$$

Thus,

$$\begin{bmatrix} G_1 + G_2 & -G_2 \\ -G_2 & G_2 + G_3 \end{bmatrix} \begin{bmatrix} v_1 \\ v_2 \end{bmatrix} = \begin{bmatrix} i_A \\ -i_B \end{bmatrix}$$

3.4

In general, the solution to Eq. (3.3) is

D3.1 For the following network, write the KCL equations for nodes 1 and 2.

ANSWER: $i_1 + \dfrac{v_1}{R_2} + \dfrac{v_1 - v_2}{R_3} + \dfrac{v_1 - v_2}{R_4} = 0$

$i_2 + \dfrac{v_1 - v_2}{R_3} + \dfrac{v_1 - v_2}{R_4} = 0$

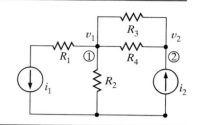

$$\mathbf{v} = \mathbf{A}^{-1}\mathbf{i} \qquad\qquad \textbf{3.5}$$

where $\mathbf{A}^{-1}$ is the inverse of the matrix $\mathbf{A}$. This calculation is both easy and straightforward for modern hand-held calculators and personal computers.

The KCL equations at nodes 1 and 2 produced two linearly independent simultaneous equations:

$$-i_A + i_1 + i_2 = 0$$
$$-i_2 + i_B + i_3 = 0$$

The KCL equation for the third node (reference) is

$$+i_A - i_1 - i_B - i_3 = 0$$

Note that if we add the first two equations, we obtain the third. Furthermore, any two of the equations can be used to derive the remaining equation. Therefore, in this $N = 3$ node circuit, only $N - 1 = 2$ of the equations are linearly independent and required to determine the $N - 1 = 2$ unknown node voltages.

Note that a nodal analysis employs KCL in conjunction with Ohm's law. Once the direction of the branch currents has been *assumed*, then Ohm's law, as illustrated by Fig. 3.2 and expressed by Eq. (3.1), is used to express the branch currents in terms of the unknown node voltages. We can assume the currents to be in any direction. However, once we assume a particular direction, we must be very careful to write the currents correctly in terms of the node voltages using Ohm's law.

EXAMPLE 3.1

Suppose that the network in Fig. 3.4 has the following parameters: $I_A = 1$ mA, $R_1 = 12$ kΩ, $R_2 = 6$ kΩ, $I_B = 4$ mA, and $R_3 = 6$ kΩ. Let us determine all node voltages and branch currents.

SOLUTION Using the parameter values Eq. (3.2) becomes

$$V_1\left[\dfrac{1}{12k} + \dfrac{1}{6k}\right] - V_2\left[\dfrac{1}{6k}\right] = 1 \times 10^{-3}$$

$$-V_1\left[\dfrac{1}{6k}\right] + V_2\left[\dfrac{1}{6k} + \dfrac{1}{6k}\right] = -4 \times 10^{-3}$$

where we employ capital letters because the voltages are constant. The equations can be written as

$$\frac{V_1}{4k} - \frac{V_2}{6k} = 1 \times 10^{-3}$$

$$-\frac{V_1}{6k} + \frac{V_2}{3k} = -4 \times 10^{-3}$$

Using Gaussian elimination, we solve the first equation for V_1 in terms of V_2:

$$V_1 = V_2\left(\frac{2}{3}\right) + 4$$

This value is then substituted into the second equation to yield

$$\frac{-1}{6k}\left(\frac{2}{3}V_2 + 4\right) + \frac{V_2}{3k} = -4 \times 10^{-3}$$

or

$$V_2 = -15 \text{ V}$$

This value for V_2 is now substituted back into the equation for V_1 in terms of V_2, which yields

$$V_1 = \frac{2}{3}V_2 + 4$$

$$= -6 \text{ V}$$

The circuit equations can also be solved using matrix analysis. In matrix form the equations are

$$\begin{bmatrix} \dfrac{1}{4k} & \dfrac{-1}{6k} \\ \dfrac{-1}{6k} & \dfrac{1}{3k} \end{bmatrix} \begin{bmatrix} V_1 \\ V_2 \end{bmatrix} = \begin{bmatrix} 1 \times 10^{-3} \\ -4 \times 10^{-3} \end{bmatrix}$$

and therefore,

$$\begin{bmatrix} V_1 \\ V_2 \end{bmatrix} = \begin{bmatrix} \dfrac{1}{4k} & \dfrac{-1}{6k} \\ \dfrac{-1}{6k} & \dfrac{1}{3k} \end{bmatrix}^{-1} \begin{bmatrix} 1 \times 10^{-3} \\ -4 \times 10^{-3} \end{bmatrix}$$

To calculate the inverse of **A**, we need the adjoint and the determinant. The adjoint is

$$\text{Adj } \mathbf{A} = \begin{bmatrix} \dfrac{1}{3k} & \dfrac{1}{6k} \\ \dfrac{1}{6k} & \dfrac{1}{4k} \end{bmatrix}$$

and the determinant is

$$|\mathbf{A}| = \left(\frac{1}{3k}\right)\left(\frac{1}{4k}\right) - \left(\frac{-1}{6k}\right)\left(\frac{-1}{6k}\right)$$

$$= \frac{1}{18k^2}$$

Therefore,

$$\begin{bmatrix} V_1 \\ V_2 \end{bmatrix} = 18k^2 \begin{bmatrix} \dfrac{1}{3k} & \dfrac{1}{6k} \\ \dfrac{1}{6k} & \dfrac{1}{4k} \end{bmatrix} \begin{bmatrix} 1 \times 10^{-3} \\ -4 \times 10^{-3} \end{bmatrix}$$

$$= 18k^2 \begin{bmatrix} \dfrac{1}{3k^2} & - \dfrac{4}{6k^2} \\ \dfrac{1}{6k^2} & - \dfrac{1}{k^2} \end{bmatrix}$$

$$= \begin{bmatrix} -6 \\ -15 \end{bmatrix}$$

Knowing the node voltages, we can determine all the currents using Ohm's law:

$$I_1 = \frac{V_1}{R_1} = \frac{-6}{12k} = -\frac{1}{2}\,\text{mA}$$

$$I_2 = \frac{V_1 - V_2}{6k} = \frac{-6 - (-15)}{6k} = \frac{3}{2}\,\text{mA}$$

and

$$I_3 = \frac{V_2}{6k} = \frac{-15}{6k} = -\frac{5}{2}\,\text{mA}$$

Figure 3.5 illustrates the results of all the calculations. Note that KCL is satisfied at every node.

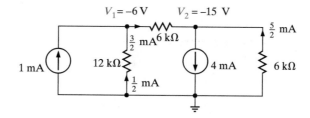

Figure 3.5 Circuit used in Example 3.1.

Let us now examine the circuit in Fig. 3.6. The current directions are assumed as shown in the figure.

At node 1 KCL yields

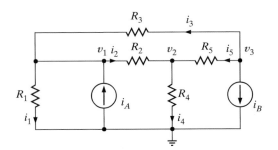

Figure 3.6 A four-node circuit.

$$i_1 - i_A + i_2 - i_3 = 0$$

or

$$\frac{v_1}{R_1} - i_A + \frac{v_1 - v_2}{R_2} - \frac{v_3 - v_1}{R_3} = 0$$

$$v_1\left(\frac{1}{R_1} + \frac{1}{R_2} + \frac{1}{R_3}\right) - v_2\frac{1}{R_2} - v_3\frac{1}{R_3} = i_A$$

At node 2 KCL yields

$$-i_2 + i_4 - i_5 = 0$$

or

$$-\frac{v_1 - v_2}{R_2} + \frac{v_2}{R_4} - \frac{v_3 - v_2}{R_5} = 0$$

$$-v_1\frac{1}{R_2} + v_2\left(\frac{1}{R_2} + \frac{1}{R_4} + \frac{1}{R_5}\right) - v_3\frac{1}{R_5} = 0$$

At node 3 the equation is

$$i_3 + i_5 + i_B = 0$$

or

$$\frac{v_3 - v_1}{R_3} + \frac{v_3 - v_2}{R_5} + i_B = 0$$

$$-v_1\frac{1}{R_3} - v_2\frac{1}{R_5} + v_3\left(\frac{1}{R_3} + \frac{1}{R_5}\right) = -i_B$$

Grouping the node equations together, we obtain

$$v_1\left(\frac{1}{R_1} + \frac{1}{R_2} + \frac{1}{R_3}\right) - v_2\frac{1}{R_2} - v_3\frac{1}{R_3} = i_A$$

$$-v_1\frac{1}{R_2} + v_2\left(\frac{1}{R_2} + \frac{1}{R_4} + \frac{1}{R_5}\right) - v_3\frac{1}{R_5} = 0$$

$$-v_1\frac{1}{R_3} - v_2\frac{1}{R_5} + v_3\left(\frac{1}{R_3} + \frac{1}{R_5}\right) = -i_B$$

3.6

Note that our analysis has produced three simultaneous equations in the three unknown node voltages $v_1, v_2,$ and v_3. The equations can also be written in matrix form as

$$
\begin{bmatrix}
\dfrac{1}{R_1}+\dfrac{1}{R_2}+\dfrac{1}{R_3} & -\dfrac{1}{R_2} & -\dfrac{1}{R_3} \\[2mm]
-\dfrac{1}{R_2} & \dfrac{1}{R_2}+\dfrac{1}{R_4}+\dfrac{1}{R_5} & -\dfrac{1}{R_5} \\[2mm]
-\dfrac{1}{R_3} & -\dfrac{1}{R_5} & \dfrac{1}{R_3}+\dfrac{1}{R_5}
\end{bmatrix}
\begin{bmatrix} v_1 \\ v_2 \\ v_3 \end{bmatrix}
=
\begin{bmatrix} i_A \\ 0 \\ -i_B \end{bmatrix}
\qquad \textbf{3.7}
$$

At this point it is important that we note the symmetrical form of the equations that describe the two previous networks. Equations (3.2) and (3.4) and Eqs. (3.6) and (3.7) exhibit the same type of symmetrical form. The **A** matrix for each network (3.4) and (3.7) is a symmetrical matrix. This symmetry is not accidental. The node equations for networks containing only resistors and independent current sources can always be written in this symmetrical form. We can take advantage of this fact and learn to write the equations by inspection. Note in the first equation of (3.2) that the coefficient of v_1 is the sum of all the conductances connected to node 1 and the coefficient of v_2 is the negative of the conductances connected between node 1 and node 2. The right-hand side of the equation is the sum of the currents entering node 1 through current sources. This equation is KCL at node 1. In the second equation in (3.2), the coefficient of v_2 is the sum of all the conductances connected to node 2, the coefficient of v_1 is the negative of the conductance connected between node 2 and node 1, and the right-hand side of the equation is the sum of the currents entering node 2 through current sources. This equation is KCL at node 2. Similarly, in the first equation in (3.6) the coefficient of v_1 is the sum of the conductances connected to node 1, the coefficient of v_2 is the negative of the conductance connected between node 1 and node 2, the coefficient of v_3 is the negative of the conductance connected between node 1 and node 3, and the right-hand side of the equation is the sum of the currents entering node 1 through current sources. The other two equations in (3.6) are obtained in a similar manner. In general, if KCL is applied to node j with node voltage v_j, the coefficient of v_j is the sum of all the conductances connected to node j and the coefficients of the other node voltages (e.g., v_{j-1}, v_{j+1}) are the negative of the sum of the conductances connected directly between these nodes and node j. The right-hand side of the equation is equal to the sum of the currents entering the node via current sources. Therefore, the left-hand side of the equation represents the sum of the currents leaving node j and the right-hand side of the equation represents the currents entering node j.

EXAMPLE 3.2

Let us apply what we have just learned to write the equations for the network in Fig. 3.7 by inspection.

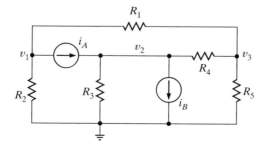

Figure 3.7 Circuit used in Example 3.2.

SOLUTION The equations are

$$v_1\left(\frac{1}{R_1} + \frac{1}{R_2}\right) - v_2(0) - v_3\left(\frac{1}{R_1}\right) = -i_A$$

$$-v_1(0) + v_2\left(\frac{1}{R_3} + \frac{1}{R_4}\right) - v_3\left(\frac{1}{R_4}\right) = i_A - i_B$$

$$-v_1\left(\frac{1}{R_1}\right) - v_2\left(\frac{1}{R_4}\right) + v_3\left(\frac{1}{R_1} + \frac{1}{R_4} + \frac{1}{R_5}\right) = 0$$

which can also be written directly in matrix form as

$$
\begin{bmatrix}
\dfrac{1}{R_1} + \dfrac{1}{R_2} & 0 & -\dfrac{1}{R_1} \\
0 & \dfrac{1}{R_3} + \dfrac{1}{R_4} & -\dfrac{1}{R_4} \\
-\dfrac{1}{R_1} & -\dfrac{1}{R_4} & \dfrac{1}{R_1} + \dfrac{1}{R_4} + \dfrac{1}{R_5}
\end{bmatrix}
\begin{bmatrix}
v_1 \\ v_2 \\ v_3
\end{bmatrix}
=
\begin{bmatrix}
-i_A \\ i_A - i_B \\ 0
\end{bmatrix}
$$

Both the equations and the **A** matrix exhibit the symmetry that will always be present in circuits that contain only resistors and current sources. ❏

EXTENSION EXERCISES

E3.1 Write the node equations for the circuit in Fig. E3.1.

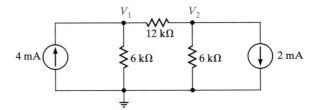

Figure E3.1

ANSWER: $\dfrac{1}{4k} V_1 - \dfrac{1}{12k} V_2 = 4 \times 10^{-3}$, $\dfrac{-1}{12k} V_1 + \dfrac{1}{4k} V_2$

$$= -2 \times 10^{-3}$$

E3.2 Find all the branch currents in the network in Fig. E3.2.

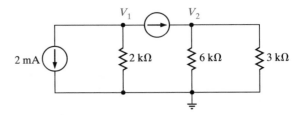

Figure E3.2

ANSWER: Note that KCL and current division yield the following currents immediately: $I_{2k\Omega} = 8\text{ mA}\uparrow$, $I_{6k\Omega} = 2\text{ mA}\downarrow$, and $I_{3k\Omega} = 4\text{ mA}\downarrow$.

Circuits Containing Dependent Current Sources

The presence of a dependent source may destroy the symmetrical form of the nodal equations that define the circuit. Consider the circuit shown in Fig. 3.8, which contains a current-controlled current source. The KCL equations for the nonreference nodes are

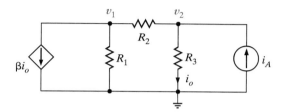

Figure 3.8 Circuit with a dependent source.

$$\beta i_o + \frac{v_1}{R_1} + \frac{v_1 - v_2}{R_2} = 0$$

and

$$\frac{v_2 - v_1}{R_2} + i_o - i_A = 0$$

where $i_o = v_2/R_3$. Simplifying the equations, we obtain

$$(G_1 + G_2)v_1 - (G_2 - \beta G_3)v_2 = 0$$
$$-G_2 v_1 + (G_2 + G_3)v_2 = i_A$$

or in matrix form

$$\begin{bmatrix} (G_1 + G_2) & -(G_2 - \beta G_3) \\ -G_2 & (G_2 + G_3) \end{bmatrix} \begin{bmatrix} v_1 \\ v_2 \end{bmatrix} = \begin{bmatrix} 0 \\ i_A \end{bmatrix}$$

Note that the presence of the dependent source has destroyed the symmetrical nature of the node equation.

EXAMPLE 3.3

Let us determine the node voltages for the network in Fig. 3.8 given the following parameters:

$$\beta = 2 \qquad R_2 = 6\text{ k}\Omega \qquad i_A = 2\text{ mA}$$
$$R_1 = 12\text{ k}\Omega \qquad R_3 = 3\text{ k}\Omega$$

SOLUTION   Using these values with the equations for the network yields

$$\frac{1}{4k} V_1 + \frac{1}{2k} V_2 = 0$$

$$-\frac{1}{6k} V_1 + \frac{1}{2k} V_2 = 2 \times 10^{-3}$$

Solving these equations using any convenient method yields $V_1 = -24/5$ V and $V_2 = 12/5$ V. We can check these answers by determining the branch currents in the network and then using that information to test KCL at the nodes. For example, the current from top to bottom through R_3 is

$$I_o = \frac{V_2}{R_3} = \frac{12/5}{3k} = \frac{4}{5k} \text{ A}$$

Similarly, the current from right to left through R_2 is

$$I_2 = \frac{V_2 - V_1}{R_2} = \frac{12/5 - (-24/5)}{6k} = \frac{6}{5k} \text{ A}$$

All the results are shown in Fig. 3.9. Note that KCL is satisfied at every node.

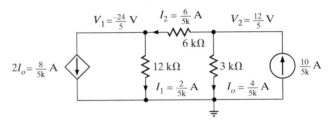

Figure 3.9 Circuit used in Example 3.3. ❏

Consider now the network in Fig. 3.10. Applying KCL at each of the nonreference nodes yields the equations

$$G_3 v_1 + G_1(v_1 - v_2) - i_A = 0$$
$$i_A + G_1(v_2 - v_1) + \alpha v_x + G_2(v_2 - v_3) = 0$$
$$G_2(v_3 - v_2) + G_4 v_3 - i_B = 0$$

where $v_x = v_2 - v_3$. Simplifying these equations, we obtain

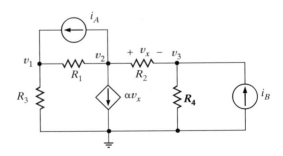

Figure 3.10 Circuit containing a voltage-controlled current source.

$$(G_1 + G_3)v_1 - G_1v_2 = i_A$$
$$-G_1v_1 + (G_1 + \alpha + G_2)v_2 - (\alpha + G_2)v_3 = -i_A$$
$$-G_2v_2 + (G_2 + G_4)v_3 = i_B$$

or in matrix form

$$\begin{bmatrix} (G_1 + G_3) & -G_1 & 0 \\ -G_1 & (G_1 + \alpha + G_2) & -(\alpha + G_2) \\ 0 & -G_2 & (G_2 + G_4) \end{bmatrix} \begin{bmatrix} v_1 \\ v_2 \\ v_3 \end{bmatrix} = \begin{bmatrix} i_A \\ -i_A \\ i_B \end{bmatrix}$$

Once again, we find that the symmetry of the equation is destroyed by the dependent source.

EXTENSION EXERCISES

E3.3 Find the node voltages in the circuit in Fig. E3.3.

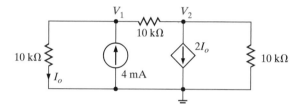

Figure E3.3

ANSWER: $V_1 = 16$ V, $V_2 = -8$ V.

E3.4 Find the voltages V_o in the network in Fig. E3.4.

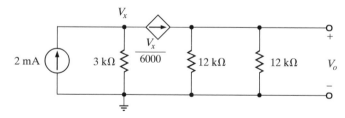

Figure E3.4

ANSWER: $V_o = 4$ V.

Circuits Containing Independent Voltage Sources

As is our practice, in our discussion of this topic we will proceed from the simplest case to those that are more complicated. The simplest case is that in which an independent voltage source is connected to the reference node. The following example illustrates this case.

EXAMPLE 3.4

Consider the circuit shown in Fig. 3.11a. Let us determine all node voltages and branch currents.

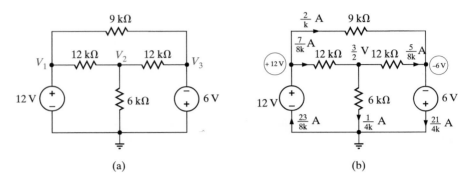

(a)   (b)

Figure 3.11 Circuit used in Example 3.4.

SOLUTION   This network has three nonreference nodes with labeled node voltages $V_1, V_2,$ and V_3. Based upon our previous discussions, we would assume that in order to find all the node voltages we would need to write a KCL equation at each of the nonreference nodes. The resulting three linearly independent simultaneous equations would produce the unknown node voltages. However, note that V_1 and V_3 are known quantities because an independent voltage source is connected directly between the nonreference node and each of these nodes. Therefore, $V_1 = 12$ V and $V_3 = -6$ V. Furthermore, note that the current through the 9-kΩ resistor is $[12 - (-6)]/9k = 2$ mA from left to right. We do not know V_2 or the current in the remaining resistors. However, since only one node voltage is unknown, a single-node equation will produce it. Applying KCL to this center node yields

$$\frac{V_2 - V_1}{12k} + \frac{V_2 - 0}{6k} + \frac{V_2 - V_3}{12k} = 0$$

or

$$\frac{V_2 - 12}{12k} + \frac{V_2}{6k} + \frac{V_2 - (-6)}{12k} = 0$$

from which we obtain

$$V_2 = \frac{3}{2} \text{ V}$$

Once all the node voltages are known, Ohm's law can be used to find the branch currents shown in Fig. 3.11b. The diagram illustrates that KCL is satisfied at every node.

Note that the presence of the voltage sources in this example has simplified the analysis, since two of the three linear independent equations are $V_1 = 12$ V and $V_3 = -6$ V. We will find that as a general rule, anytime voltage sources are present between nodes, the node voltage equations that describe the network will be simpler.

Anytime an independent voltage source is connected between the reference node and a nonreference node, the nonreference node voltage is known.

EXTENSION EXERCISES

E3.5 Find the current I_o in the network in Fig. E3.5.

ANSWER: $I_o = \dfrac{3}{4}$ mA.

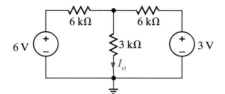

Figure E3.5

Next let us consider the case in which an independent voltage source is connected between two nonreference nodes. Once again, we will use an example to illustrate the approach.

EXAMPLE 3.5

We wish to find the currents in the two resistors in the circuit in Fig. 3.12a.

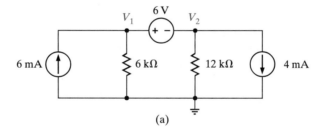

(a)

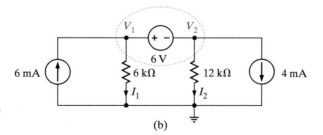

Figure 3.12 Circuits used in Example 3.5.

(b)

SOLUTION If we try to attack this problem in a brute force manner, we immediately encounter a problem. Thus far, branch currents were either known source values or could be expressed as the branch voltage divided by the branch resistance.

However, the branch current through the 6-V source is certainly not known and cannot be directly expressed using Ohm's law. We can, of course, give this current a name and write the KCL equations at the two nonreference nodes in terms of this current. However, this approach is no panacea because this technique will result in *two* linearly independent simultaneous equations in terms of *three* unknowns; that is, the two node voltages and the current in the voltage source.

To solve this dilemma, we recall that $N - 1$ linearly independent equations are required to determine the $N - 1$ nonreference node voltages in an N-node circuit. Since our network has three nodes, we need two linearly independent equations. Now note that if somehow one of the node voltages is known, we immediately know the other; that is, if V_1 is known, then $V_2 = V_1 - 6$. And if V_2 is known, then $V_1 = V_2 + 6$. Therefore, the difference in potential between the two nodes is *constrained* by the voltage source and, hence,

$$V_1 - V_2 = 6$$

This constraint equation is one of the two linearly independent equations needed to determine the node voltages.

Next consider the network in Fig. 3.12b, in which the 6-V source is completely enclosed within the dashed surface. The constraint equation governs this dashed portion of the network. The remaining equation is obtained by applying KCL to this dashed surface, which is commonly called a *supernode*. Recall that in Chapter 2 we demonstrated that KCL must hold for a surface, and this technique eliminates the problem of dealing with a current through a voltage source. KCL for the supernode is

$$-6 \times 10^{-3} + \frac{V_1}{6k} + \frac{V_2}{12k} + 4 \times 10^{-3} = 0$$

Solving these equations yields $V_1 = 10$ V and $V_2 = 4$ V and, hence, $I_1 = 5/3$ mA and $I_2 = 1/3$ mA. A quick check indicates that KCL is satisfied at every node.   ❏

EXAMPLE 3.6

Let us write the node equations for the circuit in Fig. 3.13.

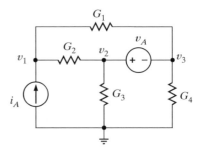

Figure 3.13 Circuit used in Example 3.6.

SOLUTION   The equation for the node labeled v_1 is

$$(v_1 - v_3)G_1 + (v_1 - v_2)G_2 - i_A = 0$$

The constraint equation for the supernode is

$$v_2 - v_3 = v_A$$

And the KCL equation for the supernode is

$$(v_2 - v_1)G_2 + v_2G_3 + (v_3 - v_1)G_1 + v_3G_4 = 0$$

The three equations will yield the node voltages.   ❏

EXAMPLE 3.7

Let us determine the current I_o in the network in Fig. 3.14a.

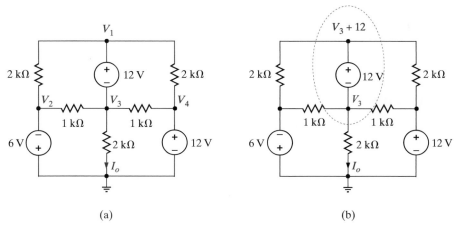

(a)   (b)

Figure 3.14 Example circuit with supernodes.

SOLUTION   Examining the network, we note that node voltages V_2 and V_4 are known and the node voltages V_1 and V_3 are constrained by the equation

$$V_1 - V_3 = 12$$

The network is redrawn in Fig. 3.14b.

Since we want to find the current I_o, V_1 (in the supernode containing V_1 and V_3) is written as $V_3 + 12$. The KCL equation at the supernode is then

$$\frac{V_3 + 12 - (-6)}{2k} + \frac{V_3 + 12 - 12}{2k} + \frac{V_3 - (-6)}{1k} + \frac{V_3 - 12}{1k} + \frac{V_3}{2k} = 0$$

Solving the equation for V_3 yields

$$V_3 = -\frac{6}{7}\,\text{V}$$

I_o can then be computed immediately as

$$I_o = \frac{-\dfrac{6}{7}}{2k} = -\frac{3}{7}\,\text{mA}$$ ❏

E3.6   Use nodal analysis to find I_o in the network in Fig. E3.6.

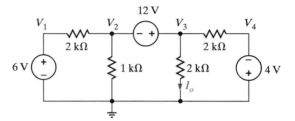

Figure E3.6

ANSWER: $I_o = 3.8\ \text{mA}$.

Circuits Containing Dependent Voltage Sources

As the following examples will indicate, networks containing dependent (controlled) sources are treated in the same manner as described earlier.

EXAMPLE 3.8

We wish to find I_o in the network in Fig. 3.15.

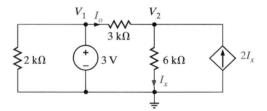

Figure 3.15 Circuit used in Example 3.8.

SOLUTION   As a general rule, when writing the circuit equations we first treat the dependent source as though it were an independent source and then write the controlling equation. Since the independent voltage source is connected between the nonreference node labeled V_1 and the reference, one of the two linearly independent equations is

$$V_1 = 3\ \text{V}$$

KCL applied to the second nonreference node yields

$$\frac{V_2 - 3}{3k} + \frac{V_2}{6k} = 2I_x$$

where the controlling equation is

$$I_x = \frac{V_2}{6k}$$

solving these equations yields $V_2 = 6$ V and, hence,

$$I_o = \frac{3 - 6}{3k} = -1 \text{ mA} \qquad \square$$

EXAMPLE 3.9

Let us find the current I_o in the network in Fig. 3.16.

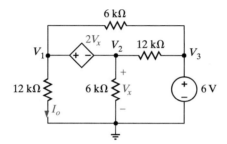

Figure 3.16 Circuit used in Example 3.9.

SOLUTION   This circuit contains both an independent voltage source and a volt-age-controlled voltage source. Note that $V_3 = 6$ V, $V_2 = V_x$, and a supernode exists between the nodes labeled V_1 and V_2.

Applying KCL to the supernode, we obtain

$$\frac{V_1 - V_3}{6k} + \frac{V_1}{12k} + \frac{V_2}{6k} + \frac{V_2 - V_3}{12k} = 0$$

where the constraint equation for the supernode is

$$V_1 - V_2 = 2V_x$$

The final equation is

$$V_3 = 6$$

Solving these equations, we find that

$$V_1 = \frac{9}{2} \text{ V}$$

and, hence,

$$I_o = \frac{V_1}{12k} = \frac{3}{8} \text{ mA} \qquad \square$$

EXTENSION EXERCISES

E3.7   Find I_o in the circuit in Fig. E3.7.

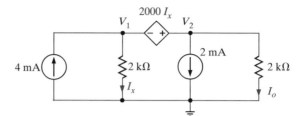

Figure E3.7

ANSWER: $I_o = \dfrac{4}{3}$ mA.

Problem-Solving Strategies Nodal Analysis

*   Select one node in the N-node circuit as the reference node. Assume that the node voltage is zero and measure all node voltages with respect to this node.

*   If only independent current sources are present in the network, write the KCL equations at the $N - 1$ nonreference nodes. If dependent current sources are present, write the KCL equations as is done for networks with only independent current sources; then write the controlling equations for the dependent sources.

*   If voltage sources are present in the network, they may be connected (1) between the reference node and a nonreference node or (2) between two nonreference nodes. In the former case, if the voltage source is an independent source, then the voltage at one of the nonreference nodes is known. If the source is dependent, it is treated as an independent source when writing the KCL equation, but an additional constraint equation is necessary, as described previously.

    In the latter case, if the source is independent, the voltage between the two nodes is constrained by the value of the voltage source, and an equation describing this constraint represents one of the $N - 1$ linearly independent equations required to determine the N-node voltages. The surface of the network described by the constraint equation (ie, the source and two connecting nodes) is called a supernode. One of the remaining $N - 1$ linearly independent equations is obtained by applying KCL at this supernode. If the voltage source is dependent, it is treated as an independent source when writing the KCL equations, but an additional constraint equation is necessary, as described previously.

3.2 LOOP ANALYSIS

In a nodal analysis the unknown parameters are the node voltages, and KCL is employed to determine them. In contrast to this approach, a loop analysis uses KVL to determine currents in the circuit. Once the currents are known, Ohm's law can be used to calculate the voltages. Recall that, in Chapter 2, we found that a single equation was sufficient to determine the current in a circuit containing a single loop. If the circuit contains N independent loops, we will show that N independent simultaneous equations will be required to describe the network. For now we will assume that the circuits are planar, which simply means that we can draw the circuit on a sheet of paper in a way such that no conductor crosses another conductor.

Our approach to loop analysis will mirror that used in nodal analysis (i.e., we will begin with simple cases and systematically proceed to those that are more difficult). Then at the end of this section we will outline a general strategy for employing loop analysis.

Circuits Containing Only Independent Voltage Sources

To begin our analysis, consider the circuit shown in Fig. 3.17. Let us also identify two loops, A-B-E-F-A and B-C-D-E-B. We now define a new set of current variables called *loop currents*, which can be used to find the physical currents in the circuit. Let us assume that current i_1 flows in the first loop and that current i_2 flows in the second loop. Then the branch current flowing from B to E through R_3 is $i_1 - i_2$. The directions of the currents have been assumed. As was the case in the nodal analysis, if the actual currents are not in the direction indicated, the values calculated will be negative.

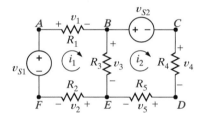

Figure 3.17 A two-loop circuit.

Applying KVL to the first loop yields

$$+v_1 + v_3 + v_2 - v_{S1} = 0$$

KVL applied to loop 2 yields

$$+v_{S2} + v_4 + v_5 - v_3 = 0$$

where $v_1 = i_1 R_1, v_2 = i_1 R_2, v_3 = (i_1 - i_2)R_3, v_4 = i_2 R_4,$ and $v_5 = i_2 R_5$.

The equations employ the passive sign convention.

DRILL

D3.2 Write the mesh equations for the following circuit.

ANSWER:
$$-v_{S1} + i_1 R_1 - v_{S2} + (i_1 - i_2)R_2 = 0$$
$$i_2 R_3 + (i_2 - i_1)R_2 + i_{2R_5} + i_2 R_4 + V_{S3} = 0$$

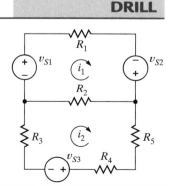

Substituting these values into the two KVL equations produces the two simultaneous equations required to determine the two loop currents; that is,

$$i_1(R_1 + R_2 + R_3) - i_2(R_3) = v_{S1}$$
$$-i_1(R_3) + i_2(R_3 + R_4 + R_5) = -v_{S2}$$

or in matrix form

$$\begin{bmatrix} R_1 + R_2 + R_3 & -R_3 \\ -R_3 & R_3 + R_4 + R_5 \end{bmatrix} \begin{bmatrix} i_1 \\ i_2 \end{bmatrix} = \begin{bmatrix} v_{S1} \\ -v_{S2} \end{bmatrix}$$

At this point it is important to define what is called a *mesh*. A mesh is a special kind of loop that does not contain any loops within it. Therefore, as we traverse the path of a mesh, we do not encircle any circuit elements. For example, the network in Fig. 3.17 contains two meshes defined by the paths *A-B-E-F-A* and *B-C-D-E-B*. The path *A-B-C-D-E-F-A* is a loop but it is not a mesh. Since the majority of our analysis in this section will involve writing KVL equations for meshes, we will refer to the currents as mesh currents and the analysis as a *mesh analysis*.

EXAMPLE 3.10

Consider the network in Fig. 3.18a. We wish to find the current I_o.

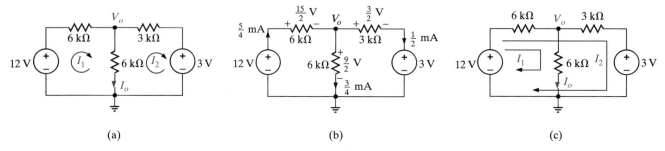

Figure 3.18 Circuits used in Example 3.10.

SOLUTION We will begin the analysis by writing mesh equations. Note that there are no + and − signs on the resistors. However, they are not needed, since we will apply Ohm's law to each resistive element as we write the KVL equations. The equation for the first mesh is

$$-12 + 6kI_1 + 6k(I_1 - I_2) = 0$$

The KVL equation for the second mesh is

$$6k(I_2 - I_1) + 3kI_2 + 3 = 0$$

where $I_o = I_1 - I_2$.

Solving the two simultaneous equations yields $I_1 = 5/4$ mA and $I_2 = 1/2$ mA. Therefore, $I_o = 3/4$ mA. All the voltages and currents in the network are shown in Fig. 3.18b. Recall from nodal analysis that once the node voltages were determined, we could check our analysis using KCL at the nodes. In this case we know the branch currents and can use KVL around any closed path to check our results. For example, applying KVL to the outer loop yields

$$-12 + \frac{15}{2} + \frac{3}{2} + 3 = 0$$

$$0 = 0$$

Since we want to calculate the current I_o, we could use loop analysis, as shown in Fig. 3.18c. Note that the loop current I_1 passes through the center leg of the network and, therefore, $I_1 = I_o$. The two loop equations in this case are

$$-12 + 6k(I_1 + I_2) + 6kI_1 = 0$$

and

$$-12 + 6k(I_1 + I_2) + 3kI_2 + 3 = 0$$

Solving these equations yields $I_1 = 3/4$ mA and $I_2 = 1/2$ mA. Since the current in the 12-V source is $I_1 + I_2 = 5/4$ mA, these results agree with the mesh analysis.

Finally, for purposes of comparison, let us find I_o using nodal analysis. The presence of the two voltage sources would indicate that this is a viable approach. Applying KCL at the top center node, we obtain

$$\frac{V_o - 12}{6k} + \frac{V_o}{6k} + \frac{V_o - 3}{3k} = 0$$

and hence,

$$V_o = \frac{9}{2} \text{ V}$$

and then

$$I_o = \frac{V_o}{6k} = \frac{3}{4} \text{ mA}$$

Note that in this case we had to solve only one equation instead of two.   ❏

Once again we are compelled to note the symmetrical form of the mesh equations that describe the circuit in Fig. 3.17. Note that the **A** matrix for this circuit is symmetrical.

Since this symmetry is generally exhibited by networks containing resistors and independent voltage sources, we can learn to write the mesh equations by inspection. In the first equation the coefficient of i_1 is the sum of the resistances through which mesh current 1 flows, and the coefficient of i_2 is the negative of the sum of the resistances common to mesh current 1 and mesh current 2. The right-hand side of the equation is the algebraic sum of the voltage sources in mesh 1. The sign of the voltage source is positive if it aids the assumed direction of the current flow and negative if it opposes the assumed flow. The first equation is KVL for mesh 1. In the second equation, the coefficient of i_2 is the sum of all the resistances in mesh 2, the coefficient of i_1 is the negative of the sum of the resistances common to mesh 1 and mesh 2, and the right-hand side of the equation is the algebraic sum of the voltage sources in mesh 2. In general, if we assume all of the mesh currents to be in the same direction (clockwise or counterclockwise), then if KVL is applied to mesh j with mesh current i_j, the coefficient of i_j is the sum of the resistances in mesh j and the coefficients of the other mesh currents $(e.g., i_{j-1}, i_{j+1})$ are the negatives of the resistances common to these meshes and mesh j. The right-hand side of the equation is equal to the algebraic sum of the voltage sources in mesh j. These voltage sources have a positive sign if they aid the current flow i_j and a negative sign if they oppose it.

EXAMPLE 3.11

Let us write the mesh equations by inspection for the network in Fig. 3.19.

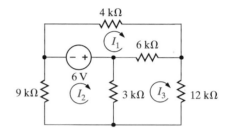

Figure 3.19 Circuit used in Example 3.11.

SOLUTION   The three linearly independent simultaneous equations are

$$(4k + 6k)I_1 - (0)I_2 - (6k)I_3 = -6$$
$$-(0)I_1 + (9k + 3k)I_2 - (3k)I_3 = 6$$
$$-(6k)I_1 - (3k)I_2 + (3k + 6k + 12k)I_3 = 0$$

or in matrix form

$$\begin{bmatrix} 10k & 0 & -6k \\ 0 & 12k & -3k \\ -6k & -3k & 21k \end{bmatrix} \begin{bmatrix} I_1 \\ I_2 \\ I_3 \end{bmatrix} = \begin{bmatrix} -6 \\ 6 \\ 0 \end{bmatrix}$$

Note the symmetrical form of the equations.   ❏

Circuits Containing Independent Current Sources

Just as the presence of a voltage source in a network simplified the nodal analysis, the presence of a current source simplifies a loop analysis. The following examples illustrate the point.

EXTENSION EXERCISES

E3.8   Use mesh equations to find V_o in the circuit in Fig. E3.8.

ANSWER: $V_o = \dfrac{33}{5} \text{ V}.$

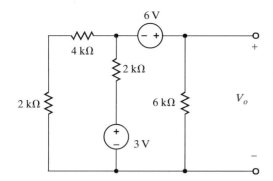

Figure E3.8

EXAMPLE 3.12

Let us find both V_o and V_1 in the circuit in Fig. 3.20.

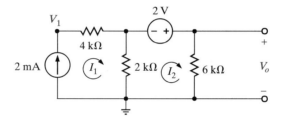

Figure 3.20 Circuit used in Example 3.12.

SOLUTION   Although it appears that there are two unknown mesh currents, the current I_1 goes directly through the current source and, therefore, I_1 is constrained to be 2 mA. Hence, only the current I_2 is unknown. KVL for the right-most mesh is

$$2k(I_2 - I_1) - 2 + 6kI_2 = 0$$

And, of course,

$$I_1 = 2 \times 10^{-3}$$

These equations yield

$$I_2 = \frac{3}{4} \text{ mA}$$

And hence,

$$V_o = 6kI_2 = \frac{9}{2} \text{ V}$$

To obtain V_1 we apply KVL around any closed path. If we use the outer loop, the KVL equation is

$$-V_1 + 4kI_1 - 2 + 6kI_2 = 0$$

And therefore,

$$V_1 = \frac{21}{2} \text{ V}$$

Note that since the current I_1 is known, the 4-kΩ resistor did not enter the equation in finding V_o. However, it appears in every loop containing the current source and, thus, is used in finding V_1.   ❏

EXAMPLE 3.13

We wish to find V_o in the network in Fig. 3.21.

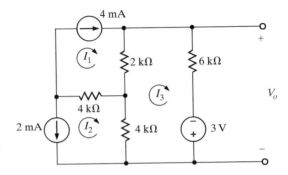

Figure 3.21 Circuit used in Example 3.13.

SOLUTION   Since the currents I_1 and I_2 pass directly through a current source, two of the three required equations are

$$I_1 = 4 \times 10^{-3}$$
$$I_2 = -2 \times 10^{-3}$$

The third equation is KVL for the mesh containing the voltage source; that is,

$$4k(I_3 - I_2) + 2k(I_3 - I_1) + 6kI_3 - 3 = 0$$

These equations yield

$$I_3 = \frac{1}{4} \text{ mA}$$

and hence,

$$V_o = 6kI_3 - 3 = \frac{-3}{2} \text{ V} \qquad ❏$$

What we have demonstrated in the previous examples is the general approach for dealing with independent current sources when writing KVL equations; that is, use one loop through each current source. The number of "window panes" in the network tells us how many equations we need. Additional KVL equations are written to cover the remaining circuit elements in the network. The following example illustrates this approach.

EXAMPLE 3.14

Let us find I_o in the network in Fig. 3.22a.

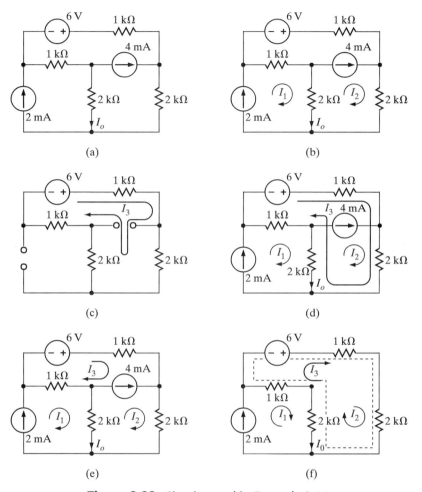

In this case the 4-mA current source is located on the boundary between two meshes. Thus we will demonstrate two techniques for dealing with this type of situation. One is a special loop technique and the other is known as the supermesh approach.

Figure 3.22 Circuits used in Example 3.14.

SOLUTION   First, we select two loop currents I_1 and I_2 such that I_1 passes directly through the 2-mA source, and I_2 passes directly through the 4-mA source, as shown in Fig. 3.22b. Therefore, two of our three linearly independent equations are

$$I_1 = 2 \times 10^{-3}$$
$$I_2 = 4 \times 10^{-3}$$

The remaining loop current I_3 must pass through the circuit elements not covered by the two previous equations and cannot, of course, pass through the current sources. The path for this remaining loop current can be obtained by open-circuiting the current sources, as shown in Fig. 3.22c. When all current are labeled on the original circuit, the KVL equation for this last loop, as shown in Fig. 3.22d, is

$$-6 + 1\mathrm{k}I_3 + 2\mathrm{k}\big(I_2 + I_3\big) + 2\mathrm{k}\big(I_3 + I_2 - I_1\big) + 1\mathrm{k}\big(I_3 - I_1\big) = 0$$

Solving the equations yields

$$I_3 = \frac{-2}{3}\,\mathrm{mA}$$

And therefore,

$$I_o = I_1 - I_2 - I_3 = \frac{-4}{3}\,\mathrm{mA}$$

Next consider the supermesh technique. In this case the three mesh currents are specified as shown in Fig. 3.22e, and since the voltage across the 4-mA current source is unknown, it is assumed to be V_x. The mesh currents constrained by the current sources are

$$I_1 = 2 \times 10^{-3}$$
$$I_2 - I_3 = 4 \times 10^{-3}$$

The KVL equations for meshes 2 and 3, respectively, are

$$2\mathrm{k}I_2 + 2\mathrm{k}\big(I_2 - I_1\big) - V_x = 0$$
$$-6 + 1\mathrm{k}I_3 + V_x + 1\mathrm{k}\big(I_3 - I_1\big) = 0$$

Adding the last two equations yields

$$-6 + 1\mathrm{k}I_3 + 2\mathrm{k}I_2 + 2\mathrm{k}\big(I_2 - I_1\big) + 1\mathrm{k}\big(I_3 - I_1\big) = 0$$

Note that the unknown voltage V_x has been eliminated. The two constraint equations, together with this latter equation, yield the desired result.

The purpose of the supermesh approach is to avoid introducing the unknown voltage V_x. The supermesh is created by mentally removing the 4-mA current source, as shown in Fig. 3.22f. Then writing the KVL equation around the dotted path, which defines the supermesh, using the original mesh currents as shown in Fig. 3.22e, yields

$$-6 + 1\mathrm{k}I_3 + 2\mathrm{k}I_2 + 2\mathrm{k}\big(I_2 - I_1\big) + 1\mathrm{k}\big(I_3 - I_1\big) = 0$$

Note that this supermesh equation is the same as that obtained earlier by introducing the voltage V_x. ❏

EXTENSION EXERCISES

E3.9 Find V_o in the network in Fig. E3.9.

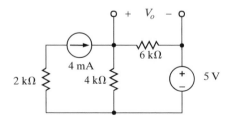

Figure E3.9

ANSWER: $V_o = \dfrac{33}{5}$ V.

E3.10   Find V_o in the network in Fig. E3.10.

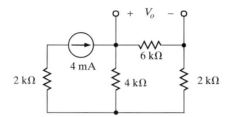

Figure E3.10

ANSWER: $V_o = 8$ V.

Circuits Containing Dependent Sources

We deal with circuits containing dependent sources just as we have in the past. First, we treat the dependent source as though it were an independent source when writing the KVL equations. Then we write the controlling equation for the dependent source. The following examples illustrate the point.

EXAMPLE 3.15

Let us find V_o in the circuit in Fig. 3.23, which contains a voltage-controlled voltage source.

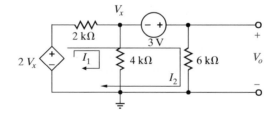

Figure 3.23 Circuit used in Example 3.15.

SOLUTION   The equations for the loop currents shown in the figure are

$$-2V_x + 2k(I_1 + I_2) + 4kI_1 = 0$$
$$-2V_x + 2k(I_1 + I_2) - 3 + 6kI_2 = 0$$

where

$$V_x = 4kI_1$$

Solving these equations yields

$$I_2 = \frac{3}{2}\,\text{mA}$$

and therefore,

$$V_o = 6kI_2 = 9 \text{ V}$$

For comparison, we will also solve the problem using nodal analysis. The presence of the voltage sources indicates that this method could be simpler. Treating the 3-V source and its connecting nodes as a supernode and writing the KCL equation for this supernode yields

$$\frac{V_x - 2V_x}{2k} + \frac{V_x}{4k} + \frac{V_x + 3}{6k} = 0$$

where

$$V_o = V_x + 3$$

These equations also yield $V_o = 9 \text{ V}$. ❏

EXAMPLE 3.16

Let us find V_o in the circuit in Fig. 3.24, which contains a voltage-controlled current source.

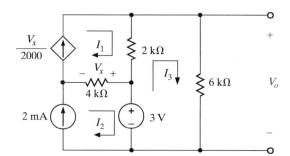

Figure 3.24 Circuit used in Example 3.16.

SOLUTION The currents I_1 and I_2 are drawn through the current sources. Therefore, two of the equations needed are

$$I_1 = \frac{V_x}{2000}$$

$$I_2 = 2 \times 10^{-3}$$

The KVL equation for the third mesh is

$$-3 + 2k(I_3 - I_1) + 6kI_3 = 0$$

where

$$V_x = 4k(I_1 - I_2)$$

Solving these equations yields

$$I_3 = \frac{11}{8} \text{ mA}$$

and hence,

$$V_o = \frac{33}{4} \text{ V}$$ ❑

As a final point, it is very important to examine the circuit carefully before selecting an analysis approach. One method could be much simpler than another, and a little time invested up front may save a lot of time in the long run.

EXTENSION EXERCISES

E3.11   Use mesh analysis to find V_o in the circuit in Fig. E3.11.

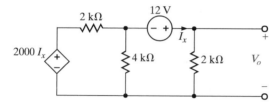

Figure E3.11

ANSWER: $V_o = 12$ V.

Problem-Solving Strategies for Loop Analysis

- One loop current is assigned to each independent loop in a circuit that contains N independent loops.

- If only independent voltage sources are present in the network, write the N linearly independent KVL equations, one for each loop. If dependent voltage sources are present, write the KVL equation as is done for circuits with only independent voltage sources; then write the controlling equations for the dependent sources.

- If current sources are present in the network, either of two techniques can be used. In the first case, one loop current is selected to pass through one of the current sources. This is done for each current source in the network. The remaining loop currents (N − the number of current sources) are determined by open-circuiting the current sources in the network and using this modified network to select them. Once all these currents are defined in the original network, the N loop equations can be written. The second approach is similar to the first with the exception that if two mesh currents pass through a particular current source, a supermesh is formed around this source. The two required equations for the meshes containing this source are the constraint equations for the two mesh currents that pass through the source and the supermesh equation. As indicated earlier, if dependent current sources are present, the controlling equations for these sources are also necessary.

E3.12 Use loop analysis to solve the network in Example 3.4 and compare the time and effort involved in the two solution techniques.

E3.13 Use nodal analysis to solve the circuit in Example 3.13 and compare the time and effort involved in the two solution strategies.

Finally, it is important to note that the techniques for nodal and loop analysis are firmly rooted in topology, and the manner in which the network equations are derived via network topology is illustrated in Appendix A.

3.3 CIRCUITS WITH OPERATIONAL AMPLIFIERS

The operational amplifier, or op-amp as it is commonly known, is an extremely important circuit. It is a versatile interconnection of devices that vastly expands our capabilities in circuit design. It is employed in everything from engine control systems to microwave ovens. The op-amp was first introduced in the 1940s for use in analog computers and now finds wide application in circuit design as a result of the advances made in integrated circuit technology. Although we will model the op-amp as a fairly simple device, its actual construction involves the use of numerous components, including resistors, capacitors, and transistors. By adding resistors and capacitors to the op-amp terminals, we can use this simple op-amp model to design networks to perform many functions, such as addition, subtraction, differentiation, and integration, as well as voltage scaling and current-to-voltage conversion. These "building blocks" can then be employed to satisfy more complex applications.

The purpose of our model is to describe the terminal characteristics (the voltage–current relationships) of the op-amp. Therefore, let us begin by considering the terminals of the commercially available LM101 op-amp. Figure 3.25a shows an outline of the LM101 DIP (dual in-line package), which has eight pins, or terminals. Pins 1, 5, and 8 are used for fine adjustments of the device's characteristics; however, a discussion of them at this point would only serve to complicate the presentation unnecessarily. Pins 4 and 7 are voltage terminals for dc power supply (typically 10 to 15 V); and, in addition, they provide a return path to ground for ac current. The op-amp has two inputs—pins 2 and 3. A positive voltage on pin 3 produces a positive output. Accordingly, pin 3 is termed the *noninverting input*. A positive voltage at pin 2 yields a negative output; hence, it is called the *inverting input*. Based on the preceding description, the op-amp, unlike the elements we have introduced thus far, is essentially a five-terminal device. The circuit symbol for the op-amp with its five terminal connections is shown in Fig. 3.25b.

Functionally, the op-amp operates like the equivalent circuit shown in Fig. 3.25c. This equivalent circuit implies a unilateral device; that is, the output is determined by the difference in input voltages, but the input is unaffected by voltages at the output. The resistance R_i is very large, the gain A is very high, and the resistance R_o is very low; typical values for these parameters are 10^5 to 10^{12} Ω, 10^5 to 10^7 V/V, and 1 to 50 Ω, respectively.

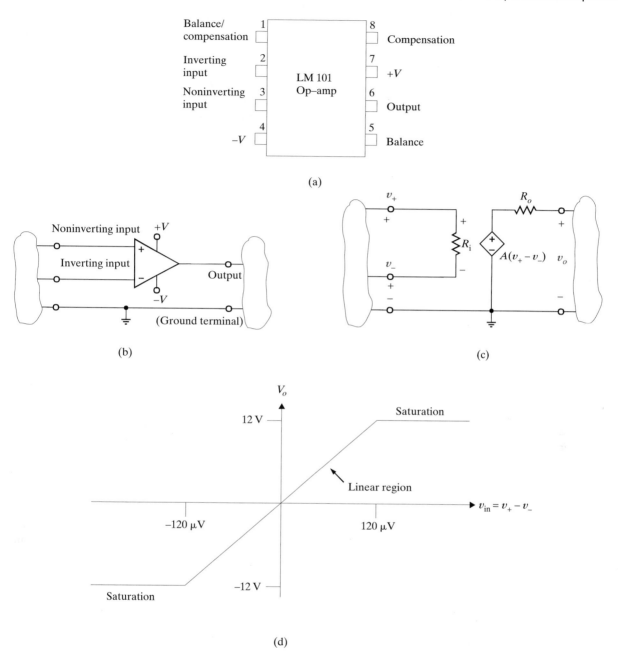

Figure 3.25 Operational amplifier representations.

The dc transfer curve (a plot of the output versus the input) for an op-amp with $A = 10^5$, $V_+ = 12\,\text{V}$ and $V_- = -12\,\text{V}$ is shown in Fig. 3.25d. The input voltage, V_{in}, is the noninverting input minus the inverting input voltages. Note first that the output voltage cannot exceed the power supply values. This is due to circuitry inside the op-amp.

Once the output reaches either power supply value, further increases in V_{in} have no effect on the output and the op-amp is said to be saturated. The portion of the curve of interest to us is the nonsaturated linear region. Here the ratio of output to input is the gain, $A = 10^5$.

The values listed previously for R_i, R_o, and A suggest an even simpler model known as the *ideal op-amp model*. R_i is very large and, therefore, can be assumed to be an open circuit. R_o is very low with respect to the recommended output load connection and, therefore, can be assumed to be zero. Finally, the gain A is a large amplification factor and can also be assumed to be infinite. At this point in our education, however, the infinite gain causes some concern. Does this mean that if the input is 1 μV the output is unlimited? Our common sense leads us to question such a condition. The reality of the situation is that the output voltages will always be between the dc power supply voltages, as shown in Fig. 3.25d. Therefore, if the op-amp is to remain in the linear region of operation, the input voltage (the difference between the noninverting and inverting input voltages) must remain within the limits.

$$\frac{V_-}{A} \leq V_{in} \leq \frac{V_+}{A}$$

For very large values of A, V_{in} approaches zero. Thus, an infinite gain yields a finite output voltage from a zero-volt input. We call this condition at the input terminals a *virtual short*.

Another important aspect of the op-amp configuration shown in Fig. 3.25c is that if the gain is assumed to be infinite, the output voltage is impossible to control. For the myriad of applications in which we need amplifiers, control of the output voltage is vital. The manner in which we accomplish this control is through feedback; that is, a portion of the output signal is fed back to the input. Later, we will show that through this mechanism the performance of the entire op-amp circuit can be controlled by external elements such as resistors. To accomplish this, the op-amp is connected to external components with a signal path from the output to the inverting input. With the op-amp connected in this manner, we will see that the high gain serves to maintain the input–output voltage relationships despite changes in load characteristics. As long as the amplifier gain A is very large, the circuit's functionality will be set by the externally connected components.

Under the assumption just stated, the ideal model for the op-amp is reduced to that shown in Fig. 3.26. The important characteristics of the model are as follows: (1) Since R_i is extremely large, the input currents to the op-amp are approximately zero (i.e., $i_+ \approx i_- \approx 0$); and (2) if the output voltage is to remain bounded, then as the gain becomes very large and approaches infinity, the voltage across the input terminals must simultaneously become infinitesimally small so that as $A \to \infty$, $v_+ - v_- \to 0$ (i.e., $v_+ - v_- = 0$ or $v_+ = v_-$). The difference between these input voltages is often called the *error signal* for the op-amp (i.e., $v_+ - v_- = v_e$).

The ground terminal $\perp$ shown on the op-amp is necessary for signal current return, and it guarantees that Kirchhoff's current law is satisfied at both the op-amp and the ground node in the circuit.

In summary, then, our ideal model for the op-amp is simply stated by the following conditions:

$$i_+ = i_- = 0$$
$$v_+ = v_-$$

3.8

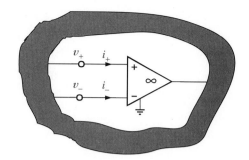

Figure 3.26 Ideal model for an operational amplifier. Model parameters: $i_+ = i_- = 0$, $v_+ = v_-$.

These simple conditions are extremely important because they form the basis of our analysis of op-amp circuits.

   The following examples will serve to illustrate the validity of our assumptions when analyzing op-amp circuits.

EXAMPLE 3.17

Let us determine the gain of the basic inverting op-amp configuration shown in Fig. 3.27a using both the nonideal and ideal op-amp models.

SOLUTION   Our model for the op-amp is shown generically in Fig. 3.27b and specifically in terms of the parameters R_i, A, and R_o in Fig. 3.27c. If the model is inserted in the network in Fig. 3.27a, we obtain the circuit shown in Fig. 3.27d, which can be redrawn as shown in Fig. 3.27e.

   The node equations for the network are

$$\frac{v_1 - v_S}{R_1} + \frac{v_1}{R_i} + \frac{v_1 - v_o}{R_2} = 0$$

$$\frac{v_o - v_1}{R_2} + \frac{v_o - Av_e}{R_o} = 0$$

where $v_e = -v_1$. The equations can be written in matrix form as

$$\begin{bmatrix} \dfrac{1}{R_1} + \dfrac{1}{R_i} + \dfrac{1}{R_2} & -\left(\dfrac{1}{R_2}\right) \\ -\left(\dfrac{1}{R_2} - \dfrac{A}{R_o}\right) & \dfrac{1}{R_2} + \dfrac{1}{R_o} \end{bmatrix} \begin{bmatrix} v_1 \\ v_0 \end{bmatrix} = \begin{bmatrix} \dfrac{v_S}{R_1} \\ 0 \end{bmatrix}$$

Solving for the node voltages, we obtain

$$\begin{bmatrix} v_1 \\ v_0 \end{bmatrix} = \frac{1}{\Delta} \begin{bmatrix} \dfrac{1}{R_2} + \dfrac{1}{R_o} & \dfrac{1}{R_2} \\ \dfrac{1}{R_2} - \dfrac{A}{R_o} & \dfrac{1}{R_1} + \dfrac{1}{R_i} + \dfrac{1}{R_o} \end{bmatrix} \begin{bmatrix} \dfrac{v_S}{R_1} \\ 0 \end{bmatrix}$$

where

$$\Delta = \left(\frac{1}{R_1} + \frac{1}{R_i} + \frac{1}{R_2}\right)\left(\frac{1}{R_2} + \frac{1}{R_o}\right) - \left(\frac{1}{R_2}\right)\left(\frac{1}{R_2} - \frac{A}{R_o}\right)$$

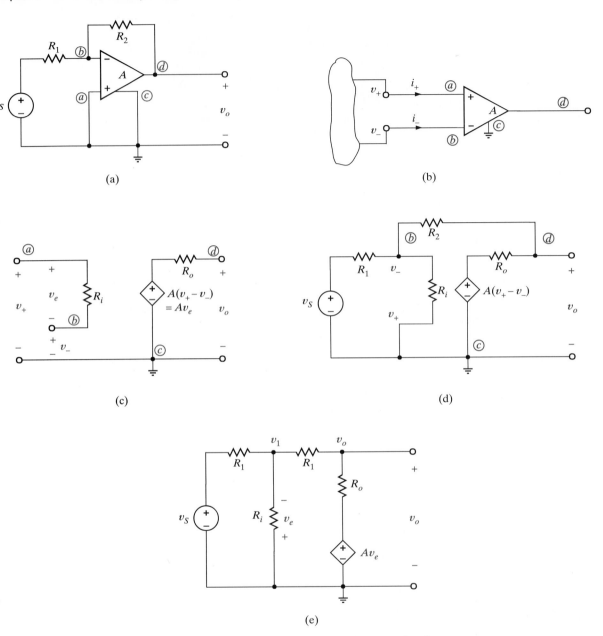

Figure 3.27 Op-amp circuit.

Hence,

$$v_o = \frac{\left(\dfrac{1}{R_2} - \dfrac{A}{R_o}\right)\left(\dfrac{v_S}{R_1}\right)}{\left(\dfrac{1}{R_1} + \dfrac{1}{R_i} + \dfrac{1}{R_2}\right)\left(\dfrac{1}{R_2} + \dfrac{1}{R_o}\right) - \left(\dfrac{1}{R_2}\right)\left(\dfrac{1}{R_2} - \dfrac{A}{R_o}\right)}$$

which can be written as

$$\frac{v_o}{v_S} = \frac{-(R_2/R_1)}{1 - \left[\left(\dfrac{1}{R_1} + \dfrac{1}{R_i} + \dfrac{1}{R_2}\right)\left(\dfrac{1}{R_2} + \dfrac{1}{R_o}\right) \Big/ \left(\dfrac{1}{R_2}\right)\left(\dfrac{1}{R_2} - \dfrac{A}{R_o}\right)\right]}$$

If we now employ typical values for the circuit parameters (e.g., $A = 10^5$, $R_i = 10^8 \ \Omega$, $R_o = 10 \ \Omega$, $R_1 = 1 \ \text{k}\Omega$, and $R_2 = 5 \ \text{k}\Omega$) the voltage gain of the network is

$$\frac{v_o}{v_S} = -4.9996994 \approx -5.000$$

However, the ideal op-amp has infinite gain and, therefore, if we take the limit of the gain equation as $A \to \infty$, we obtain

$$\lim_{A \to \infty}\left(\frac{v_o}{v_S}\right) = -\frac{R_2}{R_1} = -5.000$$

Note that the ideal op-amp yielded a result accurate to within four significant digits of that obtained from an exact solution of a typical op-amp model. These results are easily repeated for the vast array of useful op-amp circuits.

We now analyze the network in Fig. 3.27a using the ideal op-amp model. In this model

$$i_+ = i_- = 0$$
$$v_+ = v_-$$

As shown in Fig. 3.27a, $v_+ = 0$ and, therefore, $v_- = 0$. If we now write a node equation at the negative terminal of the op-amp, we obtain

$$\frac{v_S - 0}{R_1} + \frac{v_o - 0}{R_2} = 0$$

or

$$\frac{v_o}{v_S} = -\frac{R_2}{R_1}$$

and we have immediately obtained the results derived previously.

Notice that the gain is a simple resistor ratio. This fact makes the amplifier very versatile in that we can control the gain accurately and alter its value by changing only one resistor. Also, the gain is essentially independent of op-amp parameters. Since the precise values of A, R_i, and R_o are sensitive to such factors as temperature, radiation, and age, their elimination results in a gain that is stable regardless of the immediate environment. Since it is much easier to employ the ideal op-amp model rather than the nonideal model, unless otherwise stated we will use the ideal op-amp assumptions to analyze circuits that contain operational amplifiers.   ❏

EXAMPLE 3.18

Consider the op-amp circuit shown in Fig. 3.28. Let us determine an expression for the output voltage.

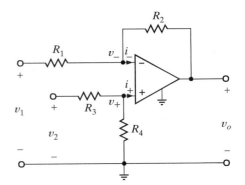

Figure 3.28 Differential amplifier operational amplifier circuit.

SOLUTION   The node equation at the inverting terminal is

$$\frac{v_1 - v_-}{R_1} + \frac{v_o - v_-}{R_2} = i_-$$

At the noninverting terminal KCL yields

$$\frac{v_2 - v_+}{R_3} = \frac{v_+}{R_4} + i_+$$

However, $i_+ = i_- = 0$ and $v_+ = v_-$. Substituting these values into the two preceding equations yields

$$\frac{v_1 - v_-}{R_1} + \frac{v_o - v_-}{R_2} = 0$$

and

$$\frac{v_2 - v_-}{R_3} = \frac{v_-}{R_4}$$

Solving these two equations for v_o results in the expression

$$v_o = \frac{R_2}{R_1}\left(1 + \frac{R_1}{R_2}\right)\frac{R_4}{R_3 + R_4}v_2 - \frac{R_2}{R_1}v_1$$

Note that if $R_4 = R_2$ and $R_3 = R_1$, the expression reduces to

$$v_o = \frac{R_2}{R_1}(v_2 - v_1)$$

Therefore, this op-amp can be employed to subtract two input voltages.   ❑

EXAMPLE 3.19

The circuit shown in Fig. 3.29a is a precision differential voltage-gain device. It is used to provide a single-ended input for an analog-to-digital converter. We wish to derive an expression for the output of the circuit in terms of the two inputs.

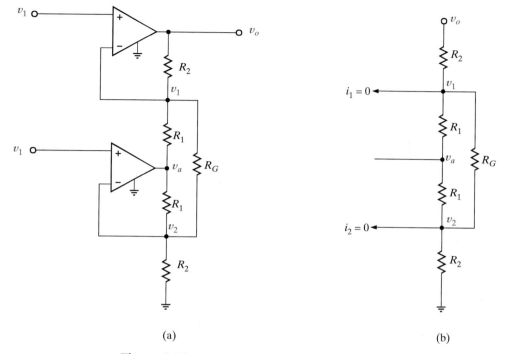

Figure 3.29 Instrumentation amplifier circuit.

SOLUTION   To accomplish this, we draw the equivalent circuit shown in Fig. 3.29b. Recall that the voltage across the input terminals of the op-amp is approximately zero and the currents into the op-amp input terminals are approximately zero. Note that we can write node equations for node voltages v_1 and v_2 in terms of v_o and v_a. Since we are interested in an expression for v_o in terms of the voltages v_1 and v_2, we simply eliminate the v_a terms from the two node equations. The node equations are

$$\frac{v_1 - v_o}{R_2} + \frac{v_1 - v_a}{R_1} + \frac{v_1 - v_2}{R_G} = 0$$

$$\frac{v_2 - v_a}{R_1} + \frac{v_2 - v_1}{R_G} + \frac{v_2}{R_2} = 0$$

Combining the two equations to eliminate v_a, and then writing v_o in terms of v_1 and v_2, yields

$$v_o = (v_1 - v_2)\left(1 + \frac{R_2}{R_1} + \frac{2R_2}{R_G}\right)$$

❑

EXAMPLE 3.20

We wish to calculate the input resistance of the op-amp circuit shown in Fig. 3.30 without any assumptions about the sizes of the parameters A, R_i, and R_o.

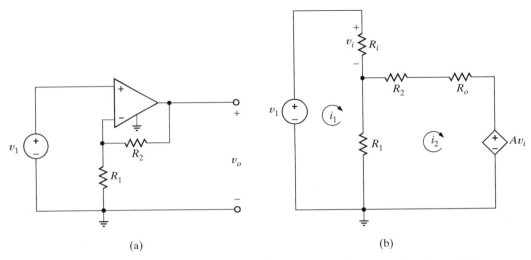

Figure 3.30 Circuit and model for a noninverting operational amplifier.

SOLUTION The input resistance is the resistance seen by the input voltage source v_1. Using the op-amp model shown in Fig. 3.25c, we obtain the circuit shown in Fig. 3.30b. The mesh equations for this circuit are

$$v_1 = i_1(R_i + R_1) - i_2 R_1$$
$$-Av_i = -i_1 R_1 + i_2(R_1 + R_2 + R_o)$$
$$v_i = R_i i_1$$

Solving these equations for i_1 yields

$$i_1 = \frac{v_1(R_1 + R_2 + R_o)}{(R_i + R_1)(R_1 + R_2 + R_o) + R_1(AR_i - R_1)}$$

The input resistance is

$$R_{in} = \frac{v_1}{i_1} = \frac{(R_i + R_1)(R_1 + R_2 + R_o) + R_1(AR_i - R_1)}{R_1 + R_2 + R_o}$$

$$= R_i + \frac{R_1(R_2 + R_o + AR_i)}{R_1 + R_2 + R_o}$$

Since A and R_i are normally very large and R_o is very small, the expression for R_{in} can be approximated by the equation

$$R_{in} \simeq \frac{AR_i}{1 + R_2/R_1}$$ ❏

EXTENSION EXERCISES

E3.14 Find I_o in the network in Fig. E3.14.

ANSWER: $I_o = 8.4$ mA.

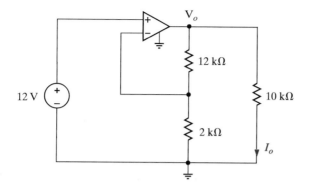

Figure E3.14

E3.15   Determine the gain of the op-amp circuit in Fig. E3.15.

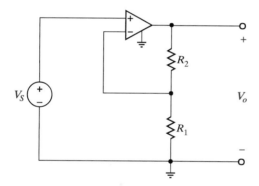

Figure E3.15

ANSWER: $\dfrac{V_o}{V_S} = 1 + \dfrac{R_2}{R_1}.$

E3.16   Determine both the gain and the output voltage of the op-amp configuration shown in Fig. E3.16.

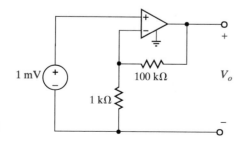

Figure E3.16

ANSWER: $V_o = 0.101$ V, gain $= 101$.

An important special case of the noninverting amplifier is the configuration shown in Fig. 3.31. Note carefully that this circuit is equivalent to that shown in Fig. E3.15 with $R_1 = \infty$ (i.e., open circuited) and $R_2 = 0$ (i.e., short circuited). Under these conditions we note that the gain of the circuit is $1 + R_2/R_1 = 1$ (i.e., $v_o = v_S$). Since v_o follows v_S, the circuit is called a *voltage follower*.

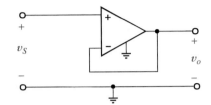

Figure 3.31 Voltage-follower operational amplifier configuration.

An obvious question at this point is this: If $v_o = v_S$, why not just connect v_S to v_o via two parallel connection wires; why do we need to place an op-amp between them? The answer to this question is fundamental and provides us with some insight that will aid us in circuit analysis and design.

Consider the circuit shown in Fig. 3.32a. In this case v_o is not equal to v_S because of the voltage drop across R_S:

$$v_o = v_S - iR_S$$

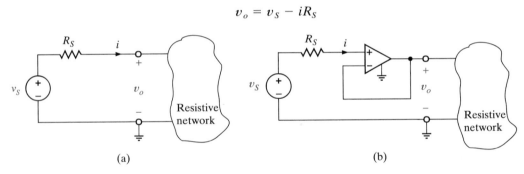

Figure 3.32 Illustration of the isolation capability of a voltage follower.

However, in Fig. 3.32b, the input current to the op-amp is zero and, therefore, v_S appears at the op-amp input. Since the gain of the op-amp configuration is $1, v_o = v_S$. In Fig. 3.32a the resistive network's interaction with the source caused the voltage v_o to be less than v_S. In other words, the resistive network loads the source voltage. However, in Fig. 3.32b the op-amp isolates the source from the resistive network, and therefore the voltage follower is referred to as a *buffer amplifier* because it can be used to isolate one circuit from another. The energy supplied to the resistive network in the first case must come from the source v_S, whereas in the second case it comes from the power supplies that supply the amplifier, and little or no energy is drawn from v_S.

EXAMPLE 3.21

Often we present amplifier gain in graphical format. Plots of the output signal versus the input are called *transfer plots* or *transfer curves*. Let us find an expression for v_o in Fig. 3.33 and plot the transfer curve.

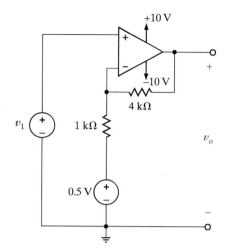

Figure 3.33 Circuit used in Example 3.21.

SOLUTION   Note that we have deviated from our normal model for the op-amp and specifically shown the dc supply voltages in the circuit. We have made an exception in this case because of their relevance to the transfer curves.

The nodal equation at the inverting input is

$$\frac{v_1 - 0.5}{1k} + \frac{v_1 - v_o}{4k} = 0$$

The output voltage is then given by the equation

$$v_o = 5v_1 - 2$$

The transfer curve for this circuit, shown in Fig. 3.34, contains several key features. First, the slope of the line,

$$\frac{dv_o}{dv_1} = 5 \text{ V/V}$$

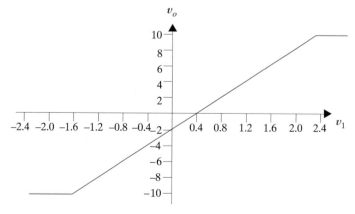

Figure 3.34 Transfer curve for an op-amp configuration.

is the gain of the circuit. Second, the output voltage is shifted by the 0.5-V source. Since v_1 is 0.4 V when v_o is 0 V, we say that the amplifier is *offset* by +0.40 V. Third, the output voltage cannot exceed the dc power supply voltage value. The output voltage obeys the transfer equation until v_o reaches either +10 V or −10 V. Beyond that point, v_o saturates at the power supply voltages. Since the ideal op-amp model does not include any information about how large the output voltage can be, it is left to the circuit designers to be conscious of the limitations of the model we are using. This is a common theme in the modeling of physical devices with mathematical expressions. Models tend to be valid only under certain conditions. It is paramount that the engineer understand the model's limitations and its implications. ❏

3.4 APPLICATIONS

At this point, we have a new element, the op-amp, which we can effectively employ in both applications and circuit design. This device is an extremely useful element that vastly expands our capability in these areas. Because of its ubiquitous nature, the addition of the op-amp to our repertoire of circuit elements permits us to deal with a wide spectrum of practical circuits. Thus, we will not only employ it here, but we will use it throughout this text.

EXAMPLE 3.22

In a light meter, a sensor produces a current proportional to the intensity of the incident radiation. We wish to obtain a voltage proportional to the light's intensity using the circuit in Fig. 3.35. Thus, we select a value of R that will produce an output voltage of 1 V for each 10μA of sensor current. Assume the sensor has zero resistance.

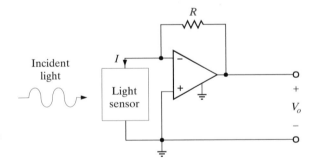

Figure 3.35 Light intensity to voltage converter.

SOLUTION   Applying KCL at the op-amp input,

$$I = V_o/R$$

Since V_o/I is 10^5,

$$R = 100 \text{ k}\Omega$$ ❏

EXAMPLE 3.23

The circuit in Fig. 3.36 is an electronic ammeter. It operates as follows: The unknown current, I through R_I produces a voltage, V_I. V_I is amplified by the op-amp to produce a voltage, V_o, which is proportional to I. The output voltage is measured with a simple voltmeter. We want to find the value of R_2 such that 10 V appears at V_o for each milliamp of unknown current.

SOLUTION   Since the current into the op-amp + terminal is zero, the relationship between V_I and I is

$$V_I = IR_I$$

The relationship between the input and output voltages is

$$V_o = V_I\left(1 + \frac{R_2}{R_1}\right)$$

or, solving the equation for V_o/I, we obtain

$$\frac{V_o}{I} = R_I\left(1 + \frac{R_2}{R_1}\right)$$

Using the required ratio V_o/I of 10^4 and resistor values from Fig. 3.36, we can find that

$$R_2 = 9\ \text{k}\Omega$$

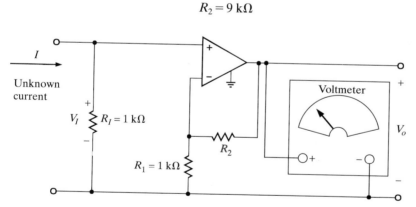

Figure 3.36 Electronic ammeter.

3.5 CIRCUIT DESIGN

EXAMPLE 3.24

A typical stereo system is shown in block form in Fig. 3.37a. The phonograph output signal is only about 2 μV. The standard input voltage for stereo power amplifiers is about 2 mV. Therefore, the phonograph signal must be amplified by a factor of 1000 before it reaches the poweramp. To accomplish this, a special-purpose amp, called the phono preamp, is used. Stereo manufacturers place them within the pre-amplifier cabinet, as shown in Fig. 3.37a.

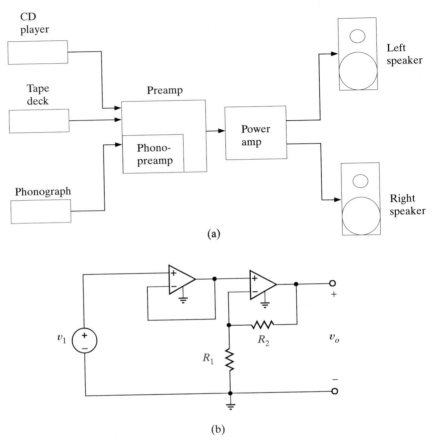

(a)

(b)

Figure 3.37 Multistage phonograph amplifier.

Let us design a phono preamp using an ideal op-amp that has an input resistance of at least 1 MΩ and a voltage gain of 1000. All resistors must be less than 10 MΩ in order to limit noise.

SOLUTION One possible network is shown in Fig. 3.37b. The input resistance requirement can be easily met with a voltage follower as the first stage of the amplifier. The second stage, or gain stage, can be a noninverting op-amp configuration. We will show in later chapters that the overall voltage gain is the product of the gains of the two stages,

$$A_V = A_{V1} A_{V2} = (1)(1 + (R_2/R_1))$$

To achieve a gain of 1000, we select $R_1 = 1\ \text{k}\Omega$ and $R_2 = 999\ \text{k}\Omega$.

EXAMPLE 3.25

A thermocouple is a two-wire element that produces a small voltage proportional to temperature. A particular thermocouple generates a voltage of 6 mV for every 1°C rise in temperature. This signal is to be used to activate a voltage-controlled switch that requires an input signal of 5 V to operate. Fig. 3.38a shows a block diagram of the system. Let us design an op-amp circuit that will trigger the switch at the boiling point of water (100°C) if the thermocouple generates 0 V at 0°C.

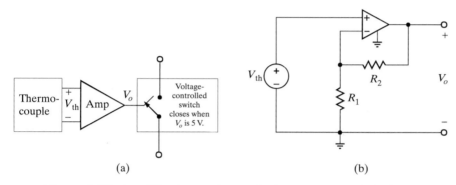

(a)   (b)

Figure 3.38 Amplifier for an electronic thermometer switch driver.

SOLUTION   The single-stage noninverting op-amp configuration shown in Fig. 3.38b should satisfy the requirement. At 100°C, the thermocouple voltage, V_{th} is

$$V_{th} = (100°)(6 \text{ mV}/°) = 600 \text{ mV}$$

This 600-mV signal must be amplified to 5 V to trigger the switch.
   Therefore, the op-amp gain is

$$A_V = \frac{5}{0.6} = 8.33$$

Selecting $R_1 = 3 \text{ k}\Omega$ and $R_2 = 22 \text{ k}\Omega$ in the op-amp circuit in Fig. 3.38b satisfies the gain requirement.   ❏

EXAMPLE 3.26

Nerve fibers carry small electrical impulses from the brain to the muscles. These impulses cause muscle contraction and body movement. Physicians can monitor voltages in a muscle as a diagnostic process to look for proper brain–nerve–muscle interaction. A simplified version of the medical instrument used for this purpose is shown in Fig. 3.39. Conductive needles are inserted into a muscle and the voltage difference is measured. The signals on these needles are modeled in Fig. 3.39 by the voltage sources V_1 and V_2. The two voltage followers are used to buffer the patient from the instrument. Let us design the amplifier such that the output voltage is $100(V_1 - V_2)$.

SOLUTION   The input–output relationship for the circuit in Fig. 3.39 is

$$V_o = (R_2/R_1)(V_1 - V_2) = 100(V_1 - V_2)$$

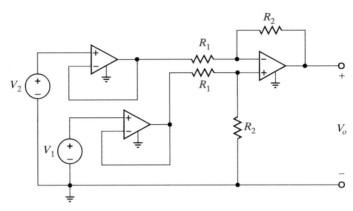

Figure 3.39   Neuromuscular voltage amplifier.

We have two unknowns and only one equation. Thus, if we select $R_1 = 1\ k\Omega$, then $R_2 = 100\ k\Omega$. The configuration with these values satisfies the requirements.   ❏

EXAMPLE 3.27

The amplitude of the signal received by the antenna of a portable AM-radio receiver is 1 μV. The signal is then amplified to supply 8 V to a speaker. What is the required voltage gain of this system? Let us design a four-stage amplifier using identical inverting op-amp configurations to meet the gain requirement.

SOLUTION   The required voltage gain is

$$A_V = 8/10^{-6} = 8 \times 10^6$$

As mentioned earlier, a four-stage amplifier has an overall gain equal to the product of the gains of the individual stages; that is,

$$A_V = A_{V1} A_{V2} A_{V3} A_{V4}$$

Since in this case the four stages are identical, we can write

$$A_{V1} = A_{V2} = A_{V3} = A_{V4} = A$$

where A represents the gain of a single stage. The inverting op-amp has a gain of

$$A = -\frac{R_2}{R_1}$$

The gain of a single stage is then

$$-\frac{R_2}{R_1} = -\sqrt[4]{(8 \times 10^6)} = -53.2$$

Therefore, the four-stage op-amp configuration shown in Fig. 3.40 will satisfy the design requirements.

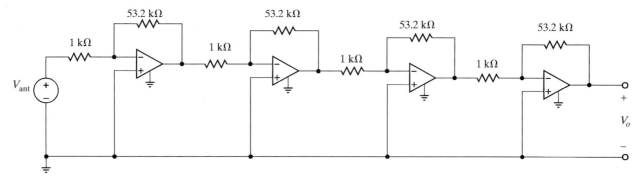

Figure 3.40 Four-stage AM-radio amplifier.   ❑

3.6 SUMMARY

Nodal analysis for an *N*-node circuit

- Select one node in the *N*-node circuit as the reference node. Assume that the node voltage is zero and measure all node voltages with respect to this node.
- If only independent current sources are present in the network, write the KCL equations as the $N - 1$ nonreference nodes. If dependent current sources are present, write the KCL equations as is done for networks with only independent current sources; then write the controlling equations for the dependent sources.
- If voltage sources are present in the network, they may be connected (1) between the reference node and a nonreference node or (2) between two nonreference nodes. In the former case, if the voltage source is an independent source, then the voltage at one of the nonreference nodes is known. If the source is dependent, it is treated as an independent source when writing the KCL equations, but an additional constraint equation is necessary.

    In the latter case, if the source is independent, the voltage between the two nodes is constrained by the value of the voltage source and an equation describing this constraint represents one of the $N - 1$ linearly independent equations required to determine the *N*-node voltages. The surface of the network described by the constraint equation (i.e., the source and two connecting nodes) is called a supernode. One of the remaining $N - 1$ linearly independent equations is obtained by applying KCL at this supernode. If the voltage source is dependent, it is treated as an independent source when writing the KCL equations, but an additional constraint equation is necessary.

Loop analysis for a N-loop circuit

- One loop current is assigned to each independent loop in a circuit that contains N independent loops.

- If only independent voltage sources are present in the network, write the N linearly independent KVL equations, one for each loop. If dependent voltage sources are present, write the KVL equations as is done for circuits with only independent voltage sources; then write the controlling equations for the dependent sources.

- If current sources are present in the network, either of two techniques can be used. In the first case, one loop current is selected to pass through one of the current sources. This is done for each current source in the network. The remaining loop currents (N − the number of current sources) are determined by open-circuiting the current sources in the network and using this modified network to select them. Once all these currents are defined in the original network, the N loop equations can be written. The second approach is similar to the first with the exception that if two mesh currents pass through a particular current source, a supermesh is formed around this source. The two required equations for the meshes containing this source are the constraint equations for the two mesh currents that pass through the source and the supermesh equation. If dependent current sources are present, the controlling equations for these sources are also necessary.

Ideal Op-Amp Model   For an ideal op-amp, $i_x + i_- = 0$ and $v_x + v_-$. Both nodal and loop analysis are useful in solving circuits containing operational amplifiers.

PROBLEMS

Section 3.1

3.1   Find I_o in the circuit in Fig. P3.1 using nodal analysis.

Write the KCL equations for nodes 1 and 2 in terms of the node voltages and solve the equations for I_o.

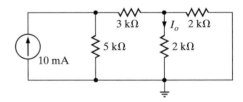

Figure P3.1

3.2 Use nodal analysis to find V_1 in the circuit in Fig. P3.2.

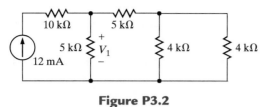

Similar to P3.1.

Figure P3.2

3.3 Find V_2 in the circuit in Fig. P3.3 using nodal analysis.

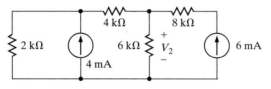

Similar to P3.1.

Figure P3.3

3.4 Use nodal analysis to find V_o in the circuit in Fig. P3.4.

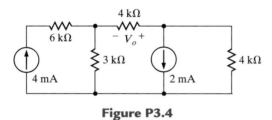

Similar to P3.1. Note that V_o, is the difference between two node voltages.

Figure P3.4

3.5 Find V_1 and V_2 in the circuit in Fig. P3.5 using nodal analysis.

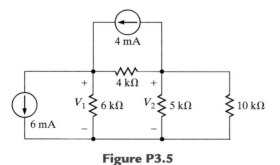

Similar to P3.1.

Figure P3.5

3.6 Use nodal analysis to find V_1 in the circuit in Fig. P3.6.

Similar to P3.1.

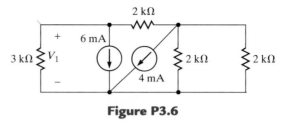

Figure P3.6

3.7 Find I_o in the circuit in Fig. P3.7 using nodal analysis.

Similar to P3.1.

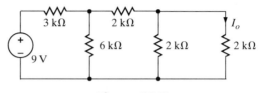

Figure P3.7

3.8 Use nodal analysis to find V_o in the circuit in Fig. P3.8.

Similar to P3.1.

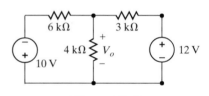

Figure P3.8

3.9 Find V_o in the circuit in Fig. P3.9 using nodal analysis.

Similar to P3.1.

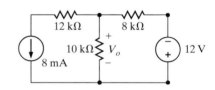

Figure P3.9

3.10 Use nodal analysis to find V_o in the circuit in Fig. P3.10.

Similar to P3.1.

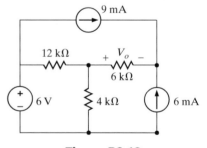

Figure P3.10

3.11 Find V_o in the circuit in Fig. P3.11 using nodal analysis.

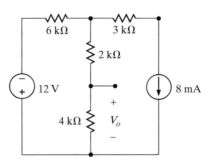

Figure P3.11

Similar to P3.1. Note that voltage division is useful in this case.

3.12 Use nodal analysis to find V_o in the circuit in Fig. P3.12.

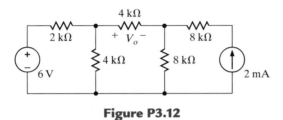

Figure P3.12

Similar to P3.4.

3.13 Find V_o in the circuit in Fig. P3.13 using nodal analysis.

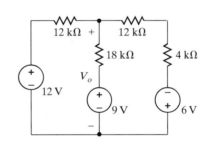

Figure P3.13

Similar to P3.1. The 12-kΩ and 4-kΩ resistors can be combined.

3.14 Use nodal analysis to find V_o in the circuit in Fig. P3.14.

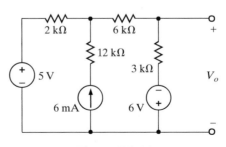

Figure P3.14

Similar to P3.13.

3.15 Find V_o in the network in Fig. P3.15 using nodal analysis.

V_o is the difference between two node voltages.

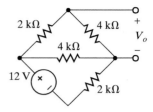

Figure P3.15

3.16 Use nodal analysis to find V_o in the circuit in Fig. P3.16.

Similar to P3.15.

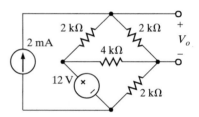

Figure P3.16

3.17 Find V_o in the network in Fig. P3.17 using nodal analysis.

Similar to P3.15.

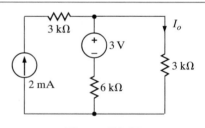

Figure P3.17

3.18 Find I_o in the circuit in Fig. P3.18 using nodal analysis.

Similar to P3.1 if the locations of the 3V source and 6KΩ resistor are exchanged.

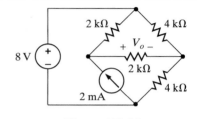

Figure P3.18

3.19 Use nodal analysis to find I_o in the circuit in Fig. P3.19.

Similar to P3.1.

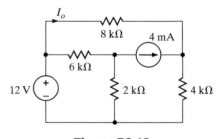

Figure P3.19

3.20 Find V_o in the circuit in Fig. P3.20 using nodal analysis.

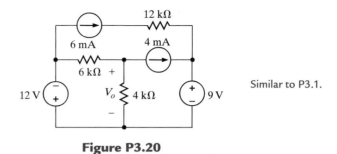

Figure P3.20

Similar to P3.1.

3.21 Use nodal analysis to find I_o and I_S in the circuit in Fig. P3.21.

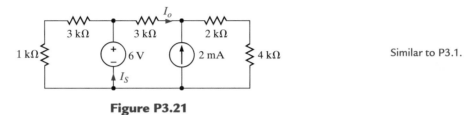

Figure P3.21

Similar to P3.1.

3.22 Find V_o in the network in Fig. P3.22 using nodal analysis.

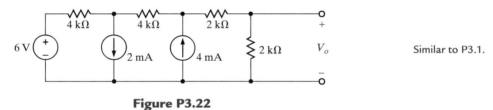

Figure P3.22

Similar to P3.1.

3.23 Find I_o in the circuit in Fig. P3.23 using nodal analysis.

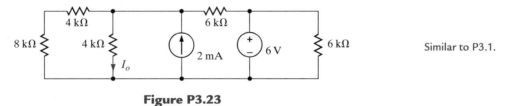

Figure P3.23

Similar to P3.1.

3.24 Use nodal analysis to find V_o in the network in Fig. P3.24.

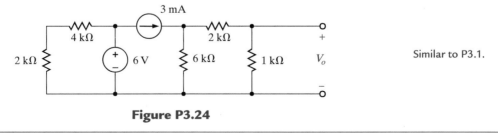

Figure P3.24

Similar to P3.1.

3.25 Use nodal analysis to find V_o in the circuit in Fig. P3.25.

Similar to P3.15.

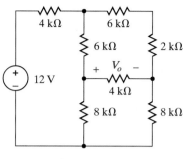

Figure P3.25

3.26 Find I_o in the circuit in Fig. P3.26 using nodal analysis.

Use the supernode technique.

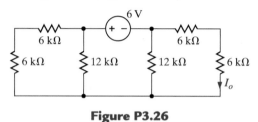

Figure P3.26

3.27 Use nodal analysis to find I_o in the circuit in Fig. P3.27.

Similar to P3.26.

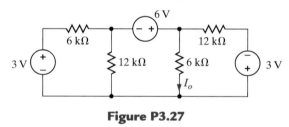

Figure P3.27

3.28 Using nodal analysis find V_o in the network in Fig. P3.28.

Similar to P3.26.

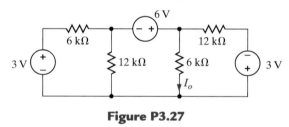

Figure P3.28

3.29 Find V_o in the network in Fig. P3.29 using nodal analysis.

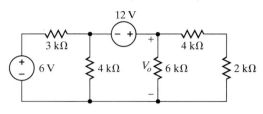

Figure P3.29

Similar to P3.26.

3.30 Use nodal analysis to find V_o in Fig. P3.30.

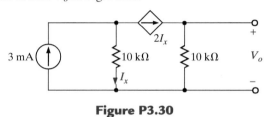

Figure P3.30

Treat the dependent source as an independent source and then include the controlling parameter.

3.31 Find I_o in the circuit in Fig. P3.31 using nodal analysis.

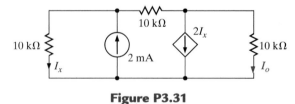

Figure P3.31

Similar to P3.30.

3.32 Find V_o in the network in Fig. P3.32 using nodal analysis.

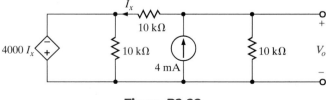

Figure P3.32

Similar to P3.30.

3.33 Use nodal analysis to find V_o in the circuit in Fig. P3.33. In addition, find all branch currents and check your answers using KCL at every node.

Similar to P3.30. Use the
supernode technique.

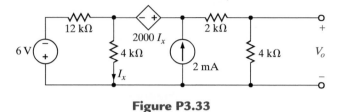

Figure P3.33

3.34 Find V_o in the network in Fig. P3.34. In addition, determine all branch currents and check KCL at every node.

Similar to P3.33.

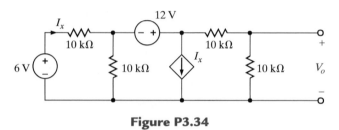

Figure P3.34

3.35 Find V_o in the circuit in Fig. P3.35.

Similar to P3.33.

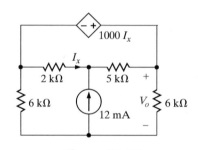

Figure P3.35

3.36 Find V_o in the circuit in Fig. P3.36.

Similar to P3.30.

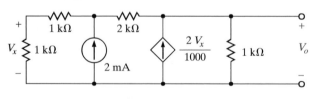

Figure P3.36

3.37 Find V_o in the circuit in Fig. P3.37.

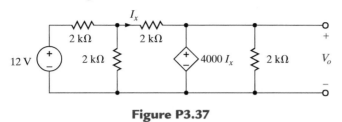

Figure P3.37

Similar to P3.30.

Section 3.2

3.38 Use mesh analysis to find V_o in the circuit in Fig. P3.38.

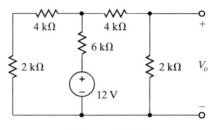

Figure P3.38

Write the mesh current equations and solve them to find V_o.

3.39 Use mesh analysis to find V_o in the network in Fig. P3.39.

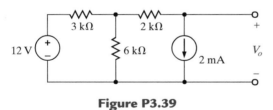

Figure P3.39

Similar to P3.38.

3.40 Use loop analysis to find V_o in the circuit in Fig. P3.40.

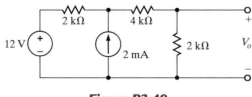

Figure P3.40

Use one loop each through the current source and the outer loop, or apply the supermesh technique.

3.41 Find I_o in the network in Fig. P3.41 using mesh analysis.

Similar to P3.39.

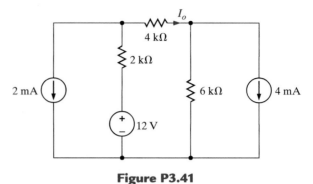

Figure P3.41

3.42 Find I_o in the network in Fig. P3.42.

Similar to P3.40.

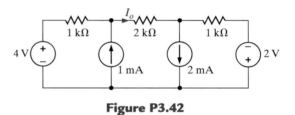

Figure P3.42

3.43 Use loop analysis to find I_o in the circuit in Fig. P3.43.

Similar to P3.40.

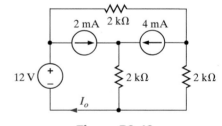

Figure P3.43

3.44 Use both nodal analysis and mesh analysis to find I_o in the circuit in Fig. P3.44.

The supernode is useful for nodal analysis, and the problem is similar to P3.39 for mesh analysis.

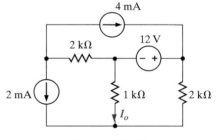

Figure P3.44

3.45 Find I_o in the network in Fig. P3.45 using mesh analysis.

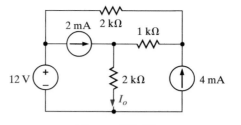

Figure P3.45

The supermesh technique is applicable here.

3.46 Find V_o in the network in Fig. P3.46 using both mesh and nodal analysis.

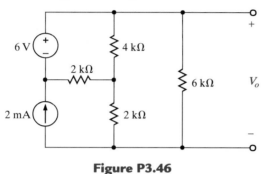

Figure P3.46

Similar to P3.44.

3.47 Use mesh analysis to find V_o in the circuit in Fig. P3.47.

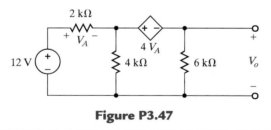

Figure P3.47

Treat the dependent source as an independent source and then include the controlling parameter.

3.48 Use loop analysis to find V_o in the circuit in Fig. P3.48.

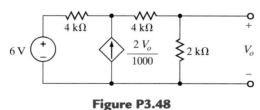

Figure P3.48

Similar to P3.40.

3.49 Find V_o in the circuit in Fig. P3.49 using mesh analysis.

Similar to P3.47.

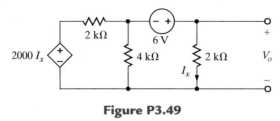

Figure P3.49

3.50 Using mesh analysis find V_o in the circuit in Fig. P3.50.

Similar to P3.44.

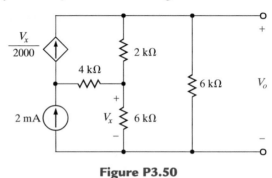

Figure P3.50

3.51 Use both nodal analysis and mesh analysis to find V_o in the circuit in Fig. P3.51.

Use a supernode for nodal analysis and a supermesh for mesh analysis.

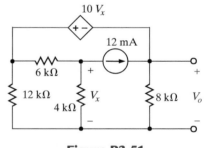

Figure P3.51

Section 3.3

3.52 Find V_o in the circuit in Fig. P3.52 assuming the op-amp is ideal.

Use $v_- = v_+ = 0$, $i_- = i_+ = 0$, and nodal analysis to find v_o.

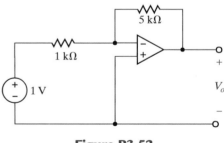

Figure P3.52

3.53 The network in Fig. P3.53 is a current-to-voltage converter or transconductance amplifier. Find v_o/i_S for this network.

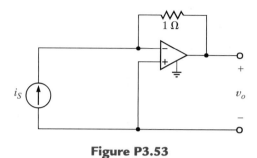

Figure P3.53

Similar to P3.52.

3.54 Find v_o in the network in Fig. P3.54 and explain what effect R_1 has on the output.

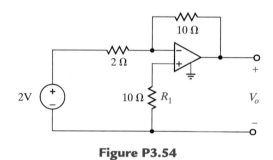

Figure P3.54

Similar to P3.52. The use of $i_+ = 0$ indicates R_1's effect.

3.55 Find V_o in the network in Fig. P3.55.

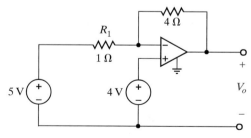

Figure P3.55

Similar to P3.52 with $v_- = v_+ = 4$.

3.56 Find the voltage gain of the op-amp circuit shown in Fig. P3.56.

Find v_+ using $i_+ = 0$. Find v_- using $i_- = 0$, and use $v_+ = v_-$ to find v_o.

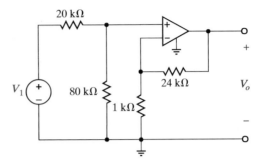

Figure P3.56

3.57 For the circuit in Fig. P3.57 find the value of R_1 that produces a voltage gain of 10.

Similar to P3.55.

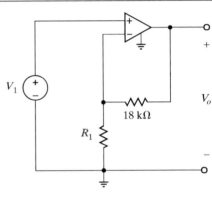

Figure P3.57

The circuit in P3.52 can generate $v_o = -2v_1$. Then $-4v_2$ can be obtained from the V_- terminal.

3.58 Given a box of 10-kΩ resistors and an op-amp, design a circuit that will have an output voltage of

$$v_o = -2v_1 - 4v_2$$

3.59 Find the expression for the output voltage in the op-amp circuit shown in Fig. P3.59.

Use $v_- = v_+ = v_1$, $i_- = i_+ = 0$, and nodal analysis to obtain v_o. Obtain v_{out} in a similar manner.

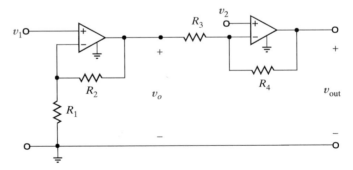

Figure P3.59

Advanced Problems

3.60 Find V_o in the circuit in Fig. P3.60 using both nodal and loop analysis.

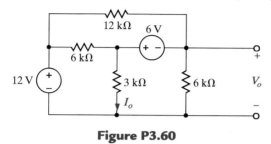

Figure P3.60

3.61 Find V_o in the circuit in Fig. P3.61 using both nodal and loop analysis.

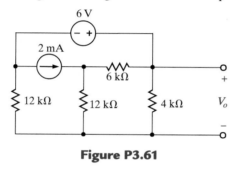

Figure P3.61

3.62 Use both nodal and loop analysis to find V_o in the circuit in Fig. P3.62.

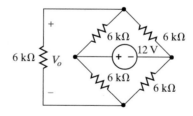

Figure P3.62

3.63 Use loop analysis to find V_o in the circuit in Fig. P3.63.

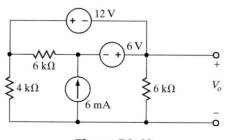

Figure P3.63

3.64 Find V_o in the circuit in Fig. P3.64.

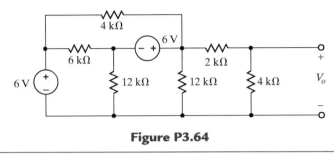

Figure P3.64

3.65 Find all branch currents and node voltages in the circuit in Fig. P3.65.

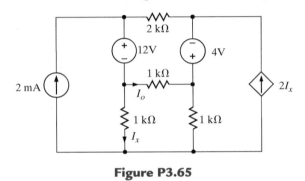

Figure P3.65

3.66 Find V_o in the circuit in Fig. P3.66 using both nodal and loop analysis.

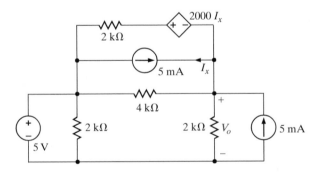

Figure P3.66

3.67 Use loop analysis to find V_o in the circuit in Fig. P3.67.

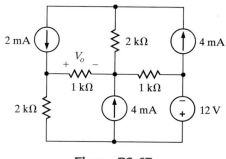

Figure P3.67

3.68 Find V_o in the network in Fig. P3.68.

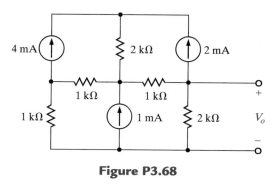

Figure P3.68

3.69 Use loop analysis to find V_o in the network in Fig. P3.69.

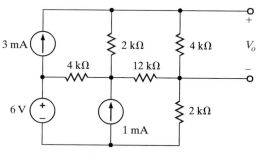

Figure P3.69

3.70 Find V_o in the circuit in Fig. P3.70.

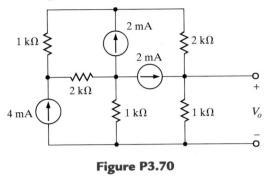

Figure P3.70

3.71 Find I_o in the circuit in Fig. P3.71.

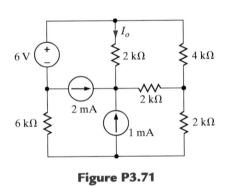

Figure P3.71

3.72 Find V_o in the network in Fig. P3.72.

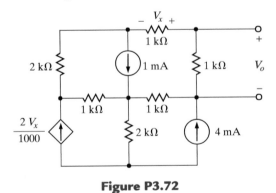

Figure P3.72

3.73 Draw a network defined by the following node equations.

$$V_1\left(\frac{1}{1\,\text{k}\Omega} + \frac{1}{2\,\text{k}\Omega} + \frac{1}{2\,\text{k}\Omega}\right) - V_2\left(\frac{1}{2\,\text{k}\Omega}\right)$$

$$= \frac{4\,\text{V}}{1\,\text{k}\Omega} + \frac{2\,\text{V}}{1\,\text{k}\Omega}; \qquad -V_1\left(\frac{1}{2\,\text{k}\Omega}\right) + V_2\left(\frac{1}{2\,\text{k}\Omega} + \frac{1}{4\,\text{k}\Omega}\right)$$

$$= -\frac{1\,\text{V}}{1\,\text{k}\Omega}$$

3.74 Draw a network defined by the following node equations.

$$V_1\left(\frac{1}{8\,\text{k}\Omega} + \frac{1}{1\,\text{k}\Omega}\right) - V_2\left(\frac{1}{1\,\text{k}\Omega}\right) = -\frac{2\,\text{V}}{1\,\text{k}\Omega} - \frac{1\,\text{V}}{1\,\text{k}\Omega}$$

$$-V_1\left(\frac{1}{1\,\text{k}\Omega}\right) + V_2\left(\frac{1}{1\,\text{k}\Omega} + \frac{1}{2\,\text{k}\Omega}\right) = \frac{2\,\text{V}}{1\,\text{k}\Omega} + \frac{4\,\text{V}}{1\,\text{k}\Omega}$$

3.75 Draw a network defined by the following mesh equations.

$$(12\,\text{k}\Omega)I_1 - (4\,\text{k}\Omega)I_3 = -28\,\text{V}$$

$$(3\,\text{k}\Omega)I_2 - (2\,\text{k}\Omega)I_3 = 16\,\text{V}$$

$$I_3 = -\frac{4\,\text{V}}{1\,\text{k}\Omega}$$

3.76 Draw a circuit defined by the following mesh equations.

$$I_1 = -\frac{6\,\text{V}}{1\,\text{k}\Omega}$$

$$-(4\,\text{k}\Omega)I_1 + (12\,\text{k}\Omega)I_2 - (2\,\text{k}\Omega)I_4 = 0$$

$$(4\,\text{k}\Omega)I_3 - (1\,\text{k}\Omega)I_4 = -12\,\text{V}$$

$$I_4 = -\frac{4\,\text{V}}{1\,\text{k}\Omega}$$

3.77 Find I_o in the circuit in Fig. P3.77.

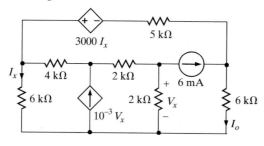

Figure P3.77

3.78 Find V_x in the circuit in Fig. P3.78.

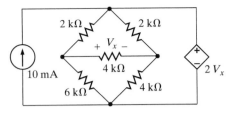

Figure P3.78

3.79 Find I_o in the circuit in Fig. P3.79.

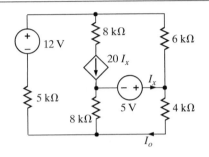

Figure P3.79

3.80 Find V_o in the circuit in Fig. P3.80.

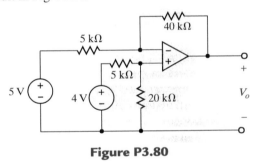

Figure P3.80

3.81 Find V_o in the circuit in Fig. P3.81.

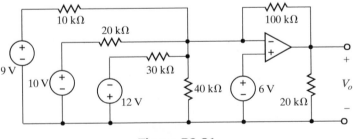

Figure P3.81

3.82 Find V_o in the circuit in Fig. P3.82.

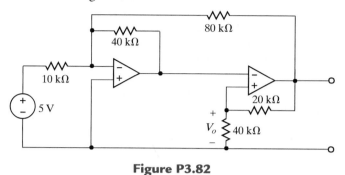

Figure P3.82

3.83 The electronic ammeter in Example 3.23 has been modified and is shown in Fig. P3.83. The selector switch allows the user to change the range of the meter. Using values for R_1 and R_2 from Example 3.23, find the values of R_A and R_B that will yield a 10-V output when the current being measured is 100 mA and 10 mA, respectively.

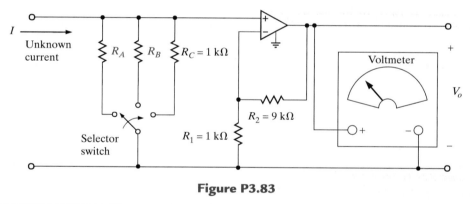

Figure P3.83

3.84 Design an op-amp circuit that has a gain of −50 using resistors no smaller than 1 kΩ.

3.85 Design a two-stage op-amp network that has a gain of −50,000 while drawing no current into its input terminal. Use no resistors smaller than 1 kΩ.

3.86 A 170°C maximum temperature digester is used in a paper mill to process wood chips that will eventually become paper. As shown in Fig. P3.86a, three electronic thermometers are placed along its length. Each thermometer outputs 0 V at 0°C and the voltage changes 25 mV/°C. We will use the average of the three thermometer voltages to find an aggregate digester temperature. Furthermore, 1 volt should appear at V_o for every 10°C of average temperature. Design such an averaging circuit using the op-amp configuration shown in Fig. 3.86 if the final output voltage must be positive.

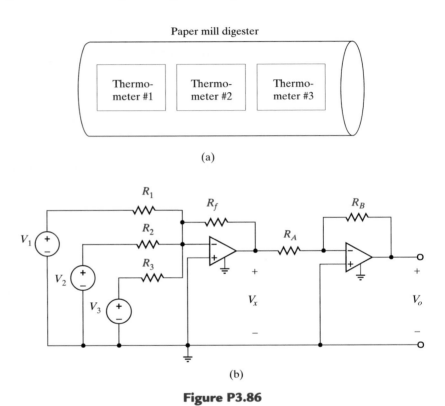

(a)

(b)

Figure P3.86

Additional Analysis Techniques

4

*A*t this point, we have mastered the ability to solve networks containing both independent and dependent sources using either nodal or loop analysis. In this chapter we introduce several new analysis techniques that bolster our arsenal of circuit analysis tools. We will find that in some situations these techniques lead to a quick solution and in other cases they do not. However, these new techniques in many cases do provide an insight into the circuit's operation that cannot be gained from a nodal or loop analysis.

In many practical situations we are interested in the analysis of some portion of a much larger network. If we can model the remainder of the network with a simple equivalent circuit, then our task will be much simpler. For example, consider the problem of analyzing some simple electronic device that is connected to the ac wall plug in our house. In this case the complete circuit includes not only the electronic device but the utility's power grid, which is connected to the device through the circuit breakers in the home. However, if we can accurately model everything outside the device with a simple equivalent circuit, then our analysis will be tractable. Two of the theorems that we will present in this chapter will permit us to do just that.

4.1 INTRODUCTION

Before introducing additional analysis techniques, let us review some of the topics we have used either explicitly or implicitly in our analyses thus far.

Equivalence

Table 4.1 is a short compendium of some of the equivalent circuits that have been employed in our analyses. This listing serves as a quick review as we begin to look at other techniques that can be used to find a specific voltage or current somewhere in a network and provide additional insight into the network's operation. In addition to the forms listed in the table, it is important to note that a series connection of current sources or a parallel connection of voltage sources is forbidden unless the sources are pointing in the same direction and have exactly the same values.

Linearity

All the circuits we have analyzed thus far and all the circuits we will treat in this book are linear circuits. We have shown earlier that the resistor is a linear element because its current–voltage relationship is a linear characteristic; that is,

$$v(t) = Ri(t)$$

Linearity requires both additivity and homogeneity (scaling). In the case of a resistive element, if $i_1(t)$ is applied, the voltage across the resistor is

$$v_1(t) = Ri_1(t)$$

Similarly, if $i_2(t)$ is applied, then

$$v_2(t) = Ri_2(t)$$

However, if $i_1(t) + i_2(t)$ is applied, the voltage across the resistor is

$$v(t) = R[i_1(t) + i_2(t)] = Ri_1(t) + Ri_2(t) = v_1(t) + v_2(t)$$

This demonstrates the additive property. In addition, if the current is scaled by a constant K_1, the voltage is also scaled by the constant K_1, since

$$RK_1i(t) = K_1Ri(t) = K_1v(t)$$

This demonstrates homogeneity.

We have shown in the preceding chapters that a circuit that contains only independent sources, linear dependent sources, and resistors is described by equations of the form

$$a_1v_1(t) + a_2v_2(t) + \cdots + a_nv_n(t) = i(t)$$

or

$$b_1i_1(t) + b_2i_2(t) + \cdots + b_ni_n(t) = v(t)$$

Note that if the independent sources are multiplied by a constant, the node voltages or loop currents are also multiplied by the same constant. Thus, we define a linear circuit as one composed of only independent sources, linear dependent sources, and linear elements. Capacitors and inductors, which we will examine in Chapter 5, are also circuit elements that have a linear input–output relationship provided that their initial storage energy is zero.

Table 4.1 Equivalent circuit forms

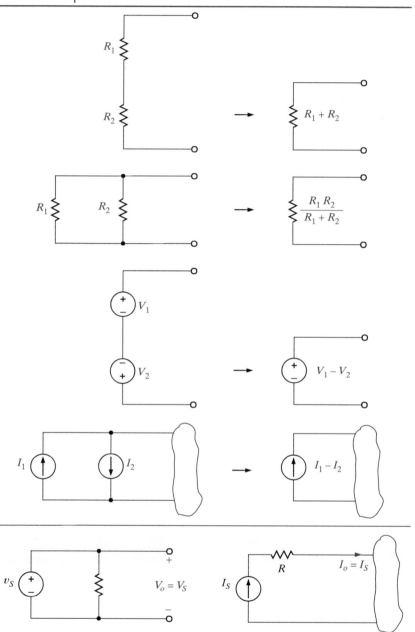

EXAMPLE 4.1

For the circuit shown in Fig. 4.1, we wish to determine the output voltage V_{out}. However, rather than approach the problem in a straightforward manner and calculate I_o, then I_1, then I_2, and so on, we will use linearity and simply assume that the output voltage is $V_{out} = 1$ V. This assumption will yield a value for the source voltage. We will then use the actual value of the source voltage and linearity to compute the actual value of V_{out}.

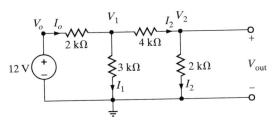

Figure 4.1 Circuit used in Example 4.1.

SOLUTION If we assume that $V_{out} = V_2 = 1$ V, then

$$I_2 = \frac{V_2}{2k} = 0.5 \text{ mA}$$

V_1 can then be calculated as

$$V_1 = 4kI_2 + V_2$$
$$= 3 \text{ V}$$

Hence,

$$I_1 = \frac{V_1}{3k} = 1 \text{ mA}$$

Now, applying KCL,

$$I_o = I_1 + I_2 = 1.5 \text{ mA}$$

Then

$$V_o = 2kI_o + V_1$$
$$= 6 \text{ V}$$

Therefore, the assumption that $V_{out} = 1$ V produced a source voltage of 6 V. However, since the actual source voltage is 12 V, the actual output voltage is 1 V(12/6) = 2 V. ❏

EXTENSION EXERCISE

E4.1 Use linearity and the assumption that $I_o = 1$ mA to compute the correct current I_o in the circuit in Fig. E4.1 if $I = 6$ mA.

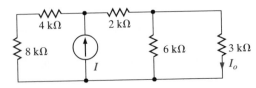

Figure E4.1

ANSWER: $I_o = 3$ mA.

4.2 SUPERPOSITION

In order to provide motivation for this subject, let us examine a simple circuit in which two sources contribute to the current in the network.

EXAMPLE 4.2

Consider the circuit in Fig. 4.2a in which the actual values of the voltage sources are left unspecified. Let us use this network to examine the concept of superposition.

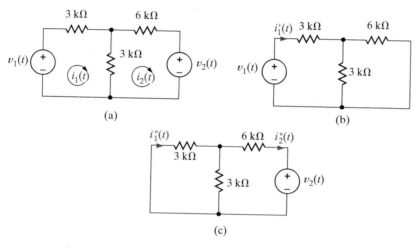

Figure 4.2 Circuits used to illustrate superposition.

SOLUTION   The mesh equations for this network are

$$6ki_1(t) - 3ki_2(t) = v_1(t)$$
$$-3ki_1(t) + 9ki_2(t) = -v_2(t)$$

Solving these equations for $i_1(t)$ yields

$$i_1(t) = \frac{v_1(t)}{5k} - \frac{v_2(t)}{15k}$$

In other words, the current $i_1(t)$ has a component due to $v_1(t)$ and a component due to $v_2(t)$. In view of the fact that $i_1(t)$ has two components, one due to each independent source, it would be interesting to examine what each source acting alone would contribute to $i_1(t)$. For $v_1(t)$ to act alone, $v_2(t)$ must be zero. As we pointed out in Chapter 2, $v_2(t) = 0$ means that the source $v_2(t)$ is replaced with a short circuit. Therefore, to determine the value of $i_1(t)$ due to $v_1(t)$ only, we employ the circuit in Fig. 4.2b and refer to this value of $i_1(t)$ as $i_1'(t)$.

$$i_1'(t) = \frac{v_1(t)}{3k + \frac{(3k)(6k)}{3k + 6k}}$$

$$= \frac{v_1(t)}{5k}$$

Let us now determine the value of $i_1(t)$ due to $v_2(t)$ acting alone and refer to this value as $i_1''(t)$. Using the network in Fig. 4.2c,

$$i_2''(t) = -\frac{v_2(t)}{6k + \frac{(3k)(3k)}{3k + 3k}} = \frac{-2v_2(t)}{15k}$$

Then, using current division, we obtain

$$i_1''(t) = \frac{-2v_2(t)}{15k}\left(\frac{3k}{3k + 3k}\right)$$

$$= \frac{-v_2(t)}{15k}$$

Now, if we add the values of $i_1'(t)$ and $i_1''(t)$, we obtain the value computed directly; that is,

$$i_1(t) = i_1'(t) + i_1''(t)$$

$$= \frac{v_1(t)}{5k} - \frac{v_2(t)}{15k}$$

Note that we have *superposed* the value of $i_1'(t)$ on $i_1''(t)$, or vice versa, to determine the unknown current.   ❏

What we have demonstrated in Example 4.2 is true in general for linear circuits and is a direct result of the property of linearity. *The principle of superposition*, which provides us with this ability to reduce a complicated problem to several easier problems—each containing only a single independent source—states that

In any linear circuit containing multiple independent sources, the current or voltage at any point in the network may be calculated as the algebraic sum of the individual contributions of each source acting alone.

When determining the contribution due to an independent source, any remaining voltage sources are made zero by replacing them with short circuits, and any remaining current sources are made zero by replacing them with open circuits.

Although superposition can be used in linear networks containing dependent sources, it is not useful in this case since the dependent source is never made zero.

It is interesting to note that, as the previous example indicates, superposition provides some insight in determining the contribution of each source to the variable under investigation.

We will now demonstrate superposition with two examples and then provide a problem-solving strategy for the use of this technique. For purposes of comparison, we will also solve the networks using both node and loop analyses. Furthermore, we will employ these same networks when demonstrating subsequent techniques, if applicable.

EXAMPLE 4.3

Let us use superposition to find V_o in the circuit in Fig. 4.3a.

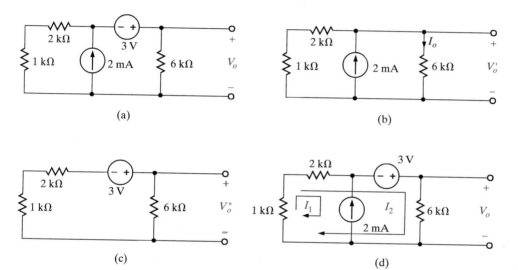

Figure 4.3 Circuits used in Example 4.3.

SOLUTION   The contribution of the 2-mA source to the output voltage is found from the network in Fig. 4.3b, using current division

$$I_o = (2 \times 10^{-3})\left(\frac{1k + 2k}{1k + 2k + 6k}\right)$$

$$= \frac{2}{3} \text{ mA}$$

and

$$V'_o = I_o(6k) = 4V$$

The contribution of the 3-V source to the output voltage is found from the circuit in Fig. 4.3c. Using voltage division,

$$V''_o = 3\left(\frac{6k}{1k + 2k + 6k}\right)$$

$$= 2 V$$

Therefore,

$$V_o = V'_o + V''_o = 6 \text{ V}$$

Although we used two separate circuits to solve the problem, both were very simple.

If we use nodal analysis and Fig. 4.3a to find V_o and recognize that the 3-V source and its connecting nodes form a supernode, V_o can be found from the node equation

$$\frac{V_o - 3}{1k + 2k} - 2 \times 10^{-3} + \frac{V_o}{6k} = 0$$

which yields $V_o = 6$ V. In addition, loop analysis applied as shown in Fig. 4.3d produces the equations

$$I_1 = -2 \times 10^{-3}$$

and

$$3k(I_1 + I_2) - 3 + 6\,kI_2 = 0$$

which yield $I_2 = 1$ mA and hence $V_o = 6$ V. ❑

DRILL

D4.1 Use superposition to find V_o in the following network.

ANSWER: $V_o = V'_o + V''_o = (9 - 3) = 6$ V

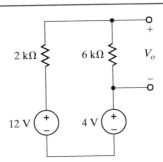

DRILL

D4.2 Use superposition to find I_o in the following network.

ANSWER: $I_o = I'_o + I''_o = (8 - 2) = 6$ mA

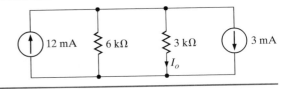

EXAMPLE 4.4

Consider now the network in Fig. 4.4a. Let us use superposition to find V_o.

SOLUTION The contribution of the 6-V source to V_o is found from the network in Fig. 4.4b, which is redrawn in Fig. 4.4c. The 2 kΩ + 6 kΩ = 8-kΩ resistor and 4-kΩ resistor are in parallel, and their combination is an 8/3-kΩ resistor. Then, using voltage division,

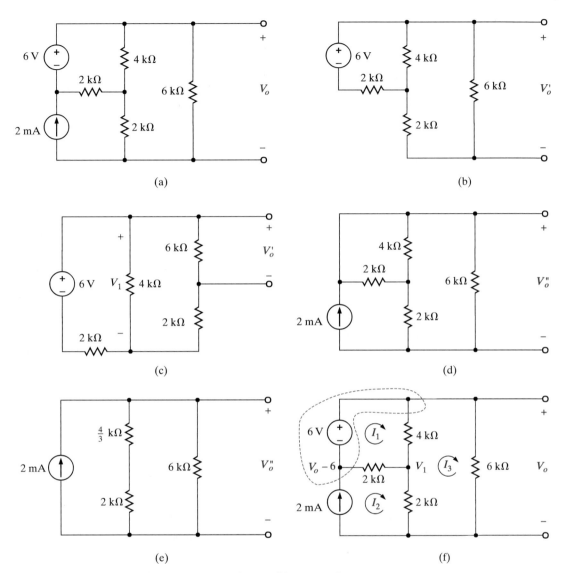

Figure 4.4 Circuits used in Example 4.4.

$$V_1 = 6\left(\dfrac{\dfrac{8}{3}\text{k}}{\dfrac{8}{3}\text{k} + 2\text{k}}\right) = \dfrac{24}{7}\,\text{V}$$

Applying voltage division again,

$$V_o' = V_1\left(\dfrac{6\text{k}}{6\text{k} + 2\text{k}}\right) = \dfrac{18}{7}\,\text{V}$$

The contribution of the 2-mA source is found from Fig. 4.4d, which is redrawn in Fig. 4.4e. V_o'' is simply equal to the product of the current source and the parallel combination of the resistors; that is,

$$V_o'' = (2 \times 10^{-3})\left(\frac{10}{3} \text{k} // 6\text{k}\right) = \frac{30}{7} \text{ V}$$

Then

$$V_o = V_o' + V_o'' = \frac{48}{7} \text{ V}$$

A nodal analysis of the network can be performed using Fig. 4.4f. The equation for the supernode is

$$-2 \times 10^{-3} + \frac{(V_o - 6) - V_1}{2\text{k}} + \frac{V_o - V_1}{4\text{k}} + \frac{V_o}{6\text{k}} = 0$$

The equation for the node labeled V_1 is

$$\frac{V_1 - V_o}{4\text{k}} + \frac{V_1 - (V_o - 6)}{2\text{k}} + \frac{V_1}{2\text{k}} = 0$$

Solving these two equations, which already contain the constraint equation for the supernode, yields $V_o = 48/7$ V.

Once again, referring to the network in Fig. 4.4f, the mesh equations for the network are

$$-6 + 4\text{k}(I_1 - I_3) + 2\text{k}(I_1 - I_2) = 0$$
$$I_2 = 2 \times 10^{-3}$$
$$2\text{k}(I_3 - I_2) + 4\text{k}(I_3 - I_1) + 6\text{k}I_3 = 0$$

Solving these equations, we obtain $I_3 = 8/7$ mA and, hence, $V_o = 48/7$ V. ❑

Problem-Solving Strategy for Applying Superposition

- In a network containing multiple independent sources, each source can be applied independently with the remaining sources turned off.

- To turn off a voltage source, replace it with a short circuit, and to turn off a current source, replace it with an open circuit.

- When the individual sources are applied to the circuit, all the circuit laws and techniques we have learned, or will soon learn, can be applied to obtain a solution.

- The results obtained by applying each source independently are then added together algebraically to obtain a solution.

Superposition can be applied to a circuit with any number of dependent and independent sources. In fact, superposition can be applied to such a network in a variety of ways. For example, a circuit with three independent sources can be solved using each source acting alone, as we have just demonstrated, or we could use two at a time and sum the result with that obtained from the third acting alone. In addition, the independent sources do not have to assume their actual value or zero. However, it is mandatory that the sum of the different values chosen add to the total value of the source.

Superposition is a fundamental property of linear equations and, therefore, can be applied to any effect that is linearly related to its cause. In this regard it is important to point out that although superposition applies to the current and voltage in a linear circuit, it cannot be used to determine power because power is a nonlinear function.

EXTENSION EXERCISE

E4.2 Compute V_o in the circuit in Fig. E4.2 using superposition.

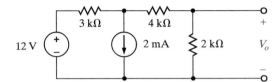

Figure E4.2

ANSWER: $V_o = \dfrac{4}{3}$ V.

4.3 SOURCE TRANSFORMATION

Before we begin discussing source transformation, it is necessary that we point out that real sources differ from the ideal models we have presented thus far. In general, a practical voltage source does not produce a constant voltage regardless of the load resistance or the current it delivers, and a practical current source does not deliver a constant current regardless of the load resistance or the voltage across its terminals.

Practical sources contain internal resistance, and this source resistance is not accessible to the user. In general, we want to minimize the resistance because it consumes power. Some typical values for internal resistance are shown in Table 4.2.

Table 4.2 Approximate Internal Resistance of Selected Sources (New Batteries) at 21°C

Battery Type	Internal Resistance
1.5-V D-cell standard carbon zinc	0.5 Ω
1.5-V D-cell heavy-duty carbon zinc	0.3 Ω
1.5-V D-cell alkaline	0.2 Ω
1.5-V D-cell nickel-cadmium	0.008 Ω
9-V standard carbon zinc	35 Ω
9-V heavy-duty carbon zinc	35 Ω
9-V alkaline	2 Ω
12-V automobile battery	0.006 Ω

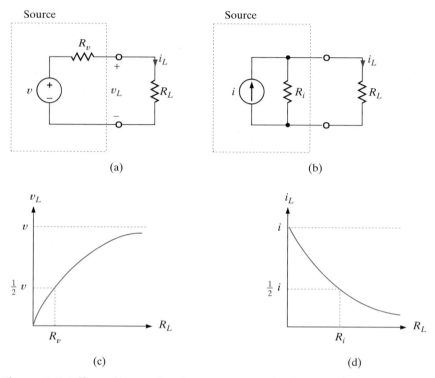

Figure 4.5 Effect of internal resistance on actual voltage and current sources.

Since practical sources contain internal resistance, the models shown in Fig. 4.5a and b more closely represent actual sources. Note that the power delivered by the practical voltage source is given by the expression

$$P_L = i_L^2 R_L$$
$$= \left(\frac{v}{R_v + R_L} \right)^2 R_L$$

which can be written as

$$P_L = \frac{v^2}{R_L} \left(\frac{1}{1 + R_v/R_L} \right)^2$$

and therefore if $R_L \gg R_v$, then

$$P_L = \frac{v^2}{R_L}$$

which is the power delivered by the ideal voltage source. Similarly, the power delivered by the practical current source is computed using current division as

$$P_L = i_L^2 R_L$$
$$= \left(\frac{iR_i}{R_i + R_L} \right)^2 R_L$$

which can be written as

$$P_L = i^2 R_L \left(\frac{1}{1 + R_L/R_i} \right)^2$$

and hence, if $R_i \gg R_L$, then

$$P_L = i^2 R_L$$

which is the power delivered by the ideal current source.

The graphs shown in Figs. 4.5c and d illustrate the effect of the internal resistance on the voltage source and current source, respectively. The graphs indicate that the output voltage approaches the ideal source voltage only for large values of load resistance and, therefore, small values of current. Similarly, the load current is approximately equal to the ideal source current only for values of the load resistance R_L that are small in comparison to the internal resistance R_i.

With this material on practical sources as background information, we now ask if it is possible to exchange one source model for another; that is, exchange a voltage source model for a current source model, or vice versa. We could exchange one source for another provided that they are equivalent; that is, each source produces exactly the same voltage and current for any load that is connected across its terminals.

Let us examine the two circuits shown in Fig. 4.6. In order to determine the conditions required for the two sources to be equivalent, let us examine the terminal conditions of each. For the network in Fig. 4.6a,

$$i = i_L + \frac{v_L}{R_i}$$

or

$$i R_i = R_i i_L + v_L$$

For the network in Fig. 4.6b,

$$v = i_L R_v + v_L$$

For the two networks to be equivalent, their terminal characteristics must be identical; that is,

$$v = i R_i \qquad \text{and} \qquad R_i = R_v \qquad\qquad \textbf{4.1}$$

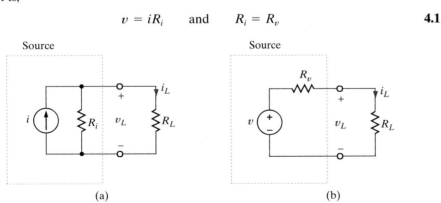

(a)   (b)

Figure 4.6 Circuits used to determine conditions for a source exchange.

The relationships specified in Eq. (4.1) and Fig. 4.6 are extremely important and the reader should not fail to grasp their significance. What these relationships tell us is that if we have embedded within a network a current source i in parallel with a resistor R, we can replace this combination with a voltage source of value $v = iR$ in series with the resistor R. The reverse is also true; that is, a voltage source v in series with a resistor R can be replaced with a current source of value $i = v/R$ in parallel with the resistor R. Parameters within the circuit (e.g., an output voltage) are unchanged under these transformations.

We must emphasize that the two equivalent circuits in Fig. 4.6, under the conditions stated in Eq. (4.1), *are equivalent only at the two external nodes*. For example, if we disconnect the load resistor, R_L, in the circuits in Fig. 4.6, the equivalent circuit in Fig. 4.6a dissipates power, but the one in Fig. 4.6b does not.

At this point let us demonstrate the utility of a source exchange using two examples.

EXAMPLE 4.5

Let us repeat Example 4.3 using source exchange.

SOLUTION   For convenience, the network is redrawn in Fig. 4.7a. Note that on the left side of the network, there is a current source in parallel with a 3-kΩ(1 kΩ + 2 kΩ) resistor, which can be transformed into a 6-V = (2 mA)(3 kΩ) source in series with the same 3-kΩ resistor. The resulting network is shown in Fig. 4.7b. The output voltage is then a simple division of the 9 volts (6 V + 3 V) between the 3-kΩ and 6-kΩ resistors. Therefore,

$$V_o = 9\left(\frac{6k}{3k + 6k}\right) = 6 \text{ V}$$

which agrees with our previous analyses. Finally, the reader is cautioned to keep the polarity of the voltage source and the direction of the current source in agreement, as shown in Fig. 4.7.

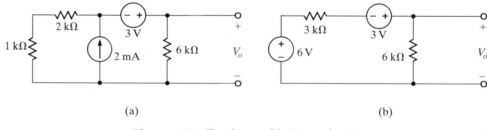

(a)   (b)

Figure 4.7 Circuits used in Example 4.5.

EXAMPLE 4.6

We will now demonstrate how to find V_o in the circuit in Fig. 4.8a using the repeated application of source transformation.

SOLUTION   If we begin at the left end of the network in Fig. 4.8a, the series combination of the 12-V source and 3-kΩ resistor is converted to a 4-mA current source in parallel with the 3-kΩ resistor. If we combine this 3-kΩ resistor with the 6-kΩ resistor, we obtain the circuit in Fig. 4.8b. Note that at this point we have eliminated one circuit element. Continuing the reduction, we convert the 4-mA source and 2-kΩ resistor into an 8-V source in series with this same 2-kΩ resistor. The two 2-kΩ resistors that are in series are now combined to produce the network in Fig. 4.8c. If we now convert the combination of the 8-V source and 4-kΩ resistor into a 2-mA source in parallel with the 4-kΩ resistor and combine the resulting current source with the other 2-mA source, we arrive at the circuit shown in Fig. 4.8d. At this point, we can simply apply current division to the two parallel resistance paths and obtain

$$I_o = (4 \times 10^{-3})\left(\frac{4k}{4k + 4k + 8k}\right) = 1 \text{ mA}$$

and hence,

$$V_o = (1 \times 10^{-3})(8k) = 8 \text{ V}$$

The reader is encouraged to consider the ramifications of working this problem using any of the other techniques we have presented.

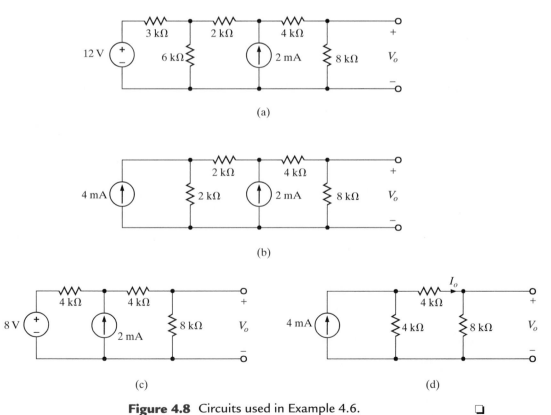

(a)

(b)

(c)

(d)

Figure 4.8 Circuits used in Example 4.6.

D4.3 Convert the following network to a single node circuit with respect to the 12 kΩ load.

ANSWER:

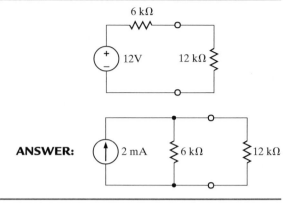

D4.4 Convert the following network to a single-loop circuit with respect to the 4-kΩ load.

ANSWER:

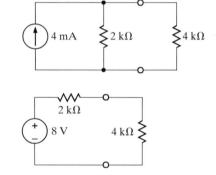

Note that this systematic, sometimes tedious, transformation allows us to reduce the network methodically to a simpler equivalent form with respect to some other circuit element. However, we should also realize that this technique is worthless for circuits of the form shown in Fig. 4.4. Furthermore, although applicable to networks containing dependent sources, it is not as useful as other techniques, and care must be taken not to transform the part of the circuit that contains the control variable.

Problem-Solving Strategy
for Applying Source Transformation

- When applicable (and, as indicated previously, this technique is not applicable in many cases), source transformation can be used to reduce the number of components in a network with respect to some particular load.

- The reduction in the number of components is achieved by systematically transforming (i.e., exchanging) a voltage source V in series with a resistor R for a current source $I = V/R$ in parallel with the resistor R, and vice versa.

- As this technique is applied, resistors in series or parallel can be combined as well as voltage sources in series and current sources in parallel. As this process proceeds, the number of elements in the network is continuously being reduced.

- When this technique is viable, it will simplify the network to at least some extent and in some cases will reduce it to a single-loop or single-node circuit containing the particular load that was present in the original circuit.

EXTENSION EXERCISE

E4.3   Compute V_o in the circuit in Fig. E4.2 using source transformation.

ANSWER: $V_o = \dfrac{4}{3}\,\text{V}.$

4.4 THÉVENIN'S AND NORTON'S THEOREMS

Thus far we have presented a number of techniques for circuit analysis. At this point we will add two theorems to our collection of tools that will prove to be extremely useful. The theorems are named after their authors, M. L. Thévenin, a French engineer, and E. L. Norton, a scientist formerly with Bell Telephone Laboratories.

Suppose that we are given a circuit and that we wish to find the current, voltage, or power that is delivered to some resistor of the network, which we will call the load. *Thévenin's theorem* tells us that we can replace the entire network, exclusive of the load, by an equivalent circuit that contains only an independent voltage source in series with a resistor in such a way that the current–voltage relationship at the load is unchanged. *Norton's theorem* is identical to the preceding statement except that the equivalent circuit is an independent current source in parallel with a resistor.

Note that this is a very important result. It tells us that if we examine any network from a pair of terminals, we know that with respect to those terminals, the entire network is equivalent to a simple circuit consisting of an independent voltage source in series with a resistor or an independent current source in parallel with a resistor.

In developing the theorems, we will assume that the circuit shown in Fig. 4.9a can be split into two parts, as shown in Fig. 4.9b. In general, circuit B is the load and may be linear or nonlinear. Circuit A is the balance of the original network exclusive of the load and must be linear. As such, circuit A may contain independent sources, dependent sources and resistors, or any other linear element. We require, however, that a dependent source and its control variable appear in the same circuit.

Circuit A delivers a current i to circuit B and produces a voltage v_o across the input terminals of circuit B. From the standpoint of the terminal relations of circuit A, we can replace circuit B by a voltage source of v_o volts (with the proper polarity), as shown in

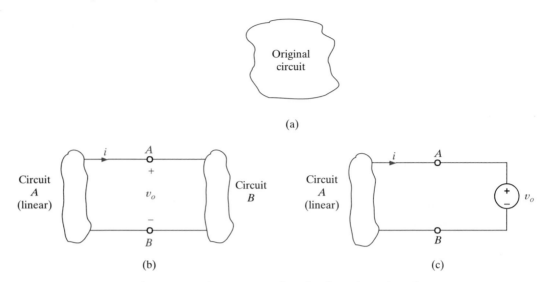

Figure 4.9 Concepts used to develop Thévenin's theorem.

Fig. 4.9c. Since the terminal voltage is unchanged and circuit A is unchanged, the terminal current i is unchanged.

Now, applying the principle of superposition to the network shown in Fig. 4.9c, the total current i shown in the figure is the sum of the currents caused by all the sources in circuit A and the source v_o that we have just added. Therefore, via superposition the current i can be written

$$i = i_o + i_{sc} \qquad \textbf{4.2}$$

where i_o is the current due to v_o with all independent sources in circuit A made zero (i.e., voltage sources replaced by short circuits and current sources replaced by open circuits), and i_{sc} is the short-circuit current due to all sources in circuit A with v_o replaced by a short circuit.

The terms i_o and v_o are related by the equation

$$i_o = \frac{-v_o}{R_{Th}} \qquad \textbf{4.3}$$

where R_{Th} is the equivalent resistance looking back into circuit A from terminals A-B with all independent sources in circuit A made zero.

Substituting Eq. (4.3) into Eq. (4.2) yields

$$i = -\frac{v_o}{R_{Th}} + i_{sc} \qquad \textbf{4.4}$$

This is a general relationship and, therefore, must hold for any specific condition at terminals A-B. As a specific case, suppose that the terminals are open circuited. For this condition, $i = 0$ and v_o is equal to the open-circuit voltage v_{oc}. Thus, Eq. (4.4) becomes

$$i = 0 = \frac{-v_{oc}}{R_{Th}} + i_{sc} \qquad \textbf{4.5}$$

Hence,

$$v_{oc} = R_{Th}i_{sc} \qquad \textbf{4.6}$$

This equation states that the open-circuit voltage is equal to the short-circuit current times the equivalent resistance looking back into circuit A with all independent sources made zero. We refer to R_{Th} as the Thévenin equivalent resistance.

Substituting Eq. (4.6) into Eq. (4.4) yields

$$i = \frac{-v_o}{R_{Th}} + \frac{v_{oc}}{R_{Th}}$$

or

$$v_o = v_{oc} - R_{Th}i \qquad \textbf{4.7}$$

Let us now examine the circuits that are described by these equations. The circuit represented by Eq. (4.7) is shown in Fig. 4.10a. The fact that this circuit is equivalent at terminals A-B to circuit A in Fig. 4.9 is a statement of *Thévenin's theorem*. The circuit represented by Eq. (4.4) is shown in Fig. 4.10b. The fact that this circuit is equivalent at terminals A-B to circuit A in Fig. 4.9 is a statement of *Norton's theorem*.

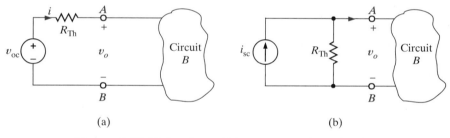

(a)   (b)

Figure 4.10 Thévenin and Norton equivalent circuits.

Note carefully that the circuits in Fig. 4.10 together with the relationship in Eq. (4.6) represent a source transformation.

The manner in which these theorems are applied depends on the structure of the original network under investigation. For example, if only independent sources are present, we can calculate the open-circuit voltage or short-circuit current and the Thévenin equivalent resistance. However, if dependent sources are also present, the Thévenin equivalent will be determined by calculating v_{oc} and i_{sc}, since this is normally the best approach for determining R_{Th} in a network containing dependent sources. Finally, if circuit A contains no *independent* sources, then both v_{oc} and i_{sc} will necessarily be zero. (Why?) Thus, we cannot determine R_{Th} by v_{oc}/i_{sc}, since the ratio is indeterminate. We must look for another approach. Notice that if $v_{oc} = 0$, then the equivalent circuit is merely the unknown resistance R_{Th}. If we apply an external source to circuit A—a test source v_t—and determine the current, i_t, which flows into circuit A from v_t, then R_{Th} can be determined from $R_{Th} = v_t/i_t$. Although the numerical value of v_t need not be specified, we could let $v_t = 1$ V and then $R_{Th} = 1/i_t$. Alternatively, we could use a current source as a test source and let $i_t = 1$ A; then $v_t = (1)R_{Th}$.

A variety of examples is now presented to demonstrate the utility of these theorems. Before beginning our analysis of several examples that will demonstrate the utility of these theorems, the reader is cautioned to remember that these theorems, in addition to an alternate method of attack, often permit us to solve several small problems rather than one large one. They allow us to replace a network, no matter how large, *at a pair of terminals* with a Thévenin or Norton equivalent circuit. In fact, we could represent the entire U.S. power grid at a pair of terminals with one of the equivalent circuits. Once this is done we can quickly analyze the effect of different loads on a network. Thus, these theorems provide us with additional insight into the operation of a specific network.

Circuits Containing Only Independent Sources

EXAMPLE 4.7

Let us use Thévenin's and Norton's theorems to find V_o in the network in Example 4.3.

SOLUTION   The circuit is redrawn in Fig. 4.11a. To determine the Thévenin equivalent, we break the network at the 6-kΩ load as shown in Fig. 4.11b. KVL indicates that the open-circuit voltage, V_{oc}, is equal to 3 V plus the voltage V_1, which is the voltage across the current source. The 2 mA from the current source flows through the two resistors (where else could it possibly go!) and, therefore, $V_1 = (2 \times 10^{-3})(1k + 2k) = 6$ V. Therefore, $V_{oc} = 9$ V. By making both sources zero, we can find the Thévenin equivalent resistance, R_{Th}, using the circuit in Fig. 4.11c. Obviously, $R_{Th} = 3$ kΩ. Now our Thévenin equivalent circuit, consisting of V_{oc} and R_{Th}, is connected back to the original terminals of the load, as shown in Fig. 4.11d. Using a simple voltage divider, we find that $V_o = 6$ V.

To determine the Norton equivalent circuit at the terminals of the load, we must find the short-circuit current as shown in Fig. 4.11e. Note that the short circuit causes the 3-V source to be directly across (i.e., in parallel with) the resistors and the current source. Therefore, $I_1 = 3/(1k + 2k) = 1$ mA. Then, using KCL, $I_{sc} = 3$ mA. We have already determined R_{Th} and, therefore, connecting the Norton equivalent to the load results in the circuit in Fig. 4.11f. Hence, V_o is equal to the source current multiplied by the parallel resistor combination, which is 6 V.   ❏

Consider for a moment some salient features of this example. Note that in applying the theorems there is no point in breaking the network to the left of the 3-V source, since the resistors in parallel with the current source are already a Norton equivalent, which can be immediately changed to a Thévenin equivalent using source transformation! Furthermore, once the network has been simplified using a Thévenin or Norton equivalent, we simply have a new network with which we can apply the theorems again. The following example illustrates this approach.

EXAMPLE 4.8

Let us use Thévenin's theorem to find V_o in the network in Fig. 4.8a, which is redrawn in Fig. 4.12a.

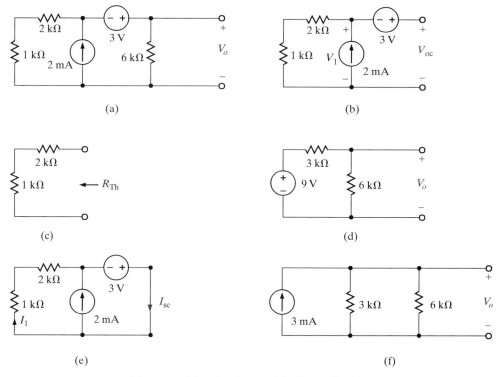

(a)

(b)

(c)

(d)

(e)

(f)

Figure 4.11 Circuits used in Example 4.7.

D4.5   Find V_o in the following network using Thévenin's theorem.

ANSWER: 12V

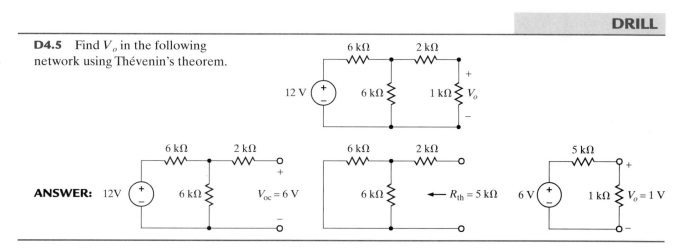

SOLUTION   If we break the network to the left of the current source, the open-circuit voltage V_{oc_1} is as shown in Fig. 4.12b. Since there is no current in the 2-kΩ resistor and therefore no voltage across it, V_{oc_1} is equal to the voltage across the 6-kΩ resistor, which can be determined by voltage division as

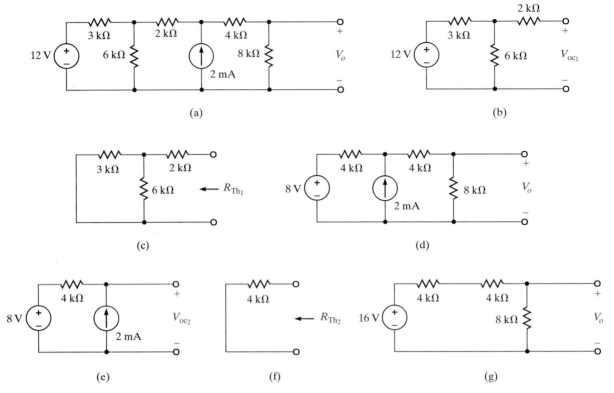

Figure 4.12 Circuits used in Example 4.8.

$$V_{oc_1} = 12\left(\frac{6k}{6k + 3k}\right) = 8 \text{ V}$$

The Thévenin equivalent resistance, R_{Th_1}, is found from Fig. 4.12c as

$$R_{Th_1} = 2k + \frac{(3k)(6k)}{3k + 6k} = 4 \text{ k}\Omega$$

Connecting this Thévenin equivalent back to the original network produces the circuit shown in Fig. 4.12d. We can now apply Thévenin's theorem again, and this time we break the network to the right of the current source as shown in Fig. 4.12e. In this case V_{oc_2} is

$$V_{oc_2} = (2 \times 10^{-3})(4k) + 8 = 16 \text{ V}$$

and R_{Th_2} obtained from Fig. 4.12f is 4 kΩ. Connecting this Thévenin equivalent to the remainder of the network produces the circuit shown in Fig. 4.12g. Simple voltage division applied to this final network yields $V_o = 8$ V. Norton's theorem can be applied in a similar manner to solve this network; however, we save that solution as an exercise for the reader. ❏

EXAMPLE 4.9

It is instructive to examine the use of Thévenin's and Norton's theorems in the solution of the network in Fig. 4.4a, which is redrawn in Fig. 4.13a.

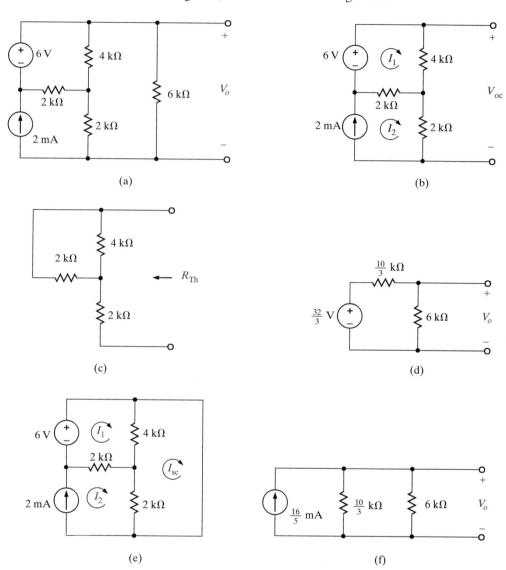

Figure 4.13   Circuits used in Example 4.9.

SOLUTION   If we break the network at the 6-kΩ load, the open-circuit voltage is found from Fig. 4.13b. The equations for the mesh currents are

$$-6 + 4kI_1 + 2k(I_1 - I_2) = 0$$

and

$$I_2 = 2 \times 10^{-3}$$

from which we easily obtain $I_1 = 5/3$ mA. Then, using KVL, V_{oc} is

$$V_{oc} = 4kI_1 + 2kI_2$$

$$= 4k\left(\frac{5}{3} \times 10^{-3}\right) + 2k(2 \times 10^{-3})$$

$$= \frac{32}{3} \text{ V}$$

R_{Th} is derived from Fig. 4.13c and is

$$R_{Th} = (2k//4k) + 2k = \frac{10}{3} k\Omega$$

Attaching the Thévenin equivalent to the load produces the network in Fig. 4.13d. Then using voltage division, we obtain

$$V_o = \frac{32}{3}\left(\frac{6k}{6k + \frac{10}{3}k}\right)$$

$$= \frac{48}{7} \text{ V}$$

In applying Norton's theorem to this problem, we must find the short-circuit current shown in Fig. 4.13e. At this point the quick-thinking reader stops immediately! Three mesh equations applied to the original circuit will immediately lead to the solution, but the three mesh equations in the circuit in Fig. 4.13e will provide only part of the answer, specifically the short-circuit current. Sometimes the use of the theorems is more complicated than a straightforward attack using node or loop analysis. This would appear to be one of those situations. Interestingly, it is not. We can find I_{sc} from the network in Fig. 4.13e without using the mesh equations. The technique is simple, but a little tricky, and so we ignore it at this time. Having said all these things, let us now finish what we have started. The mesh equations for the network in Fig. 4.13e are

$$-6 + 4k(I_1 - I_{sc}) + 2k(I_1 - 2 \times 10^{-3}) = 0$$

$$2k(I_{sc} - 2 \times 10^{-3}) + 4k(I_{sc} - I_1) = 0$$

where we have incorporated the fact that $I_2 = 2 \times 10^{-3}$ A. Solving these equations yields $I_{sc} = 16/5$ mA. R_{Th} has already been determined in the Thévenin analysis. Connecting the Norton equivalent to the load results in the circuit in Fig. 4.13f. Applying Ohm's law to this circuit yields $V_o = 48/7$ V. ❑

EXTENSION EXERCISES

E4.4 Use Thévenin theorem to find V_o in the network in Fig. E4.4.

ANSWER: $V_o = -3$ V.

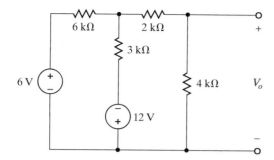

Figure E4.4

E4.5   Find V_o in the circuit in Fig. E4.2 using both Thévenin's and Norton's theorems. When deriving the Norton equivalent circuit, break the network to the left of the 4-kΩ resistor. Why?

ANSWER: $V_o = \dfrac{4}{3}$ V.

Circuits Containing Only Dependent Sources

As we have stated earlier, the Thévenin or Norton equivalent of a network containing only dependent sources is R_{Th}. The following examples will serve to illustrate how to determine this Thévenin equivalent resistance.

EXAMPLE 4.10

We wish to determine the Thévenin equivalent of the network in Fig. 4.14a at the terminals A-B.

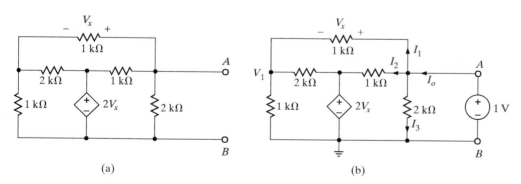

(a)   (b)

Figure 4.14 Networks employed in Example 4.10.

SOLUTION   Our approach to this problem will be to apply a 1-V source at the terminals as shown in Fig. 4.14b and then compute the current I_o and $R_{\text{Th}} = 1/I_o$.

The equations for the network in Fig. 4.14b are as follows. KVL around the outer loop specifies that

$$V_1 + V_x = 1$$

The KCL equation at the node labeled V_1 is

$$\frac{V_1}{1k} + \frac{V_1 - 2v_x}{2k} + \frac{V_1 - 1}{1k} = 0$$

Solving the equations for V_x yields $V_x = 3/7$ V. Knowing V_x, we can compute the currents $I_1, I_2,$ and I_3. Their values are

$$I_1 = \frac{V_x}{1k} = \frac{3}{7} \text{ mA}$$

$$I_2 = \frac{1 - 2V_x}{1k} = \frac{1}{7} \text{ mA}$$

$$I_3 = \frac{1}{2k} = \frac{1}{2} \text{ mA}$$

Therefore,

$$I_o = I_1 + I_2 + I_3$$

$$= \frac{15}{14} \text{ mA}$$

and

$$R_{Th} = \frac{1}{I_o}$$

$$= \frac{14}{15} \text{ k}\Omega \qquad \square$$

EXAMPLE 4.11

Let us determine R_{Th} at the terminals A-B for the network in Fig. 4.15a.

SOLUTION   Our approach to this problem will be to apply a 1-mA current source at the terminals A-B and compute the terminal voltage V_2 as shown in Fig. 4.15b. Then $R_{Th} = V_2/0.001$.

The node equations for the network are

$$\frac{V_1 - 2000I_x}{2k} + \frac{V_1}{1k} + \frac{V_1 - V_2}{3k} = 0$$

$$\frac{V_2 - V_1}{3k} + \frac{V_2}{2k} = 1 \times 10^{-3}$$

and

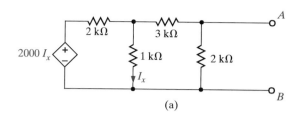

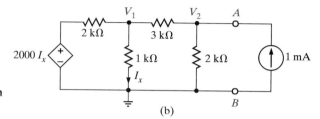

Figure 4.15 Networks used in Example 4.10.

$$I_x = \frac{V_1}{1k}$$

Solving these equations yields

$$V_2 = \frac{10}{7} \text{ V}$$

and hence,

$$R_{\text{Th}} = \frac{V_2}{1 \times 10^{-3}}$$

$$= \frac{10}{7} \text{ k}\Omega$$ ❏

EXTENSION EXERCISE

E4.6   Find the Thévenin equivalent circuit of the network in Fig. E4.6 at the terminals *A-B*.

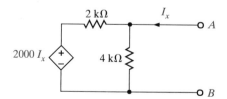

Figure E4.6

ANSWER: $R_{\text{Th}} = \frac{8}{3} \text{ k}\Omega$

Circuits Containing Both Independent and Dependent Sources

In these type of circuits we must calculate both the open-circuit voltage and short-circuit current in order to calculate the Thévenin equivalent resistance. Furthermore, we must remember that we cannot split the dependent source and its controlling variable when we break the network to find the Thévenin or Norton equivalent.

We now present two examples: one with a dependent voltage source and one with a dependent current source.

EXAMPLE 4.12

Let us use Thévenin's theorem to find V_o in the network in Fig. 4.16a.

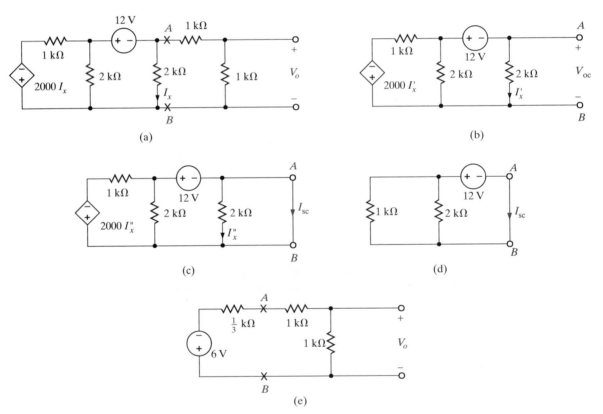

Figure 4.16 Circuits used in Example 4.12.

SOLUTION   To begin, we break the network at points A-B. Could we break it just to the right of the 12-V source? No! Why? The open-circuit voltage is calculated from the network in Fig. 4.16b. Note that we now use the source $2000I'_x$ because this circuit is different from that in Fig. 4.16a. KCL for the supernode around the 12-V source is

$$\frac{(V_{oc} + 12) - (-2000I'_x)}{1k} + \frac{V_{oc} + 12}{2k} + \frac{V_{oc}}{2k} = 0$$

where

$$I'_x = \frac{V_{oc}}{2k}$$

yielding $V_{oc} = -6$ V.

I_{sc} can be calculated from the circuit in Fig. 4.16c. Note that the presence of the short circuit forces I''_x to zero and, therefore, the network is reduced to that shown in Fig. 4.16d.

Therefore,

$$I_{sc} = \frac{-12}{\frac{2}{3}k} = -18 \text{ mA}$$

Then

$$R_{Th} = \frac{V_{oc}}{I_{sc}} = \frac{1}{3} k\Omega$$

Connecting the Thévenin equivalent circuit to the remainder of the network at terminals *A-B* produces the circuit in Fig. 4.16e. At this point, simple voltage division yields

$$V_o = (-6)\left(\frac{1k}{1k + 1k + \frac{1}{3}k}\right) = \frac{-18}{7} \text{ V} \qquad \square$$

EXAMPLE 4.13

Let us find V_o in the network in Fig. 3.24 using Thévenin's theorem.

SOLUTION   The network is redrawn in Fig. 4.17a. V_{oc} is determined from the network in Fig. 4.17b. Note that

$$I_1 = \frac{V'_x}{2k}$$

$$I_2 = 2 \text{ mA}$$

and

$$V'_x = 4k\left(\frac{V'_x}{2k} - 2 \times 10^{-3}\right)$$

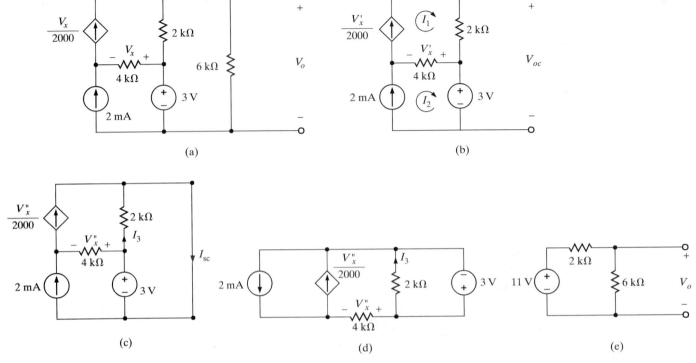

Figure 4.17 Circuits used in Example 4.13.

Solving these equations yields $I_1 = 4$ mA and, hence,

$$V_{oc} = 2kI_1 + 3 = 11 \text{ V}$$

I_{sc} is derived from the circuit in Fig. 4.17c. Note that if we collapse the short circuit, the network is reduced to that in Fig. 4.17d. Although we have temporarily lost sight of I_{sc}, we can easily find the branch currents and they, in turn, will yield I_{sc}. KCL at the node at the bottom left of the network is

$$\frac{V''_x}{4k} = \frac{V''_x}{2000} - 2 \times 10^{-3}$$

or

$$V''_x = 8 \text{ V}$$

Then since

$$I_3 = \frac{3}{2k} = \frac{3}{2} \text{ mA}$$

as shown in Fig. 4.17c

$$I_{sc} = \frac{V_x''}{2000} + I_3$$

$$= \frac{11}{2} \text{ mA}$$

Then

$$R_{Th} = \frac{V_{oc}}{I_{sc}} = 2 \text{ k}\Omega$$

Connecting the Thévenin equivalent circuit to the remainder of the original network produces the circuit in Fig. 4.17e. Simple voltage division yields

$$V_o = 11\left(\frac{6k}{2k + 6k}\right)$$

$$= \frac{33}{4} \text{ V} \qquad \square$$

Problem-Solving Strategy for Applying Thévenin's Theorem

- Remove the load and find the voltage across the open circuit terminals, V_{oc}. All the circuit analysis techniques presented here can be used to compute this voltage.
- Determine the Thévenin equivalent resistance of the network at the open terminals with the load removed. Three different types of circuits may be encountered in determining the resistance, R_{Th}.
    - If the circuit contains only independent sources, they are made zero by replacing the voltage sources with short circuits and the current sources with open circuits. R_{Th} is then found by computing the resistance of the purely resistive network at the open terminals.
    - If the circuit contains only dependent sources, an independent voltage or current source is applied at the open terminals and the corresponding current or voltage at these terminals is measured. The voltage/current ratio at the terminals is the Thévenin equivalent resistance. Since there is no energy source, the open-circuit voltage is zero in this case.
    - If the circuit contains both independent and dependent sources, the open-circuit terminals are shorted and the short-circuit current between these terminals is determined. The ratio of the open-circuit voltage to the short-circuit current is the resistance R_{Th}.
- If the load is now connected to the Thévenin equivalent circuit, consisting of V_{oc} in series with R_{Th}, the desired solution can be obtained.

The problem-solving strategy for Norton's theorem is essentially the same as that for Thévenin's theorem with the exception that we are dealing with the short-circuit current instead of the open-circuit voltage.

EXTENSION EXERCISE

E4.7 Find V_o in the circuit in Fig. E4.7 using Thévenin's theorem.

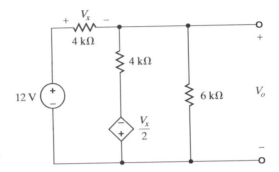

Figure E4.7

ANSWER: $V_o = \dfrac{36}{13}$ V.

At this point it is worthwhile to pause for a moment and reflect about what we have learned; that is, let us compare the use of node or loop analysis with that of the theorems discussed in this chapter. When we examine a network for analysis, one of the first things we should do is count the number of nodes and loops. Next we consider the number of sources. For example, are there a number of voltage sources or current sources present in the network? All these data, together with the information that we expect to glean from the network, give a basis for selecting the simplest approach. With the current level of computational power available to us, we can solve the node or loop equations that define the network in a flash.

With regard to the theorems, we have found that in some cases the theorems do not necessarily simplify the problem and a straightforward attack using node or loop analysis is as good an approach as any. This is a valid point provided that we are simply looking for some particular voltage or current. However, the real value of the theorems is the insight and understanding that they provide about the physical nature of the network. For example, superposition tells us what each source contributes to the quantity under investigation. However, a computer solution of the node or loop equations does not tell us the effect of changing certain parameter values in the circuit. It does not help us understand the concept of loading a network or the ramifications of interconnecting networks or the idea of matching a network for maximum power transfer. The theorems help us to understand the effect of using a transducer at the input of an amplifier with a given input resistance. They help us explain the effect of a load, such as a speaker, at the output of an amplifier. We derive none of this information from a node or loop analysis. In fact, as a simple example, suppose that a network at a specific pair of terminals has a Thévenin equivalent circuit consisting of a voltage source in series with a 2-kΩ resistor. If we connect a 2-Ω resistor to the network at these terminals, the voltage across the 2-Ω resistor will be essentially nothing. This result is fairly obvious using the Thévenin theorem approach; however, a node or loop analysis gives us no clue as to why we have obtained this result.

We have studied networks containing only dependent sources. This is a very important topic because all electronic devices, such as transistors, are modeled in this fashion. Motors in power systems are also modeled in this way. We use these amplification devices for many different purposes, such as the speed control for an automobile.

Finally, it is interesting to note that when we employ source transformation as we did in Example 4.6, we are simply converting back and forth between a Thévenin equivalent circuit and a Norton equivalent circuit.

4.5 MAXIMUM POWER TRANSFER

In circuit analysis we are sometimes interested in determining the maximum power that can be delivered to a load. By employing Thévenin's theorem, we can determine the maximum power that a circuit can supply and the manner in which to adjust the load to effect maximum power transfer.

Suppose that we are given the circuit shown in Fig. 4.18. The power that is delivered to the load is given by the expression

$$P_{\text{load}} = i^2 R_L$$

$$= \left(\frac{v}{R + R_L} \right)^2 R_L$$

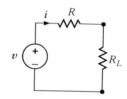

Figure 4.18 Equivalent circuit for examining maximum power transfer.

We want to determine the value of R_L that maximizes this quantity. Hence, we differentiate this expression with respect to R_L and equate the derivative to zero.

$$\frac{dP_{\text{load}}}{dR_L} = \frac{(R + R_L)^2 v^2 - 2v^2 R_L (R + R_L)}{(R + R_L)^4} = 0$$

which yields

$$R_L = R$$

In other words, maximum power transfer takes place when the load resistance $R_L = R$. Although this is a very important result, we have derived it using the simple network in Fig. 4.18. However, we should recall that v and R in Fig. 4.18 could represent the Thévenin equivalent circuit for any linear network.

EXAMPLE 4.14

Let us find the value of R_L for maximum power transfer in the network in Fig. 4.19a and the maximum power that can be transferred to this load.

SOLUTION   To begin, we derive the Thévenin equivalent circuit for the network exclusive of the load. V_{oc} can be calculated from the circuit in Fig. 4.19b. The mesh equations for the network are

$$I_1 = 2 \times 10^{-3}$$

$$3k(I_2 - I_1) + 6kI_2 + 3 = 0$$

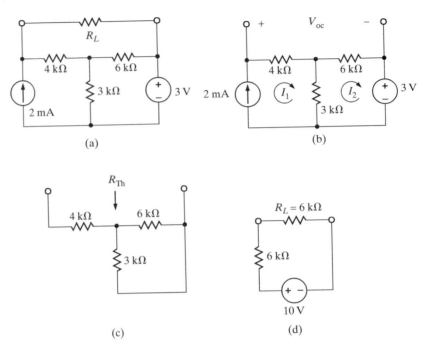

Figure 4.19 Circuits used in Example 4.14.

Solving these equations yields $I_2 = 1/3$ mA and, hence,

$$V_{oc} = 4kI_1 + 6kI_2$$

$$= 10 \text{ V}$$

R_{Th}, shown in Fig. 4.19c, is 6 kΩ; therefore, $R_L = R_{Th} = 6$ kΩ for maximum power transfer. The maximum power transferred to the load is

$$P_L = \left(\frac{10}{12k}\right)^2 (6k) = \frac{25}{6} \text{ mW} \qquad \square$$

EXAMPLE 4.15

Let us find R_L for maximum power transfer and the maximum power transferred to this load in the circuit in Fig. 4.20a.

SOLUTION We wish to reduce the network to the form shown in Fig. 4.18. We could form the Thévenin equivalent circuit by breaking the network at the load. However, a close examination of the network indicates that our analysis will be simpler if we break the network to the left of the 4-kΩ resistor. When we do this, however, we must realize, as shown in Fig. 4.18, that for maximum power transfer $R_L = R_{Th} + 4$ kΩ. V_{oc} can be calculated from the network in Fig. 4.20b. Forming a su-

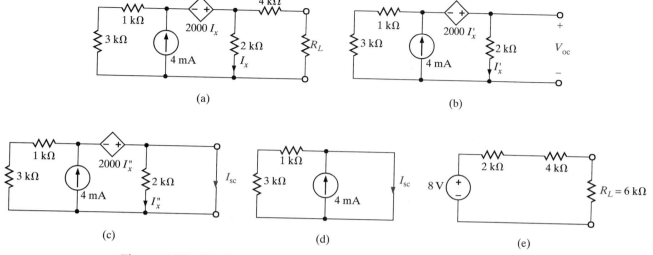

Figure 4.20 Circuits used in Example 4.15.

pernode around the dependent source and its connecting nodes, the KCL equation for this supernode is

$$\frac{V_{oc} - 2000I'_x}{1k + 3k} + (-4 \times 10^{-3}) + \frac{V_{oc}}{2k} = 0$$

where

$$I'_x = \frac{V_{oc}}{2k}$$

These equations yield $V_{oc} = 8\,\text{V}$. The short-circuit current can be found from the network in Fig. 4.20c. It is here that we find the advantage of breaking the network to the left of the 4-kΩ resistor. The short circuit shorts the 2-kΩ resistor and, therefore, $I''_x = 0$. Hence, the circuit is reduced to that in Fig. 4.20d, where clearly $I_{sc} = 4\,\text{mA}$. Then

$$R_{Th} = \frac{V_{oc}}{I_{sc}} = 2\,\text{k}\Omega$$

Connecting the Thévenin equivalent to the remainder of the original circuit produces the network in Fig. 4.20e. For maximum power transfer $R_L = R_{Th} + 4\,\text{k}\Omega = 6\,\text{k}\Omega$, and the maximum power transferred is

$$P_L = \left(\frac{8}{12k}\right)^2 (6k) = \frac{8}{3}\,\text{mW} \qquad \square$$

EXTENSION EXERCISE

E4.8   Given the circuit in Fig. E4.8, find R_L for maximum power transfer and the maximum power transferred.

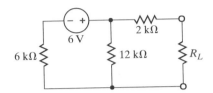

Figure E4.8

ANSWER: $R_L = 6 \text{ k}\Omega, P_L = 2/3 \text{ mW}.$

4.6 DC SPICE ANALYSIS USING SCHEMATIC CAPTURE

Introduction

Originally, SPICE (<u>S</u>imulation <u>P</u>rogram with <u>I</u>ntegrated <u>C</u>ircuit <u>E</u>mphasis) was developed at the University of California at Berkeley. It quickly became an industry standard for simulating electronic circuits. With the advent of the PC industry, several companies began selling PC and Macintosh compatible versions of SPICE. One such company, MicroSim, produced a PC compatible version called PSPICE, which we will discuss in some detail in this text. Recently, MicroSim was purchased by OrCAD. The version of PSPICE contained here is 8.0—the last MicroSim version. Version 9.0 is under development by OrCAD and will be introduced in the book's Web site when it becomes available.

In SPICE, circuit information, such as the names and values of resistors and sources as well as how they are interconnected, is input using data statements with a specific format. The particulars for each element must be typed in a precise order. This makes debugging difficult since you must know the proper format to recognize your errors.

PSPICE, as well as other SPICE-based simulators, now employs a feature known as schematic capture. The schematic editor in version 8.0 is called *Schematics*. The *Schematics* editor permits us to bypass the cryptic formatting and simply draw the circuit diagram, assigning element values via dialog boxes. *Schematics* then converts the circuit diagram to an original SPICE format for actual simulation.

When elements such as resistors and voltage sources are given values, it is convenient to use scale factors. PSPICE supports the following list:

T—tera—10^{12}	K—kilo—10^3	N—nano—10^{-9}
G—giga—10^9	M—milli—10^{-3}	P—pico—10^{-12}
MEG—mega—10^6	U—micro—10^{-6}	F—femto—10^{-15}

The component value must be immediately followed by a character—spaces are not allowed. Any text can follow the scale factor. Finally, PSPICE is case insensitive. So there is no difference between 1 MΩ and 1 mΩ in PSPICE.

Figure 4.21 shows the five-step procedure for simulating circuits using *Schematics* and PSPICE. We will demonstrate this procedure in the following example.

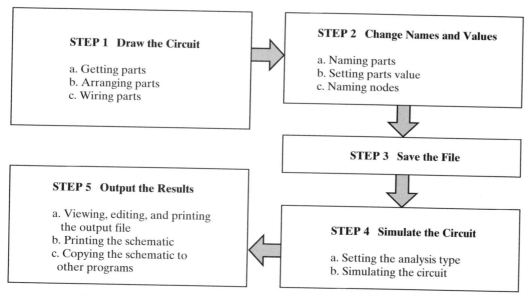

Figure 4.21 The five-step procedure for PSPICE simulations.

A dc Simulation Example

The following font conventions will be used throughout this tutorial. Uppercase text refers to PSPICE programs, menus, dialog boxes, and utilities. All boldface text denotes keyboard or mouse inputs. In each instance, boldface text's case matches that in PSPICE. Let's simulate the circuit in Fig. 4.22 using PSPICE. Following the flowchart procedure shown in Fig. 4.21, the first step is to open *Schematics* using **Start/Programs/Design-Lab 8/Schematics** sequence of pop-up menus. When *Schematics* opens, our screen will change to the Schematic Editor window shown in Fig. 4.23.

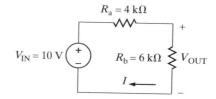

Figure 4.22 A dc circuit used for simulation.

Step 1: Drawing the Schematic

Next, we obtain and place the required parts—two resistors and a dc voltage source. To get a part, left click on the **Draw** menu and select **Get New Part**, as shown in Fig. 4.24. The **Part Browser** in Fig. 4.25 appears, listing all the parts available in PSPICE. Since we do not know the *Schematics* name for a dc voltage source, select **Libraries**, and the **Library Browser** in Fig. 4.26 appears. This browser lists all the parts libraries available to us. The dc

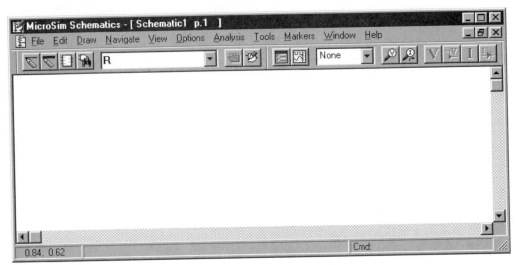

Figure 4.23 The *Schematics* editor window.

voltage source is called VDC and is located in the SOURCE.slb library that was selected and is shown in Fig. 4.26. Thus, we select the part **VDC** and click **OK**. The box in Fig. 4.25
reappears.

Figure 4.24 Getting a new part
in *Schematics*.

If we now select **Place & Close**, we revert back to the Schematic Editor in Fig. 4.23 with
one difference: the mouse pointer has become a dc voltage source symbol. The source
can be positioned within the drawing area by moving the mouse and placed by left clicking once. The result is shown in Fig. 4.27. Note that *Schematics* uses the battery symbol for
a dc source. Since the circuit contains only one source, we click the right mouse button to
stop placing sources and the mouse pointer returns to its original status.

Next we place the resistors. We repeat the process of **Draw/Get New Part** except that
this time, when the **Part Browser Basic** dialog box appears, we type in **R** and select **Place &**

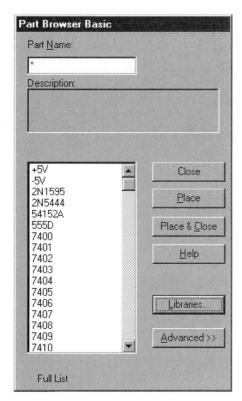

Figure 4.25 The Parts Browser window.

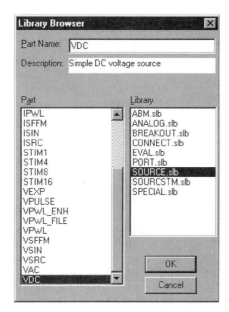

Figure 4.26 The Library Browser window.

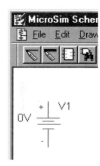

Figure 4.27 The *Schematics* editor after placing the dc voltage source.

Close. The mouse pointer then becomes a resistor. We left click once to place R1, *move the mouse some distance*, and left click again to place R2. To stop adding resistors, we right click once. Moving the mouse between part placements keeps the parts from stacking atop one another in the diagram. In order to make the schematic match the one in Fig. 4.22, we rotate R2 and place it at the right end of the diagram. This rotation is accomplished by first clicking on the **Edit** menu and selecting **Rotate**. This causes R2 to spin 90° counterclockwise. Next, we click on R2 and drag it to the desired location. A diagram similar to that shown in Fig. 4.28 should result. Note that the resistors are automatically assigned default values of 1 kΩ.

The parts can now be connected by going to the **Draw** menu and selecting **Wire**. The mouse pointer will then turn into a symbolic pencil. To connect the top of the source to R1 (i.e., R_a), use the mouse to place the pencil at the end of the wire stub protruding from Vin, click once, and release. Next, we move the mouse up and over to the left end of R1. A line is drawn up and over a 90° angle and appears dashed, as shown in Figure 4.29. *Dashed lines are not yet wires!* We must left click again to complete and cut the connection. The dashed lines become solid and the wiring connection is made. Excess wire fragments (extended dashed lines) can be removed by selecting **Redraw** from the **View** menu.

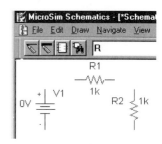

Figure 4.28 The parts after placement.

The two remaining wires can be drawn using a *Schematics* shortcut. To reactivate the wiring pencil, double right click. This shortcut reactivates the most recent mouse use. We simply repeat the steps listed previously to complete the wiring.

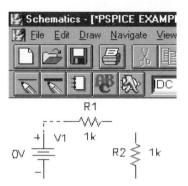

Figure 4.29 Wiring the parts.

PSPICE requires that all schematics have a ground, or reference, terminal. The voltage there will be zero and all other node voltages are referenced to it. The required part is the analog ground (AGND) in the PORT.slb library shown in Fig. 4.26. Get this part and place it at the bottom of the schematic, as shown in Fig. 4.30. Make sure that the part touches the bottom wire in the diagram so that the node dot appears in your schematic. The wiring is now completed.

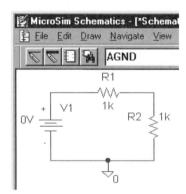

Figure 4.30 The schematic after wiring and adding the AGND part.

Step 2: Changing Component Names and Values

In order to change the voltage source's name, we double click on the text, V1. The **Edit Reference Designator** dialog box appears as shown in Fig. 4.31. Simply type in the new name for the source, in this case **Vin**, and select **OK**. Next we will change the source's voltage value by double clicking on the value, as shown in Fig. 4.30. Now the **Attributes** dialog box in Figure 4.32 appears. Type in **10V** and select **OK**. In a similar manner, the two resistors are renamed Ra and Rb and their values are set at 4 kΩ and 6 kΩ, respectively. The circuit shown in Fig. 4.33 is now ready for simulation.

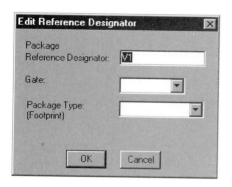

Figure 4.31 Renaming the voltage source.

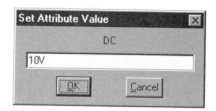

Figure 4.32 Setting the source value.

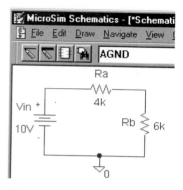

Figure 4.33 The schematic after changing the part names and values.

Step 3: Saving the Schematic

The schematic can be saved by simply going to the **File** menu and selecting **Save**. It is better to save your work in a directory other than the main PSPICE directory. All *Schematics* files are automatically given the extension .sch.

The Netlist

The netlist is actually the old-fashioned PSPICE code listing for the diagram drawn in *Schematics*. To create the netlist, go to the **Analysis** menu in Fig. 4.23 and select **Create Netlist**. What will appear at the bottom of the *Schematics* Editor window is either the message *Netlist Created* or a dialog box informing you of netlist errors. The netlist can be viewed by returning to the **Analysis** menu and selecting **Examine Netlist**, which opens the file shown in Fig. 4.34. All three elements appear along with their proper values. The text elements $N_001 $N_002 etc. are the node numbers that *Schematics* created when it converted the diagram to a netlist.

By tracing through the node numbers, we can be assured that our circuit is properly connected. The source is 10 V positive at node 2 with respect to node 0. AGND is always node number zero. Ra connects node 2 to node 1 and Rb completes the loop between nodes 0 and 1— it is perfect.

How did PSPICE generate these particular node numbers? In PSPICE, the terminals of a part symbol are called pins and are numbered. For example, the pin numbers for the symbol that represents a resistor, dc voltage, and current source symbols are shown in Figure 4.35. None of the parts in Fig. 4.35 have been rotated. When the GET NEW PART sequence generates a part with a horizontal orientation, like the resistor, pin 1 is on the left. All vertically oriented parts (the sources) have pin 1 at the top. In the netlist, the order of the node numbers is always pin 1 and then pin 2 for each component. Furthermore, when a part is rotated, the pin numbers also rotate. The most critical aspect of the pin numbers is current direction. PSPICE always yields the current flowing into pin 1 and out of pin 2. To change the node number order of a part in the netlist, we must rotate that part 180°. This can be done by clicking on the part and selecting **Rotate** in the **Edit** menu twice.

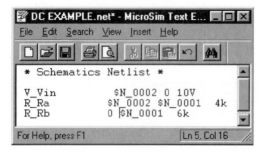

Figure 4.34 The netlist.

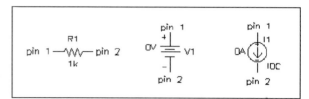

Figure 4.35 Pin numbers for resistors, voltage sources and current sources.

Step 4: Simulating the Circuit

The node voltages in the circuit are typically of interest to us, and *Schematics* permits us to identify each node with a unique name. For example, if we wish to call the output node at Rb "Vout", we simply double click on the wiring at the output node and the dialog box in Fig. 4.36 will appear. Then type **Vout** in the space shown and select **OK**. The schematic will then look like that shown in Fig. 4.37.

Figure 4.36 Labeling a node.

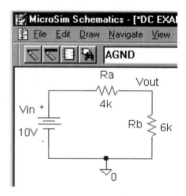

Figure 4.37 The schematic after naming the output node.

Version 8.0 has a feature that allows all node voltages and branch currents to be displayed directly on the schematic diagram. Go to the **Analysis** menu and select **Display Results on Schematic**, as shown in Fig. 4.38. To display node voltages, choose **Enable Voltage Display**. Similarly, to display branch currents, choose **Enable Current Display**. The results will be displayed on the schematic when the simulation is performed.

Simulation begins by selecting the type of analysis we wish to perform. This is done by selecting **Setup** from the **Analysis** menu. The SETUP dialog box is shown in Fig. 4.39. A *Schematics* dc analysis is requested by selecting **Bias Point Detail** and then **Close**. Since we enabled both voltage and current displays directly on the schematic, the simulation results will include all node voltages and all branch currents.

Viewing the SETUP box in Fig. 4.39, we see that a number of different analyses could be requested (for example, a DC SWEEP). In this case, *Schematics* will ask for the name of the dc source you wish to sweep, the start and stop values, and the increment. The simulation results will contain all dc node voltages and branch currents as a function of the

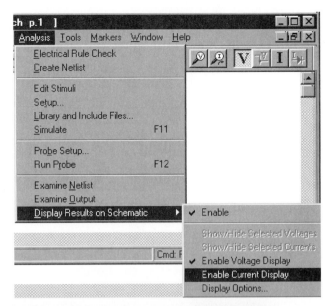

Figure 4.38 Enabling the voltage and current displays.

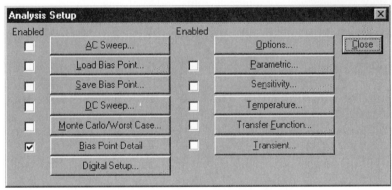

Figure 4.39 The Analysis Setup Box.

varying source value. Two additional analyses, TRANSIENT and AC SWEEP, will be discussed in subsequent chapters.

After exiting the analysis SETUP Box, we select **Run PSpice** from the **Analysis** menu. Once the simulation has run, your schematic should look much like that shown in Fig. 4.40 with node voltages and branch currents displayed. All node voltages are referenced to the AGND node. Current directions can be found be clicking on the branch current value. A small red arrow will appear, revealing the direction. The simulation results show $V_{out} = 6$ V and $I = 1$ mA.

The use of voltage division and Ohm's law quickly confirms the simulation results; that is,

$$V_{out} = V_{in}\left[\frac{R_b}{R_b + R_a}\right] = 10\left[\frac{6k}{6k + 4k}\right] = 6 \text{ V}$$

and

$$I = \frac{10}{4k + 6k} = 1 \text{ mA}$$

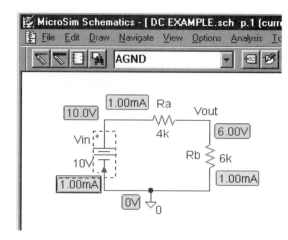

Figure 4.40 The simulated circuit.

Step 5: Viewing and Printing the Results

As seen in Fig. 4.40, simulation results can be displayed on the schematic. Alternatively, all node voltages, voltage source currents, and total power dissipation are contained in a file called the output file. Let us check the simulation results displayed on the schematic against the output file data. To view this data, select **Examine Output** from the **Analysis** menu. Within the resulting long text file is a section containing the node voltages, voltage source currents, and total power dissipation, as shown in Fig. 4.41. Indeed, V_{out} is 6.00 V and the total power dissipation is 10 mW. The current bears closer inspection. PSPICE says the current through V_{in} is −1 mA. Recall from the netlist discussion that in PSPICE, current flows from a part's pin number 1 to pin 2. As seen in Fig. 4.35, pin 1 is at the positive end of the voltage source. PSPICE is telling us that the current flowing top-down through our source (counterclockwise in Fig. 4.22) is −1 mA. This agrees with both our calculation of 1 mA flowing clockwise around the loop and the current displayed in Fig. 4.40. Note that the output file is displayed in a simple text editor. So the contents of the file can be edited, saved, printed, or copied and pasted to other programs.

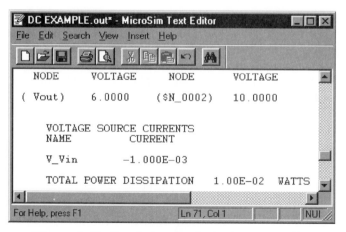

Figure 4.41 Simulation results in the output file.

To print the circuit diagram, first go to the **Configure** menu and select **Display Options**. When the **Display Options** dialog box comes up, deselect the **Grid On** checkbox to turn off the grid dots and select **OK**. Next, point the mouse pointer above and left of the upper left-hand corner of the diagram. Click left, and hold and drag the mouse beyond the lower right edge of the drawing. A box will grow as you drag, eventually surrounding the circuit. Go to the **File** menu and select **Print**. The dialog box in Fig. 4.42 will appear. Select the options **Only Print Selected Area** and **User Definable Zoom Factor**. For most of the small schematics you will create, a scale of 125 to 200% will do fine. Other options are self-explanatory. Finally, select **OK** to print.

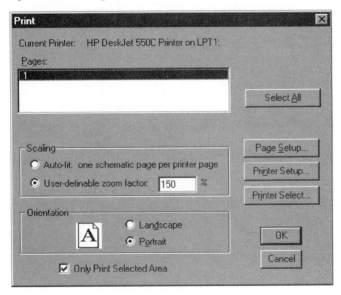

Figure 4.42 The PRINT dialog box.

To incorporate your schematic into a different program, such as Microsoft Word, draw a box around the diagram as described previously, and then under the **Edit** menu, select **Copy to Clipboard**. You can now paste the circuit into other programs.

EXAMPLE 4.16

Using the PSPICE *Schematics* editor, draw the circuit in Fig. 4.43, and use the IPROBE and VIEWPOINT parts to determine the current I and the voltage V_o.

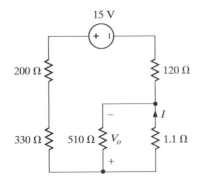

Figure 4.43

SOLUTION The *Schematics* diagram is shown in Fig. 4.44. Note how the AGND part has been judiciously placed so that the VIEWPOINT part value is indeed V_o. From Fig. 4.44, we find $V_o = 5.2348$ V and $I = 4.759$ mA.

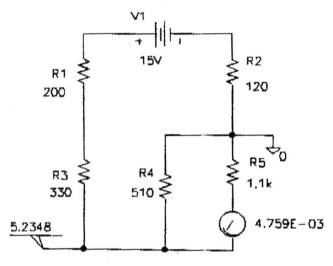

Figure 4.44 The schematics diagram for the network in Fig. 4.43. ❏

EXAMPLE 4.17

Let us use PSPICE to find the voltage V_o and the current I_x in the circuit in Fig. 4.45.

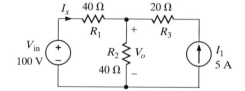

Figure 4.45

SOLUTION The PSPICE *Schematics* diagram is shown in Fig. 4.46. From this diagram, we find that $V_o = 150$ V and $I_x = -1.25$ A. ❏

2nd DC Example

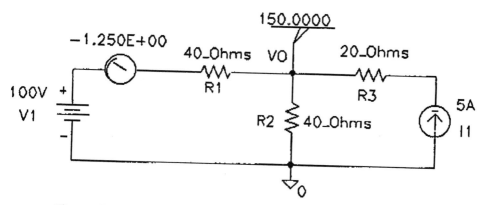

Figure 4.46 The schematics diagram for the network in Fig. 4.45.

4.7 APPLICATIONS

EXAMPLE 4.18

On Monday afternoon, Connie suddenly remembers that she has a term paper due Tuesday morning. When she sits at her computer to start typing, she discovers that the computer mouse doesn't work. After disassembly and some inspection, she finds that the mouse contains a printed circuit board that is powered by a 5-V supply contained inside the computer case. Furthermore, the board is found to contain several resistors, some op-amps and one unidentifiable device, which is connected directly to the computer's 5-V supply as shown in Fig. 4.47a. Using a voltmeter to measure the node voltages, Connie confirms that all resistors and op-amps are functioning properly and the power supply voltage reaches the mouse board. However, without knowing the mystery device's function within the circuit, she cannot determine its condition. A phone call to the manufacturer reveals that the device is indeed linear but is also proprietary. With some persuasion, the manufacturer's representative agrees that if Connie can find the Thévenin equivalent circuit for the element at nodes *A-B* with the computer on, he will tell her if it is functioning properly. Armed with a single 1-kΩ resistor and a voltmeter, Connie attacks the problem.

SOLUTION   In order to find the Thévenin equivalent for the unknown device, together with the 5-V source, Connie first isolates nodes *A* and *B* from the rest of the devices on the board to measure the open-circuit voltage. The resulting voltmeter reading is $V_{AB} = 2.4$ V. Thus, the Thévenin equivalent voltage is 2.4 V. Then she

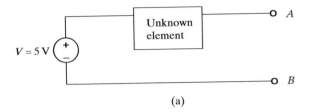

(a)

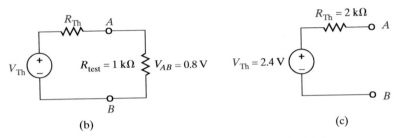

(b)   (c)

Figure 4.47 Networks used in Example 4.18.

connects the 1-kΩ resistor at nodes A-B as shown in Fig. 4.47b. The voltmeter reading is now $V_{AB} = 0.8$ V. Using voltage division to express V_{AB} in term of V_{Th}, R_{Th}
and R_{test} in Fig. 4.47b yields the expression

$$0.8 = V_{Th}\left(\frac{1k}{1k + R_{Th}}\right)$$

Solving the equations for R_{Th}, we obtain

$$R_{Th} = 2.0 \text{ k}\Omega$$

Therefore, the unknown device and the 5-V source can be represented at the terminals A-B by the Thévenin equivalent circuit shown in Fig. 4.47c. When Connie
phones the manufacturer with the data, the representative informs her that the device has indeed failed.

4.8 SUMMARY

- Linearity: This property requires both additivity and homogeneity. Using this
  property, the voltage or current somewhere in a network can be determined by
  assuming a specific value for the variable and then determining what source
  value is required to produce it. The ratio of the specified source value to that
  computed from the assumed value of the variable, together with the assumed
  value of the variable, can be used to obtain a solution.

- In a linear network containing multiple independent sources, the principle of
  superposition allows us to compute any current or voltage in the network as the
  algebraic sum of the individual contributions of each source acting alone.

- Superposition is a linear property and does not apply to nonlinear functions such as power.

- Source transformation permits us to replace a voltage source V in series with a resistance R by a current source $I = V/R$ in parallel with the resistance R. The reverse is also true.

- Using Thévenin's theorem, we can replace some portion of a network at a pair of terminals with a voltage source V_{oc} in series with a resistor R_{Th}. V_{oc} is the open-circuit voltage at the terminals, and R_{Th} is the Thévenin equivalent resistance obtained by looking into the terminals with all independent sources made zero.

- Using Norton's theorem, we can replace some portion of a network at a pair of terminals with a current source I_{sc} in parallel with a resistor R_{Th}. I_{sc} is the short-circuit current at the terminals and R_{Th} is the Thévenin equivalent resistance.

- Maximum power transfer can be achieved by selecting the load R_L to be equal to R_{Th} found by looking into the network from the load terminals.

- dc PSPICE with Schematic Capture is an effective tool in analyzing dc circuits.

PROBLEMS

Section 4.1

4.1 Find V_o in the network in Fig. P4.1 using linearity and the assumption that $V_o = 1$ mV.

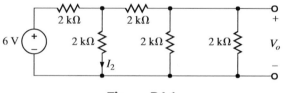

Figure P4.1

Find I_2 using $V_o = 1$ mV and then use the procedure in Example 4.1.

4.2 Find V_o in the network in Fig. P4.2 using linearity and the assumption that $V_o = 1$ mV.

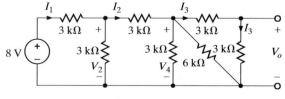

Figure P4.2

Find I_3 and use it to find V_4. Then find I_2. Follow the procedure in Example 4.1.

4.3 Find I_o in the network in Fig. P4.3 using linearity and the assumption that $I_o = 1$ mA.

Find V_2 from $I_o = 1$ mA. Find I_3 and then find I_2. Repeat the procedure until the source current is obtained. Then use the linearity to obtain I_o.

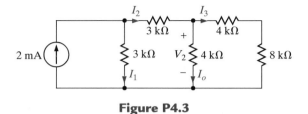

Figure P4.3

4.4 Find I_o in the network in Fig. P4.4 using linearity and the assumption that $I_o = 1$ mA.

Similar to P4.3

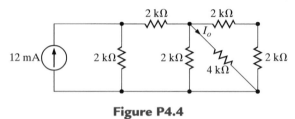

Figure P4.4

Section 4.2

4.5 Find V_o in the network in Fig. P4.5 using superposition.

Follow the procedure in Example 4.2.

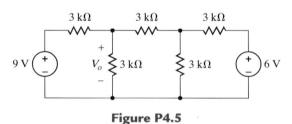

Figure P4.5

4.6 Find V_o in the network in Fig. P4.6 using superposition.

Similar to P4.5

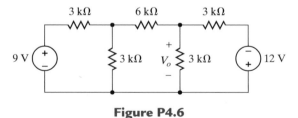

Figure P4.6

4.7   Find V_o in the network in Fig. P4.7 using superpositon.

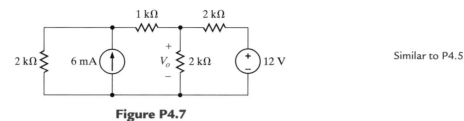

Figure P4.7

Similar to P4.5

4.8   Find I_o in the network in Fig. P4.8 using superpositon.

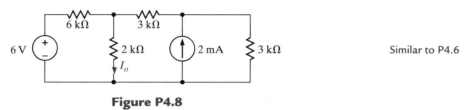

Figure P4.8

Similar to P4.6

4.9   Find V_o in the network in Fig. P4.9 using superpositon.

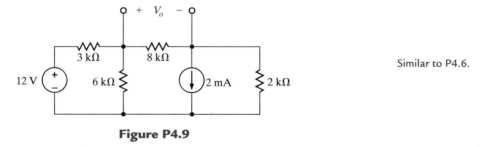

Figure P4.9

Similar to P4.6.

4.10 Find I_o in the network in Fig. P4.10 using superposition.

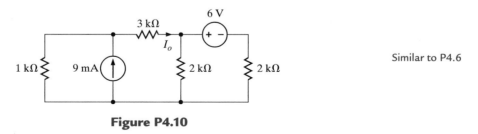

Figure P4.10

Similar to P4.6

4.11 Find V_o in the network in Fig. P4.11 using superpositon.

Similar to P4.5

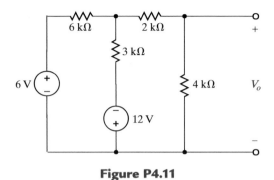

Figure P4.11

4.12 Find I_o in the network in Fig. P4.12 using superpositon.

Similar to P4.6

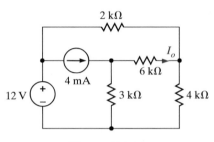

Figure P4.12

Section 4.3

4.13 Find V_o in the network in Fig. P4.13 using source transformation.

Use the procedure illustrated in Example 4.6.

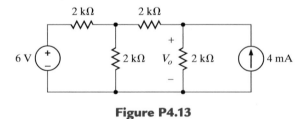

Figure P4.13

Similar to P4.13

4.14 Find V_o in the network in Fig. P4.11 using source transformation.

Similar to P4.13

4.15 Find V_o in the network in Fig. P4.9 using source transformation.

4.16 Find I_o in the network in Fig. P4.16 using source transformation.

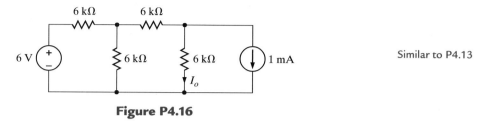

Figure P4.16

Similar to P4.13

4.17 Find V_o in the network in Fig. P4.17 using source transformation.

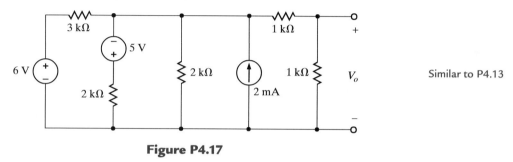

Figure P4.17

Similar to P4.13

4.18 Use source transformation to find I_o in the network in Fig. P4.18.

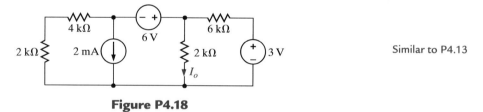

Figure P4.18

Similar to P4.13

4.19 Find I_o in the circuit in Fig. P4.18 using source transformation.

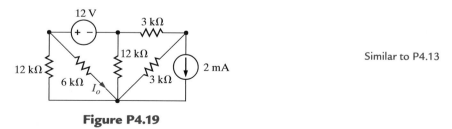

Figure P4.19

Similar to P4.13

4.20 Find V_o in the network in Fig. P4.20 using source transformation.

Similar to P4.13

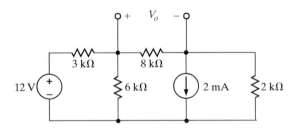

Figure P4.20

Section 4.4

4.21 Use Thévenin's theorem to find V_o in the network in Fig. P4.21.

Similar to Example 4.7

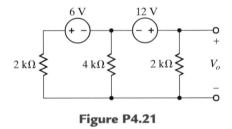

Figure P4.21

4.22 Find V_o in the network in Fig. P4.22 using Thévenin's theorem.

Similar to P4.21

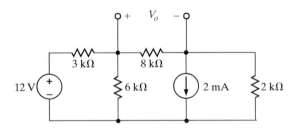

Figure P4.22

4.23 Find I_o in the network in Fig. P4.23 using Thévenin's theorem.

Similar to P4.21

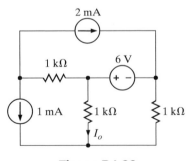

Figure P4.23

4.24 Find I_o in the network in Fig. P4.24 using Thévenin's theorem.

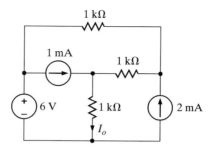

Figure P4.24

Similar to P4.22

4.25 Find I_o in the network in Fig. P4.25 using Thévenin's theorem.

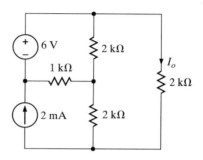

Figure P4.25

Similar to P4.22

4.26 Find I_o in the network in Fig. P4.26 using Thévenin's theorem.

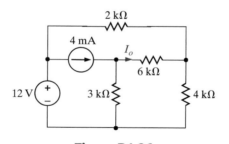

Figure P4.26

Similar to P4.22

4.27 Find V_o in the network in Fig. P4.27 using Thévenin's theorem.

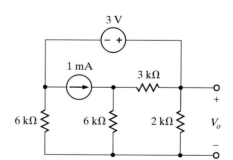

Figure P4.27

Similar to P4.22

4.28 Find V_o in the network in Fig. P4.28 using Thévenin's theorem.

Similar to P4.22

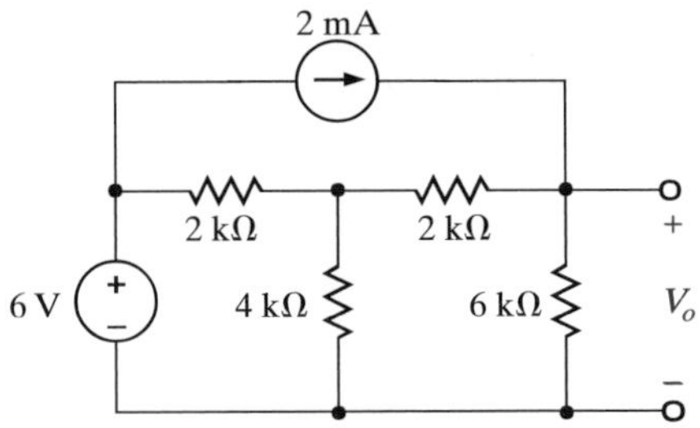

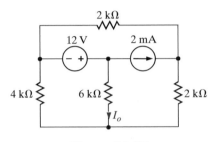

Figure P4.28

4.29 Find I_o in the network in Fig. P4.29 using Thévenin's theorem.

Similar to P4.22

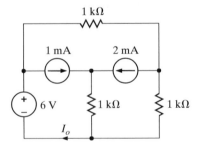

Figure P4.29

4.30 Find I_o in the network in Fig. P4.30 using Thévenin's theorem.

Similar to P4.22 with the exception that the 6-V source is the load.

Figure P4.30

4.31 Find I_o in the network in Fig. P4.31 using Thévenin's theorem.

Similar to P4.22

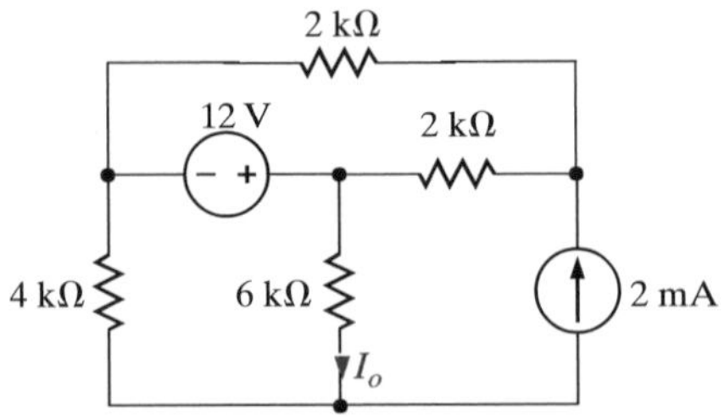

Figure P4.31

4.32 Find the Thévenin equivalent of the network in Fig. P4.32 at the terminals *A-B*.

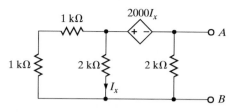

Figure P4.32

See the procedure in
Example 4.10.

4.33 Find the Thévenin equivalent of the
network in Fig. P4.33 at the terminals
A-B.

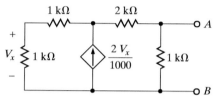

Figure P4.33

See the procedure in
Example 4.11.

4.34 Find V_o in the network in Fig. P4.34
using Thévenin's theorem.

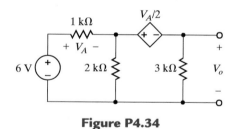

Figure P4.34

Follow the procedure used in
Example 4.12.

4.35 Use Thévenin's theorem to find I_o in the circuit in Fig. P4.35.

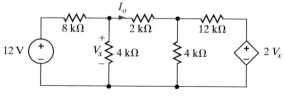

Figure P4.35

Similar to P4.34

4.36 Find V_o in the network in Fig. P4.36 using Thévenin's theorem.

Similar to P4.34

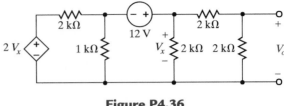

Figure P4.36

4.37 Use Thévenin's theorem to find V_o in the circuit in Fig. P4.37.

Use the procedure illustrated in
Example 4.13.

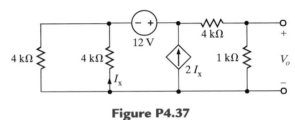

Figure P4.37

4.38 Find V_o in the network in Fig. P4.38 using Thévenin's theorem.

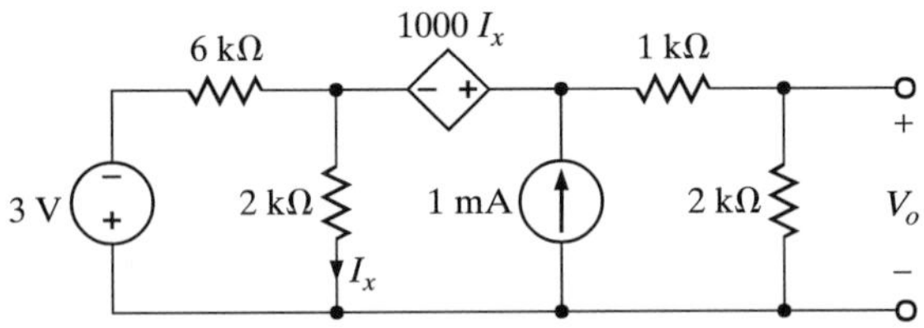

Similar to P4.34

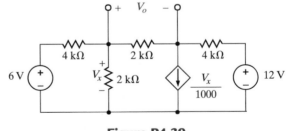

Figure P4.38

4.39 Find V_o in the circuit in Fig. P4.39 using Thévenin's theorem.

Similar to P4.37

Figure P4.39

4.40 Find V_o in the network in Fig. P4.40 using Thévenin's theorem.

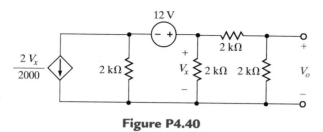

Figure P4.40

Similar to P4.37

4.41 Find I_o in the network in Fig. P4.41 using Norton's theorem.

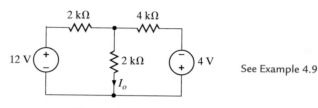

Figure P4.41

See Example 4.9

4.42 Find I_o in the network in Fig. P4.42 using Norton's theorem.

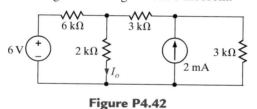

Figure P4.42

Similar to P4.41

4.43 Find I_o in the network in Fig. P4.43 using Norton's theorem.

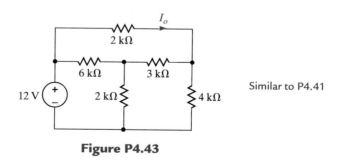

Figure P4.43

Similar to P4.41

4.44 Find V_o in the network in Fig. P4.44 using Norton's theorem.

See Example 4.9.

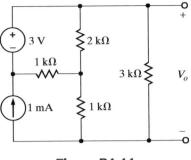

Figure P4.44

Section 4.5

4.45 Find R_L in the network in Fig. P4.45 in order to achieve maximum power transfer.

Use the procedure outlined in Example 4.14.

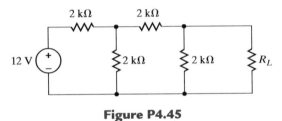

Figure P4.45

4.46 In the network in Fig. P4.46, find R_L, for maximum power transfer and the maximum power transferred to this load.

Similar to P4.45

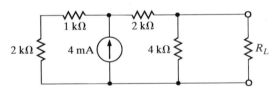

Figure P4.46

4.47 Find R_L for maximum power transfer and the maximum power that can be transferred to the load in Fig. P4.47.

Similar to P4.45

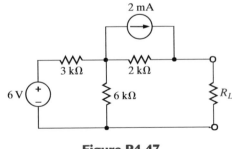

Figure P4.47

4.48 In the network in Fig. P4.48, find R_L for maximum power transfer and the maximum power transferred to this load.

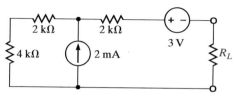

Figure P4.48

Similar to P4.45

4.49 Find R_L for maximum power transfer and the maximum power that can be transferred to the load in Fig. P4.49.

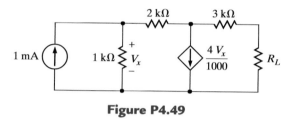

Figure P4.49

Use the procedure outlined in Example 4.15.

Section 4.6

4.50 Using the PSPICE *Schematics* editor, draw the circuit in Fig. P4.50, and use the Enable Voltage and Current Displays feature parts to determine the current I and the voltage V_o.

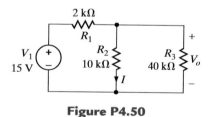

Figure P4.50

4.51 Using the PSPICE *Schematics* editor, draw the circuit in Fig. P4.51, and use the Enable Voltage and Current Displays feature parts to determine the current I and the voltages V_1 and V_2.

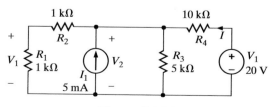

Figure P4.51

Advanced Problems

4.52 Find V_o in the circuit in Fig. P4.52 using linearity and the assumption that $V_o = 1\ \mathrm{V}$.

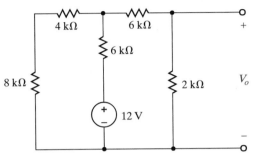

Figure P4.52

4.53 Find I_o in the network in Fig. P4.53 using superposition.

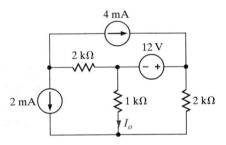

Figure P4.53

4.54 Use superposition to find V_o in the circuit in Fig. P4.54.

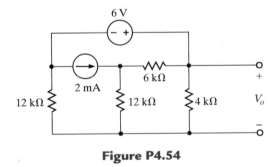

Figure P4.54

4.55 Find V_o in the circuit in Fig. P4.55 using superpositon.

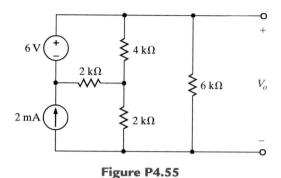

Figure P4.55

4.56 Find I_o the network in Fig. P4.56 using superpositon.

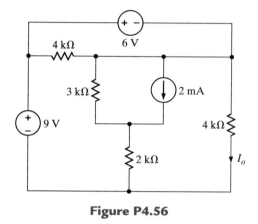

Figure P4.56

4.57 Find V_o in the network in Fig. P4.57 using superposition.

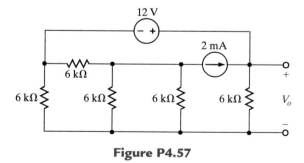

Figure P4.57

4.58 Find I_o the network in Fig. P4.58.

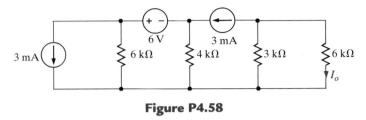

Figure P4.58

4.59 Find V_o in the network in Fig. P4.59.

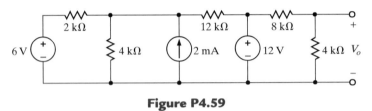

Figure P4.59

4.60 Find I_o in the network in Fig. P4.60 using source transformation.

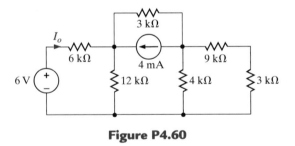

Figure P4.60

4.61 Find I_o in the network in Fig. P4.61 using source transformation.

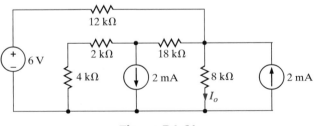

Figure P4.61

4.62 Use source transformation to find V_o in the network in Fig. P4.62.

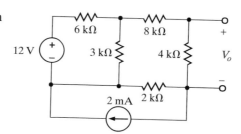

Figure P4.62

4.63 Find I_o the network in Fig. P4.63 using source transformation.

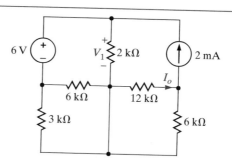

Figure P4.63

4.64 Find I_o in the network in Fig. P4.64.

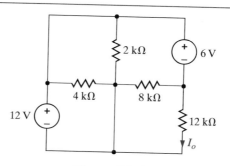

Figure P4.64

4.65 Find V_o in the network in Fig. P4.65 using source transformation.

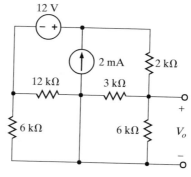

Figure P4.65

4.66 Find V_o in the network in Fig. P4.66 using Thévenin's theorem.

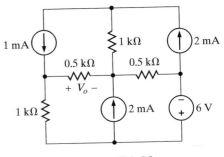

Figure P4.66

4.67 Find V_o in the network in Fig. P4.67 using Thévenin's theorem.

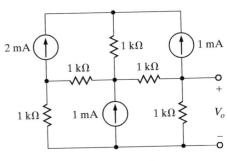

Figure P4.67

4.68 Find V_o in the network in Fig. P4.68 using Thévenin's theorem.

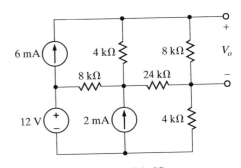

Figure P4.68

4.69 Find V_o in the network in Fig. P4.69 using Thévenin's theorem.

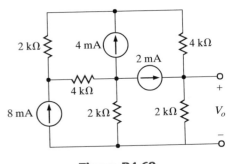

Figure P4.69

4.70 Find I_o in the network in Fig. P4.70 using Thévenin's theorem.

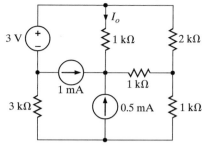

Figure P4.70

4.71 Use a combination of Thévenin's theorem and superposition to find V_o in the circuit in Fig. P4.71.

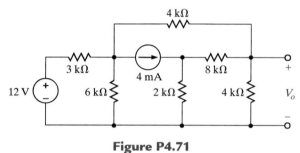

Figure P4.71

4.72 Find V_o in the network in Fig. P4.72 using Thévenin's theorem.

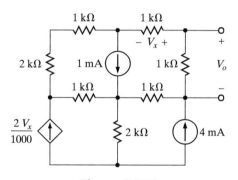

Figure P4.72

4.73 Use Norton's theorem to find V_o in the network in Fig. P4.73.

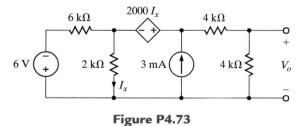

Figure P4.73

4.74 Find V_o in the network in Fig. P4.74 using Norton's theorem.

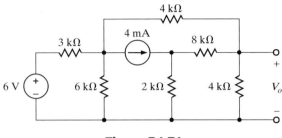

Figure P4.74

4.75 Find V_o in the network in Fig. P4.75 using Norton's theorem.

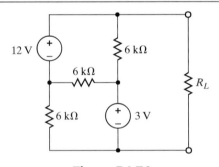

Figure P4.75

4.76 Find R_L for maximum power transfer and the maximum power that can be transferred in the network in Fig. P4.76.

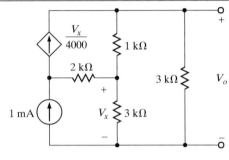

Figure P4.76

4.77 Find R_L for maximum power transfer and the maximum power that can be transferred in the network in Fig. P4.77.

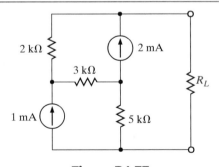

Figure P4.77

4.78 Find R_L for maximum power transfer and the maximum power that can be transferred in the network in Fig. P4.78.

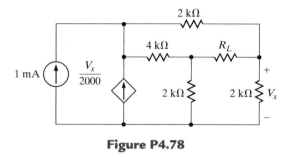

Figure P4.78

4.79 Find the Thévenin equivalent of the network in Fig. P4.79 at the terminals A-B.

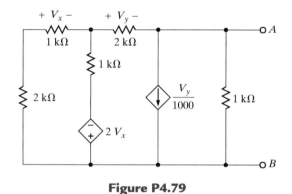

Figure P4.79

4.80 Given the linear circuit in Fig. P4.80, it is known that when a 2-kΩ load is connected to the terminals A-B, the load current is 10 mA. If a 10-kΩ load is connected to the terminals, the load current is 6 mA. Find the current in a 20-kΩ load.

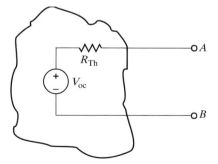

Figure P4.80

4.81 If an 8-kΩ load is connected to the terminals of the network in Fig. P4.81, $V_{AB} = 16$ V. If a 2-kΩ load is connected to the terminals, $V_{AB} = 8$ V. Find V_{AB} if a 20-kΩ load is connected to the terminals.

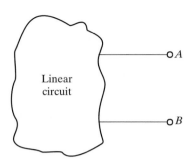

Figure P4.81

4.82 Choose R_L in Fig. P4.82 such that $I = 0.2$ mA.

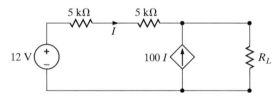

Figure P4.82

4.83 Choose R_L in Fig. P4.82 for maximum power transfer.

Capacitance and Inductance

5

*H*ave you ever wondered how a tiny camera battery is able to produce a blinding flash or how a hand-held "stun gun" can deliver 50,000 V? The answer is energy storage, and we introduce two elements that possess this property in this chapter: the capacitor and the inductor. Both capacitors and inductors are linear elements; however, unlike the resistor, their terminal characteristics are described by linear differential equations. Another distinctive feature of these elements is their ability to absorb energy from the circuit, store it temporarily, and later return it. Elements that possess this energy storage capability are referred to simply as *storage elements*.

Capacitors are capable of storing energy when a voltage is present across the element. The energy is actually stored in an electric field not unlike that produced by sliding across a car seat on a dry winter day. Conversely, inductors are capable of storing energy when a current is passing through them, causing a magnetic field to form. This phenomenon can be demonstrated by placing a needle compass in the vicinity of a current. The current causes a magnetic field whose energy deflects the compass needle.

A very important circuit, which employs a capacitor in a vital role, is also introduced. This circuit, known as an op-amp integrator, produces an output voltage that is proportional to the integral of the input voltage. The significance of this circuit is that any system (for example, electrical, mechanical, hydraulic, biological, social, economic, and so on) that can be described by a set of linear differential equations with constant coefficients can be modeled by a network consisting of op-amp integrators. Thus, very complex and costly systems can be tested safely and inexpensively prior to construction and implementation.

Finally, we examine some practical circuits where capacitors and inductors are normally found or can be effectively used in circuit design.

5.1 CAPACITORS

Note the use of the passive sign convention.

A *capacitor* is a circuit element that consists of two conducting surfaces separated by a non-conducting, or *dielectric*, material. A simplified capacitor and its electrical symbol are shown in Fig. 5.1.

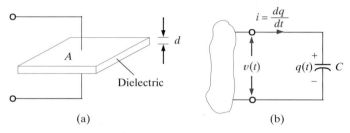

Figure 5.1 Capacitor and its electrical symbol.

There are many different kinds of capacitors, and they are categorized by the type of dielectric material that is used between the conducting plates. Although any good insulator can serve as a dielectric, each type has characteristics that make it more suitable for particular applications.

For general applications in electronic circuits (e.g., coupling between stages of amplification) the dielectric material may be paper impregnated with oil or wax, mylar, polystyrene, mica, glass, or ceramic.

Ceramic dielectric capacitors constructed of barium titanates have a large capacitance-to-volume ratio because of their high dielectric constant. Mica, glass, and ceramic dielectric capacitors will operate satisfactorily at high frequencies.

Aluminum electrolytic capacitors, which consist of a pair of aluminum plates separated by a moistened borax paste electrolyte, can provide high values of capacitance in small volumes. They are typically used for filtering, bypassing, and coupling, and in power supplies and motor-starting applications. Tantalum electrolytic capacitors have lower losses and more stable characteristics than those of aluminum electrolytic capacitors. Figure 5.2 shows a variety of typical capacitors.

In addition to these capacitors, which we deliberately insert in a network for specific applications, stray capacitance is present any time there is a difference in potential between two conducting materials separated by a dielectric. Because this stray capacitance can cause unwanted coupling between circuits, extreme care must be exercised in the layout of electronic systems on printed circuit boards.

Capacitance is measured in coulombs per volt or farads. The unit *farad* (F) is named after Michael Faraday, a famous English physicist. Capacitors may be fixed or variable and typically range from thousands of microfarads (μF) to a few picofarads (pF).

However, capacitor technology, initially driven by the modern interest in electric vehicles, is rapidly changing. For example, the capacitor in the photograph in Fig. 5.3 was designed and built at the Space Power Institute at Auburn University in order to achieve high energy and high power density capacitors for application in electric vehicles and space. This capacitor, which is specially constructed, is rated at 55 F, 500 joules (J). It is interesting to calculate the dimensions of a simple equivalent capacitor consisting of two parallel plates

Figure 5.2 Some typical capacitors (courtesy of Cornell Dubilier).

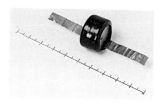

Figure 5.3 A 55-F, 500-J capacitor and a 10-in ruler for comparison.

each of area A, separated by a distance d as shown in Fig. 5.1. We learned in basic physics that the capacitance of two parallel plates of area A, separated by distance d, is

$$C = \frac{\varepsilon_o A}{d}$$

where ε_o, the permitivity of free space, is 8.85×10^{-12} F/m. If we assume the plates are separated by a distance in air of the thickness of one sheet of oil-impregnated paper, which is about 1.016×10^{-4} m, then

$$55\,\text{F} = \frac{(8.85 \times 10^{-12})A}{1.016 \times 10^{-4}}$$

$$A = 6.3141 \times 10^8 \,\text{m}^2$$

and since 1 square mile is equal to 2.59×10^6 square meters, the area is

$$A \approx 244 \text{ square miles}$$

which is the area of a medium-sized city! It would now seem that the capacitor in the photograph is much more impressive than it originally appeared. This capacitor is actually constructed using multiple layers of a combination of pressed carbon and metal fibers, which are heat treated to adhere to metal foil. The metal foils are connected to form a stack of 18 cells, within this geometry. There are literally millions of pieces of carbon and metal fibers employed to obtain the required surface area.

Suppose now that a source is connected to the capacitor shown in Fig. 5.1; then positive charges will be transferred to one plate and negative charges to the other. The charge on the capacitor is proportional to the voltage across it such that

$$q = Cv \qquad\qquad \textbf{5.1}$$

where C is the proportionality factor known as the capacitance of the element in farads.

The charge differential between the plates creates an electric field that stores energy. Because of the presence of the dielectric, the conduction current that flows in the wires that connect the capacitor to the remainder of the circuit cannot flow internally between the plates. However, via electromagnetic field theory it can be shown that this conduction current is equal to the displacement current that flows between the plates of the capacitor and is present any time that an electric field or voltage varies with time.

Our primary interest is in the current–voltage terminal characteristics of the capacitor. Since the current is

$$i = \frac{dq}{dt}$$

then for a capacitor

$$i = \frac{d}{dt}Cv$$

which for constant capacitance is

$$i = C\frac{dv}{dt} \qquad \text{5.2}$$

Equation (5.2) can be rewritten as

$$dv = \frac{1}{C}i\,dt$$

Now integrating this expression from $t = -\infty$ to some time t and assuming $v(-\infty) = 0$ yields

$$v(t) = \frac{1}{C}\int_{-\infty}^{t} i(x)\,dx \qquad \text{5.3}$$

where $v(t)$ indicates the time dependence of the voltage. Equation (5.3) can be expressed as two integrals, so that

$$v(t) = \frac{1}{C}\int_{-\infty}^{t_0} i(x)\,dx + \frac{1}{C}\int_{t_0}^{t} i(x)\,dx \qquad \text{5.4}$$

$$= v(t_0) + \frac{1}{C}\int_{t_0}^{t} i(x)\,dx$$

where $v(t_0)$ is the voltage due to the charge that accumulates on the capacitor from time $t = -\infty$ to time $t = t_0$.

The energy stored in the capacitor can be derived from the power that is delivered to the element. This power is given by the expression

$$p(t) = v(t)i(t) = Cv(t)\frac{dv(t)}{dt} \qquad \text{5.5}$$

and hence the energy stored in the electric field is

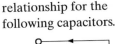

DRILL

D5.1   Write the i-v relationship for the following capacitors.

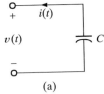

(a)

(b)

ANSWER:

(a) $i(t) = -C\dfrac{dv(t)}{dt}$

(b) $i(t) = -C\dfrac{dv(t)}{dt}$

$$w_C(t) = \int_{-\infty}^{t} Cv(x)\,\frac{dv(x)}{dx}\,dx = C\int_{-\infty}^{t} v(x)\,\frac{dv(x)}{dx}\,dx$$

$$= C\int_{v(-\infty)}^{v(t)} v(x)\,dv(x) = \tfrac{1}{2}Cv^2(x)\Big|_{v(-\infty)}^{v(t)}$$

$$= \tfrac{1}{2}Cv^2(t)\ \mathrm{J} \qquad\qquad\qquad \textbf{5.6}$$

since $v(t = -\infty) = 0$. The expression for the energy can also be written using Eq. (5.1) as

$$w_C(t) = \frac{1}{2}\frac{q^2(t)}{C} \qquad\qquad\qquad \textbf{5.7}$$

Equations (5.6) and (5.7) represent the energy stored by the capacitor, which, in turn, is equal to the work done by the source to charge the capacitor.

The polarity of the voltage across a capacitor being charged is shown in Fig. 5.1b. In the ideal case the capacitor will hold the charge for an indefinite period of time if the source is removed. If at some later time an energy-absorbing device (e.g., a flash bulb) is connected across the capacitor, a discharge current will flow from the capacitor and, therefore, the capacitor will supply its stored energy to the device.

EXAMPLE 5.1

If the charge accumulated on two parallel conductors charged to 12 V is 600 pC, what is the capacitance of the parallel conductors?

SOLUTION   Using Eq. (5.1), we find that

$$C = \frac{Q}{V} = \frac{(600)(10^{-12})}{12} = 50\ \mathrm{pF} \qquad\qquad \square$$

EXAMPLE 5.2

The voltage across a 5-μF capacitor has the waveform shown in Fig. 5.4a. Determine the current waveform.

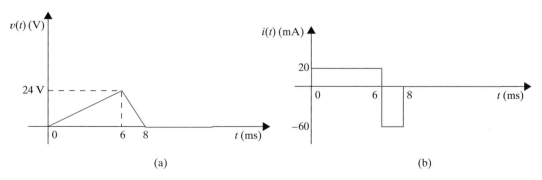

(a)   (b)

Figure 5.4 Voltage and current waveforms for a 5-μF capacitor.

SOLUTION   Note that

$$v(t) = \frac{24}{6 \times 10^{-3}} t \qquad\qquad 0 \le t \le 6 \text{ ms}$$

$$= \frac{-24}{2 \times 10^{-3}} t + 96 \qquad 6 \le t < 8 \text{ ms}$$

$$= 0 \qquad\qquad 8 \text{ ms} \le t$$

Using Eq. (5.2), we find that

$$i(t) = C \frac{dv(t)}{dt}$$

$$= 5 \times 10^{-6}(4 \times 10^{3}) \qquad 0 \le t \le 6 \text{ ms}$$

$$= 20 \text{ mA} \qquad\qquad 0 \le t \le 6 \text{ ms}$$

$$i(t) = 5 \times 10^{-6}(-12 \times 10^{3}) \qquad 6 \le t \le 8 \text{ ms}$$

$$= -60 \text{ mA} \qquad\qquad 6 \le t < 8 \text{ ms}$$

and

$$i(t) = 0 \qquad\qquad 8 \text{ ms} \le t$$

Therefore, the current waveform is as shown in Fig. 5.4b and $i(t) = 0$ for $t > 8$ ms.
❏

EXAMPLE 5.3

Determine the energy stored in the electric field of the capacitor in Example 5.2
at $t = 6$ ms.

SOLUTION   Using Eq. (5.6), we have

$$w(t) = \tfrac{1}{2} C v^{2}(t)$$

At $t = 6$ ms,

$$w(6 \text{ ms}) = \tfrac{1}{2}(5 \times 10^{-6})(24)^{2}$$

$$= 1440 \ \mu\text{J}$$
❏

EXTENSION EXERCISE

E5.1   A 10-μF capacitor has an accumulated charge of 500 nC. Determine the
voltage across the capacitor.

ANSWER: 0.05 V.

EXAMPLE 5.4

The current in an initially uncharged 4-μF capacitor is shown in Fig. 5.5a. Let us derive the waveforms for the voltage, power, and energy and compute the energy stored in the electric field of the capacitor at $t = 2$ ms.

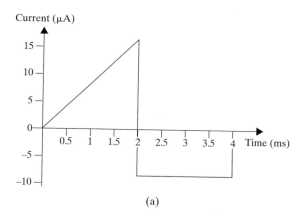

(a)

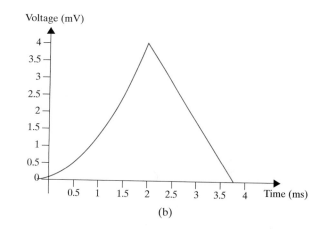

(b)

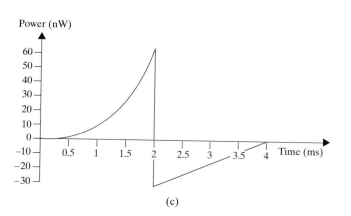

(c)

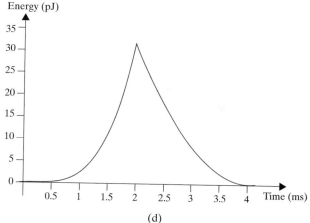

(d)

Figure 5.5 Waveforms used in Example 5.4.

SOLUTION The equations for the current waveform in the specific time intervals are

$$i(t) = \frac{16 \times 10^{-6}\, t}{2 \times 10^{-3}} \qquad 0 \le t \le 2 \text{ ms}$$
$$= -8 \times 10^{-6} \qquad 2 \text{ ms} \le t \le 4 \text{ ms}$$
$$= 0 \qquad 4 \text{ ms} < t$$

Since $v(0) = 0$, the equation for $v(t)$ in the time interval $0 \le t \le 2$ ms is

$$v(t) = \frac{1}{(4)(10^{-6})} \int_0^t 8(10^{-3}) x \, dx = 10^3 t^2$$

and hence,

$$v(2 \text{ ms}) = 10^3(2 \times 10^{-3})^2 = 4 \text{ mV}$$

In the time interval 2 ms $\le t \le$ 4 ms,

$$v(t) = \frac{1}{(4)(10^{-6})} \int_{2(10^{-3})}^{t} - (8)(10^{-6}) \, dx + (4)(10^{-3}) = -2t + 8 \times 10^{-3}$$

The waveform for the voltage is shown in Fig. 5.5b.

Since the power is $p(t) = v(t)i(t)$, the expression for the power in the time interval $0 \le t \le 2$ ms is $p(t) = 8t^3$. In the time interval 2 ms $\le t \le$ 4 ms, the equation for the power is

$$p(t) = -(8)(10^{-6})(-2t + 8 \times 10^{-3})$$
$$= 16(10^{-6})t - 64(10^{-9})$$

The power waveform is shown in Fig. 5.5c. Note that during the time interval $0 \le t \le 2$ ms, the capacitor is absorbing energy and during the interval 2 ms $\le t \le$ 4 ms, it is delivering energy.

The energy is given by the expression

$$w(t) = \int_{t_0}^{t} p(x) \, dx + w(t_0)$$

In the time interval $0 \le t \le 2$ ms,

$$w(t) = \int_{0}^{t} 8x^3 \, dx = 2t^4$$

Hence,

$$w(2 \text{ ms}) = 32 \text{ pJ}$$

In the time interval $2 \le t \le 4$ ms,

$$w(t) = \int_{2 \times 10^{-3}}^{t} [(16 \times 10^{-6})x - (64 \times 10^{-9})] \, dx + 32 \times 10^{-12}$$
$$= [(8 \times 10^{-6})x^2 - (64 \times 10^{-9})x]_{2 \times 10^{-3}}^{t} + 32 \times 10^{-12}$$
$$= (8 \times 10^{-6})t^2 - (64 \times 10^{-9})t + 128 \times 10^{-12}$$

From this expression we find that $w(2 \text{ ms}) = 32 \text{ pJ}$ and $w(4 \text{ ms}) = 0$. The energy waveform is shown in Fig. 5.5d. ❑

EXTENSION EXERCISES

E5.2 The voltage across a 2-μF capacitor is shown in Fig. E5.2. Determine the waveform for the capacitor current.

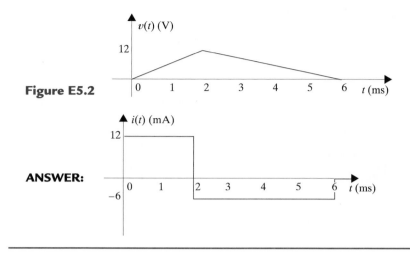

Figure E5.2

ANSWER:

E5.3   Compute the energy stored in the electric field of the capacitor in Extension Exercise E5.2 at $t = 2$ ms.

ANSWER: $w = 144$ μJ.

5.2 INDUCTORS

An *inductor* is a circuit element that consists of a conducting wire usually in the form of a coil. Two typical inductors and their electrical symbols are shown in Fig. 5.6. Inductors are typically categorized by the type of core on which they are wound. For example, the core material may be air or any nonmagnetic material, iron, or ferrite. Inductors made with air or nonmagnetic materials are widely used in radio, television, and filter circuits. Iron-core inductors are used in electrical power supplies and filters. Ferrite-core inductors are widely used in high-frequency applications. Note that in contrast to the magnetic core that confines the flux, as shown in Fig. 5.6b, the flux lines for nonmagnetic inductors extend beyond the inductor itself, as illustrated in Fig. 5.6a. Like stray capacitance, stray inductance can result from any element carrying current surrounded by flux linkages. Figure 5.7 shows a variety of typical inductors.

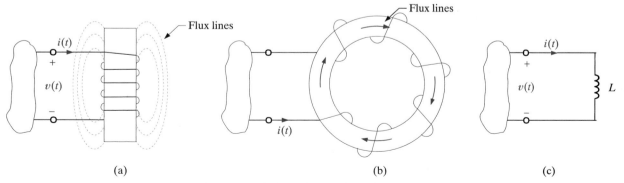

Figure 5.6 Two inductors and their electrical symbols.

Figure 5.7 Some typical inductors (courtesy of Coilcraft).

From a historical standpoint, developments that led to the mathematical model we employ to represent the inductor are as follows. It was first shown that a current-carrying conductor would produce a magnetic field. It was later found that the magnetic field and the current that produced it were linearly related. Finally, it was shown that a changing magnetic field produced a voltage that was proportional to the time rate of change of the current that produced the magnetic field; that is,

Note the use of the passive sign convention, as illustrated in Fig. 5.6.

$$v(t) = L\frac{di(t)}{dt} \qquad \textbf{5.8}$$

The constant of proportionality L is called the inductance and is measured in the unit henry, named after the American inventor Joseph Henry, who discovered the relationship. As seen in Eq. (5.8), 1 henry (H) is dimensionally equal to 1 volt-second per ampere.

Following the development of the mathematical equations for the capacitor, we find that the expression for the current in an inductor is

$$i(t) = \frac{1}{L}\int_{-\infty}^{t} v(x)\,dx \qquad \textbf{5.9}$$

which can also be written as

$$i(t) = i(t_0) + \frac{1}{L}\int_{t_0}^{t} v(x)\,dx \qquad \textbf{5.10}$$

The power delivered to the inductor can be used to derive the energy stored in the element. This power is equal to

D5.2 Write the $i\text{-}v$ relationship for the following inductors.

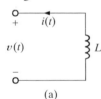

(a)

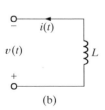

(b)

$$p(t) = v(t)i(t)$$

$$= \left[L\frac{di(t)}{dt} \right] i(t)$$

Therefore, the energy stored in the magnetic field is

$$w_L(t) = \int_{-\infty}^{t} \left[L\frac{di(x)}{dx} \right] i(t) \, dx$$

Following the development of Eq. (5.6), we obtain

$$w_L(t) = \tfrac{1}{2} Li^2(t) \text{ J} \qquad\qquad \textbf{5.12}$$

The inductor, like the resistor and capacitor, is a passive element. The polarity of the voltage across the inductor is shown in Fig. 5.6.

Practical inductors typically range from a few microhenrys to tens of henrys. From a circuit design standpoint it is important to note that inductors cannot be easily fabricated on an integrated circuit chip, and therefore chip designs typically employ only active electronic devices, resistors and capacitors, that can be easily fabricated in microcircuit form.

ANSWER:

5.11   (a) $v(t) = -L\dfrac{di(t)}{dt}$

(b) $v(t) = L\dfrac{di(t)}{dt}$

EXAMPLE 5.5

The current in a 10-mH inductor has the waveform shown in Fig. 5.8a. Determine the voltage waveform.

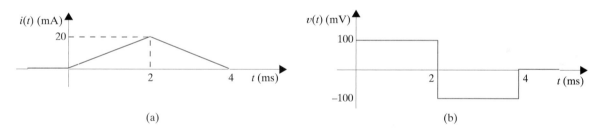

(a)   (b)

Figure 5.8 Current and voltage waveforms for a 10-mH inductor.

SOLUTION   Using Eq. (5.8) and noting that

$$i(t) = \frac{20 \times 10^{-3}t}{2 \times 10^{-3}} \qquad\qquad 0 \le t \le 2 \text{ ms}$$

$$i(t) = \frac{-20 \times 10^{-3}t}{2 \times 10^{-3}} + 40 \times 10^{-3} \quad 2 \le t \le 4 \text{ ms}$$

and

$$i(t) = 0 \qquad 4 \text{ ms} < t$$

we find that

$$v(t) = \left(10 \times 10^{-3}\right)\frac{20 \times 10^{-3}}{2 \times 10^{-3}} \quad 0 \le t \le 2 \text{ ms}$$

$$= 100 \text{ mV}$$

and

$$v(t) = (10 \times 10^{-3}) \frac{-20 \times 10^{-3}}{2 \times 10^{-3}} \quad 2 \le t \le 4 \text{ ms}$$

$$= -100 \text{ mV}$$

and $v(t) = 0$ for $t > 4$ ms. Therefore, the voltage waveform is shown in Fig. 5.8b.

❏

EXAMPLE 5.6

The current in a 2-mH inductor is

$$i(t) = 2 \sin 377t \text{ A}$$

Determine the voltage across the inductor and the energy stored in the inductor.

SOLUTION   From Eq. (5.8), we have

$$v(t) = L \frac{di(t)}{dt}$$

$$= (2 \times 10^{-3}) \frac{d}{dt} (2 \sin 377t)$$

$$= 1.508 \cos 377t \text{ V}$$

and from Eq. (5.12),

$$w_L(t) = \tfrac{1}{2} L i^2(t)$$
$$= \tfrac{1}{2} (2 \times 10^{-3})(2 \sin 377t)^2$$
$$= 0.004 \sin^2 377t \text{ J}$$

❏

EXAMPLE 5.7

The voltage across a 200-mH inductor is given by the expression

$$v(t) = (1 - 3t)e^{-3t} \text{ mV} \quad t \ge 0$$
$$= 0 \quad\quad\quad\quad\quad t < 0$$

Let us derive the waveforms for the current, energy, and power.

SOLUTION   The waveform for the voltage is shown in Fig. 5.9a. The current is derived from Eq. (5.10) as

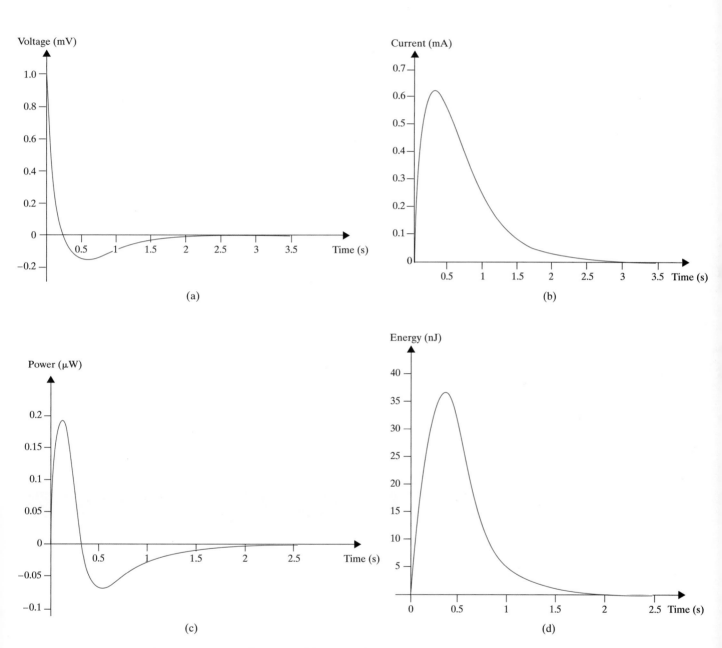

Figure 5.9 Waveforms used in Example 5.7.

$$i(t) = \frac{10^3}{200} \int_0^t (1 - 3x)e^{-3x}\, dx$$

$$= 5\left\{ \int_0^t e^{-3x}\, dx - 3 \int_0^t xe^{-3x}\, dx \right\}$$

$$= 5\left\{ \frac{e^{-3x}}{-3}\Big|_0^t - 3\left[-\frac{e^{-3x}}{9}(3x + 1) \right]_0^t \right\}$$

$$= 5te^{-3t}\,\text{mA} \qquad t \geq 0$$

$$= 0 \qquad\qquad t < 0$$

A plot of the current waveform is shown in Fig. 5.9b.
The power is given by the expression

$$p(t) = v(t)i(t)$$

$$= 5t(1 - 3t)e^{-6t}\,\mu\text{W} \qquad t \geq 0$$

$$= 0 \qquad\qquad t < 0$$

The equation for the power is plotted in Fig. 5.9c.
The expression for the energy is

$$w(t) = \tfrac{1}{2}Li^2(t)$$

$$= 2.5t^2 e^{-6t}\,\mu\text{J} \qquad t \geq 0$$

$$= 0 \qquad\qquad t < 0$$

This equation is plotted in Fig. 5.9d.   ❏

EXTENSION EXERCISES

E5.4   The current in a 5-mH inductor has the waveform shown in Fig. E5.4.
Compute the waveform for the inductor voltage.

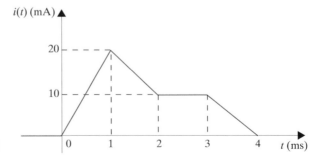

Figure E5.4

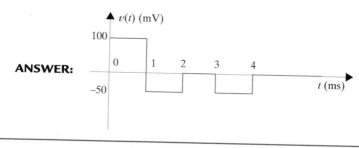

ANSWER:

E5.5 Compute the energy stored in the magnetic field of the inductor in Extension Exercise E5.4 at $t = 1.5$ ms.

ANSWER: $W = 562.5$ nJ.

5.3 FUNDAMENTAL CAPACITOR/INDUCTOR CHARACTERISTICS

Capacitors and inductors have a number of extremely important characteristics. One of the first things that the reader may have noticed is the similarity of the defining equations. Capacitors and inductors have a dual relationship; that is, their defining equations are identical if we interchange C with L and i with v, and vice versa, as shown in Table 5.1. Because of this relationship, the elements have a number of complementary features. For example, if the voltage across a capacitor is constant (i.e., non-time varying), the current through it is zero and, therefore, the capacitor looks like an open circuit to dc. Similarly, if the current in an inductor is constant, the voltage across it is zero and, therefore, the inductor looks like a short circuit to dc. An instantaneous jump in the voltage across a capacitor is not physically realizable because it requires the movement of a finite amount of charge in zero time, which is an infinite current. Similarly, an instantaneous change in the current in an inductor would require an infinite voltage. Therefore, it is not possible to change instantaneously the current in an inductor.

  Ideal capacitors and inductors only store energy; they do not dissipate energy as a resistor does. Although we model these elements as ideal devices, in practice there is some

Table 5.1 The Dual Relationship for Capacitors and Inductors

Capacitor	Inductor
$i(t) = C\dfrac{dv(t)}{dt}$	$v(t) = L\dfrac{di(t)}{dt}$
$v(t) = \dfrac{1}{C}\displaystyle\int_{t_0}^{t} i(x)\,dx + v(t_0)$	$i(t) = \dfrac{1}{L}\displaystyle\int_{t_0}^{t} v(x)\,dx + i(t_0)$
$p(t) = Cv(t)\dfrac{dv(t)}{dt}$	$p(t) = Li(t)\dfrac{di(t)}{dt}$
$w(t) = \tfrac{1}{2}Cv(t)^2$	$w(t) = \tfrac{1}{2}Li^2(t)$

E5.3 Write the i-v relationship for the practical capacitor shown below:

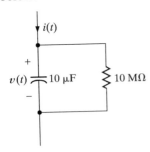

ANSWER: $i(t) =$
$$10^{-5} \frac{dv(t)}{dt} + 10^{-7} v(t)$$

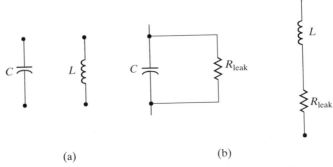

(a) (b)

Figure 5.10 (a) Ideal elements and (b) practical elements.

leakage resistance associated with each element, as shown in Fig. 5.10. In the case of the capacitor, there is normally a very large leakage resistance in parallel, which provides a conduction path between the plates. It is through this parallel resistance that the realistic capacitor slowly discharges itself. The physical coil, which forms the inductor, has significant winding resistance that quickly dissipates energy. Thus, the practical inductor is modeled as a resistor in series with an ideal inductor. Of the two elements, the capacitor is by far the better energy-storage device.

5.4 CAPACITOR AND INDUCTOR COMBINATIONS

Series Capacitors

If a number of capacitors are connected in series, their equivalent capacitance can be calculated using KVL. Consider the circuit shown in Fig. 5.11a. For this circuit

$$v(t) = v_1(t) + v_2(t) + v_3(t) + \cdots + v_N(t) \tag{5.13}$$

but

$$v_i(t) = \frac{1}{C_i} \int_{t_0}^{t} i(t)\, dt + v_i(t_0) \tag{5.14}$$

Therefore, Eq. (5.13) can be written as follows using Eq. (5.14):

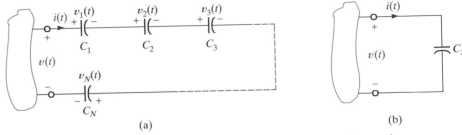

(a) (b)

Figure 5.11 Equivalent circuit for N series-connected capacitors.

$$v(t) = \left(\sum_{i=1}^{N} \frac{1}{C_i} \right) \int_{t_0}^{t} i(t) \, dt + \sum_{i=1}^{N} v_i(t_0) \qquad \textbf{5.15}$$

$$= \frac{1}{C_S} \int_{t_0}^{t} i(t) \, dt + v(t_0) \qquad \textbf{5.16}$$

where

$$v(t_0) = \sum_{i=1}^{N} v_i(t_0)$$

and

$$\frac{1}{C_S} = \sum_{i=1}^{N} \frac{1}{C_i} = \frac{1}{C_1} + \frac{1}{C_2} + \cdots + \frac{1}{C_N} \qquad \textbf{5.17}$$

Capacitors in series combine like resistors in parallel.

Thus, the circuit in Fig. 5.11b is equivalent to that in Fig. 5.11a under the conditions stated previously.

It is also important to note that since the same current flows in each of the series capacitors, each capacitor gains the same charge in the same time period. The voltage across each capacitor will depend on this charge and the capacitance of the element.

EXAMPLE 5.8

Determine the equivalent capacitance and the initial voltage for the circuit shown in Fig. 5.12.

SOLUTION   Note that these capacitors must have been charged before they were connected in series or else the charge of each would be equal and the voltages would be in the same direction.

The equivalent capacitance is

$$\frac{1}{C_S} = \frac{1}{2} + \frac{1}{3} + \frac{1}{6}$$

where all capacitance values are in microfarads.

Therefore, $C_S = 1$ μF and, as seen from the figure, $v(t_0) = -3$ V. Note that the total energy stored in the circuit is

$$w(t_0) = \tfrac{1}{2}\big[2 \times 10^{-6}(2)^2 + 3 \times 10^{-6}(-4)^2 + 6 \times 10^{-6}(-1)^2\big]$$
$$= 31 \; \mu J$$

However, the energy recoverable at the terminals is

$$w_C(t_0) = \tfrac{1}{2} C_S v^2(t)$$
$$= \tfrac{1}{2}\big[1 \times 10^{-6}(-3)^2\big]$$
$$= 4.5 \; \mu J$$

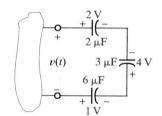

Figure 5.12 Circuit containing multiple capacitors with initial voltages.

EXAMPLE 5.9

Two previously uncharged capacitors are connected in series and then charged with a 12-V source. One capacitor is 30 μF and the other is unknown. If the voltage across the 30-μF capacitor is 8 V, find the capacitance of the unknown capacitor.

SOLUTION   The charge on the 30-μF capacitor is

$$Q = CV = (30\ \mu F)(8\ V) = 240\ \mu C$$

Since the same current flows in each of the series capacitors, each capacitor gains the same charge in the same time period.

$$C = \frac{Q}{V} = \frac{240\mu C}{4V} = 60\mu F \qquad \square$$

Parallel Capacitors

To determine the equivalent capacitance of N capacitors connected in parallel, we employ KCL. As can be seen from Fig. 5.13a,

$$i(t) = i_1(t) + i_2(t) + i_3(t) + \cdots + i_N(t) \qquad \textbf{5.18}$$

$$= C_1 \frac{dv(t)}{dt} + C_2 \frac{dv(t)}{dt} + C_3 \frac{dv(t)}{dt} + \cdots + C_N \frac{dv(t)}{dt}$$

$$= \left(\sum_{i=1}^{N} C_i \right) \frac{dv(t)}{dt}$$

$$= C_p \frac{dv(t)}{dt} \qquad \textbf{5.19}$$

where

Capacitors in parallel combine like resistors in series.

$$C_p = C_1 + C_2 + C_3 + \cdots + C_N \qquad \textbf{5.20}$$

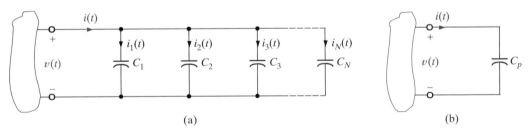

(a)   (b)

Figure 5.13 Equivalent circuit for N capacitors connected in parallel.

EXAMPLE 5.10

Determine the equivalent capacitance at terminals A-B of the circuit shown in Fig. 5.14.

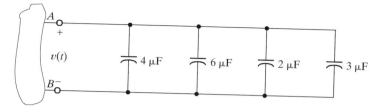

Figure 5.14 Circuit containing multiple capacitors in parallel.

SOLUTION

$$C_p = 4\mu F + 6\mu F + 2\mu F + 3\mu F$$
$$= 15\mu F$$

EXTENSION EXERCISES

E5.6 Two initially uncharged capacitors are connected as shown in Fig. E5.6. Determine the value of C_1.

ANSWER: $C_1 = 4\ \mu F$.

E5.7 Compute the equivalent capacitance of the network in Fig. E5.7.

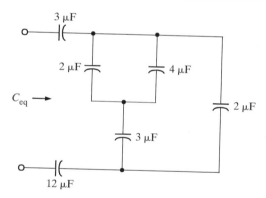

Figure E5.7

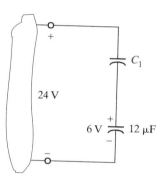

Figure E5.6

ANSWER: $C_{eq} = 1.5\ \mu F$.

E5.8 Compute the equivalent capacitance of the network in Fig. E5.8 if all the capacitors are 4 μF.

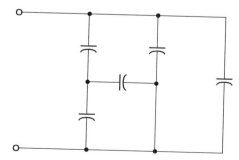

Figure E5.8

ANSWER: $C_{eq} = \dfrac{32}{3}\ \mu F.$

Series Inductors

If N inductors are connected in series, the equivalent inductance of the combination can be determined as follows. Referring to Fig. 5.15a and using KVL, we see that

$$v(t) = v_1(t) + v_2(t) + v_3(t) + \cdots + v_N(t) \qquad \textbf{5.21}$$

and therefore,

$$v(t) = L_1\frac{di(t)}{dt} + L_2\frac{di(t)}{dt} + L_3\frac{di(t)}{dt} + \cdots + L_N\frac{di(t)}{dt} \qquad \textbf{5.22}$$

$$= \left(\sum_{i=1}^{N} L_i\right)\frac{di(t)}{dt}$$

$$= L_S\frac{di(t)}{dt} \qquad \textbf{5.23}$$

where

Inductors in series combine like resistors in series.

$$L_S = \sum_{i=1}^{N} L_i \qquad \textbf{5.24}$$

Therefore, under this condition the network in Fig. 5.15b is equivalent to that in Fig. 5.15a.

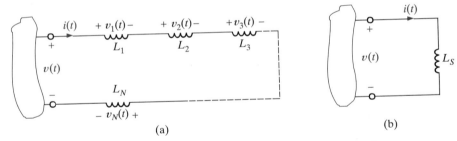

Figure 5.15 Equivalent circuit for N series-connected inductors.

EXAMPLE 5.11

Find the equivalent inductance of the circuit shown in Fig. 5.16.

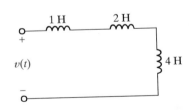

Figure 5.16 Circuit containing multiple inductors.

SOLUTION   The equivalent inductance of the circuit shown in Fig. 5.16 is

$$L_S = 1H + 2H + 4H$$
$$= 7H$$ ❑

Parallel Inductors

Consider the circuit shown in Fig. 5.17a, which contains N parallel inductors. Using KCL, we can write

$$i(t) = i_1(t) + i_2(t) + i_3(t) + \cdots + i_N(t)$$ **5.25**

However,

$$i_j(t) = \frac{1}{L_j} \int_{t_0}^{t} v(x)\, dx + i_j(t_0)$$ **5.26**

Substituting this expression into Eq. (5.25) yields

$$i(t) = \left(\sum_{j=1}^{N} \frac{1}{L_j} \right) \int_{t_0}^{t} v(x)\, dx + \sum_{j=1}^{N} i_j(t_0)$$ **5.27**

$$= \frac{1}{L_P} \int_{t_0}^{t} v(x)\, dx + i(t_0)$$ **5.28**

where

$$\frac{1}{L_P} = \frac{1}{L_1} + \frac{1}{L_2} + \frac{1}{L_3} + \cdots + \frac{1}{L_N}$$ **5.29**   Inductors in parallel combine like resistors in parallel.

and $i(t_0)$ is equal to the current in L_P at $t = t_0$. Thus, the circuit in Fig. 5.17b is equivalent to that in Fig. 5.17a under the conditions stated previously.

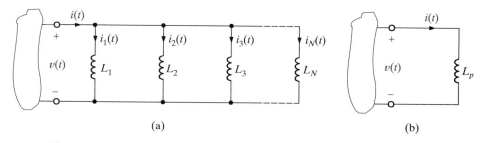

(a)   (b)

Figure 5.17 Equivalent circuits for N inductors connected in parallel.

EXAMPLE 5.12

Determine the equivalent inductance and the initial current for the circuit shown in Fig. 5.18.

SOLUTION   The equivalent inductance is

$$\frac{1}{L_P} = \frac{1}{12} + \frac{1}{6} + \frac{1}{4}$$

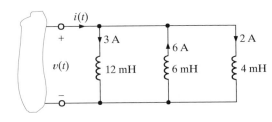

Figure 5.18 Circuit containing multiple inductors with initial currents.

where all inductance values are in millihenrys

$$L_P = 2 \text{ mH}$$

and the initial current is $i(t_0) = -1 \ A$.   ❑

The previous material indicates that capacitors combine like conductances, whereas inductances combine like resistances.

EXTENSION EXERCISES

E5.9   Compute the equivalent inductance of the network in Fig. E5.9 if all inductors are 4 mH.

ANSWER: 4.4 mH.

E5.10   Determine the equivalent inductance of the network in Fig. E5.10 if all inductors are 6 mH.

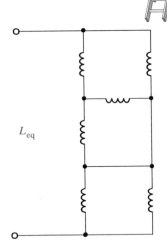

Figure E5.9

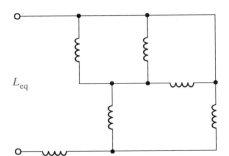

Figure E5.10

ANSWER: 9.429 mH.

5.5 *RC* OPERATIONAL AMPLIFIER CIRCUITS

The properties of the ideal op-amp are $v_+ = v_-$ and $i_+ = v_- = 0$.

Two very important *RC* op-amp circuits are the differentiator and the integrator. These circuits are derived from the circuit for an inverting op-amp by replacing the resistors R_1 and R_2, respectively, by a capacitor. Consider, for example, the circuit shown in Fig. 5.19a. The circuit equations are

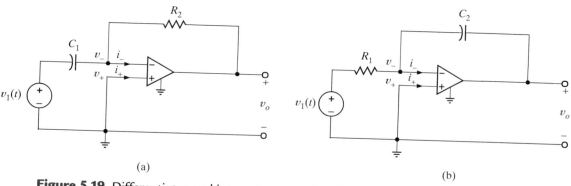

Figure 5.19 Differentiator and integrator operational amplifier circuits.

$$C_1 \frac{d}{dt}(v_1 - v_-) + \frac{v_o - v_-}{R_2} = i_-$$

However, $v_- = 0$ and $i_- = 0$. Therefore,

$$v_o(t) = -R_2 C_1 \frac{dv_1(t)}{dt} \qquad \textbf{5.30}$$

Thus, the output of the op-amp circuit is proportional to the derivative of the input. The circuit equations for the op-amp configuration in Fig. 5.19b are

$$\frac{v_1 - v_-}{R_1} + C_2 \frac{d}{dt}(v_o - v_-) = i_-$$

but since $v_- = 0$ and $i_- = 0$, the equation reduces to

$$\frac{v_1}{R_1} = -C_2 \frac{dv_o}{dt}$$

or

$$v_o(t) = \frac{-1}{R_1 C_2} \int_{-\infty}^{t} v_1(x)\, dx$$

$$= \frac{-1}{R_1 C_2} \int_{0}^{t} v_1(x)\, dx + v_o(0) \qquad \textbf{5.31}$$

If the capacitor is initially discharged, then $v_o(0) = 0$; hence,

$$v_o(t) = \frac{-1}{R_1 C_2} \int_{0}^{t} v_1(x)\, dx \qquad \textbf{5.32}$$

Thus, the output voltage of the op-amp circuit is proportional to the integral of the input voltage.

EXAMPLE 5.13

The waveform in Fig. 5.20a is applied at the input of the differentiator circuit shown in Fig. 5.19a. If $R_2 = 1$ kΩ and $C_1 = 2$ μF, determine the waveform at the output of the op-amp.

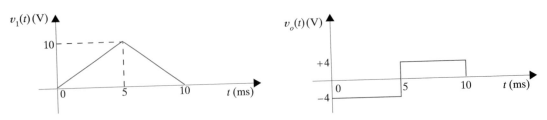

Figure 5.20 Input and output waveforms for a differentiator circuit.

SOLUTION   Using Eq. (5.30), we find that the op-amp output is

$$v_o(t) = -R_2 C_1 \frac{dv_1(t)}{dt}$$

$$= -(2)10^{-3} \frac{dv_1(t)}{dt}$$

$dv_1(t)/dt = (2)10^3$ for $0 \le t < 5$ ms, and therefore,

$$v_o(t) = -4\,\text{V} \qquad 0 \le t < 5\,\text{ms}$$

$dv_1(t)/dt = -(2)10^3$ for $5 \le t < 10$ ms, and therefore,

$$v_o(t) = 4\,\text{V} \qquad 5 \le t < 10\,\text{ms}$$

Hence, the output waveform of the differentiator is shown in Fig. 5.20b.   ❏

EXAMPLE 5.14

If the integrator shown in Fig. 5.19b has the parameters $R_1 = 5\,\text{k}\Omega$ and $C_2 = 0.2\,\mu\text{F}$, determine the waveform at the op-amp output if the input waveform is given as in Fig. 5.21a and the capacitor is initially discharged.

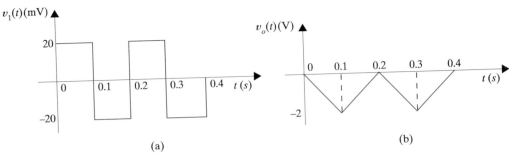

Figure 5.21 Input and output waveforms for an integrator circuit.

SOLUTION   The integrator output is given by the expression

$$v_o(t) = \frac{-1}{R_1 C_2} \int_0^t v_1(x)\,dx$$

which with the given circuit parameters is

$$v_o(t) = -10^3 \int_0^t v_1(x) \, dx$$

In the interval $0 \le t < 0.1$ s, $v_1(t) = 20$ mV. Hence,

$$v_o(t) = -10^3(20)10^{-3}t \qquad 0 \le t < 0.1 \text{ s}$$
$$= -20t$$

At $t = 0.1$ s, $v_o(t) = -2$ V. In the interval from 0.1 to 0.2 s, the integrator produces a positive slope output of $20t$ from $v_o(0.1) = -2$ V to $v_o(0.2) = 0$ V. This waveform from $t = 0$ to $t = 0.2$ s is repeated in the interval $t = 0.2$ to $t = 0.4$ s, and therefore, the output waveform is shown in Fig. 5.21b.   ❑

EXTENSION EXERCISE

E5.11   The waveform in Fig. E5.11 is applied to the input terminals of the op-amp differentiator circuit. Determine the differentiator output waveform if the op-amp circuit parameters are $C_1 = 2$ F and $R_2 = 2$ Ω.

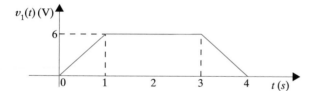

Figure E5.11

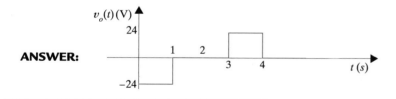

ANSWER:

5.6 APPLICATIONS

We now have two elements that have been added to our repertoire of components that can be employed in circuit applications and design—capacitors and inductors. For example, we have just found that when the capacitors are used in conjunction with op-amps, we can construct circuits that will integrate or differentiate a signal. In the remainder of the text, we will find that these two components are an integral part of our circuit analysis and design.

EXAMPLE 5.15

Integrated circuit chips, or ICs, are rectangular pieces of silicon that range in size from about 10/1000 to 200/1000 inch on each side. Chip manufacturers make electrical contact to the silicon with fine gold wire, called wire bonds, which stretch from the silicon surface to metal pins, as shown in Fig. 5.22a. The chip is eventually coated in plastic to protect it from physical damage and the pins become the access terminals to the chip.

In general, wires are not perfect conductors. They have resistance and inductance. In most cases, wiring resistance and inductance are so small that they have no measurable effect on circuit performance. However, in cases where the current changes rapidly, wire inductance can play a significant role. Let us examine the effect of the inductance of a wire bond in a high-speed microprocessor.

SOLUTION   Figure 5.22b shows an equivalent network for the situation of interest. The wire bond is modeled by the 10-nH inductor. The current source, $i(t)$, represents the current being used by the microprocessor. Finally, the 5-V source represents the chip supply voltage. As the microprocessor performs various functions, the current demand, $i(t)$, changes as shown in Fig. 5.22c. Using the element values in Fig. 5.22b, let us plot $V_{chip}(t)$ and find its minimum and maximum values.

Applying KVL in Fig. 5.22b, the chip voltage can be expressed as

$$v_{chip}(t) = 5 - L\frac{di(t)}{dt}$$

From Fig. 5.22c for $0 < t < 4$ ns,

$$\frac{di(t)}{dt} = \frac{\Delta i(t)}{\Delta t} = \frac{1}{4n} = 250 \times 10^6$$

Similarly, in the range 12 ns $< t < 20$ ns,

$$\frac{di(t)}{dt} = -125 \times 10^6$$

Based on these calculations, the resulting chip voltage, shown in Fig. 5.22d, varies widely from the 5-V requirement. The minimum value is 2.5 V and the maximum value is 6.25 V. If we desire a more stable chip voltage, we can simply put several wire bonds in parallel, as shown in Fig. 5.22e. Now $i(t)$ is divided between five different paths and so $di(t)/dt$ and the change in $V_{chip}(t)$ are reduced by a factor of 5. The resulting chip voltage, shown in Fig. 5.22f, has minimum and maximum values of 4.5 V and 5.25 V. This simple technique is employed by most IC manufacturers.   ❏

5.7 CIRCUIT DESIGN

EXAMPLE 5.16

During the dyeing process at a textile mill, dye is pumped through a pipe into a vat. As shown in Fig. 5.23a, a flowmeter mounted on the pipe monitors the flow rate

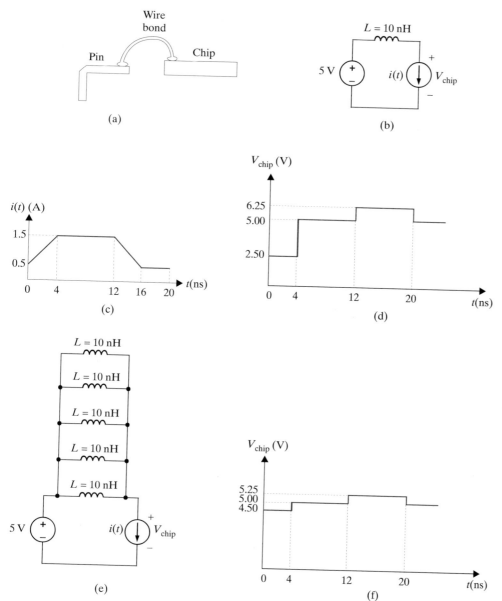

Figure 5.22 Integrated circuit wire bond example.

and outputs 1 volt at v_F for each gallon/second of flow. We want to design a circuit with an output voltage of 1 V for every 10 gallons used. Let us design a network to perform the required conversion using the inverting integrator in Fig. 5.23b, where the voltage source V_F represents the flowmeter output.

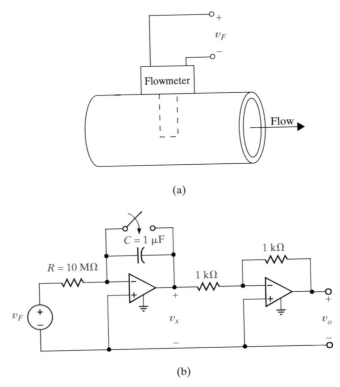

(a)

(b)

Figure 5.23 Flowmeter/volume meter conversion.

SOLUTION   The relationship between v_x and v_F in Fig. 5.23b is

$$v_x = \frac{-1}{RC} \int v_F \, dt$$

The integral of v_F will produce 1 V-s for each gallon used. Since we want 1 V at v_x for every 10 gallons, RC must be 10. If we arbitrarily choose $C = 1$ μF, then $R = 10$ MΩ. Since v_x is negative, we add the simple inverter in Fig. 5.23b to produce the voltage v_o, which has all the required characteristics. The switch around the capacitor can be closed to initialize the integrator to 0 V at v_x. Finally, since the op-amp output voltage cannot exceed its dc power supplies, as v_o reaches this limit, we will need to log the output value (normally done electronically), close the switch to reset v_o to zero, and then open the switch to begin a new integration cycle. ❏

EXAMPLE 5.17

A traffic survey is to be conducted at an intersection to determine the best traffic light timing sequence. In particular, we need to find the difference in the number of eastbound versus southbound vehicles over a 24-hour period. Sensors in the road-way as shown in Fig. 5.24a produce the voltage waveform in Fig. 5.24b each time a vehicle passes. We want to design a network that will produce a voltage propor-

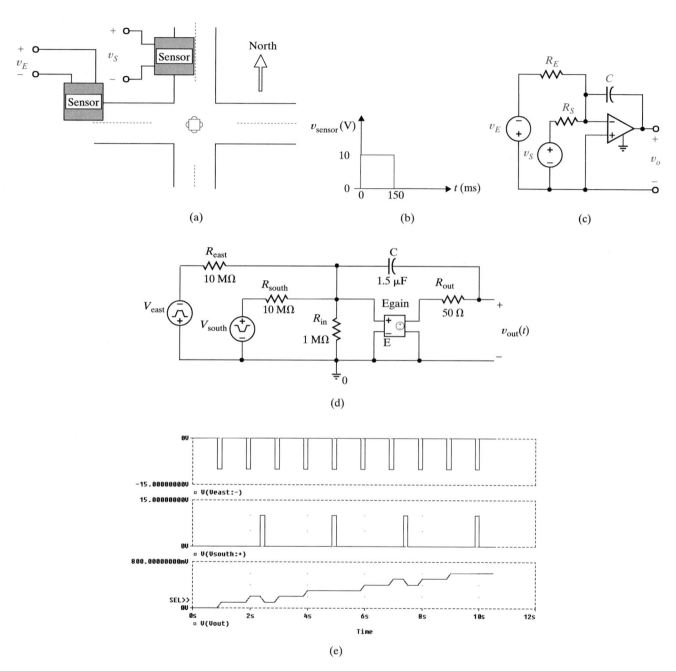

Figure 5.24 Traffic light survey.

tional to this difference where 1 V corresponds to 10 cars. Let us use an inverting integrator to produce a voltage that meets the requirements and verify its performance with PSPICE using our op-amp model with $R_i = 1\ M\Omega$, $R_o = 50\ \Omega$, and $A = 10^5$.

SOLUTION   The required network must do three things. First, it must perform a subtraction of the two sensor voltages. Second, it must integrate the difference to keep a running total of the traffic situation. And third, it must scale the voltage to meet the specification of 1 V per 10 vehicles. The circuit in Fig. 5.24c, where the voltage sources represent the sensor outputs, will meet all requirements.

KCL at the inverting op-amp input yields v_o in terms of v_E and v_S.

$$\frac{v_S}{R_S} - \frac{v_E}{R_E} = -C\left(\frac{dv_o}{dt}\right)$$

Solving for v_o, we obtain

$$v_o = \int\left[\frac{1}{R_E C}v_E - \frac{1}{R_S C}v_S\right]dt$$

Based on the sensor waveform in Fig. 5.24b, the integral of the sensor output is 1.5 V-s for each vehicle that passes. If the RC products in the preceding equation are 15, then one vehicle will produce 0.1 V at v_o as required. If we arbitrarily choose $C = 1.5\ \mu F$, then R_E and R_S are 10 MΩ. As in Example 5.16, as the integrator output voltage nears the op-amp's power supply values, the data must be logged and the capacitor discharged through a switch to reset the integrator. At the end of a 24-hour period, the recorded values of v_o can be summed to yield the daily vehicle count.

The pspice *Schematics* circuit is shown in Fig. 5.24d. To provide an inverting gain for the integrator, the gain of the dependent voltage source has been set to $-100,000$ V/V. Two VPULSE voltage sources are used to model the traffic sensors. For simulation purposes, let one eastbound vehicle pass every second and one southbound vehicle every 2.5 seconds. The required attributes for the VPULSE sources are listed in the following table.

Source	v_{east}	v_{south}
dc voltage	0	0
ac voltage	0	0
Start voltage	0	0
Pulse voltage	10	10
Delay time	0.85	2.35
Rise time	0	0
Fall time	0	0
Pulse width	0.15	0.15
Period	1	2.5

After 10 s, 10 eastbound vehicles have passed versus only four southbound. The difference of six should produce 0.6 V at v_o. From the PROBE plots of v_{east}, v_{south}, and v_o shown in Fig. 5.24e, the output voltage at 10 s is 0.608 V, which is within 1.4% of our ideal op-amp calculations.

❏

5.8 SUMMARY

- The important (dual) relationships for capacitors and inductors are as follows:

$$q = Cv$$

$$i(t) = C\frac{dv(t)}{dt} \qquad\qquad v(t) = L\frac{di(t)}{dt}$$

$$v(t) = \frac{1}{C}\int_{-\infty}^{t} i(x)\,dx \qquad\qquad i(t) = \frac{1}{L}\int_{-\infty}^{t} v(x)\,dx$$

$$p(t) = Cv(t)\frac{dv(t)}{dt} \qquad\qquad p(t) = Li(t)\frac{di(t)}{dt}$$

$$W_C(t) = 1/2Cv^2(t) \qquad\qquad W_L(t) = 1/2Li^2(t)$$

- The passive sign convention is used with capacitors and inductors.
- In dc steady state a capacitor looks like an open circuit and an inductor looks like a short circuit.
- Leakage resistance is present in practical capacitors and inductors.
- When capacitors are interconnected, their equivalent capacitance is determined as follows: Capacitors in series combine like resistors in parallel and capacitors in parallel combine like resistors in series.
- When inductors are interconnected, their equivalent inductance is determined as follows: Inductors in series combine like resistors in series and inductors in parallel combine like resistors in parallel.
- *RC* operational amplifier circuits can be used to differentiate or integrate an electrical signal.

PROBLEMS

Section 5.1

5.1 A 6-μF capacitor was charged to 12 V. Find the charge accumulated in the capacitor.

5.2 A capacitor has an accumulated charge of 600 μC with 5 V across it. What is the value of capacitance?

5.3 A 12-μF capacitor has an accumulated charge of 480 μC. Determine the voltage across the capacitor.

5.4 A 10-μF capacitor is charged by a constant current source and its voltage is increased to 2 V in 5 s. Find the value of the constant current source.

$$i = C\frac{dv}{dt}$$

Similar to P5.4

5.5 A capacitor is charged by a constant current of 2 mA and results in a voltage increase of 12 V in a 10-s interval. What is the value of the capacitance?

Similar to P5.4

5.6 An uncharged 100-μF capacitor is charged by a constant current of 1 mA. Find the voltage across the capacitor after 4 s.

$i(t) = C\dfrac{dv(t)}{dt}$ and

$W_C(t) = 1/2Cv^2(t)$

5.7 The voltage across a 100-μF capacitor is given by the expression $v(t) = 120 \sin 377t$ V. Find (a) the current in the capacitor and (b) the expression for the energy stored in the element.

$v(t) = \dfrac{1}{C}\displaystyle\int_{t_o}^{t} i(x)\,dx$

$p(t) = v(t)i(t)$

5.8 An uncharged 10-μF capacitor is charged by the current $i(t) = 10 \cos 377t$ mA. Find (a) the expression for the voltage across the capacitor and (b) the expression for the power.

5.9 The voltage across a 6-μF capacitor is shown in Fig. P5.9. Compute the waveform for the current in the capacitor.

Similar to P5.6

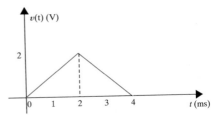

Figure P5.9

5.10 The voltage across a 100-μF capacitor is shown in Fig P5.10. Compute the waveform for the current in the capacitor.

Similar to P5.9

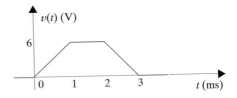

Figure P5.10

5.11 The current in a 100-μF capacitor is shown in Fig. P5.11. Determine the waveform for the voltage across the capacitor if it is initially uncharged.

Similar to P5.8

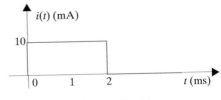

Figure P5.11

5.12 The voltage across a 2-μF capacitor is given by the waveform in Fig. P5.12. Compute the current waveform.

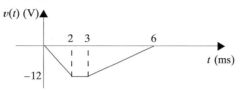

Figure P5.12

Similar to P5.9

5.13 The voltage across a 0.1-F capacitor is given by the waveform in Fig. P5.13. Find the waveform for the current in the capacitor.

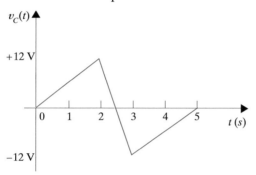

Figure P5.13

Similar to P5.9

5.14 The voltage across a 50-μF capacitor is shown in Fig. P5.14. Determine the current waveform.

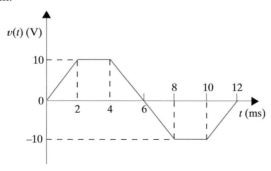

Figure P5.14

Similar to P5.9

5.15 Draw the waveform for the current in a 12-µF capacitor when the capacitor voltage is as described in Fig. P5.15.

Similar to P5.9

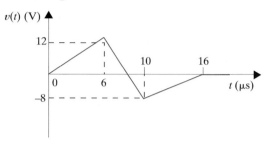

Figure P5.15

5.16 Draw the waveform for the current in a 3-µF capacitor when the voltage across the capacitor is give in Fig. P5.16.

Similar to P5.9

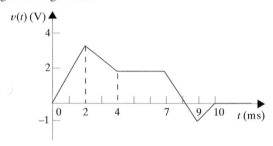

Figure P5.16

5.17 The voltage across a 6-µF capacitor is given by the waveform in Fig. P5.17. Plot the waveform for the capacitor current.

Similar to P5.9

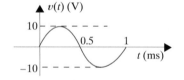

Figure P5.17

5.18 The waveform for the current in a 200-µF capacitor is shown in Fig. P5.18. Determine the waveform for the capacitor voltage.

Similar to P5.8

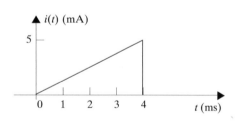

Figure P5.18

5.19 The waveform for the current in a 100-μF initially uncharged capacitor is shown in Fig. P5.19. Determine the waveform of the voltage.

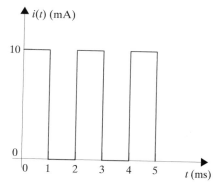

Figure P5.19

Similar to P5.8

5.20 The waveform for the current in a 100-μF initially uncharged capacitor is shown in Fig. P5.20. Determine the waveform for the capacitor's voltage.

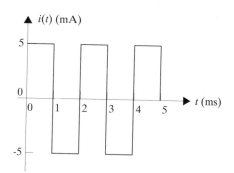

Figure P5.20

Similar to P5.8

Section 5.2

5.21 A 10-mH inductor has a sudden current change from 200 mA to 100 mA in 1 ms. Find the induced voltage.

$$v(t) = L\frac{di(t)}{dt}$$

5.22 The current in an inductor changes from 0 to 200 mA in 4 ms and induces a voltage of 100 mV. What is the value of the inductor?

Similar to P5.21

5.23 If the current $i(t) = 1.5t$ A flows through a 2-H inductor, find the energy stored at $t = 2$ s.

$$W_L(t) = 1/2Li(t)^2$$

5.24 The current in a 100-mH inductor is $i(t) = 2 \sin 377t$ A. Find (a) the voltage across the inductor and (b) the expression for the energy stored in the element.

Similar to P5.21 and P5.23

$$i(t) = \frac{1}{L} \int_{t_o}^{t} v(x)\, dx$$

$$p(t) = v(t)i(t)$$

5.25 The induced voltage across a 10-mH inductor is $v(t) = 120 \cos 377t$ V. Find (a) the expression for the inductor current and (b) the expression for the power.

5.26 The current in a 10-mH inductor is shown in Fig. P5.26. Determine the waveform for the voltage across the inductor.

Similar to P5.21

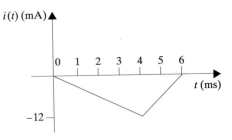

Figure P5.26

5.27 The current in a 50-mH inductor is given in Fig. P5.27. Sketch the inductor voltage.

Similar to P5.21

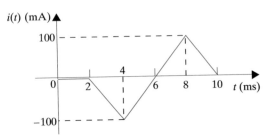

Figure P5.27

5.28 The current in a 25-mH inductor is given by the expressions

$$i(t) = 0 \qquad\qquad t < 0$$
$$i(t) = 10(1 - e^{-t})\text{mA} \qquad t > 0$$

Similar to P5.24

Find (a) the voltage across the inductor and (b) the expression for the energy stored in it.

5.29 Given the data in the previous problem, find the voltage across the inductor and the energy stored in it after 1 s.

5.30 The current in a 50-mH inductor is specified as follows:

$$i(t) = 0 \qquad\qquad t < 0$$
$$i(t) = 2te^{-4t}\,\text{A} \qquad t > 0$$

Similar to P5.28

Find (a) the voltage across the inductor, (b) the time at which the current is a maximum, and (c) the time at which the voltage is a minimum.

5.31 The current in a 10-mH inductor is shown in Fig. P5.31. Find the voltage across the inductor.

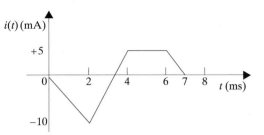

Figure P5.31

Similar to P5.21

5.32 Draw the waveform for the voltage across a 10-mH inductor when the inductor current is given by the waveform shown in Fig. P5.32.

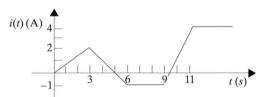

Figure P5.32

Similar to P5.21

5.33 The current in a 16-mH inductor is given by the waveform in Fig. P5.33. Find the waveform for the voltage across the inductor.

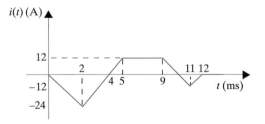

Figure P5.33

Similar to P5.21

5.34 The voltage across a 2-H inductor is given by the waveform shown in Fig. P5.34. Find the waveform for the current in the inductor.

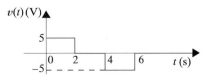

Figure P5.34

Similar to P5.25

5.35 The waveform for the voltage across a 20-mH inductor is shown in Fig. P5.35. Compute the waveform for the inductor current.

Similar to P5.25

$v(t)$ (mV)

10

0   1   2   3 t (ms)

−20

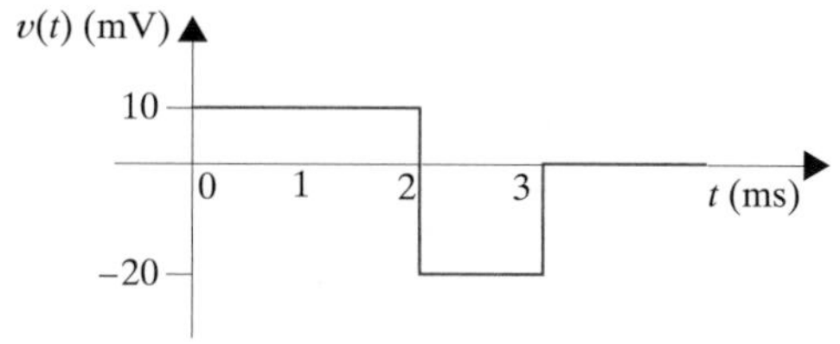

Figure P5.35

5.36 The voltage across a 10-mH inductor is shown in Fig. P5.36. Determine the waveform for the inductor current.

Similar to P5.25

$v(t)$ (mV)

10

0   1   2 t (ms)

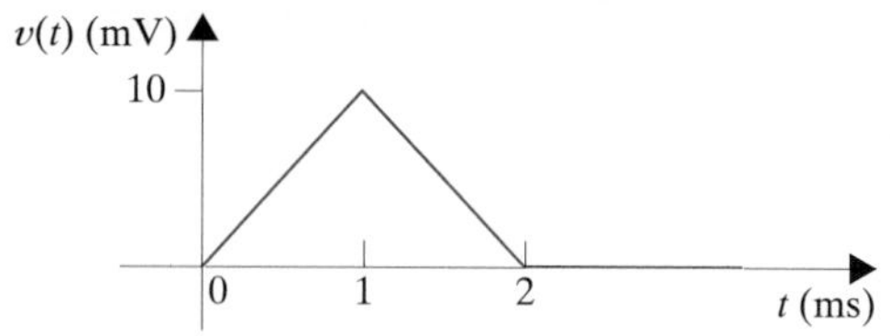

Figure P5.36

5.37 The voltage across a 2-H inductor is given by the waveform shown in Fig. P5.37. Find the waveform for the current in the inductor.

Similar to P5.25

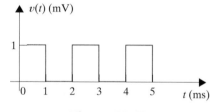

$v(t)$ (mV)

1

0   1   2   3   4   5 t (ms)

Figure P5.37

5.38 The current in a 4-mH inductor is given by the waveform in Fig. P5.38. Plot the voltage across the inductor.

Similar to P5.21

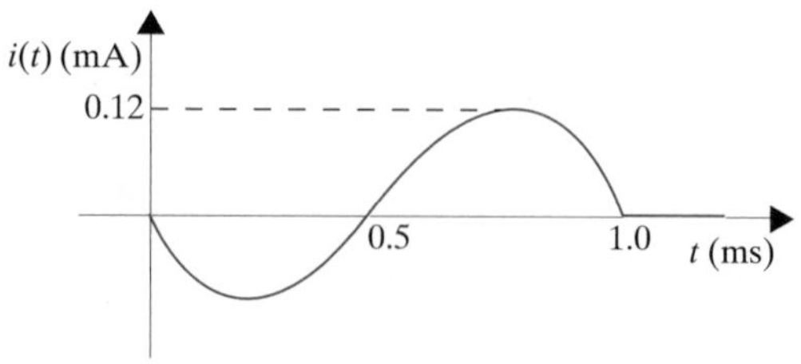

$i(t)$ (mA)

0.12

0.5   1.0 t (ms)

Figure P5.38

Section 5.3

5.39 A practical inductor, as shown in Fig. P5.39, has a current $i(t) = 10\cos 377t$ A. Find the voltage across this inductor.

10 mH   0.1 mΩ

Figure P5.39

$$v(t) = L\frac{di(t)}{dt} + Ri(t)$$

Section 5.4

5.40 A 3-μF capacitor and 6-μF capacitor are connected in parallel and charged to 12 V. Find (a) the charge stored by each capacitor and (b) the total energy stored.

1. Find C_{eq}
2. Find total voltage V
3. $W = \frac{1}{2}C_{eq}v^2$

5.41 The two capacitors in Fig. P5.41 were charged and then connected as shown. Determine the equivalent capacitance, the initial voltage at the terminals, and the total energy stored in the network.

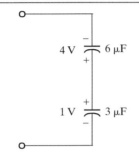

4 V   6 μF

1 V   3 μF

Figure P5.41

Similar to P5.40 except that

$$\frac{1}{C_{total}} = \frac{1}{C_1} + \frac{1}{C_2}$$

5.42 Two capacitors are connected in series as shown in Fig. P5.42. Find V_o.

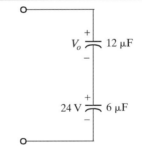

V_o 12 μF

24 V   6 μF

Figure P5.42

$$v = \frac{1}{C}\int i(t)\,dt$$

$v \propto \dfrac{1}{C}$ since $\int i(t)\,dt$ is the same for both capacitors

5.43 Three capacitors are connected as shown in Fig. P5.43. Find V_1 and V_2.

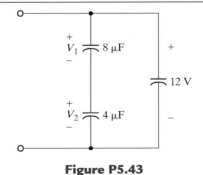

V_1 8 μF

12 V

V_2 4 μF

Figure P5.43

Similar to P5.42

Use the series and parallel relationships for combining capacitors.

5.44 What values of capacitance can be obtained by interconnecting a 4-μF capacitor, a 6-μF capacitor, and a 12-μF capacitor?

Similar to P5.44

5.45 Given a 1-, 3-, and 4-μF capacitor, can they be interconnected to obtain an equivalent 2-μF capacitor?

Similar to P5.44

5.46 Given four 2-μF capacitors, find the maximum value and minimum value that can be obtained by interconnecting the capacitors in series/parallel combinations.

5.47 Select the value of C to produce the desired total capacitance of $C_T = 2$ μF in the circuit in Fig. P5.47.

Find the expression for the total capacitance in terms of C and set it equal to 2 μF.

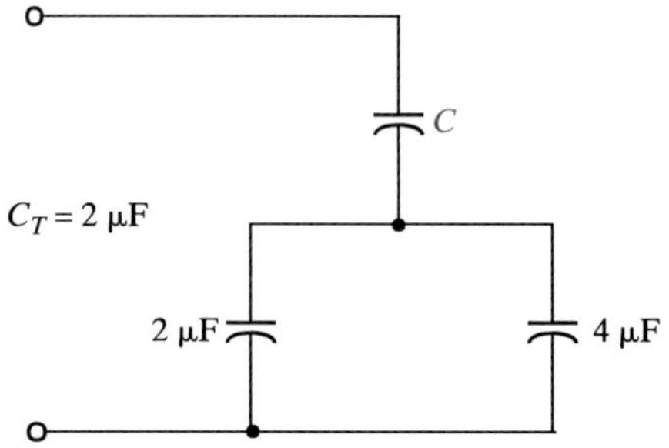

Figure P5.47

5.48 A 20-mH inductor and a 12-mH inductor are connected in series with a 1-A current source. Find (a) the equivalent inductance and (b) the total energy stored.

See the figure in Example 5.8

5.49 Two inductors are connected in parallel, as shown in Fig. P5.49. Find i.

$i = \dfrac{1}{L} \int v(t)\, dx$ and thus

$i \propto \dfrac{1}{L}$

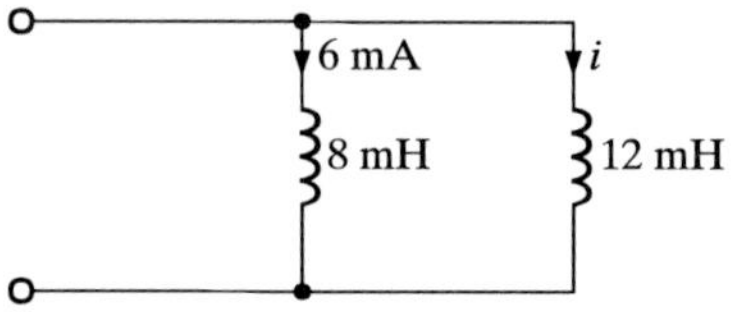

Figure P5.49

5.50 Select the value of L that produces a total inductance of $L_T = 10$ mH in the circuit in Fig. P5.50.

Similar to P5.47 with the exception that the elements are inductors.

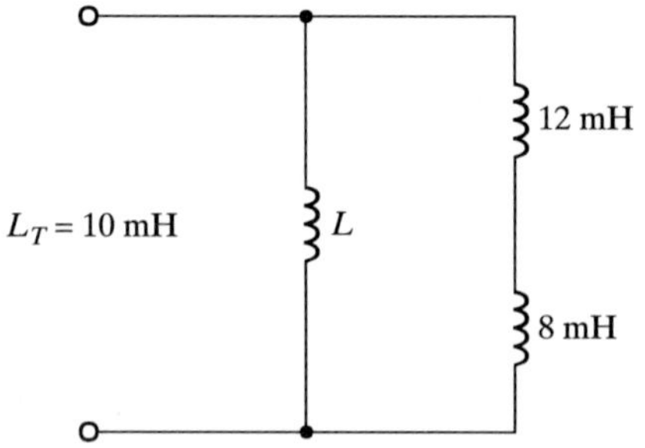

Figure P5.50

Section 5.5

5.51 For the network in Fig. P5.51 choose C such that

$$v_o = -10 \int v_S \, dt$$

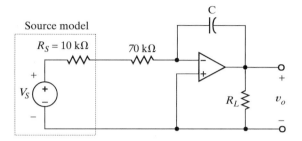

Assume a perfect op-amp and derive the relationship between v_o and v_S. Compare this equation to the derived relationship.

Figure P5.51

5.52 For the network in Fig. P5.52, $v_S(t) = 120 \cos 377t$ V. Find $v_o(t)$

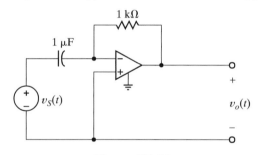

Similar to P5.51

Figure P5.52

5.53 For the network in Fig. P5.53, $v_S(t) = 115 \sin 377t$ V. Find $v_o(t)$.

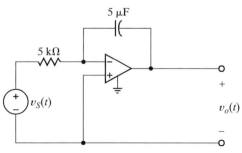

Similar to P5.52

Figure P5.53

Advanced Problems

5.54 Determine the total capacitance of the network in Fig. P5.54.

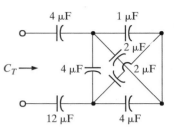

Figure P5.54

5.55 Find the equivalent capacitance at terminals *A-B* in Fig. P5.55.

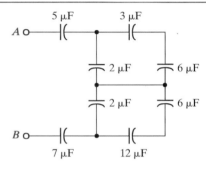

Figure P5.55

5.56 Find the total capacitance C_T of the network in Fig. P5.56.

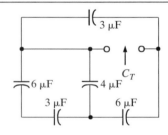

Figure P5.56

5.57 Find the total capacitance C_T of the network in Fig. P5.57.

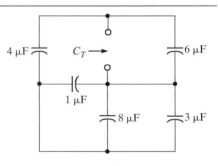

Figure P5.57

5.58 In the network in Fig. P5.58, find the capacitance C_T if (a) the switch is open and (b) the switch is closed.

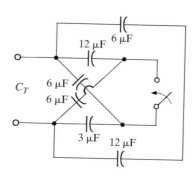

Figure P5.58

5.59 Determine the inductance at terminals A-B in the network in Fig. P5.59.

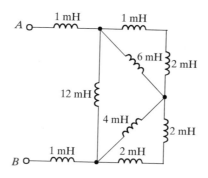

Figure P5.59

5.60 Determine the inductance at terminals A-B in the network in Fig P5.60.

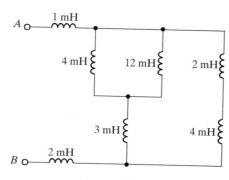

Figure P5.60

5.61 Find the total inductance at the terminals *A-B* shown in the network in Fig. P5.61.

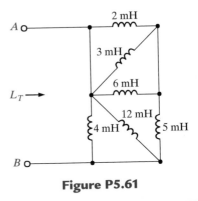

Figure P5.61

5.62 Given the network shown in Fig. P5.62, find (a) the equivalent inductance at terminals *A-B* with terminals *C-D* short circuited; and (b) the equivalent inductance at terminals *C-D* with terminals *A-B*, open circuited.

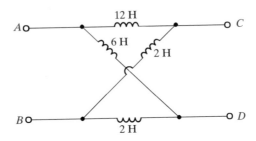

Figure P5.62

5.63 Find the total inductance at the terminals of the network in Fig. P5.63.

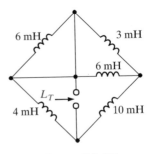

Figure P5.63

5.64 Find L_T in the network in Fig. P5.64 (a) with the switch open and (b) with the switch closed. All inductors are 12 mH.

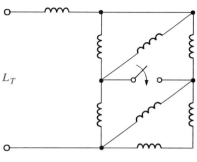

Figure P5.64

5.65 Find the value of L in the network in Fig. P5.65 so that the total inductance L_T will be 2 mH.

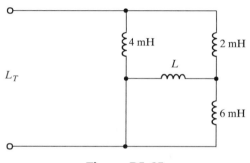

Figure P5.65

5.66 Design an op-amp circuit that will produce the output voltage

$$v_o = \int v_S \, dt - 10 v_S$$

5.67 For the network in Fig. P5.67, $v_{S_1}(t) = 80 \cos 377t$ V and $v_{S_2}(t) = 40 \cos 377t$ V. Find $v_o(t)$.

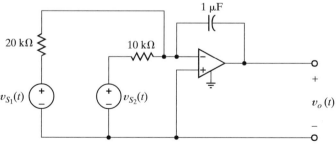

Figure P5.67

First-Order Transient Circuits

*I*n this chapter we perform what is normally referred to as a first-order transient analysis because the networks that we examine contain only a single storage element. When only a single storage element is present in the network, the network can be described by a first-order differential equation.

Our analysis involves an examination and description of the behavior of a circuit as a function of time after a sudden change in the network occurs due to switches opening or closing. Because of the presence of the storage element, the circuit response to a sudden change will go through a transition period prior to settling down to a steady-state value. It is this transition period that we will examine carefully in our transient analysis.

One of the important parameters that we will examine in our transient analysis is the circuit's time constant. This is a very important network parameter because it tells us how fast the circuit will respond to changes. We can contrast two very different systems to obtain a feel for the parameter. For example, consider the model for a room air-conditioning system and the model for a single-transistor stage of amplification in a computer chip. If we change the setting for the air conditioner from 70 degrees to 60 degrees, the unit will come on and the room will begin to cool. However, the temperature measured by a thermometer in the room will fall very slowly and, thus, the time required to reach the desired temperature is long. However, if we send a trigger signal to a transistor to change state, the action may take only a few nanoseconds. These two systems will have vastly different time constants.

Two analysis techniques are presented for performing a transient analysis: the differential equation approach, in which a differential equation is written and solved for each network, and a step-by-step approach, which takes advantage of the known form of the solution in every case. Finally, a number of application-oriented examples are presented and discussed.

6.1 INTRODUCTION

In Chapter 5 we found that capacitors and inductors were capable of storing electric energy. In the case of a charged capacitor, the energy is stored in the electric field that exists between the positively and negatively charged plates. This stored energy can be released if a circuit is somehow connected across the capacitor that provides a path through which the negative charges move to the positive charges. As we know, this movement of charge constitutes a current. The rate at which the energy is discharged is a direct function of the parameters in the circuit that is connected across the capacitor's plates.

As an example, consider the flash circuit in a camera. Recall that the operation of the flash circuit, from a user standpoint, involves depressing the push button on the camera that triggers both the shutter and the flash and then waiting a few seconds before repeating the process to take the next picture. This operation can be modeled using the circuit in Fig. 6.1a. The voltage source and resistor R_S model the batteries that power the camera and flash. The capacitor models the energy storage, the switch models the push button, and finally the resistor R models the xenon flash lamp. Thus, if the capacitor is charged, when the switch is closed, the capacitor voltage drops and energy is released through the xenon lamp, producing the flash. In practice this energy release takes about a millisecond and the discharge time is a function of the elements in the circuit. When the push button is released and the switch is then opened, the battery begins to recharge the capacitor. Once again, the time required to charge the capacitor is a function of the circuit elements. The discharge and charge cycles are graphically illustrated in Fig. 6.1b. While the discharge time is very fast, it is not instantaneous. In order to provide further insight into this phenomenon, consider what we might call a *free-body diagram* of the right half of the network in Fig. 6.1a as shown in Fig. 6.1c (that is, a charged capacitor that is discharged through a resistor). When the switch is closed, KCL for the circuit is

$$C\frac{dv_c(t)}{dt} + \frac{v_c(t)}{R} = 0$$

or

$$\frac{dv_c(t)}{dt} + \frac{1}{RC}v_c(t) = 0$$

We will demonstrate in the next section that the solution of this equation is

$$v_c(t) = V_o e^{-t/RC}$$

Note that this function is a decaying exponential and the rate at which it decays is a function of the values of R and C. The product RC is a very important parameter and we will give it a special name in the following discussions.

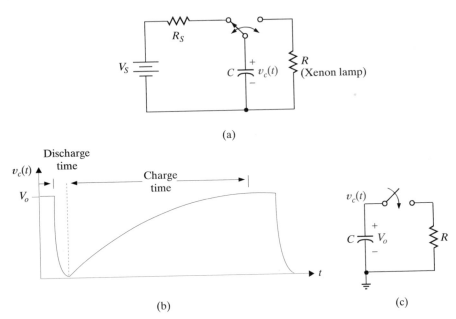

Figure 6.1 Diagrams used to describe a camera's flash circuit.

6.2 GENERAL FORM OF THE RESPONSE EQUATIONS

In our study of first-order transient circuits we will show that the solution of these circuits (i.e., finding a voltage or current) requires us to solve a first-order differential equation of the form

$$\frac{dx(t)}{dt} + ax(t) = f(t) \qquad \textbf{6.1}$$

Although there are a number of techniques for solving an equation of this type, we will obtain a general solution that we will then employ in two different approaches to transient analysis.

A fundamental theorem of differential equations states that if $x(t) = x_p(t)$ is any solution to Eq. (6.1), and $x(t) = x_c(t)$ is any solution to the homogeneous equation

$$\frac{dx(t)}{dt} + ax(t) = 0 \qquad \textbf{6.2}$$

then

$$x(t) = x_p(t) + x_c(t) \qquad \textbf{6.3}$$

is a solution to the original Eq. (6.1). The term $x_p(t)$ is called the *particular integral solution*, or forced response, and $x_c(t)$ is called the *complementary solution*, or natural response.

At the present time we confine ourselves to the situation in which $f(t) = A$ (i.e., some constant). The general solution of the differential equation then consists of two parts that are obtained by solving the two equations

$$\frac{dx_p(t)}{dt} + ax_p(t) = A \qquad\qquad \textbf{6.4}$$

$$\frac{dx_c(t)}{dt} + ax_c(t) = 0 \qquad\qquad \textbf{6.5}$$

Since the right-hand side of Eq. (6.4) is a constant, it is reasonable to assume that the solution $x_p(t)$ must also be a constant. Therefore, we assume that

$$x_p(t) = K_1 \qquad\qquad \textbf{6.6}$$

Substituting this constant into Eq. (6.4) yields

$$K_1 = \frac{A}{a} \qquad\qquad \textbf{6.7}$$

Examining Eq. (6.5), we note that

$$\frac{dx_c(t)/dt}{x_c(t)} = -a \qquad\qquad \textbf{6.8}$$

This equation is equivalent to

$$\frac{d}{dt}\left[\ln x_c(t)\right] = -a$$

Hence,

$$\ln x_c(t) = -at + c$$

and therefore,

$$x_c(t) = K_2 e^{-at} \qquad\qquad \textbf{6.9}$$

Therefore, a solution of Eq. (6.1) is

$$x(t) = x_p(t) + x_c(t)$$
$$= \frac{A}{a} + K_2 e^{-at} \qquad\qquad \textbf{6.10}$$

The constant K_2 can be found if the value of the independent variable $x(t)$ is known at one instant of time.

Equation (6.10) can be expressed in general in the form

$$x(t) = K_1 + K_2 e^{-t/\tau} \qquad\qquad \textbf{6.11}$$

Once the solution in Eq. (6.11) is obtained, certain elements of the equation are given names that are commonly employed in electrical engineering. For example, the term K_1 is referred to as the *steady-state solution*: the value of the variable $x(t)$ as $t \to \infty$ when the second term becomes negligible. The constant τ is called the *time constant* of the circuit.

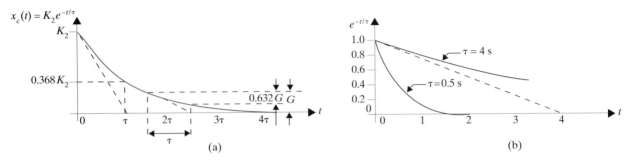

Figure 6.2 Time-constant illustrations.

Note that the second term in Eq. (6.11) is a decaying exponential that has a value, if $\tau > 0$, of K_2 for $t = 0$ and a value of 0 for $t = \infty$. The rate at which this exponential decays is determined by the time constant τ. A graphical picture of this effect is shown in Fig. 6.2a. As can be seen from the figure, the value of $x_c(t)$ has fallen from K_2 to a value of $0.368K_2$ in one time constant, a drop of 63.2%. In two time constants the value of $x_c(t)$ has fallen to $0.135K_2$, a drop of 63.2% from the value at time $t = \tau$. This means that the gap between a point on the curve and the final value of the curve is closed by 63.2% each time constant. Finally, after five time constants, $x_c(t) = 0.0067K_2$, which is less than 1%.

An interesting property of the exponential function shown in Fig. 6.2a is that the initial slope of the curve intersects the time axis at a value of $t = \tau$. In fact, we can take any point on the curve, not just the initial value, and find the time constant by finding the time required to close the gap by 63.2%. Finally, the difference between a small time constant (i.e., fast response) and a large time constant (i.e., slow response) is shown in Fig. 6.2b. These curves indicate that if the circuit has a small time constant, it settles down quickly to a steady-state value. Conversely, if the time constant is large, more time is required for the circuit to settle down or reach steady state. In any case, note that the circuit response essentially reaches steady state within five time constants (i.e., 5τ).

Note that the previous discussion has been very general in that no particular form of the circuit has been assumed—only that it results in a first-order differential equation.

6.3 ANALYSIS TECHNIQUES

The Differential Equation Approach

Equation (6.11) defines the general form of the solution of first-order transient circuits; that is, it represents the solution of the differential equation that describes an unknown current or voltage *anywhere in the network*. One of the ways that we can arrive at this solution is to solve the equations that describe the network behavior using what is often called the *state-variable approach*. In this technique we write the equation for the voltage across the capacitor and/or the equation for the current through the inductor. Recall from Chapter 5 that these quantities cannot change instantaneously. Let us first illustrate this technique in the general sense and then examine two specific examples.

Figure 6.3 *RC* and *RL* circuits.

Consider the circuit shown in Fig. 6.3a. At time $t = 0$, the switch closes. The KCL equation that describes the capacitor voltage for time $t > 0$ is

$$C \frac{dv(t)}{dt} + \frac{v(t) - V_S}{R} = 0$$

or

$$\frac{dv(t)}{dt} + \frac{v(t)}{RC} = \frac{V_S}{RC}$$

From our previous development, we assume that the solution of this first-order differential equation is of the form

$$v(t) = K_1 + K_2 e^{-t/\tau}$$

Substituting this solution into the differential equation yields

$$-\frac{K_2}{\tau} e^{-t/\tau} + \frac{K_1}{RC} + \frac{K_2}{RC} e^{-t/\tau} = \frac{V_S}{RC}$$

Equating the constant and exponential terms, we obtain

$$K_1 = V_S$$
$$\tau = RC$$

Therefore,

$$v(t) = V_S + K_2 e^{-t/RC}$$

where V_S is the steady-state value and RC is the network's time constant. K_2 is determined by the initial condition of the capacitor. For example, if the capacitor is initially uncharged (that is, the voltage across the capacitor is zero at $t = 0$), then

$$0 = V_S + K_2$$

or

$$K_2 = -V_S$$

Hence, the complete solution for the voltage $v(t)$ is

$$v(t) = V_S - V_S e^{-t/RC}$$

D6.1 If the capacitor in the network in Fig. 6.3a is initially charged to $V_S/2$, find the complete solutions for $v(t)$.

ANSWER:

$$v(t) = V_S - \frac{V_S}{2} e^{-t/RC}$$

D6.2 Find $i(t)$ for $t > 0$ in the following network:

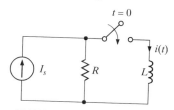

ANSWER:

$$i(t) = I_S\left(1 - e^{-\frac{R}{L}t}\right)$$

The circuit in Fig. 6.3b can be examined in a similar manner. The KVL equation that describes the inductor current for $t > 0$ is

$$L\frac{di(t)}{dt} + Ri(t) = V_S$$

A development identical to that just used yields

$$i(t) = \frac{V_S}{R} + K_2 e^{-\left(\frac{R}{L}\right)t}$$

where V_S/R is the steady-state value and L/R is the circuit's time constant. If there is no initial current in the inductor, then at $t = 0$

$$0 = \frac{V_S}{R} + K_2$$

and

$$K_2 = \frac{-V_S}{R}$$

Hence,

$$i(t) = \frac{V_S}{R} - \frac{V_S}{R}e^{-\frac{R}{L}t}$$

is the complete solution. Note that if we wish to calculate the voltage across the resistor, then

$$v_R(t) = Ri(t)$$

$$= V_S\left(1 - e^{-\frac{R}{L}t}\right)$$

EXAMPLE 6.1

Consider the circuit shown in Fig. 6.4a. Assuming that the switch has been in position 1 for a long time, at time $t = 0$ the switch is moved to position 2. We wish to calculate the current $i(t)$ for $t > 0$.

SOLUTION At $t = 0-$ the capacitor is fully charged and conducts no current since the capacitor acts like an open circuit to dc. The initial voltage across the capacitor can be found using voltage division. As shown in Fig. 6.4b,

$$v_c(0-) = 12\left(\frac{3k}{6k + 3k}\right) = 4 \text{ V}$$

The network for $t > 0$ is shown in Fig. 6.4c. The KCL equation for the voltage across the capacitor is

$$\frac{v(t)}{R_1} + C\frac{dv(t)}{dt} + \frac{v(t)}{R_2} = 0$$

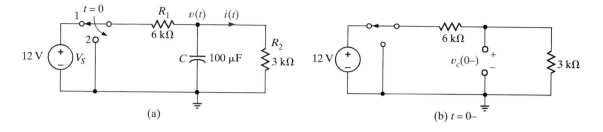

(a)

(b) $t = 0-$

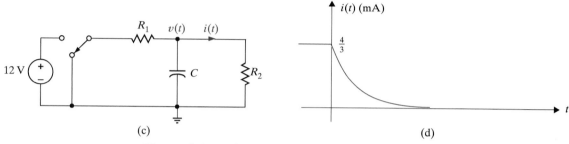

(c)

(d)

Figure 6.4 Analysis of RC circuits.

Using the component values, the equation becomes

$$\frac{dv(t)}{dt} + 5v(t) = 0$$

The form of the solution to this homogeneous equation is

$$v(t) = K_2 e^{-t/\tau}$$

If we substitute this solution into the differential equation, we find that $\tau = 0.2$ s. Thus,

$$v(t) = K_2 e^{-t/0.2} \text{ V}$$

Using the initial condition $v_c(0-) = v_c(0+) = 4$ V, we find that the complete solution is

$$v(t) = 4e^{-t/0.2} \text{ V}$$

Then $i(t)$ is simply

$$i(t) = \frac{v(t)}{R_2}$$

or

$$i(t) = \frac{4}{3} e^{-t/0.2} \text{ mA} \qquad \square$$

EXAMPLE 6.2

The switch in the network in Fig. 6.5a opens at $t = 0$. Let us find the output voltage $v_o(t)$ for $t > 0$.

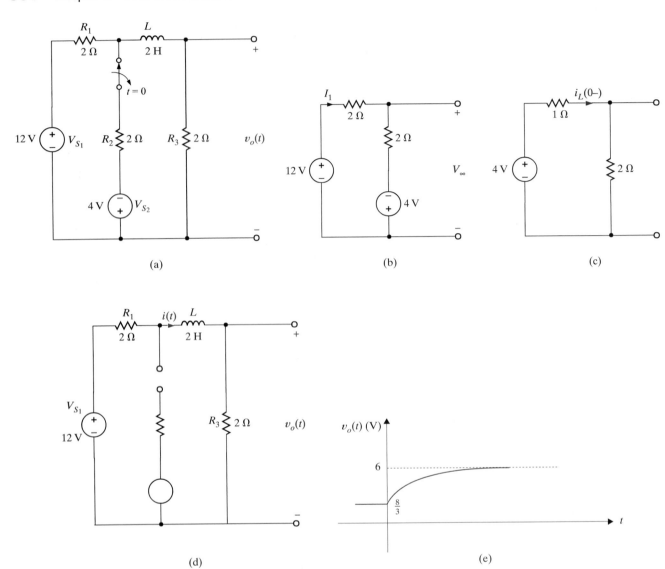

Figure 6.5 Analysis of an RL circuit.

SOLUTION   At $t = 0-$ the circuit is in steady state and the inductor acts like a short circuit. The initial current through the inductor can be found in many ways; however, we will form a Thévenin equivalent for the part of the network to the left of the inductor, as shown in Fig. 6.5b. From this network we find that $I_1 = 4$ A and $V_{oc} = 4$ V. In addition, $R_{Th} = 1$ Ω. Hence, $i_L(0-)$ obtained from Fig. 6.5c is $i_L(0-) = 4/3$ A.

The network for $t > 0$ is shown in Fig. 6.5d. The KVL equation for the circuit is

$$-V_{S_1} + R_1 i(t) + L\frac{di(t)}{dt} + R_3 i(t) = 0$$

which with the component values reduces to

$$\frac{di(t)}{dt} + 2i(t) = 6$$

The solution to this equation is of the form

$$i(t) = K_1 + K_2 e^{-t/\tau}$$

which when substituted into the differential equation yields

$$K_1 = 3$$
$$\tau = 1/2$$

Therefore,

$$i(t) = \left(3 + K_2 e^{-2t}\right) \text{A}$$

Evaluating this function at the initial condition, which is

$$i_L(0-) = i_L(0+) = i(0) = 4/3 \text{ A}$$

we find that

$$K_2 = \frac{-5}{3}$$

Hence,

$$i(t) = \left(3 - \frac{5}{3} e^{-2t}\right) \text{A}$$

and then

$$v_o(t) = 6 - \frac{10}{3} e^{-2t} \text{ V}$$

A plot of the voltage $v_o(t)$ is shown in Fig. 6.5e.   ❑

EXTENSION EXERCISES

E6.1   Find $v_C(t)$ for $t > 0$ in the circuit shown in Fig. E6.1.

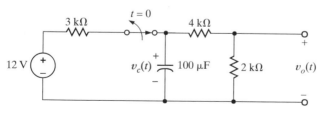

Figure E6.1

ANSWER: $v_C(t) = 8e^{-t/0.6} \text{ V}.$

E6.2 In the circuit shown in Fig. E6.2, the switch opens at $t = 0$. Find $i_1(t)$ for $t > 0$.

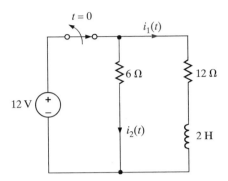

Figure E6.2

ANSWER: $i_1(t) = 1e^{-9t}$ A.

The Step-by-Step Approach

In the previous analysis technique we derived the differential equation for the capacitor voltage or inductor current, solved the differential equation, and used the solution to find the unknown variable in the network. In the very methodical technique that we will now describe we will use the fact that Eq. (6.11) is the form of the solution and employ circuit analysis to determine the constants K_1, K_2, and τ.

From Eq. (6.11) we note that as $t \to \infty$, $e^{-at} \to 0$ and $x(t) = K_1$. Therefore, if the circuit is solved for the variable $x(t)$ in steady state (i.e., $t \to \infty$) with the capacitor replaced by an open circuit [v is constant and therefore $i = C(dv/dt) = 0$] or the inductor replaced by a short circuit [i is constant and therefore $v = L(di/dt) = 0$], then the variable $x(t) = K_1$. Note that since the capacitor or inductor has been removed, the circuit is a dc circuit with constant sources and resistors, and therefore only dc analysis is required in the steady-state solution.

The constant K_2 in Eq. (6.11) can also be obtained via the solution of a dc circuit in which a capacitor is replaced by a voltage source or an inductor is replaced by a current source. The value of the voltage source for the capacitor or the current source for the inductor is a known value at one instant of time. In general, we will use the initial condition value since it is generally the one known, but the value at any instant could be used. This value can be obtained in numerous ways and is often specified as input data in a statement of the problem. However, a more likely situation is one in which a switch is thrown in the circuit and the initial value of the capacitor voltage or inductor current is determined from the previous circuit (i.e., the circuit before the switch is thrown). It is normally assumed that the previous circuit has reached steady state, and therefore the voltage across the capacitor or the current through the inductor can be found in exactly the same manner as was used to find K_1.

Finally, the value of the time constant can be found by determining the Thévenin equivalent resistance at the terminals of the storage element. Then $\tau = R_{Th}C$ for an RC circuit, and $\tau = L/R_{Th}$ for an RL circuit.

Let us now reiterate this procedure in a step-by-step fashion.

Problem-Solving Strategy for Using the Step-by-Step Approach

Step 1. We assume a solution for the variable $x(t)$ of the form $x(t) = K_1 + K_2 e^{-t/\tau}$.

Step 2. Assuming that the original circuit has reached steady state before a switch was thrown (thereby producing a new circuit), draw this previous circuit with the capacitor replaced by an open circuit or the inductor replaced by a short circuit. Solve for the voltage across the capacitor, $v_C(0-)$, or the current through the inductor, $i_L(0-)$, prior to switch action.

Step 3. Assuming that the energy in the storage element cannot change in zero time, draw the circuit, valid only at $t = 0+$. The switches are in their new positions and the capacitor is replaced by a voltage source with a value of $v_C(0+) = v_C(0-)$ or the inductor is replaced by a current source with value $i_L(0+) = i_L(0-)$. Solve for the initial value of the variable $x(0+)$.

Step 4. Assuming that steady state has been reached after the switches are thrown, draw the equivalent circuit, valid for $t > 5\tau$, by replacing the capacitor by an open circuit or the inductor by a short circuit. Solve for the steady-state value of the variable

$$x(t)|_{t > 5\tau} \doteq x(\infty)$$

Step 5. Since the time constant for all voltages and currents in the circuit will be the same, it can be obtained by reducing the entire circuit to a simple series circuit containing a voltage source, resistor, and a storage element (i.e., capacitor or inductor) by forming a simple Thévenin equivalent circuit at the terminals of the storage element. This Thévenin equivalent circuit is obtained by looking into the circuit from the terminals of the storage element. The time constant for a circuit containing a capacitor is $\tau = R_{Th} C$, and for a circuit containing an inductor it is $\tau = L/R_{Th}$.

Step 6. Using the results of steps 3, 4, and 5, we can evaluate the constants in step 1 as

$$x(0+) = K_1 + K_2$$
$$x(\infty) = K_1$$

and therefore, $K_1 = x(\infty)$, $K_2 = x(0+) - x(\infty)$, and hence the solution is

$$x(t) = x(\infty) + [x(0+) - x(\infty)]e^{-t/\tau}$$

Keep in mind that this solution form applies only to a first-order circuit having constant, dc sources. If the sources are not dc, the forced response will be different. Generally, the forced response is of the same form as the forcing functions (sources) and their derivatives.

EXAMPLE 6.3

Consider the circuit shown in Fig. 6.6a. The circuit is in steady state prior to time $t = 0$, when the switch is closed. Let us calculate the current $i(t)$ for $t > 0$.

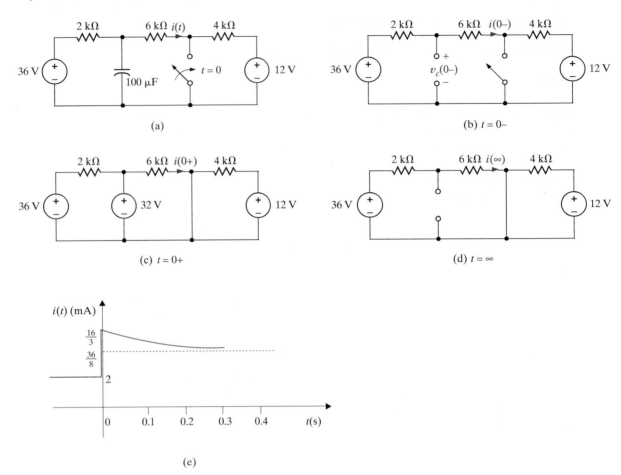

Figure 6.6 Analysis of an *RC* transient circuit with a constant forcing function.

SOLUTION

Step 1. $i(t)$ is of the form $K_1 + K_2 e^{-t/\tau}$.

Step 2. The initial voltage across the capacitor is calculated from Fig. 6.6b as

$$v_C(0-) = 36 - (2)(2)$$

$$= 32 \text{ V}$$

Step 3. The new circuit, valid only for $t = 0+$, is shown in Fig. 6.6c. The value of the voltage source that replaces the capacitor is $v_C(0-) = v_C(0+) = 32$ V. Hence,

$$i(0+) = \frac{32}{6k} = \frac{16}{3} \text{ mA}$$

Step 4. The equivalent circuit, valid for $t > 5\tau$, is shown in Fig. 6.6d. The current $i(\infty)$ caused by the 36-V source is

$$i(\infty) = \frac{36}{2k + 6k} = \frac{36}{8} \text{ mA}$$

Step 5. The Thévenin equivalent resistance, obtained by looking into the open-circuit terminals of the capacitor in Fig. 6.6d, is

$$R_{Th} = \frac{(2k)(6k)}{2k + 6k} = \frac{3}{2} \text{ k}\Omega$$

Therefore, the circuit time constant is

$$\tau = R_{Th}C$$

$$= \left(\frac{3}{2}\right)(10^3)(100)(10^{-6})$$

$$= 0.15 \text{ s}$$

Step 6.

$$K_1 = i(\infty) = \frac{36}{8} \text{ mA}$$

$$K_2 = i(0+) - i(\infty) = i(0+) - K_1$$

$$= \frac{16}{3} - \frac{36}{8}$$

$$= \frac{5}{6} \text{ mA}$$

Therefore,

$$i(t) = \frac{36}{8} + \frac{5}{6}e^{-t/0.15} \text{ mA}$$

Once again we see that although the voltage across the capacitor is continuous at $t = 0$, the current $i(t)$ in the 6-kΩ resistor jumps at $t = 0$ from 2 mA to $5\frac{1}{3}$ mA, and finally decays to $4\frac{1}{2}$ mA. ❏

EXAMPLE 6.4

The circuit shown in Fig. 6.7a is assumed to have been in a steady-state condition prior to switch closure at $t = 0$. We wish to calculate the voltage $v(t)$ for $t > 0$.

SOLUTION

Step 1. $v(t)$ is of the form $K_1 + K_2e^{-t/\tau}$.

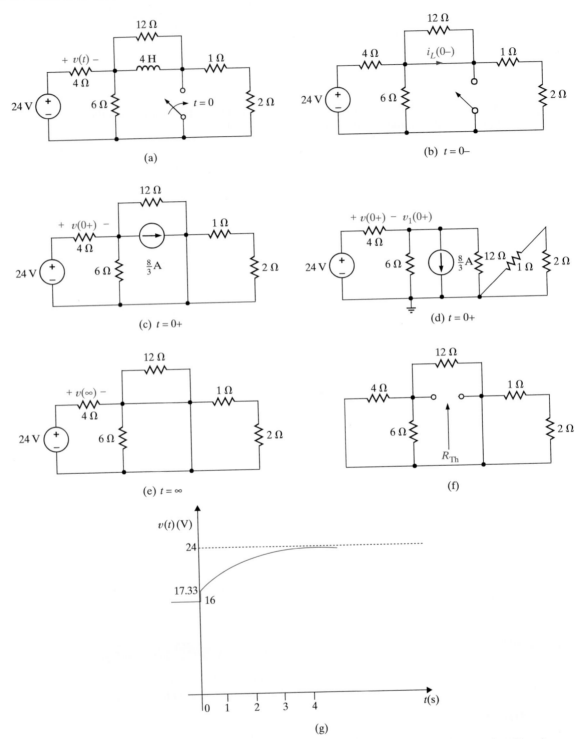

Figure 6.7 Analysis of an *RL* transient circuit with a constant forcing function.

Step 2.   In Fig. 6.7b we see that

$$i_L(0-) = \frac{24}{4 + \dfrac{(6)(3)}{6+3}} \left(\frac{6}{6+3}\right)$$

$$= \frac{8}{3} \text{ A}$$

Step 3.   The new circuit, valid only for $t = 0+$, is shown in Fig. 6.7c, which is equivalent to the circuit shown in Fig. 6.7d. The value of the current source that replaces the inductor is $i_L(0-) = i_L(0+) = \frac{8}{3}$ A. The node voltage $v_1(0+)$ can be determined from the circuit in Fig. 6.7d using a single-node equation, and $v(0+)$ is equal to the difference between the source voltage and $v_1(0+)$. The equation for $v_1(0+)$ is

$$\frac{v_1(0+) - 24}{4} + \frac{v_1(0+)}{6} + \frac{8}{3} + \frac{v_1(0+)}{12} = 0$$

or

$$v_1(0+) = \frac{20}{3} \text{ V}$$

Then

$$v(0+) = 24 - v_1(0+)$$

$$= \frac{52}{3} \text{ V}$$

Step 4.   The equivalent circuit for the steady-state condition after switch closure is given in Fig. 6.7e. Note that the 6-, 12-, 1-, and 2-Ω resistors are shorted, and therefore $v(\infty) = 24$ V.

Step 5.   The Thévenin equivalent resistance is found by looking into the circuit from the inductor terminals. This circuit is shown in Fig. 6.7f. Note carefully that R_{Th} is equal to the 4-, 6-, and 12-Ω resistors in parallel. Therefore, $R_{\text{Th}} = 2$ Ω, and the circuit time constant is

$$\tau = \frac{L}{R_{\text{Th}}} = \frac{4}{2} = 2 \text{ s}$$

Step 6.   From the previous analysis we find that

$$K_1 = v(\infty) = 24$$

$$K_2 = v(0+) - v(\infty) = -\frac{20}{3}$$

and hence that

$$v(t) = 24 - \frac{20}{3} e^{-t/2} \text{ V}$$

From Fig. 6.7b we see that the value of $v(t)$ before switch closure is 16 V. Therefore, the circuit response $v(t)$, as a function of time, is shown in Fig. 6.7g.   ❏

EXTENSION EXERCISES

E6.3   Consider the network in Fig. E6.3. The switch opens at $t = 0$. Find $v_o(t)$ for $t > 0$.

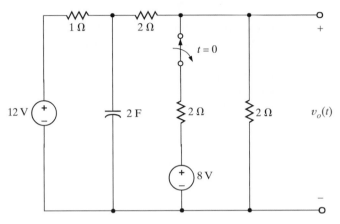

Figure E6.3

ANSWER: $v_o(t) = \dfrac{24}{5} + \dfrac{1}{5} e^{-(5/8)t}$ V.

E6.4   Consider the network in Fig. E6.4. If the switch opens at $t = 0$, find the output voltage $v_o(t)$ for $t > 0$.

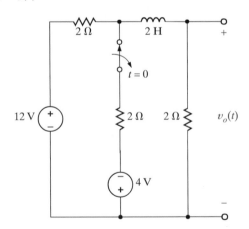

Figure E6.4

ANSWER: $v_o(t) = 6 - \dfrac{10}{3} e^{-2t}$ V.

EXAMPLE 6.5

The circuit shown in Fig. 6.8a has reached steady state with the switch in position 1. At time $t = 0$ the switch moves from position 1 to position 2. We want to calculate $v_o(t)$ for $t > 0$.

SOLUTION

Step 1. $v_o(t)$ is of the form $K_1 + K_2 e^{-t/\tau}$.

Step 2.   Using the circuit in Fig. 6.8b, we can calculate $i_L(0-)$

$$i_A = \frac{12}{4} = 3 \text{ A}$$

Then

$$i_L(0-) = \frac{12 + 2i_A}{6} = \frac{18}{6} = 3 \text{ A}$$

Step 3.   The new circuit, valid only for $t = 0+$, is shown in Fig. 6.8c. The value of the current source that replaces the inductor is $i_L(0-) = i_L(0+) = 3$ A. Because of the current source

$$v_o(0+) = (3)(6) = 18 \text{ V}$$

Step 4.   The equivalent circuit, for the steady-state condition after switch closure, is given in Fig. 6.8d. Using the voltages and currents defined in the figure, we can compute $v_o(\infty)$ in a variety of ways. For example, using node equations we can find $v_o(\infty)$ from

$$\frac{v_B - 36}{2} + \frac{v_B}{4} + \frac{v_B + 2i'_A}{6} = 0$$

$$i'_A = \frac{v_B}{4}$$

$$v_o(\infty) = v_B + 2i'_A$$

or, using loop equations,

$$36 = 2(i_1 + i_2) + 4i_1$$
$$36 = 2(i_1 + i_2) + 6i_2 - 2i_1$$
$$v_o(\infty) = 6i_2$$

Using either approach, we find that $v_o(\infty) = 27$ V.

Step 5.   The Thévenin equivalent resistance can be obtained via v_{oc} and i_{sc} because of the presence of the dependent source. From Fig. 6.8e we note that

$$i''_A = \frac{36}{2 + 4} = 6 \text{ A}$$

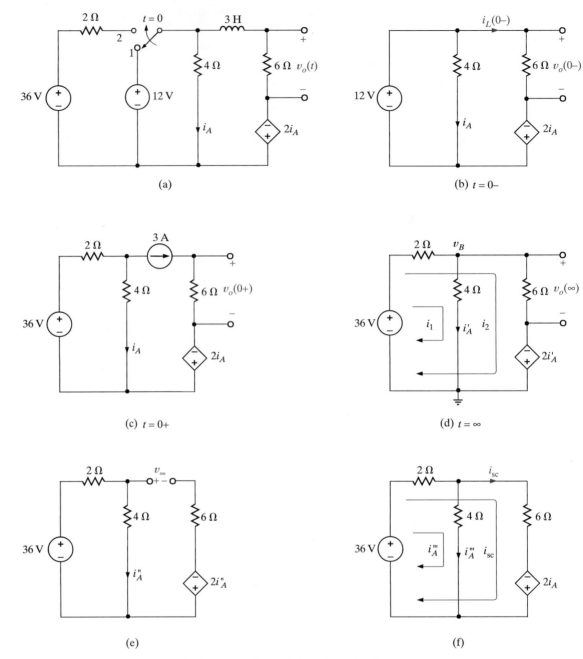

Figure 6.8 Analysis of an *RL* transient circuit containing a dependent source.

Therefore,

$$v_{oc} = (4)(6) + 2(6)$$
$$= 36 \text{ V}$$

From Fig. 6.8f we can write the following loop equations:

$$36 = 2(i'''_A + i_{sc}) + 4i'''_A$$
$$36 = 2(i'''_A + i_{sc}) + 6i_{sc} - 2i'''_A$$

Solving these equations for i_{sc} yields

$$i_{sc} = \frac{36}{8} \text{ A}$$

Therefore,

$$R_{Th} = \frac{v_{oc}}{i_{sc}} = \frac{36}{36/8} = 8 \, \Omega$$

Hence, the circuit time constant is

$$\tau = \frac{L}{R_{Th}} = \frac{3}{8} \text{ s}$$

Step 6.   Using the information just computed, we can derive the final equation for $v_o(t)$:

$$K_1 = v_o(\infty) = 27$$
$$K_2 = v_o(0+) - v_o(\infty) = 18 - 27 = -9$$

Therefore,

$$v_o(t) = 27 - 9e^{-t/(3/8)} \text{ V} \qquad \square$$

EXTENSION EXERCISE

E6.5   If the switch in the network in Fig. E6.5 closes at $t = 0$, find $v_o(t)$ for $t > 0$.

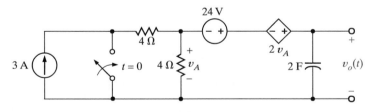

Figure E6.5

ANSWER: $v_o(t) = 24 + 36e^{-(t/12)}$ V.

At this point it is appropriate to state that not all switch action will always occur at time $t = 0$. It may occur at any time t_0. In this case the results of the step-by-step analysis yield the following equations:

$$x(t_0) = K_1 + K_2$$
$$x(\infty) = K_1$$

and

$$x(t) = x(\infty) + \left[x(t_0) - x(\infty)\right]e^{-(t-t_0)/\tau} \qquad t > t_0$$

The function is essentially time shifted by t_0 seconds.

Finally, it should be noted that if more than one independent source is present in the network, we can simply employ superposition to obtain the total response.

Thus far we have considered only sources that were constant. Although this represents a very important type of problem, it is by no means the only type nor is it necessarily the most important.

Returning to Eq. (6.1), we ask, "What is the solution if $f(t)$ is not a constant and the network equation is actually as follows?"

$$\frac{dx(t)}{dt} + ax(t) = f(t) \qquad\qquad\qquad 6.12$$

Recall that $x(t)$ consists of two parts: the natural response (complementary solution) and the forced response (particular solution). That is,

$$x(t) = x_c(t) + x_p(t) \qquad\qquad\qquad 6.13$$

where $x_c(t)$ and $x_p(t)$ must satisfy the equations

$$\frac{dx_c(t)}{dt} + ax_c(t) = 0 \qquad\qquad\qquad 6.14$$

and

$$\frac{dx_p(t)}{dt} + ax_p(t) = f(t) \qquad\qquad\qquad 6.15$$

Thus $x_c(t)$ must still be of the form

$$x_c(t) = K_2 e^{-at} \qquad\qquad\qquad 6.16$$

However, since $f(t)$ is no longer a constant, $x_p(t)$ will, in general, not be one either. Although there are many different mathematical methods for determining the appropriate form for $x_p(t)$, we will employ a deductive approach here. A method that is systematic and often more efficient will be introduced in Chapter 14.

A careful inspection of Eq. (6.15) suggests that $x_p(t)$ must consist of functional forms such as $f(t)$ and its first derivative.

EXAMPLE 6.6

Consider the circuit in Fig. 6.9. The capacitor has an initial charge and an exponentially decaying source is applied at $t = 0$. We wish to determine both $v_o(t)$ and $i_o(t)$ for $t > 0$.

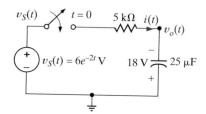

Figure 6.9 *RC* circuit with a nonconstant forcing function.

SOLUTION A nodal analysis yields the equation

$$C\frac{dv_o(t)}{dt} + \frac{v_o(t) - v_S(t)}{R} = 0 \quad t > 0$$

or

$$\frac{dv_o(t)}{dt} + 8v_o(t) = 48e^{-2t}$$

Now the natural response (complementary solution) must, as before, satisfy the homogenous equation

$$\frac{dv_{oc}(t)}{dt} + 8v_{oc}(t) = 0$$

which leads to

$$v_{oc}(t) = K_2 e^{-8t}$$

The forced response (particular solution) must satisfy the equation

$$\frac{dv_{op}(t)}{dt} + 8v_{op}(t) = 48e^{-2t}$$

We deduce that $v_{op}(t)$ must be of the form of the forcing function, e^{-2t}, and its derivative, which in this case is also e^{-2t}. Thus, we assume that

$$v_{op}(t) = K_1 e^{-2t}$$

and by substituting this assumed solution back into the differential equation we can determine the required value of K_1.

$$\frac{d}{dt}(K_1 e^{-2t}) + 8(K_1 e^{-2t}) = 48e^{-2t}$$

$$(-2 + 8)K_1 e^{-2t} = 48e^{-2t}$$

$$K_1 = 8$$

Therefore, the total solution is

$$v_o(t) = 8e^{-2t} + K_2 e^{-8t}$$

However, since $v_o(0) = -18$, K_2 can be obtained from the expression for $v_o(t)$ as

$$v_o(0) = -18 = 8e^{-2(0)} + K_2 e^{-8(0)}$$

or

$$K_2 = -26$$

Then

$$v_o(t) = 8e^{-2t} - 26e^{-8t} \text{ V} \qquad t > 0$$

To determine $i(t)$, we note that

$$i(t) = C \frac{dv_o(t)}{dt}$$

$$= 25 \times 10^{-6}(-16e^{-2t} + 208e^{-8t})$$

$$= -0.4e^{-2t} + 5.2e^{-8t} \text{ mA} \quad t > 0 \qquad \square$$

EXAMPLE 6.7

Consider the network in Fig. 6.10a. The initial inductor current is zero and a sinusoidal voltage is applied at $t = 0$. Let us find both $i_L(t)$ and $v_o(t)$ for $t > 0$.

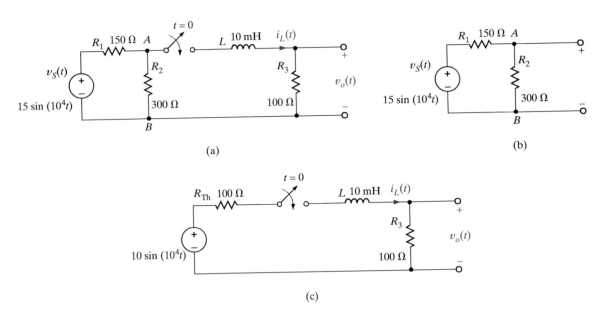

Figure 6.10 *RL* circuit with a sinusoidal forcing function.

SOLUTION To simplify the network, we will first find the Thévenin equivalent circuit looking toward the source at nodes A-B, as shown in Fig. 6.10b. The Thévenin equivalent voltage is

$$v_{Th}(t) = v_S(t)\left(\frac{300}{300 + 150}\right) = 10 \sin(10^4 t) \text{ V}$$

The equivalent resistance is the parallel combination of 150 Ω with 300 Ω or $R_{Th} = 100\ \Omega$. Now the original circuit is reduced to that in Fig. 6.10c. A loop analysis yields the expression

$$v_S(t) = (R_{Th} + R_3)i(t) + L\frac{di(t)}{dt}$$

or

$$\frac{di(t)}{dt} + 2(10^4)i(t) = 10^3 \sin(10^4 t)$$

The complementary solution must satisfy the homogeneous equation

$$\frac{di_c(t)}{dt} + 2(10^4)i_c(t) = 0$$

which requires a solution of the form

$$i_c(t) = K_1 e^{-2(10^4)t}$$

The forced response must satisfy the equation

$$\frac{di_p(t)}{dt} + 2(10^4)i_p(t) = 10^3 \sin(10^4 t)$$

The solution must be of the form of the forcing function $v_S(t)$ and its derivative. Therefore, we can write

$$i_p(t) = K_2 \sin(10^4 t) + K_3 \cos(10^4 t)$$

and by substituting this assumed solution into the differential equation we can determine K_2 and K_3.

$$\frac{d}{dt}[K_2 \sin(10^4 t) + K_3 \cos(10^4 t)] + 2(10^4)[K_2 \sin(10^4 t) + K_3 \cos(10^4 t)]$$
$$= 10^3 \sin(10^4 t)$$

Grouping all sine terms separately from the cosine terms produces two equations:

$$(2(10)^4 K_2 - 10^4 K_3)\sin(10^4 t) = 10^3 \sin(10^4 t)$$
$$(10^4 K_2 + 2(10^4 K_3)\cos(10^4 t) = 0$$

Solving for K_2 and K_3 simultaneously yields

$$K_2 = 0.040$$
$$K_3 = 0.020$$

The final expression for $i_p(t)$ is

$$i_p(t) = 40 \sin(10^4 t) - 20 \cos(10^4 t)\ \text{mA}$$

The total current is the sum of the complementary and forced solutions

$$i(t) = 0.04 \sin(10^4 t) - 0.02 \cos(10^4 t) + K_1 e^{-2(10^4)t}\ \text{A}$$

The value of K_1 can be found using the initial value of $i(t)$. At $t = 0$

$$i(0) = -0.02 + K_1 = 0$$

and

$$K_1 = 0.020$$

Therefore,

$$i(t) = 0.04 \sin(10^4 t) - 0.02 \cos(10^4 t) + 0.02 e^{-2(10^4)t} \text{ A}$$

and

$$v_o(t) = 4 \sin(10^4 t) - 2 \cos(10^4 t) + 2 e^{-2(10^4)t} \text{ V} \qquad \square$$

6.4 PULSE RESPONSE

Thus far we have examined networks in which a voltage or current source is suddenly applied. As a result of this sudden application of a source, voltages or currents in the circuit are forced to change abruptly. A forcing function whose value changes in a discontinuous manner or has a discontinuous derivative is called a *singular function*. Two such singular functions that are very important in circuit analysis are the unit impulse function and the unit step function. We will defer a discussion of the former until a later chapter and concentrate on the latter.

The *unit step function* is defined by the following mathematical relationship:

$$u(t) = \begin{cases} 0 & t < 0 \\ 1 & t > 0 \end{cases}$$

In other words, this function, which is dimensionless, is equal to zero for negative values of the argument and equal to 1 for positive values of the argument. It is undefined for a zero argument where the function is discontinuous. A graph of the unit step is shown in Fig. 6.11a. The unit step is dimensionless, and therefore a voltage step of V_o volts or a current step of I_o amperes is written as $V_o u(t)$ and $I_o u(t)$, respectively. Equivalent circuits for a voltage step are shown in Figs. 6.11b and c. Equivalent circuits for a current step are shown in Figs. 6.11d and e. If we use the definition of the unit step, it is easy to generalize this function by replacing the argument t by $t - t_0$. In this case

$$u(t - t_0) = \begin{cases} 0 & t < t_0 \\ 1 & t > t_0 \end{cases}$$

A graph of this function is shown in Fig. 6.11f. Note that $u(t = t_0)$ is equivalent to delaying $u(t)$ by t_0 seconds, so that the abrupt change occurs at time $t = t_0$.

Step functions can be used to construct one or more pulses. For example, the voltage pulse shown in Fig. 6.12a can be formulated by initiating a unit step at $t = 0$ and subtracting one that starts at $t = T$, as shown in Fig. 6.12b. The equation for the pulse is

$$v(t) = A[u(t) - u(t - T)]$$

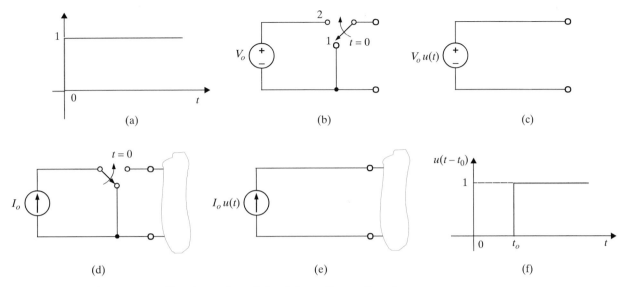

Figure 6.11 Graphs and models of the unit step function.

If the pulse is to start at $t = t_0$ and have width T, the equation would be

$$v(t) = A\{u(t - t_0) - u[t - (t_0 + T)]\}$$

Using this approach, we can write the equation for a pulse starting at any time and ending at any time. Similarly, using this approach, we could write the equation for a series of pulses, called a *pulse train*, by simply forming a summation of pulses constructed in the manner illustrated previously.

The following example will serve to illustrate many of the concepts we have just presented.

EXAMPLE 6.8

Consider the circuit shown in Fig. 6.13a. The input function is the voltage pulse shown in Fig. 6.13b. Since the source is zero for all negative time, the initial conditions for the network are zero [i.e., $v_C(0-) = 0$]. The response $v_o(t)$ for $0 < t < 0.3$ s is due to the application of the constant source at $t = 0$ and is not influenced by any source changes that will occur later. At $t = 0.3$ s the forcing function becomes zero and therefore $v_o(t)$ for $t > 0.3$ s is the source-free or natural response of the network.

Let us determine the expression for the voltage $v_0(t)$.

SOLUTION   Since the output voltage $v_o(t)$ is a voltage division of the capacitor voltage, and the initial voltage across the capacitor is zero, we know that $v_0(0+) = 0$.

If no changes were made in the source after $t = 0$, the steady-state value of $v_o(t)$ [i.e., $v_o(\infty)$] due to the application of the unit step at $t = 0$ would be

$$v_o(\infty) = \frac{9}{6k + 4k + 8k}(8k)$$

$$= 4 \text{ V}$$

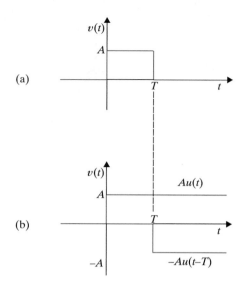

(a)

(b)

Figure 6.12 Construction of a pulse via two step functions.

D6.3   Plot following functions:

(a) $i(t) = 10\big[u(t) - u(t - 0.01)\big]\,\text{mA}$

(b) $v(t) = -10\big[u(t - 1) - u(t - 2)\big]\,\text{V}$

ANSWER: (a)   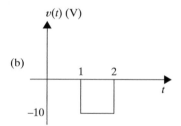   (b)

The Thévenin equivalent resistance is

$$R_{\text{Th}} = \frac{(6\text{k})(12\text{k})}{6\text{k} + 12\text{k}}$$

$$= 4\ \text{k}\Omega$$

Therefore, the circuit time constant τ is

$$\tau = R_{\text{Th}}\,C$$

$$= (4)\big(10^3\big)(100)\big(10^{-6}\big)$$

$$= 0.4\ \text{s}$$

Therefore, the response $v_o(t)$ for the period $0 < t < 0.3$ s is

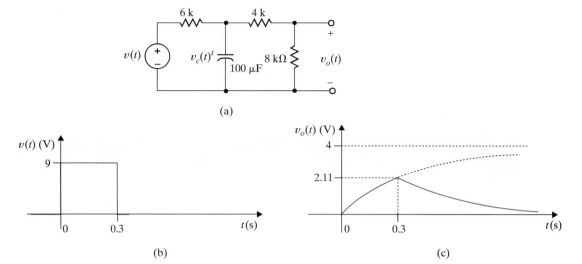

Figure 6.13 Pulse response of a network.

$$v_o(t) = 4 - 4e^{-t/0.4}\,\text{V} \qquad 0 < t < 0.3\,\text{s}$$

The capacitor voltage can be calculated by realizing that using voltage division, $v_o(t) = \frac{2}{3}v_C(t)$. Therefore,

$$v_C(t) = \frac{3}{2}(4 - 4e^{-t/0.4})\,\text{V}$$

Since the capacitor voltage is continuous,

$$v_C(0.3-) = v_C(0.3+)$$

and therefore,

$$\begin{aligned} v_o(0.3+) &= \tfrac{2}{3}v_C(0.3-) \\ &= 4(1 - e^{-0.3/0.4}) \\ &= 2.11\,\text{V} \end{aligned}$$

Since the source is zero for $t > 0.3$ s, the final value for $v_o(t)$ as $t \to \infty$ is zero. Therefore, the expression for $v_o(t)$ for $t > 0.3$ s is

$$v_o(t) = 2.11e^{-(t-0.3)/0.4}\,\text{V} \qquad t > 0.3\,\text{s}$$

The term $e^{-(t-0.3)/0.4}$ indicates that the exponential decay starts at $t = 0.3$ s. The complete solution can be written by means of superposition as

$$v_o(t) = 4(1 - e^{-t/0.4})u(t) - 4(1 - e^{-(t-0.3)/0.4})u(t - 0.3)\,\text{V}$$

or, equivalently, the complete solution is

$$v_o(t) = \begin{cases} 0 & t < 0 \\ 4(1 - e^{-t/0.4})\,\text{V} & 0 < t < 0.3\,\text{s} \\ 2.11e^{-(t-0.3)/0.4}\,\text{V} & 0.3\,\text{s} < t \end{cases}$$

which in mathematical form is

$$v_o(t) = 4(1 - e^{-t/0.4})[u(t) - u(t - 0.3)] + 2.11e^{-(t-0.3)/0.4}u(t - 0.3) \text{ V}$$

Note that the term $[u(t) - u(t - 0.3)]$ acts like a gating function that captures only the part of the step response that exists in the time interval $0 < t < 0.3$ s. The output as a function of time is shown in Fig. 6.13c. ❏

EXTENSION EXERCISE

E6.6 The voltage source in the network in Fig. E6.6a is shown in Fig. E6.6b. The initial current in the inductor must be zero. (Why?) Determine the output voltage $v_0(t)$ for $t > 0$.

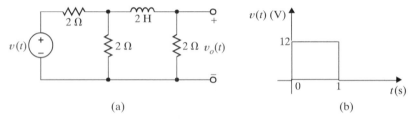

(a) (b)

Figure E6.6

ANSWER: $v_o(t) = 0$ for $t < 0, 4(1 - e^{-(3/2)t})$ V for $0 \le t \le 1$, and $3.11e^{-(3/2)(t-1)}$ V for $1 < t$.

6.5 APPLICATIONS

EXAMPLE 6.9

The network in Fig. 6.14 models the circuit for a flashing neon light. The capacitor charges toward the source voltage of 9 V, through the resistor. The neon light is designed to turn on when the capacitor voltage reaches 8 V. When the light turns on, it discharges the capacitor completely, turns off, and the cycle begins again. This is one form of what is called a relaxation oscillator. If $C = 10 \ \mu\text{F}$, let us find R so that the light flashes once per second.

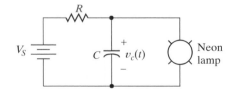

Figure 6.14 Circuit used in Example 6.9.

SOLUTION   We must first find an expression for the capacitor voltage assuming the neon light never turns on. An analogous situation exists in the network in Fig. 6.3a, where the voltage across the capacitor is

$$v_c(t) = V_S - V_S e^{-t/RC} = 9 - 9e^{-t/RC}$$

At $t = 1$ s, the capacitor voltage should be 8 V. Therefore,

$$v(1) = 8 = 9 - 9e^{-1/RC}$$

Solving for RC yields

$$RC = 0.455 \text{ s}$$

Finally,

$$R = 0.455/C = 45.5 \text{ k}\Omega \qquad \square$$

EXAMPLE 6.10

A heart pacemaker circuit is shown in Fig. 6.15. The SCR (silicon-controlled recti-fier) is a solid-state device that has two distinct modes of operation. When the volt-age across the SCR is increasing but less than 5 V, the SCR behaves like an open circuit, as shown in Fig. 6.16a. Once the voltage across the SCR reaches 5 V, the de-vice functions like a current source, as shown in Fig. 6.16b. This behavior will con-tinue as long as the SCR voltage remains above 0.2 V. At this voltage, the SCR shuts off and again becomes an open circuit.

Assume that at $t = 0$, $v_c(t)$ is 0 V and the 1-μF capacitor begins to charge toward the 6-V source voltage. Find the resistor value such that $v_c(t)$ will equal 5 V (the SCR firing voltage) at 1 s. At $t = 1$ s, the SCR fires and begins discharging the ca-pacitor. Find the time required for $v_c(t)$ to drop from 5 V to 0.2 V. Finally, plot $v_c(t)$ for the three cycles.

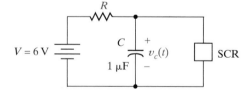

Figure 6.15 Heart pacemaker equivalent circuit.

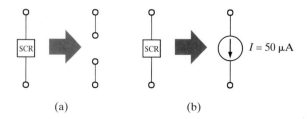

Figure 6.16 Equivalent circuits for silicon-controlled rectifier.

(a)   (b)

SOLUTION   For $t < 1$ s, the equivalent circuit for the pacemaker is shown in Fig. 6.17. As indicated in the previous example, the capacitor voltage has the form

$$v_c(t) = 6 - 6e^{-t/RC}$$

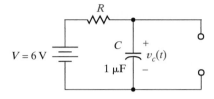

Figure 6.17 Pacemaker equivalent network during capacitor charge cycle.

A voltage of 0.2 V occurs at

$$t_1 = 0.034RC$$

while a voltage of 5 V occurs at

$$t_2 = 1.792RC$$

We desire that $t_2 - t_1 = 1$ s. Therefore,

$$t_2 - t_1 = 1.758RC = 1 \text{ s}$$

and

$$RC = 0.569 \text{ s and } R = 569 \text{ k}\Omega$$

At $t = 1$ s the SCR fires and the pacemaker is modeled by the circuit in Fig. 6.18. The form of the discharge waveform is

$$v(t) = K_1 + K_2 e^{-(t-1)/RC}$$

The term $(t - 1)$ appears in the exponential to shift the function 1 s, since during that time the capacitor was charging. Just after the SCR fires at $t = 1^+$ s, $v_c(t)$ is still 5 V, while at $t = \infty$, $v_c(t) = 6 - IR$. Therefore,

$$K_1 + K_2 = 5 \text{ and } K_1 = 6 - IR$$

Our solution, then, is of the form

$$v_c(t) = 6 - IR + (IR - 1)e^{-(t-1)/RC}$$

Let T be the time beyond 1 s necessary for $v(t)$ to drop to 0.2 V. We write

$$v_c(T + 1) = 6 - IR + (IR - 1)e^{-T/RC} = 0.2$$

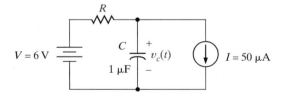

Figure 6.18 Pacemaker equivalent network during capacitor discharge cycle.

Substituting for I, R, and C, we find

$$T = 0.11 \text{ s}$$

The output waveform is shown in Fig. 6.19.   ❏

6.6 CIRCUIT DESIGN

EXAMPLE 6.11

Consider the flashing neon light discussed in Example 6.9. Let us generate the design curve for resistance value versus frequency for flash rates in the range from 0.1 to 5 flashes per second. If f represents the flash frequency, determine the value of R for flash rates of $f = 0.5$ and $f = 5$ flashes per second.

SOLUTION   From Example 6.9, we know that

$$v_c(t) = 9 - 9e^{-t/RC}$$

Let the period between flashes be denoted by the variable T. A flash should occur every time $v_c(t)$ charges to 8 V. Therefore,

$$v_c(T) = 8 = 9 - 9e^{-T/RC}$$

Solving for R yields

$$R = 0.455T/C = 45.5 \times 10^3 \, T$$

Since the period T is $1/f$,

$$R = 45.5 \times 10^3 / f$$

As a quick check, if $f = 1$ flash/second, as is the case in Example 6.9, the equation for R yields a value of 45.5 kΩ, which matches the value from Example 6.9. The plot

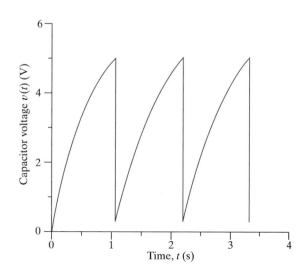

Figure 6.19 Heart pacemaker output voltage waveform.

of R versus f is shown in Fig. 6.20. For 0.5 flashes/second, $R = 91.0$ kΩ, and for 5 flashes/second, $R = 9.1$ kΩ.

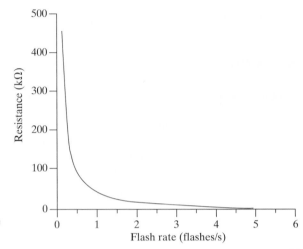

Figure 6.20 Resistor value versus flash rate for the circuit in Example 6.9.

6.7 SUMMARY

- An RC or RL transient circuit is said to be first order if it contains only a single capacitor or single inductor. The voltage or current anywhere in the network can be obtained by solving a first-order differential equation.

- The form of a first-order differential equation with a constant forcing function is

$$\frac{dx(t)}{dt} + \frac{x(t)}{\tau} = A$$

and the solution is

$$x(t) = A\tau + K_2 e^{-t/\tau}$$

where $A\tau$ is referred to as the steady-state solution and τ is called the time constant.

- The function $e^{-t/\tau}$ decays to a value that is less than 1% of its initial value after a period of 5τ. Therefore, the time constant, τ, determines the time required for the circuit to reach steady state.

- The time constant for an RC circuit is $R_{\text{Th}} C$ and for an RL circuit is L/R_{Th}, where R_{Th} is the Thévenin equivalent resistance looking into the circuit at the terminals of the storage element (i.e., capacitor or inductor).

- The two approaches proposed for solving first-order transient circuits are the differential equation approach and the step-by-step method. In the former case, the differential equation that describes the dynamic behavior of the circuit is solved to determine the desired solution. In the latter case, the initial conditions and the steady-state value of the voltage across the capacitor or current in the inductor are used in conjunction with the circuit's time constant and the known form of the desired variable to obtain a solution.

- The response of a first-order transient circuit to an input pulse can be obtained by treating the pulse as a combination of two step-function inputs.

PROBLEMS

Section 6.3

6.1   Use the differential equation approach to find $v_o(t)$ for $t > 0$ in the circuit in Fig. P6.1 and plot the response including the time interval just prior to switch action.

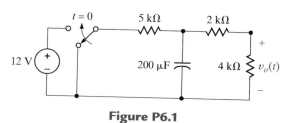

Figure P6.1

Follow the technique used in Example 6.1.

6.2   Use the differential equation approach to find $v_c(t)$ for $t > 0$ in the circuit in Fig. P6.2 and plot the response including the time interval just prior to opening the switch.

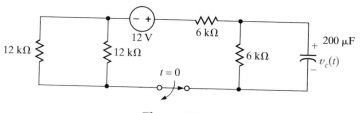

Figure P6.2

Similar to P6.1

6.3   Use the differential equation approach to find $v_c(t)$ for $t > 0$ in the circuit in Fig. P6.3 and plot the response including the time interval just prior to closing the switch.

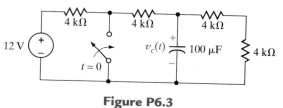

Figure P6.3

Similar to P6.1

6.4 Use the differential equation approach to find $v_o(t)$ for $t > 0$ in the circuit in Fig. P6.4 and plot the response including the time interval just prior to switch action.

Similar to P6.1

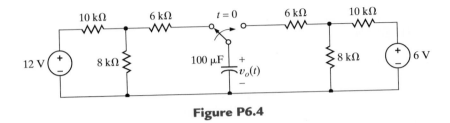

Figure P6.4

6.5 Use the differential equation approach to find $i(t)$ for $t > 0$ in the circuit in Fig. P6.5 and plot the response including the time interval just prior to switch movement.

Follow the technique outlined in Example 6.2

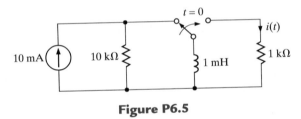

Figure P6.5

6.6 Use the differential equation approach to find $i(t)$ for $t > 0$ in the circuit in Fig. P6.6 and plot the response including the time interval just prior to opening the switch.

Similar to P6.5

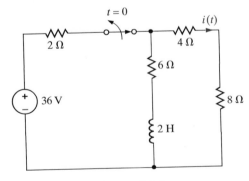

Figure P6.6

6.7 Use the differential equation approach to find $i_L(t)$ for $t > 0$ in the circuit in Fig. P6.7 and plot the response including the time interval just prior to opening the switch.

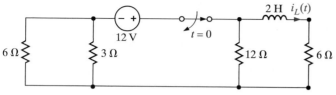

Figure P6.7

Similar to P6.5

6.8 Use the differential equation approach to find $i(t)$ for $t > 0$ in the circuit in Fig. P6.8 and plot the response including the time interval just prior to switch movement.

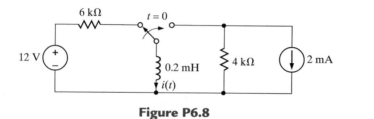

Figure P6.8

Similar to P6.5

6.9 Use the differential equation approach to find $v_o(t)$ for $t > 0$ in the circuit in Fig. P6.9 and plot the response including the time interval just prior to opening the switch.

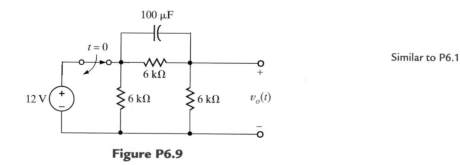

Figure P6.9

Similar to P6.1

6.10 Use the differential equation approach to find $i_o(t)$ for $t > 0$ in the circuit in Fig. P6.10 and plot the response including the time interval just prior to closing the switch.

Examine the effect of the switch on the capacitor voltage.

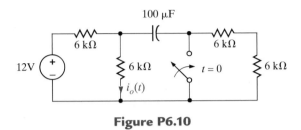

Figure P6.10

6.11 Use the differential equation approach to find $v_o(t)$ for $t > 0$ in the circuit in Fig. P6.11 and plot the response including the time interval just prior to closing the switch.

The initial condition of the capacitor must include the influence from both sources. Then the problem is similar to P6.1.

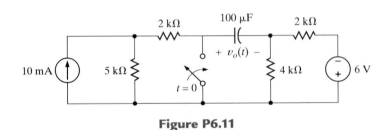

Figure P6.11

6.12 Use the differential equation approach to find $i_o(t)$ for $t > 0$ in the circuit in Fig. P6.12 and plot the response including the time interval just prior to opening the switch.

Similar to P6.1

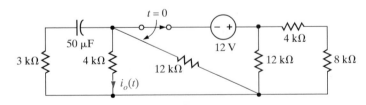

Figure P6.12

6.13 Use the differential equation approach to find $i_o(t)$ for $t > 0$ in the circuit in Fig. P6.13 and plot the response including the time interval just prior to opening the switch.

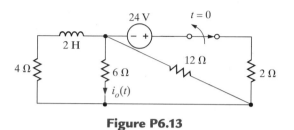

Figure P6.13

Similar to P6.5

6.14 Use the differential equation approach to find $i_o(t)$ for $t > 0$ in the circuit in Fig. P6.14 and plot the response including the time interval just prior to opening the switch.

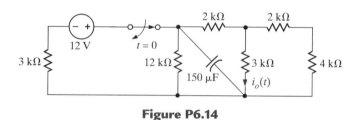

Figure P6.14

Similar to P6.1

6.15 Use the differential equation approach to find $i(t)$ for $t > 0$ in the circuit in Fig. P6.15 and plot the response including the time interval just prior to opening the switch.

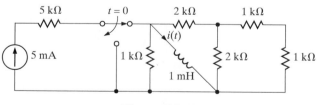

Figure P6.15

Similar to P6.5

6.16 Using the differential equation approach find $i_o(t)$ for $t > 0$ in the network in Fig. P6.16 and plot the response including the time interval just prior to opening the switch.

Similar to P6.5

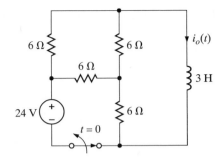

Figure P6.16

6.17 Use the differential equation approach to find $v_o(t)$ for $t > 0$ in the circuit in Fig. P6.17 and plot the response including the time interval just prior to opening the switch.

Similar to P6.1

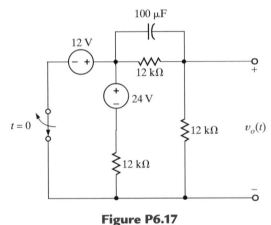

Figure P6.17

6.18 Use the differential equation approach to find $i_o(t)$ for $t > 0$ in the circuit in Fig. P6.18 and plot the response including the time interval just prior to opening the switch.

Similar to P6.1

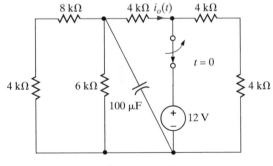

Figure P6.18

6.19 Use the differential equation approach to find $v_o(t)$ for $t > 0$ in the circuit in Fig. P6.19 and plot the response including the time interval just prior to opening the switch.

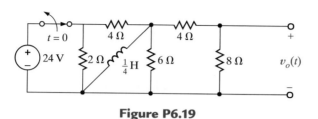

Figure P6.19

Similar to P6.5

6.20 Use the differential equation approach to find $v_o(t)$ for $t > 0$ in the circuit in Fig. P6.20 and plot the response including the time interval just prior to opening the switch.

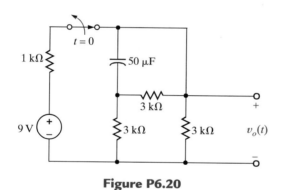

Figure P6.20

Similar to P6.1

6.21 Use the step-by step technique to find $i_o(t)$ for $t > 0$ in the network in Fig. P6.21.

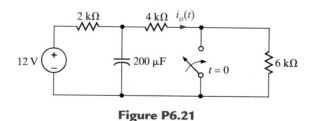

Figure P6.21

Follow the technique outlined in Example 6.3.

6.22 Use the step-by step method to find $v_o(t)$ for $t > 0$ in the network in Fig. P6.22.

Similar to P6.21

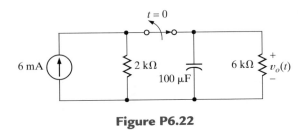

Figure P6.22

6.23 Use the step-by step method to find $i_o(t)$ for $t > 0$ in the circuit in Fig. P6.23.

Similar to P6.21

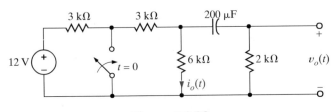

Figure P6.23

Similar to P6.21

6.24 Find $v_o(t)$ for $t > 0$ in the network in Fig. P6.23 using the step-by-step technique.

6.25 Use the step-by step method to find $i_o(t)$ for $t > 0$ in the network in Fig. P6.25.

Similar to P6.21

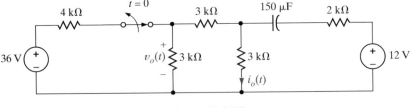

Figure P6.25

Similar to P6.21

6.26 Find $v_o(t)$ for $t > 0$ in the circuit in Fig. P6.25 using the step-by-step method.

6.27 Use the step-by step technique to find $i_o(t)$ for $t > 0$ in the network in Fig. P6.27.

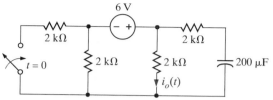

Figure P6.27

Similar to P6.21

6.28 Find $i_o(t)$ for $t > 0$ in the network in Fig. P6.28 using the step-by-step method.

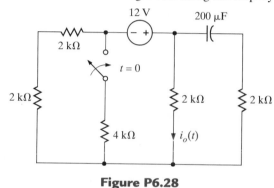

Figure P6.28

Similar to P6.21

6.29 Find $i_o(t)$ for $t > 0$ in the network in Fig. P6.29 using the step-by-step method.

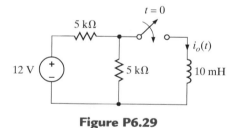

Figure P6.29

Follow the technique outlined in Example 6.4.

6.30 Find $i_o(t)$ for $t > 0$ in the network in Fig. P6.30 using the step-by-step method.

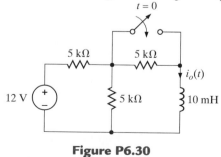

Figure P6.30

Similar to P6.29.

6.31 Find $i_o(t)$ for $t > 0$ in the network in Fig. P6.31 using the step-by-step method.

Examine the effect of opening the switch.

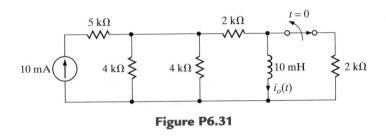

Figure P6.31

6.32 Find $v_o(t)$ for $t > 0$ in the network in Fig. P6.32 using the step-by-step method.

Similar to P6.29

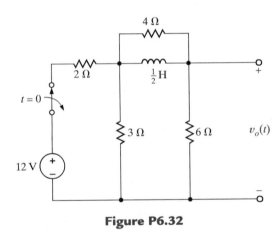

Figure P6.32

6.33 Find $v_o(t)$ for $t > 0$ in the network in Fig. P6.33 using the step-by-step technique.

Similar to P6.29

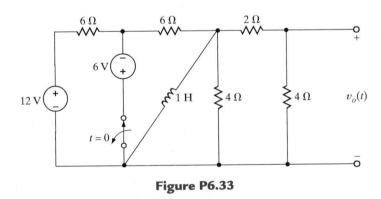

Figure P6.33

6.34 Use the step-by step method to find $v_o(t)$ for $t > 0$ in the network in Fig. P6.34.

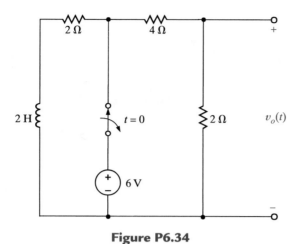

Figure P6.34

Similar to P6.29

6.35 Use the step-by step technique to find $i_o(t)$ for $t > 0$ in the circuit in Fig. P6.35.

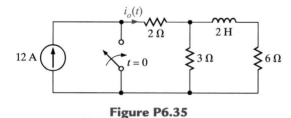

Figure P6.35

Similar to P6.29

6.36 Use the step-by step method to find $i_o(t)$ for $t > 0$ in the circuit in Fig. P6.36.

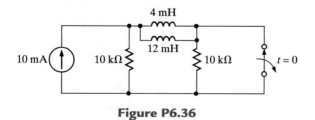

Figure P6.36

First find the equivalent inductance. Then the problem is similar to P6.29.

6.37 Find $i_o(t)$ for $t > 0$ in the circuit in Fig. P6.37 using the step-by-step method.

Similar to P6.29

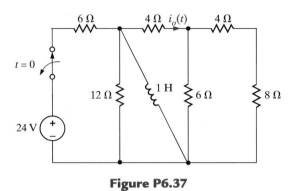

Figure P6.37

6.38 Find $v_o(t)$ for $t > 0$ in the network in Fig. P6.38 using the step-by-step technique.

Similar to P6.29

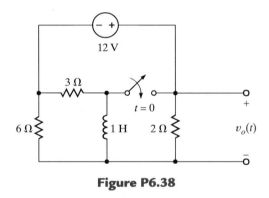

Figure P6.38

6.39 Find $i_o(t)$ for $t > 0$ in the network in Fig. P6.39 using the step-by-step method.

Follow the technique outlined
in Example 6.5.

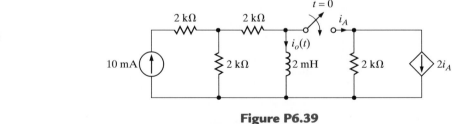

Figure P6.39

6.40 Use the step-by step method to find $v_o(t)$ for $t > 0$ in the circuit in Fig. P6.40.

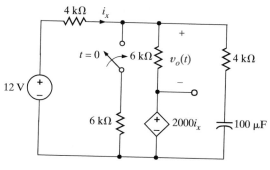

Figure P6.40

Similar to P6.39

6.41 Use the step-by step method to find $v_o(t)$ for $t > 0$ in the network in Fig. P6.41.

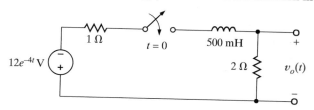

Figure P6.41

Follow the technique outlined in Example 6.6.

6.42 Find $v_o(t)$ for $t > 0$ in the circuit in Fig. P6.42 using the step-by-step technique.

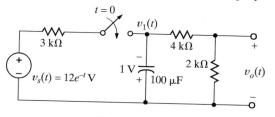

Figure P6.42

Similar to P6.41

6.43 Find $v_o(t)$ for $t > 0$ in the network in Fig. P6.43.

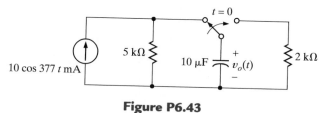

Figure P6.43

Follow the technique outlined in Example 6.7.

Section 6.4

6.44 The current source in the network in Fig. P6.44a is defined in Fig. P6.44b. The initial voltage across the capacitor must be zero. (Why?) Determine the current $i_o(t)$ for $t > 0$

Follow the technique outlined in Example 6.8.

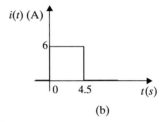

(a)

(b)

Figure P6.44

6.45 Determine the equation for the voltage $v_o(t)$ for $t > 0$, in Fig. P6.45a when subjected to the input pulse shown in Fig. P6.45b.

Similar to P6.44

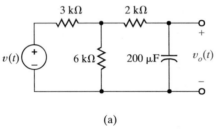

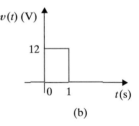

(a)

(b)

Figure P6.45

6.46 Find the output voltage $v_o(t)$ in the network in Fig. P6.46 if the input voltage is $v_i(t) = 5(u(t) - u(t - 0.05))$ V.

Similar to P6.44

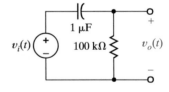

Figure P6.46

6.47 Find the output voltage $v_o(t)$ in the network in Fig. P6.47 if the input voltage is $v_i(t) = 6(u(t) - u(t - 0.05))$ V.

Similar to P6.44

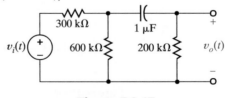

Figure P6.47

Advanced Problems

6.48 Using the differential equation approach, find $i_o(t)$ for $t > 0$ in the circuit in Fig. P6.48 and plot the response including the time interval just prior to opening the switch.

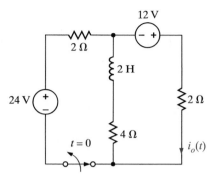

Figure P6.48

6.49 Voltages inside the picture tube of a television set can reach several thousand volts. This produces a serious hazard to technicians. Suppose a 1.2µF capacitor is connected to the tube and is charged to 20,000 V. The technician wants to discharge the capacitor through a 500-kΩ resistor. How long will it take to discharge the capacitor by 99%?

6.50 How long would it take to discharage the capacitor in Problem 6.49 to only 20 V? What is the maximum current through the resistor during discharge? What is the maximum power dissipated by the resistor during discharge?

6.51 Use the step-by-step technique to find $i_o(t)$ for $t > 0$ in the network in Fig. P6.51.

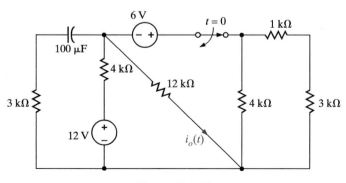

Figure P6.51

6.52 Find $v_o(t)$ for $t > 0$ in the network in Fig. P6.52 using the step-by-step method.

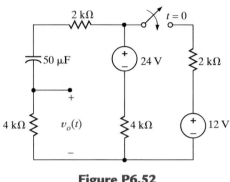

Figure P6.52

6.53 Find $i_o(t)$ for $t > 0$ in the circuit in Fig. P6.53 using the step-by-step technique.

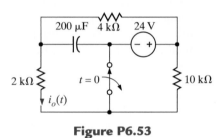

Figure P6.53

6.54 Use the step-by-step technique to find $i_o(t)$ for $t > 0$ in the network in Fig. P6.54.

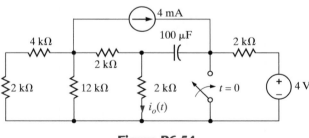

Figure P6.54

6.55 Find $v_o(t)$ for $t > 0$ in the network in Fig. P6.55 using the step-by-step method.

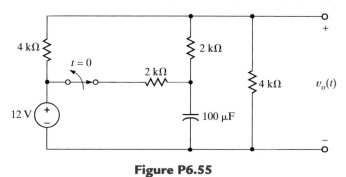

Figure P6.55

6.56 Find $v_o(t)$ for $t > 0$ in the network in Fig. P6.56 using the step-by-step method.

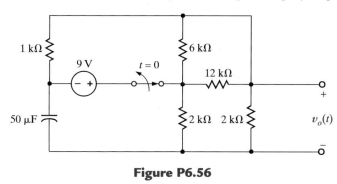

Figure P6.56

6.57 Use the step-by-step technique to find $v_o(t)$ for $t > 0$ in the circuit in Fig. P6.57.

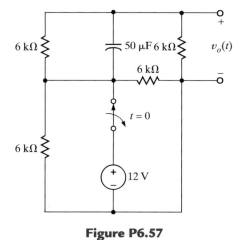

Figure P6.57

6.58 Use the step-by-step method to find $i_o(t)$ for $t > 0$ in the network in Fig. P6.58.

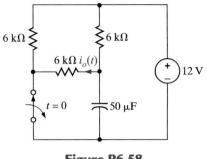

Figure P6.58

6.59 Find $v_o(t)$ for $t > 0$ in the circuit in Fig. P6.59 using the step-by-step method.

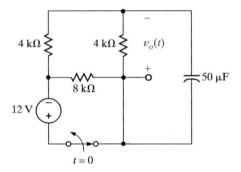

Figure P6.59

6.60 Use the step-by-step method to find $v_o(t)$ for $t > 0$ in the network in Fig. P6.60.

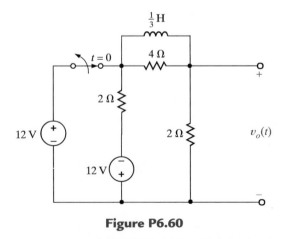

Figure P6.60

6.61 Use the step-by-step technique to find $v_o(t)$ for $t > 0$ in the circuit in Fig. P6.61.

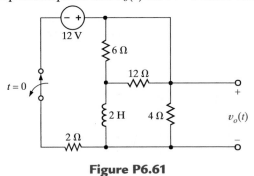

Figure P6.61

6.62 Find $v_o(t)$ for $t > 0$ in the circuit in Fig. P6.62 using the step-by-step method.

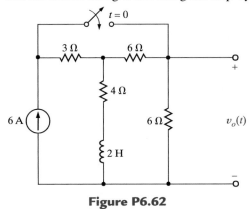

Figure P6.62

6.63 Find $v_o(t)$ for $t > 0$ in the network in Fig. P6.63 using the step-by-step technique.

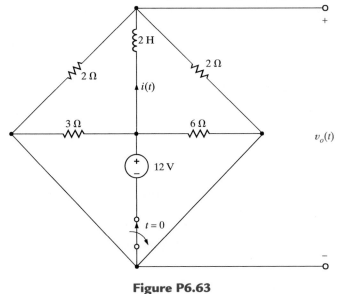

Figure P6.63

6.64 Find $v_o(t)$ for $t > 0$ in the circuit in Fig. P6.64 using the step-by-step method.

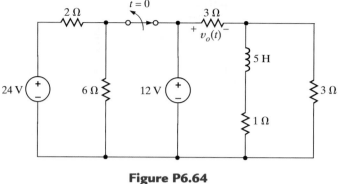

Figure P6.64

6.65 Use the step-by-step method to find $v_o(t)$ for $t > 0$ in the circuit in Fig. P6.65.

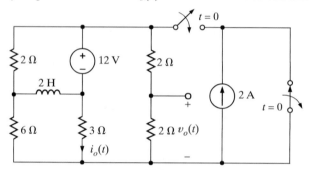

Figure P6.65

6.66 Use the step-by-step technique to find $i_o(t)$ for $t > 0$ in the network in Fig. P6.66.

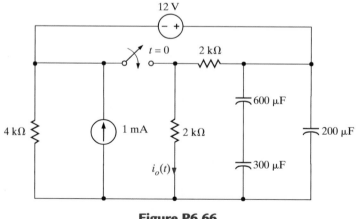

Figure P6.66

6.67 Find $i_o(t)$ for $t > 0$ in the circuit in Fig. P6.67 using the step-by-step method.

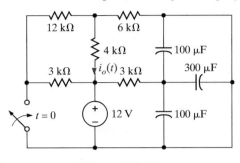

Figure P6.67

6.68 Find $i_o(t)$ for $t > 0$ in the circuit in Fig. P6.68 using the step-by-step technique.

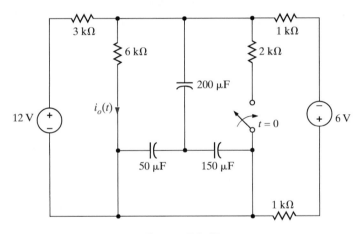

Figure P6.68

6.69 Use the step-by-step method to find $i_o(t)$ for $t > 0$ in the network in Fig. P6.69.

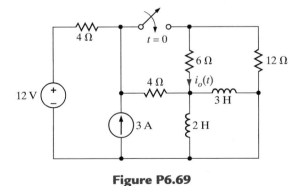

Figure P6.69

6.70 Use the step-by-step technique to find $v_o(t)$ for $t > 0$ in the network in Fig. P6.70.

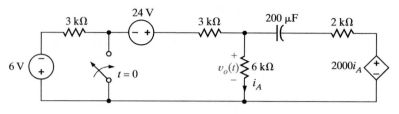

Figure P6.70

6.71 Find $v_o(t)$ for $t > 0$ in the network in Fig. P6.71 using the step-by-step method.

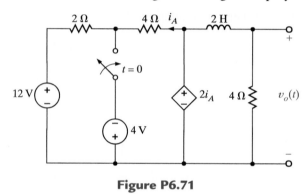

Figure P6.71

6.72 Find $v_o(t)$ for $t > 0$ in the network in Fig. P6.72 using the step-by-step method.

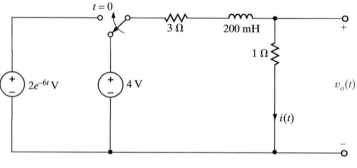

Figure P6.72

Second-Order Transient Circuits

7

In this chapter we extend our analysis of circuits that contain only a single storage element to the case where an inductor and a capacitor are present simultaneously. Although the RLC circuits are more complicated than those we have analyzed in Chapter 6, we will follow a development similar to that used earlier in obtaining a solution.

Our presentation will deal only with very simple circuits, since the analysis can quickly become complicated for networks that contain more than one loop or one nonreference node. Once we have presented the mathematical development of the response equations, we will analyze several networks in which the parameters have been chosen to illustrate the different types of response. In addition, we will extend our PSPICE analysis techniques. Finally, we will provide a number of practical applications that indicate the importance of these second-order transient circuits.

A more efficient method for providing a general mathematical solution for RLC networks will be presented in Chapter 14.

7.1 THE BASIC CIRCUIT EQUATION

To begin our development, let us consider the two basic RLC circuits shown in Fig. 7.1. We assume that energy may be initially stored in both the inductor and capacitor. The node equation for the parallel RLC circuit is

$$\frac{v}{R} + \frac{1}{L}\int_{t_0}^{t} v(x)\, dx + i_L(t_0) + C\frac{dv}{dt} = i_S(t)$$

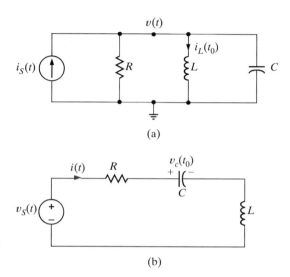

(a)

(b)

Figure 7.1 Parallel and series RLC series.

Similarly, the loop equation for the series RLC circuit is

$$Ri + \frac{1}{C}\int_{t_0}^{t} i(x)\, dx + v_C(t_0) + L\frac{di}{dt} = v_S(t)$$

Note that the equation for the node voltage in the parallel circuit is of the same form as that for the loop current in the series circuit. Therefore, the solution of these two circuits is dependent on solving one equation. If the two preceding equations are differentiated with respect to time, we obtain

$$C\frac{d^2 v}{dt^2} + \frac{1}{R}\frac{dv}{dt} + \frac{v}{L} = \frac{di_S}{dt}$$

and

$$L\frac{d^2 i}{dt^2} + R\frac{di}{dt} + \frac{i}{C} = \frac{dv_S}{dt}$$

Since both circuits lead to a second-order differential equation with constant coefficients, we will concentrate our analysis on this type of equation.

D7.1   Write the differential equation that describes the node voltage in (a) and the loop current in (b).

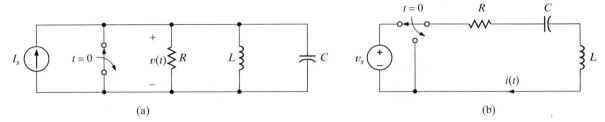

(a)   (b)

ANSWER: (a) $\dfrac{d^2 v}{dt^2} + \dfrac{1}{RC}\dfrac{dv}{dt} + \dfrac{v}{LC} = 0$

   (b) $L\dfrac{d^2 i}{dt^2} + R\dfrac{di}{dt} + \dfrac{i}{C} = 0$

7.2 MATHEMATICAL DEVELOPMENT OF THE RESPONSE EQUATIONS

In concert with our development of the solution of a first-order differential equation that results from the analysis of either an RL or an RC circuit as outlined in Chapter 6, we will now employ the same approach here to obtain the solution of a second-order differential equation that results from the analysis of RLC circuits. As a general rule, for this case we are confronted with an equation of the form

$$\frac{d^2 x(t)}{dt^2} + a_1 \frac{dx(t)}{dt} + a_2 x(t) = f(t) \qquad \text{7.1}$$

Once again we use the fact that if $x(t) = x_p(t)$ is a solution to Eq. (7.1), and if $x(t) = x_c(t)$ is a solution to the homogeneous equation

$$\frac{d^2 x(t)}{dt^2} + a_1 \frac{dx(t)}{dt} + a_2 x(t) = 0$$

then

$$x(t) = x_p(t) + x_c(t)$$

is a solution to the original Eq. (7.1). If we again confine ourselves to a constant forcing function [i.e., $f(t) = A$], the development at the beginning of Chapter 6 shows that the solution of Eq. (7.1) will be of the form

$$x(t) = \frac{A}{a_2} + x_c(t) \qquad \text{7.2}$$

Let us now turn our attention to the solution of the homogeneous equation

$$\frac{d^2 x(t)}{dt^2} + a_1 \frac{dx(t)}{dt} a_2 x(t) = 0$$

D7.2   Given the homogenous differential equation

$$4\frac{d^2 x(t)}{dt^2} + 8\frac{dx(t)}{dt} + 16x(t) = 0$$

determine the characteristic equation, the damping ratio, and the undamped resonant frequency.

ANSWER: Characteristic equation:

$$s^2 + 2s + 4 = 0$$

$$\zeta = \tfrac{1}{2}$$

$$\omega_o = 2$$

where a_1 and a_2 are constants. For simplicity we will rewrite the equation in the form

$$\frac{d^2 x(t)}{dt^2} + 2\zeta\omega_o \frac{dx(t)}{dt} + \omega_o^2 x(t) = 0 \qquad \textbf{7.3}$$

where we have made the following simple substitutions for the constants $a_1 = 2\zeta\omega_o$ and $a_2 = \omega_o^2$.

Following the development of a solution for the first-order homogeneous differential equation in Chapter 6, the solution of Eq. (7.3) must be a function whose first- and second-order derivatives have the same form, so that the left-hand side of Eq. (7.3) will become identically zero for all t. Again we assume that

$$x(t) = Ke^{st}$$

Substituting this expression into Eq. (7.3) yields

$$s^2 Ke^{st} + 2\zeta\omega_o sKe^{st} + \omega_o^2 ke^{st} = 0$$

Dividing both sides of the equation by Ke^{st} yields

$$s^2 + 2\zeta\omega_o s + \omega_o^2 = 0 \qquad \textbf{7.4}$$

This equation is commonly called the *characteristic equation*; ζ is called the exponential *damping ratio*, and ω_o is referred to as the *undamped resonant frequency*. The importance of this terminology will become clear as we proceed with the development. If this equation is satisfied, our assumed solution $x(t) = Ke^{st}$ is correct. Employing the quadratic formula, we find that Eq. (7.4) is satisfied if

$$s = \frac{-2\zeta\omega_o \pm \sqrt{4\zeta^2\omega_o^2 - 4\omega_o^2}}{2}$$

$$= -\zeta\omega_o \pm \omega_o \sqrt{\zeta^2 - 1} \qquad \textbf{7.5}$$

Therefore, there are two values of s, s_1 and s_2, that satisfy Eq. (7.4):

$$s_1 = -\zeta\omega_o + \omega_o \sqrt{\zeta^2 - 1}$$
$$s_2 = -\zeta\omega_o - \omega_o \sqrt{\zeta^2 - 1} \qquad \textbf{7.6}$$

This means that $x_1(t) = K_1 e^{s_1 t}$ is a solution of Eq. (7.3) and that $x_2(t) = K_2 e^{s_2 t}$ is also a solution of Eq. (7.3); that is,

$$\frac{d^2}{dt^2} (K_1 e^{s_1 t}) + 2\zeta\omega_o \frac{d}{dt} (K_1 e^{s_1 t}) + \omega_o^2 K_1 e^{s_1 t} = 0$$

and

$$\frac{d^2}{dt^2} (K_2 e^{s_2 t}) + 2\zeta\omega_o \frac{d}{dt} (K_2 e^{s_2 t}) + \omega_o^2 K_2 e^{s_2 t} = 0$$

The addition of these two equations produces the equality

$$\frac{d^2}{dt^2} (K_1 e^{s_1 t} + K_2 e^{s_2 t}) + 2\zeta\omega_o \frac{d}{dt} (K_1 e^{s_1 t} + K_2 e^{s_2 t}) + \omega_o^2 (K_1 e^{s_1 t} + K_2 e^{s_2 t}) = 0$$

Note that the sum of the two solutions is also a solution. Therefore, in general, the complementary solution of Eq. (7.3) is of the form

$$x_c(t) = K_1 e^{s_1 t} + K_2 e^{s_2 t} \qquad\qquad \textbf{7.7}$$

K_1 and K_2 are constants that can be evaluated via the initial conditions $x(0)$ and $dx(0)/dt$. For example, since

$$x(t) = K_1 e^{s_1 t} + K_2 e^{s_2 t}$$

then

$$x(0) = K_1 + K_2$$

and

$$\left.\frac{dx(t)}{dt}\right|_{t=0} = \frac{dx(0)}{dt} = s_1 K_1 + s_2 K_2$$

Hence, $x(0)$ and $dx(0)/dt$ produce two simultaneous equations, which when solved yield the constants K_1 and K_2.

Close examination of Eqs. (7.6) and (7.7) indicates that the form of the solution of the homogeneous equation is dependent upon the value ζ. For example, if $\zeta > 1$, the roots of the characteristic equation, s_1 and s_2, also called the *natural frequencies* because they determine the natural (unforced) response of the network, are real and unequal; if $\zeta < 1$, the roots are complex numbers; and finally, if $\zeta = 1$, the roots are real and equal. Each of these cases is very important; hence, we will now examine each one in some detail.

Case 1, $\zeta > 1$. This case is commonly called *overdamped*. The natural frequencies s_1 and s_2 are real and unequal, and therefore the natural response of the network described by the second-order differential equation is of the form

$$x_c(t) = K_1 e^{-(\zeta\omega_o - \omega_o \sqrt{\zeta^2 - 1})t} + K_2 e^{-(\zeta\omega_o + \omega_o \sqrt{\zeta^2 - 1})t} \qquad\qquad \textbf{7.8}$$

where K_1 and K_2 are found from the initial conditions. This indicates that the natural response is the sum of two decaying exponentials.

Case 2, $\zeta < 1$. This case is called *underdamped*. Since $\zeta < 1$, the roots of the characteristic equation given in Eq. (7.6) can be written as

$$s_1 = -\zeta\omega_o + j\omega_o \sqrt{1 - \zeta^2} = -\sigma + j\omega_d$$
$$s_2 = -\zeta\omega_o - j\omega_o \sqrt{1 - \zeta^2} = -\sigma - j\omega_d$$

where $j = \sqrt{-1}$, $\sigma = \zeta\omega_o$ and $\omega_d = \omega_o \sqrt{1 - \zeta^2}$. Thus, the natural frequencies are complex numbers (briefly discussed in Appendix B). The natural response is then

$$x_c(t) = K_1 e^{-(\sigma - j\omega_d)t} + K_2 e^{-(\sigma + j\omega_d)t}$$
$$x_c(t) = e^{-\sigma t}(K_1 e^{j\omega_d t} + K_2 e^{-j\omega_d t})$$

Using Euler's identities

$$e^{\pm j\theta} = \cos\theta \pm j \sin\theta$$

DRILL

D7.3 Determine the general form of the solution of the equation in Drill 7.2.

ANSWER:

$x_c(t) = e^{-t}(A_1 \cos\sqrt{3}\,t + A_2 \sin\sqrt{3}\,t)$.

we obtain

$$x_c(t) = e^{-\sigma t}\big[K_1(\cos \omega_d t + j \sin \omega_d t) + K_2(\cos \omega_d t - j \sin \omega_d t)\big]$$
$$= e^{-\sigma t}\big[(K_1 + K_2) \cos \omega_d t + (jK_1 - jK_2) \sin \omega_d t\big]$$

$$x_c(t) = e^{-\sigma t}(A_1 \cos \omega_d t + A_2 \sin \omega_d t) \qquad\qquad \textbf{7.9}$$

where A_1 and A_2, like K_1 and K_2, are constants, which are evaluated using the initial conditions $x(0)$ and $dx(0)/dt$. If $x_c(t)$ is real, K_1 and K_2 will be complex and $K_2 = K_1^*$. $A_1 = K_1 + K_2$ is therefore 2 times the real part of K_1, and $A_2 = jK_1 - jK_2$ is -2 times the imaginary part of K_1. A_1 and A_2 are the real numbers. This illustrates that the natural response is an exponentially damped oscillatory response.

Case 3, $\zeta = 1$. This case, called *critically damped*, results in

$$s_1 = s_2 = -\zeta \omega_o$$

as shown in Eq. (7.6). Therefore, Eq. (7.7) reduces to

$$x_c(t) = K_3 e^{-\zeta \omega_o t}$$

where $K_3 = K_1 + K_2$. However, this cannot be a solution to the second-order differential Eq. (7.3) because in general it is not possible to satisfy the two initial conditions $x(0)$ and $dx(0)/dt$ with the single constant K_3.

In the case where the characteristic equation has repeated roots, a solution can be obtained in the following manner. If $x_1(t)$ is known to be a solution of the second-order homogeneous equation, then via the substitution $x(t) = x_1(t)y(t)$ we can transform the given differential equation into a first-order equation in $dy(t)/dt$. Since this resulting equation is only a function of $y(t)$, it can be solved to find the general solution $x(t) = x_1(t)y(t)$.

For the present case, $s_1 = s_2 = -\zeta \omega_o$. For simplicity we let $\alpha = \zeta \omega_o$ and, hence, the basic equation is

$$\frac{d^2 x(t)}{dt^2} + 2\alpha \frac{dx(t)}{dt} + \alpha^2 x(t) = 0 \qquad\qquad \textbf{7.10}$$

and one known solution is

$$x_1(t) = K_3 e^{-\alpha t}$$

By employing the substitution

$$x_2(t) = x_1(t)y(t) = K_3 e^{-\alpha t} y(t)$$

Equation (7.10) becomes

$$\frac{d^2}{dt^2}\big[K_3 e^{-\alpha t} y(t)\big] + 2\alpha \frac{d}{dt}\big[K_3 e^{-\alpha t} y(t)\big] + \alpha^2 K_3 e^{-\alpha t} y(t) = 0$$

Evaluating the derivatives, we obtain

$$\frac{d}{dt}\left[K_3 e^{-\alpha t} y(t)\right] = -K_3 \alpha e^{-\alpha t} y(t) + K_3 e^{-\alpha t}\frac{dy(t)}{dt}$$

$$\frac{d^2}{dt^2}\left[K_3 e^{-\alpha t} y(t)\right] = K_3 \alpha^2 e^{-\alpha t} y(t) - 2K_3 \alpha e^{-\alpha t}\frac{dy(t)}{dt} + K_3 e^{-\alpha t}\frac{d^2 y(t)}{dt^2}$$

Substituting these expressions into the preceding equation yields

$$K_3 e^{-\alpha t}\frac{d^2 y(t)}{dt^2} = 0$$

Therefore,

$$\frac{d^2 y(t)}{dt^2} = 0$$

and hence,

$$y(t) = A_1 + A_2 t$$

Therefore, the general solution is

$$x_2(t) = x_1(t)y(t)$$
$$= K_3 e^{-\alpha t}(A_1 + A_2 t)$$

which can be written as

$$x_2(t) = B_1 e^{-\zeta \omega_o t} + B_2 t e^{-\zeta \omega_o t} \qquad \textbf{7.11}$$

where B_1 and B_2 are constants derived from the initial conditions.

It is informative to sketch the natural response for the three cases we have discussed: overdamped, Eq. (7.8); underdamped, Eq. (7.9); and critically damped, Eq. (7.11). Figure 7.2 graphically illustrates the three cases for the situations in which $x_c(0) = 0$. Note that the critically damped response peaks and decays faster than the overdamped response. The underdamped response is an exponentially damped sinusoid whose rate of decay is dependent on the factor ζ. Actually, the terms $\pm e^{-\zeta \omega_o t}$ define what is called the *envelope* of the response, and the damped oscillations (i.e., the oscillations of decreasing amplitude) exhibited by the waveform in Fig. 7.2b are called *ringing*.

DRILL

D7.4   Determine the general form of the solution of the equation

$$\frac{d^2 x(t)}{dt^2} + \frac{4 dx(t)}{dt} + 4 x(t) = 0$$

ANSWER:

$$x_c(t) = B_1 e^{-2t} + B_2 t e^{-2t}$$

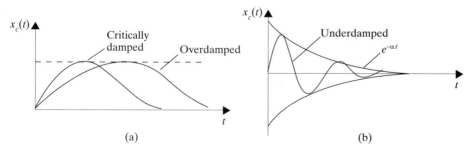

Figure 7.2 Comparison of overdamped, critically damped, and underdamped responses.

E7.1   A parallel RLC circuit has the following circuit parameters: $R = 1\,\Omega$, $L = 2$ H, and $C = 2$ F. Compute the damping ratio and the undamped natural frequency of this network.

ANSWER: $\zeta = 0.5, \omega_n = 0.5$ rad/s.

E7.2   A series RLC circuit consists of $R = 2\,\Omega$, $L = 1$ H, and a capacitor. Determine the type of response exhibited by the network if (a) $C = \frac{1}{2}$F, (b) $C = 1$ F, and (c) $C = 2$ F.

ANSWER: (a) underdamped, (b) critically damped, (c) overdamped.

7.3 THE NETWORK RESPONSE

We will now analyze a number of simple RLC networks that contain both nonzero initial conditions and constant forcing functions. Circuits that exhibit overdamped, underdamped, and critically damped responses will be considered. The approach will illustrate not only the differential equation techniques, but also the manner in which equivalent circuits, valid at $t = 0+$ and $t = \infty$, are used in the analysis.

Problem-Solving Strategy for Second-Order Transient Circuits

- Write the differential equation that describes the circuit.
- Derive the characteristic equation, which can be written in the form $s^2 + 2\zeta\omega_o s + \omega_o^2 = 0$, where ζ is the damping ratio and ω_o is the undamped resonant frequency.
- The two roots of the characteristic equation will determine the type of response. If the roots are real and unequal (i.e., $\zeta > 1$), the network response is overdamped. If the roots are real and equal (i.e., $\zeta = 1$), the network response is critically damped. If the roots are complex (i.e., $\zeta < 1$), the network response is underdamped.
- The damping condition and corresponding response for the aforementioned three cases outlined are as follows:

  Overdamped: $x(t) = K_1 e^{-(\zeta\omega_o - \omega_o\sqrt{\zeta^2-1})t} + K_2 e^{-(\zeta\omega_o + \omega_o\sqrt{\zeta^2-1})t}$

  Critically damped: $x(t) = B_1 e^{-\zeta\omega_o t} + B_2 t e^{-\zeta\omega_o t}$

  Underdamped: $x(t) = e^{-\sigma t}(A_1 \cos \omega_d t + A_2 \sin \omega_d t)$, where $\sigma = \zeta\omega_o$, and $\omega_d = \omega_o\sqrt{1 - \zeta^2}$

- Two initial conditions, either given or derived, are required to obtain the two unknown coefficients in the response equation.

The following examples will serve to demonstrate the analysis techniques.

EXAMPLE 7.1

Consider the parallel RLC circuit shown in Fig. 7.3. The second-order differential equation that describes the voltage $v(t)$ is

$$\frac{d^2v}{dt^2} + \frac{1}{RC}\frac{dv}{dt} + \frac{v}{LC} = 0$$

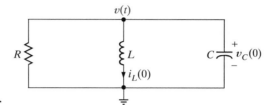

Figure 7.3 Parallel RLC circuit.

A comparison of this equation with Eqs. (7.3) and (7.4) indicates that for the parallel RLC circuit the damping term is $1/2RC$ and the resonant frequency is $1/\sqrt{LC}$. If the circuit parameters are $R = 2\,\Omega, C = \frac{1}{5}\text{F}$, and $L = 5\,\text{H}$, the equation becomes

$$\frac{d^2v}{dt^2} + 2.5\frac{dv}{dt} + v = 0$$

Let us assume that the initial conditions on the storage elements are $i_L(0) = -1\,\text{A}$ and $v_C(0) = 4\,\text{V}$.

Let us find the node voltage $v(t)$ and the inductor current.

SOLUTION   The characteristic equation for the network is

$$s^2 + 2.5s + 1 = 0$$

and the roots are

$$s_1 = -2$$
$$s_2 = -0.5$$

Since the roots are real and unequal, the circuit is overdamped, and $v(t)$ is of the form

$$v(t) = K_1e^{-2t} + K_2e^{-0.5t}$$

The initial conditions are now employed to determine the constants K_1 and K_2. Since $v(t) = v_C(t)$,

$$v_C(0) = v(0) = 4 = K_1 + K_2$$

The second equation needed to determine K_1 and K_2 is normally obtained from the expression

$$\frac{dv(t)}{dt} = -2K_1e^{-2t} - 0.5K_2e^{-0.5t}$$

However, the second initial condition is not $dv(0)/dt$. If this were the case, we would simply evaluate the equation at $t = 0$. This would produce a second equation in the unknowns K_1 and K_2. We can, however, circumvent this problem by noting that the node equation for the circuit can be written as

$$C\frac{dv(t)}{dt} + \frac{v(t)}{R} + i_L(t) = 0$$

or

$$\frac{dv(t)}{dt} = \frac{-1}{RC}v(t) - \frac{i_L(t)}{C}$$

At $t = 0$,

$$\frac{dv(0)}{dt} = \frac{-1}{RC}v(0) - \frac{1}{C}i_L(0)$$
$$= -2.5(4) - 5(-1)$$
$$= -5$$

But since

$$\frac{dv(t)}{dt} = -2K_1e^{-2t} - 0.5K_2e^{-0.5t}$$

then when $t = 0$

$$-5 = -2K_1 - 0.5K_2$$

This equation, together with the equation

$$4 = K_1 + K_2$$

produces the constants $K_1 = 2$ and $K_2 = 2$. Therefore, the final equation for the voltage is

$$v(t) = 2e^{-2t} + 2e^{-0.5t} \text{ V}$$

Note that the voltage equation satisfies the initial condition $v(0) = 4$ V. The response curve for this voltage $v(t)$ is shown in Fig. 7.4.

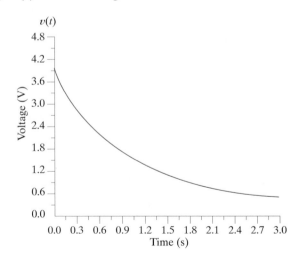

Figure 7.4 Overdamped response.

The inductor current is related to $v(t)$ by the equation

$$i_L(t) = \frac{1}{L} \int v(t)\, dt$$

Substituting our expression for $v(t)$ yields

$$i_L(t) = \frac{1}{5} \int \left[2e^{-2t} + 2e^{-0.5t} \right] dt$$

or

$$i_L(t) = -\frac{1}{5} e^{-2t} - \frac{4}{5} e^{-0.5t} \text{ A}$$

Note that in comparison with the RL and RC circuits we analyzed in Chapter 6, the response of this RLC circuit is controlled by two time constants. The first term has a time constant of $\frac{1}{2}$, and the second term has a time constant of 2.   ❑

EXAMPLE 7.2

The series RLC circuit shown in Fig. 7.5 has the following parameters: $C = 0.04$ F, $L = 1$ H, $R = 6\ \Omega$, $i_L(0) = 4$ A, and $v_C(0) = -4$ V. The equation for the current in the circuit is given by the expression

$$\frac{d^2 i}{dt^2} + \frac{R}{L} \frac{di}{dt} + \frac{i}{LC} = 0$$

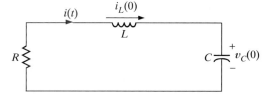

Figure 7.5 Series RLC circuit.

A comparison of this equation with Eqs. (7.3) and (7.4) illustrates that for a series RLC circuit the damping term is $R/2L$ and the resonant frequency is $1/\sqrt{LC}$. Substituting the circuit element values into the preceding equation yields

$$\frac{d^2 i}{dt^2} + 6 \frac{di}{dt} + 25i = 0$$

Let us determine the expression for both the current and the capacitor voltage.

SOLUTION   The characteristic equation is then

$$s_2 + 6s + 25 = 0$$

and the roots are

$$s_1 = -3 + j4$$
$$s_2 = -3 - j4$$

Since the roots are complex, the circuit is underdamped, and the expression for $i(t)$ is

$$i(t) = K_1 e^{-3t} \cos 4t + K_2 e^{-3t} \sin 4t$$

Using the initial conditions, we find that

$$i(0) = 4 = K_1$$

and

$$\frac{di}{dt} = -4K_1 e^{-3t} \sin 4t - 3K_1 e^{-3t} \cos 4t + 4K_2 e^{-3t} \cos 4t - 3K_2 e^{-3t} \sin 4t$$

and thus

$$\frac{di(0)}{dt} = -3K_1 + 4K_2$$

Although we do not know $di(0)/dt$, we can find it via KVL. From the circuit we note that

$$Ri(0) + L\frac{di(0)}{dt} + v_C(0) = 0$$

or

$$\frac{di(0)}{dt} = -\frac{R}{L}i(0) - \frac{v_C(0)}{L}$$

$$= -\frac{6}{1}(4) + \frac{4}{1}$$

$$= -20$$

Therefore,

$$-3K_1 + 4K_2 = -20$$

and since $K_1 = 4$, $K_2 = -2$, the expression then for $i(t)$ is

$$i(t) = 4e^{-3t} \cos 4t - 2e^{-3t} \sin 4t \text{ A}$$

Note that this expression satisfies the initial condition $i(0) = 4$. The voltage across the capacitor could be determined via KVL using this current:

$$Ri(t) + L\frac{di(t)}{dt} + v_C(t) = 0$$

or

$$v_C(t) = -Ri(t) - L\frac{di(t)}{dt}$$

Substituting the preceding expression for $i(t)$ into this equation yields

$$v_C(t) = -4e^{-3t} \cos 4t + 22e^{-3t} \sin 4t \text{ V}$$

Note that this expression satisfies the initial condition $v_C(0) = -4\,V$. A plot of the function $v_C(t)$ is shown in Fig. 7.6.

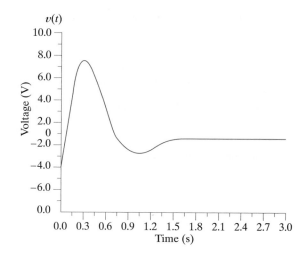

Figure 7.6 Underdamped response.

❑

EXAMPLE 7.3

Let us examine the circuit in Fig. 7.7, which is slightly more complicated than the two we have considered earlier. The two equations that describe the network are

$$L\frac{di(t)}{dt} + R_1 i(t) + v(t) = 0$$

$$i(t) = C\frac{dv(t)}{dt} + \frac{v(t)}{R_2}$$

Substituting the second equation into the first yields

$$\frac{d^2 v}{dt^2} + \left(\frac{1}{R_2 C} + \frac{R_1}{L}\right)\frac{dv}{dt} + \frac{R_1 + R_2}{R_2 LC}v = 0$$

If the circuit parameters and initial conditions are

$$R_1 = 10\,\Omega \qquad C = \tfrac{1}{8}\text{F} \qquad v_C(0) = 1\text{ V}$$
$$R_2 = 8\,\Omega \qquad L = 2\text{ H} \qquad i_L(0) = \tfrac{1}{2}\text{A}$$

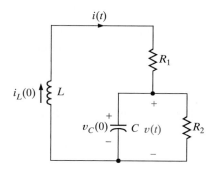

Figure 7.7 Series-parallel *RLC* circuit.

the differential equation becomes

$$\frac{d^2v}{dt^2} + 6\frac{dv}{dt} + 9v = 0$$

We wish to find expressions for the current $i(t)$ and the voltage $v(t)$.

SOLUTION The characteristic equation is then

$$s^2 + 6s + 9 = 0$$

and hence the roots are

$$s_1 = -3$$
$$s_2 = -3$$

Since the roots are real and equal, the circuit is critically damped. The term $v(t)$ is then given by the expression

$$v(t) = K_1e^{-3t} + K_2te^{-3t}$$

Since $v(t) = v_C(t)$,

$$v(0) = v_C(0) = 1 = K_1$$

In addition,

$$\frac{dv(t)}{dt} = -3K_1e^{-3t} + K_2e^{-3t} - 3K_2te^{-3t}$$

However,

$$\frac{dv(t)}{dt} = \frac{i(t)}{C} - \frac{v(t)}{R_2C}$$

Setting these two expressions equal to one another and evaluating the resultant equation at $t = 0$ yields

$$\frac{1/2}{1/8} - \frac{1}{1} = -3K_1 + K_2$$
$$3 = -3K_1 + K_2$$

Since $K_1 = 1, K_2 = 6$ and the expression for $v(t)$ is

$$v(t) = e^{-3t} + 6te^{-3t} \text{ V}$$

Note that the expression satisfies the initial condition $v(0) = 1$.

The current $i(t)$ can be determined from the nodal analysis equation at $v(t)$.

$$i(t) = C\frac{dv(t)}{dt} + \frac{v(t)}{R_2}$$

Substituting $v(t)$ from the preceding equation, we find

$$i(t) = \frac{1}{8}\left[-3e^{-3t} + 6e^{-3t} - 18te^{-3t}\right] + \frac{1}{8}\left[e^{-3t} + 6te^{-3t}\right]$$

or

$$i(t) = \tfrac{1}{2}e^{-3t} - \tfrac{3}{2}te^{-3t} \text{ A}$$

If this expression for the current is employed in the circuit equation,

$$v(t) = -L\frac{di(t)}{dt} - R_1 i(t)$$

we obtain

$$v(t) = e^{-3t} + 6te^{-3t} \text{ V}$$

which is identical to the expression derived earlier. This function is shown in Fig. 7.8.

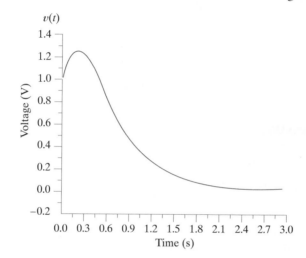

Figure 7.8 Critically damped response.

EXTENSION EXERCISES

E7.3   The switch in the network in Fig. E7.3 opens at $t = 0$. Find $i(t)$ for $t > 0$.

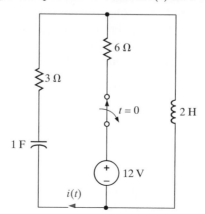

Figure E7.3

ANSWER: $i(t) = -2e^{-t/2} + 4e^{-t}$ A.

E7.4 The switch in the network in Fig. E7.4 moves from position 1 to position 2 at $t = 0$. Find $v_o(t)$ for $t > 0$.

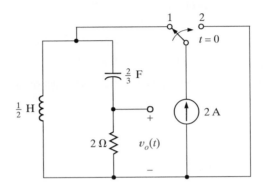

Figure E7.4

ANSWER: $v_o(t) = 2(e^{-t} - 3e^{-3t})$ V.

EXAMPLE 7.4

Consider the circuit shown in Fig. 7.9. This circuit is the same as that analyzed in Example 7.2, except that a constant forcing function is present. The circuit parameters are the same as those used in Example 7.2:

$$C = 0.04 \text{ F} \qquad i_L(0) = 4 \text{ A}$$
$$L = 1 \text{ H} \qquad v_C(0) = -4 \text{ V}$$
$$R = 6 \, \Omega$$

We want to find an expression for $v_C(t)$ for $t > 0$.

SOLUTION From our earlier mathematical development we know that the general solution of this problem will consist of a particular solution plus a complementary solution. From Example 7.2 we know that the complementary solution is of the form $K_3 e^{-3t} \cos 4t + K_4 e^{-3t} \sin 4t$. The particular solution is a constant, since the input is a constant and therefore the general solution is

$$v_C(t) = K_3 e^{-3t} \cos 4t + K_4 e^{-3t} \sin 4t + K_5$$

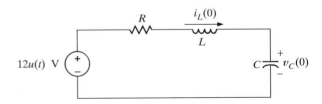

Figure 7.9 Series *RLC* circuit with a step function input.

An examination of the circuit shows that in the steady state the final value of $v_C(t)$ is 12 V, since in the steady-state condition, the inductor is a short circuit and the capacitor is an open circuit. Thus, $K_5 = 12$. The steady-state value could also be immediately calculated from the differential equation. The form of the general solution is then

$$v_C(t) = K_3 e^{-3t} \cos 4t + K_4 e^{-3t} \sin 4t + 12$$

The initial conditions can now be used to evaluate the constants K_3 and K_4.

$$v_C(0) = -4 = K_3 + 12$$
$$-16 = K_3$$

Since the derivative of a constant is zero, the results of Example 7.2 show that

$$\frac{dv_C(0)}{dt} = \frac{i(0)}{C} = 100 = -3K_3 + 4K_4$$

and since $K_3 = -16$, $K_4 = 13$. Therefore, the general solution for $v_C(t)$ is

$$v_C(t) = 12 - 16e^{-3t} \cos 4t + 13e^{-3t} \sin 4t \text{ V}$$

Note that this equation satisfies the initial condition $v_C(0) = -4$ and the final condition $v_C(\infty) = 12$ V.   ❑

EXAMPLE 7.5

Let us examine the circuit shown in Fig. 7.10. A close examination of this circuit will indicate that it is identical to that shown in Example 7.3 except that a constant forcing function is present. We assume the circuit is in steady state at $t = 0-$. The equations that describe the circuit for $t > 0$ are

$$L\frac{di(t)}{dt} + R_1 i(t) + v(t) = 24$$

$$i(t) = C\frac{dv(t)}{dt} + \frac{v(t)}{R_2}$$

Combining these equations, we obtain

$$\frac{d^2 v(t)}{dt^2} + \left(\frac{1}{R_2 C} + \frac{R_1}{L}\right)\frac{dv(t)}{dt} + \frac{R_1 + R_2}{R_2 LC} v(t) = \frac{24}{LC}$$

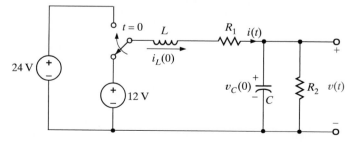

Figure 7.10 Series-parallel RLC circuit with a constant forcing function.

If the circuit parameters are $R_1 = 10\,\Omega$, $R_2 = 2\,\Omega$, $L = 2\,\text{H}$, and $C = \frac{1}{4}\text{F}$, the differential equation for the output voltage reduces to

$$\frac{d^2 v(t)}{dt^2} + 7\frac{dv(t)}{dt} + 12v(t) = 48$$

Let us determine the output voltage $v(t)$.

SOLUTION The characteristic equation is

$$s^2 + 7s + 12 = 0$$

and hence the roots are

$$s_1 = -3$$

$$s_2 = -4$$

The circuit response is overdamped, and therefore the general solution is of the form

$$v(t) = K_1 e^{-3t} + K_2 e^{-4t} + K_3$$

The steady-state value of the voltage, K_3, can be computed from Fig. 7.11a. Note that

$$v(\infty) = 4\,\text{V} = K_3$$

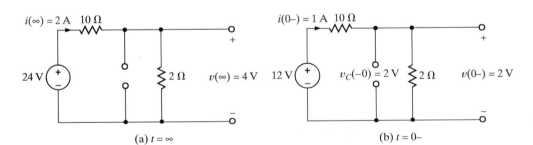

(a) $t = \infty$ (b) $t = 0-$

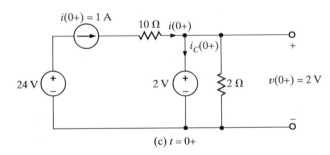

(c) $t = 0+$

Figure 7.11 Equivalent circuits at $t = \infty$, $t = 0-$, and $t = 0+$ for the circuit in Fig. 7.10.

The initial conditions can be calculated from Figs. 7.11b and c, which are valid at $t = 0-$ and $t = 0+$, respectively. Note that $v(0+) = 2$ V and, hence, from the response equation

$$v(0+) = 2 \text{ V} = K_1 + K_2 + 4$$
$$-2 = K_1 + K_2$$

Figure 7.11c illustrates that $i(0+) = 1$. From the response equation we see that

$$\frac{dv(0)}{dt} = -3K_1 - 4K_2$$

and since

$$\frac{dv(0)}{dt} = \frac{i(0)}{C} - \frac{v(0)}{R_2 C}$$
$$= 4 - 4$$
$$= 0$$

then

$$0 = -3K_1 - 4K_2$$

Solving the two equations for K_1 and K_2 yields $K_1 = -8$ and $K_2 = 6$. Therefore, the general solution for the voltage response is

$$v(t) = 4 - 8e^{-3t} + 6e^{-4t} \text{ V}$$

Note that this equation satisfies both the initial and final values of $v(t)$.   ❑

EXTENSION EXERCISE

E7.5   The switch in the network in Fig. E7.5 moves from position 1 to position 2 at $t = 0$. Compute $i_o(t)$ for $t > 0$ and use this current to determine $v_o(t)$ for $t > 0$.

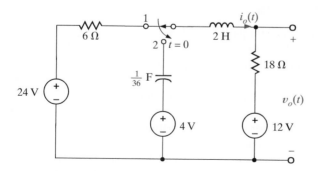

Figure E7.5

ANSWER: $i_o(t) = -\frac{11}{6}e^{-3t} + \frac{14}{6}e^{-6t}$ A, $v_o(t) = 12 + 18i_o(t)$ V.

The examples clearly indicate that the general solution of even very simple *RLC* circuits can be quite complicated. Although we can continue our current mathematical analysis and present techniques for solving more general networks, better methods exist. For example, in Chapter 14 the Laplace transform technique will be applied to analyze *RLC* circuits with a variety of forcing functions, and the reader will find that the transform technique is not only easier to use but provides some additional insight in the analysis.

7.4 TRANSIENT PSPICE ANALYSIS USING SCHEMATIC CAPTURE

Introduction

In transient analysis, we determine voltages and currents as functions of time. Typically, the time dependence is demonstrated by plotting the waveforms using time as the independent variable. PSPICE can perform this kind of analysis, called a TRANSIENT simulation, in which all voltages and currents are determined over a specified time duration. To facilitate plotting, PSPICE uses what is known as the PROBE utility, which will be described later. As an introduction to transient analysis, let us simulate the circuit in Fig. 7.12, plot the voltage $v_C(t)$ and the current $i(t)$, and extract the time constant. While we will introduce some new PSPICE topics in this section, *Schematics* fundamentals such as getting parts and wiring and editing part names and values have already been covered in Chapter 4. Also, uppercase text refers to PSPICE utilities and dialog boxes while boldface denotes keyboard or mouse inputs.

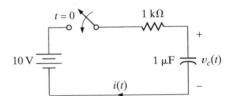

Figure 7.12 A circuit used for *Transient* simulation.

The Switch Parts

The inductor and capacitor parts are called *L* and *C*, respectively, and are in the ANALOG library. The switch, called SW_TCLOSE, is in the EVAL library. There is also a SW_TOPEN part that models an opening switch. After placing and wiring the switch along with the other parts, the *Schematics* circuit appears as that shown in Fig. 7.13.

To edit the switch's attributes, double click on the switch symbol, and the ATTRIBUTES box shown in Fig. 7.14 will appear. Deselecting the **Include Non-changeable Attributes** and **Include System-defined Attributes** fields limits the attribute list to those we can edit and is highly recommended.

The attribute **tClose** is the time at which the switch begins to close and **ttran** is the time required to complete the closure. Switch attributes **Rclosed** and **Ropen** are the switch's

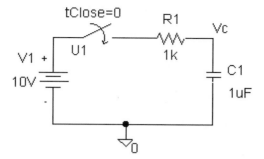

Figure 7.13 The *Schematics* circuit.

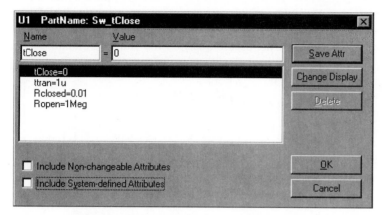

Figure 7.14 The switch's ATTRIBUTES box.

resistance in the closed and open positions, respectively. During simulations, the resistance of the switch changes linearly from **Ropen** at $t =$ **tClose** to **Rclosed** at $t =$ **tClose** + **ttran**.

When using the SW_TCLOSE and SW_TOPEN parts to simulate ideal switches, care should be taken to ensure that the values for **ttran**, **Rclosed**, and **Ropen** are appropriate for valid simulation results. In our present example, we see that the switch and R_1 are in series; thus, their resistances add. Using the default values listed in Figure 7.14, we find that when the switch is closed, the switch resistance, **Rclosed**, is 0.01 Ω, 100,000 times smaller than that of the resistor. The resulting series equivalent resistance is essentially that of the resistor. Alternatively, when the switch is open, the switch resistance is 1 MΩ, 1000 times larger than that of the resistor. Now the equivalent resistance is much larger than that of the resistor. Both are desirable scenarios.

To determine a reasonable value for **ttran**, we first estimate the duration of the transient response. The component values yield a time constant of 1 ms, and thus all voltages and currents will reach steady state in about 5 ms. For accurate simulations, **ttran** should be much less than 5 ms. Therefore, the default value of 1 μs is viable in this case.

The Importance of Pin Numbers

As mentioned in Chapter 4, each component within the various Parts libraries has two or more terminals. Within PSPICE, these terminals are called pins and are numbered sequentially starting with pin 1, as shown in Fig. 7.15 for several two-terminal parts. The significance of the pin numbers is their effect on currents plotted using the PROBE utility. PROBE always plots the current entering pin 1 and exiting pin 2. So if the current through an element is to be plotted, the part should be oriented in the *Schematics* diagram such that the defined current direction enters the part at pin 1. This can be done by using the ROTATE command in the EDIT menu. ROTATE causes the part to spin 90° counterclockwise. In our example, we will plot the current $i(t)$ by plotting the current through the capacitor, I(C1). Therefore, when the *Schematics* circuit in Fig. 7.13 was created, the capacitor was rotated 270°. As a result, pin 1 is at the top of the diagram and the assigned current direction in Fig. 7.12 matches the direction presumed by PROBE. If a component's current direction in PROBE is opposite the desired direction, simply go to the *Schematics* circuit, rotate the part in question 180°, and resimulate.

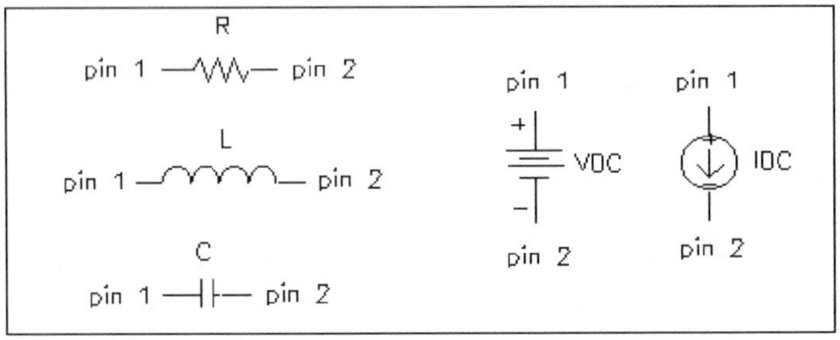

Figure 7.15 Pin numbers for common PSPICE parts.

Setting Initial Conditions

To set the initial condition of the capacitor voltage, double click on the capacitor symbol in Fig. 7.13 to open its ATTRIBUTE box, as shown in Fig. 7.16. Click on the **IC** field and set the value to the desired voltage, 0 V in this example. Setting the initial condition on an inductor current is done in a similar fashion. Be forewarned that the initial condition for a capacitor voltage is positive at pin 1 versus pin 2. Similarly, the initial condition for an inductor's current will flow into pin 1 and out of pin 2.

Setting Up a Transient Analysis

The simulation duration is selected using **SETUP** from the **ANALYSIS** menu. When the SETUP window shown in Fig. 7.17 appears, double click on the text **TRANSIENT** and the TRANSIENT window in Fig. 7.18 will appear. The simulation period described by **Final time** is selected as 6 ms. All simulations start at $t = 0$. The **No-Print Delay** field sets the time the simulation runs before data collection begins. **Print Step** is the interval used for printing data to the output file. **Print Step** has no effect on the data used to create PROBE plots. The **Detailed Bias Pt.** option is useful when simulating circuits containing transistors and diodes, and thus will not be used here. When **Skip initial transient solution** is en-

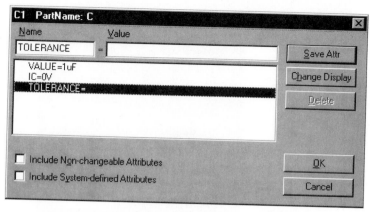

Figure 7.16 Setting the capacitor initial condition.

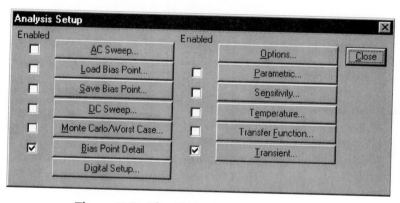

Figure 7.17 The ANALYSIS SETUP window.

abled, all capacitors and inductors that do not have specific initial condition values in their ATTRIBUTES boxes will use zero initial conditions.

Sometimes plots created in PROBE are not smooth. This is caused by an insufficient number of data points. More data points can be requested by inputting a **Step Ceiling** value. A reasonable first guess would be a hundredth of the **Final Time**. If the resulting PROBE plots are still unsatisfactory, reduce the **Step Ceiling** further. As soon as the TRANSIENT window is complete, simulate the circuit by selecting Simulate and the ANALYSIS menu.

Plotting in PROBE

When the PSPICE simulation is finished, the PROBE window shown in Fig. 7.19 will open. If not, select **Run Probe** from the **Analysis** menu. In order to plot the voltage, $v_C(t)$, select **ADD** from the **TRACE** menu. The ADD TRACES window is shown in Fig. 7.20. Notice that the options **Alias Names** and **Subcircuit Nodes** have been deselected, which

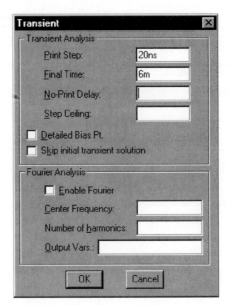

Figure 7.18 The TRANSIENT window.

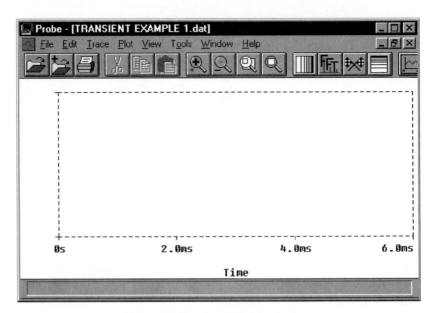

Figure 7.19 The PROBE window.

greatly simplifies the ADD TRACES window. The capacitor voltage is obtained by clicking on V(Vc) in the left column. The PROBE window should look like that shown in Fig. 7.21.

Before adding the current $i(t)$ to the plot, we note that the dc source is 10 V and the resistance is 1 kΩ, which results in a loop current of a few milliamps. Since the capacitor

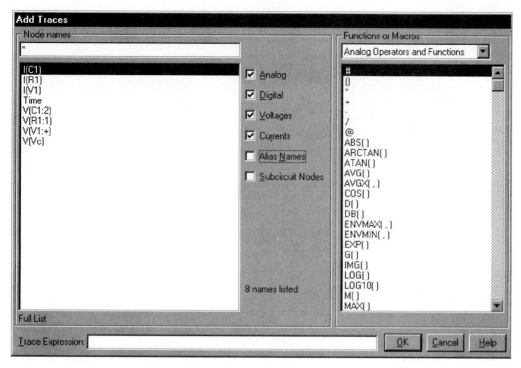

Figure 7.20 The ADD TRACES window.

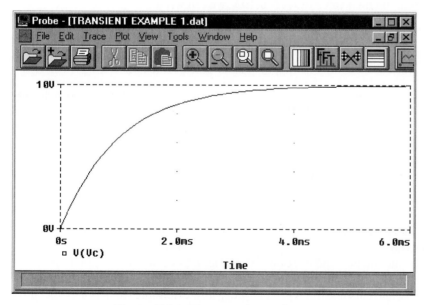

Figure 7.21 The capacitor voltage.

voltage span is much greater, we will plot the current on a second *y*-axis. From the **PLOT** menu, select **ADD Y AXIS**. To add the current to the plot, select **ADD** from the **TRACE** menu, then select I(C1). Figure 7.22 shows the PROBE plot for $v_C(t)$ and $i(t)$.

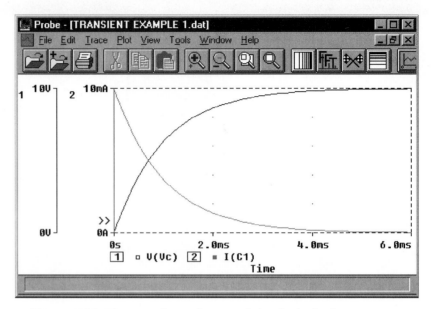

Figure 7.22 The capacitor voltage and the clockwise loop current.

Finding the Time Constant

Given that the final value of $v_C(t)$ is 10 V, we can write

$$v_C(t) = 10\left[1 - e^{-t/\tau}\right] \text{ V}$$

When $t = \tau$,

$$v_C(\tau) = 10\left(1 - e^{-1}\right) \text{ V} = 6.32 \text{ V}$$

In order to determine the time at which the capacitor voltage is 6.32 V, we activate the cursors by selecting **CURSORS/DISPLAY** in the **TOOLS** menu. There are two cursors that can be used to extract *x*-*y* data from the plots. Use the ← and → arrow keys to move the first cursor. By holding the SHIFT key down, the arrow keys move the second cursor. Moving the first cursor along the voltage plot, as shown in Fig. 7.23, we find that 6.32 V occurs at a time of 1 ms. Therefore, the time constant is 1 ms—exactly the RC product.

Saving and Printing PROBE Plots

Saving plots within PROBE requires the use of the **DISPLAY CONTROL** command in the **TOOLS** menu. Using the SAVE/RESTORE DISPLAY window shown in Fig. 7.24, we simply name the plot and click on **SAVE**. Fig. 7.24 shows that one plot has already been saved; TranExample. This procedure saves the plot attributes such as axes settings additional text, and cursor settings in a file with a .prb extension. Additionally, the .prb file contains a reference to the appropriate data file. The data file, which has a .dat extension,

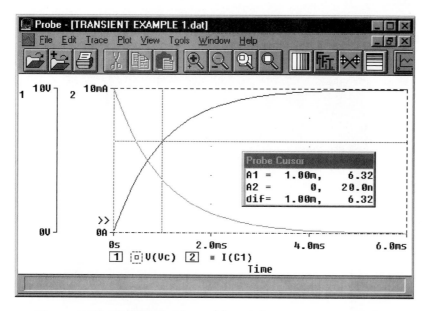

Figure 7.23 PROBE plot for $v_C(t)$ and time constant extraction.

contains the actual simulation results. Therefore, in using **DISPLAY CONTROL** to save a plot, we do not save the actual data on the plot, only the plot settings and the .dat file's name. To access an old PROBE plot, enter PROBE and select **DISPLAY CONTROL**. In the **DISPLAY CONTROL** window, select the file of interest and click on **RESTORE**. Using the **SAVE AS** option in Fig. 7.24 allows you to save the .prb file to any directory on any disk, hard or floppy.

To copy the PROBE plot to other programs such as word processors, use the **COPY TO CLIPBOARD** command in the **TOOLS** menu. To print a PROBE plot, select **PRINT**

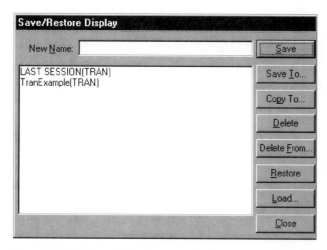

Figure 7.24 The SAVE/RESTORE DISPLAY window used in PROBE to save plots.

from the **FILE** menu and the PRINT window in Fig. 7.25 will open. The options in this window are self-explanatory.

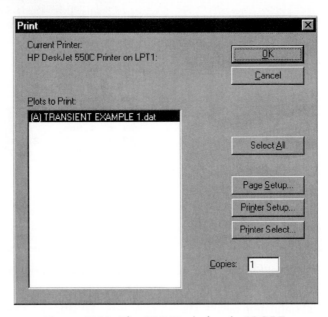

Figure 7.25 The PRINT window in PROBE.

Other PROBE Features

There are several features within PROBE for plot manipulation and data extraction. Within the **PLOT** menu reside commands for editing the plot itself. These include altering the axes, adding axes, and adding additional plots to the page. Also in the **TOOLS** menu is the **LABEL** command, which allows one to add explanatory text, lines, and shapes to the plot.

EXAMPLE 7.6

Using the PSPICE *Schematics* editor, draw the circuit in Fig. 7.26, and use the PROBE utility to find the time at which the capacitor and inductor current are equal.

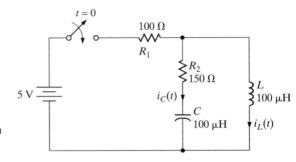

Figure 7.26 Circuit used in Example 7.6.

SOLUTION   Figure 7.27 shows the *Schematics* circuit, while the simulation results are shown in Fig. 7.28. Based on the PROBE plot, the currents are equal at 556 ns.

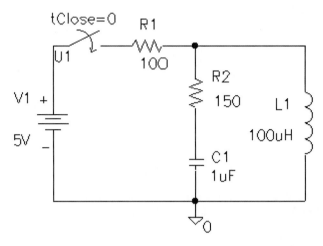

Figure 7.27

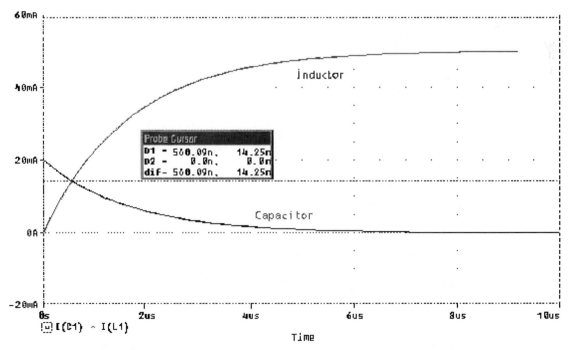

Figure 7.28 Simulation results for $i_C(t)$ and $i_L(t)$.

EXAMPLE 7.7

Using the PSPICE *Schematics* editor, draw the circuit in Fig. 7.29, and use the PROBE utility to plot $v_C(t)$ and $v_L(t)$. At what time after switching does $v_C(t)$ reach half its maximum value?

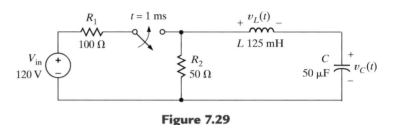

Figure 7.29

SOLUTION   Figure 7.30 shows the *Schematics* circuit, while simulation results are shown in Fig. 7.31. From the plot, we see that the maximum value of $v_C(t)$ is 40 V, which occurs at $t \leq 1$ ms. The capacitor voltage has decreased to 20 V at 4.236 ms, which is 3.236 ms after switching.

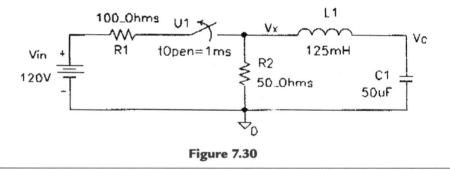

Figure 7.30

7.5 APPLICATIONS

Although there is a wide variety of practical applications for RLC circuits, we concentrate our discussion here on a particular class of circuits in which power is supplied to a network by a discharged capacitor.

EXAMPLE 7.8

Let us describe the use of a discharged capacitor in what is known as a *slapper* detonator.

SOLUTION   A device known as a capacitive discharge unit (CDU) is designed to produce very high power (e.g., megawatts in a very short time such as nanoseconds). One practical application of the CDU is its use as what is called a slapper detona-

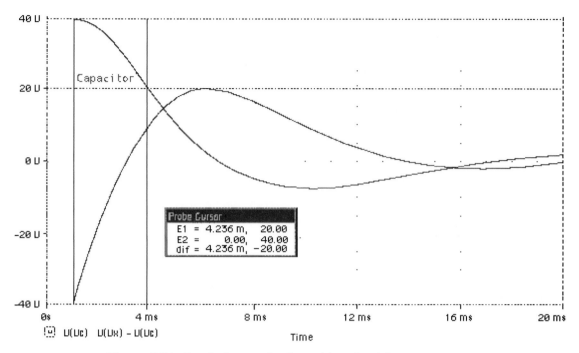

Figure 7.31 Simulation results for $v_C(t)$ and $v_L(t)$.

tor. The slapper is shown in Fig. 7.32a and consists of a polished insulating substance that is selectively coated with copper in the form of a *bowtie*. Over the center of the bowtie is placed a polyimide material called a *flyer*. The device is positioned close to an explosive pellet, as shown in Fig. 7.32a. When a large current is discharged through the bowtie, the foil heats so quickly that it explodes and the pressure produced by the explosion propels the thin piece of polyimide upward at velocities of 3 to 7 km/s. When the flyer impacts the explosive pellet, the kinetic energy of the flyer is imparted to the pellet and denotation results.

The CDU is modeled as shown in Fig. 7.32b. The resistor represents the central region of the bowtie, the inductor represents the inductance in the circuit, the switch is typically a spark-gap, and the charged capacitor is the energy source. Typical values for the components are shown in the figure. A plot of the current as a function of time following switch closure is shown in Fig. 7.33. An approximate value of the average power delivered by the device can be obtained by dividing the total energy stored in the capacitor by the total time of the discharge. The total energy stored in the capacitor is approximately 1 J and, as shown in Fig. 7.33, this energy is discharged in about 1 μs and 1 J/1μs is a megawatt.   ❏

EXAMPLE 7.9

An experimental schematic for a railgun is shown in Fig. 7.34. With switch sw-2 open, switch sw-1 is closed and the power supply charges the capacitor bank to 10 kV.

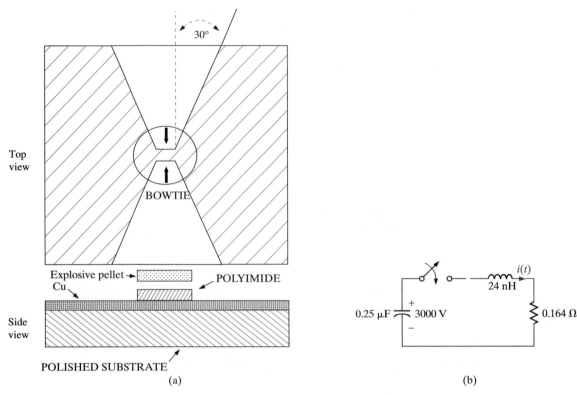

Figure 7.32 Slapper detonator and its equivalent circuit.

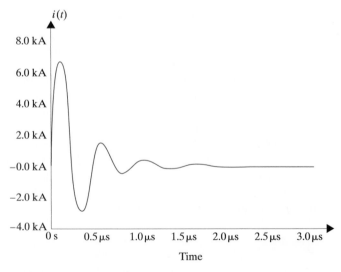

Figure 7.33 Detonator current as a function of time.

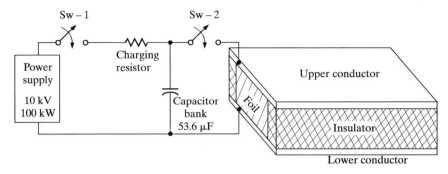

Figure 7.34 Experimental schematic for a railgun.

Then switch sw-1 is opened. The railgun is fired by closing switch sw-2. When the capacitor discharges, the current causes the foil at the end of the gun to explode, creating a hot plasma that accelerates down the tube. The voltage drop in vaporizing the foil is negligible and, therefore, more than 95% of the energy remains available for accelerating the plasma. The current flow establishes a magnetic field, and the force on the plasma caused by the magnetic field, which is proportional to the square of the current at any instant of time, accelerates the plasma. The higher the initial voltage, the more the acceleration.

The circuit diagram for the discharge circuit is shown in Fig. 7.35. The resistance of the bus (a heavy conductor) includes the resistance of the switch. The resistance of the foil and resultant plasma is negligible and, therefore, the current flowing between the upper and lower conductors is dependent upon the remaining circuit components in the closed path, as specified in Fig. 7.34.

The differential equation for the natural response of the current is

$$\frac{d^2 i(t)}{dt^2} + \frac{R_{\text{bus}}}{L_{\text{bus}}} \frac{di(t)}{dt} + \frac{i(t)}{L_{\text{bus}} C} = 0$$

Let us use the characteristic equation to describe the current waveform.

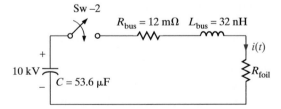

Figure 7.35 Railgun discharge circuit.

SOLUTION   Using the circuit values, the characteristic equation is

$$s^2 + 37.5 \times 10^4 s + 58.3 \times 10^{10} = 0$$

and the roots of the equation are

$$s_1, s_2 = (-18.75 \pm j74) \times 10^4$$

and hence the network is underdamped.

The roots of the characteristic equation illustrate that the damped resonant frequency is

$$\omega_d = 740 \text{ krad/s}$$

Therefore,

$$f_d = 118 \text{ kHz}$$

and the period of the waveform is

$$T = \frac{1}{f_d} = 8.5 \text{ μs}$$

An actual plot of the current is shown in Fig. 7.36, and this plot verifies that the period of the damped response is indeed 8.5 μs.

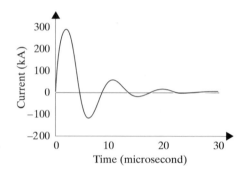

Figure 7.36 Load current with a capacitor bank charged to 10 kV.

Two additional applications for the CDU are the automotive airbag and the exploding bolts used by NASA to provide for reliable stage separation in a multistage rocket.

In the airbag application, a thin wire is welded between two pins. The wire is surrounded by a pyrotechnic mix. Upon impact, a current is passed through the wire via the pins. When the wire heats sufficiently to ignite the mix, a large volume of cool gas is produced, and the bag expands.

NASA employs a number of bolts, which are electrically insulated from the rocket, to provide the mechanical support necessary to hold two stages together. Each bolt is the resistive load in a CDU. At the time of separation, a common fire-control signal is sent to the switches in the CDUs and the energy in the capacitor is discharged through the bolt, causing it to vaporize. All bolts explode simultaneously and stage separation occurs. Clearly all bolts must release simultaneously to avoid a catastrophic accident.

7.6 CIRCUIT DESIGN

The following examples provide an introduction to some typical design strategies for second-order circuits. In these examples the problem involves the selection of the proper circuit parameters to achieve a specified transient response.

EXAMPLE 7.10

The network in Fig. 7.37 models an automobile ignition system. The voltage source represents the standard 12-V battery. The inductor is the ignition coil, which is magnetically coupled to the starter (not shown). The inductor's internal resistance is modeled by the resistor and the switch is the keyed ignition switch. Initially the switch connects the ignition circuitry to the battery, and thus the capacitor is charged to 12 V. To start the motor, we close the switch, thereby discharging the capacitor through the inductor. Assuming that optimum starter operation requires an overdamped response for $i_L(t)$ that reaches at least 1 A within 100 ms after switching, and remains above 1 A for between 1 and 1.5 s, let us find a value for the capacitor that will produce such a current waveform. In addition, let us plot the response including the time interval just prior to moving the switch and verify our design.

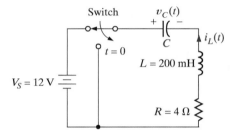

Figure 7.37 Circuit model for ignition system.

SOLUTION   Before the switch is moved at $t = 0$, the capacitor looks like an open circuit and the inductor acts like a short circuit. Thus,

$$i_L(0^-) = i_L(0^+) = 0 \text{ A} \quad \text{and} \quad v_C(0^-) = v_C(0^+) = 12 \text{ V}$$

After switching, the circuit is a series RLC unforced network described by the characteristic equation

$$s^2 + \frac{R}{L}s + \frac{1}{LC} = 0$$

with roots at $s = -s_1$ and $-s_2$. The characteristic equation is of the form

$$(s + s_1)(s + s_2) = s^2 + (s_1 + s_2)s + s_1 s_2 = 0$$

Comparing the two expressions, we see that

$$\frac{R}{L} = s_1 + s_2 = 20$$

and

$$\frac{1}{LC} = s_1 s_2$$

Since the network must be overdamped, the inductor current is of the form

$$i_L(t) = K_1 e^{-s_1 t} + K_2 e^{-s_2 t}$$

Just after switching,

$$i_L(0^+) = K_1 + K_2 = 0$$

or

$$K_2 = -K_1$$

Also, at $t = 0^+$, the inductor voltage equals the capacitor voltage because $i_L = 0$ and therefore $i_L R = 0$. Thus, we can write

$$v_L(0^+) = L\frac{di_L(0^+)}{dt} \Rightarrow -s_1 K_1 + s_2 K_1 = \frac{12}{L}$$

or

$$K_1 = \frac{60}{s_2 - s_1}$$

At this point, let us arbitrarily choose $s_1 = 3$ and $s_2 = 17$, which satisfies the condition $s_1 + s_2 = 20$, and furthermore,

$$K_1 = \frac{60}{s_2 - s_1} = \frac{60}{14} = 4.29$$

$$C = \frac{1}{Ls_1 s_2} = \frac{1}{(0.2)(3)(17)} = 98 \text{ mF}$$

Hence, $i_L(t)$ is

$$i_L(t) = 4.29\left[e^{-3t} - e^{-17t}\right] \text{ A}$$

Figure 7.38a shows a plot of $i_L(t)$. At 100 ms the current has increased to 2.39 A, which meets the initial magnitude specifications. However, one second later at $t = 1.1$ s, $i_L(t)$ has fallen to only 0.16 A—well below the magnitude-over-time requirement. Simply put, the current falls too quickly. In order to make an informed estimate for s_1 and s_2, let us investigate the effect the roots exhibit on the current waveform when $s_2 > s_1$.

Since $s_2 > s_1$, the exponential associated with s_2 will decay to zero faster than that associated with s_1. This causes $i_L(t)$ to rise—the larger the value of s_2, the faster the rise. After $5(1/s_2)$ seconds have elapsed, the exponential associated with s_2 is approximately zero and $i_L(t)$ decreases exponentially with a time constant of $\tau = 1/s_1$. Thus, to slow the fall of $i_L(t)$ we should reduce s_1. Hence, let us choose $s_1 = 1$. Since $s_1 + s_2$ must equal 20, $s_2 = 19$. Under these conditions

$$C = \frac{1}{Ls_1 s_2} = \frac{1}{(0.2)(1)(19)} = 263 \text{ mF}$$

and

$$K_1 = \frac{60}{s_2 - s_1} = \frac{60}{18} = 3.33$$

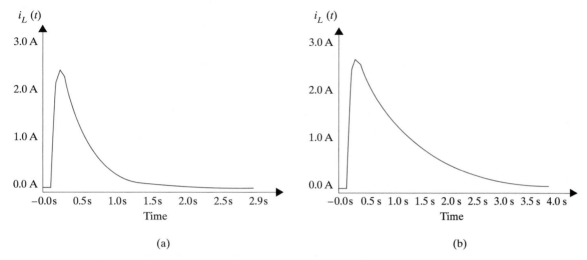

Figure 7.38 Ignition current as a function of time.

Thus, the current is

$$i_L(t) = 3.33[e^{-t} - e^{-19t}] \text{ A}$$

which is shown in Fig. 7.38b. At 100 ms the current is 2.52 A. Also, at $t = 1.1$ s, the current is 1.11 A—above the 1-A requirement. Therefore, the choice of $C = 263$ mF meets all starter specifications. ❑

EXAMPLE 7.11

A defibrillator is a device that is used to stop heart fibrillations—erratic uncoordinated quivering of the heart muscle fibers—by delivering an electric shock to the heart. The Lown defibrillator was developed by Dr. Bernard Lown in 1962. Its key feature, shown in Fig. 7.39a, is its voltage waveform. A simplified circuit diagram that is capable of producing the Lown waveform is shown in Fig. 7.39b. Let us find the necessary values for the inductor and capacitor.

SOLUTION   Since the Lown waveform is oscillatory in nature, we know that the circuit is underdamped ($\zeta < 1$) and the voltage applied to the patient is of the form

$$v_o(t) = K_1 e^{-\zeta\omega_o t} \sin[\omega t]$$

where

$$\zeta\omega_o = \frac{R}{2L}$$

$$\omega = \omega_o \sqrt{1 - \zeta^2}$$

and

$$\omega_o = \frac{1}{\sqrt{LC}}$$

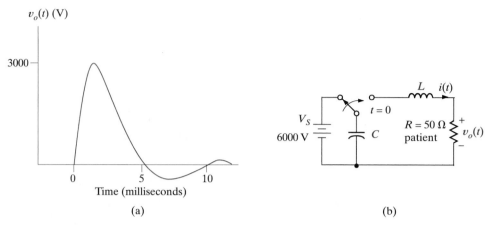

Figure 7.39 Lown defribillator waveform and simplified circuit. Reprinted with permission from John Wiley & Sons Inc., *Introduction to Biomedical Equipment Technology*.

for the series RLC circuit. From Fig. 7.39a, we see that the period of the sine function is

$$T = 10 \text{ ms}$$

Thus, we have one expression involving ω_o and ζ

$$\omega = \frac{2\pi}{T} = \omega_o \sqrt{1 - \zeta^2} = 200\pi$$

A second expression can be obtained by solving for the ratio of $v_o(t)$ at $t = T/4$ to that at $t = 3T/4$. At these two instants of time the sine function is equal to $+1$ and -1, respectively. Using values from Fig. 7.39a, we can write

$$\frac{v_o(T/4)}{-v_o(3T/4)} = \frac{K_1 e^{-\zeta\omega_o(T/4)}}{K_1 e^{-\zeta\omega_o(3T/4)}} = e^{\zeta\omega_o(T/2)} \approx \frac{3000}{250} = 12$$

or

$$\zeta\omega_o = 497.0$$

Given $R = 50 \, \Omega$, the necessary inductor value is

$$L = 50.3 \text{ mH}$$

Using our expression for ω,

$$\omega^2 = (200\pi)^2 = \omega_o^2 - (\zeta\omega_o)^2$$

or

$$(200\pi)^2 = \frac{1}{LC} - (497.0)^2$$

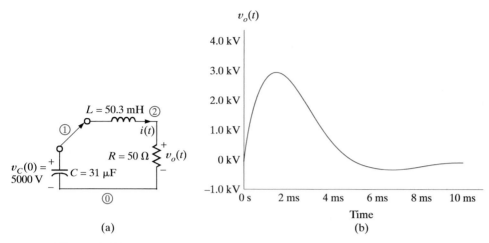

Figure 7.40 PSPICE circuit and output plot for the Lown defibrillator.

Solving for the capacitor value, we find

$$C = 31.0 \ \mu\text{F}$$

Let us verify our design using PSPICE. The PSPICE circuit is shown in Fig. 7.40a. The output voltage plot shown in Fig. 7.40b matches the Lown waveform in Fig. 7.39a; thus, we can consider the design to be a success.

It is important to note that while this solution is a viable one, it is not the only one. Like many design problems, there are often various ways to satisfy the design specifications. ❑

7.7 SUMMARY

- The voltage or current in an RLC transient circuit can be described by a constant coefficient differential equation of the form

$$\frac{d^2x(t)}{dt^2} + 2\zeta\omega_o \frac{dx(t)}{dt} + \omega_o^2 x(t) = f(t)$$

  where $f(t)$ is the network forcing function.

- The characteristic equation for a second-order circuit is $s^2 + 2\zeta\omega_o s + \omega_o^2 = 0$, where ζ is the damping ratio of ω_o is the undamped resonant frequency.

- If the two roots of the characteristic equation are

  real and unequal, then $\zeta > 1$ and the network response is overdamped

  real and equal, then $\zeta = 1$ and the network response is critically damped

  complex conjugates, then $\zeta < 1$ and the network response is underdamped

- The three types of damping together with the corresponding network response are as follows:

1. Overdamped: $x(t) = K_1 e^{-(\zeta\omega_o - \omega_o\sqrt{\zeta^2-1})t} + K_2 e^{-(\zeta\omega_o + \omega_o\sqrt{\zeta^2-1})t}$

2. Critically damped: $x(t) = B_1 e^{-\zeta\omega_o t} + B_2 t e^{-\zeta\omega_o t}$

3. Underdamped: $x(t) = e^{-\sigma t}(A_1 \cos \omega_d t + A_2 \sin \omega_d t)$, where $\sigma = \zeta\omega_o$ and $\omega_d = \omega_o \sqrt{1 - \zeta^2}$

- Two initial conditions are required to derive the two unknown coefficients in the network response equations.

PROBLEMS

Section 7.2

7.1 The differential equation that describes the current $i_o(t)$ in a network is

See Drill 7.2 and select the type of response from the damping ratio.

$$\frac{d^2 i_o(t)}{dt^2} + 6\frac{di_o(t)}{dt} + 4i_o(t) = 0$$

Find (a) the characteristic equation of the network, (b) the network's natural frequencies, and (c) the expression for $i_o(t)$.

7.2 The terminal current in a network is described by the equation

$$\frac{d^2 i_o(t)}{dt^2} + 8\frac{di_o(t)}{dt} + 16i_o(t) = 0$$

Similar to P7.1

Find (a) the characteristic equation of the network, (b) the network's natural frequencies, and (c) the equation for $i_o(t)$.

7.3 The voltage $v_1(t)$ in a network is defined by the equation

$$\frac{d^2 v_1(t)}{dt^2} + 4\frac{dv_1(t)}{dt} + 5v_1(t) = 0$$

Similar to P7.1

Find (a) the characteristic equation of the network, (b) the circuit's natural frequencies, and (c) the expression for $v_1(t)$.

7.4 The output voltage of a circuit is described by the differential equation

$$\frac{d^2 v_o(t)}{dt^2} + 8\frac{dv_o(t)}{dt} + 10v_o(t) = 0$$

Similar to P7.1

Find (a) the characteristic equation of the circuit, (b) the network's natural frequencies, and (c) the equation for $v_o(t)$.

7.5 The parameters for a parallel RLC circuit are $R = 1\,\Omega$, $L = \frac{1}{2}\,\mathrm{H}$, and $C = \frac{1}{2}\,\mathrm{F}$. Determine the type of damping exhibited by the circuit.

Derive the characteristic equation and compare it to Eq. (7.4).

7.6 A series RLC circuit contains a resistor $R = 2\,\Omega$ and a capacitor $C = \frac{1}{2}\,\mathrm{F}$. Select the value of the inductor so that the circuit is critically damped.

Similar to P7.5

7.7 A parallel RLC circuit contains a resistor $R = 1\,\Omega$ and an inductor $L = 2\,\mathrm{H}$. Select the value of the capacitor so that the circuit is critically damped.

Similar to P7.6

Section 7.3

7.8 For the underdamped circuit shown in Fig. P7.8, determine the voltage $v(t)$ if the initial conditions on the storage elements are $i_L(0) = 1\,\mathrm{A}$ and $v_C(0) = 12\,\mathrm{V}$.

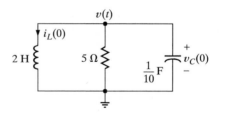

Figure P7.8

Use the problem-solving strategy for second-order transient circuits.

7.9 Given the circuit and the initial conditions of Problem 7.8, determine the current through the inductor.

The $i_L(t)$ can be obtained from $v(t)$.

7.10 In the critically damped circuit shown in Fig. P7.10, the initial conditions on the storage elements are $i_L(0) = 2\,\mathrm{A}$ and $v_c(0) = 5\,\mathrm{V}$. Determine the voltage $v(t)$.

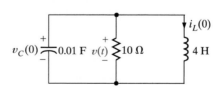

Figure P7.10

Similar to P7.8

Smilar to P7.9

7.11 Given the circuit and the initial conditions from Problem 7.10, determine the current $i_L(t)$ that is flowing through the inductor.

7.12 Find $v_o(t)$ for $t > 0$ in the circuit in Fig. P7.12 and plot the response including the time interval just prior to moving the switch.

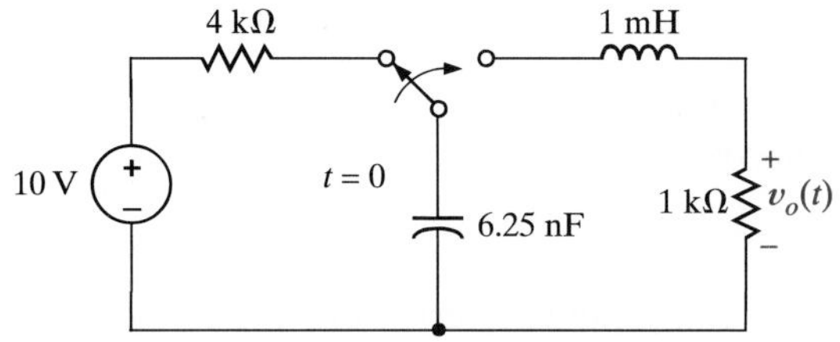

Smilar to P7.8

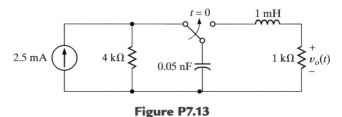

Figure P7.12

7.13 Find $v_o(t)$ for $t > 0$ in the circuit in Fig. P7.13 and plot the response including the time interval just prior to moving the switch.

Similar to P7.8

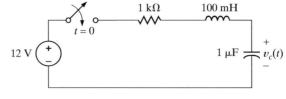

Figure P7.13

7.14 Find $v_C(t)$ for $t > 0$ in the circuit in Fig. P7.14 if $v_C(0) = 0$.

Similar to P7.8

Figure P7.14

7.15 Find $v_c(t)$ for $t > 0$ in the circuit in Fig. P7.15.

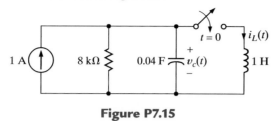

Figure P7.15

Similar to P7.8

7.16 Find $i_L(t)$ for $t > 0$ in the circuit in Fig. P7.15.

Similar to P7.8

7.17 Given the circuit in Fig. P7.17, find the equation for $i(t), t > 0$.

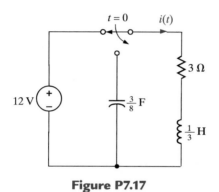

Figure P7.17

Similar to P7.8

7.18 Find $v_o(t)$ for $t > 0$ in the circuit in Fig. P7.18 and plot the response including the time interval just prior to closing the switch.

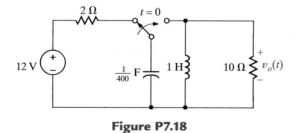

Figure P7.18

Similar to P7.8

7.19 In the circuit shown in Fig. P7.19, find $v(t), t > 0$.

Similar to P7.8

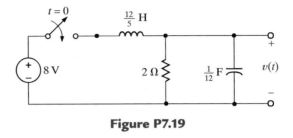

Figure P7.19

Section 7.4

7.20 Using the PSPICE *Schematics* editor, draw the circuit in Figure P7.20, and use the PROBE utility to plot $v_C(t)$ and determine the time constants for $0 < t < 1$ ms and 1 ms $< t < \infty$. Also, find the maximum voltage on the capacitor.

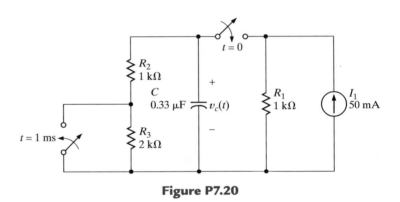

Figure P7.20

7.21 Using the PSPICE *Schematics* editor, draw the circuit in Figure P7.21, and use the PROBE utility to find the maximum values of $v_L(t), i_C(t)$, and $i(t)$.

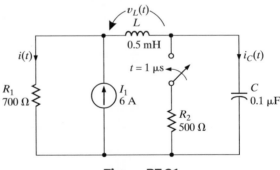

Figure P7.21

Section 7.6

7.22 Design a series RCL circuit with $R \geq 1\ k\Omega$ that has the characteristic equation

$$s^2 + 4 \times 10^7 s + 4 \times 10^{14} = 0$$

Similar to Example 7.10

7.23 Design a parallel RLC circuit with $R \geq 1\ k\Omega$ that has the characteristic equation

$$s^2 + 4 \times 10^7 s + 3 \times 10^{14} = 0$$

Similar to P7.2

Advanced Problems

7.24 Find $i_o(t)$ for $t > 0$ in the circuit in Fig. P7.24 and plot the response including the time interval just prior to closing the switch.

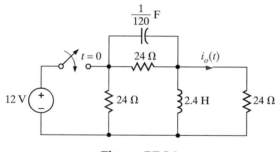

Figure P7.24

7.25 Find $v_o(t)$ for $t > 0$ in the circuit in Fig. P7.25 and plot the response including the time interval just prior to $t = 0$.

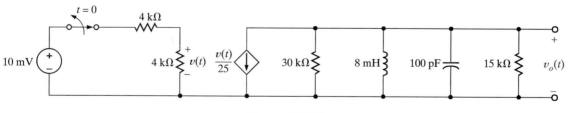

Figure P7.25

7.26 Find $i_o(t)$ for $t > 0$ in the circuit in Fig. P7.26 and plot the response including the time interval just prior to opening the switch.

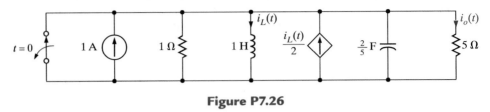

Figure P7.26

7.27 Find $v_o(t)$ for $t > 0$ in the circuit in Fig. P7.27 and plot the response including the time interval just prior to closing the switch.

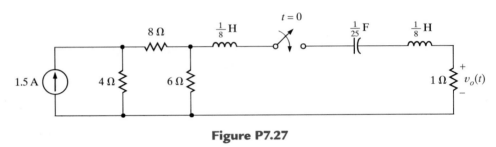

Figure P7.27

7.28 Find $v_o(t)$ for $t > 0$ in the circuit in Fig. P7.28 and plot the response including the time interval just prior to closing the switch.

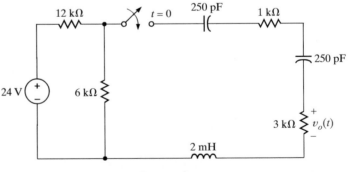

Figure P7.28

7.29 In the circuit shown in Fig. P7.29, switch action occurs at $t = 0$. Determine the voltage $v_o(t), t > 0$.

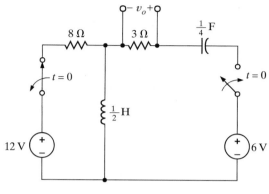

Figure P7.29

7.30 Find $v_o(t)$ for $t > 0$ in the circuit in Fig. P7.30 and plot the response including the time interval just prior to moving the switch.

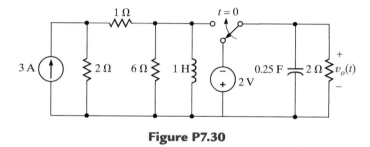

Figure P7.30

7.31 Find $v_o(t)$ for $t > 0$ in the circuit in Fig. P7.31 and plot the response including the time interval just prior to moving the switch.

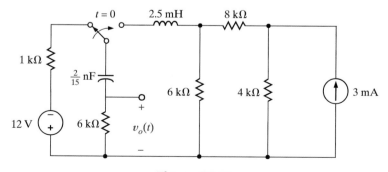

Figure P7.31

7.32 Find $v_o(t)$ for $t > 0$ in the circuit in Fig. P7.32 and plot the response including the time interval just prior to moving the switch.

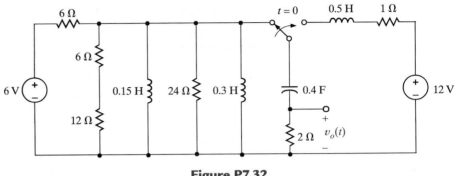

Figure P7.32

7.33 For the network in Fig. P7.33, choose C and L such that for $t > 0$, the characteristic equation is

$$s^2 + 7 \times 10^5 s + 10^{10} = 0$$

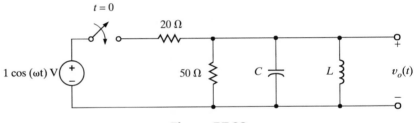

Figure P7.33

AC Steady-State Analysis

8

*I*n the preceding chapters we have considered in some detail both the natural and forced response of a network. We found that the natural response was a characteristic of the network and was independent of the forcing function. The forced response, however, depends directly on the type of forcing function, which until now has generally been a constant. At this point we will diverge from this tack to consider an extremely important excitation: the *sinusoidal forcing function*. Nature is replete with examples of sinusoidal phenomena, and although this is important to us as we examine many different types of physical systems, one reason that we can appreciate at this point for studying this forcing function is that it is the dominant waveform in the electric power industry. The signal present at the ac outlets in our home, office, laboratory, and so on is sinusoidal. In addition, we will demonstrate in Chapter 15 that via Fourier analysis we can represent any periodic electrical signal by a sum of sinusoids.

In this chapter we concentrate on the steady-state forced response of networks with sinusoidal driving functions. We will ignore the initial conditions and the transient or natural response, which will eventually vanish for the type of circuits with which we will be dealing. We refer to this as an *ac steady-state analysis*.

Our approach will be to begin by first studying the characteristics of a sinusoidal function as a prelude to using it as a forcing function for a circuit. We will mathematically relate this sinusoidal forcing function to a complex forcing function, which will lead us to define a phasor. By employing phasors we effectively transform a set of differential equations with sinusoidal forcing functions in the time domain into a set of algebraic equations containing complex numbers in the frequency domain. We will show that in this frequency domain Kirchhoff's laws are valid and, thus, all the analysis techniques that we have learned for dc analysis are applicable in ac steady-state analysis. Finally, we demonstrate the power of PSPICE in the solution of ac steady-state circuits.

8.1 SINUSOIDS

Let us begin our discussion of sinusoidal functions by considering the sine wave

$$x(\omega t) = X_M \sin \omega t \qquad \textbf{8.1}$$

where $x(t)$ could represent either $v(t)$ or $i(t)$. X_M is the *amplitude* or *maximum value*, ω is the *radian* or *angular frequency*, and ωt is the *argument* of the sine function. A plot of the function in Eq. (8.1) as a function of its argument is shown in Fig. 8.1a. Obviously, the function repeats itself every 2π radians. This condition is described mathematically as $x(\omega t + 2\pi) = x(\omega t)$ or in general for period T as

$$x\big[\omega(t + T)\big] = x(\omega t) \qquad \textbf{8.2}$$

meaning that the function has the same value at time $t + T$ as it does at time t.

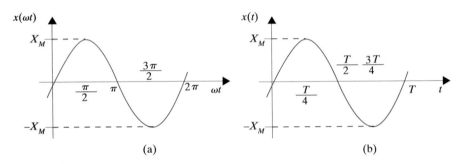

Figure 8.1 Plots of a sine wave as a function of both ωt and t.

The waveform can also be plotted as a function of time, as shown in Fig. 8.1b. Note that this function goes through one period every T seconds, or in other words, in 1 second it goes through $1/T$ periods or cycles. The number of cycles per second, called hertz, is the frequency f, where

The relationship between frequency and period.

$$f = \frac{1}{T} \qquad \textbf{8.3}$$

Now since $\omega T = 2\pi$, as shown in Fig. 8.1a, we find that

$$\omega = \frac{2\pi}{T} = 2\pi f \qquad \textbf{8.4}$$

The relationship among frequency, period, and radian frequency.

which is, of course, the general relationship among period in seconds, frequency in hertz, and radian frequency.

Now that we have discussed some of the basic properties of a sine wave, let us consider the following general expression for a sinusoidal function:

$$x(t) = X_M \sin(\omega t + \theta) \qquad \textbf{8.5}$$

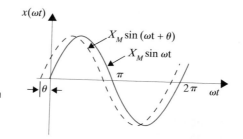

Figure 8.2 Graphical illustration of $X_M \sin(\omega t + \theta)$ leading $X_M \sin \omega t$ by θ radians.

In this case $(\omega t + \theta)$ is the argument of the sine function, and θ is called the *phase angle*. A plot of this function is shown in Fig. 8.2, together with the original function in Eq. (8.1) for comparison. Because of the presence of the phase angle, any point on the waveform $X_M \sin(\omega t + \theta)$ occurs θ radians earlier in time than the corresponding point on the waveform $X_M \sin \omega t$. Therefore, we say that $X_M \sin \omega t$ *lags* $X_M \sin(\omega t + \theta)$ by θ radians. In the more general situation, if

Phase lag defined.

$$x_1(t) = X_{M_1} \sin(\omega t + \theta)$$

and

$$x_2(t) = X_{M_2} \sin(\omega t + \phi)$$

then $x_1(t)$ leads $x_2(t)$ by $\theta - \phi$ radians and $x_2(t)$ lags $x_1(t)$ by $\theta - \phi$ radians. If $\theta = \phi$, the waveforms are identical and the functions are said to be *in phase*. If $\theta \neq \phi$, the functions are *out of phase*.

The phase angle is normally expressed in degrees rather than radians, and therefore we will simply state at this point that we will use the two forms interchangeably; that is,

In phase and out of phase defined.

$$x(t) = X_M \sin\left(\omega t + \frac{\pi}{2}\right) = X_M \sin(\omega t + 90°) \qquad \textbf{8.6}$$

Rigorously speaking, since ωt is in radians, the phase angle should be also. However, it is common practice and convenient to use degrees for phase, and therefore, that will be our practice in this text.

In addition, it should be noted that adding to the argument integer multiples of either 2π radians or $360°$ does not change the original function. This can easily be shown mathematically but is visibly evident when examining the waveform, as shown in Fig. 8.2.

Although our discussion has centered the sine function, we could just as easily have used the cosine function, since the two waveforms differ only by a phase angle; that is,

$$\cos \omega t = \sin\left(\omega t + \frac{\pi}{2}\right) \qquad \textbf{8.7}$$

$$\sin \omega t = \cos\left(\omega t - \frac{\pi}{2}\right) \qquad \textbf{8.8}$$

Some trigonometric identities that are useful in phase angle calculations.

It should be noted that when comparing one sinusoidal function with another *of the same frequency* to determine the phase difference, it is necessary to express both functions as either sines or cosines with positive amplitudes. Once in this format, the phase angle between the functions can be computed as outlined previously. Two other trigonometric identities that normally prove useful in phase angle determination are

$$-\cos(\omega t) = \cos(\omega t \pm 180°) \qquad \text{8.9}$$
$$-\sin(\omega t) = \sin(\omega t \pm 180°) \qquad \text{8.10}$$

Finally, the angle-sum and angle-difference relationships for sines and cosines may be useful in the manipulation of sinusoidal functions. These relations are

$$\begin{aligned}
\sin(\alpha + \beta) &= \sin\alpha\cos\beta + \cos\alpha\sin\beta \\
\cos(\alpha + \beta) &= \cos\alpha\cos\beta - \sin\alpha\sin\beta \\
\sin(\alpha - \beta) &= \sin\alpha\cos\beta - \cos\alpha\sin\beta \\
\cos(\alpha - \beta) &= \cos\alpha\cos\beta + \sin\alpha\sin\beta
\end{aligned} \qquad \text{8.11}$$

Our interest in these relations will revolve about the cases where $\alpha = \omega t$ and $\beta = \theta$. Since θ is a constant, the preceding equations indicate that either a sinusoidal or cosinusoidal function with a phase angle can be written as the sum of a sine function and a cosine function with the same arguments. For example, using Eq. (8.11), Eq. (8.5) can be written as

$$\begin{aligned}
x(t) &= X_M \sin(\omega t + \theta) \\
&= X_M(\sin\omega t \cos\theta + \cos\omega t \sin\theta) \\
&= A \sin\omega t + B \cos\omega t
\end{aligned}$$

where

$$A = X_M \cos\theta$$
$$B = X_M \sin\theta$$

or

$$x(t) = \sqrt{A^2 + B^2} \sin(\omega t + \tan^{-1}(B/A))$$

EXAMPLE 8.1

We wish to plot the waveforms for the following functions: (a) $v(t) = 1\cos(\omega t + 45°)$, (b) $v(t) = 1\cos(\omega t + 225°)$, and (c) $v(t) = 1\cos(\omega t - 315°)$.

SOLUTION   Figure 8.3a shows a plot of the function $v(t) = 1\cos\omega t$. Figure 8.3b is a plot of the function $v(t) = 1\cos(\omega t + 45°)$. Figure 8.3c is a plot of the function $v(t) = 1\cos(\omega t + 225°)$. Note that since

$$v(t) = 1\cos(\omega t + 225°) = 1\cos(\omega t + 45° + 180°)$$

this waveform is 180° out of phase with the waveform in Fig. 8.3b; that is, $\cos(\omega t + 225°) = -\cos(\omega t + 45°)$, and Fig. 8.3c is the negative of Fig. 8.3b. Finally, since the function

$$v(t) = 1\cos(\omega t - 315°) = 1\cos(\omega t - 315° + 360°) = 1\cos(\omega t + 45°)$$

this function is identical to that shown in Fig. 8.3b.   ❑

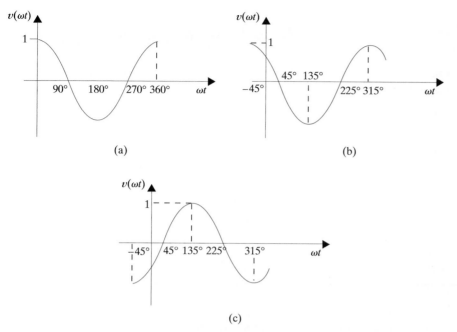

Figure 8.3 Cosine waveforms with various phase angles.

EXAMPLE 8.2

Determine the frequency and the phase angle between the two voltages $v_1(t) = 12 \sin(1000t + 60°)$ V and $v_2(t) = -6 \cos(1000t + 30°)$ V.

SOLUTION   The frequency in hertz (Hz) is given by the expression

$$f = \frac{\omega}{2\pi} = \frac{1000}{2\pi} = 159.2 \text{ Hz}$$

Using Eq. (8.9), $v_2(t)$ can be written as

$$v_2(t) = -6\cos(\omega t + 30°) = 6\cos(\omega t + 210°) \text{ V}$$

Then employing Eq. (8.7), we obtain

$$6\sin(\omega t + 300°) \text{ V} = 6\sin(\omega t - 60°) \text{ V}$$

Now that both voltages of the same frequency are expressed as sine waves with positive amplitudes, the phase angle between $v_1(t)$ and $v_2(t)$ is $60° - (-60) = 120°$; that is, $v_1(t)$ leads $v_2(t)$ by 120° or $v_2(t)$ lags $v_1(t)$ by 120°.   ❏

EXTENSION EXERCISES

E8.1   Given the voltage $v(t) = 120 \cos(314t + \pi/4)$V, determine the frequency of the voltage in hertz and the phase angle in degrees.

ANSWER: $f = 50$ Hz, $\theta = 45°$.

E8.2   Three branch currents in a network are known to be

$$i_1(t) = 2 \sin(377t + 45°)\text{A}$$

$$i_2(t) = 0.5 \cos(377t + 10°)\text{A}$$

$$i_3(t) = -0.25 \sin(377t + 60°)\text{A}$$

Determine the phase angles by which $i_1(t)$ leads $i_2(t)$ and $i_1(t)$ leads $i_3(t)$.

ANSWER: i_1 leads i_2 by $-55°$, i_1 leads i_3 by $165°$.

8.2 SINUSOIDAL AND COMPLEX FORCING FUNCTIONS

In the preceding chapters we applied a constant forcing function to a network and found that the steady-state response was also constant.

In a similar manner, if we apply a sinusoidal forcing function to a linear network, the steady-state voltages and currents in the network will also be sinusoidal. This should also be clear from the KVL and KCL equations. For example, if one branch voltage is a sinusoid of some frequency, the other branch voltages must be sinusoids of the same frequency if KVL is to apply around any closed path. This means, of course, that the forced solutions of the differential equations that describe a network with a sinusoidal forcing function are sinusoidal functions of time. For example, if we assume that our input function is a voltage $v(t)$ and our output response is a current $i(t)$ as shown in Fig. 8.4, then if $v(t) = A \sin(\omega t + \theta)$, $i(t)$ will be of the form $i(t) = B \sin(\omega t + \phi)$. The critical point here is that we know the form of the output response, and therefore the solution involves simply determining the values of the two parameters B and ϕ.

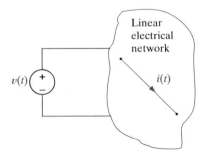

Figure 8.4 Current response to an applied voltage in an electrical network.

EXAMPLE 8.3

Consider the circuit in Fig. 8.5. Let us derive the expression for the current.

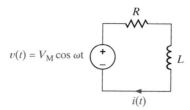

Figure 8.5 A simple *RL* circuit.

SOLUTION   The KVL equation for this circuit is

$$L\frac{di(t)}{dt} + Ri(t) = V_M \cos \omega t$$

Since the input forcing function is $X_M \cos \omega t$, we assume that the forced response component of the current $i(t)$ is of the form

$$i(t) = A \cos(\omega t + \phi)$$

which can be written using Eq. (8.11) as

$$i(t) = A \cos \phi \cos \omega t - A \sin \phi \sin \omega t$$
$$= A_1 \cos \omega t + A_2 \sin \omega t$$

Note that this is, as we observed in Chapter 6, of the form of the forcing function $\cos \omega t$ and its derivative $\sin \omega t$. Substituting this form for $i(t)$ into the preceding differential equation yields

$$L\frac{d}{dt}(A_1 \cos \omega t + A_2 \sin \omega t) + R(A_1 \cos \omega t + A_2 \sin \omega t) = V_M \cos \omega t$$

Evaluating the indicated derivative produces

$$-A_1 \omega L \sin \omega t + A_2 \omega L \cos \omega t + RA_1 \cos \omega t + RA_2 \cos \omega t = V_M \cos \omega t$$

By equating coefficients of the sine and cosine functions, we obtain

$$-A_1 \omega L + A_2 R = 0$$
$$A_1 R + A_2 \omega L = V_M$$

that is, two simultaneous equations in the unknowns A_1 and A_2. Solving these two equations for A_1 and A_2 yields

$$A_1 = \frac{RV_M}{R^2 + \omega^2 L^2}$$

$$A_2 = \frac{\omega L V_M}{R^2 + \omega^2 L^2}$$

Therefore,

$$i(t) = \frac{RV_M}{R^2 + \omega^2 L^2} \cos \omega t + \frac{\omega L V_M}{R^2 + \omega^2 L^2} \sin \omega t$$

which, using the last identity in Eq. (8.11), can be written as

$$i(t) = A \cos(\omega t + \phi)$$

where A and ϕ are determined as follows:

$$A \cos \phi = \frac{RV_M}{R^2 + \omega^2 L^2}$$

$$A \sin \phi = \frac{-\omega L V_M}{R^2 + \omega^2 L^2}$$

Hence,

$$\tan \phi = \frac{A \sin \phi}{A \cos \phi} = -\frac{\omega L}{R}$$

and therefore,

$$\phi = -\tan^{-1} \frac{\omega L}{R}$$

and since

$$(A \cos \phi)^2 + (A \sin \phi)^2 = A^2(\cos^2 \phi + \sin^2 \phi) = A^2$$

$$A^2 = \frac{R^2 V_M^2}{\left(R^2 + \omega^2 L^2\right)^2} + \frac{(\omega L)^2 V_M^2}{\left(R^2 + \omega^2 L^2\right)^2} = \frac{V_M^2}{R^2 + \omega^2 L^2}$$

$$A = \frac{V_M}{\sqrt{R^2 + \omega^2 L^2}}$$

Hence, the final expression for $i(t)$ is

$$i(t) = \frac{V_M}{\sqrt{R^2 + \omega^2 L^2}} \cos\left(\omega t - \tan^{-1} \frac{\omega L}{R}\right)$$

The preceding analysis indicates that ϕ is zero if $L = 0$ and hence $i(t)$ is in phase with $v(t)$. If $R = 0$, $\phi = 90°$ and the current lags the voltage by 90°. If L and R are both present, the current lags the voltage by some angle between 0° and 90°.   ❏

This example illustrates an important point—solving even a simple one-loop circuit containing one resistor and one inductor is very complicated when compared to the solution of a single-loop circuit containing only two resistors. Imagine for a moment how laborious it would be to solve a more complicated circuit using the procedure we have employed in Example 8.3. To circumvent this approach, we will establish a correspondence between sinusoidal time functions and complex numbers. We will then show that this relationship leads to a set of algebraic equations for currents and voltages in a network (e.g., loop currents or node voltages) in which the coefficients of the variables are complex numbers. Hence, once again we will find that determining the currents or voltages in a circuit can be accomplished by solving a set of algebraic equations; however, in this case, their solution is complicated by the fact that variables in the equations have complex, rather than real, coefficients.

The vehicle we will employ to establish a relationship between time-varying sinusoidal functions and complex numbers is Euler's equation, which for our purposes is written as

$$e^{j\omega t} = \cos \omega t + j \sin \omega t \qquad \textbf{8.12}$$

This complex function has a real part and an imaginary part:

$$\text{Re}(e^{j\omega t}) = \cos \omega t$$
$$\text{Im}(e^{j\omega t}) = \sin \omega t \qquad \textbf{8.13}$$

where $\text{Re}(\cdot)$ and $\text{Im}(\cdot)$ represent the real part and the imaginary part, respectively, of the function in the parentheses.

Now suppose that we select as our forcing function in Fig. 8.4 the nonrealizable voltage

$$v(t) = V_M e^{j\omega t} \qquad \textbf{8.14}$$

which because of Euler's identity can be written as

$$v(t) = V_M \cos \omega t + j V_M \sin \omega t \qquad \textbf{8.15}$$

The real and imaginary parts of this function are each realizable. We think of this complex forcing function as two forcing functions, a real one and an imaginary one, and as a consequence of linearity, the principle of superposition applies and thus the current response can be written as

$$i(t) = I_M \cos(\omega t + \phi) + j I_M \sin(\omega t + \phi) \qquad \textbf{8.16}$$

where $I_M \cos(\omega t + \phi)$ is the response due to $V_M \cos \omega t$ and $j I_M \sin(\omega t + \phi)$ is the response due to $j V_M \sin \omega t$. This expression for the current containing both a real and an imaginary term can be written via Euler's equation as

$$i(t) = I_M e^{j(\omega t + \phi)} \qquad \textbf{8.17}$$

Because of the preceding relationships we find that rather than applying the forcing function $V_M \cos \omega t$ and calculating the response $I_M \cos(\omega t + \phi)$, we can apply the complex forcing function $V_M e^{j\omega t}$ and calculate the response $I_M e^{j(\omega t + \phi)}$, the real part of which is the desired response $I_M \cos(\omega t + \phi)$. Although this procedure may initially appear to be more complicated, it is not. It is via this technique that we will convert the differential equation to an algebraic equation that is much easier to solve.

EXAMPLE 8.4

Once again, let us determine the current in the *RL* circuit examined in Example 8.3. However, rather than applying $V_M \cos \omega t$ we will apply $V_M e^{j\omega t}$.

SOLUTION   The forced response will be of the form

$$i(t) = I_M e^{j(\omega t + \phi)}$$

where only I_M and ϕ are unknown. Substituting $v(t)$ and $i(t)$ into the differential equation for the circuit, we obtain

$$RI_M\, e^{j(\omega t+\phi)} + L\frac{d}{dt}\left(I_M\, e^{j(\omega t+\phi)}\right) = V_M\, e^{j\omega t}$$

Taking the indicated derivative, we obtain

$$RI_M\, e^{j(\omega t+\phi)} + j\omega L I_M\, e^{j(\omega t+\phi)} = V_M\, e^{j\omega t}$$

Dividing each term of the equation by the common factor $e^{j\omega t}$ yields

$$RI_M\, e^{j\phi} + j\omega L I_M\, e^{j\phi} = V_M$$

which is an algebraic equation with complex coefficients. This equation can be written as

$$I_M\, e^{j\phi} = \frac{V_M}{R + j\omega L}$$

Converting the right-hand side of the equation to exponential or polar form produces the equation

$$I_M\, e^{j\phi} = \frac{V_M}{\sqrt{R^2 + \omega^2 L^2}}\, e^{j[-\tan^{-1}(\omega L/R)]}$$

Summary of complex number relationships:

$$x + jy = re^{j\theta}$$
$$r = \sqrt{x^2 + y^2}$$
$$\theta = \tan^{-1}\frac{y}{x}$$
$$x = r\cos\theta$$
$$y = r\sin\theta$$
$$\frac{1}{e^{j\theta}} = e^{-j\theta}$$

(A quick refresher on complex numbers is given in Appendix C for readers who need to sharpen their skills in this area.) The preceding form clearly indicates that the magnitude and phase of the resulting current are

$$I_M = \frac{V_M}{\sqrt{R^2 + \omega^2 L^2}}$$

and

$$\phi = -\tan^{-1}\frac{\omega L}{R}$$

However, since our actual forcing function was $V_M\cos\omega t$ rather than $V_M\, e^{j\omega t}$, our actual response is the real part of the complex response:

$$i(t) = I_M\cos(\omega t + \phi)$$

$$= \frac{V_M}{\sqrt{R^2 + \omega^2 L^2}}\cos\left(\omega t - \tan^{-1}\frac{\omega L}{R}\right)$$

Note that this is identical to the response obtained in the previous example by solving the differential equation for the current $i(t)$.   ❏

8.3 PHASORS

Once again let us assume that the forcing function for a linear network is of the form

$$v(t) = V_M e^{j\omega t}$$

Then every steady-state voltage or current in the network will have the same form and the same frequency ω; for example, a current $i(t)$ will be of the form $i(t) = I_M e^{j(\omega t + \phi)}$.

As we proceed in our subsequent circuit analyses, we will simply note the frequency and then drop the factor $e^{j\omega t}$ since it is common to every term in the describing equations. Dropping the term $e^{j\omega t}$ indicates that every voltage or current can be fully described by a magnitude and phase. For example, a voltage $v(t)$ can be written in exponential form as

$$v(t) = V_M \cos(\omega t + \theta) = \text{Re}\left[V_M e^{j(\omega t + \theta)}\right] \qquad \textbf{8.18}$$

or as a complex number

$$v(t) = \text{Re}\left(V_M \underline{/\theta}\ e^{j\omega t}\right) \qquad \textbf{8.19}$$

Since we are working with a complex forcing function, the real part of which is the desired answer, and each term in the equation will contain $e^{j\omega t}$, we can drop $\text{Re}(\cdot)$ and $e^{j\omega t}$ and work only with the complex number $V_M \underline{/\theta}$. This complex representation is commonly called a *phasor*. As a distinguishing feature, phasors will be written in boldface type. In a completely identical manner a voltage $v(t) = V_M \cos(\omega t + \theta) = \text{Re}\left[V_M e^{j(\omega t + \theta)}\right]$ and a current $i(t) = I_M \cos(\omega t + \phi) = \text{Re}\left[I_M e^{j(\omega t + \phi)}\right]$ are written in phasor notation as $\mathbf{V} = V_M \underline{/\theta}$ and $\mathbf{I} = V_M \underline{/\phi}$, respectively.

If $v(t) = V_m \cos(\omega t + \theta)$ and $i(t) = I_m \cos(\omega t + \phi)$, then in phasor notation

$$\mathbf{V} = V_m \underline{/\theta} \quad \text{and}$$

$$\mathbf{I} = I_m \underline{/\phi}$$

EXAMPLE 8.5

Again, we consider the *RL* circuit in Example 8.3. Let us use phasors to determine the expression for the current.

SOLUTION   The differential equation is

$$L \frac{di(t)}{dt} Ri(t) = V_M \cos \omega t$$

The forcing function can be replaced by a complex forcing function that is written as $\mathbf{V}e^{j\omega t}$ with phasor $\mathbf{V} = V_M \underline{/0°}$. Similarly, the forced response component of the current $i(t)$ can be replaced by a complex function that is written as $\mathbf{I}e^{j\omega t}$ with phasor $\mathbf{I} = I_M \underline{/\phi}$. From our previous discussions we recall that the solution of the differential equation is the real part of this current.

Using the complex forcing function, we find that the differential equation becomes

$$L \frac{d}{dt}\left(\mathbf{I}e^{j\omega t}\right) + R\mathbf{I}e^{j\omega t} = \mathbf{V}e^{j\omega t}$$

$$j\omega L\mathbf{I}e^{j\omega t} + R\mathbf{I}e^{j\omega t} = \mathbf{V}e^{j\omega t}$$

Note that $e^{j\omega t}$ is a common factor and, as we have already indicated, can be eliminated, leaving the phasors; that is,

The differential equation is reduced to a phasor equation.

$$j\omega L\mathbf{I} + R\mathbf{I} = \mathbf{V}$$

Therefore,

$$\mathbf{I} = \frac{\mathbf{V}}{R + j\omega L} = I_M \underline{/\phi} = \frac{V_M}{\sqrt{R^2 + \omega^2 L^2}} \underline{\bigg/ -\tan^{-1}\frac{\omega L}{R}}$$

Thus,

$$i(t) = \frac{V_M}{\sqrt{R^2 + \omega^2 L^2}} \cos\left(\omega t - \tan^{-1}\frac{\omega L}{R}\right)$$

which once again is the function we obtained earlier. ❏

Phasor analysis

1 Using phasors, transform a set of differential equations in the time domain into a set of algebraic equations in the frequency domain.

2 Solve the algebric equations for the unknown phasors.

3 Transform the now-known phasors back to the time domain.

We define relations between phasors after the $e^{j\omega t}$ term has been eliminated as "phasor, or frequency domain, analysis." Thus we have transformed a set of differential equations with sinusoidal forcing functions in the time domain into a set of algebraic equations containing complex numbers in the frequency domain. In effect, we are now faced with solving a set of algebraic equations for the unknown phasors. The phasors are then simply transformed back to the time domain to yield the solution of the original set of differential equations. In addition, we note that the solution of sinusoidal steady-state circuits would be relatively simple if we could write the phasor equation directly from the circuit description. In Section 8.4 we will lay the groundwork for doing just that.

Note that in our discussions we have tacitly assumed that sinusoidal functions would be represented as phasors with a phase angle based on a cosine function. Therefore, if sine functions are used, we will simply employ the relationship in Eq. (8.7) to obtain the proper phase angle.

In summary, while $v(t)$ represents a voltage in the time domain, the phasor $\mathbf{V}$ represents the voltage in the frequency domain. The phasor contains only magnitude and phase information, and the frequency is implicit in this representation. The transformation from the time domain to the frequency domain, as well as the reverse transformation, is shown in Table 8.1. Recall that the phase angle is based on a cosine function and, therefore, if a sine function is involved, a 90° shift factor must be employed, as shown in the table.

Table 8.1 Phasor Representation

Time Domain	Frequency Domain
$A \cos(\omega t \pm \theta)$	$A \underline{/\pm\theta}$
$A \sin(\omega t \pm \theta)$	$A \underline{/\pm\theta - 90°}$

The following examples illustrate the use of the phasor transformation.

It is important to note, however, that if a network contains only sine sources, there is no need to perform the 90° shift. We simply perform the normal phasor analysis and then the *imaginary* part of the time-varying complex solution is the desired response. Simply put, cosine sources generate a cosine response and sine sources generate a sine response.

EXAMPLE 8.6

Convert the time functions $v(t) = 24 \cos(337t - 45°)$ V and $i(t) = 12 \sin(337t + 120°)$ A to phasors.

SOLUTION   Using the phasor transformation shown previously, we have

$$\mathbf{V} = 24 \underline{/-45°} \text{ V}$$

$$\mathbf{I} = 12 \underline{/120° - 90°} = 12 \underline{/30°} \text{ A} \qquad \square$$

EXAMPLE 8.7

Convert the phasors $\mathbf{V} = 16 \underline{/20°}$ V and $\mathbf{I} = 10 \underline{/-75°}$ A from the frequency domain to the time domain if the frequency is 1 kHz.

SOLUTION   Employing the reverse transformation for phasors, we find that

$$v(t) = 16 \cos(2000\pi t + 20°)\text{V}$$

$$i(t) = 10 \cos(2000\pi t - 75°)\text{A} \qquad \square$$

EXTENSION EXERCISES

E8.3   Convert the following voltage functions to phasors.

$$v_1(t) = 12 \cos(377t - 425°)\text{V}$$

$$v_2(t) = 18 \sin(2513t + 4.2°)\text{V}$$

ANSWER: $\mathbf{V}_1 = 12 \underline{/-425°}$ V, $\mathbf{V}_2 = 18 \underline{/-85.8°}$ V.

E8.4   Convert the following phasors to the time domain if the frequency is 400 Hz.

$$\mathbf{V}_1 = 10 \underline{/20°}\text{V}$$

$$\mathbf{V}_2 = 12 \underline{/-60°}\text{V}$$

ANSWER: $v_1(t) = 10 \cos(800\pi t + 20°)$ V, $v_2(t)) = 12 \cos(800\pi t - 60°)$ V.

8.4 PHASOR RELATIONSHIPS FOR CIRCUIT ELEMENTS

As we proceed in our development of the techniques required to analyze circuits in the sinusoidal steady state, we are now in a position to establish the phasor relationships between voltage and current for the three passive elements R, L, and C.

In the case of a resistor as shown in Fig. 8.6a, the voltage–current relationship is known to be

$$v(t) = Ri(t) \qquad\qquad \textbf{8.20}$$

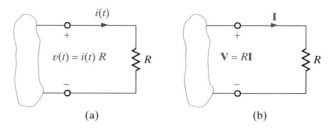

Current and voltage are in phase.

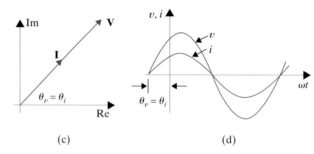

Figure 8.6 Voltage–current relationships for a resistor.

Applying the complex voltage $V_M e^{j(\omega t + \theta_v)}$ results in the complex current $I_M e^{j(\omega t + \theta_i)}$, and therefore Eq. (8.20) becomes

$$V_M e^{j(\omega t + \theta_v)} = R I_M e^{j(\omega t + \theta_i)}$$

which reduces to

$$V_M e^{j\theta_v} = R I_M e^{j\theta_i} \qquad \textbf{8.21}$$

Equation (8.21) can be written in phasor form as

$$\mathbf{V} = R\mathbf{I} \qquad \textbf{8.22}$$

where

$$\mathbf{V} = V_M e^{j\theta_v} = V_M \underline{/\theta_v} \quad \text{and} \quad \mathbf{I} = I_M e^{j\theta_i} = I_M \underline{/\theta_i}$$

From Eq. (8.21) we see that $\theta_v = \theta_i$ and thus the current and voltage for this circuit are *in phase*.

Historically, complex numbers have been represented as points on a graph in which the x-axis represents the real axis and the y-axis the imaginary axis. The line segment connecting the origin with the point provides a convenient representation of the magnitude and angle when the complex number is written in a polar form. A review of Appendix C will indicate how these complex numbers or line segments can be added, subtracted, and so on. Since phasors are complex numbers, it is convenient to represent the phasor voltage and current graphically as line segments. A plot of the line segments representing the phasors is called a *phasor diagram*. This pictorial representation of phasors provides immediate information on the relative magnitude of one phasor with another, the angle between two phasors, and the relative position of one phasor with respect to another (i.e., leading or lagging). A phasor diagram and the sinusoidal waveforms for the resistor are shown in Figs. 8.6c and d, respectively. A phasor diagram will be drawn for each of the other circuit elements in the remainder of this section.

EXAMPLE 8.8

If the voltage $v(t) = 24 \cos(377t + 75°)$ V is applied to a 6-Ω resistor as shown in Fig. 8.6a, we wish to determine the resultant current.

SOLUTION   Since the phasor voltage is

$$\mathbf{V} = 24 \;\underline{/75°}\; \text{V}$$

the phasor current from (Eq. (8.22)) is

$$\mathbf{I} = \frac{24 \;\underline{/75°}}{6} = 4 \;\underline{/75°}\; \text{A}$$

which in the time domain is

$$i(t) = 4 \cos(377t + 75°) \text{ A}$$

❏

EXTENSION EXERCISE

E8.5   The current in a 4-Ω resistor is known to be $\mathbf{I} = 12 \;\underline{/60°}$ A. Express the voltage across the resistor as a time function if the frequency of the current is 4 kHz.

ANSWER: $v(t) = 48 \cos(8000\pi t + 60°)$ V.

The voltage–current relationship for an inductor, as shown in Fig. 8.7a, is

$$v(t) = L\frac{di(t)}{dt} \qquad\qquad \textbf{8.23}$$

Substituting the complex voltage and current into this equation yields

$$V_M e^{j(\omega t + \theta_v)} = L\frac{d}{dt} I_M e^{j(\omega t + \theta_i)}$$

which reduces to

$$V_M e^{j\theta_v} = j\omega L I_M e^{j\theta_i} \qquad\qquad \textbf{8.24}$$

Equation (8.24) in phasor notation is

$$\mathbf{V} = j\omega L\mathbf{I} \qquad\qquad \textbf{8.25}$$

Note that the differential equation in the time domain (8.23) has been converted to an algebraic equation with complex coefficients in the frequency domain. This relationship is

The derivative process yields a frequency-dependent impedance.

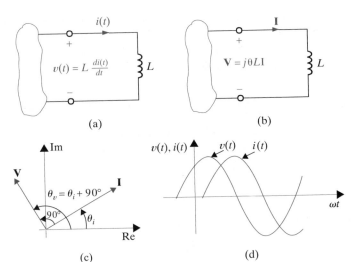

Figure 8.7 Voltage–current relationships for an inductor.

The voltage leads the current or the current lags the voltage.

shown in Fig. 8.7b. Since the imaginary operator $j = 1e^{j90°} = 1\underline{/90°} = \sqrt{-1}$, Eq. (8.24) can be written as

$$V_M e^{j\theta_v} = \omega L I_M e^{j(\theta_i + 90°)} \qquad \textbf{8.26}$$

Therefore, the voltage and current are *90° out of phase* and in particular the voltage leads the current by 90° or the current lags the voltage by 90°. The phasor diagram and the sinusoidal waveforms for the inductor circuit are shown in Figs. 8.7c and d, respectively.

EXAMPLE 8.9

The voltage $v(t) = 12\cos(377t + 20°)$ V is applied to a 20-mH inductor as shown in Fig. 8.7a. Find the resultant current.

SOLUTION   The phasor current is

Applying $\mathbf{V} = j\omega L\mathbf{I}$

$$\frac{x_1\underline{/\theta_1}}{x_2\underline{/\theta_2}} = \frac{x_1}{x_2}\underline{/\theta_1 - \theta_2}$$

$$\mathbf{I} = \frac{\mathbf{V}}{j\omega L} = \frac{12\underline{/20°}}{\omega L\underline{/90°}}$$

$$= \frac{12\underline{/20°}}{(377)(20\times10^{-3})\underline{/90°}}$$

$$= 1.59\underline{/-70°}\ \text{A}$$

or

$$i(t) = 1.59\cos(377t - 70°)\ \text{A} \qquad \square$$

EXTENSION EXERCISE

E8.6   The current in a 0.05-H inductor is $\mathbf{I} = 4\underline{/-30°}$ A. If the frequency of the current is 60 Hz, determine the voltage across the inductor.

ANSWER: $v_L(t) = 75.4\cos(377t + 60°)$ V.

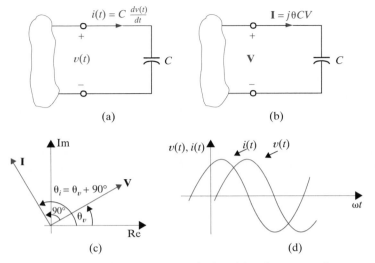

Figure 8.8 Voltage–current relationships for a capacitor.

The voltage–current relationship for our last passive element, the capacitor, as shown in Fig. 8.8a, is

$$i(t) = C\frac{dv(t)}{dt} \qquad\qquad 8.27$$

Once again employing the complex voltage and current, we obtain

$$I_M e^{j(\omega t + \theta_i)} = C\frac{d}{dt}V_M e^{j(\omega t + \theta_v)}$$

which reduces to

$$I_M e^{j\theta_i} = j\omega C V_M e^{j\theta_v} \qquad\qquad 8.28$$

In phasor notation this equation becomes

$$\mathbf{I} = j\omega C\mathbf{V} \qquad\qquad 8.29$$

The derivative process yields a frequency-dependent impedance.

Equation (8.27), a differential equation in the time domain, has been transformed into Eq. (8.29), an algebraic equation with complex coefficients in the frequency domain. The phasor relationship is shown in Fig. 8.8b. Substituting $j = 1e^{j90°}$ into Eq. (8.28) yields

$$I_M e^{j\theta_i} = \omega C V_M e^{j(\theta_v + 90°)} \qquad\qquad 8.30$$

Note that the voltage and current are *90° out of phase*. Equation (8.30) states that the current leads the voltage by 90° or the voltage lags the current by 90°. The phasor diagram and the sinusoidal waveforms for the capacitor circuit are shown in Figs. 8.8c and d, respectively.

EXAMPLE 8.10

The voltage $v(t) = 100\cos(314t + 15°)$ V is applied to a 100-μF capacitor as shown in Fig. 8.8a. Find the current.

Applying $\mathbf{I} = j\omega C\mathbf{V}$

SOLUTION   The resultant phasor current is

$$\mathbf{I} = j\omega C(100\ \underline{/15°})$$
$$= (314)(100 \times 10^{-6}\ \underline{/90°})(100\ \underline{/15°})$$
$$= 3.14\ \underline{/105°}\ A$$

Therefore, the current written as a time function is

$$i(t) = 3.14\cos(314t + 105°)\ A \qquad \square$$

EXTENSION EXERCISE

E8.7   The current in a 150-μF is $\mathbf{I} = 3.6\ \underline{/-145°}\ A$. If the frequency of the current is 60 Hz, determine the voltage across the capacitor.

ANSWER: $\mathbf{V}_C = 63.66\ \underline{/-235°}\ V.$

8.5 IMPEDANCE AND ADMITTANCE

We have examined each of the circuit elements in the frequency domain on an individual basis. We now wish to treat these passive circuit elements in a more general fashion. We now define the two-terminal input *impedance* **Z**, also referred to as the driving point impedance, in exactly the same manner in which we defined resistance earlier. Later we will examine another type of impedance, called transfer impedance.

Impedance is defined as the ratio of the phasor voltage **V** to the phasor current **I**:

$$\mathbf{Z} = \frac{\mathbf{V}}{\mathbf{I}} \qquad\qquad \textbf{8.31}$$

at the two terminals of the element related to one another by the passive sign convention, as illustrated in Fig 8.9. Since **V** and **I** are complex, the impedance **Z** is complex and

$$\mathbf{Z} = \frac{V_M\ \underline{/\theta_v}}{I_M\ \underline{/\theta_i}} = \frac{V_M}{I_M}\ \underline{/\theta_v - \theta_i} = Z\ \underline{/\theta_z} \qquad \textbf{8.32}$$

Since **Z** is the ratio of **V** to **I**, the units of **Z** are ohms. Thus, impedance in an ac circuit is analogous to resistance in a dc circuit. In rectangular form, impedance is expressed as

$$\mathbf{Z}(\omega) = R(\omega) + jX(\omega) \qquad \textbf{8.33}$$

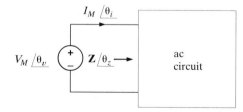

Figure 8.9 General impedance relationship.

where $R(\omega)$ is the real, or resistive, component and $X(\omega)$ is the imaginary, or reactive, component. In general, we simply refer to R as the resistance and X as the reactance. It is important to note that R and X are real functions of ω and therefore $\mathbf{Z}(\omega)$ is frequency dependent. Equation (8.33) clearly indicates that $\mathbf{Z}$ is a complex number; however, it is not a phasor, since phasors denote sinusoidal functions.

Equations (8.32) and (8.33) indicate that

$$Z \underline{/\theta_z} = R + jX \qquad\qquad \textbf{8.34}$$

Therefore,

$$Z = \sqrt{R^2 + X^2} \qquad\qquad \textbf{8.35}$$

$$\theta_z = \tan^{-1}\frac{X}{R}$$

where

$$R = Z\cos\theta_z$$

$$X = Z\sin\theta_z$$

For the individual passive elements the impedance is as shown in Table 8.2. However, just as it was advantageous to know how to determine the equivalent resistance in dc circuits, we want to learn how to determine the equivalent impedance in ac circuits..

Table 8.2 Passive element impedance

Passive element	Impedance
R	$\mathbf{Z} = R$
L	$\mathbf{Z} = j\omega L = jX_L = \omega L \underline{/90°},\, X_L = \omega L$
C	$\mathbf{Z} = \dfrac{1}{j\omega C} = jX_C = -\dfrac{1}{\omega C}\underline{/90°},\, X_C = -\dfrac{1}{\omega C}$

KCL and KVL are both valid in the frequency domain and we can use this fact, as was done in Chapter 2 for resistors, to show that impedances can be combined using the same rules that we established for resistor combinations. That is, if $\mathbf{Z}_1, \mathbf{Z}_2, \mathbf{Z}_3, \dots, \mathbf{Z}_n$ are connected in series, the equivalent impedance $\mathbf{Z}_s$ is

$$\mathbf{Z}_s = \mathbf{Z}_1 + \mathbf{Z}_2 + \mathbf{Z}_3 + \cdots + \mathbf{Z}_n \qquad\qquad \textbf{8.36}$$

and if $\mathbf{Z}_1, \mathbf{Z}_2, \mathbf{Z}_3, \dots, \mathbf{Z}_n$ are connected in parallel, the equivalent impedance is given by

$$\frac{1}{\mathbf{Z}_p} = \frac{1}{\mathbf{Z}_1} + \frac{1}{\mathbf{Z}_2} + \frac{1}{\mathbf{Z}_3} + \cdots + \frac{1}{\mathbf{Z}_n} \qquad\qquad \textbf{8.37}$$

EXAMPLE 8.11

Technique

1 Express $v(t)$ as a phasor and determine the impedance of each passive element.

2 Combine impedances and solve for the phasor **I**.

3 Convert the phasor **I** to $i(t)$.

Determine the equivalent impedance of the network shown in Fig. 8.10 if the frequency is $f = 60$ Hz. Then compute the current $i(t)$ if the voltage source is $v(t) = 50 \cos(\omega t + 30°)$ V. Finally, calculate the equivalent impedance if the frequency is $f = 400$ Hz.

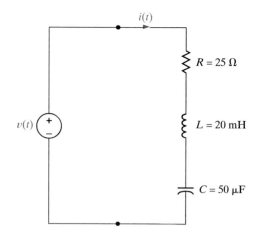

Figure 8.10 Series ac circuit.

SOLUTION   The impedances of the individual elements at 60 Hz are

$$\mathbf{Z}_R = 25\Omega$$
$$\mathbf{Z}_L = j\omega L = j(2\pi \times 60)(20 \times 10^{-3}) = j7.54 \ \Omega$$
$$\mathbf{Z}_C = \frac{-j}{\omega C} = \frac{-j}{(2\pi \times 60)(50 \times 10^{-6})} = -j53.05 \ \Omega$$

Since the elements are in series,

$$\mathbf{Z} = \mathbf{Z}_R + \mathbf{Z}_L + \mathbf{Z}_C$$
$$= 25 - j45.51 \ \Omega$$

The current in the circuit is given by

$$\mathbf{I} = \frac{\mathbf{V}}{\mathbf{Z}} = \frac{50 \ \underline{/30°}}{25 - j45.51} = \frac{50 \ \underline{/30°}}{51.93 \ \underline{/-61.22°}} = 0.96 \ \underline{/91.22°} \ \text{A}$$

or in the time domain, $i(t) = 0.96 \cos(377t + 91.22°)$ A.

If the frequency is 400 Hz, the impedance of each element is

$$\mathbf{Z}_R = 25 \ \Omega$$
$$\mathbf{Z}_L = j\omega L = j50.27 \ \Omega$$
$$\mathbf{Z}_C = \frac{-j}{\omega C} = -j7.96 \ \Omega$$

The total impedance is then

$$\mathbf{Z} = 25 + j42.31 = 49.14 \ \underline{/59.42°} \ \Omega$$

It is important to note that at the frequency $f = 60$ Hz, the reactance of the circuit is capacitive; that is, if the impedance is written as $R + jX$, $X < 0$; however, at $f = 400$ Hz the reactance is inductive since $X > 0$.   ❏

EXTENSION EXERCISE

E8.8   Find the current $i(t)$ in the network in Fig. E8.8.

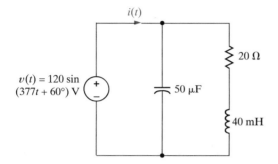

Figure E8.8

ANSWER: $i(t) = 3.88 \cos(377t - 39.2°)$ A.

Another quantity that is very useful in the analysis of ac circuits is the two-terminal input *admittance*, which is the reciprocal of impedance; that is,

$$\mathbf{Y} = \frac{1}{\mathbf{Z}} = \frac{\mathbf{I}}{\mathbf{V}}$$ 8.38

The units of $\mathbf{Y}$ are siemens, and this quantity is analogous to conductance in resistive dc circuits. Since $\mathbf{Z}$ is a complex number, $\mathbf{Y}$ is also a complex number.

$$\mathbf{Y} = Y_M \underline{/\theta_y}$$ 8.39

which is written in rectangular form as

$$\mathbf{Y} = G + jB$$ 8.40

where G and B are called *conductance* and *susceptance*, respectively. Because of the relationship between $\mathbf{Y}$ and $\mathbf{Z}$, we can express the components of one quantity as a function of the components of the other. From the expression

$$G + jB = \frac{1}{R + jX}$$ 8.41

we can show that

$$G = \frac{R}{R^2 + X^2}, \quad B = \frac{-X}{R^2 + X^2}$$ 8.42

Technique for taking the reciprocal:

$$\frac{1}{R + jX} = \frac{R - jX}{(R + jX)(R - jX)}$$
$$= \frac{R - jX}{R^2 + X^2}$$

D8.2 Find $\mathbf{Y}_p$

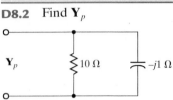

ANSWER: $\mathbf{Y}_p = 0.1 + j1$ S.

D8.3 Find $\mathbf{Y}_s$

ANSWER:

$\mathbf{Y}_s = 0.05 - j0.05$ S.

and in a similar manner, we can show that

$$R = \frac{G}{G^2 + B^2}, \qquad X = \frac{-B}{G^2 + B^2} \qquad \text{8.43}$$

It is very important to note that in general R and G are *not* reciprocals of one another. The same is true for X and B. The purely resistive case is an exception. In the purely reactive case the quantities are negative reciprocals of one another.

The admittance of the individual passive elements is

$$\mathbf{Y}_R = \frac{1}{R} = G$$

$$\mathbf{Y}_L = \frac{1}{j\omega L} = -\frac{1}{\omega L}\underline{/90^\circ} \qquad \text{8.44}$$

$$\mathbf{Y}_C = j\omega C = \omega C\underline{/90^\circ}$$

Once again, since KCL and KVL are valid in the frequency domain, we can show, using the same approach outlined in Chapter 2 for conductance in resistive circuits, that the rules for combining admittances are the same as those for combining conductances; that is, if $\mathbf{Y}_1, \mathbf{Y}_2, \mathbf{Y}_3, \ldots, \mathbf{Y}_n$ are connected in parallel, the equivalent admittance is

$$\mathbf{Y}_p = \mathbf{Y}_1 + \mathbf{Y}_2 + \cdots + \mathbf{Y}_n \qquad \text{8.45}$$

and if $\mathbf{Y}_1, \mathbf{Y}_2, \ldots, \mathbf{Y}_n$ are connected in series, the equivalent admittance is

$$\frac{1}{\mathbf{Y}_S} = \frac{1}{\mathbf{Y}_1} + \frac{1}{\mathbf{Y}_2} + \cdots + \frac{1}{\mathbf{Y}_n} \qquad \text{8.46}$$

EXAMPLE 8.12

Calculate the equivalent admittance $\mathbf{Y}_p$ for the network in Fig. 8.11 and use it to determine the current $\mathbf{I}$ if $\mathbf{V}_S = 60\underline{/45^\circ}$ V.

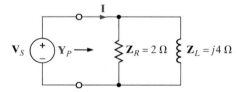

Figure 8.11 An Example parallel circuit.

SOLUTION   From Fig. 8.11 we note that

$$\mathbf{Y}_R = \frac{1}{\mathbf{Z}_R} = \frac{1}{2}\,\text{S}$$

$$\mathbf{Y}_L = \frac{1}{\mathbf{Z}_L} = \frac{-j}{4}\,\text{S}$$

Therefore,

$$\mathbf{Y}_p = \frac{1}{2} - j\frac{1}{4}\,\text{S}$$

Admittances add in parallel.

and hence,

$$\mathbf{I} = \mathbf{Y}_p\mathbf{V}_S$$
$$= \left(\frac{1}{2} - j\frac{1}{4}\right)(60\,\underline{/45°})$$
$$= 33.5\,\underline{/18.43°}\text{ A} \qquad \square$$

EXTENSION EXERCISE

E8.9   Find the current **I** in the network in Fig. E8.9.

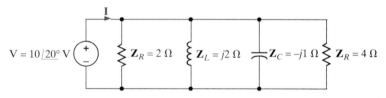

$V = 10\,\underline{/20°}\text{ V}$ $\mathbf{Z}_R = 2\,\Omega$ $\mathbf{Z}_L = j2\,\Omega$ $\mathbf{Z}_C = -j1\,\Omega$ $\mathbf{Z}_R = 4\,\Omega$

Figure E8.9

ANSWER: $\mathbf{I} = 9\,\underline{/53.7°}\text{ A}$.

As a prelude to our analysis of more general ac circuits, let us examine the techniques for computing the impedance or admittance of circuits in which numerous passive elements are interconnected. The following example illustrates that our technique is analogous to our earlier computations of equivalent resistance.

EXAMPLE 8.13

Consider the network shown in Fig. 8.12a. The impedance of each element is given in the figure. We wish to calculate the equivalent impedance of the network $\mathbf{Z}_{eq}$ at terminals A-B.

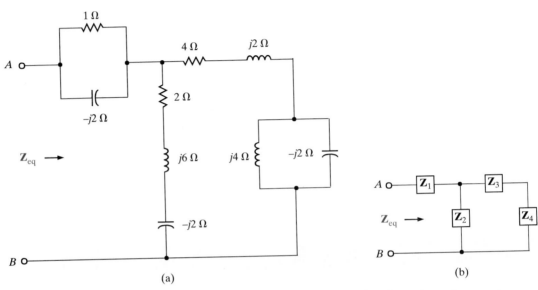

(a)

(b)

Figure 8.12 Example circuit for determining equivalent impedance in two steps.

Technique:

1 Add the admittances of elements in parallel.

2 Add the impedances of elements in series.

3 Convert back and forth between admittance and impedance in order to combine neighboring elements.

SOLUTION The equivalent impedance $\mathbf{Z}_{eq}$ could be calculated in a variety of ways; we could use only impedances, or only admittances, or a combination of the two. We will use the latter. We begin by noting that the circuit in Fig. 8.12a can be represented by the circuit in Fig. 8.12b.

Note that

$$\mathbf{Y}_4 = \mathbf{Y}_L + \mathbf{Y}_C$$
$$= \frac{1}{j4} + \frac{1}{-j2}$$
$$= j\frac{1}{4}\ \text{S}$$

Therefore,

$$\mathbf{Z}_4 = -j4\ \Omega$$

Now

$$\mathbf{Z}_{34} = \mathbf{Z}_3 + \mathbf{Z}_4$$
$$= (4 + j2) + (-j4)$$
$$= 4 - j2\ \Omega$$

and hence,

$$\mathbf{Y}_{34} = \frac{1}{\mathbf{Z}_{34}}$$

$$= \frac{1}{4 - j2}$$

$$= 0.20 + j0.10 \text{ S}$$

Since

$$\mathbf{Z}_2 = 2 + j6 - j2$$

$$= 2 + j4 \; \Omega$$

then

$$\mathbf{Y}_2 = \frac{1}{2 + j4}$$

$$= 0.10 - j0.20 \text{ S}$$

$$\mathbf{Y}_{234} = \mathbf{Y}_2 + \mathbf{Y}_{34}$$

$$= 0.30 - j0.10 \text{ S}$$

The reader should note carefully our approach—we are adding impedances in series and adding admittances in parallel.

From $\mathbf{Y}_{234}$ we can compute $\mathbf{Z}_{234}$ as

$$\mathbf{Z}_{234} = \frac{1}{\mathbf{Y}_{234}}$$

$$= \frac{1}{0.30 - j0.10}$$

$$= 3 + j1 \; \Omega$$

Now

$$\mathbf{Y}_1 = \mathbf{Y}_R + \mathbf{Y}_C$$

$$= \frac{1}{1} + \frac{1}{-j2}$$

$$= 1 + j\frac{1}{2} \text{ S}$$

and then

$$\mathbf{Z}_1 = \frac{1}{1 + j\dfrac{1}{2}}$$

$$= 8.0 - j0.4 \; \Omega$$

Therefore,

$$\mathbf{Z}_{eq} = \mathbf{Z}_1 + \mathbf{Z}_{234}$$
$$= 0.8 - j0.4 + 3 + j1$$
$$= 3.8 + j0.6 \ \Omega$$ ❏

EXTENSION EXERCISE

E8.10   Compute the impedance $\mathbf{Z}_T$ in the network in Fig. E8.10.

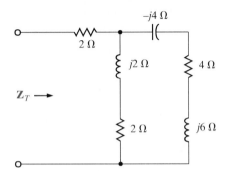

Figure E8.10

ANSWER: $\mathbf{Z}_T = 3.38 + j1.08 \ \Omega$.

8.6 PHASOR DIAGRAMS

Impedance and admittance are functions of frequency, and therefore their values change as the frequency changes. These changes in $\mathbf{Z}$ and $\mathbf{Y}$ have a resultant effect on the current–voltage relationships in a network. This impact of changes in frequency on circuit parameters can be easily seen via a phasor diagram. The following examples will serve to illustrate these points.

EXAMPLE 8.14

Let us sketch the phasor diagram for the network shown in Fig. 8.13.

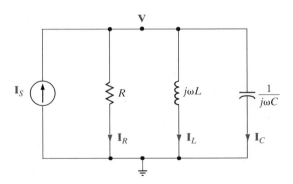

Figure 8.13 Example parallel circuit.

SOLUTION   The pertinent variables are labeled on the figure. For convenience in forming a phasor diagram, we select **V** as a reference phasor and arbitrarily assign it a 0° phase angle. We will, therefore, measure all currents with respect to this phasor. We suffer no loss of generality by assigning **V** a 0° phase angle, since if it is actually, for example, 30°, we will simply rotate the entire phasor diagram by 30° because all the currents are measured with respect to this phasor.

At the upper node in the circuit KCL is

$$\mathbf{I}_S = \mathbf{I}_R + \mathbf{I}_L + \mathbf{I}_C = \frac{\mathbf{V}}{R} + \frac{\mathbf{V}}{j\omega L} + \frac{\mathbf{V}}{1/j\omega C}$$

Since $\mathbf{V} = V_M\,\underline{/0°}$, then

$$\mathbf{I}_S = \frac{V_M\,\underline{/0°}}{R} + \frac{V_M\,\underline{/-90°}}{\omega L} + V_M\,\omega C\,\underline{/90°}$$

The phasor diagram that illustrates the phase relationship between $\mathbf{V}, \mathbf{I}_R, \mathbf{I}_L$, and $\mathbf{I}_C$ is shown in Fig. 8.14a. For small values of ω such that the magnitude of $\mathbf{I}_L$ is greater than that of $\mathbf{I}_C$, the phasor diagram for the currents is shown in Fig. 8.14b. In the case of large values of ω, that is, those for which $\mathbf{I}_C$ is greater than $\mathbf{I}_L$, the phasor diagram for the currents is shown in Fig. 8.14c. Note that as ω increases, the phasor $\mathbf{I}_S$ moves from $\mathbf{I}_{S_1}$ to $\mathbf{I}_{S_n}$ along a locus of points specified by the dashed line shown in Fig. 8.14d.

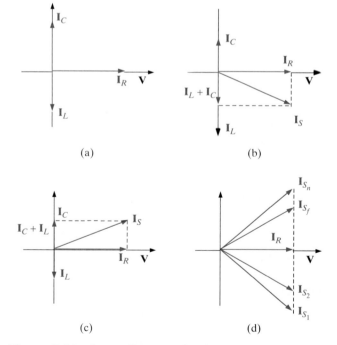

(a)

(b)

(c)

(d)

From a graphical standpoint, phasors can be manipulated like vectors.

Figure 8.14 Phasor diagrams for the circuit in Fig. 8.13.

D8.4   Find the frequency at which $v(t)$ and $i(t)$ are in phase.

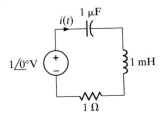

ANSWER:

$f = 5.03 \times 10^3$ Hz.

Note that $\mathbf{I}_S$ is in phase with $\mathbf{V}$ when $\mathbf{I}_C = \mathbf{I}_L$ or, in other words, when $\omega L = 1/\omega C$. Hence, the node voltage $\mathbf{V}$ is in phase with the current source $\mathbf{I}_S$ when

$$\omega = \frac{1}{\sqrt{LC}}$$

This can also be seen from the KCL equation

$$\mathbf{I} = \left[\frac{1}{R} + j\left(\omega C - \frac{1}{\omega L}\right)\right]\mathbf{V} \qquad \square$$

EXAMPLE 8.15

Let us determine the phasor diagram for the series circuit shown in Fig. 8.15a.

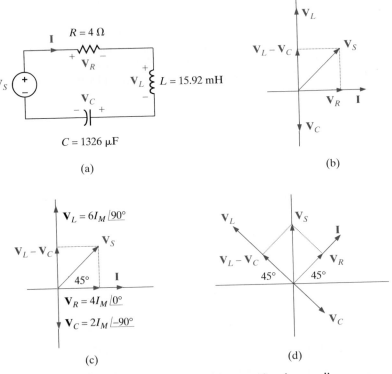

Figure 8.15   Series circuit and certain specific phasor diagrams.

SOLUTION   KVL for this circuit is of the form

$$\mathbf{V}_S = \mathbf{V}_R + \mathbf{V}_L + \mathbf{V}_C$$

$$= \mathbf{I}R + \omega L\mathbf{I}\ \underline{/90°} + \frac{\mathbf{I}}{\omega C}\ \underline{/-90°}$$

If we select $\mathbf{I}$ as a reference phasor so that $\mathbf{I} = I_M \underline{/0°}$, then if $\omega L I_M > I_M/\omega C$, the phasor diagram will be of the form shown in Fig. 8.15b. Specifically, if $\omega = 377$ rad/s (i.e., $f = 60$ Hz), then $\omega L = 6$ and $1/\omega C = 2$. Under these conditions the phasor diagram is as shown in Fig. 8.15c. If, however, we select $\mathbf{V}_S$ as reference with, for example,

$$v_S(t) = 12\sqrt{2}\cos(377t + 90°)\text{ V}$$

then

$$\mathbf{I} = \frac{\mathbf{V}}{\mathbf{Z}} = \frac{12\sqrt{2}\ \underline{/90°}}{4 + j6 - j2}$$

$$= \frac{12\sqrt{2}\ \underline{/90°}}{4\sqrt{2}\ \underline{/45°}}$$

$$= 3\ \underline{/45°}\text{ A}$$

and the entire phasor diagram, as shown in Figs. 8.15b and c, is rotated 45°, as shown in Fig. 8.15d.   ❏

EXTENSION EXERCISE

E8.11   Draw a phasor diagram illustrating all currents and voltages for the network in Fig. E8.11.

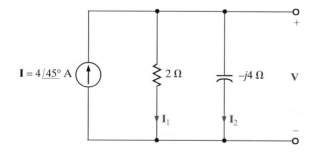

Figure E8.11

ANSWER:

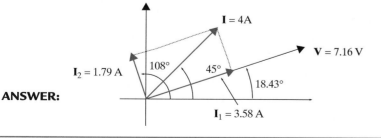

8.7 BASIC ANALYSIS USING KIRCHOFF'S LAWS

We have shown that Kirchhoff's laws apply in the frequency domain, and therefore they can be used to compute steady-state voltages and currents in ac circuits. This approach involves expressing these voltages and currents as phasors, and once this is done, the ac steady-state analysis employing phasor equations is performed in an identical fashion to that used in the dc analysis of resistive circuits. Complex number algebra is the tool that is used for the mathematical manipulation of the phasor equations, which, of course, have complex coefficients. We will begin by illustrating that the techniques we have applied in the solution of dc resistive circuits are valid in ac circuit analysis also—the only difference being that in steady-state ac circuit analysis the algebraic phasor equations have complex coefficients.

Problem-solving Strategies for AC Steady-state Analysis

- For relatively simple circuits (e.g., those with a single source), use
  - Ohm's law for ac analysis, i.e., $\mathbf{V} = \mathbf{IZ}$
  - The rules for combining $\mathbf{Z}_s$ and $\mathbf{Y}_s$
  - KCL and KVL
  - Current and voltage division
- For more complicated circuits with multiple sources, use
  - Nodal analysis
  - Loop of mesh analysis
  - Superposition
  - Source exchange
  - Thévenin's and Norton's theorems

Technique

1. Compute $\mathbf{I}_1$.

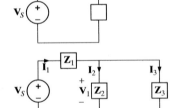

2. Determine $\mathbf{V}_1 = \mathbf{V}_S - \mathbf{I}_1 \mathbf{Z}_1$.

   Then $\mathbf{I}_2 = \dfrac{\mathbf{V}_1}{\mathbf{Z}_2}$ and

   $\mathbf{I}_3 = \dfrac{\mathbf{V}_1}{\mathbf{Z}_3}$

Current and voltage division are also applicable.

At this point, it is important for the reader to understand that in our manipulation of algebraic phasor equations with complex coefficients we will, for the sake of simplicity, normally carry only two digits to the right of the decimal point. In doing so, we will introduce round-off errors in our calculations. Nowhere are these errors more evident than when two or more approaches are used to solve the same problem, as is done in the following example.

EXAMPLE 8.16

We wish to calculate all the voltages and currents in the circuit shown in Fig. 8.16a.

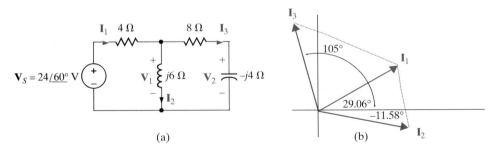

(a)   (b)

Figure 8.16 (a) Example ac circuit, (b) phasor diagram for the currents.

SOLUTION   Our approach will be as follows. We will calculate the total impedance seen by the source $\mathbf{V}_S$. Then we will use this to determine $\mathbf{I}_1$. Knowing $\mathbf{I}_1$, we can compute $\mathbf{V}_1$ using KVL. Knowing $\mathbf{V}_1$, we can compute $\mathbf{I}_2$ and $\mathbf{I}_3$, and so on.

The total impedance seen by the source $\mathbf{V}_S$ is

$$
\begin{aligned}
\mathbf{Z}_{eq} &= 4 + \frac{(j6)(8 - j4)}{j6 + 8 - j4} \\
&= 4 + \frac{24 + j48}{8 + j2} \\
&= 4 + 4.24 + j4.94 \\
&= 9.61 \; \underline{/30.94°} \; \Omega
\end{aligned}
$$

Then

$$
\begin{aligned}
\mathbf{I}_1 &= \frac{\mathbf{V}_S}{\mathbf{Z}_{eq}} = \frac{24 \; \underline{/60°}}{9.61 \; \underline{/30.94°}} \\
&= 2.5 \; \underline{/29.06°} \; \text{A}
\end{aligned}
$$

$\mathbf{V}_1$ can be determined using KVL:

$$
\begin{aligned}
\mathbf{V}_1 &= \mathbf{V}_S - 4\mathbf{I}_1 \\
&= 24 \; \underline{/60°} - 10 \; \underline{/29.06°} \\
&= 3.26 + j15.93 = 16.26 \; \underline{/78.43°} \; \text{V}
\end{aligned}
$$

Note that $\mathbf{V}_1$ could also be computed via voltage division:

$$
\mathbf{V}_1 = \frac{\mathbf{V}_S \dfrac{(j6)(8 - j4)}{j6 + 8 - j4}}{4 + \dfrac{(j6)(8 - j4)}{j6 + 8 - 4}} \; \text{V}
$$

which from our previous calculation is

$$
\begin{aligned}
\mathbf{V}_1 &= \frac{(24 \; \underline{/60°})(6.51 \; \underline{/49.36°})}{9.61 \; \underline{/30.94°}} \\
&= 16.26 \; \underline{/78.42°} \; \text{V}
\end{aligned}
$$

Knowing $\mathbf{V}_1$, we can calculate both $\mathbf{I}_2$ and $\mathbf{I}_3$:

$$
\begin{aligned}
\mathbf{I}_2 &= \frac{\mathbf{V}_1}{j6} = \frac{16.26 \; \underline{/78.43°}}{6 \; \underline{/90°}} \\
&= 2.71 \; \underline{/-11.58°} \; \text{A}
\end{aligned}
$$

and

$$
\begin{aligned}
\mathbf{I}_3 &= \frac{\mathbf{V}_1}{8 - j4} \\
&= 1.82 \; \underline{/105°} \; \text{A}
\end{aligned}
$$

Note that $\mathbf{I}_2$ and $\mathbf{I}_3$ could have been calculated by current division. For example, $\mathbf{I}_2$ could be determined by

$$
\begin{aligned}
\mathbf{I}_2 &= \frac{\mathbf{I}_1(8 - j4)}{8 - j4 + j6} \\
&= \frac{(2.5 \ \underline{/29.06°})(8.94 \ \underline{/-26.57°})}{8 + j2} \\
&= 2.71 \ \underline{/-11.55°} \text{ A}
\end{aligned}
$$

Finally, $\mathbf{V}_2$ can be computed as

$$
\begin{aligned}
\mathbf{V}_2 &= \mathbf{I}_3(-j4) \\
&= 7.28 \ \underline{/15°} \text{ V}
\end{aligned}
$$

This value could also have been computed by voltage division. The phasor diagram for the currents $\mathbf{I}_1$, $\mathbf{I}_2$, and $\mathbf{I}_3$ is shown in Fig. 8.16b and is an illustration of KCL.

Finally, the reader is encouraged to work the problem in reverse; that is, given $\mathbf{V}_2$, find $\mathbf{V}_S$. Note that if $\mathbf{V}_2$ is known, $\mathbf{I}_3$ can be computed immediately using the capacitor impedance. Then $\mathbf{V}_2 + \mathbf{I}_3(8)$ yields $\mathbf{V}_1$. Knowing $\mathbf{V}_1$ we can find $\mathbf{I}_2$. Then $\mathbf{I}_2 + \mathbf{I}_3 = \mathbf{I}_1$, and so on. Note that this analysis, which is the subject of Extension Exercise E8.12, involves simply a repeated application of Ohm's law, KCL, and KVL.❏

EXTENSION EXERCISE

E8.12   In the network in Fig. E8.12, $\mathbf{V}_o$ is known to be $8 \ \underline{/45°}$ V. Compute $\mathbf{V}_S$.

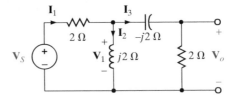

Figure E8.12

ANSWER: $\mathbf{V}_S = 17.89 \ \underline{/-18.43°}$ V.

8.8 ANALYSIS TECHNIQUES

In this section we revisit the circuit analysis methods that were successfully applied earlier to dc circuits and illustrate their applicability to ac steady-state analysis. The vehicle we employ to present these techniques is examples in which all the theorems, together with nodal analysis and loop analysis, are used to obtain a solution.

EXAMPLE 8.17

Let us determine the current $\mathbf{I}_o$ in the network in Fig. 8.17a using nodal analysis, loop analysis, superposition, source exchange, Thévenin's theorem, and Norton's theorem.

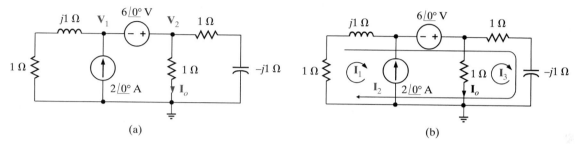

Figure 8.17 Circuits used in Example 8.17 for node and loop analysis.

SOLUTION

(1) Nodal Analysis   We begin with a nodal analysis of the network. The KCL equation for the supernode that includes the voltage source is

Summing the current, leaving the supernode. Outbound currents have a positive sign.

$$\frac{\mathbf{V}_1}{1 + j} - 2\,\underline{/0^\circ} + \frac{\mathbf{V}_2}{1} + \frac{\mathbf{V}_2}{1 - j} = 0$$

and the associated KVL constraint equation is

$$\mathbf{V}_1 + 6\,\underline{/0^\circ} = \mathbf{V}_2$$

Solving for $\mathbf{V}_1$ in the second equation and using this value in the first equation yields

$$\frac{\mathbf{V}_2 - 6\,\underline{/0^\circ}}{1 + j} - 2\,\underline{/0^\circ} + \mathbf{V}_2 + \frac{\mathbf{V}_2}{1 - j} = 0$$

or

$$\mathbf{V}_2\left[\frac{1}{1 + j} + 1 + \frac{1}{1 - j}\right] = \frac{6 + 2 + 2j}{1 + j}$$

Solving for $\mathbf{V}_2$, we obtain

$$\mathbf{V}_2 = \left(\frac{4 + j}{1 + j}\right)\mathbf{V}$$

DRILL

D8.5   Use nodal analysis to find I_o.

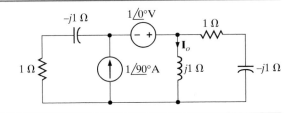

ANSWER: $I_o = 1.581 \; \underline{/-18.4°}$ A.

Therefore,

$$I_o = \frac{4 + j}{1 + j} = \left(\frac{5}{2} - \frac{3}{2}j\right)A$$

(2) Loop Analysis   The network in Fig. 8.17b is used to perform a loop analysis. Note that one loop current is selected that passes through the independent current source. The three loop equations are

> *Just as in a dc analysis, the loop equations assume a decrease in potential level is + and an increase is −.*

$$I_1 = -2 \; \underline{/0°}$$

$$1(I_1 + I_2) + j1(I_1 + I_2) - 6 \; \underline{/0°} + 1(I_2 + I_3) - j1(I_2 + I_3) = 0$$

$$1I_3 + 1(I_2 + I_3) - j1(I_2 + I_3) = 0$$

Combining the first two equations yields

$$I_2(2) + I_3(1 - j) = 8 + 2j$$

The third loop equation can be simplified to the form

$$I_2(1 - j) + I_3(2 - j) = 0$$

Solving this last equation for I_2 and substituting the value into the previous equation yields

$$I_3\left[\frac{-4 + 2j}{1 - j} + 1 - j\right] = 8 + 2j$$

or

$$I_3 = \frac{-10 + 6j}{4}$$

DRILL

D8.6   Find I_o using the loop currents shown in the circuit.

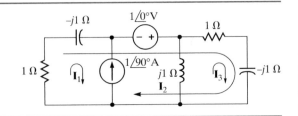

ANSWER: $I_o = I_3 = 1.581 \; \underline{/-18.40}$ A.

and finally

$$\mathbf{I}_o = -\mathbf{I}_3 = \left(\frac{5}{2} - \frac{3}{2}j\right)\mathrm{A}$$

(3) Superposition   In using superposition, we apply one independent source at a time. The network in which the current source acts alone is shown in Fig. 8.18a. By combining the two parallel impedances on each end of the network, we obtain the circuit in Fig. 8.18b, where

In applying superposition in this case, each source is applied independently and the results are added to obtain the solution.

$$\mathbf{Z}' = \frac{(1+j)(1-j)}{(1+j)+(1-j)} = 1\ \Omega$$

Therefore, using current division

$$\mathbf{I}_o' = 1\ \underline{/0°}\ \mathrm{A}$$

The circuit in which the voltage source acts alone is shown in Fig. 8.18c. The voltage $\mathbf{V}_1''$ obtained using voltage division is

D8.7   Find $\mathbf{I}_o$ in Drill 8.6 using superposition.

$$\mathbf{V}_1'' = \frac{(6\ \underline{/0°})\left[\dfrac{1(1-j)}{1+1-j}\right]}{1+j+\left[\dfrac{1(1-j)}{1+1-j}\right]}$$

$$= \frac{6(1-j)}{4}\ \mathrm{V}$$

and hence,

$$\mathbf{I}_0'' = \frac{6}{4}(1-j)\mathrm{A}$$

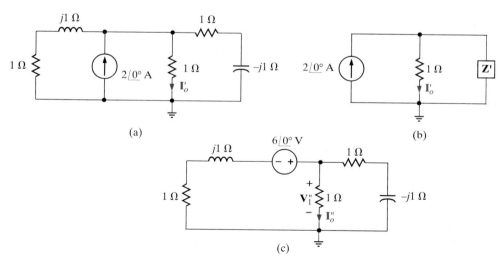

Figure 8.18 Circuits used in Example 8.17 for a superposition analysis.

Then

$$\mathbf{I}_o = \mathbf{I}'_o + \mathbf{I}''_o = 1 + \frac{6}{4}(1 - j) = \left(\frac{5}{2} - \frac{3}{2}j\right)\mathrm{A}$$

In source exchange, a voltage source in series with an impedance can be exchanged for a current source in parallel with the impedance, and vice versa. Repeated application systematically reduces the number of circuit elements.

(4) Source exchange As a first step in the source exchange approach, we exchange the current source and parallel impedance for a voltage source in series with the impedance, as shown in Fig. 8.19a.

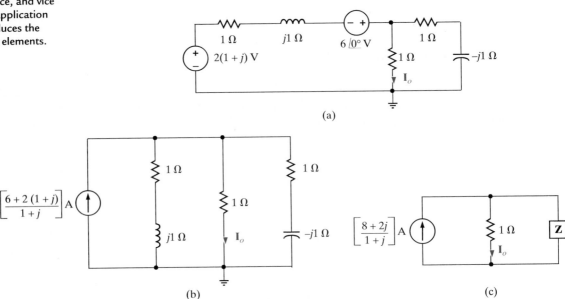

(a)

(b)   (c)

Figure 8.19 Circuits used in Example 8.17 for a source exchange analysis.

Adding the two voltage sources and transforming them and the series impedance into a current source in parallel with that impedance is shown in Fig. 8.19b. Combining the two impedances that are in parallel with the 1-Ω resistor produces the network in Fig. 8.19c, where

DRILL

D8.8 Find $\mathbf{V}_o$ using source exchange.

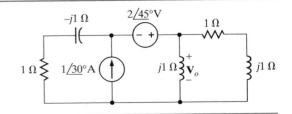

ANSWER: $\mathbf{V}_o = 1.486 \,\underline{/110.7°}$ V.

$$\mathbf{Z} = \frac{(1 + j)(1 - j)}{1 + j + 1 - j} = 1 \, \Omega$$

Therefore, using current division,

$$\mathbf{I}_o = \left(\frac{8 + 2j}{1 + j}\right)\left(\frac{1}{1 + j}\right) = \frac{4 + j}{1 + j}$$

$$= \left(\frac{5}{2} - \frac{3}{2}j\right)\mathrm{A}$$

(5) Thévenin Analysis   In applying Thévenin's theorem to the circuit in Fig. 8.20a, we first find the open-circuit voltage, $\mathbf{V}_{oc}$, as shown in Fig. 8.20a. In order to simplify the analysis, we perform a source exchange on the left end of the network, which results in the circuit in Fig. 8.20b. Now using voltage division

$$\mathbf{V}_{oc} = [6 + 2(1 + j)]\left[\frac{1 - j}{1 - j + 1 + j}\right]$$

or

$$\mathbf{V}_{oc} = (5 - 3j)\mathrm{V}$$

The Thévenin equivalent impedance, $\mathbf{Z}_{Th}$, obtained at the open-circuit terminals when the current source is replaced with an open circuit and the voltage source is replaced with a short circuit is shown in Fig. 8.20c and calculated to be

$$\mathbf{Z}_{Th} = \frac{(1 + j)(1 - j)}{1 + j + 1 - j} = 1 \, \Omega$$

In this Thévenin analysis

1   Remove the 1-Ω load and find the voltage across the open terminals, $\mathbf{V}_{oc}$.

2   Determine the impedance $\mathbf{Z}_{Th}$ at the open terminals with all sources made zero.

3   Construct the following circuit and determine $\mathbf{I}_o$.

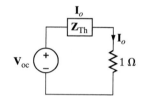

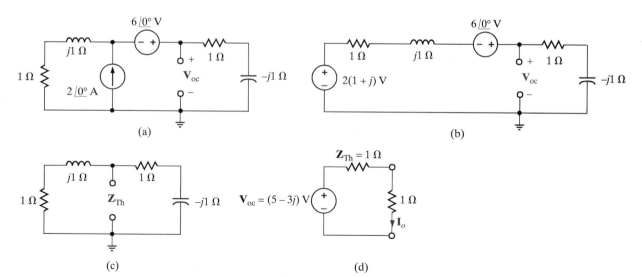

Figure 8.20 Circuits used in Example 8.17 for a Thévenin analysis.

DRILL

D8.9 Use Thévenin's theorem to find $\mathbf{V}_o$ in the following circuit.

ANSWER: 1. Removing just the resistor produces an unsolvable (impossible) circuit.

2. Removing just the capaitor produces $\mathbf{V}_{oc} = 1.5 + 0.5j$ V, $\mathbf{Z}_{Th} = 0.5 + 0.5j$ Ω.

3. Removing both resistor and capacitor produces $\mathbf{V}_{oc} = 1 + 2j$ V, $\mathbf{Z}_{Th} = j$ Ω. Either approach 2 or 3 produces $\mathbf{V}_o = 2 - j$ V.

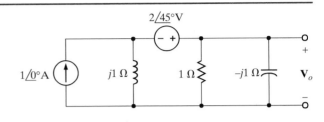

In this Norton analysis

1 Remove the 1-Ω load and find the current $\mathbf{I}_{sc}$ through the short-circuited terminals.

2 Determine the impedance $\mathbf{Z}_{Th}$ at the open load terminals with all sources made zero.

3 Construct the following circuit and determine $\mathbf{I}_o$.

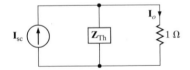

Connecting the Thévenin equivalent circuit to the 1-Ω resistor containing $\mathbf{I}_o$ in the original network yields the circuit in Fig. 8.20d. The current $\mathbf{I}_o$ is then

$$\mathbf{I}_o = \left(\frac{5}{2} - \frac{3}{2}j\right)\text{A}$$

(6) Norton Analysis Finally, in applying Norton's theorem to the circuit in Fig. 8.17a, we calculate the short-circuit current, $\mathbf{I}_{sc}$, using the network in Fig. 8.21a. Note that because of the short circuit, the voltage source is directly across the impedance in the left-most branch. Therefore,

$$\mathbf{I}_1 = \frac{6 \; \underline{/0^\circ}}{1 + j}$$

Then using KCL

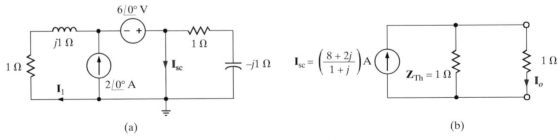

(a)   (b)

Figure 8.21 Circuits used in Example 8.17 for a Norton analysis.

$$\mathbf{I}_{sc} = \mathbf{I}_1 + 2 \underline{/0^\circ} = 2 + \frac{6}{1 + j}$$

$$= \left(\frac{8 + 2j}{1 + j}\right)\mathbf{A}$$

The Thévenin equivalent impedance, $\mathbf{Z}_{Th}$, is known to be 1 Ω and, therefore, connecting the Norton equivalent to the 1-Ω resistor containing $\mathbf{I}_o$ yields the network in Fig. 8.21b. Using current division, we find that

$$\mathbf{I}_o = \frac{1}{2}\left(\frac{8 + 2j}{1 + j}\right)$$

$$= \left(\frac{5}{2} - \frac{3}{2}j\right)\mathbf{A} \qquad \square$$

Let us now consider an example containing a dependent source.

EXAMPLE 8.18

Let us determine the voltage $\mathbf{V}_o$ in the circuit in Fig. 8.22a. In this example we will use node equations, loop equations, Thévenin's theorem, and Norton's theorem. We will omit the techniques of superposition and source transformation. Why?

How does the presence of a dependent source affect superposition and source exchange?

SOLUTION

(1) Nodal Analysis   In order to perform a nodal analysis, we label the node voltages and identify the supernode as shown in Fig. 8.22b. The constraint equation for the supernode is

$$\mathbf{V}_3 + 12 \underline{/0^\circ} = \mathbf{V}_1$$

and the KCL equations for the nodes of the network are

$$\frac{\mathbf{V}_1 - \mathbf{V}_2}{-j1} + \frac{\mathbf{V}_3 - \mathbf{V}_2}{1} - 4 \underline{/0^\circ} + \frac{\mathbf{V}_3 - \mathbf{V}_o}{1} + \frac{\mathbf{V}_3}{j1} = 0$$

$$\frac{\mathbf{V}_2 - \mathbf{V}_1}{-j1} + \frac{\mathbf{V}_2 - \mathbf{V}_3}{1} - 2\left(\frac{\mathbf{V}_3 - \mathbf{V}_o}{1}\right) = 0$$

$$4 \underline{/0^\circ} + \frac{\mathbf{V}_o - \mathbf{V}_3}{1} + \frac{\mathbf{V}_o}{1} = 0$$

At this point we can solve the foregoing equations using a matrix analysis or, for example, substitute the first and last equations into the remaining two equations, which yields

$$3\mathbf{V}_o - (1 + j)\mathbf{V}_2 = -(4 + j12)$$
$$-(4 + j2)\mathbf{V}_o + (1 + j)\mathbf{V}_2 = 12 + j16$$

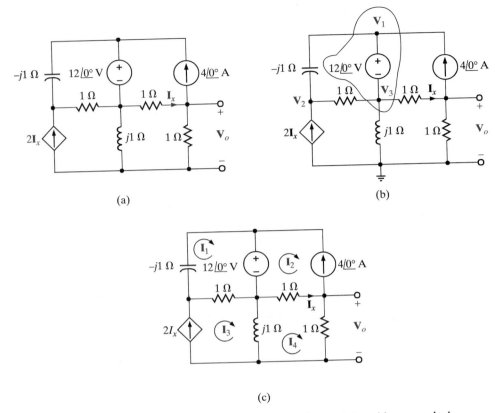

Figure 8.22 Circuits used in Example 8.18 for nodal and loop analysis.

Solving these equations for $\mathbf{V}_o$ yields

$$\mathbf{V}_o = \frac{-(8 + j4)}{1 + j2}$$

$$= -4\ \underline{/143.13°}\ \text{V}$$

(2) Loop Analysis The mesh currents for the network are defined in Fig. 8.22c. The constraint equations for the circuit are

$$\mathbf{I}_2 = -4\ \underline{/0°}$$
$$\mathbf{I}_x = \mathbf{I}_4 - \mathbf{I}_2 = \mathbf{I}_4 + 4\ \underline{/0°}$$
$$\mathbf{I}_3 = 2\mathbf{I}_x = 2\mathbf{I}_4 + 8\ \underline{/0°}$$

The KVL equations for mesh 1 and mesh 4 are

$$-j1\mathbf{I}_1 + 1(\mathbf{I}_1 - \mathbf{I}_3) = -12\ \underline{/0°}$$
$$j1(\mathbf{I}_4 - \mathbf{I}_3) + 1(\mathbf{I}_4 - \mathbf{I}_2) + 1\mathbf{I}_4 = 0$$

Note that if the constraint equations are substituted into the second KVL equation, the only unknown in the equation is $\mathbf{I}_4$. This substitution yields

$$\mathbf{I}_4 = -4\ \underline{/143.13^\circ}\ \text{A}$$

and hence,

$$\mathbf{V}_o = -4\ \underline{/143.13^\circ}\ \text{V}$$

(3) Thévenin's Theorem   In applying Thévenin's theorem, we will find the open-circuit voltage and then determine the Thévenin equivalent impedance using a test source at the open-circuit terminals. We could determine the Thévenin equivalent impedance by calculating the short-circuit current; however, we will determine this current when we apply Norton's theorem.

The open-circuit voltage is determined from the network in Fig. 8.23a. Note that $\mathbf{I}_x' = 4\ \underline{/0^\circ}$ A and since $2\mathbf{I}_x'$ flows through the inductor, the open-circuit voltage $\mathbf{V}_{oc}$ is

$$\mathbf{V}_{oc} = -1(4\ \underline{/0^\circ}) + j1(2\mathbf{I}_x')$$
$$= -4 + j8\ \text{V}$$

To determine the Thévenin equivalent impedance, we turn off the independent sources, apply a test voltage source to the output terminals, and compute the current leaving the test source. As shown in Fig. 8.23b, since $\mathbf{I}_x''$ flows in the test source, KCL requires that the current in the inductor be $\mathbf{I}_x''$ also. KVL around the mesh containing the test source indicates that

$$j1\mathbf{I}_x'' - 1\mathbf{I}_x'' - \mathbf{V}_{\text{test}} = 0$$

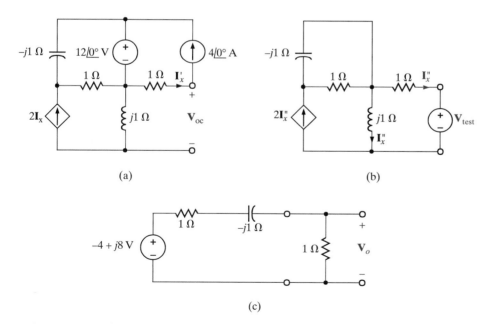

(a)   (b)

(c)

Figure 8.23 Circuits used in Example 8.18 when applying Thévenin's theorem.

Therefore,

$$\mathbf{I}_x'' = \frac{-\mathbf{V}_{\text{test}}}{1 - j}$$

Then

$$\mathbf{Z}_{\text{Th}} = \frac{\mathbf{V}_{\text{test}}}{-\mathbf{I}_x''}$$

$$= 1 - j \, \Omega$$

If the Thévenin equivalent network is now connected to the load, as shown in Fig. 8.23c, the output voltage $\mathbf{V}_o$ is found to be

$$\mathbf{V}_o = \frac{-4 + 8j}{2 - j1} \quad (1)$$

$$= -4 \, \underline{/143.13°} \, \text{V}$$

(4) Norton's Theorem In using Norton's theorem, we will find the short-circuit current from the network in Fig. 8.24a. Once again, using the supernode, the constraint and KCL equations are

$$\mathbf{V}_3 + 12 \, \underline{/0°} = \mathbf{V}_1$$

$$\frac{\mathbf{V}_2 - \mathbf{V}_1}{-j1} + \frac{\mathbf{V}_2 - \mathbf{V}_3}{1} - 2\mathbf{I}_x''' = 0$$

$$\frac{\mathbf{V}_1 - \mathbf{V}_2}{-j1} + \frac{\mathbf{V}_3 - \mathbf{V}_2}{1} - 4 \, \underline{/0°} + \frac{\mathbf{V}_3}{j1} + \mathbf{I}_x''' = 0$$

$$\mathbf{I}_x''' = \frac{\mathbf{V}_3}{1}$$

Substituting the first and last equations into the remaining equations yields

$$(1 + j)\mathbf{V}_2 - (3 + j)\mathbf{I}_x''' = j12$$
$$-(1 + j)\mathbf{V}_2 + (2)\mathbf{I}_x''' = 4 - j12$$

Solving these equations for $\mathbf{I}_x'$ yields

$$\mathbf{I}_x''' = \frac{-4}{1 + j} \, \mathbf{A}$$

The KCL equation at the right-most node in the network in Fig. 8.24a is

$$\mathbf{I}_x''' = 4 \, \underline{/0°} + \mathbf{I}_{\text{sc}}$$

Solving for $\mathbf{I}_{\text{sc}}$, we obtain

$$\mathbf{I}_{\text{sc}} = \frac{-(8 + j4)}{1 + j} \, \mathbf{A}$$

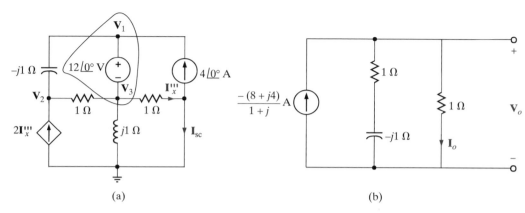

Figure 8.24 Circuits used in Example 8.18 when applying Norton's theorem.

The Thévenin equivalent impedance was found earlier to be

$$\mathbf{Z}_{Th} = 1 - j\,\Omega$$

Using the Norton equivalent network, the original network is reduced to that shown in Fig. 8.24b. The voltage $\mathbf{V}_o$ is then

$$\mathbf{V}_o = \frac{-(8 + j4)}{1 + j}\left[\frac{(1)(1 - j)}{1 + 1 - j}\right]$$

$$= -4\left[\frac{3 - j}{3 + j}\right]$$

$$= -4\,\underline{/-36.87°}\ \text{V}$$

❏

EXTENSION EXERCISES

E8.13   Use nodal analysis to find $\mathbf{V}_o$ in the network in Fig. E8.13.

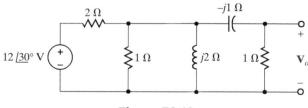

Figure E8.13

ANSWER: $\mathbf{V}_o = 2.12\,\underline{/75°}\ \text{V}.$

E8.14 Use mesh equations to find $\mathbf{V}_o$ in the network in Fig. E8.14.

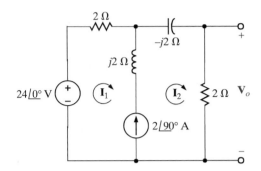

Figure E8.14

ANSWER: $\mathbf{V}_o = 10.88 \;/36°$ V.

E8.15 Using superposition, find $\mathbf{V}_o$ in the network in Fig. E8.15.

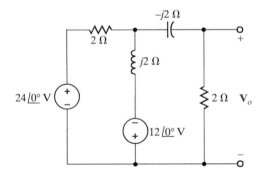

Figure E8.15

ANSWER: $\mathbf{V}_o = 12 \;/90°$ V.

E8.16 Find $\mathbf{V}_o$ in the network in Fig. E8.15 using source transformation.

ANSWER: $\mathbf{V}_o = 12 \;/90°$ V.

E8.17 Find $\mathbf{V}_o$ in the network in Fig. E8.14 using Thévenin's theorem.

ANSWER: $\mathbf{V}_o = 10.88 \;/36.3°$ V.

E8.18 Use Norton's theorem to find $\mathbf{V}_o$ in the network in Fig. E8.15.

ANSWER: $\mathbf{V}_o = 12 \;/90°$ V.

8.9 AC PSPICE ANALYSIS USING SCHEMATIC CAPTURE

Introduction

In this chapter we have found that an ac steady-state analysis is facilitated by the use of phasors. PSPICE can perform ac steady-state simulations, outputting magnitude and phase data for any voltage or current phasors of interest. Additionally, PSPICE can perform an AC SWEEP in which the frequency of the sinusoidal sources is varied over a user-defined range. In this case, the simulation results are the magnitude and phase of every node voltage and branch current as a function of frequency.

We will introduce five new *Schematics*/PSPICE topics in this section: defining AC sources, simulating at a single frequency, simulating over a frequency range, using the PROBE feature to create plots and finally saving and printing these plots. *Schematics* fundamentals such as getting parts, wiring and changing part names and values have already been discussed in Chapter 4. As in Chapter 4, we will use the following font conventions. Uppercase text refers to programs and utilities within PSPICE such as the AC SWEEP feature and PROBE graphing utility. All boldface text, whether upper or lowercase, denotes keyboard or mouse entries. For example, when placing a resistor into a circuit schematic, one must specify the resistor **VALUE** using the keyboard. The case of the boldface text matches that used in PSPICE.

Defining AC Sources

Figure 8.25 shows the circuit we will simulate at a frequency of 60 Hz. We will continue to follow the flowchart shown in Figure 4.21 in performing this simulation. Inductor and capacitor parts are in the ANALOG library and are called L and C, respectively. The AC source, VAC, is in the SOURCE library. Figure 8.26 shows the resulting *Schematics* diagram after wiring and editing the part's names and values.

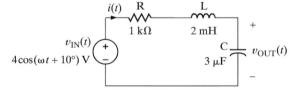

Figure 8.25 Circuit for ac simulation.

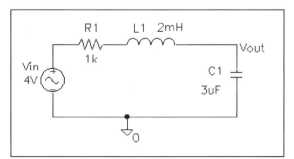

Figure 8.26 The *Schematics* diagram for the circuit in Figure 8.25.

To set up the AC source for simulation, double-click on the source symbol to open its ATTRIBUTES box which is shown, after editing, in Figure 8.27. As indicated earlier, we deselect the fields **Include Non-changeable Attributes and Include System-defined Attributes**. Each line in the ATTRIBUTES box is called an attribute of the ac source. Each attribute has a name and a value. The **DC** attribute is the dc value of the source for dc analyses. The **ACMAG** and **ACPHASE** attributes set the magnitude and phase of the phasor representing V_{in} for ac analyses. Each of these attributes defaults to zero. The value of the **ACMAG** attribute was set to 4 V when we created the schematic in Fig. 8.26. To set the **ACPHASE** attribute to 10°, click on the **ACPHASE** attribute line, enter 10 in the **Value** field, press **Save Attr** and **OK**. When the ATTRIBUTE box looks like that shown in Fig. 8.27, the source is ready for simulation.

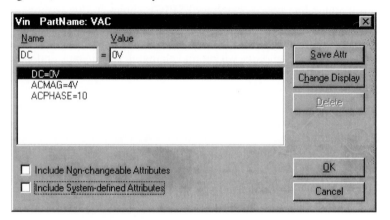

Figure 8.27 Setting the ac source phase angle.

Single Frequency AC Simulations

Next, we must specify the frequency for simulation. This is done by selecting **Setup** from the **Analysis** menu. The SETUP box in Fig. 8.28 should appear. If we double-click on the text **AC Sweep**, the AC SWEEP AND NOISE ANALYSIS window in Fig. 8.29 will open. All of the fields in Fig. 8.29 have been set for our 60 Hz simulation.

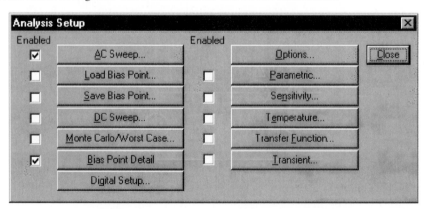

Figure 8.28 The ANALYSIS SETUP window.

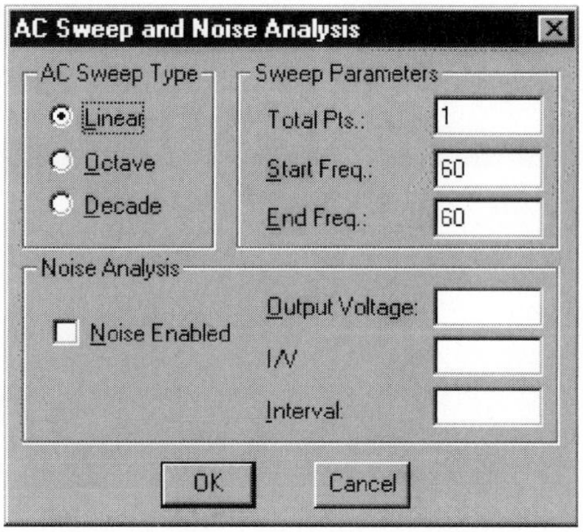

Figure 8.29 Setting the frequency range for a single frequency simulation.

Since the simulation will be performed at only one frequency, 60 Hz, graphing the simulation results is not an attractive option. Instead, we will write the magnitude and phase of the phasors V_{OUT} and I to the output file using the VPRINT1 and IPRINT parts, from the SPECIAL library, which have been added to the circuit diagram, as shown in Fig. 8.30. The VPRINT1 part acts as a voltmeter, measuring the voltage at any single node with respect to the ground node. There is also a VPRINT2 part that measures the voltage between any two nonreference nodes. Similarly, the IPRINT part acts as an ammeter and must be placed in series with the branch current of interest. By convention, current in the IPRINT part is assumed to exit from the negatively marked terminal. To find the clockwise loop current, as defined in Fig. 8.25, the IPRINT part has been flipped. The FLIP command is in the EDIT menu.

After placing the VPRINT1 part, double click on it to open its ATTRIBUTES box, shown in Fig. 8.31. The VPRINT1 part can be configured to meter the node voltage in any kind of simulation; dc, ac, or transient. Since an ac analysis was specified in the SETUP window in Fig. 8.28, the values of the **AC**, **MAG**, and **PHASE** attributes are set to **Y**, where "Y" stands for YES. This process is repeated for the IPRINT part. When we return to *Schematics*, the simulation is ready to run.

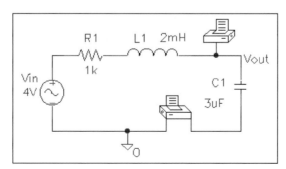

Figure 8.30 A *Schematics* diagram ready for single frequency ac simulation.

When an AC SWEEP is performed, PSPICE, unless instructed otherwise, will attempt to plot the results using the PROBE plotting program. To turn off this feature, select **Probe Setup** in the **Analysis** menu. When the PROBE SETUP window shown in Fig. 8.32 appears, select **Do Not Auto-Run Probe** and **OK**.

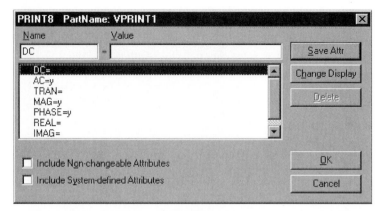

Figure 8.31 Setting the VPRINT1 measurements for ac magnitude and phase.

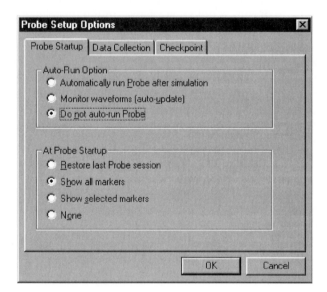

Figure 8.32 The PROBE SETUP window.

The circuit is simulated by selecting **Simulate** from the **Analysis** menu. Since the results are in the output file, select **Examine Output** from the **Analysis** menu to view the data. At the bottom of the file, we find the results as seen in Fig. 8.33: $\mathbf{V}_{OUT} = 2.651 \; \underline{/-38.54°}$ V and $\mathbf{I} = 2.998 \; \underline{/51.46°}$ mA.

Variable Frequency AC Simulations

In order to sweep the frequency over a range, 1 Hz to 10 MHz, for example, return to the AC SWEEP AND NOISE ANALYSIS box shown in Fig. 8.29. Change the fields to that shown in Fig. 8.34. Since the frequency range is so large, we have chosen a log axis for frequency with 50 data points in each decade. We can now plot the data using the PROBE utility. This procedure requires two steps. First, we remove the VPRINT1 and IPRINT

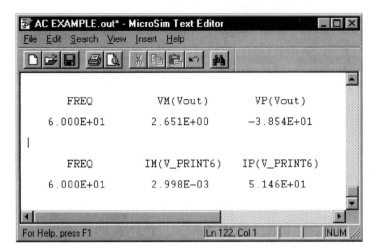

Figure 8.33 Magnitude and phase data for V_{OUT} and I are at the bottom of the output file.

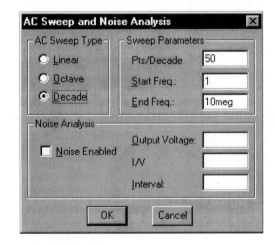

Figure 8.34 Setting the frequency range for a swept frequency simulation.

parts in Fig. 8.30. Second, we return to the PROBE SETUP window shown in Fig. 8.32, and select **Automatically Run Probe After Simulation** and **OK**.

Creating Plots in PROBE

When the PSPICE simulation is finished, the PROBE window shown in Fig. 8.35 will open. Note that the frequency is displayed on a log axis, as requested. In order to plot the magnitude and phase of V_{OUT}, select **Add** from the **Trace** menu. The ADD TRACES window shown in Fig. 8.36 will appear. To display the magnitude of V_{OUT}, we select **V(V_{OUT})** from the left column. When a voltage or current is selected, the magnitude of the phasor will be plotted. Now the PROBE window should look like that shown in Fig. 8.37.

Before adding the phase to the plot, we note that V_{OUT} spans a small range (i.e., 0 to 4 V). Since the phase change could span a much greater range, we will plot the phase on

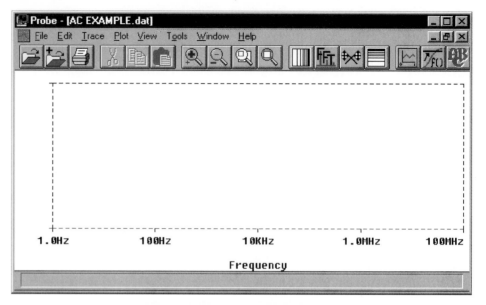

Figure 8.35 The PROBE window.

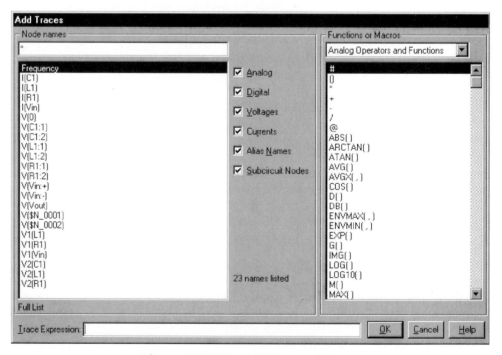

Figure 8.36 The Add Traces window.

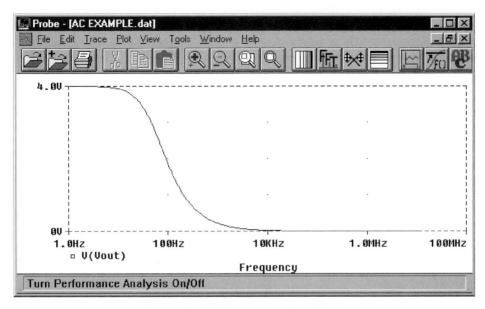

Figure 8.37 The magnitude of $\mathbf{V}_{OUT}$.

a second *y*-axis. From the **Plot** menu, select **Add Y Axis**. To add the phase to the plot, select **Add** from the **Trace** menu. On the right side of the ADD TRACES window in Fig. 8.36, scroll down to the entry, **P()**. Click on that, and then click on **V(Vout)** in the left column. The TRACE EXPRESSION line at the bottom of the window will contain the expression P(V(Vout))—the phase of Vout. Figure 8.38 shows the PROBE plot for both magnitude and phase of $\mathbf{V}_{OUT}$.

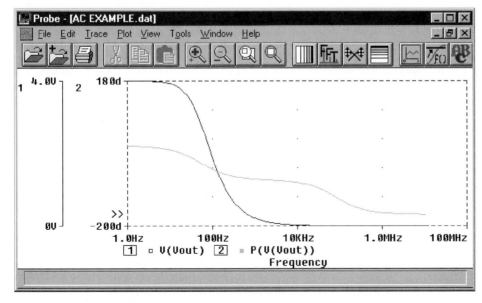

Figure 8.38 The magnitude and phase of $\mathbf{V}_{OUT}$.

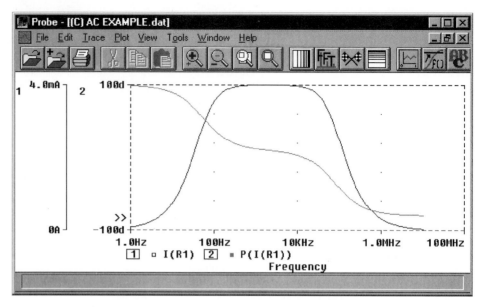

Figure 8.39 The magnitude and phase of the current, I.

In order to plot the current, I, on a new plot, we select **New** from the **Window** menu. Then, add the traces for the magnitude and phase of the current through R1 (PSPICE calls it I(R1)) using the process described above for plotting the magnitude and phase of Vout. The results are shown in Fig. 8.39.

The procedures for saving and printing PROBE plots, as well as the techniques for plot manipulation and data extraction, are described in Chapter 6.

EXAMPLE 8.19

Using the PSPICE *Schematics* editor, draw the circuit in Figure 8.40, and use the PROBE utility to create plots for the magnitude and phase of $\mathbf{V_{OUT}}$ and **I**. At what frequency is $|\mathbf{V_{OUT}}| = 1$ V? What are the phases of V_{OUT} and **I** and the magnitude of **I** at that frequency?

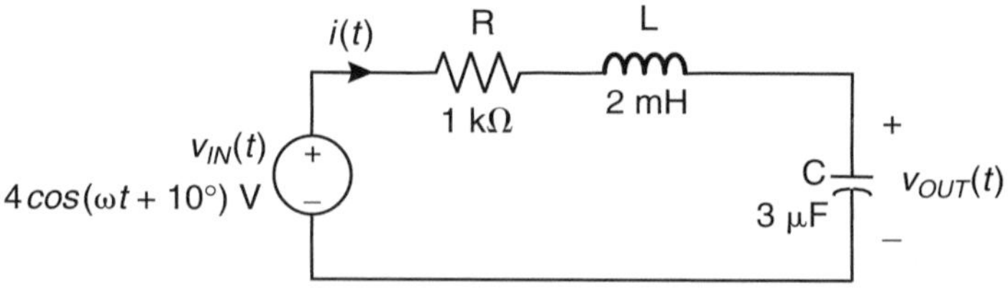

Figure 8.40

i(t)   R   L

$v_{IN}(t)$

$4\cos(\omega t + 10°)$ V

1 kΩ   2 mH

C $v_{OUT}(t)$

3 μF

SOLUTION The *Schematics* diagram for the simulation is shown in Figure 8.41 where the ac source voltage and phase have been set to 4 V and 10° respectively. An AC SWEEP analysis was run over the frequency range 1 Hz to 10 MHz at 50 data points per decade. Simulation results for $\mathbf{V_{OUT}}$ magnitude and phase are shows in Figure 8.42. Using the CURSOR tool in PROBE, $\mathbf{V_{OUT}}$ was found to be 1 $\underline{/-65.6°}$ V

AC Analysis Example

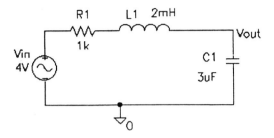

Figure 8.41

at $f = 205.65$ Hz. A second plot, shown in Figure 8.43, was created for the current **I** by plotting I(R1). Using the CURSOR, the current at 205.65 Hz was found to be 3.88 $\underline{/24.3°}$ mA.

When plotting currents, remember that current direction is determined by the netlist as shown below.

```
*   Schematics Netlist   *
R_R1           $N_0002 $N_0001   1k
L_L1           $N_0001 Vout    2mH
V_Vin          SN_0002 0 DC OV AC AV 10
C_C1           0 Vout    3uF
```

Note that R_R1 is connected between nodes 2 and 1 (node 0 is always ground). So, PROBE thinks the current through R_1 flows left-to-right, the same direction we want to plot. If the netlist current direction were opposite our assumption, we would have to plot negative I(RI) with PROBE.   ❏

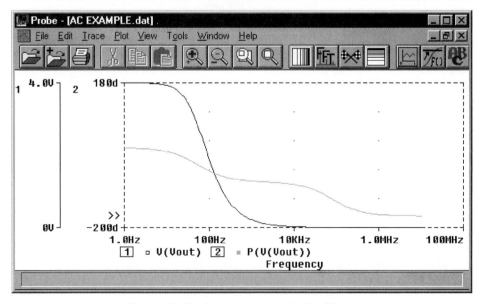

Figure 8.42 Simulation results for $\mathbf{V}_{OUT}$.

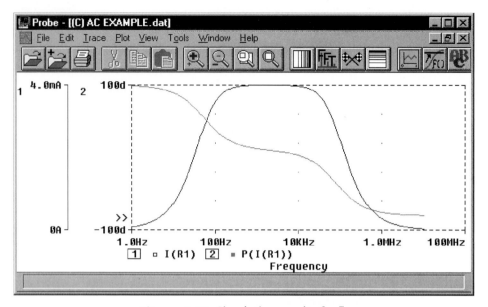

Figure 8.43 Simulation results for **I**.

EXAMPLE 8.20

Using the PSPICE Schematics editor, draw the circuit in Figure 8.44, and use the PROBE utility to create plots for the magnitude and phase of $\mathbf{V_{OUT}}$ and **I**. At what frequency does maximum |**I**| occur? What are the phasors $\mathbf{V_{OUT}}$ and **I** at that frequency?

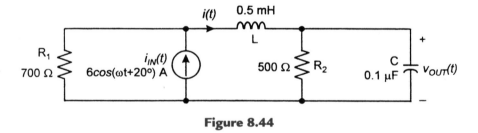

Figure 8.44

SOLUTION The Schematics diagram for the simulation is shown in Figure 8.45 where an AC SWEEP has been setup for the frequency range 10 Hz to 10 MHz at 100 data points per decade. The phase angle of the ac current source part, IAC, is zero degrees and is not changeable in PSPICE. Therefore, we will perform the simulation with the phase of LAC at zero and add 20° to all simulation data. Plots for $\mathbf{V_{OUT}}$ and **I** magnitudes and phases are given in Figures 8.46 and 8.47 respectively. From Figure 8.47 we see that maximum inductor current magnitude occurs at my TI-86 says 31.35, using the fmax features kHz. The corresponding phasors are $\mathbf{I} = 5.94 \underline{/16.09°}$ and $\mathbf{V_{OUT}} = \underline{/-68.11°}$ V. ❑

AC Example 2

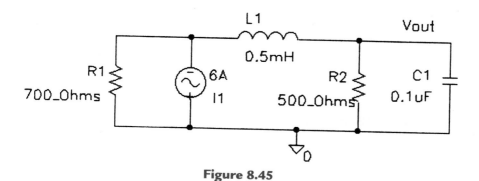

Figure 8.45

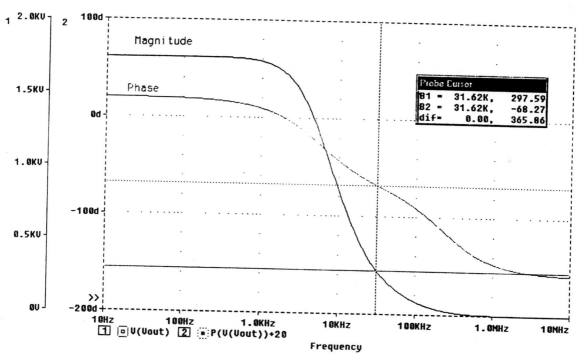

Figure 8.46 Simulation results for $\mathbf{V}_{OUT}$.

EXTENSION PROBLEM

E8.19 Using the PSPICE Schematics editor, draw the circuit in Figure E8.19, and use the VPRINT1 and IPRINT parts to determine the phasors $\mathbf{V_{OUT}}$ and $\mathbf{I}$ at a frequency of 60 Hz?

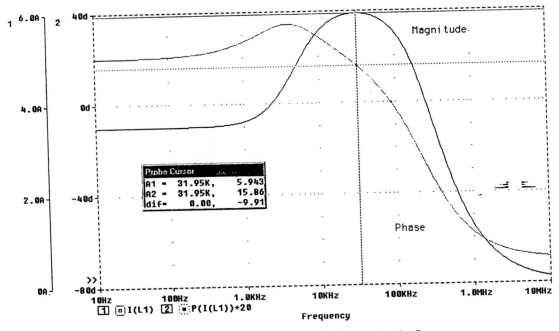

Figure 8.47 Simulation results for **I**.

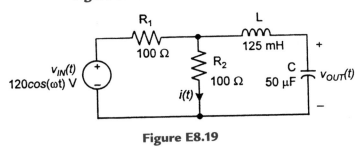

Figure E8.19

ANSWER: $\mathbf{V}_{OUT} = 63.22 \angle{-83.24°}$ V and $\mathbf{I} = 70.64 \angle{-83.24°}$ mA.

8.10 SUMMARY

The sinusoidal function definition The sinusoidal function $x(t) = X_M \sin(\omega t + \theta)$ has an amplitude of X_M, a radian frequency of ω, a period of $2\pi/\omega$, and a phase angle of θ.

The phase lead and phase lag definitions If $x_1(t) = X_{M_1} \sin(\omega t + \theta)$ and $x_2(t) = X_{M_2} \sin(\omega t + \phi)$, $x_1(t)$ leads $x_2(t)$ by $\theta - \phi$ radians and $x_2(t)$ lags $x_1(t)$ by $\theta - \phi$ radians.

The phasor definition The sinusoidal voltage $v(t) = V_M \cos(\omega t + \theta)$ can be written in exponential form as $v(t) = \text{Re}[V_M e^{j(\omega t + \theta)}]$ and in phasor form as $\mathbf{V} = V_M \angle{\theta}$.

The phase relationship in θ_v and θ_i for elements R, L and C   If θ_v and θ_i represent the phase angles of the voltage across and the current through a circuit element, then $\theta_i = \theta_v$ if the element is a resistor, θ_i lags θ_v by 90° if the element is an inductor, θ_i leads θ_v by 90° if the element is a capacitor.

The impedances of R, L and C   Impedance, **Z**, is defined as the ratio of the phasor voltage, **V**, to the phasor current, **I**, where $\mathbf{Z} = R$ for a resistor, $\mathbf{Z} = j\omega L$ for an inductor, and $\mathbf{Z} = 1/j\omega C$ for a capacitor.

The phasor diagrams   Phasor diagrams can be used to display the magnitude and phase relationships of various voltages and currents in a network.

Frequency domain analysis:

1. Represent all voltages, $v_i(t)$, and all currents, $i_j(t)$, as phasors and represent all passive elements by their impedance or admittance.
2. Solve for the unknown phasors in the frequency (ω) domain.
3. Transform the now-known phasors back to the time domain.

Solution techniques for ac steady-state problems:

> Ohm's law
> KCL and KVL
> PSPICE
> Nodal and loop analysis
> Superposition and source exchange
> Thévenin's and Norton's theorems

PROBLEMS

Section 8.1

8.1   Given $v(t) = 10 \cos(377t - 60°)$ V, determine the period and the frequency in hertz.

$$f = \frac{\omega}{2\pi}$$

$$T = \frac{1}{f}$$

8.2   Determine the relative phase relationship of the two waves.

$$v_1(t) = 10 \cos(377t - 30°) \text{ V}$$
$$v_2(t) = 10 \cos(377t + 90°) \text{ V}$$

Use the phase lead or lag relationship.

8.3   Given the following voltage and current

$$i(t) = 5 \sin(377t - 20°) \text{ V}$$

$$v(t) = 10 \cos(377t + 30°) \text{ V}$$

determine the phase relationship between $i(t)$ and $v(t)$.

Convert $i(t)$ to a cosine function and find the phase relationship.

8.4   Determine the phase angles by which $v_1(t)$ leads $i_1(t)$ and $v_1(t)$ leads $i_2(t)$, where

Similar to P8.3.

$$v_1(t) = 4 \sin(377t + 25°)$$
$$i_1(t) = 0.05 \cos(377t - 20°)$$
$$i_2(t) = -0.1 \sin(377t + 45°)$$

Sections 8.2, 8.3, 8.4, 8.5

$i(t) = \dfrac{v(t)}{R}$

$\mathbf{I} = \mathbf{V}/R$

if $v(t) = V_m \cos(\omega t + \theta)$,

$\mathbf{V} = V_m e^{j\phi}$

8.5   Calculate the current in the resistor in Fig. P8.5 if the voltage input is

(a) $v_1(t) = 10 \cos(377t + 80°)$ V

(b) $v_2(t) = 5 \sin(377t + 30°)$ V

Give the answers in both the time and frequency domains.

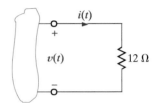

Figure P8.5

$v(t) = L\dfrac{di(t)}{dt}$

$i(t) = \dfrac{1}{L}\int v(t)\,dt, \mathbf{I} = \dfrac{\mathbf{V}}{j\omega L}$

8.6   Calculate the current in the inductor shown in Fig. P8.6 if the voltage input is

(a) $v_1(t) = 10 \cos(377t + 45°)$ V

(b) $v_2(t) = 5 \sin(377t - 90°)$ V

Give the answers in both the time and frequency domains.

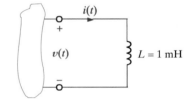

Figure P8.6

$i(t) = C\dfrac{dv(t)}{dt}$

$\mathbf{I} = j\omega C\mathbf{V}$

8.7   Calculate the current in the capacitor shown in Fig. P8.7 if the voltage input is

(a) $v_1(t) = 10 \cos(377t - 30°)$ V

(b) $v_2(t) = 5 \sin(377t + 60°)$ V

Give the answers in both the time and frequency domains.

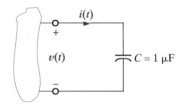

Figure P8.7

8.8   Find the frequency domain impedance, $\mathbf{Z}$, for the network in Fig. P8.8.

Impedances add in series.

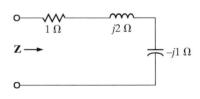

Figure P8.8

8.9   Find the frequency domain impedance, **Z**, for the circuit in Fig. P8.9.

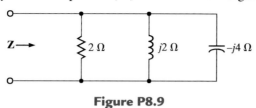

Figure P8.9

Admittances add in parallel.

8.10   Find the frequency domain impedance, **Z**, in the network in Fig. P8.10.

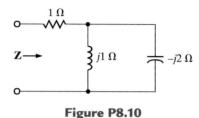

Figure P8.10

A series parallel combination.

8.11   Find the impedance, **Z**, shown in Fig. P8.11 at a frequency of 60 Hz.

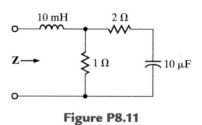

Figure P8.11

Use $j\omega L$ and $1/j\omega C$ for L and C, respectively. Then combine the impedances using the series and parallel relationships.

8.12   Find the impedance, **Z**, shown in Fig. P8.11 at a frequency of 400 Hz.

8.13   Find the frequency domain impedance, **Z**, shown in Fig. P8.13.

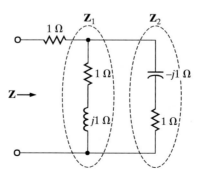

Figure P8.13

Find $\mathbf{Z}_1$ and $\mathbf{Z}_2$, and then combine them since they are in parallel.

8.14 Find the frequency domain impedance, **Z**, shown in Fig. P8.14.

There are two sets of parallel combinations, and the remaining impedances are in series with them.

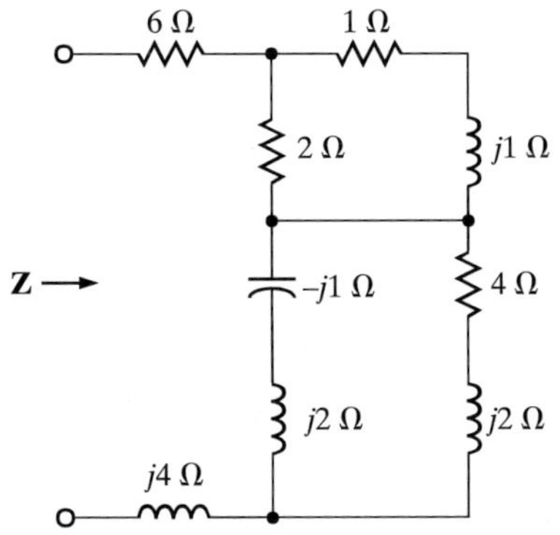

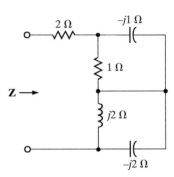

Figure P8.14

8.15 Find the frequency domain impedance, **Z**, shown in P8.15.

Similar to P8.14.

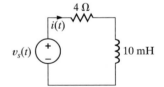

Figure P8.15

8.16 Draw the frequency domain circuit and calculate $i(t)$ for the circuit shown in Fig. P8.16 if $v_s(t) = 2 \cos(377t)$ V.

Replace L by $j\omega L$ and $v_s(t)$ by

$$\mathbf{V} = 2 \underline{/0°} \text{ V}, \mathbf{I} = \frac{V}{R + j\omega L}.$$

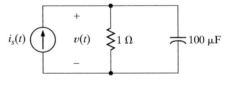

Figure P8.16

8.17 Draw the frequency domain circuit and calculate $v(t)$ for the circuit shown in Fig. P8.17 if $i_s(t) = 10 \cos(377t + 30°)$ A.

Replace C by $j\omega C$ and $i_s(t)$ by

$$\mathbf{I}_s = 10 \underline{/30°} \text{ A},$$

$$\mathbf{V} = \mathbf{I}_s \left(R \Big/\!\!\Big/ \frac{1}{j\omega C} \right).$$

Figure P8.17

8.18 Draw the frequency domain circuit and calculate $i(t)$ for the circuit shown in Fig. P8.18 if $v_s(t) = 10 \cos(377t + 30°)$ V.

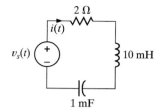

Figure P8.18

$$I = \frac{V}{2 + j\omega L + \dfrac{1}{j\omega C}}$$

8.19 Draw the frequency domain circuit and calculate $v(t)$ for the circuit shown in Fig. P8.19 if $i_s(t) = 20 \cos(377t + 120°)$ A.

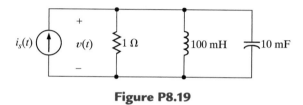

Figure P8.19

Similar to P8.17.

8.20 Draw the frequency domain circuit and calculate $i(t)$ for the circuit shown in Fig. P8.20 if $v_s(t) = 10 \cos(1000t + 30°)$ V.

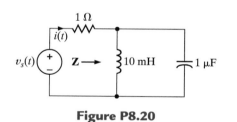

Figure P8.20

Find Z and $I = V_s/Z$.

8.21 Draw the frequency domain circuit and calculate $v(t)$ for the circuit shown in Fig. P8.21 if $i_s(t) = 2 \cos(1000t + 120°)$ A.

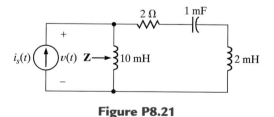

Figure P8.21

Find Z and $V = I_s Z$.

Section 8.6

Convert $v_R(t)$, $v_L(t)$ and $v_S(t)$ to phasors. Then use vector techniques to show that the sum of $v_R(t)$ and $v_L(t)$ is $v_S(t)$.

8.22 The voltages $v_R(t)$ and $v_L(t)$ in the circuit shown in Fig. P8.16, can be drawn as phasors in a phasor diagram. Use a phasor diagram to show that $v_R(t) + v_L(t) = v_s(t)$.

Similar to P8.22.

8.23 The currents $i_R(t)$ and $i_C(t)$ in the circuit shown in Fig. P8.17 can be drawn as phasors in a phasor diagram. Use the diagram to shown that $i_R(t) + v_C(t) = i_s(t)$.

Similar to P8.22.

8.24 The voltages $v_R(t)$, $v_L(t)$ and $v_C(t)$ in the circuit shown in Fig. P8.18, can be drawn as phasors in a phasor diagram. Show that $v_R(t) + v_L(t) + v_C(t) = v_s(t)$.

Similar to P8.22.

8.25 The currents $i_R(t)$, $i_L(t)$, and $i_C(t)$ in the circuit shown in Fig. P8.19, can be drawn as phasors in a phasor diagram. Show that $i_R(t) + i_L(t) + i_C(t) = i_s(t)$.

Similar to P8.22.

8.26 The currents $i_L(t)$ and $i_C(t)$ of the inductor and capacitor in the circuit shown in Fig. P8.20 can be drawn as phasors in a phasor diagram. Show that $i_L(t) + i_C(t) = i(t)$.

When $\omega L = \dfrac{1}{\omega C}$, $\mathbf{Z} = 1\,\Omega$.

8.27 Find the value of C in the circuit shown in Fig. P8.27 so that $\mathbf{Z}$ is purely resistive at the frequency of 60 Hz.

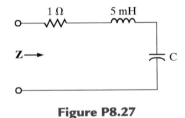

Figure P8.27

8.28 Find the frequency at which the circuit shown in Fig. P8.28 is purely resistive.

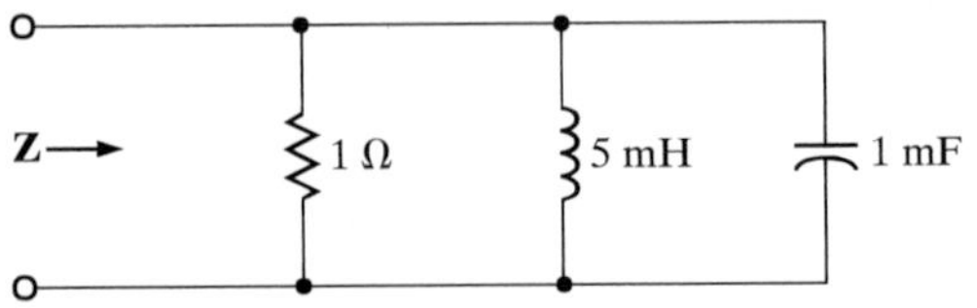

Similar to P8.27.

Figure P8.28

8.29 The impedance of the circuit in Fig. P8.29 is real at $f = 60$ Hz. What is the value of L?

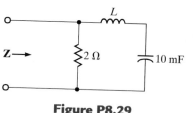

Figure P8.29

Similar to P8.27.

8.30 In the circuit shown in Fig. P8.30, determine the frequency at which $i(t)$ is in phase with $v_s(t)$.

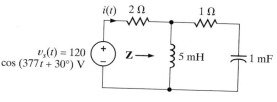

$v_s(t) = 120 \cos(377t + 30°)$ V

Figure P8.30

The quantities are in phase when the imaginary part of Z is zero.

8.31 In the circuit shown in Fig. P8.31, determine the value of the inductance such that $v(t)$ is in phase with $i_s(t)$.

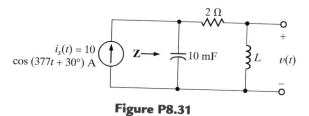

$i_s(t) = 10 \cos(377t + 30°)$ A

Figure P8.31

Similar to P8.30.

8.32 In the circuit shown in Fig. P8.32, determine the frequency at which $i(t)$ is in phase with $v_s(t)$.

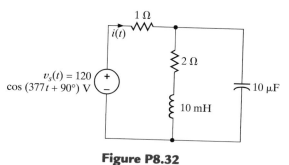

$v_s(t) = 120 \cos(377t + 90°)$ V

Figure P8.32

Similar to P8.30.

Section 8.7

8.33 Find the current **I** shown in Fig. P8.33.

Use the current division rule.

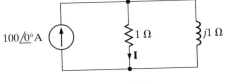

Figure P8.33

8.34 Find the voltage **V** shown in Fig. P8.34.

Use the voltage division rule.

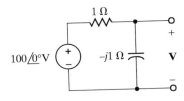

Figure P8.34

8.35 Find the voltage **V** shown in Fig. P8.35.

Find **Z**. $\mathbf{V} = (\mathbf{Z})1 \angle 30°$.

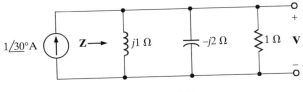

Figure P8.35

8.36 Find the frequency domain voltage, $\mathbf{V}_o$, as shown in Fig. P8.36.

Find $\mathbf{I}_o$ using the current division rule. Then $\mathbf{V}_o = \mathbf{I}_o(-j12)$V.

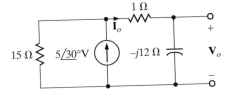

Figure P8.36

8.37 Find the voltage $\mathbf{V}_o$ as shown in Fig. P8.37.

Use the voltage division rule.

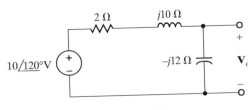

Figure P8.37

8.38 Find the voltage, $\mathbf{V}_o$, shown in Fig. P8.38.

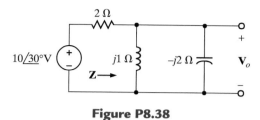

Figure P8.38

Find **Z**. Then use the voltage division rule.

8.39 Find the frequency domain current, $\mathbf{I}_o$, shown in Fig. P8.39.

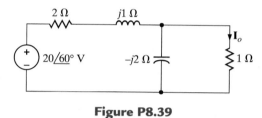

Figure P8.39

Similar to P8.38.

8.40 Given the network in Fig. P8.40, determine the value of $\mathbf{V}_o$ if $\mathbf{V}_s = 24 \underline{/0°}$ V.

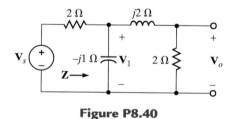

Figure P8.40

Find **Z**. Find $\mathbf{V}_1$ using the voltage division rule. Find $\mathbf{V}_o$ using the voltage division rule again.

8.41 Determine $\mathbf{I}_o$ in the network shown in Fig. P8.41 if $\mathbf{V}_s = 12 \underline{/0°}$ V.

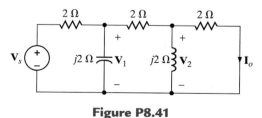

Figure P8.41

Similar to P8.40.

8.42 Determine I_o in the network in Fig. P8.42 if $I_s = 12 \underline{/0°}$ A.

Find **Z**. Use the current division rule to find I_1. Then use the current division rule to find I_o.

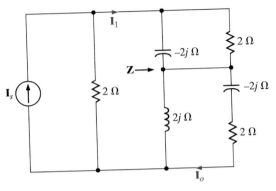

Figure P8.42

8.43 Find V_o in the network in Fig. P8.43.

Use the voltage division rule to find V_o.

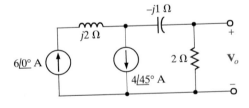

Figure P8.43

Section 8.8

8.44 Find V_o in the circuit in Fig. P8.44.

Use nodal analysis, superposition or source exchange.

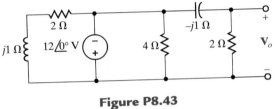

Figure P8.44

8.45 Determine V_o in the circuit in Fig. P8.45.

Use source exchange or superposition.

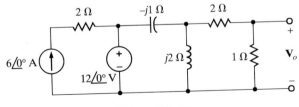

Figure P8.45

8.46 Draw the frequency domain network and calculate $v_o(t)$ in the circuit shown in Fig. P8.46 if $v_S(t)$ is $10\cos(10^4 t + 30°)$ V. Also, use a phasor diagram to determine $v_L(t)$.

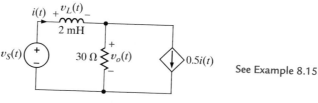

See Example 8.15

Figure P8.46

8.47 Find $\mathbf{V}_o$ in the network in Fig. P8.47.

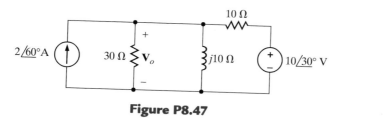

Use source exchange or superposition.

Figure P8.47

8.48 Find $\mathbf{V}_o$ in the network in Fig. P8.48.

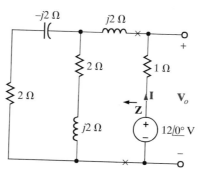

Find **Z**. Then **I**. KVL will then yield $\mathbf{V}_o$.

Figure P8.48

8.49 Use nodal analysis to find $\mathbf{I}_o$ in the circuit in Fig. P8.49.

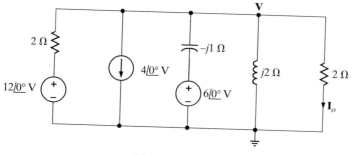

Assume that the node voltage of the top node is **V** and write the KCL equation for that node.

Figure P8.49

8.50 Using nodal analysis, find $\mathbf{I}_o$ in the circuit in Fig. P8.50.

Write the nodal equation for the two nodes with voltages $\mathbf{V}_1$ and $\mathbf{V}_2$.

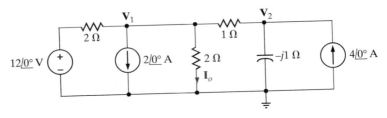

Figure P8.50

8.51 Find $\mathbf{I}_o$ in the network in Fig. P8.51.

Use nodal, loop, or mesh analysis.

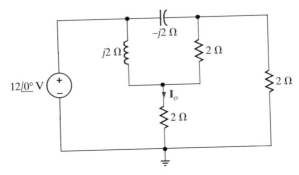

Figure P8.51

8.52 Find $\mathbf{V}_o$ in the network in Fig. P8.52 using nodal analysis.

Write the KCL equation for the node with voltage $\mathbf{V}$. Then use voltage division to find $\mathbf{V}_o$.

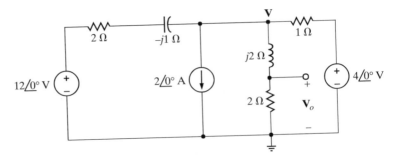

Figure P8.52

8.53 Find $\mathbf{I}_o$ in the circuit in Fig. P8.53 using nodal analysis.

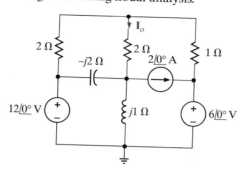

Write the KCL equation for the nodes that are not connected to a voltage source.

Figure P8.53

8.54 Use the supernode technique to find $\mathbf{I}_o$ in the circuit in Fig. P8.54.

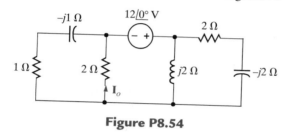

Construct a supernode for the $12\ \underline{/0^\circ}$ V source and write the node equation.

Figure P8.54

8.55 Find $\mathbf{I}_o$ in the network in Fig. P8.55.

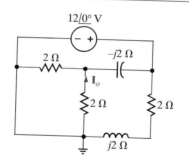

Construct a supernode for the $12\ \underline{/0^\circ}$ V source and write the node equations for the supernode and the remaining nonreference node.

Figure P8.55

8.56 Find $\mathbf{I}_o$ in the network in Fig. P8.56.

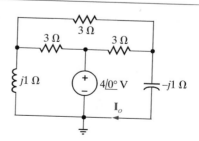

Similar to P8.55.

Figure P8.56

8.57 Find $\mathbf{V}_o$ in the network in Fig. P8.57.

Similar to P8.55.

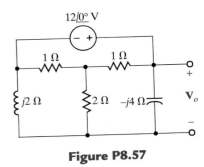

Figure P8.57

8.58 Find $\mathbf{V}_o$ in the network in Fig. P8.58.

Use Thévenin's theorem and remove the 1-Ω load .

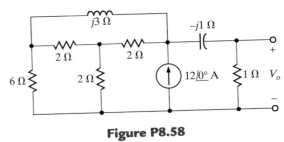

Figure P8.58

8.59 Find $\mathbf{V}_o$ in the network in Fig. P8.59.

Similar to P8.58.

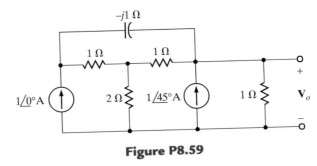

Figure P8.59

8.60 Use nodal analysis to find $\mathbf{V}_o$ in the circuit in Fig. P8.60.

Use the supernode technique.

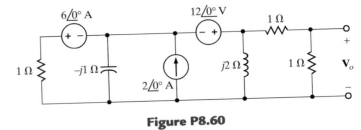

Figure P8.60

8.61 Use nodal analysis to determine I_o in the network in Fig. P8.61.

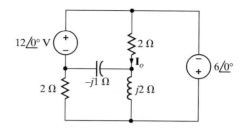

Figure P8.61

Only one nodal equation is necessary.

8.62 Find V_o in the network in Fig. P8.62.

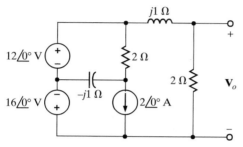

Figure P8.62

Use voltage division.

8.63 Find the voltage across the inductor in the circuit shown in Fig. P8.63 using nodal analysis.

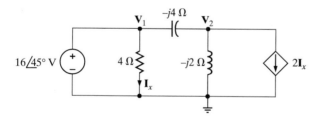

Figure P8.63

V_2 is the only unknown voltage.

8.64 The low-frequency equivalent circuit for a common-emitter transistor amplifier is shown in Fig. P8.64. Compute the voltage gain V_o/V_S.

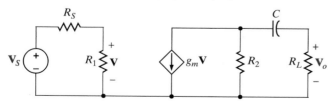

Figure P8.64

V can be obtained by voltage division and then V_o by current division.

8.65 Use nodal analysis to find $\mathbf{V}_o$ in the circuit in Fig. P8.65.

Use the supernode technique.

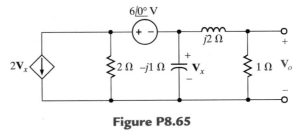

Figure P8.65

8.66 Use mesh analysis to find $\mathbf{V}_o$ in the circuit shown in Fig. P8.66.

$\mathbf{V}_o = 2\mathbf{I}_2$.

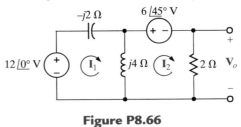

Figure P8.66

8.67 Use mesh analysis to find $\mathbf{V}_o$ in the circuit shown in Fig. P8.67.

$\mathbf{V}_o = 2(\mathbf{I}_1 - \mathbf{I}_2)$.

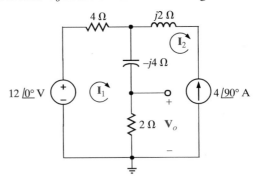

Figure P8.67

8.68 Find $\mathbf{V}_o$ in the circuit in Fig. P8.68 using mesh analysis.

The current sources make the mesh analysis straightforward.

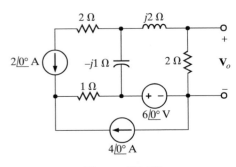

Figure P8.68

8.69 Use mesh analysis to find $\mathbf{V}_o$ in the
circuit in Fig. P8.69.

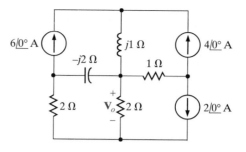

Similar to P8.68.

Figure P8.69

8.70 Using loop analysis, find $\mathbf{I}_o$ in the
network in Fig. P8.70.

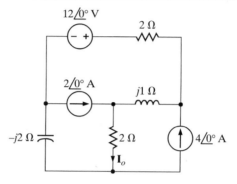

Use loop analysis or the
supermesh technique to avoid
a current source.

Figure P8.70

8.71 Using superposition, find $\mathbf{V}_o$ in the circuit in Fig. P8.71.

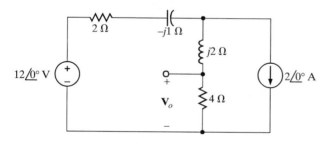

Use the $12 \underline{/0°}$ V source and
open the $2 \underline{/0°}$ A source to
find $\mathbf{V}_0'$.
Use the $2 \underline{/0°}$ A source and
replace the voltage source with
a short circuit to find
$\mathbf{V}_o''. \mathbf{V}_o = \mathbf{V}_o' + \mathbf{V}_o''$.

Figure P8.71

8.72 Use superposition to determine $\mathbf{V}_o$ in
the circuit in Fig. P8.72.

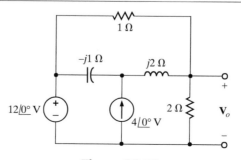

Use superposition and current
division.

Figure P8.72

8.73 Find $\mathbf{V}_o$ in the network in Fig. P8.73 using superposition.

Similar to P8.72.

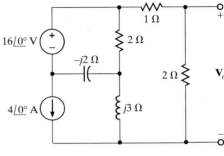

Figure P8.73

See Example 8.17.

8.74 Use source transformation to determine $\mathbf{I}_o$ in the network in Fig. P8.49.

Similar to P8.74.

8.75 Use source transformation to determine $\mathbf{I}_o$ in the network in Fig. P8.50.

8.76 Use source exchange to determine $\mathbf{V}_o$ in the network in Fig. P8.76.

Similar to P8.74.

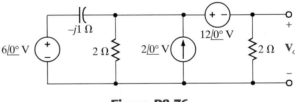

Figure P8.76

See the steps in Section 8.7.

8.77 Solve Problem 8.45 using Thévenin's theorem.

Similar to P8.77.

8.78 Solve Problem 8.60 using Thévenin's theorem.

Similar to P8.77.

8.79 Solve Problem 8.62 using Thévenin's theorem.

8.80 Find $\mathbf{V}_o$ in the network in Fig. P8.80 using Thévenin's theorem.

See Example 8.18.

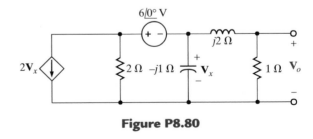

Figure P8.80

8.81 Solve Problem 8.72 using Norton's theorem.

See the steps in Section 8.7.

8.82 Solve Problem 8.73 using Norton's theorem.

Similar to P8.81.

8.83 Apply Thévenin's theorem to find $\mathbf{V}_o$ in the circuit in Fig. P8.83.

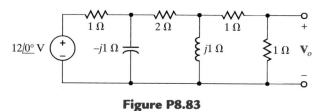

Figure P8.83

Use Thévenin's theorem twice to simplify the circuit.

8.84 Apply Norton's theorem to find $\mathbf{V}_o$ in the network in Fig. P8.84.

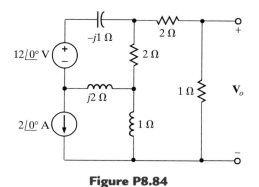

Figure P8.84

Similar to P8.81.

8.85 Find $\mathbf{V}_x$ in the circuit in Fig. P8.85 using Norton's theorem.

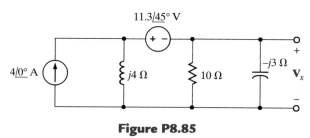

Figure P8.85

Similar to P8.81.

Advanced Problems

8.86 Find **Z** in the network in Fig. P8.86.

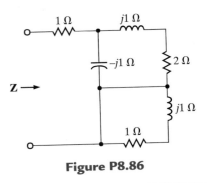

Figure P8.86

8.87 Find **Z** in the network in Fig. P8.87.

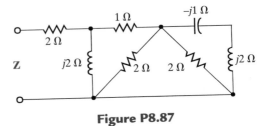

Figure P8.87

8.88 Find **Z** in the network in Fig. P8.88.

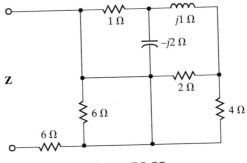

Figure P8.88

8.89 Determine the values of the two elements R_p and L_p or C_p in the network in Fig. P8.89 so that **Z** is $10 + j2 \ \Omega$ at a frequency of 60 Hz.

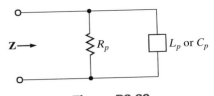

Figure P8.89

8.90 Determine the values of the two elements (R, L, or C) connected in parallel as shown in Fig. P8.90 so that $\mathbf{Z}$ is $5 - j5\ \Omega$ at 60 Hz.

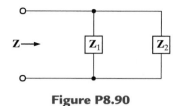

Figure P8.90

8.91 Draw the frequency domain network and calculate $v_o(t)$ in the circuit shown in Fig. P8.91 if $v_S(t)$ is $4 \sin(500t + 45°)$ V and $i_S(t)$ is $1 \cos(500t + 45°)$ A. Also, use a phasor diagram to determine $v_1(t)$.

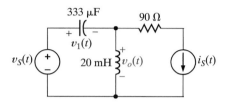

Figure P8.91

8.92 Draw the frequency domain network and calculate $v_o(t)$ in the circuit shown in Fig. P8.92 if $i_1(t)$ is $200 \cos(10^5 t + 60°)$ mA, $i_2(t)$ is $100 \sin(10^5 t + 90°)$ mA, and $v_S(t) = 10 \sin(10^5 t)$ V. Also, use a phasor diagram to determine $v_C(t)$.

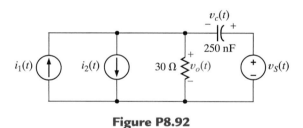

Figure P8.92

8.93 Draw the frequency domain network and calculate $v_o(t)$ in the circuit shown in Fig. P8.93 if $v_S(t)$ is $10 \cos(2\pi ft)$ V, where $f = 400$ Hz. Also, use a phasor diagram to determine $v(t)$.

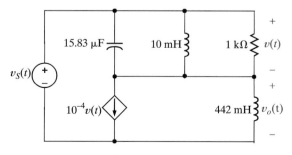

Figure P8.93

8.94 Find $v_o(t)$ in Problem 8.93 using Thévenin's theorem.

8.95 Find $\mathbf{V}_o$ using Norton's theorem for the circuit in Fig. P8.95.

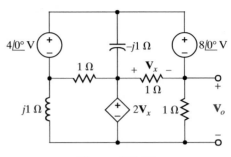

Figure P8.95

8.96 Given the network in Fig. P8.96, find the Thévenin's equivalent of the network at the terminals A-B.

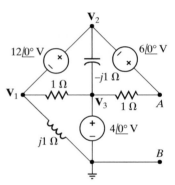

Figure P8.96

8.97 Use Thévenin's theorem to find $\mathbf{V}_o$ in the network in Fig. P8.97.

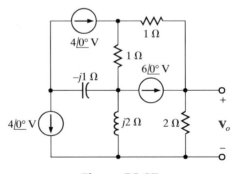

Figure P8.97

8.98 Use Norton's theorem to find $\mathbf{V}_o$ in the network in Fig. P8.98.

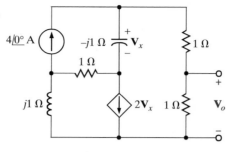

Figure P8.98

8.99 Using the PSPICE Schematics editor, draw the circuit in Figure P8.99. At what frequency are the magnitudes of $i_C(t)$ and $i_L(t)$ equal?

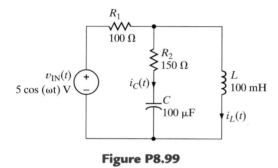

Figure P8.99

8.100 Using the PSPICE Schematics editor, draw the circuit in Figure P8.100. At what frequency are the phases of $i_1(t)$ and $v_X(t)$ equal?

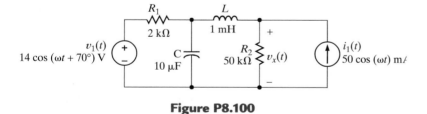

Figure P8.100

Steady-State Power Analysis

9

Do you understand the electrical wiring circuit in your home that is used to power all the appliances and lights? Do you know the hazards associated with working with electrical power circuits? These are just two of the topics that will be examined in this chapter.

We have been primarily concerned in the preceding chapters with determining the voltage or current at some point in a network. Of equal importance to us in many situations is the power that is supplied or absorbed by some element. Typically, electrical and electronic devices have peak power or maximum instantaneous power ratings that cannot be exceeded without damaging the devices.

In electrical and electronic systems, power comes in all sizes. The power absorbed by some device on an integrated circuit chip may be in picowatts, whereas the power supplied by a large generating station may be in gigawatts. Note that the range between these two examples is phenomenally large (10^{21}).

In our previous work we defined instantaneous power to be the product of voltage and current. Average power, obtained by averaging the instantaneous power, is the average rate at which energy is absorbed or supplied. In the dc case, where both current and voltage are constant, the instantaneous power is equal to the average power. However, as we will demonstrate, this is not the case when the currents and voltages are sinusoidal functions of time.

In this chapter we explore the many ramifications of power in ac circuits. We examine instantaneous power, average power, maximum power transfer, average power for periodic nonsinusoidal waveforms, the power factor, complex power, and power measurement.

Some very practical safety considerations will be introduced and discussed through a number of examples.

Finally, a wide variety of application-oriented examples is presented. The origin of many of these examples is the common household wiring circuit.

9.1 INSTANTANEOUS POWER

By employing the sign convention adopted in the earlier chapters, we can compute the instantaneous power supplied or absorbed by any device as the product of the instantaneous voltage across the device and the instantaneous current through it.

Consider the circuit shown in Fig. 9.1. In general, the steady-state voltage and current for the network can be written as

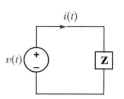

Figure 9.1 Simple ac network.

$$v(t) = V_M \cos(\omega t + \theta_v) \qquad \text{9.1}$$

$$i(t) = I_M \cos(\omega t + \theta_i) \qquad \text{9.2}$$

The instantaneous power is then

$$
\begin{aligned}
p(t) &= v(t)i(t) \\
&= V_M I_M \cos(\omega t + \theta_v) \cos(\omega t + \theta_i) \qquad \text{9.3}
\end{aligned}
$$

Employing the following trigonometric identity,

$$\cos \phi_1 \cos \phi_2 = \tfrac{1}{2}\left[\cos(\phi_1 - \phi_2) + \cos(\phi_1 + \phi_2)\right] \qquad \text{9.4}$$

we find that the instantaneous power can be written as

$$p(t) = \frac{V_M I_M}{2}\left[\cos(\theta_v - \theta_i) + \cos(2\omega t + \theta_v + \theta_i)\right] \qquad \text{9.5}$$

Note that the instantaneous power consists of two terms. The first term is a constant (i.e., it is time independent), and the second term is a cosine wave of twice the excitation frequency. We will examine this equation in more detail in Section 9.2.

EXAMPLE 9.1

The circuit in Fig. 9.1 has the following parameters: $v(t) = 4 \cos(\omega t + 60°)$ V and $\mathbf{Z} = 2\ /30°\ \Omega$. We wish to determine equations for the current and the instantaneous power as a function of time and plot these functions with the voltage on a single graph for comparison.

SOLUTION Since

$$
\begin{aligned}
\mathbf{I} &= \frac{4\ /60°}{2\ /30°} \\
&= 2\ /30°\ \text{A}
\end{aligned}
$$

then

$$i(t) = 2 \cos(\omega t + 30°)\ \text{A}$$

From Eq. (9.5),

$$
\begin{aligned}
p(t) &= 4[\cos(30°) + \cos(2\omega t + 90°)] \\
&= 3.46 + 4 \cos(2\omega t + 90°)\ \text{W}
\end{aligned}
$$

Note that $p(t)$ contains a dc term and a cosine wave with twice the frequency of $v(t)$ and $i(t)$.

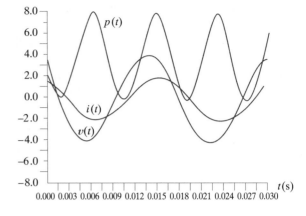

Figure 9.2 Plots of $v(t)$, $i(t)$, and $p(t)$ for the circuit in Example 9.1 using $f = 60$ Hz.

A plot of this function, together with plots of the voltage and current, is shown in Fig. 9.2. As can be seen in the figure, the instantaneous power has an average value, and the frequency is twice that of the voltage or current. ❏

9.2 AVERAGE POWER

The average value of any periodic waveform (e.g., a sinusoidal function) can be computed by integrating the function over a complete period and dividing this result by the period. Therefore, if the voltage and current are given by Eqs. (9.1) and (9.2), respectively, the average power is

$$P = \frac{1}{T} \int_{t_0}^{t_0+T} p(t) \, dt$$

$$= \frac{1}{T} \int_{t_0}^{t_0+T} V_M I_M \cos(\omega t + \theta_v) \cos(\omega t + \theta_i) \, dt \qquad \textbf{9.6}$$

where t_0 is arbitrary, $T = 2\pi/\omega$ is the period of the voltage or current, and P is measured in watts. Actually, we may average the waveform over any integral number of periods so that Eq. (9.6) can also be written as

$$P = \frac{1}{nT} \int_{t_0}^{t_0+nT} V_M I_M \cos(\omega t + \theta_v) \cos(\omega t + \theta_i) \, dt \qquad \textbf{9.7}$$

where n is a positive integer.

Employing Eq. (9.5) for the expression in (9.6), we obtain

$$P = \frac{1}{T} \int_{t_0}^{t_0+T} \frac{V_M I_M}{2} \left[\cos(\theta_v - \theta_i) + \cos(2\omega t + \theta_v + \theta_i) \right] dt \qquad \textbf{9.8}$$

We could, of course, plod through the indicated integration; however, with a little fore-thought we can determine the result by inspection. The first term is independent of t, and therefore a constant in the integration. Integrating the constant over the period and dividing by the period simply results in the original constant. The second term is a cosine wave. It is well known that the average value of a cosine wave over one complete period or an integral number of periods is zero, and therefore the second term in Eq. (9.8) vanishes. In view of this discussion, Eq. (9.8) reduces to

$$P = \tfrac{1}{2} V_M I_M \cos(\theta_v - \theta_i) \qquad \textbf{9.9}$$

Note that since $\cos(-\theta) = \cos(\theta)$, the argument for the cosine function can be either $\theta_v - \theta_i$ or $\theta_i - \theta_v$. In addition, note that $\theta_v - \theta_i$ is the angle of the circuit impedance as shown in Fig. 9.1. Therefore, *for a purely resistive circuit,*

A frequently used equation for calculating the average power

$$P = \tfrac{1}{2} V_M I_M \qquad \textbf{9.10}$$

and *for a purely reactive circuit,*

$$P = \tfrac{1}{2} V_M I_M \cos(90°)$$
$$= 0$$

Because purely reactive impedances absorb no average power, they are often called *lossless elements.* The purely reactive network operates in a mode in which it stores energy over one part of the period and releases it over another.

EXAMPLE 9.2

We wish to determine the average power absorbed by the impedance shown in Fig. 9.3.

SOLUTION   From the figure we note that

$$\mathbf{I} = \frac{\mathbf{V}}{\mathbf{Z}} = \frac{V_M \angle \theta_v}{2 + j2} = \frac{10 \angle 60°}{2.83 \angle 45°} = 3.53 \angle 15° \text{ A}$$

Therefore,

$$I_M = 3.53 \text{ A} \qquad \text{and} \qquad \theta_i = 15°$$

Hence,

$$P = \tfrac{1}{2} V_M I_M \cos(\theta_v - \theta_i)$$
$$= \tfrac{1}{2}(10)(3.53) \cos(60° - 15°)$$
$$= 12.5 \text{ W}$$

Since the inductor absorbs no power, we can employ Eq. (9.10) provided that V_M in that equation is the voltage across the resistor. Using voltage division, we obtain

$$\mathbf{V}_R = \frac{(10 \angle 60°)(2)}{2 + j2} = 7.07 \angle 15° \text{ V}$$

DRILL

D9.1   Find the average power for the following $\mathbf{V}$ and $\mathbf{I}$:

(1) $\mathbf{V} = 2 \angle 30°$ V,
    $\mathbf{I} = 2 \angle 30°$ A
(2) $\mathbf{V} = 2 \angle 30°$ V,
    $\mathbf{I} = 2 \angle 120°$ A
(3) $\mathbf{V} = 2 \angle 30°$ V,
    $\mathbf{I} = 2 \angle 60°$ A

ANSWER: (1) 2 W
              (2) 0 W
              (3) 1.732 W

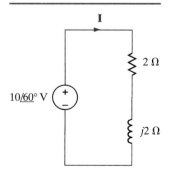

Figure 9.3 Example *RL* circuit.

and therefore,

$$P = \tfrac{1}{2}(7.07)(3.53)$$
$$= 12.5 \text{ W}$$

In addition, using Ohm's law, we could also employ the expressions

$$P = \frac{1}{2}\frac{V_M^2}{R}$$

or

$$P = \tfrac{1}{2}I_M^2 R$$

where once again we must be careful that the V_M and I_M in these equations refer to the voltage across the resistor and the current through it, respectively.   ❑

EXAMPLE 9.3

For the circuit shown in Fig. 9.4, we wish to determine both the total average power absorbed and the total average power supplied.

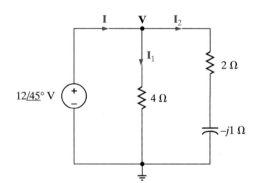

Figure 9.4 Example circuit for illustrating a power balance.

SOLUTION   From the figure we note that

$$\mathbf{I}_1 = \frac{12\ \underline{/45^\circ}}{4} = 3\ \underline{/45^\circ}\ \text{A}$$

$$\mathbf{I}_2 = \frac{12\ \underline{/45^\circ}}{2 - j1} = \frac{12\ \underline{/45^\circ}}{2.24\ \underline{/-26.57^\circ}} = 5.36\ \underline{/71.57^\circ}\ \text{A}$$

and therefore,

$$\mathbf{I} = \mathbf{I}_1 + \mathbf{I}_2$$
$$= 3\ \underline{/45^\circ} + 5.36\ \underline{/71.57^\circ}$$
$$= 8.15\ \underline{/62.10^\circ}\ \text{A}$$

The average power absorbed in the 4-Ω resistor is

$$P_4 = \tfrac{1}{2}V_M I_M = \tfrac{1}{2}(12)(3) = 18 \text{ W}$$

The average power absorbed in the 2-Ω resistor is

$$P_2 = \tfrac{1}{2}I_M^2 R = \tfrac{1}{2}(5.34)^2(2) = 28.7 \text{ W}$$

Therefore, the total average power absorbed is

$$P_A = 18 + 28.7 = 46.7 \text{ W}$$

Note that we could have calculated the power absorbed in the 2-Ω resistor using $\tfrac{1}{2}V_M^2/R$ if we had first calculated the voltage across the 2-Ω resistor.

The total average power supplied by the source is

$$\begin{aligned}
P_S &= \tfrac{1}{2}V_M I_M \cos(\theta_v - \theta_i) \\
&= \tfrac{1}{2}(12)(8.15)\cos(45° - 62.10°) \\
&= 46.7 \text{ W}
\end{aligned}$$

Thus, the total average power supplied is, of course, equal to the total average power absorbed. ❏

EXTENSION EXERCISES

E9.1   Find the average power adsorbed by each resistor in the network in Fig. E9.1.

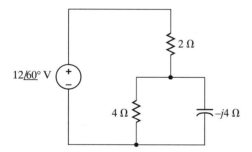

Figure E9.1

ANSWER: $P_{2\Omega} = 7.20$ W, $P_{4\Omega} = 7.20$ W.

E9.2   Given the network in Fig. E9.2, find the average power absorbed by each passive circuit element and the total average power supplied by the current source.

ANSWER: $P_{3\Omega} = 56.60$ W, $P_{4\Omega} = 33.96$ W, $P_L = 0$, $P_{cs} = 90.50$ W.

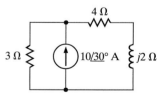

Figure E9.2

When determining average power, if more than one source is present in a network, we can use any of our network analysis techniques to find the necessary voltage and/or current to compute the power. However, we must remember that in general we cannot apply

Superposition is not applicable to power. Why?

superposition to power. We should, however, mention that there is a special case in which superposition does apply to power. If the current is of the form

$$i(t) = I_1 \cos(\omega_1 t + \theta_1) + I_2 \cos(\omega_2 t + \theta_2)$$

and if $n\omega_1 = m\omega_2$, where n and m are two different integers, then $n \cdot 2\pi/T_1 = m \cdot 2\pi/T_2$ so that $nT_2 = mT_1 = T$, where $\cos \omega_1 t$ has m full periods in time T and $\cos \omega_2 t$ has n full periods in time T. These sinusoids are said to be harmonically related. This subject will be discussed in detail in Chapter 16.

The average power that is absorbed by resistor R, in a time interval T, is given by

$$P = \frac{1}{T} \int_0^T [I_1 \cos(\omega_1 t + \theta_1) + I_2 \cos(\omega_2 t + \theta_2)]^2 R \, dt$$

where both ω_1 and ω_2 are periodic in T. The preceding equation can be written as

$$
\begin{aligned}
P = \frac{1}{T} \int_0^T &[I_1^2 \cos^2(\omega_1 t + \theta_1) + I_2^2 \cos^2(\omega_2 t + \theta_2) \\
&+ 2I_1 I_2 \cos(\omega_1 t + \theta_1)\cos(\omega_2 t + \theta_2)] R \, dt \\
= \frac{I_1^2}{2} R &+ \frac{I_2^2}{2} R + \frac{1}{T} \int_0^T \{I_1 I_2 \cos[(\omega_1 + \omega_2)t + \theta_1 + \theta_2] \\
&+ I_1 I_2 \cos[(\omega_1 - \omega_2)t + \theta_1 - \theta_2]\} R \, dt
\end{aligned}
$$

Since $(\omega_1 + \omega_2) = (m + n)(2\pi/T)$ and $(\omega_1 - \omega_2) = (m - n)(2\pi/T)$, the integral will be zero. Therefore,

$$P = \frac{I_1^2}{2} R + \frac{I_2^2}{2} R$$

Superposition also applies if one, and only one, of the sources is dc.

EXAMPLE 9.4

Consider the network shown in Fig. 9.5. We wish to determine the total average power absorbed and supplied by each element.

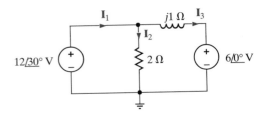

Figure 9.5 Example *RL* circuit with two sources.

SOLUTION   From the figure we note that

$$\mathbf{I}_2 = \frac{12 \,\underline{/30°}}{2} = 6 \,\underline{/30°} \text{ A}$$

and

$$\mathbf{I}_3 = \frac{12\ \underline{/30^\circ} - 6\ \underline{/0^\circ}}{j1} = \frac{4.39 + j6}{j1} = 7.44\ \underline{/-36.21^\circ}\ \text{A}$$

The power absorbed by the 2-Ω resistor is

$$P_2 = \tfrac{1}{2} V_M I_M = \tfrac{1}{2}(12)(6) = 36\ \text{W}$$

According to the direction of $\mathbf{I}_3$ the 6 $\underline{/0^\circ}$-V source is absorbing power. The power it absorbs is given by

$$\begin{aligned} P_{6\ \underline{/0^\circ}} &= \tfrac{1}{2} V_M I_M \cos(\theta_v - \theta_i) \\ &= \tfrac{1}{2}(6)(7.44)\cos[0^\circ - (-36.21^\circ)] \\ &= 18\ \text{W} \end{aligned}$$

At this point an obvious question arises: How do we know if the 6 $\underline{/0^\circ}$-V source is supplying power to the remainder of the network or absorbing it? The answer to this question is actually straightforward. If we employ our passive sign convention that was adopted in the earlier chapters—that is, if the current reference direction enters the positive terminal of the source and the answer is positive—the source is absorbing power. If the answer is negative, the source is supplying power to the remainder of the circuit. A generator sign convention could have been used, and under this condition the interpretation of the sign of the answer would be reversed. Note that once the sign convention is adopted and used, the sign for average power will be negative only if the angle difference is greater than 90° (i.e., $|\theta_v - \theta_i| > 90^\circ$). To obtain the power supplied to the network, we compute $\mathbf{I}_1$ as

$$\begin{aligned} \mathbf{I}_1 &= \mathbf{I}_2 + \mathbf{I}_3 \\ &= 6\ \underline{/30^\circ} + 7.44\ \underline{/-36.21^\circ} \\ &= 11.29\ \underline{/-7.10^\circ}\ \text{A} \end{aligned}$$

Therefore, the power supplied by the 12 $\underline{/30^\circ}$-V source using the generator sign convention is

$$\begin{aligned} P_S &= \tfrac{1}{2}(12)(11.29)\cos(30^\circ + 7.10^\circ) \\ &= 54\ \text{W} \end{aligned}$$

and hence the power absorbed is equal to the power supplied.   ❏

Under the following condition

If $P = \mathbf{IV} > 0$, power is being absorbed.
If $P = \mathbf{IV} < 0$, power is being generated.

EXAMPLE 9.5

The current in a 2-Ω resistor (12 $\underline{/30^\circ}$-V source) is of the form

$$i(t) = 4\cos(377t + 30^\circ) + 2\cos(754t + 60^\circ)$$

We wish to find the average power absorbed by the resistor.

SOLUTION   Since $\omega_1 = 377$ and $\omega_2 = 754\ (\text{i.e., } \omega_2 = 2\omega_1)$, then

$$P = \tfrac{1}{2}(4^2 + 2^2)2 = 20\ \text{W}$$ ❏

EXTENSION EXERCISES

E9.3 Determine the total average power absorbed and supplied by each element in the network in Fig. E9.3.

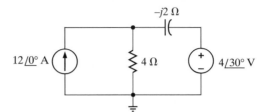

Figure E9.3

ANSWER: $P_{\text{cs}} = -69.4$ W, $P_{VS} = 19.8$ W, $P_{4\Omega} = 49.6$ W, $P_C = 0$.

E9.4 Given the network in Fig. E9.4, determine the total average power absorbed or supplied by each element.

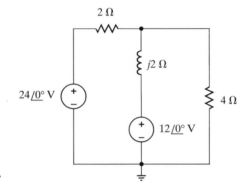

Figure E9.4

ANSWER: $P_{24\underline{/0^\circ}} = -55.4$ W, $P_{12\underline{/0^\circ}} = 5.5$ W, $P_{2\Omega} = 22.2$ W, $P_{4\Omega} = 27.7$ W, $P_L = 0$.

This impedance matching concept is an important issue in the design of high-speed computer chips and motherboards. For today's high-speed chips with internal clocks running at about 1 GHz and motherboards with a bus speed above 100 MHz, impedance matching is necessary in order to obtain the required speed for signal propagation. Although this high-speed transmission line is based on a distributed circuit (discussed later in electrical engineering courses), the impedance matching technique for the transmission line is the same as that of the lumped parameter circuit for maximum average power transfer.

9.3 MAXIMUM AVERAGE POWER TRANSFER

In our study of resistive networks we addressed the problem of maximum power transfer to a resistive load. We showed that if the network excluding the load was represented by a Thévenin equivalent circuit, maximum power transfer would result if the value of the load resistor was equal to the Thévenin equivalent resistance (i.e., $R_L = R_{\text{Th}}$). We will now reexamine this issue within the present context to determine the load impedance for the network shown in Fig. 9.6 that will result in maximum average power being absorbed by the load impedance $\mathbf{Z}_L$.

The equation for average power at the load is

$$P_L = \tfrac{1}{2} V_L I_L \cos(\theta_{V_L} - \theta_{i_L})$$ **9.11**

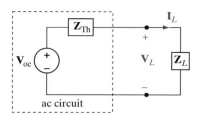

Figure 9.6 Circuit used to examine maximum power transfer.

The phasor current and voltage at the load are given by the expressions

$$\mathbf{I}_L = \frac{\mathbf{V}_{oc}}{\mathbf{Z}_{Th} + \mathbf{Z}_L} \qquad \textbf{9.12}$$

$$\mathbf{V}_L = \frac{\mathbf{V}_{oc}\,\mathbf{Z}_L}{\mathbf{Z}_{Th} + \mathbf{Z}_L} \qquad \textbf{9.13}$$

where

$$\mathbf{Z}_{Th} = R_{Th} + jX_{Th} \qquad \textbf{9.14}$$

and

$$\mathbf{Z}_L = R_L + jX_L \qquad \textbf{9.15}$$

The magnitude of the phasor current and voltage are given by the expressions

$$I_L = \frac{V_{oc}}{\left[\left(R_{Th} + R_L\right)^2 + \left(X_{Th} + X_L\right)^2\right]^{1/2}} \qquad \textbf{9.16}$$

$$V_L = \frac{V_{oc}\left(R_L^2 + X_L^2\right)^{1/2}}{\left[\left(R_{Th} + R_L\right)^2 + \left(X_{Th} + X_L\right)^2\right]^{1/2}} \qquad \textbf{9.17}$$

The phase angles for the phasor current and voltage are contained in the quantity $\left(\theta_{V_L} - \theta_{i_L}\right)$. Note also that $\theta_{V_L} - \theta_{i_L} = \theta_{\mathbf{Z}_L}$ and, in addition,

$$\cos\theta_{\mathbf{Z}_L} = \frac{R_L}{\left(R_L^2 + X_L^2\right)^{1/2}} \qquad \textbf{9.18}$$

Substituting Eqs. (9.16) to (9.18) into Eq. (9.11) yields

$$P_L = \frac{1}{2}\frac{V_{oc}^2 R_L}{\left(R_{Th} + R_L\right)^2 + \left(X_{Th} + X_L\right)^2} \qquad \textbf{9.19}$$

which could, of course, be obtained directly from Eq. (9.16) using $P_L = \frac{1}{2}I_L^2 R_L$. Once again, a little forethought will save us some work. From the standpoint of maximizing P_L, V_{oc} is a constant. The quantity $\left(X_{Th} + X_L\right)$ absorbs no power, and therefore any nonzero value of this quantity only serves to reduce P_L. Hence, we can eliminate this term by selecting $X_L = -X_{Th}$. Our problem then reduces to maximizing

$$P_L = \frac{1}{2}\frac{V_{oc}^2 R_L}{\left(R_L + R_{Th}\right)^2} \qquad \textbf{9.20}$$

DRILL

D9.2   Determine $\mathbf{Z}_L$ for maximum average power transfer to the load and the maximum average power transferred to the load.

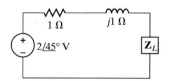

ANSWER: $\mathbf{Z}_L = 1 - j1\ \Omega$

Maximum average power transfer $= \frac{1}{2}$ W.

However, this is the same quantity we maximized in the purely resistive case by selecting $R_L = R_{Th}$. Therefore, for maximum average power transfer to the load shown in Fig.(ff6) 9.6, $\mathbf{Z}_L$ should be chosen so that

$$\mathbf{Z}_L = R_L + jX_L = R_{Th} - jX_{Th} = \mathbf{Z}_{Th}^* \qquad \textbf{9.21}$$

Finally, if the load impedance is purely resistive (i.e., $X_L = 0$), the condition for maximum average power transfer can be derived via the expression

$$\frac{dP_L}{dR_L} = 0$$

where P_L is the expression in Eq. (9.19) with $X_L = 0$. *The value of R_L that maximizes P_L under the condition $X_L = 0$ is*

$$R_L = \sqrt{R_{Th}^2 + X_{Th}^2} \qquad \textbf{9.22}$$

Problem-solving Strategy for Maximum Average Power Transfer

1. Remove the load $\mathbf{Z}_L$ and find the Thévenin equivalent for the remainder of the circuit.
2. Construct the circuit shown in Fig. 9.6.
3. Select $\mathbf{Z}_L = \mathbf{Z}_{Th}^* = R_{Th} - jX_{Th}$, and then $\mathbf{I}_L = \mathbf{V}_{oc}/2\,R_{Th}$ and the maximum average power transfer $= \dfrac{1}{2}\mathbf{I}_L^2\,R_{Th} = \mathbf{V}_{oc}^2/8\,R_{Th}$.

EXAMPLE 9.6

Given the circuit in Fig. 9.7a, we wish to find the value of $\mathbf{Z}_L$ for maximum average power transfer. In addition, we wish to find the value of the maximum average power delivered to the load.

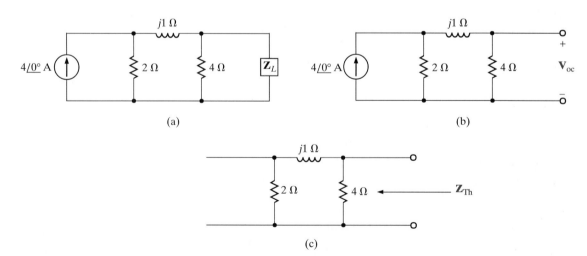

Figure 9.7 Circuits for illustrating maximum average power transfer.

SOLUTION   In order to solve the problem, we form a Thévenin equivalent at the load. The circuit in Fig. 9.7b is used to compute the open-circuit voltage

$$\mathbf{V}_{oc} = \frac{4 \underline{/0°} (2)}{6 + j1} (4) = 5.26 \underline{/-9.46°} \text{ V}$$

The Thévenin equivalent impedance can be derived from the circuit in Fig. 9.7c. As shown in the figure,

$$\mathbf{Z}_{Th} = \frac{4(2 + j1)}{6 + j1} = 1.41 + j0.43 \ \Omega$$

Therefore, $\mathbf{Z}_L$ for maximum average power transfer is

$$\mathbf{Z}_L = 1.41 - j0.43 \ \Omega$$

With $\mathbf{Z}_L$ as given previously, the current in the load is

$$\mathbf{I} = \frac{5.26 \underline{/-9.46°}}{2.82} = 1.87 \underline{/-9.46°} \text{ A}$$

Therefore, the maximum average power transferred to the load is

$$P_L = \tfrac{1}{2} I_M^2 R_L = \tfrac{1}{2}(1.87)^2(1.41) = 2.47 \text{ W} \qquad \square$$

In this Thévenin analysis,

1 Remove $\mathbf{Z}_L$ and find the voltage across the open terminals, $\mathbf{V}_{oc}$.

2 Determine the impedance $\mathbf{Z}_{Th}$ at the open terminals with all independent sources made zero.

3 Construct the following circuit and determine $\mathbf{I}$ and P_L.

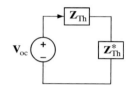

EXAMPLE 9.7

For the circuit shown in Fig. 9.8a, we wish to find the value of $\mathbf{Z}_L$ for maximum average power transfer. In addition, let us determine the value of the maximum average power delivered to the load.

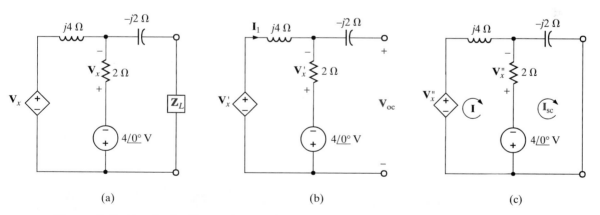

(a)   (b)   (c)

Figure 9.8 Circuits for illustrating maximum average power transfer.

SOLUTION   We will first reduce the circuit, with the exception of the load, to a Thévenin equivalent circuit. The open-circuit voltage can be computed from Fig. 9.8b. The equations for the circuit are

$$\mathbf{V}'_x + 4 = (2 + j4)\mathbf{I}_1$$
$$\mathbf{V}'_x = -2\mathbf{I}_1$$

When there is a dependent source, both $\mathbf{V}_{oc}$ and $\mathbf{I}_{sc}$ must be found and $\mathbf{Z}_{Th}$ computed from the equation

$$\mathbf{Z}_{Th} = \frac{\mathbf{V}_{oc}}{\mathbf{I}_{sc}}$$

Solving for $\mathbf{I}_1$, we obtain

$$\mathbf{I}_1 = \frac{1 \,\underline{/-45^\circ}}{\sqrt{2}}$$

The open-circuit voltage is then

$$\begin{aligned}
\mathbf{V}_{oc} &= 2\mathbf{I}_1 - 4 \,\underline{/0^\circ} \\
&= \sqrt{2} \,\underline{/-45^\circ} - 4 \,\underline{/0^\circ} \\
&= -3 - j1 \\
&= -3.16 \,\underline{/-161.57^\circ} \text{ V}
\end{aligned}$$

The short-circuit current can be derived from Fig. 9.8c. The equations for this circuit are

$$\begin{aligned}
\mathbf{V}_x'' + 4 &= (2 + j4)\mathbf{I} - 2\mathbf{I}_{sc} \\
-4 &= -2\mathbf{I} + (2 - j2)\mathbf{I}_{sc} \\
\mathbf{V}_x'' &= -2(\mathbf{I} - \mathbf{I}_{sc})
\end{aligned}$$

Solving these equations for $\mathbf{I}_{sc}$ yields

$$\mathbf{I}_{sc} = -(1 + j2) \text{ A}$$

The Thévenin equivalent impedance is then

$$\mathbf{Z}_{Th} = \frac{\mathbf{V}_{oc}}{\mathbf{I}_{sc}} = \frac{3 + j1}{1 + j2} = 1 - j1 \ \Omega$$

Therefore, for maximum average power transfer the load impedance should be

$$\mathbf{Z}_L = 1 + j1 \ \Omega$$

The current in this load $\mathbf{Z}_L$ is then

$$\mathbf{I}_L = \frac{\mathbf{V}_{oc}}{\mathbf{Z}_{Th} + \mathbf{Z}_L} = \frac{-3 - j1}{2} = 1.58 \,\underline{/-161.57^\circ} \text{ A}$$

Hence, the maximum average power transferred to the load is

$$\begin{aligned}
P_L &= \tfrac{1}{2}(1.58)^2(1) \\
&= 1.25 \text{ W} \qquad\qquad \square
\end{aligned}$$

EXTENSION EXERCISES

E9.5   Given the network in Fig. E9.5, find $\mathbf{Z}_L$ for maximum average power transfer and the maximum average power transferred to the load.

ANSWER: $\mathbf{Z}_L = 1 + j1 \ \Omega, P_L = 45$ W.

E9.6   Find $\mathbf{Z}_L$ for maximum average power transfer and the maximum average power transferred to the load in the network in Fig. E9.6.

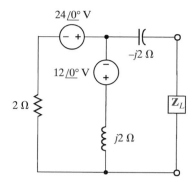

Figure E9.5

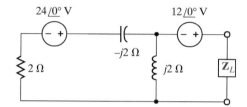

Figure E9.6

ANSWER: $\mathbf{Z}_L = 2 - j2\ \Omega, P_L = 45$ W.

9.4 EFFECTIVE OR RMS VALUES

In the preceding sections of this chapter we have shown that the average power absorbed by a resistive load is directly dependent on the type, or types, of sources that are delivering power to the load. For example, if the source was dc, the average power absorbed was $I^2 R$, and if the source was sinusoidal, the average power was $\frac{1}{2} I_M^2 R$. Although these two types of waveforms are extremely important, they are by no means the only waveforms we will encounter in circuit analysis. Therefore, a technique by which we can compare the *effectiveness* of different sources in delivering power to a resistive load would be quite useful.

In order to accomplish this comparison, we define what is called the *effective value of a periodic waveform*, representing either voltage or current. Although either quantity could be used, we will employ current in the definition. Hence, we define the effective value of a periodic current as a constant or dc value, which as current would deliver the same average power to a resistor R. Let us call the constant current I_{eff}. Then the average power delivered to a resistor as a result of this current is

$$P = I_{\text{eff}}^2 R$$

Similarly, the average power delivered to a resistor by a periodic current $i(t)$ is

$$P = \frac{1}{T} \int_{t_0}^{t_0+T} i^2(t) R \, dt$$

Equating these two expressions, we find that

$$I_{\text{eff}} = \sqrt{\frac{1}{T} \int_{t_0}^{t_0+T} i^2(t)\, dt} \qquad\qquad \textbf{9.23}$$

Note that this effective value is found by first determining the *square* of the current, then computing the average or *mean* value, and finally taking the square *root*. Thus in "reading" the mathematical Eq. (9.23), we are determining the root mean square, which we abbreviate as rms, and therefore I_{eff} is called I_{rms}.

Since dc is a constant, the rms value of dc is simply the constant value. Let us now determine the rms value of other waveforms. The most important waveform is the sinusoid and, therefore, we address this particular one in the following example.

EXAMPLE 9.8

We wish to compute the rms value of the waveform $i(t) = I_M \cos(\omega t - \theta)$, which has a period of $T = 2\pi/\omega$.

SOLUTION Substituting these expressions into Eq. (9.23) yields

$$I_{\text{rms}} = \left[\frac{1}{T} \int_0^T I_M^2 \cos^2(\omega t - \theta)\, dt \right]^{1/2}$$

Using the trigonometric identity

$$\cos^2 \phi = \tfrac{1}{2} + \tfrac{1}{2} \cos 2\phi$$

we find that the preceding equation can be expressed as

$$I_{\text{rms}} = I_M \left\{ \frac{\omega}{2\pi} \int_0^{2\pi/\omega} \left[\tfrac{1}{2} + \tfrac{1}{2} \cos(2\omega t - 2\theta) \right] dt \right\}^{1/2}$$

Since we know that the average or mean value of a cosine wave is zero,

$$I_{\text{rms}} = I_M \left(\frac{\omega}{2\pi} \int_0^{2\pi/\omega} \frac{1}{2}\, dt \right)^{1/2}$$

$$= I_M \left[\frac{\omega}{2\pi} \left(\frac{t}{2} \right) \Big|_0^{2\pi/\omega} \right]^{1/2} = \frac{I_M}{\sqrt{2}} \qquad\qquad \textbf{9.24}$$

Therefore, the rms value of a sinusoid is equal to the maximum value divided by the $\sqrt{2}$. Hence, a sinusoidal current with a maximum value of I_M delivers the same average power to a resistor R as a dc current with a value of $I_M/\sqrt{2}$. Recall that earlier a phasor $\mathbf{X}$ was defined as $X_M \,\underline{/\theta}$ for a sinusoidal wave of the form $X_m \cos(\omega t + \theta)$. This phasor can also be represented as $X_m/\sqrt{2} \,\underline{/\theta}$ if the units are given in rms. For example, 120 $\underline{/30°}$ V rms is equivalent to 170 $\underline{/30°}$ V. ❏

Upon using the rms values for voltage and current, the average power can be written, in general, as

$$P = V_{\text{rms}} I_{\text{rms}} \cos(\theta_v - \theta_i) \qquad\qquad \textbf{9.25}$$

The power absorbed by a resistor R is

$$P = I_{rms}^2 R = \frac{V_{rms}^2}{R} \qquad \textbf{9.26}$$

In dealing with voltages and currents in numerous electrical applications, it is important to know whether the values quoted are maximum, average, rms, or what. For example, the normal 120-V ac electrical outlets have an rms value of 120 V, an average value of 0 V, and a maximum value of $120\sqrt{2}$ V.

Finally, if the current in a resistor R is composed of a sum of sinusoidal waves of different frequencies, the power absorbed by the resistor can be expressed as

$$P = \left(I_{1rms}^2 + I_{2rms}^2 + \cdots + I_{nrms}^2\right)R \qquad \textbf{9.27}$$

where the rms value of the total current is

$$I_{rms} = \sqrt{I_{1rms}^2 + I_{2rms}^2 + \cdots + I_{nrms}^2} \qquad \textbf{9.28}$$

Each component represents a current of different frequency.

EXAMPLE 9.9

We wish to compute the rms value of the voltage waveform shown in Fig. 9.9.

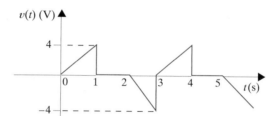

Figure 9.9 Waveform used to illustrate rms values.

SOLUTION   The waveform is periodic with period $T = 3$ s. The equation for the voltage in the time frame $0 \le t \le 3$ s is

$$v(t) = \begin{cases} 4t \text{ V} & 0 < t \le 1\text{ s} \\ 0\text{ V} & 1 < t \le 2\text{ s} \\ -4t + 8\text{ V} & 2 < t \le 3\text{ s} \end{cases}$$

The rms value is

$$\begin{aligned}
V_{rms} &= \left\{\frac{1}{3}\left[\int_0^1 (4t)^2\, dt + \int_1^2 (0)^2\, dt + \int_2^3 (8 - 4t)^2\, dt\right]\right\}^{1/2} \\
&= \left[\frac{1}{3}\left(\frac{16t^3}{3}\bigg|_0^1 + \left(64t - \frac{64t^2}{2} + \frac{16t^3}{3}\right)\bigg|_2^3\right)\right]^{1/2} \\
&= 1.89 \text{ V}
\end{aligned}$$

❑

EXAMPLE 9.10

Determine the rms value of the current waveform in Fig. 9.10 and use this value to compute the average power delivered to a 2-Ω resistor through which this current is flowing.

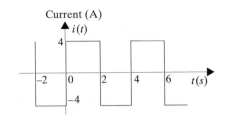

Figure 9.10 Waveform used to illustrate rms values.

SOLUTION   The current waveform is periodic with a period of $T = 4$ s. The rms value is

$$I_{\text{rms}} = \left\{ \frac{1}{4} \left[\int_0^2 (4)^2 \, dt + \int_2^4 (-4)^2 \, dt \right] \right\}^{1/2}$$

$$= \left[\frac{1}{4} \left(16t \Big|_0^2 + 16t \Big|_2^4 \right) \right]^{1/2}$$

$$= 4 \text{ A}$$

The average power delivered to a 2-Ω resistor with this current is

$$P = I_{\text{rms}}^2 R = (4)^2(2) = 32 \text{ W}$$ ❏

EXAMPLE 9.11

We wish to find the rms value of the current

$$i(t) = 12 \sin 377t + 6 \sin(754t + 30°) \text{ A}$$

SOLUTION   Since the frequencies of the two sinusoidal waves are different,

$$I_{\text{rms}}^2 = I_{1\text{rms}}^2 + I_{2\text{rms}}^2$$

$$= \left(\frac{12}{\sqrt{2}} \right)^2 + \left(\frac{6}{\sqrt{2}} \right)^2$$

Therefore,

$$I_{\text{rms}} = 9.49 \text{ A}$$ ❏

EXTENSION EXERCISES

E9.7   Compute the rms value of voltage waveform shown in Fig. E9.7.

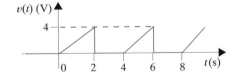

Figure E9.7

ANSWER: $V_{\text{rms}} = 1.633$ V.

E9.8   The current waveform in Fig. E9.8 is flowing through a 4-Ω resistor. Compute the average power delivered to the resistor.

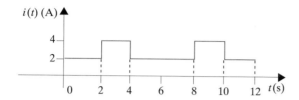

Figure E9.8

ANSWER: $P = 32$ W.

E9.9   The current waveform in Fig. E9.9 is flowing through a 10-Ω resistor. Determine the average power delivered to the resistor.

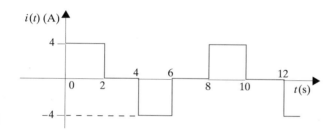

Figure E9.9

ANSWER: $P = 80$ W.

E9.10   The current $i(t)$ is flowing in a 4-Ω resistor. Determine the power absorbed by the resistor if

$$i(t) = 4 \sin(377t - 30°) + 6 \cos(754t - 45°) \text{ A}$$

ANSWER: $P = 104$ W.

9.5 THE POWER FACTOR

The power factor is a very important quantity. Its importance stems in part from the economic impact it has on industrial users of large amounts of power. In this section we carefully define this term and then illustrate its significance via some practical examples.

In Section 9.4 we showed that a load operating in the ac steady state is delivered an average power of

$$P = V_{\text{rms}} I_{\text{rms}} \cos(\theta_v - \theta_i)$$

We will now further define the terms in this important equation. The product $V_{\text{rms}} I_{\text{rms}}$ is referred to as the *apparent power*. Although the term $\cos(\theta_v - \theta_i)$ is a dimensionless quantity, and the units of P are watts, apparent power is normally stated in volt-amperes (VA) or kilovolt-amperes (kVA) in order to distinguish it from average power.

We now define the *power factor* (pf) as the ratio of the average power to the apparent power; that is,

$$\text{pf} = \frac{P}{V_{rms} I_{rms}} = \cos(\theta_v - \theta_i) \qquad \textbf{9.29}$$

where

$$\cos(\theta_v - \theta_i) = \cos\theta_{\mathbf{Z}_L} \qquad \textbf{9.30}$$

DRILL

D9.3 If a load impedance is 10 /30° Ω, find the power factor.

ANSWER:
pf = $\cos 30° = \sqrt{3}/2$.

The angle $\theta_v - \theta_i = \theta_{\mathbf{Z}_L}$ is the phase angle of the load impedance and is often referred to as the *power factor angle*. The two extreme positions for this angle correspond to a purely resistive load where $\theta_{\mathbf{Z}_L} = 0$ and the pf is 1, and the purely reactive load where $\theta_{\mathbf{Z}_L} = \pm 90°$ and the pf is 0. It is, of course, possible to have a unity pf for a load containing R, L, and C elements if the values of the circuit elements are such that a zero phase angle is obtained at the particular operating frequency.

There is, of course, a whole range of power factor angles between $\pm 90°$ and $0°$. If the load is an equivalent RC combination, then the pf angle lies between the limits $-90° < \theta_{\mathbf{Z}_L} < 0°$. On the other hand, if the load is an equivalent RL combination, then the pf angle lies between the limits $0 < \theta_{\mathbf{Z}_L} < 90°$. Obviously, confusion in identifying the type of load could result, due to the fact that $\cos\theta_{\mathbf{Z}_L} = \cos(-\theta_{\mathbf{Z}_L})$. To circumvent this problem, the pf is said to be either *leading* or *lagging*, where these two terms *refer to the phase of the current with respect to the voltage*. Since the current leads the voltage in an RC load, the load has a leading pf. In a similar manner, an RL load has a lagging pf; therefore, load impedances of $\mathbf{Z}_L = 1 - j1$ and $\mathbf{Z}_L = 2 + j1$ have power factors of $\cos(-45°) = 0.707$ leading and $\cos(26.57°) = 0.894$ lagging, respectively.

EXAMPLE 9.12

An industrial load consumes 88 kW at a pf of 0.707 lagging from a 480-V rms line. The transmission line resistance from the power company's transformer to the plant is 0.08 Ω. Let us determine the power that must be supplied by the power company (a) under present conditions and (b) if the pf is somehow changed to 0.90 lagging. (It is economically advantageous to have a power factor as close to one as possible.)

Technique

1 Given P_L, pf and V rms, determine I rms.

2 Then
$P_s = P_L + I^2 \text{ rms } R_{line}$,
where R_{line} is the line resistance.

SOLUTION (a) The equivalent circuit for these conditions is shown in Fig. 9.11. Using Eq. (9.29), we obtain the magnitude of the rms current into the plant:

$$I_{rms} = \frac{P_L}{(\text{pf})(V_{rms})}$$

Figure 9.11 Example circuit for examining changes in power factor.

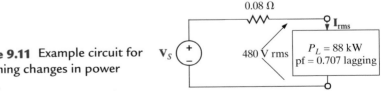

$$= \frac{(88)(10^3)}{(0.707)(480)}$$

$$= 259.3 \text{ A rms}$$

The power that must be supplied by the power company is

$$P_S = P_L + (0.08)I_{rms}^2$$

$$= 88,000 + (0.08)(259.3)^2$$

$$= 93.38 \text{ kW}$$

(b) Suppose now that the pf is somehow changed to 0.90 lagging but the voltage remains constant at 480 V. The rms load current for this condition is

$$I_{rms} = \frac{P_L}{(pf)(V_{rms})}$$

$$= \frac{(88)(10^3)}{(0.90)(480)}$$

$$= 203.7 \text{ A rms}$$

Under these conditions the power company must generate

$$P_S = P_L + (0.08)I_{rms}^2$$

$$= 88,000 + (0.08)(203.7)^2$$

$$= 91.32 \text{ kW}$$

Note carefully the difference between the two cases. A simple change in the pf of the load from 0.707 lagging to 0.90 lagging has had an interesting effect. Note that in the first case the power company must generate 93.38 kW in order to supply the plant with 88 kW of power because the low power factor means that the line losses will be high—5.38 kW. However, in the second case the power company need only generate 91.32 kW in order to supply the plant with its required power, and the corresponding line losses are only 3.32 kW.   ❏

The example clearly indicates the economic impact of the load's power factor. A low power factor at the load means that the utility generators must be capable of carrying more current at constant voltage, and they must also supply power for higher $I_{rms}^2 R$ line losses than would be required if the load's power factor were high. Since line losses represent energy expended in heat and benefit no one, the utility will insist that a plant maintain a high pf, typically 0.90 lagging, and adjust its rate schedule to penalize plants that do not conform to this requirement. In the next section we will demonstrate a simple and economical technique for achieving this power factor correction.

EXTENSION EXERCISE

E9.11   An industrial load consumes 100 kW at 0.707 pf lagging. The 60-Hz line voltage at the load is 480 $\underline{/0°}$ V rms. The transmission-line resistance between the power company transformer and the load is 0.1 Ω. Determine the power savings that could be obtained if the pf is changed to 0.94 lagging.

ANSWER: Power saved is 3.771 kW.

9.6 COMPLEX POWER

In our study of ac steady-state power, it is convenient to introduce another quantity, which is commonly called *complex power*. To develop the relationship between this quantity and others we have presented in the preceding sections, consider the circuit shown in Fig. 9.12.

The complex power is defined to be

$$\mathbf{S} = \mathbf{V}_{rms}\mathbf{I}_{rms}^* \hspace{3cm} 9.31$$

where $\mathbf{I}_{rms}^*$ refers to the complex conjugate of $\mathbf{I}_{rms}$; that is, if $\mathbf{I}_{rms} = I_{rms}\underline{/\theta_i} = I_R + jI_I$, then $\mathbf{I}_{rms}^* = I_{rms}\underline{/-\theta_i} = I_R - jI_I$. Complex power is then

$$\mathbf{S} = V_{rms}\underline{/\theta_v}\, I_{rms}\underline{/-\theta_i} = V_{rms}I_{rms}\underline{/\theta_v - \theta_i} \hspace{2cm} 9.32$$

or

$$\mathbf{S} = V_{rms}I_{rms}\cos(\theta_v - \theta_i) + jV_{rms}I_{rms}\sin(\theta_v - \theta_i) \hspace{1cm} 9.33$$

where, of course, $\theta_v - \theta_i = \theta_{\mathbf{z}}$. We note from Eq. (9.33) that the real part of the complex power is simply the *real* of *average power*. The imaginary part of **S** we call the *reactive* or *quadrature power*. Therefore, complex power can be expressed in the form

$$\mathbf{S} = P + jQ \hspace{3cm} 9.34$$

where

$$P = \mathrm{Re}(\mathbf{S}) = V_{rms}I_{rms}\cos(\theta_v - \theta_i) \hspace{1.5cm} 9.35$$
$$Q = \mathrm{Im}(\mathbf{S}) = V_{rms}I_{rms}\sin(\theta_v - \theta_i) \hspace{1.5cm} 9.36$$

As shown in Eq. (9.33), the magnitude of the complex power is what we have called the *apparent power*, and the phase angle for complex power is simply the power factor angle. Complex power, like apparent power, is measured in volt-amperes, real power is measured in watts, and in order to distinguish Q from the other quantities, which in fact have the same dimensions, it is measured in volt-amperes reactive, or var.

In addition to the relationships expressed previously, we note that since

$$\cos(\theta_v - \theta_i) = \frac{\mathrm{Re}(\mathbf{Z})}{Z}$$

$$\sin(\theta_v - \theta_i) = \frac{\mathrm{Im}(\mathbf{Z})}{Z}$$

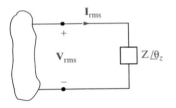

Figure 9.12 Circuit used to explain power relationships.

and

$$I_{\mathrm{rms}} = \frac{V_{\mathrm{rms}}}{Z}$$

Eqs. (9.35) and (9.36) can be written as

$$P = I_{\mathrm{rms}}^2 \, \mathrm{Re}(\mathbf{Z}) \qquad \text{9.37}$$
$$Q = I_{\mathrm{rms}}^2 \, \mathrm{Im}(\mathbf{Z}) \qquad \text{9.38}$$

and therefore, Eq. (9.34) can be expressed as

$$\mathbf{S} = I_{\mathrm{rms}}^2 \, \mathbf{Z} \qquad \text{9.39}$$

The diagrams in Fig. 9.13 serve to explain further the relationships among the various quantities of power. As shown in Fig. 9.13a, the phasor current can be split into two components: one that is in phase with $\mathbf{V}_{\mathrm{rms}}$ and one that is 90° out of phase with $\mathbf{V}_{\mathrm{rms}}$. Equations (9.35) and (9.36) illustrate that the in-phase component produces the real power, and the 90° component, called the *quadrature component*, produces the reactive or quadrature power. In addition, Eqs. (9.34) to (9.36) indicate that

$$\tan(\theta_v - \theta_i) = \frac{Q}{P} \qquad \text{9.40}$$

which relates the pf angle to P and Q in what is called the *power triangle*.

The relationships among $\mathbf{S}$, P, and Q can be expressed via the diagrams shown in Figs. 9.13b and c. In Fig. 9.13b we note the following conditions. If Q is positive, the load is inductive, the power factor is lagging, and the complex number $\mathbf{S}$ lies in the first quadrant. If Q is negative, the load is capacitive, the power factor is leading, and the complex number $\mathbf{S}$ lies in the fourth quadrant. If Q is zero, the load is resistive, the power factor is unity, and the complex number $\mathbf{S}$ lies along the positive real axis. Figure 9.13c illustrates the relationships expressed by Eqs. (9.37) to (9.39) for an inductive load.

Finally, it is important to state that complex power, like energy, is conserved; that is, the total complex power delivered to any number of individual loads is equal to the sum of the complex powers delivered to each individual load, regardless of how the loads are interconnected.

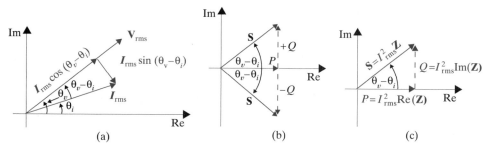

Figure 9.13 Diagram for illustrating power relationships.

Problem-solving strategies involving P or S

If $v(t)$ and $i(t)$ are known and we wish to find P given an impedance $\mathbf{Z}\,\underline{/\theta} = R + jX$, two viable approaches are as follows:

1 Determine $\mathbf{V}$ and $\mathbf{I}$ and then calculate

$$P = VI \cos\theta \qquad \text{or} \qquad P = VI \cos(\underline{/\mathbf{V}} - \underline{/\mathbf{I}})$$

2 Use $\mathbf{I}$ to calculate the real part of $\mathbf{S}$; that is,

$$P = R_e(\mathbf{S}) = I^2 R$$

The latter method may be easier to calculate than the former. However, if the imaginary part of the impedance, X, is not zero, then

$$P \neq \frac{V^2}{R}$$

which is a common mistake. Furthermore, the P and Q portions of $\mathbf{S}$ are directly related to $Z\,\underline{/\theta}$ and provide a convenient way in which to relate power, current, and impedance. That is,

$$\tan\theta = \frac{Q}{P}$$

$$\mathbf{S} = I^2 \mathbf{Z}$$

The following example illustrates the usefulness of $\mathbf{S}$.

EXAMPLE 9.13

A load operates at 20 kW, 0.8 pf lagging. The load voltage is $220\,\underline{/0°}$ V rms at 60 Hz. The impedance of the line is $0.09 + j0.3\ \Omega$. We wish to determine the voltage and power factor at the input to the line.

SOLUTION   The circuit diagram for this problem is shown in Fig. 9.14. As illustrated in Fig. 9.13,

$$S = \frac{P}{\cos\theta} = \frac{P}{pf} = \frac{20{,}000}{0.8} = 25{,}000 \text{ VA}$$

Therefore, at the load

$$\mathbf{S}_L = 25{,}000\,\underline{/\theta} = 25{,}000\,\underline{/36.87°} = 20{,}000 + j15{,}000 \text{ VA}$$

Technique

1 Use the given P_L, cos θ, and V_L rms to obtain S_L and I_L based on Eqs. (9.35) and (10.31), respectively.

2 Use I_L and Z_{line} to obtain S_{line} using Eq. (9.39).

3 Use $\mathbf{S}_s = \mathbf{S}_{line} + \mathbf{S}_L$.

4 $\mathbf{V}_s = \mathbf{S}_s / \mathbf{I}_L^*$ yields V_S and θ_v. Since $\mathbf{V}_s = V_s\,\underline{/\theta_v}$ and θ_v is the phase of $\mathbf{I}_L$, pf $= \cos(\theta_v - \theta_i)$.

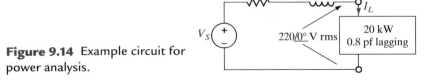

Figure 9.14 Example circuit for power analysis.

Since $\mathbf{S}_L = \mathbf{V}_L \mathbf{I}_L^*$

$$\mathbf{I}_L = \left[\frac{25{,}000\ \underline{/36.87°}}{220\ \underline{/0°}} \right]^*$$

$$= 113.64\ \underline{/36.87°}\ \text{A rms}$$

The complex power losses in the line are

$$\mathbf{S}_{\text{line}} = I_L^2 \mathbf{Z}_{\text{line}}$$

$$= (113.64)^2 (0.09 + j03)$$

$$= 1162.26 + j3874.21\ \text{VA}$$

As stated earlier, complex power is conserved and, therefore, the complex power at the generator is

$$\mathbf{S}_S = \mathbf{S}_L + \mathbf{S}_{\text{line}}$$

$$= 21{,}162.26 + j18{,}874.21$$

$$= 28{,}356.25\ \underline{/41.73°}\ \text{VA}$$

Hence, the generator voltage is

$$V_L = \frac{|\mathbf{S}_L|}{I_L} = \frac{28{,}356.25}{113.64}$$

$$= 249.53\ \text{V rms}$$

and the generator power factor is

$$\cos(41.73°) = 0.75\ \text{lagging}$$

We could have solved this problem using KVL. For example, we calculated the load current as

$$\mathbf{I}_L = 113.64\ \underline{/-36.87°}\ \mathbf{A}\ \text{rms}$$

Hence, the voltage drop in the transmission line is

$$\mathbf{V}_{\text{line}} = \left(113.64\ \underline{/-36.87°}\right)(0.09 + j0.3)$$

$$= 35.59\ \underline{/36.43°}\ \text{V rms}$$

Therefore, the generator voltage is

$$\mathbf{V}_S = 220\ \underline{/0°} + 35.59\ \underline{/36.43°}$$

$$= 249.53\ \underline{/4.86°}\ \text{V rms}$$

Hence, the generator voltage is 249.53 V rms. In addition,

$$\theta_v - \theta_i = 4.86° - (-36.87°) = 41.73°$$

and therefore,

$$\text{pf} = \cos(41.73°) = 0.75\ \text{lagging} \qquad \square$$

EXAMPLE 9.14

Two networks A and B are connected by two conductors having a net impedance of $\mathbf{Z} = 0 + j1\Omega$, as shown in Fig. 9.15. The voltages at the terminals of the networks

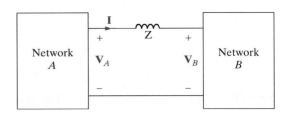

Figure 9.15 Network used in Example 9.14.

are $\mathbf{V}_A = 120 \underline{/30°}$ V rms and $\mathbf{V}_B = 120 \underline{/0°}$ V rms. We wish to determine the average power flow between the networks and identify which is the source and which is the load.

SOLUTION As shown in Fig. 9.15,

$$\mathbf{I} = \frac{\mathbf{V}_A - \mathbf{V}_B}{\mathbf{Z}}$$

$$= \frac{120 \underline{/30°} - 120 \underline{/0°}}{j1}$$

$$= 62.12 \underline{/15°} \text{ A rms}$$

The power delivered by network A is

$$P_A = |\mathbf{V}_A||\mathbf{I}| \cos(\theta_{\mathbf{V}_A} - \theta_{\mathbf{I}})$$
$$= (120)(62.12) \cos(30° - 15°)$$
$$= 7200.4 \text{ W}$$

The power absorbed by network B is

$$P_B = |\mathbf{V}_B||\mathbf{I}| \cos(\theta_{\mathbf{V}_B} - \theta_{\mathbf{I}})$$
$$= (120)(62.12) \cos(0° - 15°)$$
$$= 7200.4 \text{ W}$$

If the power flow had actually been from network B to network A, the resultant signs on P_A and P_B would have been negative. ❏

EXTENSION EXERCISES

E9.12 An industrial load requires 40 kW at 0.84 pf lagging. The load voltage is $220 \underline{/0°}$ V rms at 60 Hz. The transmission-line impedance is $0.1 + j0.25$ Ω. Determine the real and reactive power losses in the line and the real reactive power required at the input to the transmission line.

ANSWER: $P_{\text{line}} = 4.685$ kW, $Q_{\text{line}} = 11.713$ kvar, $P_S = 44.685$ kW, $Q_S = 45.31$ kvar.

E9.13 A load requires 60 kW at 0.85 pf lagging. The 60-Hz line voltage at the load is $220 \underline{/0°}$ V rms. If the transmission-line impedance is $0.12 + j0.18$ Ω, determine the line voltage and power factor at the input.

ANSWER: $\mathbf{V}_{\text{in}} = 284.6 \underline{/5.8°}$ V rms, $\text{pf}_{\text{in}} = 0.792$ lagging.

9.7 POWER FACTOR CORRECTION

Industrial plants that require large amounts of power have a wide variety of loads. However, by nature the loads normally have a lagging power factor. In view of the results obtained in Example 9.12, we are naturally led to ask if there is any convenient technique for raising the power factor of a load. Since a typical load may be a bank of induction motors or other expensive machinery, the technique for raising the pf should be an economical one in order to be feasible.

To answer the question we pose, let us examine the circuit shown in Fig. 9.16. The circuit illustrates a typical industrial load. In parallel with this load we have placed a capacitor. The original complex power for the load $\mathbf{Z}_L$, which we will denote as $\mathbf{S}_{old}$, is

$$\mathbf{S}_{old} = P_{old} + jQ_{old} = |\mathbf{S}_{old}|\; \underline{/\theta_{old}}$$

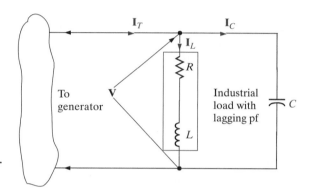

Figure 9.16 Circuit for power factor correction.

The complex power for the capacitor is

$$\mathbf{S}_{cap} = 0 + jQ_{cap} = |\mathbf{S}_{cap}|\; \underline{/-90°}$$

The new complex power that results from adding a capacitor is

$$\mathbf{S}_{old} + \mathbf{S}_{cap} = \mathbf{S}_{new} = P_{old} + jQ_{new} = |\mathbf{S}_{new}|\; \underline{/\theta_{new}} \qquad\qquad \textbf{9.41}$$

where θ_{new} is specified by the required power factor. The difference between the new and old complex powers is caused by the addition of the capacitor and since the capacitor is purely reactive,

$$\mathbf{S}_{cap} = +jQ_{cap} = -j\omega C V_{rms}^2 \qquad\qquad \textbf{9.42}$$

Equation (9.42) can be used to find the required value of C in order to achieve the new specified power factor. The procedure we have described is illustrated in Fig. 9.17, where $\theta_{old} = \theta_{v_L} - \theta_{i_L}$ and $\theta_{new} = \theta_{v_T} - \theta_{i_T}$. Hence, we can obtain a particular power factor for the total load simply by a judicious selection of a capacitor and placing it in parallel with the original load. In general, we want the power factor to be large, and therefore the power factor angle must be small (i.e., the larger the desired power factor, the smaller the angle $\theta_{v_T} - \theta_{i_T}$).

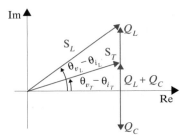

Figure 9.17 Illustration of the technique for power factor correction.

EXAMPLE 9.15

Plastic kayaks are manufactured using a process called rotomolding, which is diagrammed in Fig. 9.18. Molten plastic is injected into a mold, which is then spun on the long axis of the kayak until the plastic cools, resulting in a hollow one-piece craft. Suppose that the induction motors used to spin the molds consume 50 kW at a pf of 0.8 lagging from a 220 $\underline{/0°}$-V rms, 60-Hz line. We wish to raise the pf to 0.95 lagging by placing a bank of capacitors in parallel with the load.

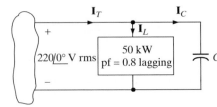

Figure 9.18 Rotomolding manufacturing process.

Technique

1. Find Q_{old} from P_L and θ_{old}, or the equivalent pf_{old}.
2. Find θ_{new} from the desired pf_{new}.
3. Determine $Q_{new} = P_{old} \tan \theta_{new}$.
4. $Q_{new} - Q_{old} = Q_{cap} = -\omega C V^2$ rms.

SOLUTION The circuit diagram for this problem is shown in Fig. 9.19. $P_L = 50$ kW and since $\cos^{-1} 0.8 = 36.87°$, $\theta_{old} = 36.87°$. Therefore,

$$Q_{old} = P_{old} \tan \theta_{old} = (50)(10^3)(0.75) = 37.5 \text{ kvar}$$

Hence,

$$S_{old} = P_{old} + jQ_{old} = 50{,}000 + j37{,}500$$

and

$$S_{cap} = 0 + jQ_{cap}$$

Since the required power factor is 0.95.

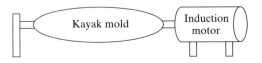

Figure 9.19 Example circuit for power factor correction.

Furthermore,

$$\theta_{new} = \cos^{-1}(pf_{new}) = \cos^{-1}(0.95)$$
$$= 18.19°$$

Then

$$Q_{new} = P_{old} \tan \theta_{new}$$
$$= 50,000 \tan(18.19°)$$
$$= 16,430 \text{ var}$$

Hence

$$Q_{new} - Q_{old} = Q_{cap} = -\omega C V^2 \text{ rms}$$
$$16,430 - 37,500 = -\omega C V^2 \text{ rms}$$

Solving the equation for C yields

$$C = \frac{21,070}{(377)(220)^2}$$
$$= 1155 \ \mu F$$

By using a capacitor of this magnitude in parallel with the industrial load, we create, from the utility's perspective, a load pf of 0.95 lagging. However, the parameters of the actual load remain unchanged. Under these conditions, the current supplied by the utility to the kayak manufacturer is less and therefore they can use smaller conductors for the same amount of power. Or, if the conductor size is fixed, the line losses will be less since these losses are a function of the square of the current. ❏

EXTENSION EXERCISE

E9.14 Compute the value of the capacitor necessary to change the power factor in Extension Exercise E9.11 to 0.95 lagging.

ANSWER: $C = 773 \ \mu F.$

9.8 POWER MEASUREMENTS

In the preceding sections we have illustrated techniques for computing power. We will now show how power is actually measured in an electrical network. An instrument used to measure average power is the wattmeter. This instrument contains a low-impedance current coil (which ideally has zero impedance) that is connected in series with the load, and a high-impedance voltage coil (which ideally has infinite impedance) that is connected

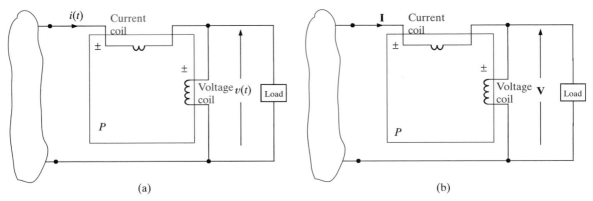

Figure 9.20 (a) and (b) Wattmeter connections for power measurement.

across the load. If the voltage and current are periodic and the wattmeter is connected as shown in Fig. 9.20a, it will read

$$P = \frac{1}{T} \int_0^T v(t)i(t)\, dt$$

where $v(t)$ and $i(t)$ are defined on the figure. Note that $i(t)$ is referenced entering the $\pm$ terminal of the current coil and $v(t)$ is referenced positive at the $\pm$ terminal of the voltage coil.

In addition, the $\pm$ terminal of the current coil in Fig. 9.20a is usually connected to the source of electric power and the $\pm$ terminal of the voltage coil is normally connected to the line that links the current terminal with no sign. When the current and voltage coils are connected in the configuration shown, the wattmeter will provide an upscale reading.

In the frequency domain the equivalent circuit is shown in Fig. 9.20b, where the current and voltage are referenced the same as in the time domain and the wattmeter reading will be

$$P = \mathrm{Re}(\mathbf{V}\mathbf{I}^*) = VI \cos(\theta_v - \theta_i)$$

If $v(t)$ and $i(t)$ or $\mathbf{V}$ and $\mathbf{I}$ are correctly chosen, the reading will be average power. In Fig. 9.20 the connections will produce a reading of power delivered to the load. Since the two coils are completely isolated from one another within the meter, they could be connected anywhere in the circuit and the reading may or may not have meaning.

Note that if one of the coils on the wattmeter is reversed, the equations for the power are the negative of what they were before the coil reversal due to the change in the variable reference as related to the $\pm$ terminal. Due to the physical construction of wattmeters, the $\pm$ terminal of the potential coil should always be connected to the same line as the current coil, as shown in Fig. 9.21a. If it becomes necessary to reverse a winding to produce an upscale reading, either coil can be reversed. For example, reversing the current coil results in the network shown in Fig. 9.21b. Note that if the $\pm$ terminal of the

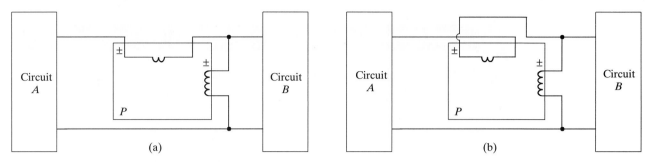

Figure 9.21 Wattmeter connections.

potential coil is connected to the line containing the current coil and the meter is reading upscale, the power, P, is flowing through the wattmeter from circuit A to circuit B.

EXAMPLE 9.16

Given the network shown in Fig. 9.22, we wish to determine the wattmeter reading.

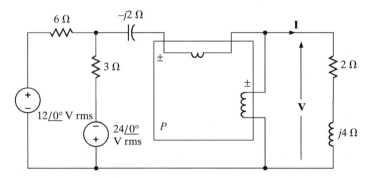

Figure 9.22 Use of the wattmeter in power measurement.

SOLUTION Using any one of a number of techniques (e.g., Thévenin's theorem), we can show that

$$\mathbf{I} = 2.68 \underline{/153.43°} \text{ A rms}$$

$$\mathbf{V} = 11.99 \underline{/216.86°} \text{ V rms}$$

and therefore the wattmeter reads

$$P = \text{Re}(\mathbf{VI}^*) = |\mathbf{V}||\mathbf{I}| \cos(\theta_v - \theta_i)$$

$$= (11.99)(2.68) \cos(216.86° - 153.43°)$$

$$= 14.37 \text{ W}$$

the average power delivered to the impedance $(2 + j4)$ Ω. ❑

EXTENSION EXERCISE

E9.15   Determine the wattmeter reading in the network in Fig. E9.15.

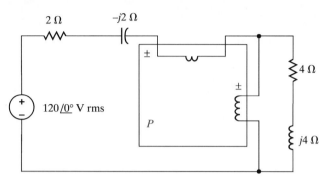

Figure E9.15

ANSWER: $P = 1440$ W.

Modern electronics technology has reduced the size and enhanced the capabilities of all types of measurement equipment. For example, the handheld power meter shown in Fig. 9.23 can be used to measure and display voltage, current, and power.

9.9 SINGLE-PHASE THREE-WIRE CIRCUITS

The single-phase three-wire ac circuit shown in Figure 9.24 is an important topic because it is the typical ac power network found in households. Note that the voltage sources are equal; that is, $\mathbf{V}_{an} = \mathbf{V}_{nb} = \mathbf{V}$. Thus, the magnitudes are equal and the phases are equal (single phase). The line-to-line voltage $\mathbf{V}_{ab} = 2\mathbf{V}_{an} = 2\mathbf{V}_{nb} = 2\mathbf{V}$. Within a household, lights and small appliances are connected from one line to *neutral n*, and large appliances such as hot water heaters and air conditioners are connected line to line. Lights operate at about 120 V rms and large appliances operate at approximately 240 V rms.

Let us now attach two identical loads to the single-phase three-wire voltage system using perfect conductors as shown in Fig. 9.24b. From the figure we note that

$$\mathbf{I}_{aA} = \frac{\mathbf{V}}{\mathbf{Z}_L}$$

and

$$\mathbf{I}_{bB} = -\frac{\mathbf{V}}{\mathbf{Z}_L}$$

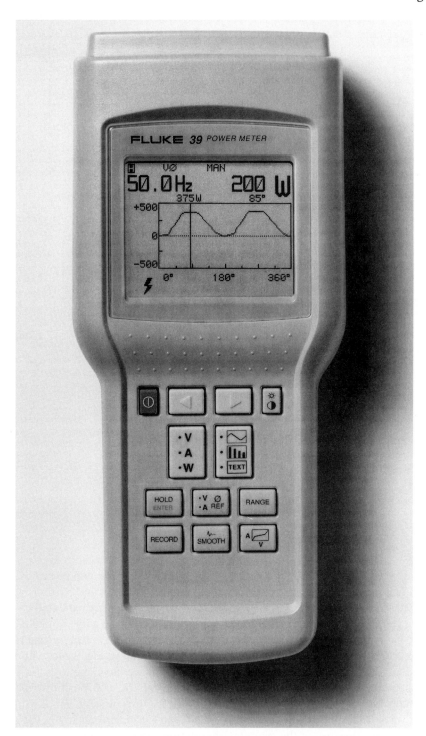

Figure 9.23 A Handheld Power Meter (Courtesy of Fluke Corporation).

D9.5 In Fig. 9.24b, if
$\mathbf{V} = 120\ \underline{/0^\circ}$ V and
$\mathbf{Z}_L = 100\ +\ j1\ \Omega$, find $\mathbf{I}_{nN}$.

ANSWER: $\mathbf{I}_{nN} = 0$.

D9.6 In Fig. 9.24d, if

$$\mathbf{V} = 120\ \underline{/0^\circ}\ \text{V}$$
$$\mathbf{Z}_L = 100\ +\ j1\ \Omega$$
$$\mathbf{Z}_{LL} = 10\ +\ j0.1\ \Omega$$
$$\mathbf{Z}_{\text{line}} = 0$$
$$\mathbf{Z}_n = 0$$

find $\mathbf{I}_a$.

ANSWER: $13.2\ \underline{/0.57^\circ}$ A.

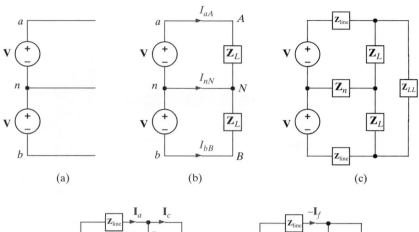

(a) (b) (c)

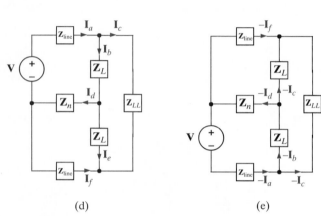

(d) (e)

Figure 9.24 Single-phase three-wire system.

KCL at point N is

$$\mathbf{I}_{nN} = -(\mathbf{I}_{aA} + \mathbf{I}_{bB})$$
$$= -\left(\frac{\mathbf{V}}{\mathbf{Z}_L} - \frac{\mathbf{V}}{\mathbf{Z}_L}\right)$$
$$= 0$$

Note that there is no current in the neutral wire, and therefore it could be removed without affecting the remainder of the system; that is, all the voltages and currents would be unchanged. One is naturally led to wonder just how far the simplicity exhibited by this system will extend. For example, what would happen if each line had a line impedance, if the neutral conductor had an impedance associated with it, and if there were a load tied from line to line? To explore these questions, consider the circuit in Fig. 9.24c. Although we could examine this circuit using many of the techniques we have employed in previous chapters, the symmetry of the network suggests that perhaps superposition may lead us to some conclusions without having to resort to a brute-force assault. Employing superposition, we consider the two circuits in Figs. 9.24d and e. The currents in Fig. 9.24d are labeled arbitrarily. Because of the symmetrical relationship between Figs. 9.24d and e, the currents in Fig. 9.24e correspond directly to those in Fig. 9.24d. If we add the two *phasor* currents in each branch, we find that the neutral current is again zero. A neutral current of zero is a direct result of the symmetrical nature of the network. If either the line impedances $\mathbf{Z}_{line}$ or the load impedances $\mathbf{Z}_L$ are unequal, the neutral current will be nonzero. We will make direct use of these concepts when we study three-phase networks in Chapter 10.

EXAMPLE 9.17

A three-wire single-phase household circuit is shown in Fig. 9.25a. The use of the lights, stereo, and range for a 24-hour period is demonstrated in Fig. 9.25b. Let us calculate the energy use over the 24 hours in kilowatt-hours. Assuming that this represents a typical day and that our utility rate is \$0.08/kWh, let us also estimate the power bill for a 30-day month.

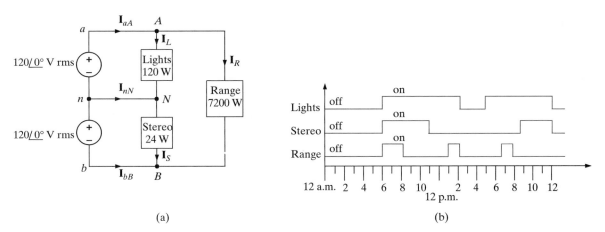

(a)   (b)

Figure 9.25 Household three-wire network and appliance usage.

SOLUTION   Applying nodal analysis to Fig. 9.25a yields

$$\mathbf{I}_{aA} = \mathbf{I}_L + \mathbf{I}_R$$

$$\mathbf{I}_{bB} = -\mathbf{I}_S - \mathbf{I}_R$$

$$\mathbf{I}_{nN} = \mathbf{I}_S - \mathbf{I}_L$$

The current magnitudes for each load can be found from the corresponding power levels as follows.

$$I_L = \frac{P_L}{V_{an}} = \frac{120}{120} = 1 \text{ A rms}$$

$$I_S = \frac{P_S}{V_{nb}} = \frac{24}{120} = 0.2 \text{ A rms}$$

$$I_R = \frac{P_R}{V_{ab}} = \frac{7200}{240} = 30 \text{ A rms}$$

The energy used is simply the integral of the power delivered by the two sources over the 24-hour period. Since the voltage magnitudes are constants, we can express the energy delivered by the sources as

$$E_{an} = V_{an} \int I_{aA} \, dt$$

$$E_{nb} = V_{nb} \int -I_{bB} \, dt$$

The integrals of I_{aA} and I_{bB} can be determined graphically from Fig. 9.25b.

$$\int_{12\text{a.m.}}^{12\text{a.m.}} I_{aA} \, dt = 4I_R + 15I_L = 135$$

$$\int_{12\text{a.m.}}^{12\text{a.m.}} -I_{bB} \, dt = 8I_S + 4I_R = 121.6$$

Therefore, the daily energy for each source and the total energy is

$$E_{an} = 16.2 \text{ kWh}$$
$$E_{nb} = 14.6 \text{ kWh}$$
$$E_{\text{total}} = 30.8 \text{ kWh}$$

Over a 30-day month, a \$0.08/kWh utility rate results in a power bill of

$$\text{Cost} = (30.8)(30)(0.08) = \$73.92$$ ❑

9.10 SAFETY CONSIDERATIONS

Although this book is concerned primarily with the theory of circuit analysis, we recognize that most students, by this point in their study, will have begun to relate the theory to the electrical devices and systems that they encounter in the world around them. Thus, it seems advisable to depart briefly from the theoretical and spend some time discussing the very practical and important subject of safety. Electrical safety is a very broad and diverse topic that would require several volumes for a comprehensive treatment. Instead, we will limit our discussion to a few introductory concepts and illustrate them with examples.

It would be difficult to imagine that anyone in our society could have reached adolescence without having experienced some form of electrical shock. Whether that shock was from a harmless electrostatic discharge or from accidental contact with an energized electrical circuit, the response was probably the same—an immediate and involuntary muscular reaction. In either case, the cause of the reaction is current flowing through the body. The severity of the shock depends on several factors, the most important of which are the magnitude, the duration, and the pathway of the current through the body.

The effect of electrical shock varies widely from person to person. Figure 9.26 shows the general reactions that occur as a result of 60-Hz ac current flow through the body from hand to hand, with the heart in the conduction pathway. Observe that there is an intermediate range of current, from about 0.1 to 0.2 A, which is most likely to be fatal. Current levels in this range are apt to produce ventricular fibrillation, a disruption of the orderly contractions of the heart muscle. Recovery of the heartbeat generally does not occur without immediate medical intervention. Current levels above that fatal range tend to cause the heart muscle to contract severely, and if the shock is removed soon enough, the heart may resume beating on its own.

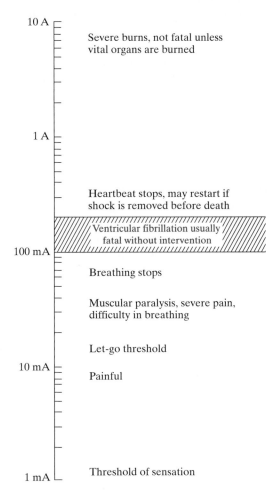

Figure 9.26 Effects of electrical shock. (From C. F. Dalziel and W. R. Lee, "Lethal Electric Currents," *IEEE Spectrum*, February 1969, pp. 44–50, and C. F. Dalziel, "Electric Shock Hazard," *IEEE Spectrum*, February 1972, pp. 41–50.)

The voltage required to produce a given current depends on the quality of the contact to the body and the impedance of the body between the points of contact. The electrostatic voltage such as might be produced by sliding across a car seat on a dry winter day may be on the order of 20,000 to 40,000 V, and the current surge upon touching the door handle, on the order of 40 A. However, the pathway for the current flow is mainly over the body surface, and its duration is for only a few microseconds. Although that shock could be disastrous for some electronic components, it causes nothing more than mild discomfort and aggravation to a human being.

Electrical appliances found about the home typically require 120 or 240 V rms for operation. Although the voltage level is small compared with that of the electrostatic shock, the potential for harm to the individual and to property is much greater. Accidental contact is more apt to result in current flow either from hand to hand or from hand to foot—either of which will subject the heart to shock. Moreover, the relatively slowly changing (low-frequency) 60-Hz current tends to penetrate more deeply into the body as opposed to remaining on the surface as a rapidly changing (high-frequency) current would tend to do. In addition, the energy source has the capability of sustaining a current flow without depletion. Thus, subsequent discussion will concentrate primarily on hazards associated with the 60-Hz ac power system.

The single-phase three-wire system introduced earlier is commonly, although not exclusively, used for electrical power distribution in residences. Two important aspects of this, or any, system that relate to safety were not mentioned earlier: circuit fusing and grounding.

Each branch circuit, regardless of the type of load it serves, is protected from excessive current flow by circuit breakers or fuses. Receptacle circuits are generally limited to 20 amps and lighting circuits to 15 amps. Clearly, these cannot protect persons from lethal shock. The primary purpose of these current-limiting devices is to protect equipment.

The neutral conductor of the power system is connected to ground (earth) at a multitude of points throughout the system and, in particular, at the service entrance to the residence. The connection to earth may be by way of a driven ground rod or by contact to a cold water pipe of a buried metallic water system. The 120-V branch circuits radiating from the distribution panel (fuse box) generally consist of three conductors rather than only two, as was shown in Fig. 9.24. The third conductor is the ground wire, as shown in Fig. 9.27.

The ground conductor may appear to be redundant, since it plays no role in the normal operation of a load that might be connected to the receptacle. Its role is illustrated by the following example.

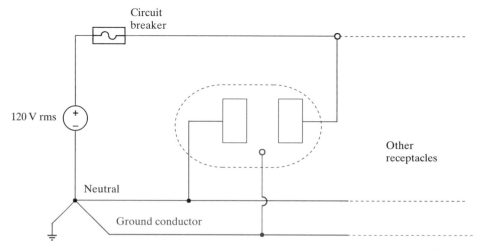

Figure 9.27 A household receptacle.

EXAMPLE 9.18

Joe has a workshop in his basement where he uses a variety of power tools such as drills, saws, and sanders. The basement floor is concrete, and being below ground level, it is usually damp. Damp concrete is a relatively good conductor. Unknown to Joe, the insulation on a wire in his electric drill has been nicked and the wire is in contact with (or shorted to) the metal case of the drill, as shown in Fig. 9.28. Is Joe in any danger when using the drill?

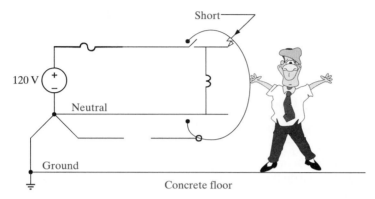

Figure 9.28 Faulty circuit, when the tool does not have the case grounded through the power cord.

SOLUTION   Without the ground conductor connected to the metal case of the tool, Joe would receive a severe, perhaps fatal, shock when he attempts to use the drill. The voltage between his hand and his feet would be 120 V, and the current through his body would be limited by the resistance of his body and of the concrete floor. Typically, the circuit breakers would not operate. However, if the ground conductor is present and properly connected to the drill case, the case will remain at ground potential, the 120-V source becomes shorted to ground, the circuit breaker operates, and Joe lives to drill another hole.   ❏

It was mentioned earlier that the circuit breaker or fuse cannot provide effective protection against shock. There is, however, a special type of device called a ground-fault interrupter (GFI) that can provide protection for personnel. This device detects current flow outside the normal circuit. Consider the circuit of Fig. 9.28. In the normal safe operating condition, the current in the neutral conductor must be the same as that in the line conductor. If at anytime the current in the line does not equal the current in the neutral, then a secondary path has somehow been established, creating an unsafe condition. This secondary path is called a fault. For example, the fault path in Fig. 9.28 is through Joe and the concrete floor. The GFI detects this fault and opens the circuit in response. Its principle of operation is illustrated by the following example.

EXAMPLE 9.19

Let us describe the operation of a GFI.

SOLUTION   Consider the action of the magnetic circuit in Fig. 9.29. Under normal operating conditions, i_1 and i_2 are equal, and if the coils in the neutral and line conductors are identical, as we learned in basic physics, the magnetic flux in the core will be zero. Consequently, no voltage will be induced in the sensing coil.

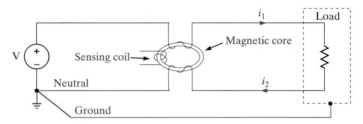

Figure 9.29 Ground-fault interrupter circuit.

If a fault should occur at the load, current will flow in the ground conductor and perhaps in the earth; thus, i_1 and i_2 will no longer be equal, the magnetic flux will not be zero, and a voltage will be induced in the sensing coil. That voltage can be used to activate a circuit breaker. This is the essence of the GFI device.   ❏

Ground-fault interrupters are available in the form of circuit breakers and also as receptacles. They are now required in branch circuits that serve outlets in areas such as bathrooms, basements, garages, and outdoor sites. The devices will operate at ground-fault currents on the order of a few milliamperes. Unfortunately, the GFI is a relatively new device and electrical code requirements are generally not retroactive. Thus few older residences have them.

Requirements for the installation and maintenance of electrical systems are meticulously defined by various codes that have been established to provide protection of personnel and property. Installation, alteration, or repair of electrical devices and systems should be undertaken only by qualified persons. The subject matter that we study in circuit analysis does not provide that qualification.

The following examples illustrate the potential hazards that can be encountered in a variety of everyday situations. We begin by revisiting a situation described in a previous example.

EXAMPLE 9.20

Suppose that a man is working on the roof of a mobile home with a hand drill. It is early in the day, the man is barefoot, and dew covers the mobile home. The ground prong on the electrical plug of the drill has been removed. Will the man be shocked if the "hot" electrical line shorts to the case of the drill?

SOLUTION   To analyze this problem, we must construct a model that adequately represents the situation described. In his book *Medical Instrumentation* (Boston: Houghton Mifflin Company, 1978), John G. Webster suggests the following values for resistance of the human body: R_{skin} (dry) = 15 kΩ, R_{skin} (wet) = 150 Ω, R_{limb} (arm or leg) = 100 Ω, and R_{trunk} = 200 Ω.

The network model is shown in Fig. 9.30. Note that since the ground line is open circuited, a closed path exists from the hot wire through the short, the human body, the mobile home, and the ground. For the conditions stated previously, we assume that the surface contact resistances R_{sc_1} and R_{sc_2} are 150 Ω each. The body resistance, R_{body}, consisting of arm, trunk, and leg, is 400 Ω. The mobile home resistance is assumed to be zero, and the ground resistance, R_{gnd}, from the mobile home ground to the actual source ground is assumed to be 1 Ω. Therefore, the magnitude of the current through the body from hand to foot would be

$$\mathbf{I}_{body} = \frac{120}{R_{sc_1} + R_{body} + R_{sc_2} + R_{gnd}}$$

$$= \frac{120}{701}$$

$$= 171 \text{ mA}$$

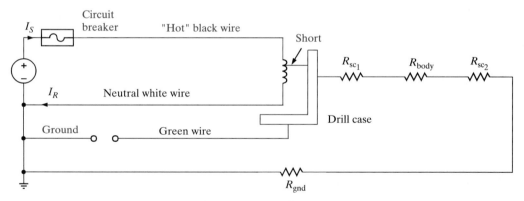

Figure 9.30 Model for Example 9.20.

A current of this magnitude can easily cause heart failure.

It is important to note that additional protection would be provided if the circuit breaker were a ground-fault interrupter. ❏

Two boys are playing basketball in their backyard. In order to cool off, they decide to jump into their pool. The pool has a vinyl lining, so the water is electrically insulated from the earth. Unknown to the boys, there is a ground fault in one of the pool lights. One boy jumps in and while standing in the pool with water up to his

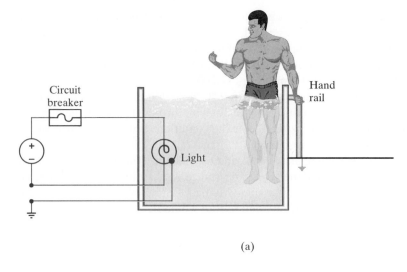

(a)

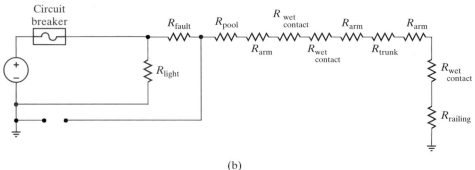

(b)

Figure 9.31 Diagrams used in Example 9.21.

chest, reaches up to pull in the other boy, who is holding onto a grounded hand rail, as shown in Fig. 9.31a. What is the impact of this action?

SOLUTION   The action in Fig. 9.31a is modeled as shown in Fig. 9.31b. Note that since a ground fault has occurred, there exists a current path through the two boys. Assuming that the fault, pool, and railing resistances are approximately zero, the magnitude of the current through the two boys would be

$$\mathbf{I} = \frac{120}{(3R_{arm}) + 3(3R_{wet\ contact}) + R_{trunk}}$$

$$= \frac{120}{950}$$

$$= 126\ \text{mA}$$

This current level would cause severe shock in both boys. The boy outside the pool could experience heart failure. ❏

EXAMPLE 9.22

A patient in a medical laboratory has a muscle stimulator attached to her left fore-arm. Her heart rate is being monitored by an EKG machine with two differential electrodes over the heart and the ground electrode attached to her right ankle. This activity is illustrated in Fig. 9.32a. The stimulator acts as a current source that dri-ves 150 mA through the muscle from the active electrode to the passive electrode. If the laboratory technician mistakenly decides to connect the passive electrode of the stimulator to the ground electrode of the EKG system to achieve a common ground, is there any risk?

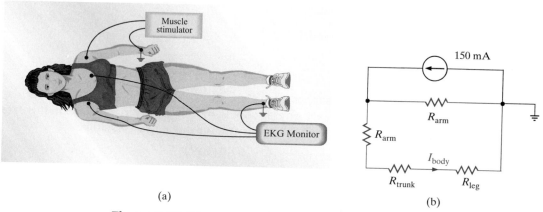

(a)   (b)

Figure 9.32 Diagrams used in Example 9.22.

SOLUTION   When the passive electrode of the stimulator is connected to the ground electrode of the EKG system, the equivalent network in Fig. 9.32b illustrates the two paths for the stimulator current: one through half an arm and the other through half an arm and the body. Using current division, the body current is

$$I_{body} = \frac{(150)(10^{-3})(50)}{50 + 50 + 200 + 100}$$

$$= 19 \text{ mA}$$

Therefore, a dangerously high level of current will flow from the stimulator through the body to the EKG ground.   ❏

EXAMPLE 9.23

A cardiac care patient with a pacing electrode has ignored the hospital rules and is listening to a cheap stereo. The stereo has an amplified 60-Hz hum that is very an-noying. The patient decides to dismantle the stereo partially in an attempt to elim-inate the hum. In the process, while he is holding one of the speaker wires, the other touches the pacing electrode. What are the risks in this situation?

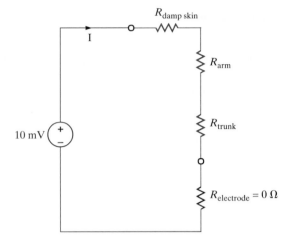

Figure 9.33 Circuit model
for Example 9.23.

SOLUTION   Let us suppose that the patient's skin is damp and that the 60-Hz voltage across the speaker wires is only 10 mV. Then the circuit model in this case would be shown in Fig. 9.33. The current through the heart would be

$$I = \frac{(10)(10^{-3})}{150 + 100 + 200}$$

$$= 22.2 \ \mu A$$

It is known that 10 μA delivered directly to the heart is potentially lethal.   ❏

EXAMPLE 9.24

While maneuvering in a muddy area, a crane operator accidentally touched a high-voltage line with the boom of the crane, as illustrated in Fig. 9.34a. The line potential was 7200 V. The neutral conductor was grounded at the pole. When the crane operator realized what had happened, he jumped from the crane and ran in the direction of the pole, which was approximately 10 m away. He was electrocuted as he ran. Can we explain this very tragic accident?

SOLUTION   The conditions depicted in Fig. 9.34a can be modeled as shown in Fig. 9.34b. The crane was at 7200 V with respect to earth. Therefore, a gradient of 720 V/m existed along the earth between the crane and the power pole. This earth between the crane and the pole is modeled as a resistance. If the man's stride was about 1 m, the difference in potential between his feet was approximately 720 V. A man standing in the same area with his feet together was unharmed.   ❏

The examples of this section have been provided in an attempt to illustrate some of the potential dangers that exist when working or playing around electric power. In the worst case, failure to prevent an electrical accident can result in death. However, even

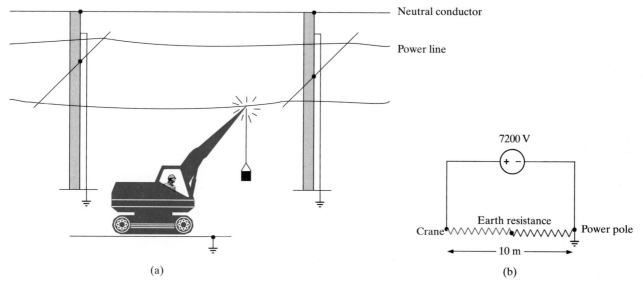

Figure 9.34 Illustrations used in Example 9.24.

nonlethal electrical contacts can cause such things as burns or falls. Therefore, we must always be alert to ensure not only our own safety, but also that of others who work and play with us.

The following guidelines will help us minimize the chances of injury.

SAFETY Guidelines

1. Avoid working on energized electrical systems.

2. Always assume that an electrical system is energized unless you can absolutely verify that it is not.

3. Never make repairs or alterations that are not in compliance with the provisions of the prevailing code.

4. Do not work on potentially hazardous electrical systems alone.

5. If another person is "frozen" to an energized electrical circuit, deenergize the circuit, if possible. If that cannot be done, use nonconductive material such as dry wooden boards, sticks, belts, and articles of clothing to separate the body from the contact. Act quickly but take care to protect yourself.

6. When handling long metallic equipment, such as ladders, antennas, and so on, outdoors, be continuously aware of overhead power lines and avoid any possibility of contact with them.

EXTENSION EXERCISE

E9.16   A woman is driving her car in a violent rainstorm. While she is waiting at an intersection, a power line falls on her car and makes contact. The power line voltage is 7200 V.

(a) Assuming that the resistance of the car is negligible, what is the potential current-through her body if, while holding the door handle with a dry hand, she steps out onto the wet ground?

(b) If she remained in the car, what would happen?

ANSWER: (a) $\mathbf{I}$ = 463 mA, extremely dangerous; (b) she should be safe.

Safety when working with electric power must always be a primary consideration. Regardless of how efficient or expedient an electrical network is for a particular application, it is worthless if it is also hazardous to human life.

The safety device shown in Fig. 9.35, which is also used for troubleshooting, is a proximity-type sensor that will indicate if a circuit is energized by simply touching the conductor on the outside of the insulation. This device is typically carried by all electricians and is helpful when working on electric circuits.

In addition to the numerous deaths that occur each year due to electrical accidents, fire damage that results from improper use of electrical wiring and distribution equipment amounts to millions of dollars per year.

To prevent the loss of life and damage to property, very detailed procedures and specifications have been established for the construction and operation of electrical systems to ensure their safe operation. The *National Electrical Code* ANSI C1 (ANSI—American National Standards Institute) is the primary guide. There are other codes, however: for example, the national electric safety code, ANSI C2, which deals with safety requirements for public utilities. The Underwriters' Laboratory (UL) tests all types of devices and systems to ensure that they are safe for use by the general public. We find the UL label on all types of electrical equipment that is used in the home, such as appliances and extension cords.

Electric energy plays a very central role in our lives. It is extremely important to our general health and well-being. However, if not properly used, it can be lethal.

9.11 APPLICATIONS

The following application-oriented examples illustrate a practical use of the material studied in this chapter.

EXAMPLE 9.25

An electric lawn mower requires 12 A rms at 120 V rms but will operate down to 110 V rms without damage. At 110 V rms, the current draw is 13.1 A rms, as shown in Fig. 9.36. Give the maximum length extension cord that can be used with a 120-V rms power source if the extension cord is made from

1. 16-gauge wire ($4 \text{ m}\Omega/\text{ft}$)
2. 14-gauge wire ($2.5 \text{ m}\Omega/\text{ft}$)

SOLUTION The voltage drop across the extension cord is

$$V_{\text{cord}} = (2)(13.1)R_{\text{cord}} = 10 \text{ V rms}$$

Figure 9.35 A modern safety or troubleshooting device (courtesy of Fluke Corporation).

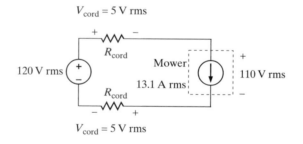

$V_{\text{cord}} = 5$ V rms

R_{cord}

120 V rms

R_{cord}

Mower

13.1 A rms

110 V rms

$V_{\text{cord}} = 5$ V rms

Figure 9.36 Circuit model in Example 9.25.

or

$$R_{\text{cord}} = 0.382 \ \Omega$$

If ℓ_{cord} is the length of the extension cord, then for 16-gauge wire we find

$$\ell_{\text{cord}} = \frac{R_{\text{cord}}}{0.004} = 95.5 \text{ feet}$$

and for 14-gauge wire

$$\ell_{\text{cord}} = \frac{R_{\text{cord}}}{0.0025} = 152.8 \text{ feet}$$

EXAMPLE 9.26

The dc power supply shown in Fig. 9.37 is designed to supply the power requirement for a computer. The dc supply has an efficiency of 92%, and at unity input power factor provides 5 V at 40 A dc. Let us determine the current requirement at the ac wall plug.

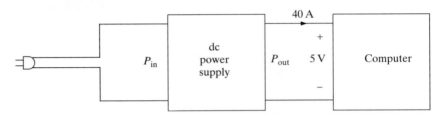

Figure 9.37 Circuit model in Example 9.26.

SOLUTION   The output power of the dc supply is

$$P_{out} = (40)(5) = 200 \text{ W}$$

The input power to the supply is

$$P_{in} = \frac{P_{out}}{0.92} = 217.4 \text{ W}$$

The ac power requirement is

$$P_{ac} = P_{in} = V_{rms} I_{rms} \, pf$$

and hence,

$$I_{rms} = \frac{P_{in}}{V_{rms} \, pf} = 1.81 \text{ A rms} \qquad \square$$

EXAMPLE 9.27

After stopping for coffee after dark, Joe starts his car and notices that his alternator light fails to turn off, indicating total alternator failure. Highway conditions permit driving at 55 mph after dark and Joe is 82.5 miles from home. He recalls that he had purchased a 120 Ah 12-V car battery just last month, advertised to produce "600 cold cranking amps." Joe estimates that it had taken about 6 seconds to start the car.

Joe immediately turns off the radio and heater to conserve power. His two headlights each draw 50 watts on low beam. Assuming the rest of the car's electrical load totals to 50 watts, will Joe make it home before the battery is completely discharged?

SOLUTION   If Joe assumes that the battery is completely charged, then charge = 120 Ah. Since the exact current required to start the engine is unknown,

Joe decides to use the worst case value of 600 A. Thus the charge required to start the engine is estimated to be

$$\text{charge}_{\text{start}} = (600 \text{ A}) \frac{(6 \text{ s})}{3600 \text{ s/h}}$$

$$= 1 \text{ Ah}$$

The electrical load on the battery caused by the two lights and other elements within the car is

$$I = \frac{150 \text{ W}}{12 \text{ V}}$$

$$= 12.5 \text{ A}$$

The time required to drive home at 55 mph is

$$t = \frac{82.5}{55}$$

$$= 1.5 \text{ h}$$

The total charge drain on the battery including starting is

$$\text{charge}_{\text{total}} = 1 + 12.5(1.5) = 19.75 \text{ Ah}$$

Therefore, Joe should make it home with charge to spare.   ❑

EXAMPLE 9.28

The circuit in Fig. 9.38a models an electric field probe amplifier including capacitive effects. As drawn, the average power delivered to the load is not optimized. However, the compensating impedance $\mathbf{Z}_C$ can be used to modify the output power. Let us design $\mathbf{Z}_C$ for maximum power transfer if the input signal $\mathbf{V}_S$ is sinusoidal at 10^6 rad/s.

SOLUTION   For maximum power transfer, the impedance of $\mathbf{Z}_C$ in parallel with R_L must be the complex conjugate of the Thévenin equivalent impedance at nodes A-B, looking back toward the source. Therefore, let us find the Thévenin impedance for the network as redrawn in Fig. 9.38b. Nodal analysis yields

$$(\mathbf{V}_S - \mathbf{V}_{\text{oc}})j\omega C_2 = g_m \mathbf{V}_S + \frac{\mathbf{V}_{\text{oc}}}{R_2}$$

or

$$\mathbf{V}_{\text{oc}} = \mathbf{V}_S(j\omega C_2 - g_m)\mathbf{Z}_2$$

where

$$\mathbf{Z}_2 = R_2 // \frac{1}{j\omega C_2}$$

To find $\mathbf{I}_{\text{sc}}$ we short nodes A and B as shown in Fig. 9.38c. All of the dependent source current flows through the short; therefore,

$$\mathbf{I}_{\text{sc}} = -g_m \mathbf{V}_S$$

Use the problem-solving strategy for maximum average power transfer delineated in Section 9.3.

1 Remove the load $\mathbf{Z}_L$ and find the Thévenin equivalent for the remainder of the circuit.

2 Construct a circuit consisting of the Thévenin equivalent circuit and the load.

3 Select a load $\mathbf{Z}_L$ such that $\mathbf{Z}_L = \mathbf{Z}_{\text{Th}}^*$ and determine the values of the load elements.

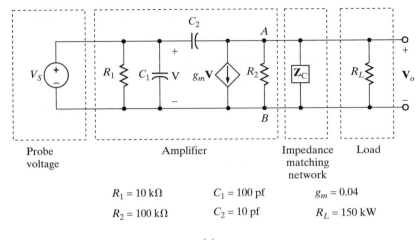

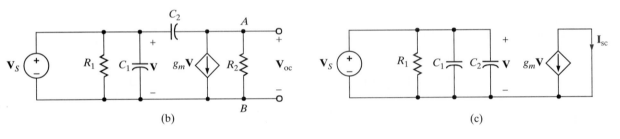

$R_1 = 10\ \text{k}\Omega$ $C_1 = 100\ \text{pf}$ $g_m = 0.04$

$R_2 = 100\ \text{k}\Omega$ $C_2 = 10\ \text{pf}$ $R_L = 150\ \text{kW}$

(a)

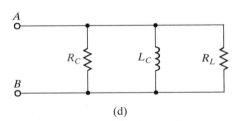

(b)   (c)

(d)

Figure 9.38 Circuits used in Example 9.28.

and $\mathbf{Z}_{\text{Th}}$ is

$$\mathbf{Z}_{\text{Th}} = \frac{\mathbf{V}_{\text{oc}}}{\mathbf{I}_{\text{sc}}} = \mathbf{Z}_2\left(1 - \frac{j\omega C_2}{g_m}\right)$$

Using element values from Fig. 9.38a yields

$$\mathbf{Z}_{\text{Th}} = \frac{10^5}{1 + j1}\left(1 - j2.5 \times 10^{-4}\right) \approx \frac{10^5}{1 + j1}$$

or

$$\mathbf{Z}_{\text{Th}} = 70.7\ \underline{/-45^\circ}\ \text{k}\Omega$$

Maximum power transfer requires

$$\mathbf{Z}_C//R_L = \mathbf{Z}_{Th}^*$$

Since $\mathbf{Z}_{Th}$ is capacitive, it is sensible that $\mathbf{Z}_C$ should be inductive. Let us use the network in Fig. 9.38d for the compensating impedance and find the values of L_C and R_C for maximum average power transfer. The equivalent parallel impedance is

$$\mathbf{Z}_{eq} = R_C//R_L//j\omega L_C$$

or

$$\mathbf{Z}_{eq} = \frac{R_C//R_L}{1 - j\dfrac{R_C//R_L}{\omega L_C}}$$

Based on the requirements for maximum power transfer, i.e.

$$\frac{R_C//R_L}{1 - j\dfrac{R_C//R_L}{\omega L_C}} = \left(\frac{10^5}{1 - j1}\right)^*$$

we find that

$$R_C//R_L = 100 \text{ k}\Omega$$

or

$$R_C = 300 \text{ k}\Omega$$

and

$$\frac{R_C//R_L}{\omega L_C} = 1$$

or

$$L_C = 100 \text{ mH}$$

These values ensure that the amplifier sees an equivalent load, $\mathbf{Z}_{eq}$, which produces maximum average power transfer.   ❏

EXAMPLE 9.29

For safety reasons the National Electrical Code restricts the circuit-breaker rating in a 120-V household lighting branch circuit to no more than 20 A. Furthermore, the code also requires a 25% safety margin for continuous-lighting loads. Under these conditions let us determine the number of 100-W lighting fixtures that can be placed in one branch circuit.

SOLUTION   The model for the branch circuit is shown in Fig. 9.39. The current drawn by each 100-W bulb is

$$I_{bulb} = 100/120 = 0.833 \text{ A rms}$$

Using the safety margin recommendation, the estimated current drawn by each bulb is 25% higher or

$$I_{bulb} = (1.25)(0.83) = 1.04 \text{ A rms}$$

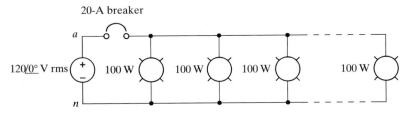

Figure 9.39 20-A branch circuit for household lighting.

Therefore, the maximum number of fixtures on one circuit breaker is

$$n = 20/1.04 = 19 \text{ fixtures}$$

EXAMPLE 9.30

While sitting in a house reading a book, we notice that every time the air conditioner comes on, the lights momentarily dim. Let us investigate this phenomenon using the single-phase three-wire circuit shown in Fig. 9.40a and some typical current requirements for a 10,000-Btu/h air conditioner, assuming a line resistance of 0.5 Ω.

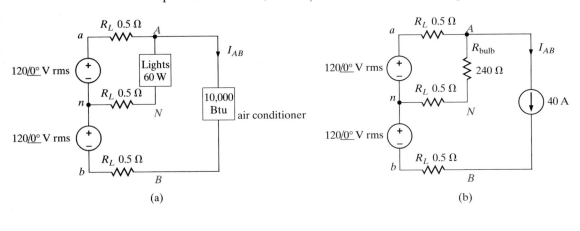

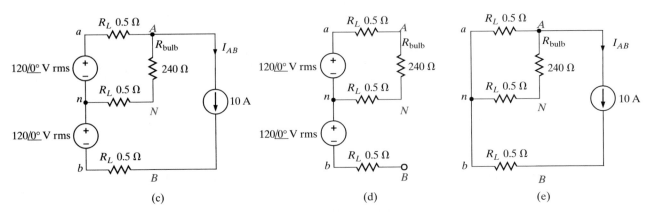

Figure 9.40 Networks used in Example 9.30.

SOLUTION   The 60-W light bulb can be roughly modeled by its equivalent resistance:

$$P_{bulb} = \frac{V_{an}^2}{R_{bulb}}$$

or

$$R_{bulb} = 240 \ \Omega$$

When the air conditioner unit first turns on, the current requirement is 40 A, as shown in Fig. 9.40b. As the compressor motor comes up to speed, the current requirement drops quickly to a steady-state value of 10 A, as shown in Fig. 9.40c. We will compare the voltage across the light fixture, V_{AN}, both at turn on and in steady state.

Using superposition, let us first find that portion of V_{AN} caused by the voltage sources. The appropriate circuit is shown in Fig. 9.40d. Using voltage division, we find that

$$V_{AN1} = V_{AN}\left(\frac{R_{bulb}}{R_{bulb} + 2R_L}\right)$$

or

$$V_{AN1} = 119.50 \text{ V rms}$$

Figure 9.40e will yield the contribution to V_{AN} caused by the 10-A steady-state current. Using current division to calculate the current through the light bulb, we find that

$$V_{AN2} = -\left\{I_{AB}\left(\frac{R_L}{R_{bulb} + 2R_L}\right)\right\}R_{bulb}$$

or

$$V_{AN2} = -4.98 \text{ V rms}$$

Therefore, the steady-state value of V_{AN} is

$$V_{AN} = V_{AN1} + V_{AN2} = 114.52 \text{ V rms}$$

At start-up, our expression for V_{AN2} can be used with $I_{AB} = 40$ A, which yields $V_{AN2} = -19.92$ V rms. The resulting value for V_{AN} is

$$V_{AN} = V_{AN1} + V_{AN2} = 119.50 - 19.92 = 99.58 \text{ V rms}$$

The voltage delivered to the light fixture at start-up is 13% lower than the steady-state value, resulting in a momentary dimming of the lights.

Technique

1 Find the resistance of the light bulb.

2 Use a large current source to represent the transient current of the air conditioner and a small current source to represent the steady-state current.

3 Find the voltage drop across the light bulb during both the transient and steady-state operations.

9.12 CIRCUIT DESIGN

The following design example is presented in the context of a single-phase three-wire household circuit and serves as a good introduction to the material in the next chapter.

EXAMPLE 9.31

A light-duty commercial single-phase three-wire 60-Hz circuit serves lighting, heating, and motor loads, as shown in Fig. 9.41a. Lighting and heating loads are essentially pure resistance and, hence, unity power factor (pf), whereas motor loads have lagging pf.

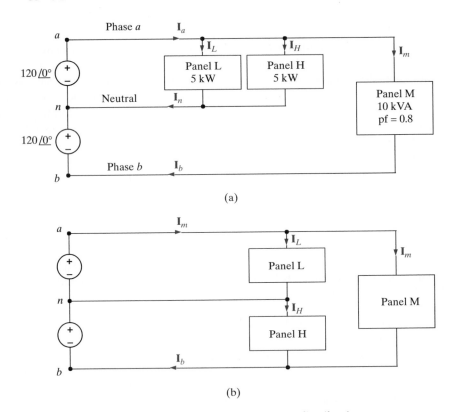

(a)

(b)

Figure 9.41 Single-phase three-wire power distribution system.

We wish to design a balanced configuration for the network and determine its economic viability using the following procedure.

(a) Compute the phase and neutral currents, and the complex power and pf for each source.

(b) Now move the heating load (panel H) to phase b, as shown in Fig. 9.41b. This is called "balancing" the load. Repeat the analysis of (a).

(c) Assume that the phase and neutral conductor resistances are each 0.05 Ω and have negligible effect on the results of (a). Evaluate the system line losses for (a) and (b). If the loads in question operate 24 hours per day and 365 days per year, at $0.08 kWh, how much energy (and money) is saved by operating in the balanced mode?

SOLUTION   (a) The magnitudes of the rms currents are

$$I_L = I_H = \frac{P}{V} = \frac{5000}{120} = 41.67 \text{ A}$$

and

$$I_m = \frac{10{,}000}{240} = 41.67 \text{ A}$$

In addition,

$$\theta_m = \cos^{-1}(0.8) = -36.9°$$

Therefore,

$$\mathbf{I}_a = \mathbf{I}_L + \mathbf{I}_H + \mathbf{I}_m$$
$$= 41.67 \,\underline{/0°} + 41.67 \,\underline{/0°} + 41.67 \,\underline{/-36.9°}$$
$$= 119.4 \,\underline{/-12.1°} \text{ A}$$

The currents in the neutral and phase b lines are

$$\mathbf{I}_n = \mathbf{I}_L + \mathbf{I}_H = 83.34 \,\underline{/0°} \text{ A}$$
$$\mathbf{I}_b = 41.67 \,\underline{/-36.9°} \text{ A}$$

The complex power and power factor for each source are

$$\mathbf{S}_a = \mathbf{V}_{an}\mathbf{I}_a^* = \left(120 \,\underline{/0°}\right)\left(119.4 \,\underline{/ + 12.1°}\right) = 14 + j3 \text{ kVA}$$
$$\text{pf}_a = \cos(12.1°) = 0.9778 \text{ lagging}$$

and in a similar manner

$$\mathbf{S}_b = \mathbf{V}_{bn}\mathbf{I}_b^* = 4 + j3 \text{ kVA}$$
$$\text{pf}_b = 0.8 \text{ lagging}$$

(b) Under the balanced condition

$$\mathbf{I}_a = \mathbf{I}_L + \mathbf{I}_m = 41.67 \,\underline{/0°} + 41.67 \,\underline{/-36.9°}$$
$$= 79.06 \,\underline{/-18.4°} \text{ A}$$

and

$$\mathbf{I}_b = 79.06 \,\underline{/-18.4°} \text{ A}$$
$$\mathbf{I}_n = 0$$

Therefore,

$$\mathbf{S}_a = \mathbf{V}_{na}\mathbf{I}_a^* = 9 + j3 \text{ kVA}$$
$$\text{pf}_a = 0.9487 \text{ lagging}$$

and

$$\mathbf{S}_b = \mathbf{V}_{bn}\mathbf{I}_b^* = 9 + j3 \text{ kVA}$$
$$\text{pf}_b = 0.9487 \text{ lagging}$$

(c) The power loss in the lines in kW is

$$P_{\text{loss}} = R_a I_a^2 + R_b I_b^2 + R_n I_c^2$$
$$= 0.05\left(I_a^2 + I_b^2 + I_n^2\right)/1000$$

The total energy loss for a year is

$$W_{\text{loss}} = (24)(365)P_{\text{loss}} = 8760\,P_{\text{loss}}$$

and the annual cost is

$$\text{Cost} = \$0.08\,W_{\text{loss}}$$

A comparison of the unbalanced and balanced cases is shown in the following table.

	Unbalanced Case	Balanced Case
P_{loss} (kW)	1.147	0.625
W_{loss} (kW-hr)	10,034	5,475
Cost ($)	804	438

Therefore, the annual savings obtained using the balanced configuration is

$$\text{Annual energy savings} = 10{,}048 - 5{,}475 = 4{,}573\ \text{kWh}$$
$$\text{Annual cost savings} = 804 - 438 = \$366 \qquad \square$$

9.13 SUMMARY

Instantaneous power   If the current and voltage are sinusoidal functions of time, the instantaneous power is equal to a time-independent average value plus a sinusoidal term that has a frequency twice that of the voltage or current.

Average power $P = \frac{1}{2}VI \cos(\theta_v - \theta_i) = \frac{1}{2}VI \cos\theta$, where θ is the phase of the impedance.

Resistive load $P = \frac{1}{2}I^2 R = \frac{1}{2}VI$ since $\mathbf{V}$ and $\mathbf{I}$ are in phase.

Reactive load $P = \frac{1}{2}VI \cos(\pm 90°) = 0$

Superposition   When multiple sources are present in a network, superposition cannot be used in computing power unless each source is of a different frequency.

Maximum average power transfer   To obtain the maximum average power transfer to a load, the load impedance should be chosen equal to the conjugate of the Thévenin equivalent impedance representing the remainder of the network.

rms or effective value of a periodic waveform   The effective, or rms, value of a periodic waveform was introduced as a means of measuring the effectiveness of a source in delivering power to a resistive load. The effective value of a periodic waveform is found by determining the root-mean-square value of the waveform.

The rms value of a sinusoidal function is equal to the maximum value of the sinusoid divided by $\sqrt{2}$.

Power factor Apparent power is defined as the product $V_{rms}I_{rms}$. The power factor is defined as the ratio of the average power to the apparent power and is said to be leading when the phase of the current leads the voltage, and lagging when the phase of the current lags the voltage. The power factor of a load with a lagging power factor can be corrected by placing a capacitor in parallel with the load.

Complex power The complex power, $\mathbf{S}$, is defined as the product $\mathbf{V}_{rms}\mathbf{I}_{rms}^*$. The complex power $\mathbf{S}$ can be written as $\mathbf{S} = P + jQ$, where P is the real or average power and Q is the imaginary or quadrature power.

$$\mathbf{S} = I^2\mathbf{Z} = I^2R + jI^2X$$

The single-phase three-wire circuit The single-phase three-wire circuit is the one commonly used in households. Large appliances are connected line to line and small appliances and lights are connected line to neutral.

Safety Safety must be a primary concern in the design and use of any electrical circuit. The National Electric Code is the primary guide for the construction and operation of electrical systems.

PROBLEMS

Section 9.2

9.1 The voltage and current at the input of a circuit are given by the expressions

$$v(t) = 170 \cos(\omega t + 30°) \text{ V}$$
$$i(t) = 5 \cos(\omega t + 45°) \text{ A}$$

$P = \frac{1}{2}VI \cos \theta$

Determine the average power absorbed by the circuit.

9.2 The voltage and current at the input of a network are given by the expressions

$$v(t) = 6 \cos \omega t \text{ V}$$
$$i(t) = 4 \sin \omega t \text{ A}$$

Similar to P9.1.

Determine the average power absorbed by the network.

9.3 Find the average power absorbed by the resistor in the circuit shown in Fig. P9.3 if $v_1(t) = 10 \cos(377t + 60°)$ V and $v_2(t) = 20 \cos(377t + 120°)$ V.

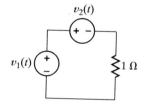

$P = \frac{1}{2}\dfrac{V^2}{R}$

Figure P9.3

9.4   Find the average power absorbed by the resistor in the circuit shown in Fig. P9.4.
Let $i_1(t) = 4\cos(377t + 60°)$ A, $i_2(t) = 6\cos(754t + 10°)$ A, and $i_3(t) = 4\cos(377t - 30°)$ A.

$P = \frac{1}{2}I^2R$

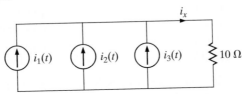

Figure P9.4

9.5   Compute the average power absorbed by each of the elements to the right of the dashed line in the circuit shown in Fig. P9.5.

Only the resistor absorbs
average power.

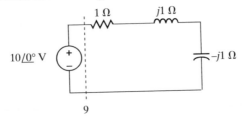

Figure P9.5

9.6   Compute the average power absorbed by the elements to the right of the dashed line in the network shown in Fig. P9.6.

Similar to P9.5.

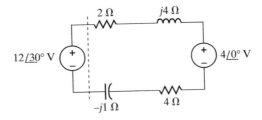

Figure P9.6

9.7   Determine the average power supplied by each source in the network shown in Fig. P9.7.

KCL can be used to find all
voltages and currents needed
to determine the unknowns.

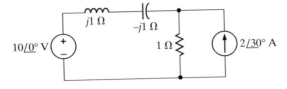

Figure P9.6

9.8   Given the network in Fig. P9.8, show that the power supplied by the sources is equal to the power absorbed by the passive elements.

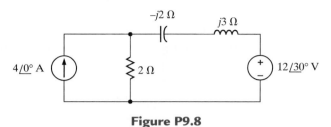

Similar to P9.7.

Figure P9.8

9.9   Find the average power absorbed by the network shown in Fig. P9.9.

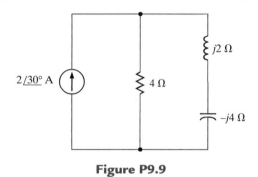

Similar to P9.7.

Figure P9.9

9.10   Calculate the average power absorbed by the 1-Ω resistor in the network shown in Fig. P9.10.

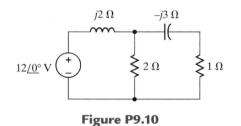

Use the voltage division rule to find the voltage across the 1-Ω resistor. Then employ $P = \frac{1}{2}V^2/R$.

Figure P9.10

9.11   Find the average power supplied and/or absorbed by each element in Fig. P9.11.

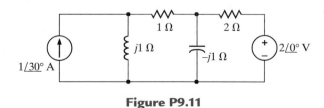

Use nodal analysis to find voltages. Find the current in the voltage source. Then use $P = \frac{1}{2}VI \cos \theta$ for the sources and $P = \frac{1}{2}V^2/R$ for the resistor.

Figure P9.11

9.12 Given the circuit in Fig. P9.12, find the average power supplied and the average power absorbed by each element.

The current division rule will yield the currents, and they, in turn, will produce the source voltage.

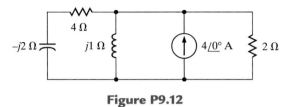

Figure P9.12

9.13 Determine the average power absorbed by the 4-Ω resistor in the network shown in Fig. P9.13.

Using nodal analysis, find the voltage across the 4-Ω resistor.

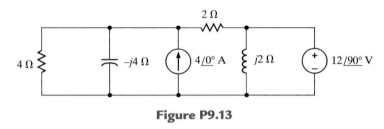

Figure P9.13

9.14 Find the average power absorbed by each element in the network shown in Fig. P9.14.

Nodal analysis will yield the node voltages and then the element currents.

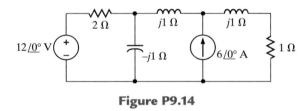

Figure P9.14

9.15 Given the network in Fig. P9.15, find the average power supplied to the circuit.

Use nodal analysis to find the voltage across the current source. Then all currents can be calculated.

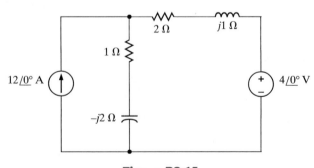

Figure P9.15

9.16 Determine the average power absorbed by the 4-Ω in the network shown in Fig. P9.16.

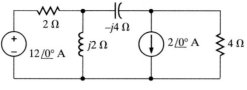

Figure P9.16

Similar to P9.13.

9.17 Determine the average power supplied to the network shown in Fig. P9.17.

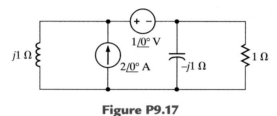

Figure P9.17

Use the supernode technique to find the node voltages. Then all currents can be obtained immediately.

9.18 Given the circuit in Fig. P9.18, determine the amount of average power supplied to the network.

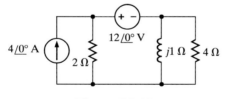

Figure P9.18

Similar to P9.17.

9.19 Determine the average power supplied to the network in Fig. P9.19.

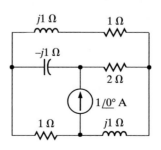

Figure P9.19

Similar to P9.17.

Section 9.3

Find the Thévenin equivalent circuit for the network with the load $\mathbf{Z}_L$ removed. Then let $\mathbf{Z}_L = \mathbf{Z}_{Th}^*$ and find the maximum power.

9.20 Determine the impedance $\mathbf{Z}_L$ for maximum average power transfer and the value of the maximum power transferred to the $\mathbf{Z}_L$ for the circuit shown in Fig. P9.20.

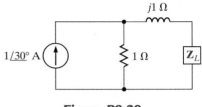

Figure P9.20

Find the current for maximum average power transfer. Then use the average power equation.

9.21 Show that the maximum average power transferred to a load is given by the following expression, where V_{oc} and R_{Th} are defined in the chapter.

$$P_L = \frac{V_{oc}^2}{8R_{Th}}$$

Similar to P9.20.

9.22 Determine the impedance $\mathbf{Z}_L$ for maximum average power transfer and the value of the maximum average power absorbed by the load in the network shown in Fig. P9.22.

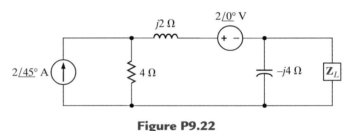

Figure P9.22

Similar to P9.20.

9.23 Determine the impedance $\mathbf{Z}_L$ for maximum average power transfer and the value of the maximum average power transferred to $\mathbf{Z}_L$ for the circuit shown in Fig. P9.23.

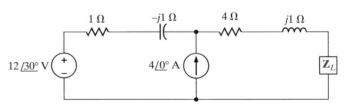

Figure P9.23

9.24 Repeat Problem 9.23 for the network in Fig. P9.24.

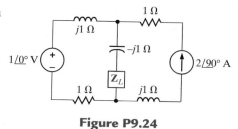

Figure P9.24

9.25 Repeat Problem 9.23 for the network in Fig. P9.25.

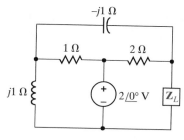

Figure P9.25

9.26 Repeat Problem 9.23 for the network in Fig. P9.26.

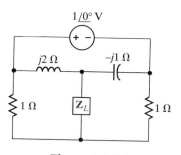

Figure P9.26

9.27 Repeat Problem 9.23 for the network in Fig. P9.27.

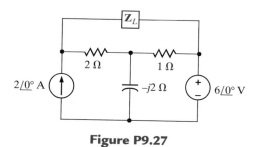

Figure P9.27

Similar to P9.20.

9.28 Repeat Problem 9.23 for the network in Fig. P9.28.

Similar to P9.20.

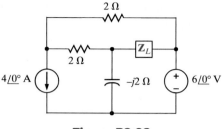

Figure P9.28

9.29 Repeat Problem 9.23 for the network in Fig. P9.29.

Similar to P9.20.

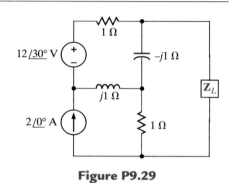

Figure P9.29

Section 9.4

$$V_{rms} = \sqrt{\frac{1}{T} \int_0^T v^2(t) \, dt}$$

9.30 Compute the rms value of the voltage given by the expression $v(t) = 10 + 20 \cos(377t + 30°)$ V.

9.31 Compute the rms value of the voltage given by the waveform shown in Fig. P9.31.

Similar to P9.30; carefully select the bounds of the integral.

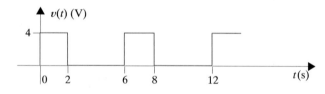

Figure P9.31

9.32 Calculate the rms value of the waveform shown in Fig. P9.32.

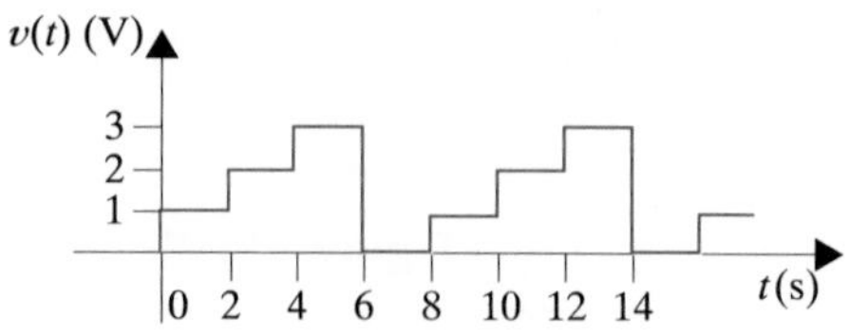

Similar to P9.31.

Figure P9.32

9.33 Calculate the rms value of the waveform in Fig. P9.33.

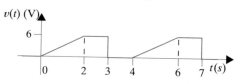

Figure P9.33

The equation of the voltage between 0 and 2 s is 3t.

9.34 Calculate the rms value of the waveform in Fig. P9.34.

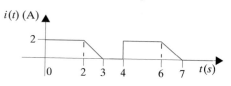

Figure P9.34

Similar to P9.33.

9.35 Calculate the rms value of the waveform in Fig. P9.35.

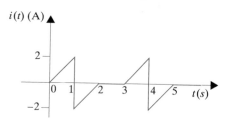

Figure P9.35

Find the equation of the functions in the two intervals $0 \leq t < 1$ s, and $0 \leq t < 2$ s. Integrate to find rms value.

9.36 Find the rms value of the exponential waveform shown in Fig. P9.36.

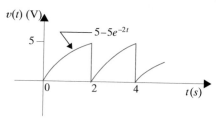

Figure P9.36

Use $T = 2$ seconds.

9.37 Determine the rms value of the rectified sine wave shown in Fig. P9.37. Use the fact that $\sin^2 \theta = \frac{1}{2} - \frac{1}{2} \cos 2\theta$.

The integral's upper bound = 1, the lower bound = 0, T = 2, and the function is $\sin \pi t$.

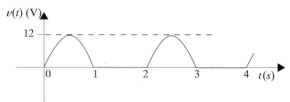

Figure P9.37

9.38 Find the rms value of the current defined by the expression

See Example 9.11.

$$i(t) = 10 \cos (377t) + 12 \sin (754t + 80°) \text{ A}$$

9.39 Find the rms value of the voltage defined by the expression

Shortcut: Note that $\cos (t + 240°) + \cos t + \cos (t + 120°) = 0$.

$$v(t) = \cos t + \cos (t + 120°).$$

9.40 Write the phasor with rms magnitude in polar form for the following waveforms.

(a) $v_1(t) = 170 \cos (377t + 30°) \text{ V}$
(b) $i_2(t) = 10 \cos (377t + 45°) \text{ A}$
(c) $i_3(t) = 20 \cos (377t + 120°) \text{ A}$
(d) $v_4(t) = 32 \sqrt{2} \cos (377t - 120°) \text{ V}$
(e) $v_5(t) = 20 \cos (377t + 30°) \text{ V}$

Section 9.5

9.41 A plant consumes 20 kW of power from a 240-V rms line. If the load power factor is 0.9, what is the angle by which the load voltage leads the load current? What is the load current phasor if the line voltage has a phasor of 240 $\underline{/0°}$ V rms?

Find θ from pf = $\cos \theta$.

9.42 A plant consumes 100 kW of power at 0.9 pf lagging. If the load current is 200 A rms, find the load voltage.

Find V using $P = VI \cos \theta$.

9.43 A plant draws 250 A rms from a 240-V rms line to supply a load with 100 kW. What is the power factor of the load?

pf = $\cos \theta$

9.44 An industrial load consumes 100 kW at 0.8 pf lagging. If an ammeter in the transmission line indicates that the load current is 200 A rms, find the load voltage.

$P = VI \cos \theta$.

9.45 An industrial load that consumes 40 kW is supplied by the power company through a transmission line with 0.1 Ω resistance, with 44 kW. If the voltage at the load is 240 V rms, find the power factor at the load.

$P_s = P_{line} + P_{Load}$ will yield the line current.

9.46 The power company supplies 40 kW to an industrial load. The load draws 200 A rms from the transmission line. If the load voltage is 240 V rms and the load power factor is 0.8 lagging, find the losses in the transmission line.

Similar to P9.45.

Section 9.6

9.47 Determine the real power, the reactive power, the complex power, and the power factor for a load having the following characteristics.
  (a) $\mathbf{I} = 2\ \underline{/40°}$ A rms, $\mathbf{V} = 450\ \underline{/70°}$ V rms.
  (b) $\mathbf{I} = 1.5\ \underline{/-20°}$ A rms, $\mathbf{Z} = 5000\ \underline{/15°}$ Ω.
  (c) $\mathbf{V} = 200\ \underline{/+35°}$ V rms, $\mathbf{Z} = 1500\ \underline{/-15°}$ Ω.

The complex power can be expressed as

$$S = VI^* = I^2 Z = \left(\frac{V}{Z}\right)^2 Z$$

9.48 An industrial load operates at 30 kW, 0.8 pf lagging. The load voltage is $240\ \underline{/0°}$ V rms. The real and reactive power losses in the transmission-line feeder are 1.8 kW and 2.4 kvar, respectively. Find the impedance of the trasmission line and the input voltage to the line.

Recall that

$$R_{Line} = \frac{P_{Line}}{I_L^2}$$

$$X_{Line} = \frac{Q_{Line}}{I_L^2}$$

9.49 A transmission line with impedance $0.08 + j0.25$ Ω is used to deliver power to a load. The load is inductive and the load voltage is $240\ \underline{/0°}$ V rms at 60 Hz. If the load requires 12 kW and the real power loss in the line is 560 W, determine the power factor angle of the load.

The line losses can be used to find the line current.

9.50 Calculate the voltage $\mathbf{V}_S$ that must be supplied to obtain 2 kW, $240\ \underline{/0°}$ V rms, and a power factor of 0.9 lagging at the load $\mathbf{Z}_L$ in the network in Fig. P9.50.

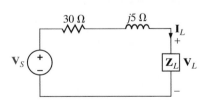

Figure P9.50

I_L can be found using the expression $P = VI \cos \theta$.

9.51 Use Kirchhoff's laws to compute the source voltage of the network shown in Fig. P9.51.

$\mathbf{V}_S = \mathbf{I}_S \mathbf{Z}_{\text{Line}} + 240 \underline{/0°}$,
where $\mathbf{I}_S$ is the source current.

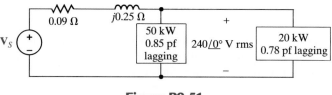

Figure P9.51

9.52 Given the network in Fig. P9.52, determine the input voltage $\mathbf{V}_S$.

Similar to P9.51.

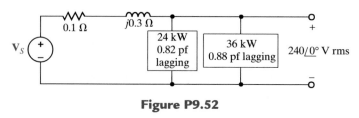

Figure P9.52

9.53 Given the network in Fig. P9.53, determine the input voltage $\mathbf{V}_S$.

Similar to P9.51.

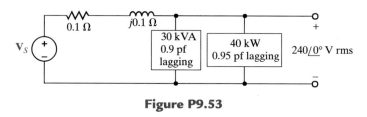

Figure P9.53

9.54 Find the input source voltage and the power factor of the source for the network shown in Fig. P9.54.

Find the current in the 24-kW load. Then find the voltage across the 12-kW load. Find the current in the 12-kW load. Then determine $\mathbf{V}_S$.

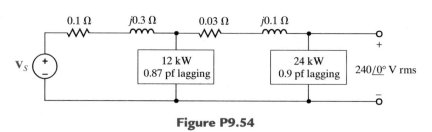

Figure P9.54

Section 9.7

9.55 A plant consumes 60 kW at a power factor of 0.75 lagging from a 240-V rms 60-Hz line. Determine the value of the capacitor that when placed in parallel with the load will change the load power factor to 0.9 lagging.

$Q_{new} - Q_{old} = -\omega C V_{rms}^2$

9.56 A particular load has a pf of 0.8 lagging. The power delivered to the load is 40 kW from a 270-V rms 60-Hz line. What value of capacitance placed in parallel with the load will raise the pf to 0.9 lagging?

Similar to P9.55.

9.57 A small plant has a bank of induction motors that consume 64 kW at a pf of 0.68 lagging. The 60-Hz line voltage across the motors is 240 $\underline{/0°}$ V rms. The local power company has told the plant to raise the pf to 0.92 lagging. What value of capacitance is required?

Similar to P9.55.

9.58 The 60-Hz line voltage for a 60-kW, 0.76-pf lagging industrial load is 240 $\underline{/0°}$ V rms. Find the value of capacitance that when placed in parallel with the load will raise the power factor to 0.9 lagging.

Similar to P9.55.

9.59 An industrial load consumes 44 kW at 0.82 pf lagging from a 240 $\underline{/0°}$-V-rms 60-Hz line. A bank of capacitors totaling 600 μF is available. If these capacitors are placed in parallel with the load, what is the new power factor of the total load?

Similar to P9.55.

9.60 A bank of induction motors consumes 36 kW at 0.78 pf lagging from a 60-Hz 240 $\underline{/0°}$-V-rms line. If 200 μF of capacitors are placed in parallel with the load, what is the new power factor of the total load?

Similar to P9.55.

Section 9.8

9.61 Given the network in Fig. P9.61, determine the wattmeter reading

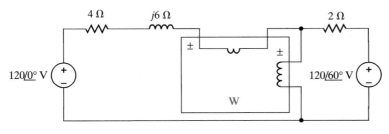

Assume the current coil is replaced with a short circuit and the voltage coil is open circuited. Find the current and voltage of the watt meter. $P = \frac{1}{2} IV \cos \theta$.

Figure P9.61

9.62 Given the network in Fig. P9.62, determine the wattmeter reading.

Similar to P9.61.

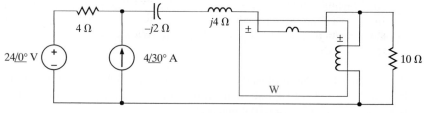

Figure P9.62

Section 9.9

9.63 A single-phase three-wire 60-Hz circuit serves three loads, as shown in Fig. P9.63. Determine $\mathbf{I}_{aA}$, $\mathbf{I}_{nN}$, $\mathbf{I}_c$, and the energy use over a 24-hour period in kilowatt-hours.

See Example 9.17.

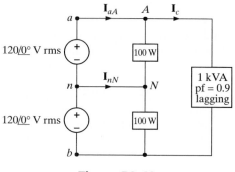

Figure P9.63

Section 9.10

9.64 A 5.1-kW household range is designed to operate on a 240-V rms sinusoidal voltage, as shown in Fig. P9.64a. However, the electrician has mistakenly connected the range to 120 V rms, as shown in Fig. P9.64b. What is the effect of this error?

Find the power consumed by the range by assuming that I_L is the maximum current allowed in both connections. Check to see if the power causes any damage.

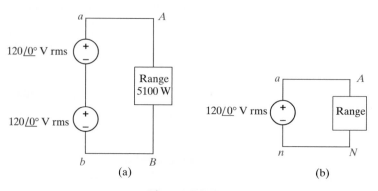

Figure P9.64

9.65 A man and his son are flying a kite. The kite becomes entangled in a 7200-V power line close to a power pole. The man crawls up the pole to remove the kite. While trying to remove the kite, the man accidentally touches the 7200-V line. Assuming the power pole is well grounded, what is the potential current through the man's body?

> Use the resistances of an arm or leg, trunk, and skin defined in Problem 9.20 to find the current.

9.66 A number of 120-V household fixtures are to be used to provide lighting for a large room. The total lighting load is 8 kW. The National Electric Code requires that no circuit breaker be larger than 20 A with a 25% safety margin. Determine the number of identical branch circuits needed for this requirement.

> Find I_{load} in the 8 kW load. $I_{breaker} = (1.25)I_{load}$, $I_{breaker}/(20\text{ A})$ is used to determine the number of circuits.

9.67 In order to test a light socket, a woman, while standing on cushions that insulate her from the ground, sticks her finger into the socket, as shown in Fig. P9.67. The tip of her finger makes contact with one side of the line, and the side of her finger makes contact with the other side of the line. Assuming that any portion of a limb has a resistance of 95 Ω, is there any current in the body? Is there any current in the vicinity of the heart?

> Construct an equivalent circuit similar to that in P9.65. Determine if the current will go through the heart.

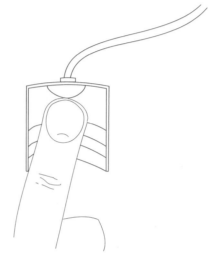

Figure P9.67

9.68 An inexperienced mechanic is installing a 12-V battery in a car. The negative terminal has been connected. He is currently tightening the bolts on the positive terminal. With a tight grip on the wrench, he turns it so that the gold ring on his finger makes contact with the frame of the car. This situation is modeled in Fig. P9.68, where we assume that the resistance of the wrench is negligible and the resistance of the contact is as follows:

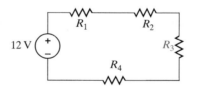

Figure P9.68

> Find the current in the circuit. Then find the power.

$$R_1 = R_{bolt\ to\ wrench} = 0.012\ \Omega$$
$$R_2 = R_{wrench\ to\ ring} = 0.012\ \Omega$$
$$R_3 = R_{ring} = 0.012\ \Omega$$
$$R_4 = R_{ring\ to\ frame} = 0.012\ \Omega$$

What power is quickly dissipated in the gold ring, and what is the impact of this power dissipation?

Advanced Problems

9.69 Find the average power absorbed by the 2-Ω resistor in the circuit shown in Fig P9.69.

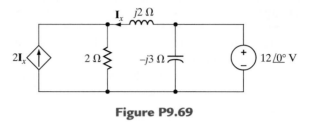

Figure P9.69

9.70 Determine the average power absorbed by a 2-Ω resistor connected at the output terminals of the network shown in Fig. P9.70.

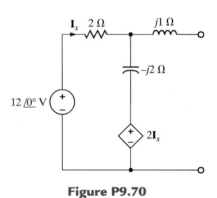

Figure P9.70

9.71 Determine the average power absorbed by the 2-kΩ output resistor in Fig. P9.71.

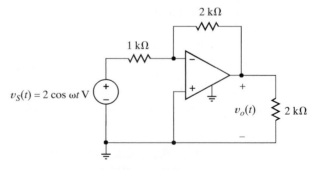

Figure P9.71

9.72 Determine the average power absorbed by the 4-kΩ resistor in Fig. P9.72.

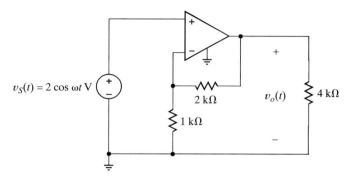

$v_S(t) = 2 \cos \omega t$ V

2 kΩ

1 kΩ

$v_o(t)$

4 kΩ

Figure P9.72

9.73 Given the network in Fig. P9.73, determine which elements are supplying power, which ones are absorbing power, and how much power is being supplied and absorbed.

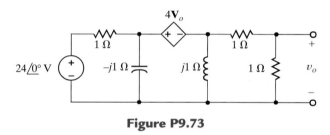

$4\mathbf{V}_o$

1 Ω

1 Ω

$24\underline{/0°}$ V

$-j1$ Ω

$j1$ Ω

1 Ω

v_o

Figure P9.73

9.74 Given the network in Fig. P9.74, find the total average power supplied and the average power absorbed in the 4-Ω resistor.

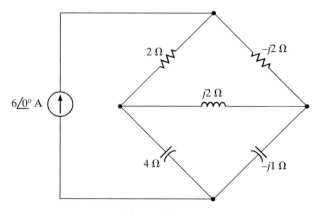

2 Ω

$-j2$ Ω

$j2$ Ω

$6\underline{/0°}$ A

4 Ω

$-j1$ Ω

Figure P9.74

9.75 Given the network in Fig. P9.75, find the average power supplied and the total average power absorbed.

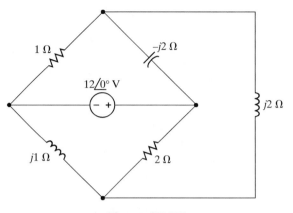

Figure P9.75

9.76 Find the impedance $\mathbf{Z}_L$ for maximum average power transfer and the value of the maximum average power transferred to $\mathbf{Z}_L$ for the circuit shown in Fig. P9.76.

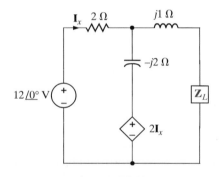

Figure P9.76

9.77 Determine $\mathbf{Z}_L$ for maximum average power transfer and the value of the maximum average power transferred for the network shown in Fig. P9.77.

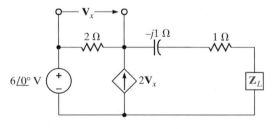

Figure P9.77

9.78 Find the impedance $\mathbf{Z}_L$ for maximum average power transfer and the value of the maximum average power transferred to $\mathbf{Z}_L$ for the circuit shown in Fig. P9.78.

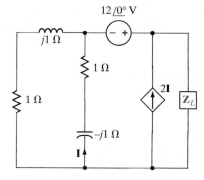

Figure P9.78

9.79 Repeat Problem 9.78 for the network in Fig. P9.79.

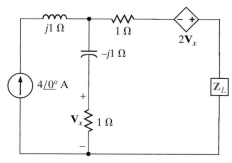

Figure P9.79

9.80 Given the network in Fig. P9.80, find $\mathbf{Z}_L$ for maximum average power transfer and the maximum average power transferred.

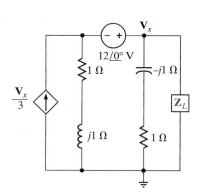

Figure P9.80

9.81 Determine the impedance $\mathbf{Z}_L$ for maximum average power transfer and the value of the maximum average power absorbed by the load in the network shown in Fig. P9.81.

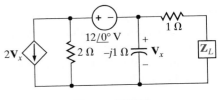

Figure P9.81

9.82 Determine the impedance $\mathbf{Z}_L$ for maximum average power transfer and the value of the maximum average power absorbed by the load in the network shown in Fig. P9.82.

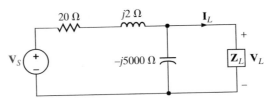

Figure P9.82

9.83 Determine the voltage $\mathbf{V}_S$ that must be supplied to obtain 1.5 kW, $0.75\ \underline{/0°}$ A rms, and a power factor of 0.85 leading at the load $\mathbf{Z}_L$ in the network in Fig. P9.83.

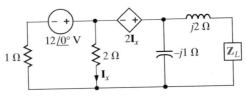

Figure P9.83

9.84 Given the network in Fig. P9.84, compute the wattmeter reading.

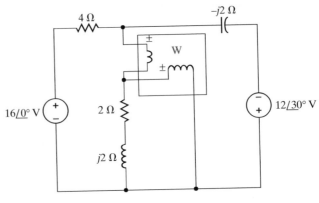

Figure P9.84

9.85 For the network in Fig. P9.85, find the impedance $\mathbf{Z} = R + jX$ that is required for maximum average power transfer.

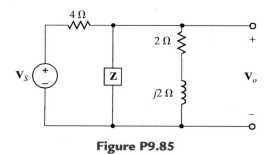

Figure P9.85

Polyphase Circuits

10

While traveling throughout our city or across the countryside, we often encounter power transmission lines overhead. If we look carefully, we find that there are typically three large conductors. Depending upon the configuration, there may be one or two smaller conductors that are placed above the power conductors and used for lightning protection. This three-phase power distribution line is employed to transmit power from the generating station to the user. The importance of this three-phase circuit stems from the fact that we can show that (1) power distributed in this manner is steady, rather than pulsating, as is the case in single-phase distribution and, therefore, there is less wear and tear on mechanical conversion equipment; and (2) by using a three-phase set of voltages we can deliver the same amount of power as that delivered in the single-phase case with fewer conductors and therefore much less material.

10.1 THREE-PHASE CIRCUITS

In this chapter we add a new dimension to our study of ac steady-state circuits. Up to this point we have dealt with what we refer to as single-phase circuits. Now we extend our analysis techniques to polyphase circuits or, more specifically, three-phase circuits (that is, circuits containing three voltage sources that are one-third of a cycle apart in time).

There are a number of important reasons why we study three-phase circuits. It is more advantageous and economical to generate and transmit electric power in the polyphase mode rather than with single-phase systems. As a result, most electric power is transmitted in polyphase circuits. In the United States the power system frequency is 60 Hz, whereas in other parts of the world 50 Hz is common.

Power transmission is most efficiently accomplished at very high voltage. Since this voltage can be extremely high in comparison to the level at which it is normally used (e.g., in the household), there is a need to raise and lower the voltage. This can be accomplished in ac systems using transformers, which we will study in Chapter 11.

As the name implies, three-phase circuits are those in which the forcing function is a three-phase system of voltages. If the three sinusoidal voltages have the same magnitude and frequency and each voltage is 120° out of phase with the other two, the voltages are said to be *balanced*. If the loads are such that the currents produced by the voltages are also balanced, the entire circuit is referred to as a *balanced three-phase circuit*.

A balanced set of three-phase voltages can be represented in the frequency domain as shown in Fig. 10.1a, where we have assumed that their magnitudes are 120 V rms. From the figure we note that

$$
\begin{aligned}
\mathbf{V}_{an} &= 120 \ \underline{/0°} \ \text{V rms} \\
\mathbf{V}_{bn} &= 120 \ \underline{/-120°} \ \text{V rms} \\
\mathbf{V}_{cn} &= 120 \ \underline{/-240°} \ \text{V rms} \\
&= 120 \ \underline{/120°} \ \text{V rms}
\end{aligned}
$$

10.1

Note that our double-subscript notation is exactly the same as that employed in the earlier chapters; that is, $\mathbf{V}_{an}$ means the voltage at point a with respect to the point n. We will

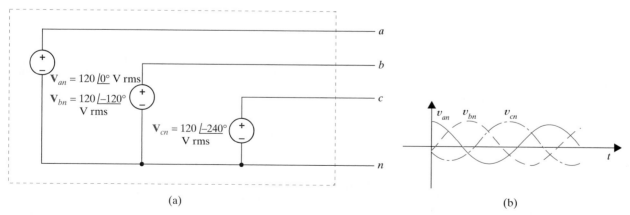

(a)   (b)

Figure 10.1 Balanced three-phase voltages.

also employ the double-subscript notation for currents; that is, $\mathbf{I}_{an}$ is used to represent the current from a to n. However, we must be very careful in this case to describe the precise path, since in a circuit there will be more than one path between the two points. For example, in the case of a single loop the two possible currents in the two paths will be 180° out of phase with each other.

The preceding phasor voltages can be expressed in the time domain as

$$v_{an}(t) = 120\sqrt{2}\cos\omega t \text{ V}$$
$$v_{bn}(t) = 120\sqrt{2}\cos(\omega t - 120°) \text{ V} \qquad \textbf{10.2}$$
$$v_{cn}(t) = 120\sqrt{2}\cos(\omega t - 240°) \text{ V}$$

These time functions are shown in Fig. 10.1b.

Finally, let us examine the instantaneous power generated by a three-phase system. Assume that the voltages in Fig. 10.1 are

$$v_{an}(t) = V_m\cos\omega t \text{ V}$$
$$v_{bn}(t) = V_m\cos(\omega t - 120°) \text{ V} \qquad \textbf{10.3}$$
$$v_{cn}(t) = V_m\cos(\omega t - 240°) \text{ V}$$

If the load is balanced, the currents produced by the sources are

$$i_a(t) = I_m\cos(\omega t - \theta) \text{ A}$$
$$i_b(t) = I_m\cos(\omega t - \theta - 120°) \text{ A} \qquad \textbf{10.4}$$
$$i_c(t) = I_m\cos(\omega t - \theta - 240°) \text{ A}$$

The instantaneous power produced by the system is

$$\begin{aligned}
p(t) &= p_a(t) + p_b(t) + p_c(t) \\
&= V_m I_m[\cos\omega t\cos(\omega t - \theta) + \cos(\omega t - 120°)\cos(\omega t - \theta - 120°) \\
&\quad + \cos(\omega t - 240°)\cos(\omega t - \theta - 240°)]
\end{aligned} \qquad \textbf{10.5}$$

Using the trigonometric identity

$$\cos\alpha\cos\beta = \tfrac{1}{2}[\cos(\alpha - \beta) + \cos(\alpha + \beta)] \qquad \textbf{10.6}$$

Eq. (10.5) becomes

$$\begin{aligned}
p(t) &= \frac{V_m I_m}{2}[\cos\theta + \cos(2\omega t - \theta) + \cos\theta \\
&\quad + \cos(2\omega t - \theta - 240°) + \cos\theta + \cos(2\omega t - \theta - 480°)]
\end{aligned} \qquad \textbf{10.7}$$

which can be written as

$$\begin{aligned}
p(t) &= \frac{V_m I_m}{2}[3\cos\theta + \cos(2\omega t - \theta) \\
&\quad + \cos(2\omega t - \theta - 120°) + \cos\theta(2\omega t - \theta + 120°)]
\end{aligned} \qquad \textbf{10.8}$$

There exists a trigonometric identity that allows us to simplify the preceding expression. The identity, which we will prove later using phasors, is

$$\cos\phi + \cos(\phi - 120°) + \cos(\phi + 120°) = 0 \qquad \textbf{10.9}$$

If we employ this identity, the expression for the power becomes

$$p(t) = 3 \frac{V_m I_m}{2} \cos \theta \ \text{W} \qquad\qquad \textbf{10.10}$$

Note that this equation indicates that the instantaneous power is always constant in time, rather than pulsating as in the single-phase case. Therefore, power delivery from a three-phase voltage source is very smooth, which is another important reason why power is generated in three-phase form.

10.2 THREE-PHASE CONNECTIONS

By far the most important polyphase voltage source is the balanced three-phase source. This source, as illustrated by Fig. 10.2, has the following properties. The phase voltages—that is, the voltage from each line a, b, and c to the neutral n—are given by

$$\mathbf{V}_{an} = V_p \ \underline{/0°}$$
$$\mathbf{V}_{bn} = V_p \ \underline{/-120°} \qquad\qquad \textbf{10.11}$$
$$\mathbf{V}_{cn} = V_p \ \underline{/+120°}$$

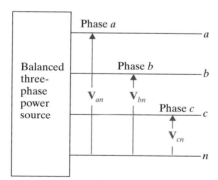

Figure 10.2 Balanced three-phase voltage source.

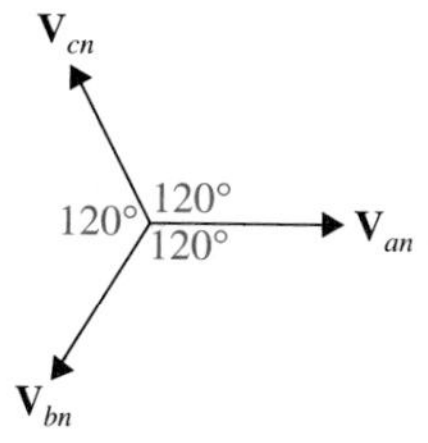

Figure 10.3 Phasor diagram for a balanced three-phase voltage source.

The phasor diagram for these voltages is shown in Fig. 10.3. The phase sequence of this set is said to be *abc* (called positive phase sequence), meaning that $\mathbf{V}_{bn}$ *lags* $\mathbf{V}_{an}$ by 120°.

We will standardize our notation so that we always label the voltages $\mathbf{V}_{an}$, $\mathbf{V}_{bn}$, and $\mathbf{V}_{cn}$ and observe them in the order *abc*. Furthermore, we will normally assume with no loss of generality that $\underline{/\mathbf{V}_{an}} = 0°$.

An important property of the balanced voltage set is that

$$\mathbf{V}_{an} + \mathbf{V}_{ab} + \mathbf{V}_{cn} = 0 \qquad\qquad \textbf{10.12}$$

This property can easily be seen by resolving the voltage phasors into components along the real and imaginary axes. It can also be demonstrated via Eq. (10.9).

From the standpoint of the user who connects a load to the balanced three-phase voltage source, it is not important how the voltages are generated. It is important to note,

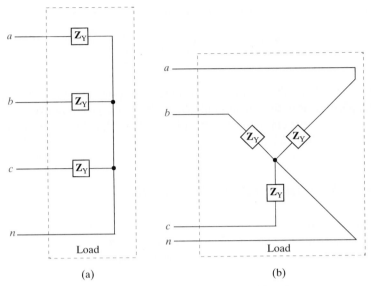

Figure 10.4 Wye (Y)-connected loads.

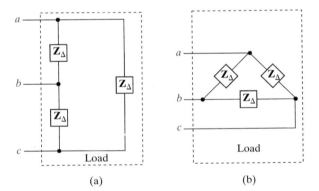

Figure 10.5 Delta (Δ)-connected loads.

however, that if the load currents generated by connecting a load to the power source shown in Fig. 10.2 are also *balanced*, there are two possible equivalent configurations for the load. The equivalent load can be considered as being connected in either a *wye* (Y) or a *delta* (Δ) configuration. The balanced wye configuration is shown in Fig. 10.4a and equivalently in Fig. 10.4b. The delta configuration is shown in Fig. 10.5a and equivalently in Fig. 10.5b. Note that in the case of the delta connection, there is no neutral line. The actual function of the neutral line in the wye connection will be examined and it will be shown that in a balanced system the neutral line carries no current and, for purposes of analysis, may be omitted.

10.3 SOURCE/LOAD CONNECTIONS

Since the source and the load can each be connected in either Y or Δ, three-phase balanced circuits can be connected Y–Y, Y–Δ, Δ–Y, or Δ–Δ. Our approach to the analysis of all of these circuits will be "Think Y" and, therefore, we will analyze the Y–Y connection first.

Balanced Wye–Wye Connection

Suppose now that the source and load are both connected in a wye, as shown in Fig. 10.6. The phase voltages with positive phase sequence are

$$\mathbf{V}_{an} = V_p \, \underline{/0°}$$
$$\mathbf{V}_{bn} = V_p \, \underline{/-120°}$$
$$\mathbf{V}_{cn} = V_p \, \underline{/+120°}$$

10.13

where V_p, the phase voltage, is the magnitude of the phasor voltage from the neutral to any line. The *line-to-line* or, simply, *line voltages* can be calculated using KVL; for example,

$$
\begin{aligned}
\mathbf{V}_{ab} &= \mathbf{V}_{an} - \mathbf{V}_{bn} \\
&= V_p \, \underline{/0°} - V_p \, \underline{/-120°} \\
&= V_p - V_p \left[-\frac{1}{2} - j\frac{\sqrt{3}}{2} \right] \\
&= V_p \left[\frac{3}{2} + j\frac{\sqrt{3}}{2} \right] \\
&= \sqrt{3} \, V_p \, \underline{/30°}
\end{aligned}
$$

The phasor addition is shown in Fig. 10.7a. In a similar manner, we obtain the set of line-to-line voltages as

$$\mathbf{V}_{ab} = \sqrt{3} \, V_p \, \underline{/30°}$$
$$\mathbf{V}_{bc} = \sqrt{3} \, V_p \, \underline{/-90°}$$
$$\mathbf{V}_{ca} = \sqrt{3} \, V_p \, \underline{/-210°}$$

10.14

All the line voltages together with the phase voltages are shown in Fig. 10.7b. We will denote the magnitude of the line voltages as V_L, and therefore, for a balanced system,

$$V_L = \sqrt{3} \, V_p$$

10.15

Hence, in a wye-connected system, the line voltage is equal to $\sqrt{3}$ times the phase voltage.

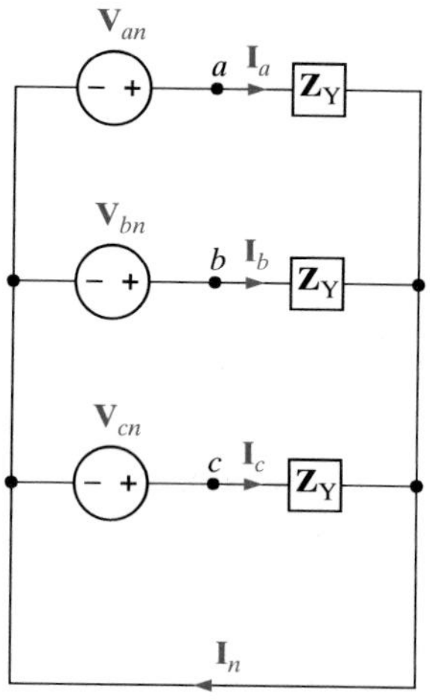

Figure 10.6 Balanced three-phase wye–wye connection.

Conversion rules

$$\underline{/\mathbf{V}_{ab}} = \underline{/\mathbf{V}_{an}} + 30°$$
$$\mathbf{V}_{ab} = \sqrt{3} \, V_{an}$$

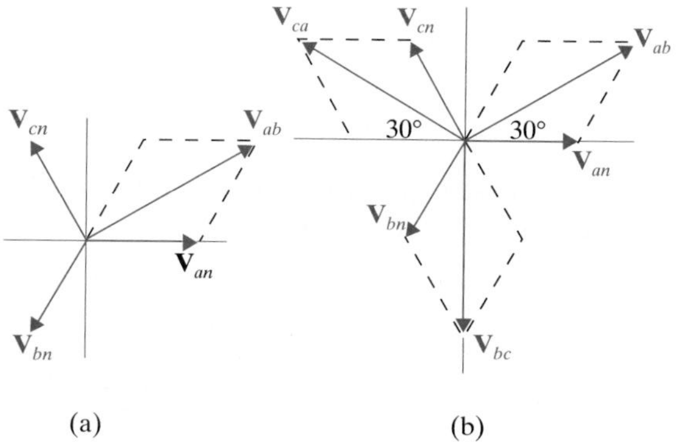

Figure 10.7 Phasor representation of phase and line voltages in a balanced wye–wye system.

(a)   (b)

DRILL

D10.1   If $\mathbf{V}_{an} = 120 \, \underline{/30°}$ V rms, determine $\mathbf{V}_{ab}$.

ANSWER:
$\mathbf{V}_{ab} = 208 \, \underline{/60°}$ V rms

DRILL

D10.2   If $\mathbf{V}_{ab} = 199.2 \, \underline{/45°}$ V rms, determine $\mathbf{V}_{an}$.

ANSWER:
$\mathbf{V}_{an} = 115 \, \underline{/15°}$ V rms

As shown in Fig. 10.6, the line current for the a phase is

$$\mathbf{I}_a = \frac{\mathbf{V}_{an}}{\mathbf{Z}_Y} = \frac{V_p \underline{/0°}}{\mathbf{Z}_Y} \qquad \textbf{10.16}$$

where $\mathbf{I}_b$ and $\mathbf{I}_c$ have the same magnitude but lag $\mathbf{I}_a$ by 120° and 240°, respectively. The neutral current $\mathbf{I}_n$ is then

$$\mathbf{I}_n = (\mathbf{I}_a + \mathbf{I}_b + \mathbf{I}_c) = 0 \qquad \textbf{10.17}$$

Since there is no current in the neutral, this conductor could contain any impedance or it could be an open or a short circuit, without changing the results found previously.

As illustrated by the wye–wye connection in Fig. 10.6, the current in the line connecting the source to the load is the same as the phase current flowing through the impedance $\mathbf{Z}_Y$. Therefore, in a *wye–wye connection*,

$$I_L = I_Y \qquad \textbf{10.18}$$

where I_L is the magnitude of the line current and I_Y is the magnitude of the current in a wye-connected load.

It is important to note that although we have a three-phase system composed of three sources and three loads, we can analyze a single phase and use the phase sequence to obtain the voltages and currents in the other phases. This is, of course, a direct result of the balanced condition. We may even have impedances present in the lines; however, as long as the system remains balanced, we need analyze only one phase. If the line impedances in lines a, b, and c are equal, the system will be balanced. Recall that the balance of the system is unaffected by whatever appears in the neutral line, and since the neutral line impedance is arbitrary, we assume that it is zero (i.e., a short circuit).

EXAMPLE 10.1

An *abc*-sequence three-phase voltage source connected in a balanced wye has a line voltage of $\mathbf{V}_{ab} = 208 \underline{/-30°}$ V rms. We wish to determine the phase voltages.

SOLUTION The magnitude of the phase voltage is given by the expression

$$V_p = \frac{208}{\sqrt{3}}$$
$$V_p = 120 \text{ V rms}$$

The phase relationships between the line and phase voltages are shown in Fig. 10.7. From this figure we note that

The phase of
$\mathbf{V}_{bn} = \underline{/\mathbf{V}_{bn}} = \underline{/\mathbf{V}_{an}} - 120°$
The phase of
$\mathbf{V}_{cn} = \underline{/\mathbf{V}_{cn}} = \underline{/\mathbf{V}_{an}} + 120°$

$$\mathbf{V}_{an} = 120 \underline{/-60°} \text{ V rms}$$
$$\mathbf{V}_{bn} = 120 \underline{/-180°} \text{ V rms}$$
$$\mathbf{V}_{cn} = 120 \underline{/+60°} \text{ V rms}$$

The magnitudes of these voltages are quite common, and one often hears that the electric service in a building, for example, is three-phase 208/120 V.   ❏

EXAMPLE 10.2

A three-phase wye-connected load is supplied by an *abc*-sequence balanced three-phase wye-connected source with a phase voltage of 120 V rms. If the line impedance and load impedance per phase are $1 + j1\ \Omega$ and $20 + j10\ \Omega$, respectively, we wish to determine the value of the line currents and the load voltages.

SOLUTION   The phase voltages are

$$\mathbf{V}_{an} = 120\ \underline{/0°}\ \text{V rms}$$
$$\mathbf{V}_{bn} = 120\ \underline{/-120°}\ \text{V rms}$$
$$\mathbf{V}_{cn} = 120\ \underline{/+120°}\ \text{V rms}$$

The per phase circuit diagram is shown in Fig. 10.8. The line current for the *a* phase is

$$\mathbf{I}_{aA} = \frac{120\ \underline{/0°}}{21 + j11}$$
$$= 5.06\ \underline{/-27.65°}\ \text{A rms}$$

The load voltage for the *a* phase, which we call $\mathbf{V}_{AN}$, is

$$\mathbf{V}_{AN} = \left(5.06\ \underline{/-27.65°}\right)(20 + j10)$$
$$= 113.15\ \underline{/1.08°}\ \text{V rms}$$

The corresponding line currents and load voltages for the *b* and *c* phases are

$$\mathbf{I}_{bB} = 5.06\ \underline{/-147.65°}\ \text{A rms} \qquad \mathbf{V}_{BN} = 113.15\ \underline{/-121.08°}\ \text{V rms} \qquad \underline{/\mathbf{I}_{bB}} = \underline{/\mathbf{I}_{aA}} - 120°$$
$$\mathbf{I}_{cC} = 5.06\ \underline{/-267.65°}\ \text{A rms} \qquad \mathbf{V}_{CN} = 113.15\ \underline{/-241.08°}\ \text{V rms} \qquad \underline{/\mathbf{I}_{cC}} = \underline{/\mathbf{I}_{aA}} + 120°$$

To reemphasize and clarify our terminology, phase voltage, V_p, is the magnitude of the phasor voltage from the neutral to any line, while line voltage, V_L, is the magnitude of the phasor voltage between any two lines. Thus, the values of V_L and V_p will depend on the point at which they are calculated in the system.   ❏

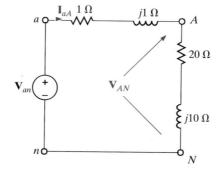

Figure 10.8 Per phase circuit diagram for the problem in Example 10.2.

EXTENSION EXERCISES

E10.1   The voltage for the *a* phase of an *abc*-phase-sequence balanced wye-connected source is $\mathbf{V}_{an} = 120 \underline{/90°}$ V rms. Determine the line voltages for this sources

ANSWER: $\mathbf{V}_{ab} = 208 \underline{/120°}$ V rms, $\mathbf{V}_{bc} = 208 \underline{/0°}$ V rms, $\mathbf{V}_{ca} = 208 \underline{/-120°}$ V rms.

E10.2   An *abc*-phase-sequence three-phase voltage source connected in a balanced wye has a line voltage of $\mathbf{V}_{ab} = 208 \underline{/0°}$ V rms. Determine the phase voltages of the source.

ANSWER: $\mathbf{V}_{an} = 120 \underline{/-30°}$ V rms, $\mathbf{V}_{bn} = 120 \underline{/-150°}$ V rms, $\mathbf{V}_{cn} = 120 \underline{/-270°}$ V rms.

E10.3   A three-phase wye load is supplied by an *abc*-sequence balanced three-phase wye-connected source through a transmission line with an impedance of $1 + j1 \, \Omega$ per phase. The load impedance is $8 + j3 \, \Omega$ per phase. If the load voltage for the *a* phase is $104.02 \underline{/26.6°}$ V rms (i.e., $V_p = 104.02$ V rms at the load end), determine the phase voltages of the source.

ANSWER: $\mathbf{V}_{an} = 120 \underline{/30°}$ V rms, $\mathbf{V}_{bn} = 120 \underline{/-90°}$ V rms, $\mathbf{V}_{cn} = 120 \underline{/-210°}$ V rms.

The previous analysis indicates that we can simply treat a three-phase balanced circuit on a per phase basis and use the phase relationship to determine all voltages and currents. Let us now examine the situations in which either the source or the load is connected in Δ.

Delta-Connected Source

Consider the delta-connected source shown in Fig. 10.9a. Note that the sources are connected line to line. We found earlier that the relationship between line-to-line and line-to-neutral voltages was given by Eq. (10.14) and illustrated in Fig. 10.7 for an *abc*-phase sequence of voltages. Therefore, if the delta sources are

$$\mathbf{V}_{ab} = V_L \underline{/0°}$$

$$\mathbf{V}_{bc} = V_L \underline{/-120°}$$ **10.19**

$$\mathbf{V}_{ca} = V_L \underline{/+120°}$$

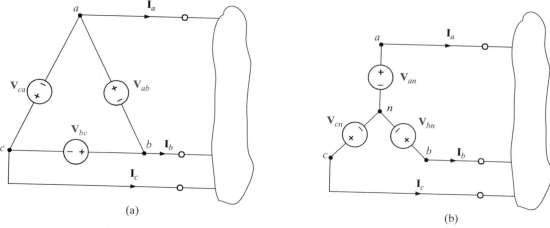

Figure 10.9 Sources connected in delta and wye.

where V_L is the magnitude of the phase voltage, the equivalent wye sources shown in Fig. 10.9b are

$$\mathbf{V}_{an} = \frac{V_L}{\sqrt{3}} \underline{/-30°} = V_p \underline{/-30°}$$

$$\mathbf{V}_{bn} = \frac{V_L}{\sqrt{3}} \underline{/-150°} = V_p \underline{/-150°}$$ **10.20**

$$\mathbf{V}_{cn} = \frac{V_L}{\sqrt{3}} \underline{/-270°} = V_p \underline{/+90°}$$

where V_p is the magnitude of the phase voltage of an equivalent wye-connected source. Therefore, if we encounter a network containing a delta-connected source, we can easily convert the source from delta to wye so that all the techniques we have discussed previously can be applied in an analysis.

Problem-Solving Strategies for Three-Phase Balanced AC Power Circuits

1. Convert the source/load connection to a wye–wye connection if either the source, load, or both are connected in delta since the wye–wye connection can be easily used to obtain the unknown phasors.

2. Only the unknown phasors for the a-phase of the circuit need be determined since the three-phase system is balanced.

3. Finally, convert the now-known phasors to the corresponding phasors in the original system.

EXAMPLE 10.3

Consider the network shown in Fig. 10.10a. We wish to determine the line currents and the magnitude of the line voltage at the load.

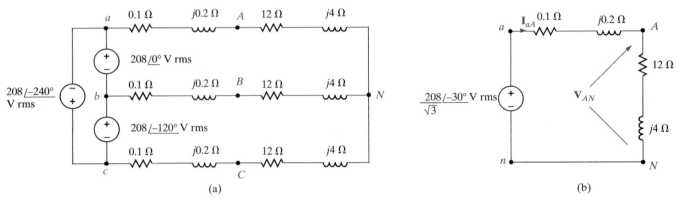

Figure 10.10 Delta–wye network and an equivalent single-phase (*a*-phase) diagram.

SOLUTION The single-phase diagram for the network is shown in Fig. 10.10b. The line current $\mathbf{I}_{aA}$ is

$$\mathbf{I}_{aA} = \frac{(208/\sqrt{3})\,\underline{/-30°}}{12.1 + j4.2}$$

$$= 9.38\,\underline{/-49.14°}\ \text{A rms}$$

and thus $\mathbf{I}_{bB} = 9.38\,\underline{/-169.14°}$ V rms and $\mathbf{I}_{cC} = 9.38\,\underline{/70.86°}$ V rms. The voltage $\mathbf{V}_{AN}$ is then

$$V_{AN} = (9.38\,\underline{/-49.14°})(12 + j4)$$

$$= 118.65\,\underline{/-30.71°}\ \text{V rms}$$

Therefore, the magnitude of the line voltage at the load is

$$V_L = \sqrt{3}\,(118.65)$$

$$= 205.51\ \text{V rms}$$

The phase voltage at the source is $V_p = 208/\sqrt{3} = 120$ V rms, while the phase voltage at the load is $V_p = 205.51/\sqrt{3} = 118.65$ V rms. Clearly, we must be careful with our notation and specify where the phase or line voltage is taken.   ❏

EXTENSION EXERCISE

E10.4 Consider the network shown in Fig. E10.4. Compute the magnitude of the line voltages at the load.

ANSWER: $V_L = 205.2$ V rms.

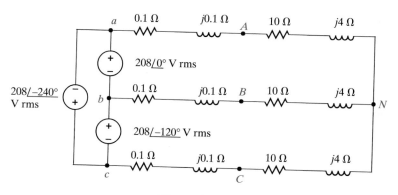

Figure E10.4

Delta-Connected Load

Consider now the Δ-connected load shown in Fig. 10.11. Note that in this connection the line-to-line voltage is the voltage across each load impedance.

If the phase voltages of the source are

$$\mathbf{V}_{an} = V_P \underline{/0°}$$
$$\mathbf{V}_{bn} = V_P \underline{/-120°}$$
$$\mathbf{V}_{cn} = V_P \underline{/+120°}$$

10.21

then the line voltages are

$$\mathbf{V}_{ab} = \sqrt{3}\,V_P \underline{/30°} = V_L \underline{/30°} = \mathbf{V}_{AB}$$
$$\mathbf{V}_{bc} = \sqrt{3}\,V_P \underline{/-90°} = V_L \underline{/-90°} = \mathbf{V}_{BC}$$
$$\mathbf{V}_{ca} = \sqrt{3}\,V_P \underline{/-210°} = V_L \underline{/-210°} = \mathbf{V}_{CA}$$

10.22

where V_L is the magnitude of the line voltage at both the delta-connected load and at the source since there is no line impedance present in the network.

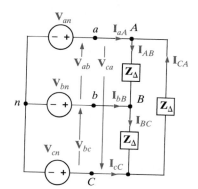

Figure 10.11 Balanced three-phase wye–delta system.

From Fig. 10.11 we note that if $\mathbf{Z}_\Delta = Z_\Delta \underline{/\theta}$, the phase currents at the load are

$$\mathbf{I}_{AB} = \frac{\mathbf{V}_{AB}}{\mathbf{Z}_\Delta} \qquad \text{10.23}$$

where $\mathbf{I}_{BC}$ and $\mathbf{I}_{CA}$ have the same magnitude but lag $\mathbf{I}_{AB}$ by 120° and 240°, respectively. KCL can now be employed in conjunction with the phase currents to determine the line currents. For example,

$$\mathbf{I}_{aA} = \mathbf{I}_{AB} + \mathbf{I}_{AC}$$
$$= \mathbf{I}_{AB} - \mathbf{I}_{CA}$$

However, it is perhaps easier to simply convert the balanced Δ-connected load to a balanced Y-connected load using the Δ-Y transformation.

Consider the networks in Fig. 10.12. The transformation that relates the impedances $\mathbf{Z}_a$, $\mathbf{Z}_b$, and $\mathbf{Z}_c$ to the impedances $\mathbf{Z}_1$, $\mathbf{Z}_2$, and $\mathbf{Z}_3$ is derived in the following manner. For the networks to be equivalent at each corresponding pair of terminals, it is necessary that the impedance at the corresponding terminals be equal (e.g., the impedances at terminals a and b with c open circuited must be the same for both networks). Therefore, if we equate the impedances for each corresponding set of terminals, we obtain the following equations:

$$\mathbf{Z}_{ab} = \mathbf{Z}_a + \mathbf{Z}_b = \frac{\mathbf{Z}_1(\mathbf{Z}_2 + \mathbf{Z}_3)}{\mathbf{Z}_1 + \mathbf{Z}_2 + \mathbf{Z}_3}$$

$$\mathbf{Z}_{bc} = \mathbf{Z}_b + \mathbf{Z}_c = \frac{\mathbf{Z}_3(\mathbf{Z}_1 + \mathbf{Z}_2)}{\mathbf{Z}_3 + \mathbf{Z}_1 + \mathbf{Z}_2}$$

$$\mathbf{Z}_{ca} = \mathbf{Z}_c + \mathbf{Z}_a = \frac{\mathbf{Z}_2(\mathbf{Z}_1 + \mathbf{Z}_3)}{\mathbf{Z}_2 + \mathbf{Z}_1 + \mathbf{Z}_3}$$

Solving this set of equations for $\mathbf{Z}_a$, $\mathbf{Z}_b$, and $\mathbf{Z}_c$ yields

$$\mathbf{Z}_a = \frac{\mathbf{Z}_1\mathbf{Z}_2}{\mathbf{Z}_1 + \mathbf{Z}_2 + \mathbf{Z}_3}$$

$$\mathbf{Z}_b = \frac{\mathbf{Z}_1\mathbf{Z}_3}{\mathbf{Z}_1 + \mathbf{Z}_2 + \mathbf{Z}_3} \qquad \text{10.24}$$

$$\mathbf{Z}_c = \frac{\mathbf{Z}_2\mathbf{Z}_3}{\mathbf{Z}_1 + \mathbf{Z}_2 + \mathbf{Z}_3}$$

DIRLL

D10.5 A three-phase balanced delta-connected load has the following parameters:
$\mathbf{V}_{AB} = 208 \underline{/30°}$ V rms and $\mathbf{I}_{AB} = 5 \underline{/10°}$ A rms. Determine the load impedance.

ANSWER:
$\mathbf{Z}_\Delta = 41.6 \underline{/20°}$ Ω.

DRILL

D10.6 A three-phase balanced delta-connected load has a phase current of $\mathbf{I}_{AB} = 10 \underline{/15°}$ A rms. Find the line current $\mathbf{I}_{aA}$.

ANSWER:
$\mathbf{I}_{aA} = 17.32 \underline{/-15°}$ A rms.

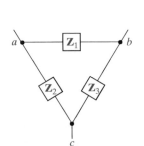

Figure 10.12 Wye and delta network configurations.

Similarly, if we solve this set of equations for $\mathbf{Z}_1$, $\mathbf{Z}_2$, and $\mathbf{Z}_3$, we obtain

$$\mathbf{Z}_1 = \frac{\mathbf{Z}_a\mathbf{Z}_b + \mathbf{Z}_b\mathbf{Z}_c + \mathbf{Z}_c\mathbf{Z}_a}{\mathbf{Z}_c}$$

$$\mathbf{Z}_2 = \frac{\mathbf{Z}_a\mathbf{Z}_b + \mathbf{Z}_b\mathbf{Z}_c + \mathbf{Z}_c\mathbf{Z}_a}{\mathbf{Z}_b}$$ **10.25**

$$\mathbf{Z}_3 = \frac{\mathbf{Z}_a\mathbf{Z}_b + \mathbf{Z}_b\mathbf{Z}_c + \mathbf{Z}_c\mathbf{Z}_a}{\mathbf{Z}_a}$$

These equations are general relationships and therefore apply to any set of impedances connected in a wye or delta configuration.

For a balanced load,

$$\mathbf{Z}_Y = \frac{1}{3}\mathbf{Z}_\Delta$$

and then the line current $\mathbf{I}_{aA}$ is simply

$$\mathbf{I}_{aA} = \frac{\mathbf{V}_{an}}{\mathbf{Z}_Y}$$

Finally, using the same approach as that employed earlier to determine the relationship between the line voltages and phase voltages in a Y–Y connection, we can show that the relationship between the *magnitudes* of the phase currents in the Δ-connected load and the line currents is

$$I_L = \sqrt{3}\,I_\Delta$$ **10.26**

EXAMPLE 10.4

A balanced delta-connected load contains a 10-Ω resistor in series with a 20-mH inductor in each phase. The voltage source is an *abc*-sequence three-phase 60-Hz, balanced wye with a voltage $\mathbf{V}_{an} = 120\ \underline{/30°}$ V rms. We wish to determine all Δ currents and line currents.

SOLUTION The impedance per phase in the delta load is $\mathbf{Z}_\Delta = 10 + j7.54\ \Omega$. The line voltage $\mathbf{V}_{ab} = 120\ \sqrt{3}\ \underline{/60°}$ V rms. Since there is no line impedance, $\mathbf{V}_{AB} = \mathbf{V}_{ab} = 120\ \sqrt{3}\ \underline{/60°}$ V rms. Hence,

$$\mathbf{I}_{AB} = \frac{120\ \sqrt{3}\ \underline{/60°}}{10 + j7.54}$$

$$= 16.60\ \underline{/+22.98°}\ \text{A rms}$$

If $\mathbf{Z}_\Delta = 10 + j7.54\ \Omega$, then

$$\mathbf{Z}_Y = \frac{1}{3}\mathbf{Z}_\Delta$$

$$= 3.33 + j2.51\ \Omega$$

Then the line current

$$\mathbf{I}_{aA} = \frac{\mathbf{V}_{an}}{\mathbf{Z}_Y} = \frac{120 \,\underline{/30°}}{3.33 + j2.51}$$

$$= \frac{120 \,\underline{/30°}}{4.17 \,\underline{/37.01°}}$$

$$= 28.78 \,\underline{/-7.01°} \text{ A rms}$$

Therefore, the remaining phase and line currents are

$$\mathbf{I}_{BC} = 16.60 \,\underline{/-97.02°} \text{ A rms} \qquad \mathbf{I}_{bB} = 28.78 \,\underline{/-127.01°} \text{ A rms}$$
$$\mathbf{I}_{CA} = 16.60 \,\underline{/+142.98°} \text{ A rms} \qquad \mathbf{I}_{cC} = 28.78 \,\underline{/-112.99°} \text{ A rms} \qquad ❑$$

In summary, the relationship between the line voltage and phase voltage and the line current and phase current for both the Y and Δ configurations are shown in Fig. 10.13. The currents and voltages are shown for one phase. The two remaining phases have the same magnitude but lag by 120° and 240°, respectively.

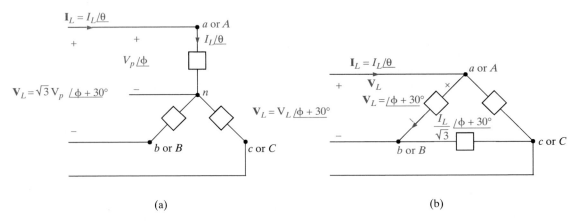

Figure 10.13 Voltage and current relationships for Y and Δ configurations.

Careful observation of Table 10.1 indicates that the following rules apply when solving problems in balanced three-phase systems:

Table 10.1 The voltage, current, and impedance relationships for Y and Δ configurations.

	Y	**Δ**
Line Voltage $(\mathbf{V}_{ab}$ or $\mathbf{V}_{AB})$	$\sqrt{3}\,V_p \,\underline{/\phi + 30°}$ $= V_L \,\underline{/\phi + 30°}$	$V_L \,\underline{/\phi + 30°}$
Line current $\mathbf{I}_{aA}$	$I_L \,\underline{/\theta}$	$I_L \,\underline{/\theta}$
Phase voltage	$V_p \,\underline{/\phi}(\mathbf{V}_{an}$ or $\mathbf{V}_{AN})$	$\sqrt{3}\,V_p \,\underline{/\phi + 30°}$
Phase current	$I_L \,\underline{/\theta}$	$\dfrac{I_L}{\sqrt{3}} \,\underline{/\theta + 30°}$
Load impedance	$\mathbf{Z}_Y \,\underline{/\phi - \theta}$	$3\mathbf{Z}_Y \,\underline{/\phi - \theta}$

- The phase of the voltages and currents in a Δ connection is 30° ahead of those in a Y connection.

- The magnitude of the line voltage or, equivalently, the Δ-connection phase voltage, is $\sqrt{3}$ times that of the Y-connection phase voltage.

- The magnitude of the line current or, equivalently, the Y-connection phase current, is $\sqrt{3}$ times that of the Δ-connection phase current.

- The load impedance in the Y connection is one-third of that in the Δ-connection and the phase is identical.

EXTENSION EXERCISES

E10.5   An *abc*-sequence three-phase voltage source connected in a balanced wye supplies power to a balanced delta-connected load. The line current for the *a* phase is $\mathbf{I}_{aA} = 12 \; \underline{/40°}$ A rms. Find the phase currents in the delta-connected load.

ANSWER: $\mathbf{I}_{AB} = 6.93 \; \underline{/70°}$ A rms, $\mathbf{I}_{BC} = 6.93 \; \underline{/-50°}$ A rms, $\mathbf{I}_{CA} = 6.93 \; \underline{/-170°}$ A rms.

E10.6   An *abc*-sequence balanced three-phase wye-connected source supplies power to a balanced delta-connected load. The load impedance per phase is $12 + j8 \; \Omega$. If the current $\mathbf{I}_{AB}$ in one phase of the delta is $14.42 \; \underline{/86.31°}$ A rms, determine the line currents and phase voltages at the source.

ANSWER: $\mathbf{I}_a = 24.98 \; \underline{/56.31°}$ A rms, $\mathbf{I}_b = \mathbf{I}_a \; \underline{/-120°}, \mathbf{I}_c = \mathbf{I}_a \; \underline{/-240°}$, $\mathbf{V}_{an} = 120 \; \underline{/90°}$ V rms, $\mathbf{V}_{bn} = \mathbf{V}_{an} \; \underline{/-120°}, \mathbf{V}_{cn} = \mathbf{V}_{an} \; \underline{/-240°}$.

Consider now two examples in which the loads consist of a parallel combination of a delta-connected load and a wye-connected load.

EXAMPLE 10.5

Consider the network shown in Fig. 10.14a. We wish to determine all the load currents.

SOLUTION   The circuit diagram for the *a* phase is shown in Fig. 10.14b. Since the parallel combination of 30 Ω and 20 Ω is 12 Ω,

$$\mathbf{I}_{aA} = \frac{120 \; \underline{/0°}}{16}$$
$$= 7.5 \; \underline{/0°} \text{ A rms}$$

and hence

$$\mathbf{V}_{AN} = (7.5 \; \underline{/0°}) \, (12)$$
$$= 90 \; \underline{/0°} \text{ V rms}$$

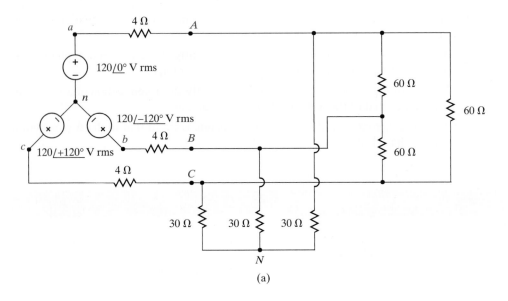

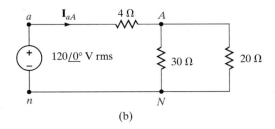

Figure 10.14 Networks used in Example 10.5: (a) original network; (b) *a*-phase equivalent network.

In the original wye-connected load,

$$\mathbf{I}_{AN} = \frac{90 \,\underline{/0°}}{30}$$

$$= 3 \,\underline{/0°} \text{ A rms}$$

and therefore $\mathbf{I}_{BN} = 3 \,\underline{/-120°}$ A rms and $\mathbf{I}_{CN} = 3 \,\underline{/+120°}$ A rms. For the original delta-connected load

$$\mathbf{V}_{AB} = 90 \,\sqrt{3} \,\underline{/0° + 30°}$$

$$= 155.88 \,\underline{/30°} \text{ V rms}$$

and hence,

$$\mathbf{I}_{AB} = \frac{155.88 \,\underline{/30°}}{60}$$

$$= 2.60 \,\underline{/30°} \text{ A rms}$$

Therefore, $\mathbf{I}_{BC} = 2.60 \,\underline{/-90°}$ A rms and $\mathbf{I}_{CA} = 2.60 \,\underline{/+150°}$ A rms.

As shown in Fig. 10.14a, the line current must be equal to the sum of the load currents. Hence,

$$\mathbf{I}_{aA} = \mathbf{I}_{AN} + \mathbf{I}_{AB} + \mathbf{I}_{AC}$$
$$= \mathbf{I}_{AN} + \mathbf{I}_{AB} - \mathbf{I}_{CA}$$
$$= 3 \underline{/0°} + 2.6 \underline{/30°} - 2.6 \underline{/150°}$$
$$= 7.5 \underline{/0°} \text{ A rms}$$

which is a simple check on our calculations.   ❏

EXAMPLE 10.6

A balanced three-phase system has a load consisting of a balanced wye in parallel with a balanced delta. The impedance per phase for the wye is $10 + j6 \; \Omega$ and for the delta $24 + j9 \; \Omega$. The source is a balanced delta and $\mathbf{V}_{ab} = 208 \underline{/30°}$ V rms. If the line impedance per phase is $1 + j0.5 \; \Omega$, we want to determine the line currents, the load phase voltages when the load is converted to an equivalent wye, and the currents in the delta-connected load.

SOLUTION   Converting the delta load to an equivalent wye load, we obtain

$$\mathbf{Z}_{Y1} = \tfrac{1}{3}\mathbf{Z}_{\Delta 1} = 8 + j3\Omega$$

The per phase equivalent circuit is shown in Fig. 10.15. Since the voltage at the source is $\mathbf{V}_{ab} = 208 \underline{/30°}$ V rms, the phase voltage at the source is $\mathbf{V}_{an} = 120 \underline{/0°}$ V rms.

Recall that since the loads are balanced, the neutral points can be connected. The equivalent wye load impedance is

$$\mathbf{Z}_Y = \frac{\mathbf{Z}_{Y1}\mathbf{Z}_{Y2}}{\mathbf{Z}_{Y1} + \mathbf{Z}_{Y2}}$$
$$= \frac{(10 + j6)(8 + j3)}{10 + j6 + 8 + j3} = 4.95 \underline{/24.95°}$$
$$= 4.49 + j2.09 \; \Omega$$

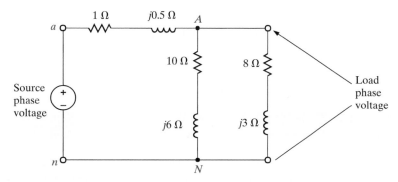

Figure 10.15 Per phase equivalent circuit for Example 10.6.

The line current $\mathbf{I}_a$ is then

$$\mathbf{I}_a = \frac{\mathbf{V}_{an}}{\mathbf{Z}_{line} + \mathbf{Z}_Y}$$

$$= \frac{120 \;/0°}{1 + j0.5 + 4.49 + j2.09}$$

$$= 19.77 \;/-25.26° \text{ A rms}$$

The phase voltage at the load is then $V_p = |\mathbf{V}_{AN}|$, where

$$\mathbf{V}_{AN} = \mathbf{I}_a \mathbf{Z}_Y$$

$$= (19.77 \;/-25.26°)(4.95 \;/24.95°)$$

$$= 97.86 \;/-0.31° \text{ V rms}$$

Therefore, the line currents and load phase voltages are

$$\mathbf{I}_{aA} = 19.77 \;/-25.26° \text{ A rms} \qquad \mathbf{V}_{AN} = 97.86 \;/-0.31° \text{ V rms}$$

$$\mathbf{I}_{bB} = 19.77 \;/-145.26° \text{ A rms} \qquad \mathbf{V}_{BN} = 97.86 \;/-120.31° \text{ V rms}$$

$$\mathbf{I}_{cC} = 19.77 \;/+94.74° \text{ A rms} \qquad \mathbf{V}_{CN} = 97.86 \;/+119.69° \text{ V rms}$$

In order to determine the phase currents in the delta-connected load, we revisit the original circuit. Although the wye-connected load, with which we replaced the delta-connected load, is equivalent *at the external terminals*, we must use the original circuit to determine *internal* voltages or currents.

Since

$$\mathbf{V}_{AN} = 97.86 \;/-0.31° \text{ V rms}$$

then

$$\mathbf{V}_{AB} = 97.86 \sqrt{3} \;/-0.31° + 30°$$

$$= 169.5 \;/29.69° \text{ V rms}$$

Hence,

$$\mathbf{I}_{AB} = \frac{\mathbf{V}_{AB}}{\mathbf{Z}_\Delta} = \frac{169.5 \;/29.69°}{24 + j9}$$

$$= 6.61 \;/-9.13° \text{ A rms}$$

The remaining phase currents in the delta are

$$\mathbf{I}_{BC} = 6.61 \;/-110.87° \text{ A rms}$$

$$\mathbf{I}_{CA} = 6.61 \;/-129.13° \text{ A rms}$$ ❏

EXTENSION EXERCISE

E10.7   In a balanced three-phase system the load consists of a balanced wye in parallel with a balanced delta. The impedance per phase for the wye is $8 + j4 \; \Omega$ and for the delta is $18 + j6 \; \Omega$. The source is an *abc*-phase-sequence balanced wye and

$\mathbf{V}_{an} = 120 \,\underline{/60°}$ V rms. If the line impedance per phase is $1 + j1\ \Omega$, determine the magnitude of the phase currents in each load.

ANSWER: $I_\Delta = 8.07$ A rms, $I_Y = 9.88$ A rms.

10.4 POWER RELATIONSHIPS

Whether the load is connected in a wye or a delta, the real and reactive power per phase is

$$P_p = V_p I_p \cos\theta$$
$$Q_p = V_p I_p \sin\theta \qquad \textbf{10.27}$$

where θ is the angle between the phase voltage and the line current, or

$$P_p = \frac{V_L I_L}{\sqrt{3}} \cos\theta$$

$$Q_p = \frac{V_L I_L}{\sqrt{3}} \sin\theta \qquad \textbf{10.28}$$

For wye-connected system
$$P_p = (V_L/\sqrt{3})I_L \cos\theta$$

For delta-connected system
$$P_p = V_L(I_L/\sqrt{3})\cos\theta$$

The total real and reactive power for all three phases is then

$$P_T = \sqrt{3}\, V_L I_L \cos\theta$$
$$Q_T = \sqrt{3}\, V_L I_L \sin\theta \qquad \textbf{10.29}$$

and, therefore, the magnitude of the complex power (apparent power) is

$$S_T = \sqrt{P_T^2 + Q_T^2}$$
$$= \sqrt{3}\, V_L I_L$$

and

$$\underline{/\mathbf{S}_T} = \theta$$

EXAMPLE 10.7

A three-phase balanced wye–delta system has a line voltage of 208 V rms. The total real power absorbed by the load is 1200 W. If the power factor angle of the load is 20° lagging, we wish to determine the magnitude of the line current and the value of the load impedance per phase in the delta.

SOLUTION   The line current can be obtained from Eq. (10.28). Since the real power per phase is 400 W,

$$400 = \frac{208 I_L}{\sqrt{3}} \cos 20°$$

$$I_L = 3.54 \text{ A rms}$$

DRILL

D10.9 A three-phase balanced delta-connected load has a phase voltage of 208 V rms and a phase current of 10 A rms. Determine the complex power if the pf is 0.9.

ANSWER:
2080 $\underline{/25.8°}$ VA.

The magnitude of the current in each leg of the delta-connected load is

$$I_\Delta = \frac{I_L}{\sqrt{3}}$$

$$= 2.05 \text{ A rms}$$

Therefore, the magnitude of the delta impedance in each phase of the load is

$$|\mathbf{Z}_\Delta| = \frac{V_L}{I_\Delta}$$

$$= \frac{208}{2.05}$$

$$= 101.46 \ \Omega$$

Since the power factor angle is 20° lagging, the load impedance is

$$\mathbf{Z}_\Delta = 101.46 \ \underline{/20°}$$

$$= 95.34 + j34.70 \ \Omega \qquad \qquad \square$$

EXAMPLE 10.8

For the circuit in Example 10.2 we wish to determine the real and reactive power per phase at the load and the total real power, reactive power, and complex power at the source.

SOLUTION   From the data in Example 10.2 the complex power per phase at the load is

$$\mathbf{S}_{\text{load}} = \mathbf{VI}^*$$

$$= \left(113.15 \ \underline{/-108°}\right)\left(5.06 \ \underline{/27.65°}\right)$$

$$= 572.54 \ \underline{/26.57°}$$

$$= 512.07 + j256.09 \text{ VA}$$

Therefore, the real and reactive power per phase at the load are 512.07 W and 256.09 var, respectively.

The complex power per phase at the source is

$$\mathbf{S}_{\text{source}} = \mathbf{VI}^*$$

$$= \left(120 \ \underline{/-0°}\right)\left(5.06 \ \underline{/27.65°}\right)$$

$$= 607.2 \ \underline{/27.65°}$$

$$= 537.86 + j281.78 \text{ VA}$$

Therefore, total real power, reactive power, and apparent power at the source are 1613.6 W, 845.2 var, and 1821.6 VA, respectively. $\qquad \square$

EXAMPLE 10.9

A balanced three-phase source serves three loads as follows:

Load 1: 24 kW at 0.6 lagging power factor
Load 2: 10 kW at unity power factor
Load 3: 12 kVA at 0.8 leading power factor

If the line voltage at the loads is 208 V rms at 60 Hz, we wish to determine the line current and the combined power factor of the loads.

SOLUTION   From the data we find that

$$\mathbf{S}_1 = 24,000 + j32,000$$
$$\mathbf{S}_2 = 10,000 + j0$$
$$\mathbf{S}_3 = 12,000 \,\underline{/-36.9°} = 9600 - j7200$$

Therefore,

$$\mathbf{S}_{\text{load}} = 43,600 + j24,800$$
$$= 50,160 \,\underline{/29.63°} \text{ VA}$$
$$I_L = \frac{|\mathbf{S}_{\text{load}}|}{\sqrt{3}\, V_L}$$
$$= \frac{50,160}{208 \sqrt{3}}$$
$$I_L = 139.23 \text{ A rms}$$

and the combined power factor is

$$\text{pf}_{\text{load}} = \cos 29.63°$$
$$= 0.869 \text{ lagging} \qquad \square$$

The sum of three complex powers
$$\mathbf{S}_{\text{load}} = \mathbf{S}_1 + \mathbf{S}_2 + \mathbf{S}_3$$

EXAMPLE 10.10

Given the three-phase system in Example 10.9, let us determine the line voltage and power factor at the source if the line impedance is $\mathbf{Z}_{\text{line}} = 0.05 + j0.02 \ \Omega$.

SOLUTION   The complex power absorbed by the line impedances is

$$\mathbf{S}_{\text{line}} = 3(R_{\text{line}} I_L^2 + jX_{\text{line}} I_L^2)$$
$$= 2908 + j1163 \text{ VA}$$

The complex power delivered by the source is then

$$\mathbf{S}_S = \mathbf{S}_{\text{load}} + \mathbf{S}_{\text{line}}$$
$$= 43,600 + j24,800 + 2908 + j1163$$
$$= 53,264 \,\underline{/29.17°} \text{ VA}$$

The line voltage at the source is then

$$V_{L_S} = \frac{S_S}{\sqrt{3}\, I_L}$$
$$= 220.87 \text{ V rms}$$

Recall that the complex power for all three line, is
$$\mathbf{S}_{\text{line}} = 3I_L^2 \mathbf{Z}_{\text{line}}$$

and the power factor at the source is

$$\text{pf}_S = \cos 29.17°$$
$$= 0.873 \text{ lagging} \qquad \square$$

EXTENSION EXERCISES

E10.8   A three-phase balanced wye–wye system has a line voltage of 208 V rms. The total real power absorbed by the load is 12 kW at 0.8 pf lagging. Determine the per-phase impedance of the load.

ANSWER: $\mathbf{Z} = 2.88 \,\underline{/36.87°} \,\Omega$.

E10.9   For the balanced wye-wye system described in Extension Exercise E10.3, determine the real and reactive power and the complex power at both the source and the load.

ANSWER: $\mathbf{S}_{\text{load}} = 1186.77 + j444.66 \text{ VA}; \mathbf{S}_{\text{source}} = 1335.65 + j593.55 \text{ VA}$.

E10.10   A 480-V rms line feeds two balanced three-phase loads. If the two loads are rated as follows,

Load 1: 5 kVA at 0.8 pf lagging

Load 2: 10 kVA at 0.9 pf lagging

determine the magnitude of the line current from the 480-V rms source.

ANSWER: $I_L = 17.97 \text{ A rms}$.

10.5 THREE-PHASE MEASUREMENT

Power Measurement

The measurement of three-phase power would seem to be a simple task, since it would appear that in order to accomplish the measurement we would simply replicate the single-phase measurement for each phase. For a balanced system this task would become even simpler, since we need only measure the power in one phase and multiply the wattmeter reading by 3. However, the neutral terminal in the wye-connected load may be inaccessible, and of course, in a delta-connected load there is no neutral point. Our measurement technique must, therefore, deal only with external lines to the load.

The method we will describe is applicable for both wye and delta connections; however, we will present the measurement method using a balanced wye-connected load. Consider the circuit shown in Fig. 10.16. The current coil of each wattmeter is connected in series with the line current, and the voltage coil of each wattmeter is connected between a line and what we will call the virtual neutral N^*, which is nothing more than an arbitrary point.

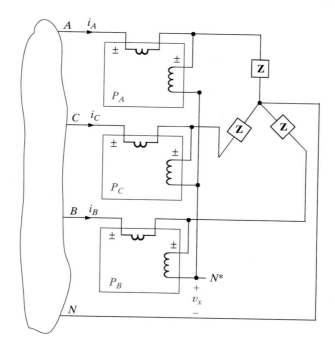

Figure 10.16 Wattmeter connections for power measurement.

The average power measured by wattmeter A is

$$P_A = \frac{1}{T} \int_0^T v_{AN^*} i_A \, dt \qquad \textbf{10.31}$$

where T is the period of all voltages and currents in the system. In a similar manner,

$$P_B = \frac{1}{T} \int_0^T v_{BN^*} i_B \, dt \qquad \textbf{10.32}$$

$$P_C = \frac{1}{T} \int_0^T v_{CN^*} i_C \, dt \qquad \textbf{10.33}$$

The sum of all the wattmeter measurements is

$$P = \frac{1}{T} \int_0^T \left(v_{AN^*} i_A + v_{BN^*} i_B + v_{CN^*} i_C \right) dt \qquad \textbf{10.34}$$

As shown in Fig. 10.16, the voltages v_{AN^*}, v_{BN^*}, and v_{CN^*} can be expressed as

$$\begin{aligned}
v_{AN^*} &= v_{AN} - v_x \\
v_{BN^*} &= v_{BN} - v_x \\
v_{CN^*} &= v_{CN} - v_x
\end{aligned} \qquad \textbf{10.35}$$

Substituting Eq. (10.35) into Eq. (10.34) yields

$$P = \frac{1}{T} \int_0^T \left(v_{AN} i_A + v_{BN} i_B + v_{CN} i_C \right) dt - \frac{1}{T} \int_0^T v_x (i_A + i_B + i_C) \, dt$$

However,

$$i_A + i_B + i_C = 0$$

and hence the expression for the power reduces to

$$P = \frac{1}{T}\int_0^T v_{AN} i_A \, dt + \frac{1}{T}\int_0^T v_{BN} i_B \, dt + \frac{1}{T}\int_0^T v_{CN} i_C \, dt \qquad \textbf{10.36}$$

which we recognize to be the total power absorbed by the three-phase wye load. There-fore, using the three wattmeters shown in Fig. 10.16, we can measure the power absorbed by a three-phase load, and in addition, this method applies whether the system is bal-anced or unbalanced, and whether the load is wye or delta connected.

If we connect the virtual neutral to line C, the voltage coil on wattmeter C will have zero volts across it; the wattmeter will read zero watts and can thus be removed. Hence, the two wattmeters A and B will then measure all the power in the three-phase load. This configuration is shown in Fig. 10.17 and is known as the *two-wattmeter method* for power measurement. The two-wattmeter method can always be used when the load is balanced, but in the unbalanced case, is valid only for a three-wire load (that is, a load with no neu-tral conductor). In general, if there are n wires from the source to the load, $n - 1$ wattmeters are required. As shown in Fig. 10.17, the total power measured by the two-wattmeter method is

$$P_T = P_A + P_B = V_{AC} I_{aA} \cos\left(\underline{/\mathbf{V}_{AC}} - \underline{/\mathbf{I}_{aA}}\right)$$
$$+ V_{BC} I_{bB} \cos\left(\underline{/\mathbf{V}_{BC}} - \underline{/\mathbf{I}_{bB}}\right) \qquad \textbf{10.37}$$

where, for example, $\underline{/\mathbf{V}_{AC}}$ represents the phase angle of the voltage $\mathbf{V}_{AC}$.

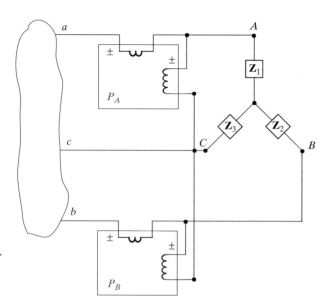

Figure 10.17 Two-wattmeter method for power measurement.

EXAMPLE 10.11

A balanced wye–delta system has an *abc*-phase-sequence source with $\mathbf{V}_{an} = 120\ \underline{/0°}$ V rms. The balanced load has a phase impedance of $10 + j5\ \Omega$. We wish to find the power absorbed by the load using the two-wattmeter method.

SOLUTION   If $\mathbf{V}_{an} = 120\ \underline{/0°}$ V rms, then

$$\mathbf{V}_{AB} = 208\ \underline{/30°}\ \text{V rms}$$
$$\mathbf{V}_{BC} = 208\ \underline{/-90°}\ \text{V rms}$$
$$\mathbf{V}_{CA} = 208\ \underline{/-210°}\ \text{V rms}$$

and hence $\mathbf{V}_{AC} = 208\ \underline{/-30°}$ V rms. Since the phase impedance is $10 + j5 = 11.18\ \underline{/26.57°}\ \Omega$, the magnitude of the delta current is

$$I_\Delta = \frac{208}{11.18} = 18.60\ \text{A rms}$$

Power for each phase
$P_p = I_p^2 R_p$

The average power per phase is, therefore,

$$P_p = (18.60)^2(10)$$
$$= 3461\ \text{W}$$

and hence the total power is

$$P_T = 10{,}383\ \text{W}$$

Using the two-wattmeter method, we note that

$$\mathbf{I}_{aA} = 18.60\ \sqrt{3}\ \underline{/-26.57°}\ \text{A rms}$$
$$\mathbf{I}_{bB} = 18.60\ \sqrt{3}\ \underline{/-120 - 26.57°}\ \text{A rms}$$

where $\underline{/\mathbf{I}_{aA}} = \underline{/\mathbf{V}_{an}} - \underline{/Z_Y}$. Therefore, using Eq. (10.37), we have

$$P_T = (208)(32.22) \cos(-30° + 26.57°)$$
$$\qquad + (208)(32.22) \cos(-90° + 120° + 26.57°)$$
$$= 10{,}383\ \text{W} \qquad\qquad\qquad □$$

EXTENSION EXERCISE

E10.11   Given the data in Extension Exercise E10.6, if the two-wattmeter method is used to measure power, find the reading of each wattmeter.

ANSWER: $P_A = 5185$ W, $P_B = 2303$ W.

Power Factor Measurement

The two-wattmeter method can also be used to compute the power factor angle if the load is balanced. To illustrate how this can be accomplished, consider the wye-connected load shown in Fig. 10.17, where $\mathbf{Z} = Z_Y \, \underline{/\theta}$. If the source has an $\underline{/\mathbf{V}_{an}} = 0°$, then as shown in Eq. (10.37),

$$P_A = V_{AC} I_{aA} \cos(\underline{/\mathbf{V}_{AC}} - \underline{/\mathbf{I}_{aA}})$$
$$V_{AC} = V_L \text{ and } \underline{/\mathbf{V}_{AC}} - \underline{/\mathbf{I}_{CA}} + 180° = -30°$$

Similarly,

$$I_{aA} = I_L \text{ and } \underline{/\mathbf{I}_{aA}} = -\theta$$

Therefore,

$$P_A = V_L I_L \cos(-30° + \theta)$$

In a similar manner we can show that

$$P_B = V_L I_L \cos(30° + \theta)$$

Then the ratio of the two wattmeter readings is

$$\frac{P_A}{P_B} = \frac{\cos(\theta - 30°)}{\cos(\theta + 30°)}$$

If we now employ the trigonometric identities in Eq. (8.11) and recall that $\cos 30° = \sqrt{3}/2$ and $\sin 30° = 1/2$, it is straightforward to show that the preceding equation can be reduced to

$$\tan \theta = \frac{(P_A - P_B)\sqrt{3}}{P_A + P_B}$$

and since $P_T = P_A + P_B$,

$$\theta = \tan^{-1} \frac{(P_A - P_B)\sqrt{3}}{P_T} \tag{10.38}$$

The preceding equations indicate that if $P_A = P_B$, the load is resistive; if $P_A > P_B$, the load is inductive; and if $P_A < P_B$, the load is capacitive. Finally, this technique is valid whether the load is wye or delta connected.

EXAMPLE 10.12

In a balanced wye–delta system, two wattmeters are connected to measure the total power. We wish to determine the power factor of the load if the wattmeter readings are $P_A = 1200$ W and $P_B = 480$ W.

SOLUTION   Using Eq. (10.38), we have

$$\theta = \tan^{-1} \frac{(1200 - 480)\sqrt{3}}{1680}$$

$$= 36.59°$$

Therefore,

$$\cos \theta = \text{pf} = 0.80 \text{ lagging} \qquad \square$$

EXAMPLE 10.13

If the source in Example 10.12 has a line voltage of 208 V rms, we wish to find the delta load impedance.

SOLUTION   The total power $P_T = P_A + P_B = 1680$ W; therefore, the power per phase is $P_p = 1680/3 = 560$ W. Hence, using the equation

$$V_L I_\Delta \cos \theta = P_p$$

$$I_\Delta = \frac{560}{(208)(0.8)}$$

$$= 3.37 \text{ A rms}$$

Power for each phase

$$P_p = V_L \left(\frac{I_L}{\sqrt{3}} \right) \cos \theta$$

$$= V_L I_\Delta \cos \theta$$

Therefore,

$$\mathbf{Z} = \frac{208}{3.37} \underline{/36.59°} = 61.72 \underline{/36.59°}$$

$$= 49.56 + j36.79 \; \Omega \qquad \square$$

EXAMPLE 10.14

In a balanced wye–delta system, two wattmeters are used to measure total power. Wattmeter A reads 800 W and wattmeter B reads 400 W after the current coil terminals are reversed. If the line voltage is 208 V rms, we wish to determine the total average power, the power factor, and the impedance of the load.

SOLUTION   From the data given $P_A = 800$ and $P_B = -400$; therefore, the total power P_T is

$$P_T = P_A + P_B$$

$$= 400 \text{ W}$$

The power factor is computed from

$$\theta = \tan^{-1} \frac{[800 - (-400)]\sqrt{3}}{400}$$

$$\theta = 79.11°$$

and therefore,

$$\cos \theta = \text{pf} = 0.19 \text{ lagging}$$

The delta impedance is determined from

$$V_L I_\Delta \cos \theta = P_P$$

$$(208) I_\Delta (0.19) = \frac{400}{3}$$

$$I_\Delta = 3.37 \text{ A rms}$$

Therefore,

$$\mathbf{Z}_\Delta = \frac{208}{3.37} \underline{/79.11°} = 61.72 \underline{/79.11°} \ \Omega \qquad \qquad ❑$$

EXTENSION EXERCISES

E10.12   Two wattmeters are used to measure the total power in the load of a balanced wye–wye system. The line voltage is 208 V rms and the wattmeter readings are $P_A = 1600$ W and $P_B = 840$ W. Compute the impedance per phase of the load.

ANSWER: Z = 15.59 $\underline{/28.35°}$ Ω.

E10.13   Two wattmeters are used to measure the total power in the load of a balanced wye–wye system. The line voltage is 208 V rms. Wattmeter A reads 1280 W and wattmeter B reads 540 W when the current coil terminals are reversed. Determine the power factor of the load and the impedance per phase of the load.

ANSWER: pf = 0.23 lagging, **Z** = 13.34 $\underline{/76.79°}$ Ω.

10.6 POWER FACTOR CORRECTION

Major precautions for three-phase power factor correction:

Must distinguish P_T and P_P.
Must use appropriate V rms for Y- and Δ-connections.

In Section 9.7 we illustrated a simple technique for raising the power factor of a load. The method involved judiciously selecting a capacitor and placing it in parallel with the load. In a balanced three-phase system, power factor correction is performed in exactly the same manner. It is important to note, however, that the $\mathbf{S}_{cap}$ specified in Eq. (9.42) is provided by three capacitors, and in addition, V_{rms} in the equation is the voltage across each capacitor. The following example illustrates the technique.

EXAMPLE 10.15

In the balanced three-phase system shown in Fig. 10.18, the line voltage is 34.5 kV rms at 60 Hz. We wish to find the values of the capacitors C such that the total load has a power factor of 0.94 leading.

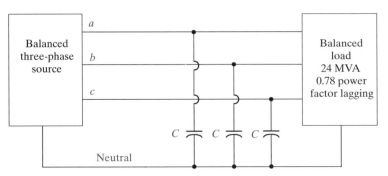

Figure 10.18 Network used in Example 10.15.

SOLUTION   Following the development outlined in Section 10.7 for single-phase power factor correction, we obtain

$$\mathbf{S}_{\text{old}} = 24 \; \underline{/\cos^{-1} 0.78} \text{ MVA}$$
$$= 18.72 + j15.02 \text{ MVA}$$

and

$$\theta_{\text{new}} = -\cos^{-1} 0.94$$
$$= -19.95°$$

Therefore,

$$\mathbf{S}_{\text{new}} = 18.72 + j18.72 \tan(-19.95°)$$
$$= 18.72 - j6.80 \text{ MVA}$$

and

$$\mathbf{S}_{\text{cap}} = \mathbf{S}_{\text{new}} - \mathbf{S}_{\text{old}}$$
$$= -j21.82 \text{ MVA}$$

However,

$$-j\omega C \, V_{\text{rms}}^2 = -j21.82 \text{ MVA}$$

and since the line voltage is 34.5 kV rms, then

$$(377)\left(\frac{34.5\text{k}}{\sqrt{3}}\right)^2 C = \frac{21.82}{3}$$

Hence,

$$C = 48.6 \; \mu\text{F} \qquad \Box$$

> The reactive power to be supplied by C is derived from the expression
> $$jQ_{\text{cap}} = -j\omega C \, V_{\text{rms}}^2$$
> The phase voltage for the Y connection is
> $$V_Y = \frac{34.5\text{k}}{\sqrt{3}}$$

EXTENSION EXERCISE

E10.14   Find C in Example 10.15 such that the load has a power factor of 0.90 lagging.

ANSWER: $C = 13.26 \; \mu\text{F}$.

Finally, the reader will recall that our entire discussion in this chapter has focused on balanced systems. It is extremely important, however, to point out that in the unbalanced three-phase system the problem is much more complicated because of the mutual inductive coupling between phases in power apparatus.

10.7 APPLICATIONS

The first of the following two examples illustrates the manner in which power flow is measured when utilities are interconnected, answering the question of who is supplying power to whom. The second example demonstrates the actual method in which capacitors are specified by the manufacturer for power factor correction.

EXAMPLE 10.16

Two balanced three-phase systems, X and Y, are interconnected with lines having impedance $\mathbf{Z}_{line} = 1 + j2\ \Omega$. The line voltages are $\mathbf{V}_{ab} = 12\ \underline{/0^\circ}$ kV rms and $\mathbf{V}_{AB} = 12\ \underline{/5^\circ}$ kV rms, as shown in Fig. 10.19a. We wish to determine which system is the source, which is the load, and the average power supplied by the source and absorbed by the load.

SOLUTION   When we draw the per phase circuit for the system as shown in Fig. 10.19b, the analysis will be essentially the same as that of Example 9.15.

The network in Fig. 10.19b indicates that

$$\mathbf{I}_{aA} = \frac{\mathbf{V}_{an} - \mathbf{V}_{AN}}{\mathbf{Z}_{line}}$$

$$= \frac{\dfrac{12{,}000}{\sqrt{3}}\ \underline{/-30^\circ} - \dfrac{12{,}000}{\sqrt{3}}\ \underline{/-25^\circ}}{\sqrt{5}\ \underline{/63.43^\circ}}$$

$$= 270.30\ \underline{/-180.93^\circ}\ \text{A rms}$$

The average power absorbed by system Y is

$$P_Y = \sqrt{3}\, V_{AB} I_{aA} \cos(\theta_{\mathbf{V}_{an}} - \theta_{\mathbf{I}_{aA}})$$
$$= \sqrt{3}\,(12{,}000)(270.30)\cos(-25^\circ + 180.93^\circ)$$
$$= -5.130\ \text{MW}$$

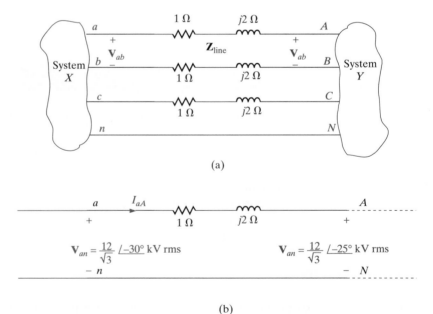

(a)

(b)

Figure 10.19 Circuits used in Example 10.16: (a) original three-phase system; (b) per phase circuit.

Note that system Y is not the load, but rather the source and supplies 5.130 MW.

System X absorbs the following average power:

$$P_X = \sqrt{3}\, V_{ab} I_{Aa} \cos\left(\theta_{V_{an}} - \theta_{I_{aA}}\right)$$

where

$$\mathbf{I}_{Aa} = -\mathbf{I}_{aA} = 270.30 \; \underline{/-0.93°} \; \text{A rms}$$

Therefore,

$$P_X = \sqrt{3}\,(12{,}000)(270.30) \cos(-30° + 0.93°)$$

$$= 4.910 \; \text{MW}$$

and hence system X is the load.

The difference in the power supplied by system Y and that absorbed by system X is, of course, the power absorbed by the resistance of the three lines. ❏

The preceding example illustrates an interesting point. Note that the phase difference between the two ends of the power line determines the direction of the power flow. Since the numerous power companies throughout the United States are tied together to form the U.S. power grid, the phase difference across the interconnecting transmission lines reflects the manner in which power is transferred between power companies.

Capacitors for power factor correction are usually specified by the manufacturer in vars rather than in farads. Of course, the supplier must also specify the voltage at which the capacitor is designed to operate and a frequency of 60 Hz is assumed. The relationship between capacitance and the vars rating is

$$Q_R = \frac{V^2}{Z_c}$$

where Q_R is vars rating, V is the voltage rating, and Z_c is the capacitor's impedance at 60 Hz. Thus, a 500-V, 600-var capacitor has a capacitance of

$$C = \frac{Q_R}{\omega V^2} = \frac{600}{(377)(500)^2}$$

or

$$C = 6.37 \; \mu\text{F}$$

and can be used in any application where the voltage across the capacitor does not exceed the rated value of 500 V.

EXAMPLE 10.17

Table 10.2 lists the voltage and power ratings for three power factor correction capacitors. Let us determine which of them, if any, can be employed in Example 10.15.

Table 10.2 Rated voltage and vars for power factor correction capacitors

Capacity	Rated Voltage (kV)	Rated Q (Muans)
1	10.0	4.0
2	50.0	25.0
3	20.0	7.5

SOLUTION   From Fig. 10.18 we see that the voltage across the capacitors is the line-to-neutral voltage, which is

$$V_{an} = \frac{V_{ab}}{\sqrt{3}} = \frac{34,500}{\sqrt{3}}$$

or

$$V_{an} = 19.9 \text{ kV}$$

Therefore, only those capacitors with rated voltages greater than or equal to 19.9 kV can be used in this application, which eliminates capacitor 1. Let us now determine the capacitance of capacitors 2 and 3. For capacitor 2,

$$C_2 = \frac{Q}{\omega V^2} = \frac{25 \times 10^6}{(377)(50,000)^2}$$

or

$$C_2 = 26.63 \ \mu\text{F}$$

which is much smaller than the required 48.6 μF. The capacitance of capacitor 3 is

$$C_3 = \frac{Q}{\omega V^2} = \frac{7.5 \times 10^6}{(377)(20,000)^2}$$

or

$$C_3 = 49.7 \ \mu\text{F}$$

which is within 2.5% of the required value. Obviously, capacitor 3 is the best choice. ❏

10.8 CIRCUIT DESIGN

In the following example we examine the selection of both the conductor and the capacitor in a practical power factor correction problem.

EXAMPLE 10.18

Two stores, as shown in Fig. 10.20, are located at a busy intersection. The stores are fed from a balanced three-phase 60-Hz source with a line voltage of 13.8 kV rms.

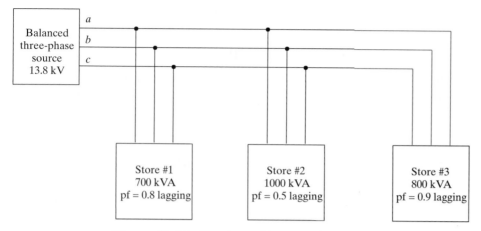

Figure 10.20 Circuit used in Example 10.18.

The power line is constructed of a #4ACSR (aluminum cable steel reinforced) conductor that is rated at 170 A rms.

A third store, shown in Fig. 10.20, wishes to locate at the intersection. Let us determine (1) if the #4ACSR conductor will permit the addition of this store, and (2) the value of the capacitors connected in wye that are required to change the overall power factor for all three stores to 0.92 lagging.

SOLUTION   (1) The complex power for each of the three loads is

$$\mathbf{S}_1 = 700 \ \underline{/36.9°} = 560 + j420 \ \text{kVA}$$

$$\mathbf{S}_2 = 1000 \ \underline{/60°} = 500 + j866 \ \text{kVA}$$

$$\mathbf{S}_3 = 800 \ \underline{/25.8°} = 720 + j349 \ \text{kVA}$$

Therefore, the total complex power is

$$\mathbf{S}_T = \mathbf{S}_1 + \mathbf{S}_2 + \mathbf{S}_3$$
$$= 1780 + j1635$$
$$= 2417 \ \underline{/42.57°} \ \text{kVA}$$

Since

$$\mathbf{S}_T = \sqrt{3} \, V_L I_L$$

the line current is

$$I_L = \frac{(2417)(10^3)}{\sqrt{3} \,(13.8)(10^3)}$$
$$= 101.1 \ \text{A rms}$$

Since this value is well below the rated value of 170 A rms, the conductor is sized properly and we can safely add the third store.

(2) The combined power factor for the three loads is found from the expression

$$\cos \theta = \text{pf} = \frac{1780}{2417} = 0.7365 \text{ lagging}$$

By adding capacitors we wish to change this power factor to 0.92 lagging. This new power factor corresponds to a θ_{new} of 23.07°. Therefore, the new complex power is

$$\mathbf{S}_{\text{new}} = 1780 + j1780 \tan(23.07°)$$
$$= 1780 + j758.28 \text{ KVA}$$

As illustrated in Fig. 9.17, the difference between $\mathbf{S}_{\text{new}}$ and $\mathbf{S}_T$ is that supplied by the purely reactive capacitor and, therefore,

$$\mathbf{S}_{\text{cap}} = jQ_c = \mathbf{S}_{\text{new}} - \mathbf{S}_T$$

or

$$jQ_c = j(758.28 - 1635)$$
$$= -j876.72 \text{ kVA}$$

Thus,

$$-j\omega C V_{\text{rms}}^2 = \frac{-j876.72\text{k}}{3}$$

and

$$377 \left(\frac{13.8 \times 10^3}{\sqrt{3}} \right)^2 C = \frac{876.72}{3} \times 10^3$$

Therefore,

$$C = 12.2 \text{ μF}$$

Hence, three capacitors of this value connected in wye at the load will yield a total power factor of 0.92 lagging.   ❏

10.9 SUMMARY

- An important advantage of the balanced three-phase system is that it provides very smooth power delivery.
- Because of the balanced condition, it is possible to analyze a circuit on a per phase basis, thereby providing a significant computational shortcut to a solution.
- A balanced three-phase source and load system: A balanced three-phase voltage source has three sinusoidal voltages of the same magnitude and frequency, and each voltage is 120° out of phase with the others. A positive-phase-sequence balanced voltage source is one in which $\mathbf{V}_{bn}$ lags $\mathbf{V}_{an}$ by 120° and $\mathbf{V}_{cn}$ lags $\mathbf{V}_{bn}$ by 120°.
- The relationships between wye- and delta-connected sources are as follows:

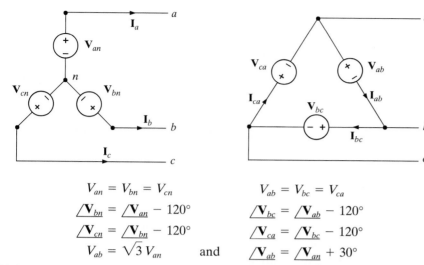

$$V_{an} = V_{bn} = V_{cn} \qquad\qquad V_{ab} = V_{bc} = V_{ca}$$

$$\underline{/V_{bn}} = \underline{/V_{an}} - 120° \qquad \underline{/V_{bc}} = \underline{/V_{ab}} - 120°$$

$$\underline{/V_{cn}} = \underline{/V_{bn}} - 120° \qquad \underline{/V_{ca}} = \underline{/V_{bc}} - 120°$$

$$V_{ab} = \sqrt{3}\, V_{an} \qquad \text{and} \qquad \underline{/V_{ab}} = \underline{/V_{an}} + 30°$$

If the load currents generated by connecting a load to a balanced three-phase voltage source are also balanced, the load is connected in either a balanced wye or a balanced delta configuration. The currents in the balanced system are

$$I_a = I_b = I_c \qquad\qquad I_{ab} = I_{bc} = I_{ca}$$

$$\underline{/I_b} = \underline{/I_a} - 120° \qquad \underline{/I_{bc}} = \underline{/I_{ab}} - 120°$$

$$\underline{/I_c} = \underline{/I_b} - 120° \qquad \underline{/I_{ca}} = \underline{/I_{bc}} - 120°$$

$$I_a = \sqrt{3}\, I_{ab} \qquad \text{and} \qquad \underline{/I_{ab}} = \underline{/I_a} + 30°$$

Note that the load voltages and currents are similar to those outlined for the source.

- The three-phase terminology is as shown in Table 10.3.

Table 10.3 Three-phase terminology

Quantity	Wye		Delta
I_a, I_b, I_c	Line current (I_L)		
	Phase current (I_0)		
V_{an}, V_{bn}, V_{cn}	Line-to-neutral voltage		
	Phase voltage (V_0)		
	Line-to-line, phase-to-phase, line voltage (V_L)		
V_{ab}, V_{bc}, V_{ca}			Phase voltage (V_0)
I_{ab}, I_{bc}, I_{ca}			Phase current (I_0)

- In a balanced system the voltages and currents sum to zero.

$$\mathbf{V}_{an} + \mathbf{V}_{bn} + \mathbf{V}_{cn} = 0$$

$$\mathbf{I}_a + \mathbf{I}_b + \mathbf{I}_c = 0 \qquad\qquad \text{(no current in the neutral line)}$$

and

$$\mathbf{V}_{ab} + \mathbf{V}_{bc} + \mathbf{V}_{ca} = 0$$

$$\mathbf{I}_{ab} + \mathbf{I}_{bc} + \mathbf{I}_{ca} = 0$$

- The steps recommended for solving balanced three-phase ac circuits are as follows:
  1. If the source/load connection is not wye–wye, then transform the system to a wye–wye connection.
  2. Determine the unknown phasors in the wye–wye connection and deal only with the phase a.
  3. Convert the now-known phasors back to the corresponding phasors in the original connection.

- The two-wattmeter method is as follows:
  The two-wattmeter method is a technique for measuring real power in a three-phase, three-wire system using only two wattmeters.
  The power factor angle of a load in a balanced three-phase system can be computed using two wattmeters.

- Power factor correction in a balanced three-phase environment is performed in the same manner as in the single-phase case. Three capacitors are put in parallel with the load to reduce the lagging phase caused by the three-phase load.

PROBLEMS

Sections 10.1 and 10.2

10.1 Sketch a phasor representation of an abc-sequence balanced three-phase Y-connected source, including $\mathbf{V}_{an}$, $\mathbf{V}_{bn}$, and $\mathbf{V}_{cn}$ if $\mathbf{V}_{an} = 120 \: \underline{/15°}$ V rms.

10.2 Sketch a phasor representation of an abc-sequence balanced three-phase Δ-connected source, including $\mathbf{V}_{ab}$, $\mathbf{V}_{bc}$, and $\mathbf{V}_{ca}$ if $\mathbf{V}_{ab} = 208 \: \underline{/0°}$ V rms.

Convert the phase voltage phasors to line voltage phasors and plot both sets.

10.3 Sketch a phasor representation of a balanced three-phase system containing both phase voltages and line voltages if $\mathbf{V}_{an} = 120 \: \underline{/60°}$ V rms. Label all magnitudes and assume an abc-phase sequence.

Similar to Example 10.1.

10.4 A positive-sequence three-phase balanced wye voltage source has a phase voltage of $\mathbf{V}_{an} = 240 \: \underline{/90°}$ V rms. Determine the line voltages of the source.

Convert the line voltage phasors to the phase voltage phasors and plot both sets.

10.5 Sketch a phasor representation of a balanced three-phase system containing both phase voltages and line voltages if $\mathbf{V}_{ab} = 208 \: \underline{/45°}$ V rms. Label all phasors and assume an abc-phase sequence.

Section 10.3

10.6   A positive-sequence balanced three-phase wye-connected source with a phase voltage of 120 V rms supplies power to a balanced wye-connected load. The per phase load impedance is $40 + j10\ \Omega$. Determine the line currents in the circuit if $\underline{/\mathbf{V}_{an}} = 0°$.

Use a single phase analysis to obtain the line current.

10.7   A positive-sequence balanced three-phase wye-connected source supplies power to a balanced wye-connected load. The magnitude of the line voltages is 208 V rms. If the load impedance per phase is $36 + j12\ \Omega$, determine the line currents if $\underline{/\mathbf{V}_{an}} = 0°$.

Similar to P10.6.

10.8   An *abc*-sequence balanced three-phase wye-connected source supplies power to a balanced wye-connected load. The line impedance per phase is $1 + j0\ \Omega$, and the load impedance per phase is $20 + j20\ \Omega$. If the source line voltage $\mathbf{V}_{ab}$ is $208\ \underline{/0°}$ V rms, find the line currents.

Similar to P10.6 except that the impedance includes line impedance.

10.9   In a three-phase balanced wye–wye system, the source is an *abc*-sequence set of voltages with $\mathbf{V}_{an} = 120\ \underline{/60°}$ V rms. The per phase impedance of the load is $12 + j10\ \Omega$. If the line impedance per phase is $0.8 + j0.5\ \Omega$, find the line currents and the load voltages.

Similar to P10.8 with the use of Ohm's law.

10.10   An *abc*-sequence set of voltages feeds a balanced three-phase wye–wye system. The line and load impedances are $1 + j1\ \Omega$ and $10 + j10\ \Omega$, respectively. If the load voltage on the *a* phase is $\mathbf{V}_{AN} = 110\ \underline{/30°}$ V rms, determine the line voltages of the input.

Similar to P10.9.

10.11   In a balanced three-phase wye–wye system, the source is an *abc*-sequence set of voltages. The load voltage on the *a* phase is $\mathbf{V}_{AN} = 110\ \underline{/80°}$ V rms, $\mathbf{Z}_{line} = 1 + j1.4\ \Omega$, and $\mathbf{Z}_{load} = 10 + j13\ \Omega$. Determine the input sequence of line-to-neutral voltages.

Similar to P10.10.

10.12   Find the equivalent impedances $\mathbf{Z}_{ab}$, $\mathbf{Z}_{bc}$, and $\mathbf{Z}_{ca}$ in the network in Fig. P10.12.

Table 10.1 is helpful in this case.

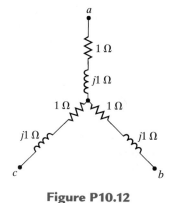

Figure P10.12

Table 10.1 is helpful in this case.

10.13 A balanced *abc*-sequence of voltages feeds a balanced three-phase wye–wye system. The line and load impedances are $0.6 + j0.9\ \Omega$ and $8 + j12\ \Omega$ respectively. The load voltage on the a phase is $\mathbf{V}_{AN} = 110\ \underline{/10°}$ V rms. Find the line voltage $\mathbf{V}_{ab}$.

Similar to P10.13.

10.14 In a balanced three-phase wye–wye system, the source is an *abc*-sequence set of voltages. $\mathbf{Z}_{\text{line}} = 1 + j1.8\ \Omega$, $\mathbf{Z}_{\text{load}} = 14 + j12\ \Omega$, and the load voltage on the *a* phase is $\mathbf{V}_{AN} = 440\ \underline{/20°}$ V rms. Find the line voltage $\mathbf{V}_{ab}$.

The line current can be determined from the power and impedance.

10.15 An *abc*-phase sequence balanced three-phase source feeds a balanced load. The system is connected wye–wye and $\underline{/\mathbf{V}_{an}} = 30°$. The line impedance is $0.5 + j0.2\ \Omega$, the load impedance is $16 + j10\ \Omega$, and the total power absorbed by the load is 1836.54 W. Determine the magnitude of the source voltage $\mathbf{V}_{an}$.

The power loss can be used to derive the line current.

10.16 In a balanced three-phase wye–wye system, the total power loss in the lines is 272.57 W. $\mathbf{V}_{AN} = 115.28\ \underline{/31.65°}$ V rms and the power factor of the load is 0.8 lagging. If the line impedance is $2 + j1\ \Omega$, determine the load impedance.

Voltage division is helpful in this case.

10.17 In a balanced three-phase wye–wye system the load impedance is $8 + j4\ \Omega$. The source has phase sequence *abc* and $\mathbf{V}_{an} = 120\ \underline{/0°}$ V rms. If the load voltage $\mathbf{V}_{AN} = 111.62\ \underline{/-1.33°}$ V rms, determine the line impedance.

Similar to P10.17.

10.18 In a balanced three-phase wye–wye system the load impedance is $10 + j1\ \Omega$. The source has phase sequence *abc* and the line voltage $\mathbf{V}_{ab} = 220\ \underline{/30°}$ V rms. If the load voltage $\mathbf{V}_{AN} = 120\ \underline{/0°}$ V rms, determine the line impedance.

Find the line impedance first.

10.19 In a balanced three-phase wye–wye system the load impedance is $20 + j12\ \Omega$. The source has an *abc*-phase sequence and $\mathbf{V}_{an} = 120\ \underline{/0°}$ V rms. If the load voltage is $\mathbf{V}_{AN} = 110\ \underline{/-0.2°}$ V rms, determine the magnitude of the line current if the load is suddenly short circuited.

The line current can be obtained from the load power. Then Ohm's law can be used to obtain the desired quantities.

10.20 In a balanced three-phase wye–wye system, the source is an *abc*-sequence set of voltages and $\mathbf{V}_{an} = 120\ \underline{/30°}$ V rms. The power absorbed by the load is 3435 W and the load impedance is $10 + j2\ \Omega$. Find the two possible line impedances if the power generated by the sources is 3774 W. Which line impedance is more likely to occur in an actual power transmission system?

10.21 In a balanced three-phase wye–wye system, the source is an *abc*-sequence set of voltages and $\mathbf{V}_{an} = 120 \underline{/50°}$ V rms. The load voltage on the *a* phase is110 $\underline{/470°}$ V rms and the load impedance is $16 + j20 \; \Omega$. Find the line impedance.

Similar to P10.20.

10.22 In a balanced three-phase wye–wye system, the source is an *abc*-sequence set of voltages and $\mathbf{V}_{an} = 120 \underline{/40°}$ V rms. If the *a*-phase line current and line impedance are known to be 7.10 $\underline{/-10.28°}$ A rms and $1 + j1 \; \Omega$, respectively, find the load impedance.

Ohm's law and KVL are applicable in this case.

10.23 An *abc*-sequence set of voltages feeds a balanced three-phase wye–wye system. If $\mathbf{V}_{an} = 440 \underline{/30°}$ V rms, $\mathbf{V}_{AN} = 413.28 \underline{/29.78°}$ V rms, and $\mathbf{Z}_{line} = 2 + j1.5 \; \Omega$, find the load impedance.

The line current can be obtained using $\mathbf{V}_{an} - \mathbf{V}_{AN}$ and $\mathbf{Z}_{line}$. The load impedance can be derived from Ohm's law.

10.24 In a three-phase balanced system a delta-connected source supplies power to a wye-connected load. If the line impedance is $0.2 + j0.4 \; \Omega$, the load impedance $6 + j4 \; \Omega$, and the source phase voltage $\mathbf{V}_{ab} = 210 \underline{/40°}$ V rms, find the magnitude of the line voltage at the load.

Once the system is converted to wye–wye, the voltage division rule can be easily applied.

10.25 Given the network in Fig. 10.25, compute the line currents and the magnitude of the phase voltage at the load.

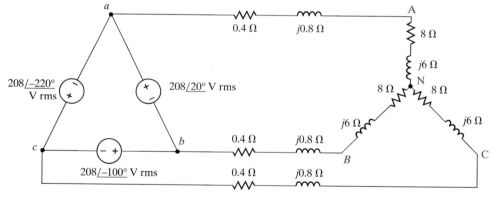

Figure P10.25

Similar to P10.24.

10.26 In a balanced three-phase delta–wye system the source has an *abc*-phase sequence. The line and load impedances are $0.6 + j0.3 \; \Omega$ and $12 + j3 \; \Omega$, respectively. If the line current $\mathbf{I}_{aA} = 9.6 \underline{/-20°}$ A rms, determine the phase voltages of the source.

One possible approach would be to solve for the voltages in a wye–wye system and then convert back to a delta–wye configuration.

10.27 An *abc*-phase-sequence three-phase balanced wye-connected 60-Hz source supplies a balanced delta-connected load. The phase impedance in the load consists of a 20-Ω resistor in series with a 20-mH inductor, and the phase voltage at the source is $\mathbf{V}_{an} = 120 \:\underline{/30°}$ V rms. If the line impedance is zero, find the line currents in the system.

Convert to a wye–wye system.

10.28 An *abc*-phase-sequence three-phase balanced wye-connected source supplies a balanced delta-connected load. The impedance per phase in the delta load is $12 + j6 \:\Omega$. The line voltage at the source is $\mathbf{V}_{ab} = 120 \sqrt{3} \:\underline{/40°}$ V rms. If the line impedance is zero, find the line currents in the balanced wye–delta system.

Similar to P10.27.

10.29 An *abc*-phase-sequence three-phase balanced wye-connected source supplies power to a balanced delta-connected load. The impedance per phase in the load is $14 + j7 \:\Omega$. If the source voltage for the *a* phase is $\mathbf{V}_{an} = 120 \:\underline{/80°}$ V rms and the line impedance is zero, find the phase currents in the wye-connected source.

Similar to P10.27.

10.30 An *abc*-phase-sequence three-phase balanced wye-connected source supplies a balanced delta-connected load. The impedance per phase of the delta load is $20 + j4 \:\Omega$. If $\mathbf{V}_{AB} = 115 \:\underline{/35°}$ V rms, find the line current.

Similar to P10.27.

10.31 An *abc*-phase-sequence three-phase balanced wye-connected source supplies a balanced delta-connected load. The impedance per phase of the delta load is $10 + j5 \:\Omega$. If the line impedance is zero and the line current in the *a* phase is known to be $\mathbf{I}_{aA} = 5 \:\underline{/-30°}$ A rms, find the load voltage $\mathbf{V}_{AB}$.

Use the per phase circuit to find $\mathbf{V}_{AN}$ and then convert to $\mathbf{V}_{AB}$.

10.32 An *abc*-phase-sequence three-phase balanced wye-connected source supplies power to a balanced delta-connected load. The impedance per phase of the delta load is $14 + j7 \:\Omega$. If the line impedance is zero and the line current in the *a* phase is $\mathbf{I}_{aA} = 20 \:\underline{/31°}$ A rms, find the load voltages of the balanced source.

Similar to P10.30.

10.33 In a balanced three-phase wye–delta system, the source has an *abc*-phase sequence and $\mathbf{V}_{an} = 120 \:\underline{/0°}$ V rms. If the line impedance is zero and the line current $\mathbf{I}_{aA} = 5 \:\underline{/20°}$ A rms, find the load impedance per phase in the delta.

Similar to P10.30.

10.34 In a balanced three-phase wye–delta system, the source has an *abc*-phase sequence and $\mathbf{V}_{an}$ = 120 $\underline{/40°}$ V rms. The line and load impedance are 0.5 + *j*0.4 Ω and 36 + *j*18 Ω, respectively. Find the delta currents in the load.

Use a wye–wye system to find the line currents. Then convert the line currents to currents in the delta-corrected load.

10.35 In a three-phase balanced delta–delta system, the source has an *abc*-phase sequence. The line and load impedances are 0.3 + *j*0.2 Ω and 9 + *j*3 Ω respectively. If the load current in the delta is $\mathbf{I}_{AB}$ = 15 $\underline{/40°}$ A rms, find the phase voltages of the source.

Similar to P10.34.

10.36 In a three-phase balanced delta–delta system, the source has an *abc*-phase sequence. The line and load impedances are 0.5 + 0.1*j* Ω and 10 + *j*5 Ω, respectively. If $\mathbf{V}_{AB}$ = 115 $\underline{/30°}$ V rms, find the phase voltage of the sources.

Find the line current in the equivalent wye–wye system. Determine the source voltage and convert it to its delta equivalent.

10.37 In a balanced three-phase delta–delta system, the source has an *abc*-phase sequence. The phase angle for the source voltage is $\underline{/\mathbf{V}_{ab}}$ = 40° and $\mathbf{I}_{ba}$ = 4 $\underline{/15°}$ A rms. If the total power absorbed by the load is 1400 W, find the load impedance.

The total power can be used to determine the real part of the load impedance.

10.38 A three-phase load impedance consists of a balanced wye in parallel with a balanced delta. What is the equivalent wye load and what is the equivalent delta load if the phase impedances of the wye and delta are 6 + *j*3 Ω and 15 + *j*10 Ω, respectively?

Similar to P10.5.

10.39 In a balanced three-phase system, the *abc*-phase-sequence source is wye connected and $\mathbf{V}_{an}$ = 120 $\underline{/20°}$ V rms. The load consists of two balanced wyes with phase impedances of 8 + *j*2 Ω and 12 + *j*3 Ω. If the line impedance is zero, find the line currents and the phase current in each load.

The two load currents can be summed to yield the line current.

10.40 In a balanced three-phase system, the *abc*-phase-sequence source is delta connected and $\mathbf{V}_{ab}$ = 120 $\underline{/30°}$ V rms. The load consists of two balanced wyes with phase impedances of 10 + *j*1 Ω and 20 + *j*5 Ω. If the line impedance is zero, find the line currents and the load phase voltage.

Similar to P10.30.

Section 10.4

10.41 An *abc*-sequence wye-connected source having a phase-*a* voltage of 120 $\underline{/0°}$ V rms is attached to a wye-connected load having a per phase impedance of 100 $\underline{/70°}$ Ω. If the line impedance is 1 $\underline{/20°}$ Ω, determine the total complex power produced by the voltage sources and the real and reactive power dissipated by the load.

The line current, obtained from Ohm's law, yields the unknowns.

The complex power expression leads to the unknowns.

10.42 The magnitude of the complex power (apparent power) supplied by a three-phase balanced wye–wye system is 3600 VA. The line voltage is 208 V rms. If the line impedance is negligible and the power factor angle of the load is 75°, determine the load impedance.

The load pf yields the power factor angle, which, in turn, yields the loads complex power and then the load impedance.

10.43 A three-phase *abc*-sequence wye-connected source supplies 14 kVA with a power factor of 0.75 lagging to a delta load. If the delta load consumes 12 kVA at a power factor of 0.7 lagging and has a phase current of $10 \,/\!-30°$ A rms, determine the phase impedance of the load and the line.

The expressions for real power and Ohm's law are applicable here.

10.44 A three-phase balanced wye–wye system has a line voltage of 208 V rms. The line current is 6 A rms and the total real power absorbed by the load is 1800 W. Determine the load impedance phase, if the line impedance is negligible.

The total complex power can be used to find the line current.

10.45 A balanced three-phase wye–wye system has two parallel loads. Load 1 is rated at 3000 VA, 0.7 pf lagging, and load 2 is rated at 2000 VA, 0.75 pf leading. If the line voltage is 208 V rms, find the magnitude of the line current.

Similar to P10.45.

10.46 Two smallindustrial plants represent balanced three-phase loads. The plants receive their power from a balanced three-phase source with a line voltage of 4.6 kV rms. Plant 1 is rated at 300 kVA, 0.8 pf lagging, and plant 2 is rated at 350 kVA, 0.8 pf lagging. Determine the power line current.

Similar to P10.45.

10.47 A balanced three-phase source serves two loads:
Load 1: 36 kVA at 0.8 pf lagging
Load 2: 18 kVA at 0.6 pf lagging
The line voltage at the load is 208 V rms at 60 Hz. Find the line current and the combined power factor at the load.

Similar to P10.45.

10.48 A balanced three-phase source serves the following loads:
Load 1: 48 kVA at 0.9 pf lagging
Load 2: 24 kVA at 0.75 pf lagging
The line voltage at the load is 208 V rms at 60 Hz. Determine the line current and the combined power factor at the load.

10.49 A small shopping center contains three stores that represent three balanced three-phase loads. The power lines to the shopping center represent a three-phase source with a line voltage of 13.8 kV rms. The three loads are

Load 1: 500 kVA at 0.8 pf lagging
Load 2: 400 kVA at 0.85 pf lagging
Load 3: 300 kVA at 0.90 pf lagging
Find the power line current.

Similar to P10.45.

10.50 The following loads are served by a balanced three-phase source:

Load 1: 18 kVA at 0.8 pf lagging
Load 2: 8 kVA at 0.8 pf leading
Load 3: 12 kVA at 0.75 pf lagging

The load voltage is 208 V rms at 60 Hz. If the line impedance is negligible, find the power factor of the source.

Similar to P10.45.

Section 10.5

10.51 The source in the network in Fig. P10.51 has an *abc*-phase sequence and $\mathbf{V}_{ab} = 208 \underline{/60°}$ V rms. Determine the two-wattmeter readings.

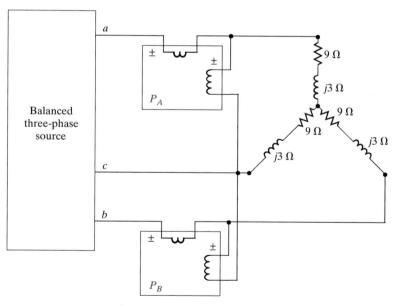

The equation for the two-wattmeter measurement is applicable here.

Figure P10.51

10.52 The source in the network in Fig. P10.52 has an *abc*-phase sequence and $\mathbf{V}_{ab} = 208 \,/10°$ V rms. Determine the two-wattmeter readings.

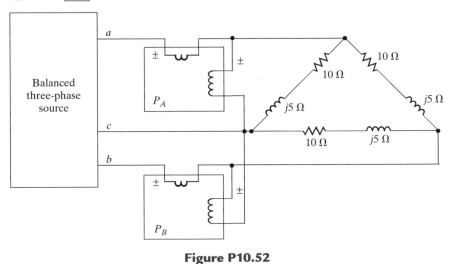

Similar to P10.51.

Figure P10.52

10.53 A balanced wye–wye three-phase system has an *abc*-sequence source with $\mathbf{V}_{an} = 120 \,/30°$ V rms. The balanced load has a per phase impedance of $24 + j16 \,\Omega$. Calculate the readings of the wattmeters if the two-wattmeter method is used to measure the total three-phase power.

Similar to P10.51.

10.54 In a balanced three-phase system the source has an *abc*-phase sequence, is connected in wye, and $\mathbf{V}_{an} = 120 \,/0°$ V rms. There are two parallel loads connected in wye. Load 1 has a phase impedance of $4 + j3 \,\Omega$ and load 2 has a phase impedance of $12 + j8 \,\Omega$. Find the total power absorbed by the loads using the two-wattmeter method.

Similar to P10.51.

10.55 A balanced three-phase system has an *abc*-phase sequence wye-connected source and $\mathbf{V}_{an} = 120 \,/60°$ V rms. Two loads are connected in parallel. Load 1 is connected in wye, and the phase impedance is $8 + j2 \,\Omega$. Load 2 is connected in delta, and the phase impedance is $8 + j3 \,\Omega$. Find the total power absorbed by the loads using the two-wattmeter method.

Similar to P10.51.

10.56 Two wattmeters are employed to measure the total power in a balanced three-phase wye–delta system. The phase voltage of the source is 120 V rms. The meter readings are $P_A = 1200$ W and $P_B = 800$ W. Determine the impedance per phase of the delta.

See Example 10.14.

10.57 In a balanced wye–delta three-phase system two wattmeters are used to measure total power. The line voltages are 208 V rms. If wattmeter A reads 1200 W and wattmeter B reads 600 W, determine the power factor and the load impedance per phase.

Similar to 11.56.

Section 10.6

10.58 A three-phase *abc*-sequence wye-connected source with $\mathbf{V}_{an} = 220\ \underline{/0°}$ V rms supplies power to a wye-connected load that consumes 50 kW of power in each phase at a pf of 0.8 lagging. Three capacitors are found that each have an impedance of $-j2.0\ \Omega$, and they are connected in parallel with the previous load in a wye configuration. Determine the power factor of the combined load as seen by the source.

See Example 10.15.

10.59 If the three capacitors in the network in Problem 10.58 are connected in a delta configuration, determine the power factor of the combined load as seen by the source.

Similar to P10.58.

10.60 Find C in the network in Fig. P10.60 such that the total load has a power factor of 0.9 lagging.

See Example 10.15.

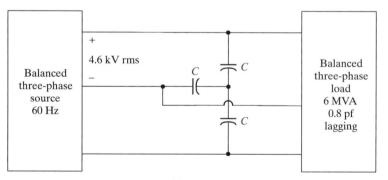

Figure P10.60

Similar to 10.60.

10.61 Find C in the network in Fig. P10.60 so that the total load has a power factor of 0.9 leading.

Similar to P10.60.

10.62 Find C in the network in Fig. P10.62 such that the total load has a power factor of 0.92 leading.

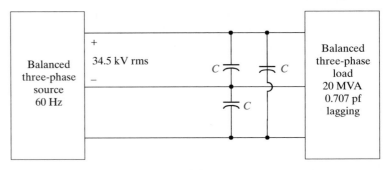

Figure P10.62

Similar P10.60.

10.63 Find the value of C in Problem 10.62 such that the total load has a power factor of 0.92 lagging.

Advanced Problems

10.64 Find the equivalent $\mathbf{Z}$ of the network in Fig. P10.64.

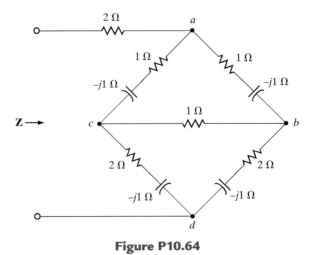

Figure P10.64

10.65 Find the equivalent **Z** of the network in Fig. P10.65.

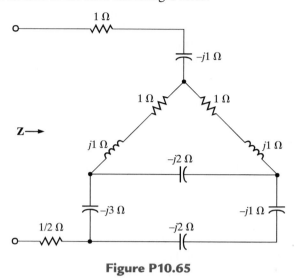

Figure P10.65

10.66 Find the equivalent **Z** of the network in Fig. P10.66.

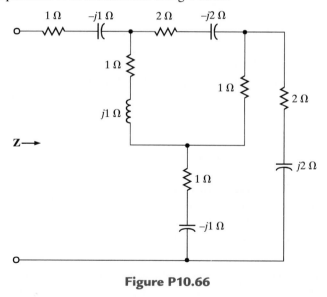

Figure P10.66

10.67 In a balanced three-phase system, the source is a balanced wye with an *abc*-phase sequence and $\mathbf{V}_{ab} = 208\ \underline{/60°}$ V rms. The load consists of a balanced wye with a phase impedance of $8 + j5\ \Omega$ in parallel with a balanced delta with a phase impedance of $21 + j12\ \Omega$. If the line impedance is $1.2 + j1\ \Omega$, find the phase currents in the balanced wye load.

10.68 In a balanced three-phase system, the source is a balanced wye with an *abc*-phase sequence and $\mathbf{V}_{an} = 120 \;\underline{/0°}$ V rms. The load is a balanced wye with a phase impedance of $6 + j4 \; \Omega$ in parallel with a balanced delta having a phase impedance of $12 + j6 \; \Omega$. If the line impedance is $1 + j1 \; \Omega$, find the magnitude of the phase voltage of the delta.

10.69 In a balanced three-phase system, the source is a balanced wye with an *abc*-phase sequence and $\mathbf{V}_{ab} = 215 \;\underline{/50°}$ V rms. The load is a balanced wye in parallel with a balanced delta. The phase impedance of the wye is $5 + j3 \; \Omega$ and the phase impedance of the delta is $18 + j12 \; \Omega$. If the line impedance is $1 + j0.8 \; \Omega$, find the line currents and the phase currents in the loads.

10.70 In a balanced three-phase system the source, which has *abc*-phase sequence, is connected in delta and $\mathbf{V}_{ab} = 220 \;\underline{/55°}$ V rms. There are two loads connected in parallel. Load 1 is connected in wye and the phase impedance is $4 + j3 \; \Omega$. Load 2 is connected in wye and the phase impedance is $8 + j6 \; \Omega$. Compute the delta currents in the source if the line impedance connecting the source to the loads is $0.2 + j0.1 \; \Omega$.

10.71 In a balanced three-phase system the source has an *abc*-phase sequence and is connected in delta. There are two parallel wye-connected loads. The phase impedance of load 1 and load 2 is $4 + j4 \; \Omega$ and $10 + j4 \; \Omega$, respectively. The line impedance connecting the source to the loads is $0.3 + j0.2 \; \Omega$. If the current in the *a* phase of load 1 is $\mathbf{I}_{AN_1} = 10 \;\underline{/20°}$ A rms, find the delta currents in the source.

10.72 In a balanced three-phase system the source has an *abc*-phase sequence and is connected in delta. There are two loads connected in parallel. The line connecting the source to the loads has an impedance of $0.2 + j0.1 \; \Omega$. Load 1 is connected in wye and the phase impedance is $4 + j2 \; \Omega$. Load 2 is connected in delta and the phase impedance is $12 + j6 \; \Omega$. The current $\mathbf{I}_{AB}$ in the delta load is $16 \;\underline{/45°}$ A rms. Find the phase voltages of the source.

10.73 In a balanced three-phase system the source has an *abc*-phase sequence and is connected in delta. There are two loads connected in parallel. Load 1 is connected in wye and has a phase impedance of $6 + j2 \; \Omega$. Load 2 is connected in delta and has a phase impedance of $9 + j3 \; \Omega$. The line impedance is $0.6 + j0.2 \; \Omega$. Determine the phase voltages of the source if the current in the *a* phase of load 1 is $\mathbf{I}_{AN_1} = 10 \;\underline{/30°}$ A rms.

10.74 A balanced three-phase delta-connected source supplies power to a load consisting of a balanced delta in parallel with a balanced wye. The phase impedance of the delta is $24 + j12 \ \Omega$, and the phase impedance of the wye is $12 + j8 \ \Omega$. The abc-phase-sequence source voltages are $\mathbf{V}_{ab} = 440 \ \underline{/60°}$ V rms, $\mathbf{V}_{bc} = 440 \ \underline{/-60°}$ V rms, and $\mathbf{V}_{ca} = 440 \ \underline{/-180°}$ V rms, the line impedance per phase is $1 + j0.8 \ \Omega$. Find the line currents and the power absorbed by the wye-connected load.

10.75 A cluster of loads are served by a balanced three-phase source with a line voltage of 4160 V rms. Load 1 is 240 kVA at 0.8 pf lagging and load 2 is 160 kVA at 0.92 pf lagging. A third load is unknown except that it has a power factor of unity. If the line current is measured and found to be 62 A rms, find the complex power of the unknown load.

10.76 A three-phase abc-sequence wye-connected source supplies 14 kVA with a power factor of 0.75 lagging to a parallel combination of a wye load and a delta load. If the wye load consumes 9 kVA at a power factor of 0.6 lagging and has a phase current of $10 \ \underline{/-30°}$ A rms, determine the phase impedance of the delta load.

10.77 In a balanced three-phase system, the source has an abc-phase sequence, is Y connected, and $\mathbf{V}_{an} = 120 \ \underline{/20°}$ V rms. The source feeds two parallel loads, both of which are Y connected. The impedance of load 1 is $8 + j6 \ \Omega$. The complex power for the a phase of load 2 is $600 \ \underline{/36°}$ VA. Find the line current for the a phase and the total complex power of the source.

10.78 A balanced three-phase source supplies power to three loads. The loads are
Load 1: 30 kVA at 0.8 pf lagging
Load 2: 24 kW at 0.6 pf leading
Load 3: unknown
The line voltage at the load and line current at the source are 208 V rms and 166.8 A rms, respectively. If the combined power factor at the load is unity, find the unknown load.

10.79 A balanced three-phase source serves the following loads:
Load 1: 18 kVA at 0.8 pf lagging
Load 2: 10 kVA at 0.7 pf leading
Load 3: 12 kW at unity pf
Load 4: 16 kVA at 0.6 pf lagging
The line voltage at the load is 208 V rms at 60 Hz and the line impedance is $0.02 + j0.04 \ \Omega$. Find the line voltage and power factor at the source.

10.80 A balanced three-phase source supplies power to three loads. The loads are
Load 1: 18 kW at 0.8 pf lagging
Load 2: 10 kVA at 0.6 pf leading
Load 3: unknown
If the line voltage at the loads is 208 V rms, the line current at the source is 116.39 A rms, and the combined power factor at the load is 0.86 lagging, find the unknown load.

10.81 A balanced three-phase source supplies power to three loads. The loads are
Load 1: 20 kW at 0.6 pf lagging
Load 2: 12 kVA at 0.75 pf lagging
Load 3: unknown
If the line voltage at the loads is 208 V rms, the line current at the source is 98.60 A rms, and the combined power factor at the load is 0.88 lagging, find the unknown load.

10.82 A standard practice for utility companies is to divide its customers into single-phase users and three-phase users. The utility must provide three-phase users, typically industries, with all three phases. However, single-phase users, residential, and light commercial are connected to only one phase. To reduce cable costs, all single-phase users in a neighborhood are connected together. This means that even if the three-phase users present perfectly balanced loads to the power grid, the single-phase loads will never be in balance, resulting in current flow in the neutral connection.

Consider the 60-Hz, *abc*-sequence network in Fig. P10.82. With a line voltage of 416 $\underline{/30°}$ V rms, phase *a* supplies the single-phase users on A Street, phase *b* supplies B Street, and phase *c* supplies C Street. Furthermore, the three-phase industrial load, which is connected in delta, is balanced. Find the neutral current.

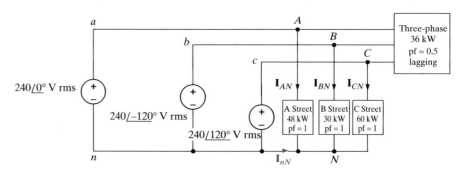

Figure P10.82

Magnetically Coupled Networks

11

High-voltage power lines crisscross the country to supply electric power to us in every facet of our lives. Many of the electric utilities are tied together to form a power grid. As indicated earlier, in order to deliver power in an efficient manner, it is transmitted at high voltage. Five-hundred-kilovolt lines are typical transmission facilities. However, the power that is supplied to our homes, for example, is typically 208/120 V rms. Thus, it is natural for us to question how the high voltage supplied through the transmission lines is stepped down to this level. Hence, one of the topics that we will address in this chapter is the transformer, which is the primary component in this application.

The recent advances in integrated circuits have caused designers to avoid the use of transformers (and inductors too) in a whole host of situations. For example, to avoid the use of a transformer, we may use a large number of other components such as resistors, capacitors, op-amps, and the like. However, while the trade-off in number of components may seem unreasonable, it is not. It is very easy to fabricate such components as resistors and capacitors, and they require much less space than a single inductor.

Nevertheless, transformers remain an important electrical component. In addition to power systems, where the transformers play a fundamental role, there are other applications. For example, transformers are used to reject high-frequency noise in audio and industrial control systems, and they are built into special wall plugs that are used to step down the voltage for the purpose of recharging batteries for calculators and hand tools.

11.1 MUTUAL INDUCTANCE

To begin our description of two coupled coils, we employ *Faraday's law*, which can be stated as follows: The induced voltage in a coil is proportional to the time rate of change of flux and the number of turns, N, in the coil. The two coupled coils are shown in Fig. 11.1 together with the following flux components.

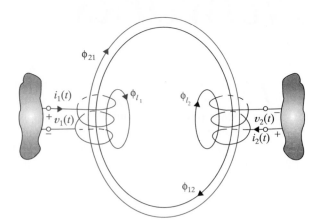

Figure 11.1 Flux relationships for mutually coupled coils.

ϕ_{l_1}	The flux in coil 1 that is produced by the current in coil 1 and does not link coil 2
ϕ_{l_2}	The flux in coil 2 that is produced by the current in coil 2 and does not link coil 1
ϕ_{12}	The flux in coil 1 produced by the current in coil 2
ϕ_{21}	The flux in coil 2 produced by the current in coil 1
$\phi_{11} = \phi_{l_1} + \phi_{21}$	The flux in coil 1 produced by the current in coil 1
$\phi_{22} = \phi_{l_2} + \phi_{12}$	The flux in coil 2 produced by the current in coil 2
ϕ_1	The total flux in coil 1
ϕ_2	The total flux in coil 2

In order to write the equations that describe the coupled coils, we define the voltages and currents, using the passive sign convention, at each pair of terminals as shown in Fig. 11.1.

Mathematically, Faraday's law can be written as

$$v_1(t) = N_1 \frac{d\phi_1}{dt} \tag{11.1}$$

The flux ϕ_1 will be equal to ϕ_{11}, the flux in coil 1 caused by current in coil 1, plus or minus the flux in coil 1 caused by current in coil 2; that is,

$$\phi_1 = \phi_{11} + \phi_{12} \tag{11.2}$$

If the current in coil 2 is such that the fluxes add, then the plus sign is used; if the current in coil 2 is such that the fluxes oppose one another, a minus sign is used. The equation for the voltage can now be written

$$v_1(t) = N_1 \frac{d\phi_1}{dt} \tag{11.3}$$

$$= N_1 \frac{d\phi_{11}}{dt} + N_1 \frac{d\phi_{12}}{dt}$$

From basic physics we know that

$$\phi_{11} = N_1 i_1 \mathcal{P}_{11} \tag{11.4}$$

$$\phi_{12} = N_2 i_2 \mathcal{P}_{12}$$

where the $\mathcal{P}$'s are constants (permeances) that depend on the magnetic paths taken by the flux components. The voltage equation can then be written

$$v_1(t) = N_1^2 \mathcal{P}_{11} \frac{di_1}{dt} + N_1 N_2 \mathcal{P}_{12} \frac{di_2}{dt} \tag{11.5}$$

The constant $N_1^2 \mathcal{P}_{11} = L_{11}$ (the same L that we used before) is now called the *self-inductance*, and the constant $N_1 N_2 \mathcal{P}_{12} = L_{12}$ is called the *mutual inductance*. Therefore,

$$v_1(t) = L_{11} \frac{di_1}{dt} + L_{12} \frac{di_2}{dt} \tag{11.6}$$

Using the same technique, we can write

$$v_2(t) = N_2^2 \mathcal{P}_{22} \frac{di_2}{dt} + N_1 N_2 \mathcal{P}_{21} \frac{di_1}{dt} \tag{11.7}$$

which can be written

$$v_2(t) = L_{22} \frac{di_2}{dt} + L_{21} \frac{di_1}{dt} \tag{11.8}$$

$$v_1(t) = L_{11} \frac{di_1(t)}{dt} + M \frac{di_2(t)}{dt}$$

$$v_2(t) = L_{22} \frac{di_2(t)}{dt} + M \frac{di_1(t)}{dt}$$

If the medium through which the magnetic flux passes is linear, then $\mathcal{P}_{12} = \mathcal{P}_{21}$. Hence, $L_{12} = L_{21} = M$. For convenience, let us define $L_1 = L_{11}$ and $L_2 = L_{22}$.

We now need to examine the physical details of coupled coils. In basic physics we learned the *right-hand rule*, which states that if we curl the fingers of our right hand around the coil in the direction of the current, the flux produced by the current is in the direction of our thumb.

In order to indicate the physical relationship of the coils and, therefore, simplify the sign convention for the mutual terms, we employ what is commonly called the *dot convention*. Dots are placed beside each coil so that if currents are entering both dotted terminals or leaving both dotted terminals, the fluxes produced by these currents will add. In order to place the dots on a pair of coupled coils, we arbitrarily select one terminal of either coil and place a dot there. Using the right-hand rule, we determine the direction of the flux produced by this coil when current is entering the dotted terminal. We then examine the other coil to determine which terminal the current would have to enter to produce a flux that would add to the flux produced by the first coil. Place a dot on this terminal. The dots have been placed on the two coupled circuits in Fig. 11.2; verify that they are correct.

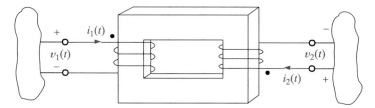

Figure 11.2 Illustration of the dot convention for coupled coils.

When the equations for the terminal voltages are written, the dots can be used to define the sign of the mutually induced voltages. If the currents $i_1(t)$ and $i_2(t)$ are both entering or both leaving dots, the sign of the mutual voltage $M(di_2/dt)$ will be the same in an equation as the self-induced voltage $L_1(di_1/dt)$. If one current enters a dot and the other current leaves a dot, the mutual induced voltage and self-induced voltage terms will have opposite signs.

Table 11.1 summarizes the key self- and mutual-inductance relationships.

Treating the induced voltage as a dependent source makes it easier to visualize the coupling effect and serves as a better model for deriving the Thévenin equivalent impedance when employing Thévenin's or Norton's theorem.

EXAMPLE 11.1

Determine the expressions for $v_1(t)$ and $v_2(t)$ in the circuit shown in Fig. 11.3.

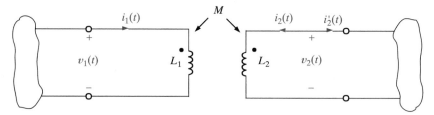

Figure 11.3 Two mutually coupled inductor circuits.

SOLUTION For the circuit in Fig. 11.3 the voltage equations for the variables as assigned on the figure are

$$v_1(t) = L_1 \frac{di_1}{dt} + M \frac{di_2}{dt}$$

$$v_2(t) = L_2 \frac{di_2}{dt} + M \frac{di_1}{dt}$$ ❑

EXAMPLE 11.2

Determine the equations for $v_1(t)$ and $v_2(t)$ in the circuits shown in Fig. 11.3 if we replace $i_2(t)$ with $i_2'(t)$ (i.e., secondary current leaving the dot).

D11.1   Determine which rule in Table 11.1 must be used to determine the self- or mutual-inductance relationship for each of the following circuits.

1.

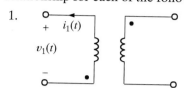

ANSWER: Self-inductance rule 2; mutual-inductance rule 1

2.

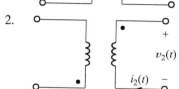

ANSWER: Self-inductance rule 2; mutual-inductance counterpart of rule 2.

3.

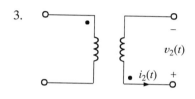

ANSWER: Self-inductance rule 2; mutual-inductance counterpart of rule 2.

4.

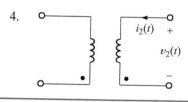

ANSWER: Self-inductance rule 1; mutual-inductance counterpart of rule 2.

SOLUTION   The equations are

$$v_1(t) = L_1 \frac{di_1}{dt} - M \frac{di_2'}{dt}$$

$$v_2(t) = -L_2 \frac{di_2'}{dt} + M \frac{di_1}{dt} \qquad \square$$

EXTENSION EXERCISE

E11.1   Write the equations for $v_1(t)$ and $v_2(t)$ in the circuit in Fig. E11.1.

ANSWER: $v_1(t) = -L_1 \frac{di_1}{dt} - M \frac{di_2}{dt}, \ v_2(t) = L_2 \frac{di_2}{dt} + M \frac{di_1}{dt}.$

Table 11.1 Self- and mutual-inductance relationships

Self-inductance

Rule 1

Rule 2

$$v(t) = L\frac{di(t)}{dt}$$

$$v(t) = -L\frac{di(t)}{dt}$$

Mutual inductance

Rule 1

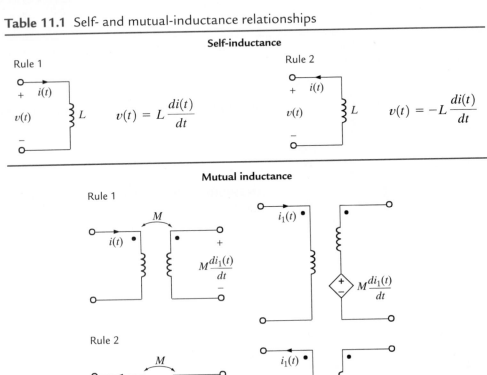

$$M\frac{di_1(t)}{dt}$$

Rule 2

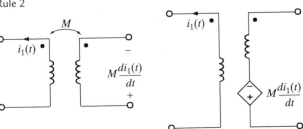

$$M\frac{di_1(t)}{dt}$$

Counterpart of rule 1

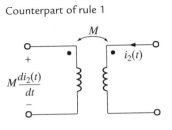

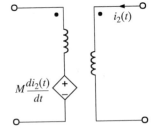

$$M\frac{di_2(t)}{dt}$$

Counterpart of rule 2

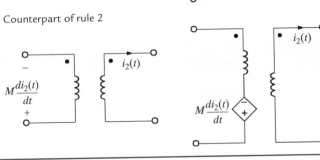

$$M\frac{di_2(t)}{dt}$$

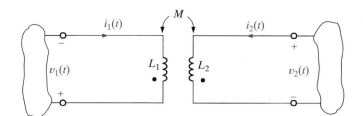

Figure E11.1

Assume the coupled circuit in Fig. 11.3 is excited with a sinusoidal source. The voltages will be of the form $\mathbf{V}_1 e^{j\omega t}$ and $\mathbf{V}_2 e^{j\omega t}$, and the currents will be of the form $\mathbf{I}_1 e^{j\omega t}$ and $\mathbf{I}_2 e^{j\omega t}$, where $\mathbf{V}_1$, $\mathbf{V}_2$, $\mathbf{I}_1$, and $\mathbf{I}_2$, are phasors. Substituting these voltages and currents into Eqs. (11.6) and (11.8), we obtain

$$\mathbf{V}_1 = j\omega L_1 \mathbf{I}_1 + j\omega M \mathbf{I}_2$$
$$\mathbf{V}_2 = j\omega L_2 \mathbf{I}_2 + j\omega M \mathbf{I}_1 \qquad \textbf{11.9}$$

The model of the coupled circuit in the frequency domain is identical to that in the time domain except for the way the elements and variables are labeled. The sign on the mutual terms is handled in the same manner as is done in the time domain.

Problem-Solving Strategy for Circuits With Coupled Inductors

1. Treat induced voltages as dependent sources using Table 11.1.
2. Write the KVL equations(s) for the network.
3. Solve the equation(s) to obtain the desired answer.

DRILL

D11.2   Determine the equations for $v_1(t)$ and $v_2(t)$ in the following circuit.

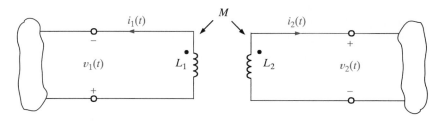

ANSWER:

$$v_1(t) = L_1 \frac{di_1(t)}{dt} + M \frac{di_2(t)}{dt}$$

$$v_2(t) = -L_2 \frac{di_2(t)}{dt} - M \frac{di_1(t)}{dt}$$

EXAMPLE 11.3

The two mutually coupled coils in Fig. 11.4a can be interconnected in four possible ways. We wish to determine the equivalent inductance of each of the four possible interconnections.

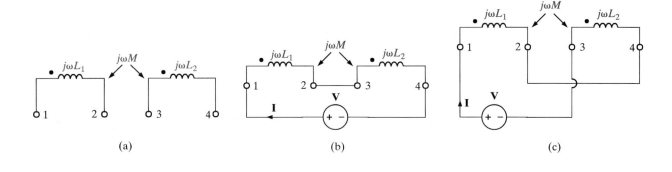

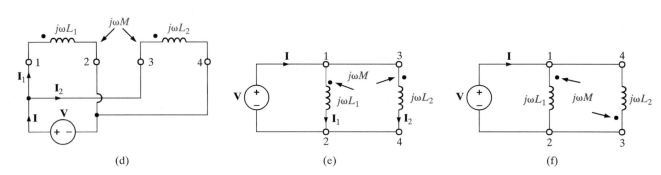

Figure 11.4 Circuits used in Example 11.3.

SOLUTION   Case 1 is shown in Fig. 11.4b. In this case

$$\mathbf{V} = j\omega L_1 \mathbf{I} + j\omega M \mathbf{I} + j\omega L_2 \mathbf{I} + j\omega M \mathbf{I}$$
$$= j\omega L_{eq} \mathbf{I}$$

where $L_{eq} = L_1 + L_2 + 2M$.

Case 2 is shown in Fig. 11.4c. Using KVL, we obtain

$$\mathbf{V} = j\omega L_1 \mathbf{I} - j\omega M \mathbf{I} + j\omega L_2 \mathbf{I} - j\omega M \mathbf{I}$$
$$= j\omega L_{eq} \mathbf{I}$$

where $L_{eq} = L_1 + L_2 - 2M$.

Case 3 is shown in Fig. 11.4d and redrawn in Fig. 11.4e. The two KVL equations are

$$\mathbf{V} = j\omega L_1 \mathbf{I}_1 + j\omega M \mathbf{I}_2$$
$$\mathbf{V} = j\omega M \mathbf{I}_1 + j\omega L_2 \mathbf{I}_2$$

Solving these equations for $\mathbf{I}_1$ and $\mathbf{I}_2$ yields

$$I_1 = \frac{\mathbf{V}(L_2 - M)}{j\omega(L_1 L_2 - M^2)}$$

$$I_2 = \frac{\mathbf{V}(L_1 - M)}{j\omega(L_1 L_2 - M^2)}$$

Using KCL gives us

$$\mathbf{I} = \mathbf{I}_1 + \mathbf{I}_2 = \frac{\mathbf{V}(L_1 + L_2 - 2M)}{j\omega(L_1 L_2 - M^2)} = \frac{\mathbf{V}}{j\omega L_{eq}}$$

where

$$L_{eq} = \frac{L_1 L_2 - M^2}{L_1 + L_2 - 2M}$$

Case 4 is shown in Fig. 11.4f. The voltage equations in this case will be the same as those in Case 3 except that the signs of the mutual terms will be negative. Therefore,

$$L_{eq} = \frac{L_1 L_2 - M^2}{L_1 + L_2 + 2M} \qquad \square$$

EXAMPLE 11.4

We wish to determine the output voltage $\mathbf{V}_o$ in the circuit in Fig. 11.5.

The network in Fig. 11.5 is redrawn in Fig. 11.6 to employ dependent sources to represent mutual inductance.

Figure 11.5 Example of a magnetically coupled circuit.

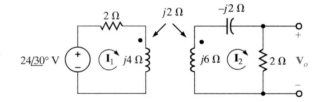

Figure 11.6 The network in Fig. 11.5 with dependent source.

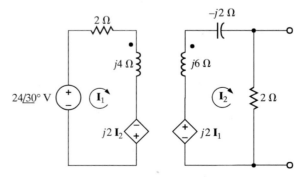

SOLUTION The two KVL equations for the network are

$$(2 + j4)\mathbf{I}_1 - j2\mathbf{I}_2 = 24 \underline{/30°}$$

$$-j2\mathbf{I}_1 + (2 + j6 - j2)\mathbf{I}_2 = 0$$

Solving the equations yields

$$\mathbf{I}_2 = 2.68 \,\underline{/3.43°}\ \text{A}$$

Therefore,

$$\mathbf{V}_o = 2\mathbf{I}_2$$
$$= 5.36 \,\underline{/3.43°}\ \text{V} \qquad \square$$

Let us now consider a more complicated example involving mutual inductance.

EXAMPLE 11.5

Consider the circuit in Fig. 11.7. We wish to write the mesh equations for this network.

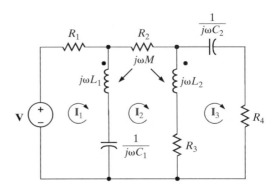

Figure 11.7 is redrawn with dependent sources in Fig. 11.8.

Figure 11.7 Example of a magnetically coupled circuit.

SOLUTION   Because of the multiple currents that are present in the coupled inductors, we must be very careful in writing the circuit equations.

The mesh equations for the phasor network are

$$\mathbf{I}_1 R_1 + j\omega L_1 (\mathbf{I}_1 - \mathbf{I}_2) + j\omega M (\mathbf{I}_2 - \mathbf{I}_3) + \frac{1}{j\omega C_1}(\mathbf{I}_1 - \mathbf{I}_2) = \mathbf{V}$$

$$\frac{1}{j\omega C_1}(\mathbf{I}_2 - \mathbf{I}_1) + j\omega L_1 (\mathbf{I}_2 - \mathbf{I}_1) + j\omega M (\mathbf{I}_3 - \mathbf{I}_2) + R_2 \mathbf{I}_2 + j\omega L_2 (\mathbf{I}_2 - \mathbf{I}_3)$$
$$+ j\omega M (\mathbf{I}_2 - \mathbf{I}_1) + R_3 (\mathbf{I}_2 - \mathbf{I}_3) = 0$$

$$R_3 (\mathbf{I}_3 - \mathbf{I}_2) + j\omega L_2 (\mathbf{I}_3 - \mathbf{I}_2) + j\omega M (\mathbf{I}_2 - \mathbf{I}_1) + \frac{1}{j\omega C_2}\mathbf{I}_3 + R_4 \mathbf{I}_3 = 0$$

which can be rewritten in the form

$$\left(R_1 + j\omega L_1 + \frac{1}{j\omega C_1}\right)\mathbf{I}_1 - \left(j\omega L_1 + \frac{1}{j\omega C_1} - j\omega M\right)\mathbf{I}_2 - j\omega M \mathbf{I}_3 = \mathbf{V}$$

$$-\left(j\omega L_1 + \frac{1}{j\omega C_1} - j\omega M\right)\mathbf{I}_1$$
$$+\left(\frac{1}{j\omega C_1} + j\omega L_1 + R_2 + j\omega L_2 + R_3 - j2\omega M\right)\mathbf{I}_2$$
$$-\left(j\omega L_2 + R_3 - j\omega M\right)\mathbf{I}_3 = 0$$

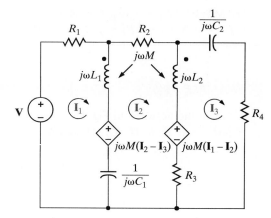

Figure 11.8 The network in Fig. 11.7 with dependent sources.

$$-j\omega M\mathbf{I}_1 - (R_3 + j\omega L_2 - j\omega M)\mathbf{I}_2 + \left(R_3 + j\omega L_2 + \frac{1}{j\omega C_2} + R_4\right)\mathbf{I}_3 = 0$$

Note the symmetrical form of these equations.   ❏

EXTENSION EXERCISES

E11.2   Find the currents $\mathbf{I}_1$ and $\mathbf{I}_2$ and the output voltage $\mathbf{V}_o$ in the network in Fig. E11.2.

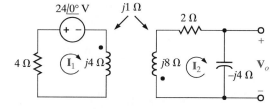

Figure E11.2

ANSWER: $\mathbf{I}_1 = -4.29 \,\underline{/137.2°}$ A, $\mathbf{I}_2 = 0.96 \,\underline{/-16.26°}$ A, $\mathbf{V}_o = 3.84 \,\underline{/-106.26°}$ V.

E11.3   Write the KVL equations in standard form for the network in Fig. E11.3.

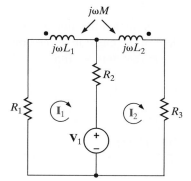

Figure E11.3

ANSWER: $(R_1 + j\omega L_1 + R_2)\mathbf{I}_1 - (R_2 + j\omega M)\mathbf{I}_2 = -\mathbf{V}_1,$
$-(R_2 + j\omega M)\mathbf{I}_1 + (R_2 + j\omega L_2 + R_3)\mathbf{I}_2 = \mathbf{V}_1.$

EXAMPLE 11.6

Given the network in Fig. 11.9 with the parameters $\mathbf{Z}_S = 3 + j1\ \Omega$, $j\omega L_1 = j2\ \Omega$, $j\omega L_2 = j2\ \Omega$, $j\omega M = j1\ \Omega$, and $\mathbf{Z}_L = 1 - j1\ \Omega$, determine the impedance seen by the source $\mathbf{V}_S$.

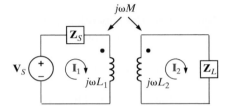

Figure 11.9 Circuit employed in Example 11.6.

SOLUTION   The mesh equations for the network are

$$\mathbf{V}_S = (\mathbf{Z}_S + j\omega L_1)\mathbf{I}_1 - j\omega M\mathbf{I}_2$$
$$0 = -j\omega M\mathbf{I}_1 + (j\omega L_2 + \mathbf{Z}_L)\mathbf{I}_2$$

If we now define $\mathbf{Z}_{11} = \mathbf{Z}_S + j\omega L_1$ and $\mathbf{Z}_{22} = j\omega L_2 + \mathbf{Z}_L$, then the second equation yields

$$\mathbf{I}_2 = \frac{j\omega M}{\mathbf{Z}_{22}}\mathbf{I}_1$$

If this secondary mesh equation is substituted into the primary mesh equation, we obtain

$$\mathbf{V}_S = \mathbf{Z}_{11}\mathbf{I}_1 + \frac{\omega^2 M^2}{\mathbf{Z}_{22}}\mathbf{I}_1$$

and therefore

$$\frac{\mathbf{V}_S}{\mathbf{I}_1} = \mathbf{Z}_{11} + \frac{\omega^2 M^2}{\mathbf{Z}_{22}}$$

which is the impedance seen by $\mathbf{V}_S$. Note that the mutual term is squared and therefore the impedance is independent of the location of the dots.

Using the values of the circuit parameters, we find that

$$\frac{\mathbf{V}_S}{\mathbf{I}_1} = (3 + j1 + j2) + \frac{1}{j2 + 1 - j1}$$
$$= 3 + j3 + 0.5 - j0.5$$
$$= 3.5 + j2.5\ \Omega$$ ❑

EXTENSION EXERCISE

E11.4   Find the impedance seen by the source in the circuit in Fig. E11.4.

ANSWER: $\mathbf{Z}_S = 2.25 \, \underline{/20.9°} \, \Omega$

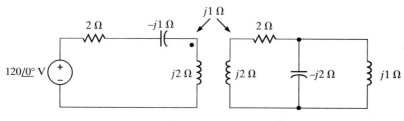

Figure E11.4

11.2 ENERGY ANALYSIS

We now perform an energy analysis on a pair of mutually coupled inductors, which will yield some interesting relationships for the circuit elements. Our analysis will involve the performance of an experiment on the network shown in Fig. 11.10. Before beginning the experiment, we set all voltages and currents in the circuit equal to zero. Once the circuit is quiescent, we begin by letting the current $i_1(t)$ increase from zero to some value I_1 with the right-side terminals open circuited. Since the right-side terminals are open, $i_2(t) = 0$, and therefore the power entering these terminals is zero. The instantaneous power entering the left-side terminals is

$$p(t) = v_1(t)i_1(t) = \left[L_1 \frac{di_1(t)}{dt} \right] i_1(t)$$

The energy stored within the coupled circuit at t_1 when $i_1(t) = I_1$ is then

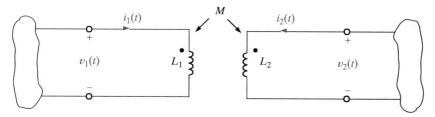

Figure 11.10 Magnetically coupled circuit.

$$\int_0^{t_1} v_1(t) i_1(t)\, dt = \int_0^{I_1} L_1 i_1(t)\, di_1(t) = \tfrac{1}{2} L_1 I_1^2$$

Continuing our experiment, starting at time t_1, we let the current $i_2(t)$ increase from zero to some value I_2 at time t_2 while holding $i_1(t)$ constant at I_1. The energy delivered through the right-side terminals is

$$\int_{t_1}^{t_2} v_2(t) i_2(t)\, dt = \int_0^{I_2} L_2 i_2(t)\, di_2(t) = \tfrac{1}{2} L_2 I_2^2$$

However, during the interval t_1 to t_2 the voltage $v_1(t)$ is

$$v_1(t) = L_1 \frac{di_1(t)}{dt} + M \frac{di_2(t)}{dt}$$

Since $i_1(t)$ is a constant I_1, the energy delivered through the left-side terminals is

$$\int_{t_1}^{t_2} v_1(t) i_1(t)\, dt = \int_{t_1}^{t_2} M \frac{di_2(t)}{dt} I_1\, dt = M I_1 \int_0^{I_2} di_2(t)$$
$$= M I_1 I_2$$

Therefore, the total energy stored in the network for $t > t_2$ is

$$w = \tfrac{1}{2} L_1 I_1^2 + \tfrac{1}{2} L_2 I_2^2 + M I_1 I_2 \tag{11.10}$$

We could, of course, repeat our entire experiment with either the dot on L_1 or L_2, but not both, reversed, and in this case the sign on the mutual inductance term would be negative, producing

$$w = \tfrac{1}{2} L_1 I_1^2 + \tfrac{1}{2} L_2 I_2^2 - M I_1 I_2$$

It is very important for the reader to realize that in our derivation of the preceding equation, by means of the experiment, the values I_1 and I_2 could have been any values at any *time*; therefore, the energy stored in the magnetically coupled inductors at any instant of time is given by the expression

$$w(t) = \tfrac{1}{2} L_1 [i_1(t)]^2 + \tfrac{1}{2} L_2 [i_2(t)]^2 \pm M i_1(t) i_2(t) \tag{11.11}$$

The two coupled inductors represent a passive network, and therefore, the energy stored within this network must be nonnegative for any values of the inductances and currents.

The equation for the instantaneous energy stored in the magnetic circuit can be written as

$$w(t) = \tfrac{1}{2} L_1 i_1^2 + \tfrac{1}{2} L_2 i_2^2 \pm M i_1 i_2$$

Adding and subtracting the term $\tfrac{1}{2}(M^2/L_2) i_1^2$ and rearranging the equation yields

$$w(t) = \frac{1}{2}\left(L_1 - \frac{M^2}{L_2}\right) i_1^2 + \frac{1}{2} L_2 \left(i_2 + \frac{M}{L_2} i_1\right)^2$$

From this expression we recognize that the instantaneous energy stored will be nonnegative if

$$M \leq \sqrt{L_1 L_2} \tag{11.12}$$

Note that this equation specifies an upper limit on the value of the mutual inductance. We define the coefficient of coupling between the two inductors L_1 and L_2 as

$$k = \frac{M}{\sqrt{L_1 L_2}}$$ **11.13**

and we note from Eq. (11.12) that its range of values is

$$0 \leq k \leq 1$$ **11.14**

This coefficient is an indication of how much flux in one coil is linked with the other coil; that is, if all the flux in one coil reaches the other coil, then we have 100% coupling and $k = 1$. For large values of k (i.e., $k > 0.5$), the inductors are said to be tightly coupled, and for small values of k (i.e., $k \leq 0.5$), the coils are said to be loosely coupled. If there is no coupling, $k = 0$. The previous equations indicate that the value for the mutual inductance is confined to the range

$$0 \leq M \leq \sqrt{L_1 L_2}$$ **11.15**

and that the upper limit is the geometric mean of the inductances L_1 and L_2.

EXAMPLE 11.7

The coupled circuit in Fig. 11.11a has a coefficient of coupling of 1 (i.e., $k = 1$). We wish to determine the energy stored in the mutually coupled inductors at time $t = 5$ ms. $L_1 = 2.653$ mH and $L_2 = 10.61$ mH.

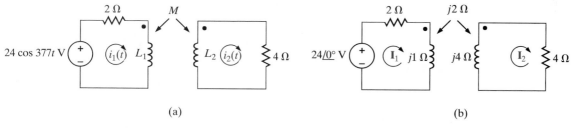

(a)   (b)

Figure 11.11 Example of a magnetically coupled circuit drawn in the time and frequency domains.

SOLUTION   From the data the mutual inductance is

$$M = \sqrt{L_1 L_2} = 5.31 \text{ mH}$$

The frequency domain equivalent circuit is shown in Fig. 11.11b, where the impedance values for X_{L_1}, X_{L_2}, and X_M are 1, 4, and 2, respectively. The mesh equations for the network are then

$$(2 + j1)\mathbf{I}_1 - j2\mathbf{I}_2 = 24 \,\underline{/0°}$$
$$-j2\mathbf{I}_1 + (4 + j4)\mathbf{I}_2 = 0$$

Solving these equations for the two mesh currents yields

$$\mathbf{I_1} = 9.41\;\underline{/-11.31°}\;\text{A} \quad \text{and} \quad \mathbf{I_2} = 3.33\;\underline{/+33.69°}\;\text{A}$$

and therefore,

$$i_1(t) = 9.41 \cos(377t - 11.31°)\;\text{A}$$
$$i_2(t) = 3.33 \cos(377t + 33.69°)\;\text{A}$$

At $t = 5$ ms, $377t = 1.885$ rad or $108°$, and therefore,

$$i_1(t = 5\text{ ms}) = 9.41 \cos(108° - 11.31°) = -1.10\;\text{A}$$
$$i_2(t = 5\text{ ms}) = 3.33 \cos(108° + 33.69°) = -2.61\;\text{A}$$

Therefore, the energy stored in the coupled inductors at $t = 5$ ms is

$$
\begin{aligned}
w(t)|_{t=0005\,s} &= \tfrac{1}{2}(2.653)(10^{-3})(-1.10)^2 + \tfrac{1}{2}(10.61)(10^{-3})(-2.61)^2 \\
&\quad -(5.31)(10^{-3})(-1.10)(-2.61) \\
&= (1.61)(10^{-3}) + (36.14)(10^{-3}) - (15.25)(10^{-3}) \\
&= 2.25\;\text{mJ}
\end{aligned}
$$

❏

EXTENSION EXERCISE

E11.5   The network in Fig. E11.5 operates at 60 Hz. Compute the energy stored in the mutually coupled inductors at time $t = 10$ ms.

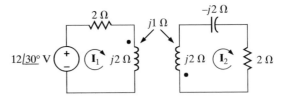

Figure E11.5

ANSWER: $w(10\text{ ms}) = 39\;\text{mJ}$.

11.3 THE IDEAL TRANSFORMER

Consider the situation illustrated in Fig. 11.12, showing two coils of wire wound on a single closed magnetic core. The magnetic core concentrates the flux so that all the flux links all the turns of both coils. In the ideal case we also neglect wire resistance. Let us now examine the coupling equations under the condition that the same flux goes through each winding and therefore,

$$v_1(t) = N_1 \frac{d\phi_1}{dt} = N_1 \frac{d\phi}{dt}$$

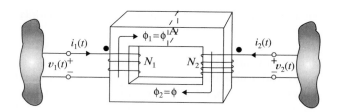

Figure 11.12 Transformer employing a magnetic core.

and

$$v_2(t) = N_2 \frac{d\phi_2}{dt} = N_2 \frac{d\phi}{dt}$$

and therefore,

$$\frac{v_1}{v_2} = \frac{N_1 \dfrac{d\phi}{dt}}{N_2 \dfrac{d\phi}{dt}} = \frac{N_1}{N_2} \qquad\qquad \textbf{11.16}$$

Another relationship can be developed between the currents $i_1(t)$ and $i_2(t)$ and the number of turns in each coil. To develop this relationship, we employ, from electromagnetic field theory, Ampère's law, which is written in mathematical form as

$$\oint H \cdot dl = i_{enclosed} = N_1 i_1 + N_2 i_2 \qquad\qquad \textbf{11.17}$$

where H is the magnetic field intensity and the integral is over the closed path traveled by the flux around the transformer core. For the ideal core material, μ (permeability) $= \infty$, $H = 0$. Therefore,

$$N_1 i_1 + N_2 i_2 = 0 \qquad\qquad \textbf{11.18}$$

or

$$\frac{i_1}{i_2} = -\frac{N_2}{N_1} \qquad\qquad \textbf{11.19}$$

Note that if we divide Eq. (11.18) by N_1 and multiply it by v_1, we obtain

$$v_1 i_1 + \frac{N_2}{N_1} v_1 i_2 = 0$$

However, since $v_1 = (N_1/N_2)v_2$,

$$v_1 i_1 + v_2 i_1 = 0$$

and hence the total power into the device is zero, which means that an ideal transformer is lossless.

Therefore, to summarize the dot convention for an ideal transformer,

$$v_1 = \frac{N_1}{N_2} v_2 \qquad\qquad \textbf{11.20}$$

where both voltages are referenced positive at the dots and

$$N_1 i_1 + N_2 i_2 = 0 \qquad \textbf{11.21}$$

where both currents are defined entering the dots.

Consider now the circuit shown in Fig. 11.13, where the symbol used for the transformer indicates that it is an iron-core transformer. Because of the relationship between the dots and both the assigned currents and voltages, the phasor voltages $\mathbf{V}_1$ and $\mathbf{V}_2$ are related by the expression

$$\frac{\mathbf{V}_1}{\mathbf{V}_2} = \frac{N_1}{N_2}$$

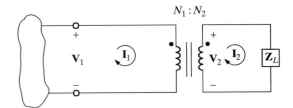

Figure 11.13 Ideal transformer circuit used to illustrate input impedance.

and the phasor currents, from Eq. (11.21), are related by

$$\frac{\mathbf{I}_1}{\mathbf{I}_2} = \frac{N_2}{N_1}$$

The sign in Eq. (11.21) is reversed, since the direction of $\mathbf{I}_2$ is reversed. These two preceding equations can be rewritten as

$$\mathbf{V}_1 = \frac{N_1}{N_2} \mathbf{V}_2$$

$$\mathbf{I}_1 = \frac{N_2}{N_1} \mathbf{I}_2 \qquad \textbf{11.22}$$

Also note that

$$\mathbf{S}_1 = \mathbf{V}_1 \mathbf{I}_1^* = \left(\frac{N_1}{N_2} \mathbf{V}_2 \right) \left(\frac{N_2}{N_1} \mathbf{I}_2 \right)^*$$

$$= \mathbf{V}_2 \mathbf{I}_2^* = \mathbf{S}_2$$

From the figure we note that $\mathbf{Z}_L = \mathbf{V}_2 / \mathbf{I}_2$, and therefore the input impedance

$$\mathbf{Z}_1 = \frac{\mathbf{V}_1}{\mathbf{I}_1} = \left(\frac{N_1}{N_2} \right)^2 \mathbf{Z}_L \qquad \textbf{11.23}$$

where $\mathbf{Z}_L$ is reflected into the primary side by the turns ratio.

If we now define the turns ratio as

$$n = \frac{N_2}{N_1} \qquad \textbf{11.24}$$

then the defining equations for the *ideal transformer* are

$$\mathbf{V}_1 = \frac{\mathbf{V}_2}{n^2}$$

$$\mathbf{I}_1 = n\mathbf{I}_2$$

$$\mathbf{S}_1 = \mathbf{S}_2 \qquad\qquad \textbf{11.25}$$

$$\mathbf{Z}_1 = \frac{\mathbf{Z}_L}{n^2}$$

Equations (11.25) define the important relationships for an ideal transformer. Care must be exercised in using these relationships because the signs on the voltages and currents are dependent on the assigned references and how they are related to the dots.

Just as the voltage-current relationship is specified in Ohm's law for a resistor, Fig. 11.14 and the following equations summarize the voltage and current relationships for the ideal transformer:

$$\frac{\mathbf{V}_1}{\mathbf{V}_2} = \frac{N_1}{N_2} \qquad\qquad \textbf{11.26}$$

$$N_1 \mathbf{I}_1 = N_2 \mathbf{I}_2$$

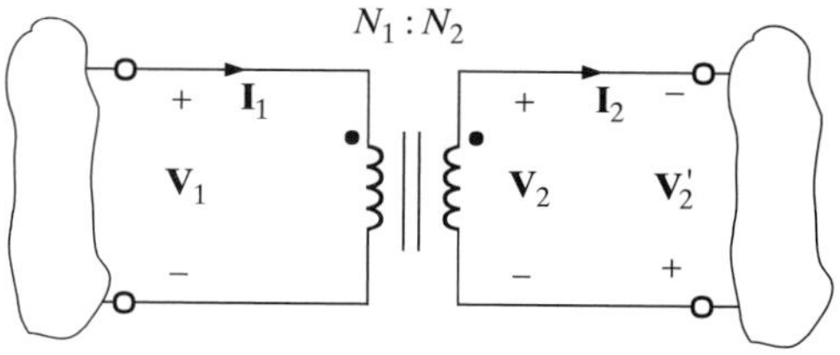

Figure 11.14 Ideal transformer current–voltage relationship.

Note carefully the directions of the currents and the positive direction of the voltages with respect to the dots. If any one of these directions change, then the variables in the equations are replaced by negative quantities. For example,

$$\frac{\mathbf{V}_1}{-\mathbf{V}_2'} = \frac{N_1}{N_2} \qquad\qquad \textbf{11.27}$$

Although practical transformers do not use dots, per se, they use markings specified by the National Electrical Manufacturers Association (NEMA) that are conceptually equivalent to the dots.

Tables 11.2 and 11.3 summarize the rules for determining the signs on the voltage and current in the ideal transformer equations.

The rules outlined in Tables 11.2 and 11.3 can be readily applied to analyze circuits containing ideal transformers. Other combinations not listed in the tables can be generalized by the reader.

Table 11.2 Rules for determining the signs in voltage equations

Equation	Voltage Assignment
$\dfrac{v_1}{v_2} = \dfrac{N_1}{N_2}$	
$\dfrac{v_1}{v_2} = -\dfrac{N_1}{N_2}$	

Table 11.3 Rules for determining the signs in current equations

Equation	Current Assignment
$\dfrac{i_1}{i_2} = \dfrac{N_2}{N_1}$	
$\dfrac{i_1}{i_2} = -\dfrac{N_2}{N_1}$	

Problem-Solving Strategy for Ideal Transformer Circuits

1. Use Tables 11.2 and 11.3 to determine the relationships among the currents and voltages based upon the location of the dots.

2. If the unknown(s) of interest is on the primary side, then reflect the circuit on the secondary side of the transformer to the primary; otherwise reflect the primary circuit to the secondary side. The resultant circuit does not contain a transformer.

3. Use the appropriate techniques to analyze the reflected circuit and determine the required unknown(s).

D11.3   Determine the voltage relationship for the following configurations.

1.

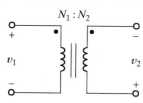

$N_1 : N_2$

ANSWER: $\dfrac{v_1}{v_2} = -\dfrac{N_1}{N_2}$

2.

$N_1 : N_2$

ANSWER: $\dfrac{v_1}{v_2} = -\dfrac{N_1}{N_2}$

3.

$N_1 : N_2$

ANSWER: $\dfrac{v_1}{v_2} = \dfrac{N_1}{N_2}$

D11.4   Determine the current relationship for the following configurations.

1.

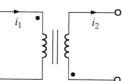

ANSWER: $\dfrac{i_1}{i_2} = -\dfrac{N_2}{N_1}$

2.

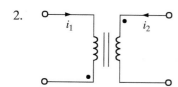

ANSWER: $\dfrac{i_1}{i_2} = \dfrac{N_2}{N_1}$

3.

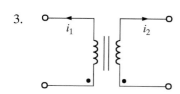

ANSWER: $\dfrac{i_1}{i_2} = -\dfrac{N_2}{N_1}$

EXAMPLE 11.8

Given the circuit shown in Fig. 11.15, we wish to determine all indicated voltages and currents.

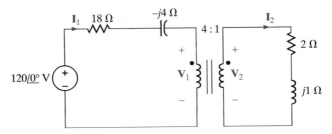

Figure 11.15 Ideal transformer circuit.

SOLUTION　Because of the relationships between the dots and the currents and voltages, the transformer equations are

$$\mathbf{V}_1 = -\frac{\mathbf{V}_2}{n} \quad \text{and} \quad \mathbf{I}_1 = -n\mathbf{I}_2$$

where $n = \frac{1}{4}$. The reflected impedance at the input to the transformer is

$$\mathbf{Z}_1 = 4^2 \mathbf{Z}_L = 16(2 + j1) = 32 + j16 \ \Omega$$

Therefore, the current in the source is

$$\mathbf{I}_1 = \frac{120 \ \underline{/0^\circ}}{18 - j4 + 32 + j16} = 2.33 \ \underline{/-13.5^\circ} \ A$$

The voltage across the input to the transformer is then

$$\begin{aligned}
\mathbf{V}_1 &= \mathbf{I}_1 \mathbf{Z}_1 \\
&= (2.33 \ \underline{/-13.5^\circ})(32 + j16) \\
&= 83.49 \ \underline{/13.07^\circ} \ V
\end{aligned}$$

Hence, $\mathbf{V}_2$ is

$$\begin{aligned}
\mathbf{V}_2 &= -n\mathbf{V}_1 \\
&= -\tfrac{1}{4}(83.49 \ \underline{/13.07^\circ}) \\
&= 20.87 \ \underline{/193.07^\circ} \ V
\end{aligned}$$

The current $\mathbf{I}_2$ is

$$\begin{aligned}
\mathbf{I}_2 &= -\frac{\mathbf{I}_1}{n} \\
&= -4(2.33 \ \underline{/-13.5^\circ}) \\
&= 9.33 \ \underline{/166.50^\circ} \ A
\end{aligned}$$

E11.6   Compute the current $\mathbf{I}_1$ in the network in Fig. E11.6.

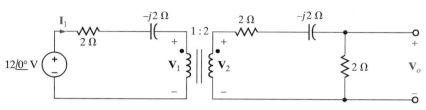

Figure E11.6

ANSWER: $\mathbf{I}_1 = 3.07 \underline{/39.81°}$ A.

E11.7   Find $\mathbf{V}_o$ in the network in Fig. E11.6.

ANSWER: $\mathbf{V}_o = 3.07 \underline{/39.81°}$ V.

E11.8   In the network in Fig. E11.8 the voltage $\mathbf{V}_o = 10 \underline{/0°}$ V. Find the input voltage $\mathbf{V}_S$.

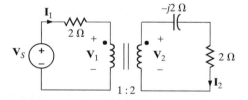

Figure E11.8

ANSWER: $V_S = 25.5 \underline{/-11.31°}$ V.

Another technique for simplifying the analysis of circuits containing an ideal transformer involves the use of either Thévenin's or Norton's theorems to obtain an equivalent circuit that replaces the transformer and either the primary or secondary circuit. This technique usually requires more effort, however, than the approach presented thus far. Let us demonstrate this approach by employing Thévenin's theorem to derive an equivalent circuit for the transformer and primary circuit of the network shown in Fig. 11.16a. The equations for the transformer in view of the direction of the currents and voltages and the position of the dots are

$$\mathbf{I}_1 = n\mathbf{I}_2$$

$$\mathbf{V}_1 = \frac{\mathbf{V}_2}{n}$$

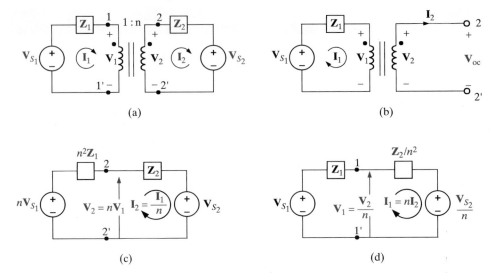

Figure 11.16 Circuit containing an ideal transformer and some of its equivalent networks.

Forming a Thévenin equivalent at the secondary terminals $2 = 2'$, as shown in Fig. 11.16b, we note that $\mathbf{I}_2 = 0$ and therefore $\mathbf{I}_1 = 0$. Hence

$$\mathbf{V}_{oc} = \mathbf{V}_2 = n\mathbf{V}_1 = n\mathbf{V}_{S_1}$$

The Thévenin equivalent impedance obtained by looking into the open-circuit terminals with $\mathbf{V}_{S_1}$ replaced by a short circuit is $\mathbf{Z}_1$, which when reflected into the secondary by the turns ratio is

$$\mathbf{Z}_{Th} = n^2\mathbf{Z}_1$$

Therefore, one of the resulting equivalent circuits for the network in Fig. 11.16a is as shown in Fig. 11.16c. In a similar manner we can show that replacing the transformer and its secondary circuit by an equivalent circuit results in the network shown in Fig. 11.16d.

It can be shown in general that when developing an equivalent circuit for the transformer and its primary circuit, each primary voltage is multiplied by n, each primary current is divided by n, and each primary impedance is multiplied by n^2. Similarly, when developing an equivalent circuit for the transformer and its secondary circuit, each secondary voltage is divided by n, each secondary current is multiplied by n, and each secondary impedance is divided by n^2. Powers are the same, whether calculated on the primary or secondary side.

The reader should recall from our previous analysis that if either dot on the transformer is reversed, then n is replaced by $-n$ in the equivalent circuits. In addition, it should be noted that the development of these equivalent circuits is predicated on the assumption that removing the transformer will divide the network into two parts; that is, there are no connections between the primary and secondary other than through the transformer. If any external connections exist, the equivalent circuit technique cannot in general be used. Finally, it should be pointed out that if the primary or secondary circuits are more complicated than those shown in Fig. 11.16a, Thévenin's theorem may be applied to reduce the network to that shown in Fig. 11.16a. Also, we can simply reflect the complicated circuit component by component from one side of the transformer to the other.

EXAMPLE 11.9

Given the circuit in Fig. 11.17a, we wish to draw the two networks obtained by replacing the transformer and the primary, and the transformer and the secondary, with equivalent circuits.

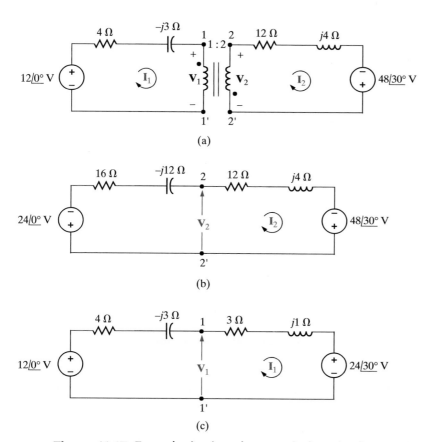

Figure 11.17 Example circuit and two equivalent circuits.

SOLUTION   Due to the relationship between the assigned currents and voltages and the location of the dots, the network containing an equivalent circuit for the primary, and the network containing an equivalent circuit for the secondary, are shown in Figs. 11.17b and c, respectively. The reader should note carefully the polarity of the voltage sources in the equivalent networks.   ❏

EXAMPLE 11.10

Let us determine the output voltage $\mathbf{V}_o$ in the circuit in Fig. 11.18a.

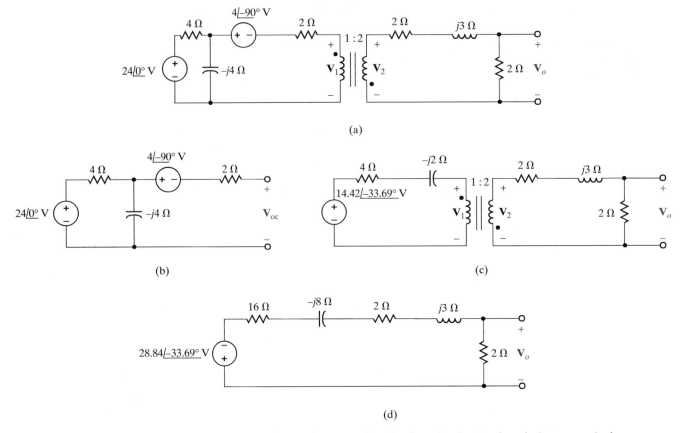

Figure 11.18 Example network and other circuits used to derive an equivalent network.

SOLUTION We begin our attack by forming a Thévenin equivalent for the primary circuit. From Fig. 11.18b we can show that the open-circuit voltage is

$$\mathbf{V}_{oc} = \frac{24\ \underline{/0^\circ}}{4 - j4}(-j4) - 4\ \underline{/-90^\circ}$$

$$= 12 - j8 = 14.42\ \underline{/-33.69^\circ}\ \text{V}$$

The Thévenin equivalent impedance looking into the open-circuit terminals with the voltage sources replaced by short circuits is

$$\mathbf{Z}_{\text{Th}} = \frac{(4)(-j4)}{4 - j4} + 2$$

$$= 4 - j2\ \Omega$$

The circuit in Fig. 11.18a thus reduces to that shown in Fig. 11.18c. Forming an equivalent circuit for the transformer and primary results in the network shown in Fig. 11.18d. Therefore, the voltage $\mathbf{V}_o$ is

$$\mathbf{V}_o = \frac{-28.84 \; \underline{/-33.69°}}{20 - j5} \quad (2)$$
$$= 2.80 \; \underline{/160.35°} \text{ V} \qquad \square$$

EXTENSION EXERCISES

E11.9   Given the network in Fig. E11.9, form an equivalent circuit for the transformer and secondary, and use the resultant network to compute $\mathbf{I}_1$.

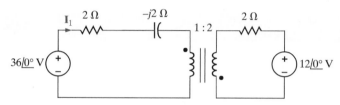

Figure E11.9

ANSWER: $\mathbf{I}_1 = 13.12 \; \underline{/38.66°}$ A.

E11.10   Given the network in Fig. E11.10, form an equivalent circuit for the transformer and primary, and use the resultant network to find $\mathbf{V}_o$.

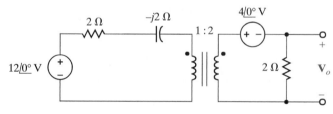

Figure E11.10

ANSWER: $\mathbf{V}_o = 13.12 \; \underline{/38.66°}$ V.

EXAMPLE 11.11

Determine $\mathbf{I}_1, \mathbf{I}_2, \mathbf{V}_1$, and $\mathbf{V}_2$ in the network in Fig. 11.19.

SOLUTION   The nodal equations at nodes 1 and 2 are

$$\frac{10 - \mathbf{V}_1}{2} = \frac{\mathbf{V}_1 - \mathbf{V}_2}{2} + \mathbf{I}_1$$

$$\mathbf{I}_2 + \frac{\mathbf{V}_1 - \mathbf{V}_2}{2} = \frac{\mathbf{V}_2}{2j}$$

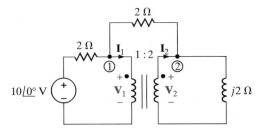

Figure 11.19

The transformer relationships are $\mathbf{V}_2 = 2\mathbf{V}_1$ and $\mathbf{I}_1 = 2\mathbf{I}_2$. The first nodal equation yields $\mathbf{I}_1 = 5\,A$ and therefore $\mathbf{I}_2 = 2.5\,A$. The second nodal equation, together with the contraint equations specified by the transformer, yields $\mathbf{V}_1 = \sqrt{5}\,\underline{/63°}\,V$ and $\mathbf{V}_2 = 2\sqrt{5}\,\underline{/63°}\,V$. ❏

EXTENSION EXERCISE

E11.11 Determine $\mathbf{I}_1, \mathbf{I}_2, \mathbf{V}_1$, and $\mathbf{V}_2$ in the network in Fig. E11.11.

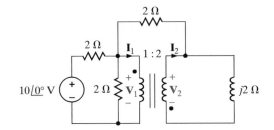

Figure E11.11

ANSWER: $\mathbf{I}_1 = 3.08\,\underline{/-13.71°}\,A,$ $\qquad \mathbf{I}_2 = 1.54\,\underline{/166.29°}\,A$

$\mathbf{V}_1 = 0.85\,\underline{/-19.98°}\,V,$ $\qquad \mathbf{V}_2 = 1.71\,\underline{/-160.02°}\,V.$

11.4 SAFETY CONSIDERATIONS

Transistors are used extensively in modern electronic equipment to provide a low-voltage power supply. As examples, in computer systems a common voltage level is 5 V dc, portable radios use 9 V dc, and military and airplane equipment operates at 28 V dc. When transformers are used to connect these low-voltage transistor circuits to the power line, there is generally less danger of shock within the system because the transformer provides electrical isolation from the line voltage. However, from a safety standpoint, a transformer, although helpful in many situations, is not an absolute solution. We must always be vigilant when working with any electrical equipment to minimize the dangers of electrical shock.

In power electronics equipment or power systems the danger is severe. The problem in these cases is that of high voltage from a low-impedance source, and we must constantly remember that the line voltage in our homes can be lethal.

Consider now the following example, which illustrates a hidden danger that could surprise even the experienced professional, with devastating consequences.

EXAMPLE 11.12

Two adjacent homes, A and B, are fed from different transformers, as shown in Fig.
11.20a. A surge on the line feeding house B has caused the circuit breaker X-Y to
open. House B is now left without power. In an attempt to help his neighbor, the res-
ident of house A volunteers to connect a long extension cord between a wall plug
in house A and a wall plug in house B, as shown in Fig. 11.20b. Later, the line technician from the utility company comes to reconnect the circuit breaker. Is the line
technician in any danger in this situation?

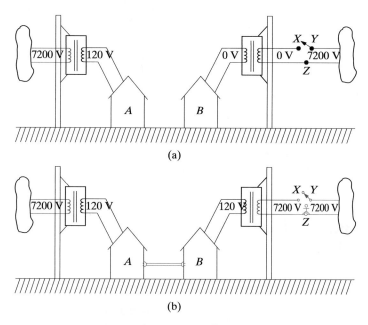

Figure 11.20 Diagrams used in Example 11.12.

SOLUTION   Unaware of the extension cord connection, the line techinician be-
lieves that there is no voltage between points X and Z. However, because of the
electrical connection between the two homes, 7200 V exists between the two points,
and the line techinician could be seriously injured or even killed if he comes in contact with this high voltage.   ❏

11.5 APPLICATIONS

The following examples demonstrate several applications for transformers. We will first
revisit the problem of transmitting power over long distances and illustrate the important
role the transformer plays in that application. We then examine the use of a transformer

in distributing power to our homes and, finally, consider its use in recharging the battery in a hand-held calculator.

EXAMPLE 11.13

Let us investigate two ways to deliver 120-MW three-phase power from a generating station a distance of 200 km to a 240-V single-phase load with at least 97% efficiency. We will consider two cases: without and with a transformer.

SOLUTION

Case 1.   Since the single-phase power is 40 MW, the current per phase is

$$I = \frac{P_{1\phi}}{V} = \frac{40M}{240} = 166.67 \text{ kA}$$

as shown in Fig. 11.21. Given a load voltage of 240 V, the generator voltage must be no more than

$$V_S = \frac{V_{\text{load}}}{0.97} = 247.42 \text{ V}$$

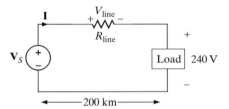

Figure 11.21 Single-phase model for Case 1.

and the voltage drop due to line resistance no more than

$$V_{\text{line}} = (0.03)V_S = 7.42 \text{ V}$$

Therefore, the line resistance is

$$R_{\text{line}} = \frac{V_{\text{line}}}{I} = 44.54 \ \mu\Omega$$

over the 200-km distance. The line resistance can be written as

$$R_{\text{line}} = \rho \frac{L}{A}$$

where ρ is the resistivity of the cable, L is its length (200 km), and A is its cross-sectional area. Typical values of ρ are approximately 8 $\mu\Omega$-cm. Given these values, the required cable has a cross-sectional area of

$$A = \rho \frac{L}{R_{\text{line}}}$$

or

$$A = 359.2 \text{ m}^2$$

which corresponds to a diameter of 21.4 m or 70 feet! Obviously, this is not feasible.

Case 2.   Let us try transmitting the power at 200 kV and reducing the voltage to 240 V with a step-down transformer at the load, as shown in Fig. 11.22. Now the per phase current is only

$$I = \frac{P_{1\phi}}{V} = \frac{40\text{M}}{200\text{k}} = 200 \text{ A}$$

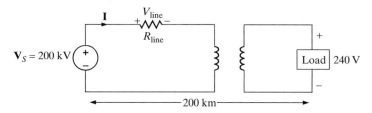

Figure 11.22 Single-phase model for Case 2.

If we employ a standard transmission cable that has a cross-sectional area of 1.75 in^2 and a resistance per unit length of 0.07 Ω/km, the voltage drop on the line is

$$V_{\text{line}} = R_{\text{line}} I = (0.07)(200)(200) = 2800 \text{ V}$$

The efficiency of the transmission is

$$\text{Efficiency} = \left(\frac{200\text{k} - 2.8\text{k}}{200\text{k}} \right) 100\% = 98.6\%$$

Now we have a very efficient (economical) transmission of power using a relatively small and therefore inexpensive cable. This is why power utilities use high voltages for the distribution of power. ❑

EXAMPLE 11.14

The local transformer in Fig. 11.23 provides the last voltage step-down in a power distribution system. A common sight on utility poles in residential areas, it is a single-phase transformer that typically has 13.8 kV line to neutral on its primary coil, and a center tap secondary coil provides both 120 V and 240 V to service several residences. Let us find the turns ratio necessary to produce the 240-V secondary voltage. Assuming that the transformer provides 200-A service to each of 10 houses, let us determine the minimum power rating for the transformer and the maximum current in the primary.

SOLUTION   The turns ratio is given by

$$n = \frac{V_2}{V_1} = \frac{240}{13,800} = \frac{1}{57.5}$$

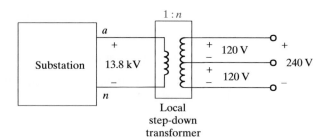

Figure 11.23 Local transformer subcircuit with center tap.

If I_H is the maximum current per household, then the maximum primary current is

$$I_1 = nI_2 = n(10I_H) = 34.78 \text{ A}$$

The maximum power delivered to the primary is then

$$S_1 = V_1 I_1 = (13{,}800)(34.78) = 480 \text{ kVA}.$$

Therefore, the transformer must have a power rating of at least 480 kVA.   ❏

EXAMPLE 11.15

A simplified circuit diagram for a calculator recharger is shown in Fig. 11.24. It operates on 120 V rms at 60 Hz and produces a dc output of 9 V. The dc conversion is done by the peak detector subcircuit, which simply outputs a dc voltage equal to the peak value of the ac signal entering it. Let us determine the required turns ratio for the transformer.

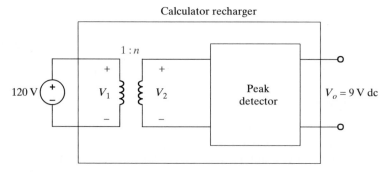

Figure 11.24 Simplified diagram for calculator recharger.

SOLUTION   For an output of 9 V, the input to the peak detector must be 9 V, or

$$v_2(t) = 9 \sin(120\pi t)$$

which has an rms value of

$$V_2 = \frac{V_P}{\sqrt{2}} = 6.36 \text{ V rms}$$

Therefore, the required turns ratio is

$$n = \frac{V_2}{V_1} = \frac{6.36}{120} = \frac{1}{18.86}$$

❏

11.6 CIRCUIT DESIGN

The following simple example illustrates a technique for employing a transformer in a configuration that will extend the life of a set of Christmas tree lights.

EXAMPLE 11.16

The bulbs in a set of Christmas tree lights normally operate at 120 V rms. However, they last much longer if they are instead connected to 108 V rms. Using a 120-V–12-V transformer, let us design an autotransformer that will provide 108 V rms to the bulbs.

SOLUTION   The two-winding transformers we have presented thus far provide electrical isolation between primary and secondary windings, as shown in Fig. 11.25a.

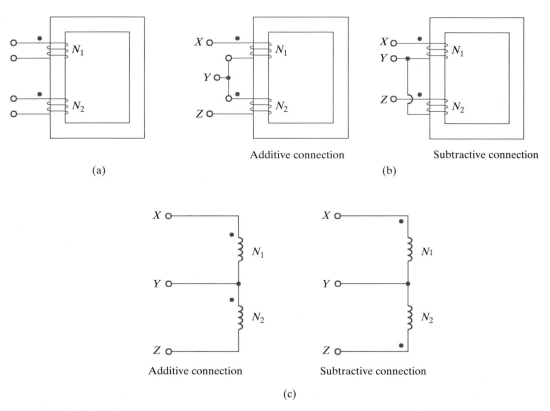

Figure 11.25 Autotransformer: (a) normal two-winding transformer with adjacent windings; (b) two-winding transformer interconnected to create a single-winding, three-terminal autotransformer; (c) symbolic representation of (b).

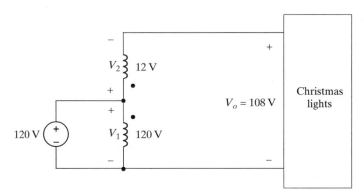

Figure 11.26 Autotransformer for low-voltage Christmas tree lights.

It is possible, however, to interconnect primary and secondary windings serially, creating a three-terminal device, known as an autotransformer, as shown in Fig. 11.25b and represented in Fig. 11.25c. As well shall see, this arrangement offers certain practical advantages over the isolated case. Note that the three-terminal arrangement is essentially one continuous winding with an internal tap.

To reduce the voltage from 120 V to 108 V requires that the two coils be connected such that their voltages are in opposition to each other, corresponding to a substractive connector (in Fig. 11.25b) as shown in Fig. 11.26. In this arrangement, the voltage across both coils is

$$V_o = V_1 - V_2 = 120 - 12 = 108 \text{ V rms}$$

and the lights are simply connected across both coils.

11.7 SUMMARY

Mutual inductance   Mutual inductance occurs when inductors are placed in close proximity to one another and share a common magnetic flux.

The dot convention for mutual inductance   The dot convention governs the sign of the induced voltage in one coil based upon the current direction in another. Table 11.1 indicates that if a current enters a coil at the dotted terminal, the induced voltage has a positive sign on the dotted end of the other coil. Likewise, if the current leaves a coil from the dotted end, the induced voltage has a negative sign on the dotted end of the other coil.

The relationship between the mutual inductance and self-inductance of two coils   An energy analysis indicates that $M = k\sqrt{L_1 L_2}$, where k, the coefficient of coupling, has a value between 0 and 1.

The ideal transformer   An ideal transformer has infinite core permeability and winding conductance. The voltage and current can be transformed between the primary and secondary ends based on the ratio of the number of winding turns between the primary and secondary.

The dot convention for an ideal transformer   Tables 11.2 and 11.3 summarize the rules for determining the signs on the voltage and current in ideal transformer equations. If both voltages have the same polarity on the dotted ends, a positive sign is used in the equation; otherwise a negative sign is used. Likewise, if both currents enter or leave the dotted ends simultaneously, a negative sign should be used; otherwise a positive sign is used.

Equivalent circuits involving ideal transformers   Based upon the location of the circuits unknowns, either the primary or secondary can be reflected to the other side of the transformer in order to form a single circuit containing the desired unknown. The reflected voltages, currents, and impedances are a function of the dot convention and turn ratio.

PROBLEMS

Section 11.1

11.1   Given the network in Fig. P11.1, (a) write the equations for $v_a(t)$ and $v_b(t)$, and (b) write the equations for $v_c(t)$ and $v_d(t)$.

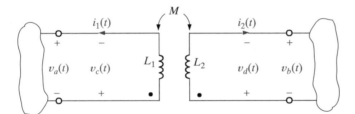

Table 11.1 is helpful in problems of this type.

Figure P11.1

11.2   Given the network in Fig. P11.2, (a) write the equations for $v_a(t)$ and $v_b(t)$, and (b) write the equations for $v_c(t)$ and $v_d(t)$.

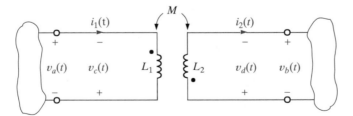

Similar to P11.1

Figure P11.2

11.3 Given the network in Fig. P11.3, (a) write the equations for $v_a(t)$ and $v_b(t)$, and (b) write the equations for $v_c(t)$ and $v_d(t)$.

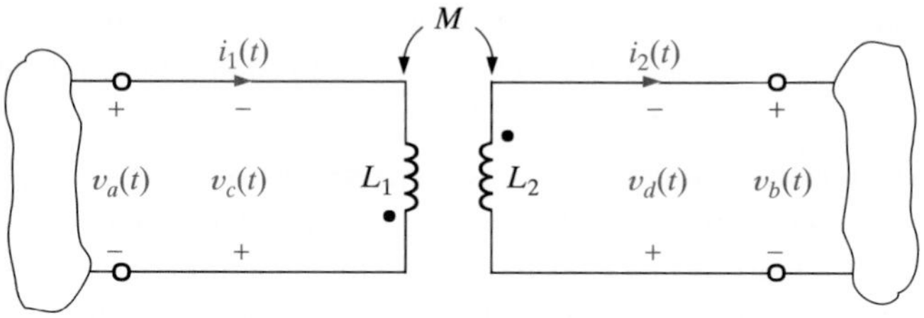

Similar to P11.1

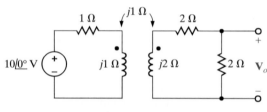

Figure P11.3

11.4 Find $\mathbf{V}_o$ in the network in Fig. P11.4.

Loop equations in conjunction with Table 11.1 will yield the solution.

Figure P11.4

11.5 Find $\mathbf{V}_o$ in the circuit in Fig. P11.5.

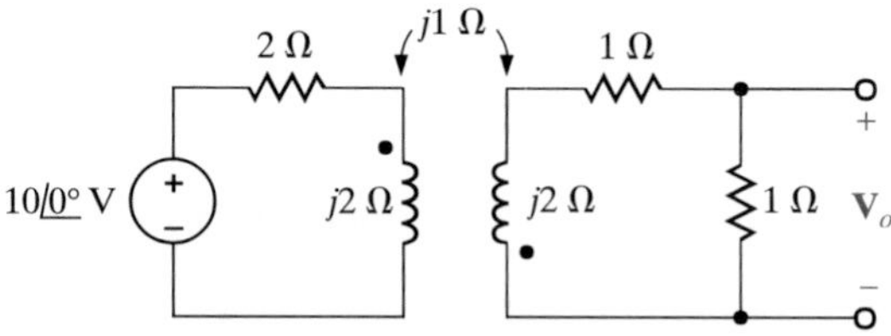

Similar to P11.4

Figure P11.5

11.6 Given the network in Fig. P11.6, find $\mathbf{V}_o$.

Similar to P11.4

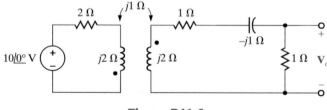

Figure P11.6

11.7   Find $\mathbf{V}_o$ in the network in Fig. P11.7.

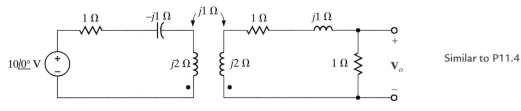

Figure P11.7

Similar to P11.4

11.8   Find $\mathbf{V}_o$ in the circuit in Fig. P11.8.

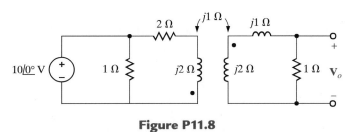

Figure P11.8

Similar to P11.4. The use of source exchange is also helpful.

11.9   Write the mesh equations for the network in Fig. 11.9.

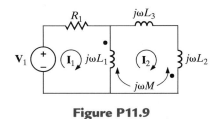

Figure P11.9

The use of Table 11.1 is helpful.

11.10   Write the mesh equations for the network in Fig. 11.10.

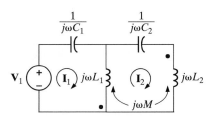

Figure P11.10

Similar to P11.9

11.11 Write the mesh equations for the network in Fig. 11.11 and determine $\mathbf{V}_o/\mathbf{V}_1$.

Similar to P11.9

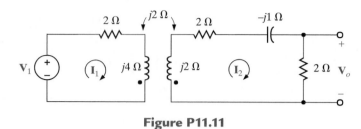

Figure P11.11

11.12 Write the mesh equations for the network shown in Fig. P11.12.

Similar to P11.9

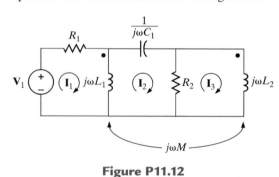

Figure P11.12

11.13 Write the mesh equations for the network shown in Fig. P11.13.

Similar to P11.9

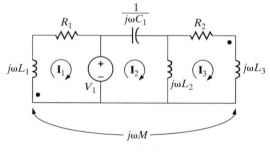

Figure P11.13

11.14 Find $\mathbf{V}_o$ in the network in Fig. P11.14.

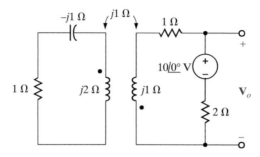

Figure P11.14

Similar to P11.8

11.15 Find $\mathbf{V}_o$ in the network in Fig. P11.15.

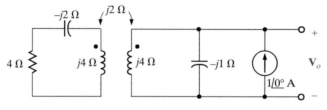

Figure P11.15

Similar to P11.8

11.16 Find $\mathbf{V}_o$ in the network in Fig. P11.16.

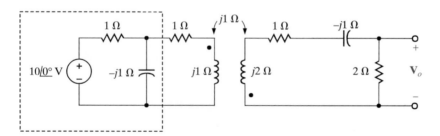

Figure P11.16

Similar to P11.4 if the dashed block is replaced by its Thévenin equivalent network.

11.17 Find $\mathbf{V}_o$ in the network in Fig. P11.17.

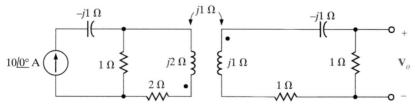

Figure P11.17

Similar to P11.16

11.18 Find $\mathbf{V}_o$ in the network in Fig. P11.18.

Similar to P11.16

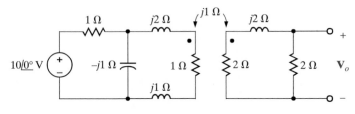

Figure P11.18

11.19 Find $\mathbf{V}_o$ in the network in Fig. P11.19.

Similar to P11.16

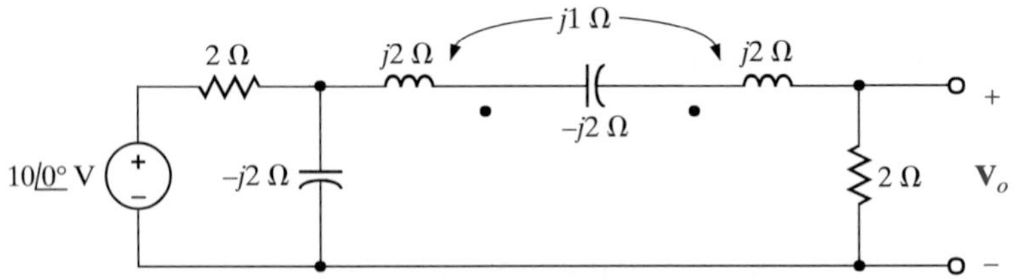

Figure P11.19

11.20 Find $\mathbf{V}_o$ in the network in Fig. P11.20.

Similar to P11.16

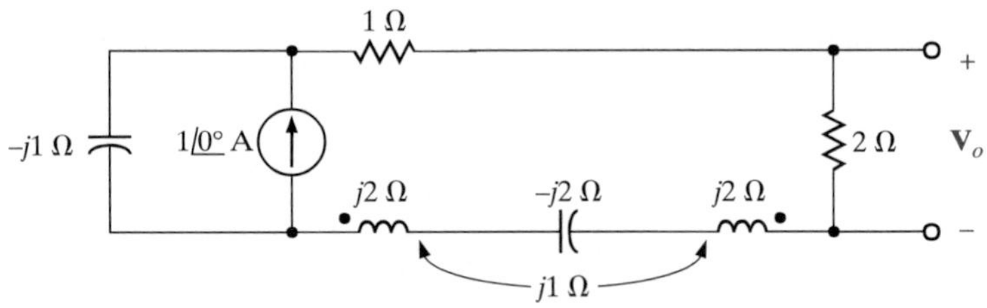

Figure P11.20

11.21 Find $\mathbf{V}_o$ in the network in Fig. P11.21.

Similar to P11.16

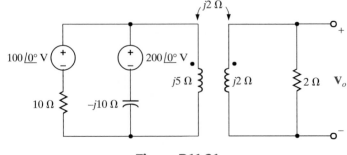

Figure P11.21

11.22 Find $\mathbf{V}_o$ in the network in Fig. P11.22.

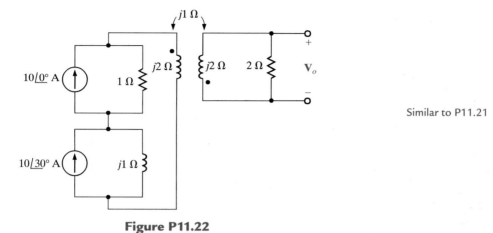

Similar to P11.21

Figure P11.22

11.23 Find $\mathbf{V}_o$ in the network in Fig. P11.23.

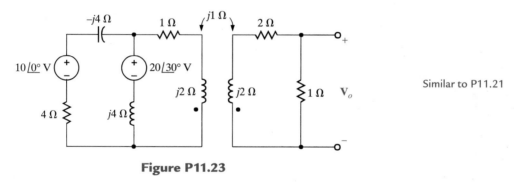

Similar to P11.21

Figure P11.23

11.24 Find $\mathbf{V}_o$ in the network in Fig. P11.24.

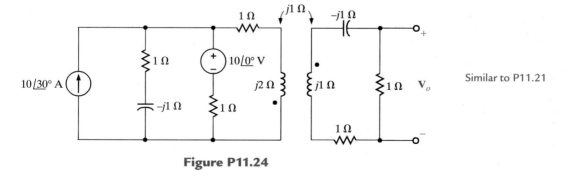

Similar to P11.21

Figure P11.24

11.25 Find $\mathbf{V}_o$ in the network in Fig. P11.25.

Similar to P11.21

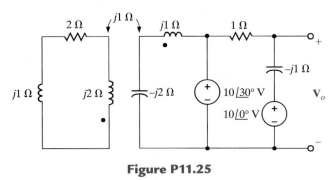

Figure P11.25

11.26 Find $\mathbf{V}_o$ in the network in Fig. P11.26.

Similar to P11.4

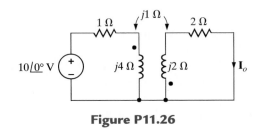

Figure P11.26

11.27 Find $\mathbf{I}_o$ in the circuit in Fig. P11.27.

Similar to P11.8

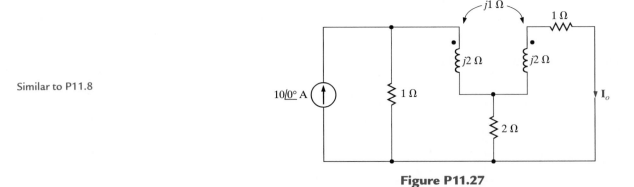

Figure P11.27

11.28 Find $\mathbf{V}_o$ in the circuit in Fig. P11.28.

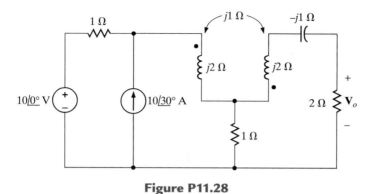

Figure P11.28

Similar to P11.8

Section 11.2

11.29 Two coils in a network are positioned such that there is 100% coupling between them. If the inductance of one coil is 10 mH and the mutual inductance is 6 mH, compute the inductance of the other coil.

Equation 11.13 is useful in this case.

11.30 Two coils are positioned such that there is 90% coupling between them. If the inductances of the coils are 10 mH and 20 mH, respectively, find the mutual inductance.

Similar to P11.29

11.31 The currents in the network in Fig. P11.31 are known to be $i_1(t) = 10\cos(377t - 30°)$ mA and $i_2(t) = 20\cos(377t - 45°)$ mA. The inductances are $L_1 = 2$ H and $L_2 = 2$ H and $k = 0.8$. Determine $v_1(t)$ and $v_2(t)$.

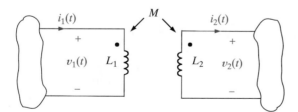

Figure P11.31

Once M is known, the KVL equations can be written.

11.32 If $L_1 = L_2 = 4\,\text{H}$ and $k = 0.8$, find $i_1(t)$ and $i_2(t)$ in the circuit in Fig. P11.32.

Similar to P11.31

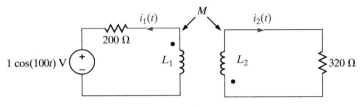

Figure P11.32

See Example 11.7.

11.33 Determine the energy stored in the coupled inductors in the circuit in P11.31 at $t = 1\,\text{ms}$.

Similar to P11.33

11.34 Determine the energy stored in the coupled inductors in the network in P11.32 at $t = 2\,\text{ms}$.

Section 11.3

11.35 Determine the input impedance seen by the source in the circuit in Fig. P11.35.

If the secondary is reflected to the primary, the input impedance can be obtained easily.

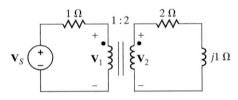

Figure P11.35

11.36 Determine the input impedance seen by the source in the circuit in Fig. P11.36.

Similar to P11.35

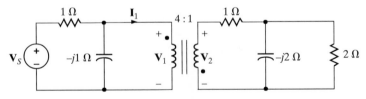

Figure P11.36

11.37 Determine I_1, I_2, V_1, and V_2 in the network in Fig. P11.37.

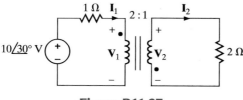

Figure P11.37

V_1 and I_1 can be determined by reflecting the secondary to the primary. Then reflecting the primary to the secondary will yield I_2 and V_2. The formulas that relate V_1 and I_1 to V_2 and I_2 can also be used.

11.38 Determine I_1, I_2, V_1, and V_2 in the network in Fig. P11.38.

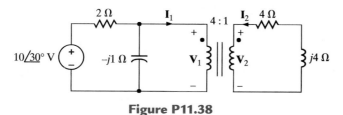

Figure P11.38

Similar to P11.37

11.39 Determine I_1, I_2, V_1, and V_2 in the network in Fig. P11.39.

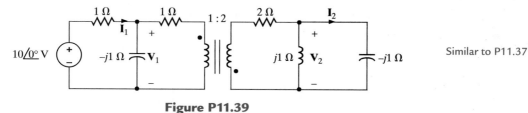

Figure P11.39

Similar to P11.37

11.40 Determine I_1, I_2, V_1, and V_2 in the network in Fig. P11.40.

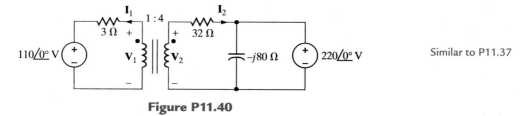

Figure P11.40

Similar to P11.37

11.41 Determine $\mathbf{I}_1$, $\mathbf{I}_2$, $\mathbf{V}_1$, and $\mathbf{V}_2$ in the network in Fig. P11.41.

Similar to P11.37 with the use
of source exchange

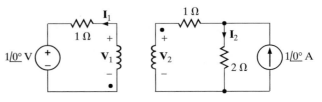

Figure P11.41

11.42 Determine $\mathbf{V}_o$ in the circuit in Fig. P11.42.

Similar to P11.37

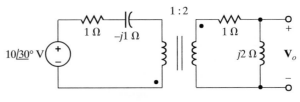

Figure P11.42

11.43 Determine $\mathbf{V}_S$ in the circuit in Fig. P11.43.

Similar to P11.37

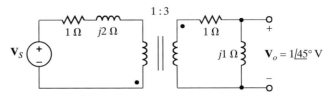

Figure P11.43

11.44 Determine $\mathbf{I}_S$ in the circuit in Fig. P11.44.

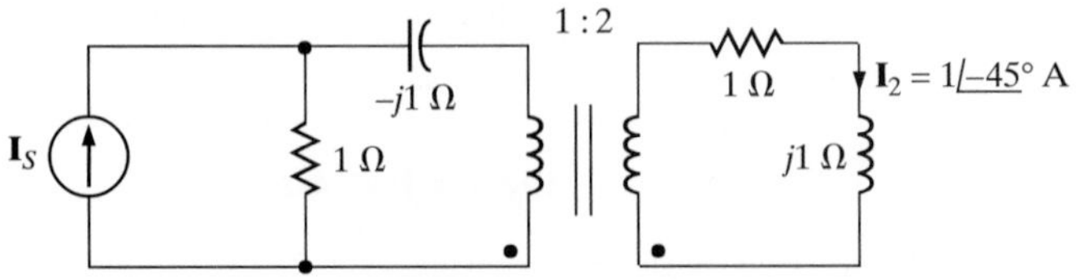

Similar to P11.37

Figure P11.44

11.45 The output stage of an amplifier in an old radio is to be matched to the impedance of a speaker, as shown in Fig. P11.45. If the impedance of the speaker is 8 Ω and the amplifier requires a load impedance of 3.2 kΩ, determine the turns ratio of the ideal transformer.

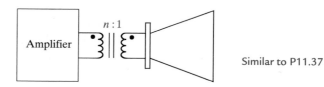

Similar to P11.37

Figure P11.45

11.46 Derive an equivalent network for the circuit in Fig. P11.46 and use it to determine the current I_1.

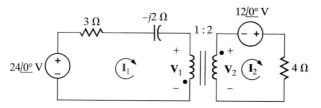

Similar to P11.37

Figure P11.46

11.47 Form an equivalent circuit for the transformer and the primary in the network shown in Fig. P11.47 and use it to determine I_2.

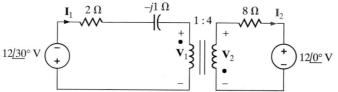

Similar to P11.37

Figure P11.47

11.48 Find **I** in the network in Fig. P.11.48.

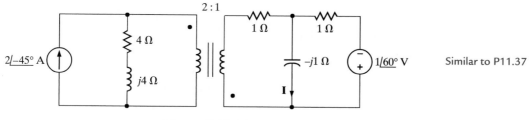

Similar to P11.37

Figure P11.48

Advanced Problems

11.49 Write the mesh equations for the network shown in Fig. P11.49.

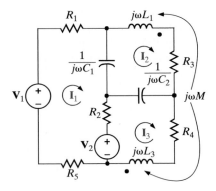

Figure P11.49

11.50 Write the mesh equations for the network shown in Fig. P11.50.

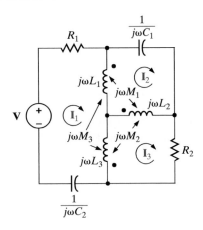

Figure P11.50

11.51 Find I_o in the circuit in Fig. P11.51.

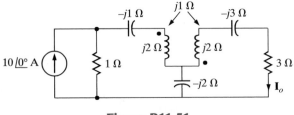

Figure P11.51

11.52 Find V_o in the network in Fig. P11.52.

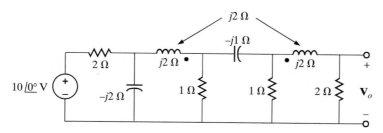

Figure P11.52

11.53 Find V_o in the network in Fig. P11.53.

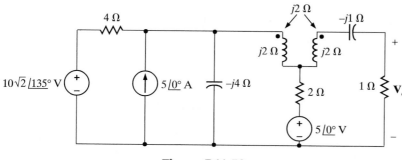

Figure P11.53

11.54 Find V_o in the network in Fig. P11.54.

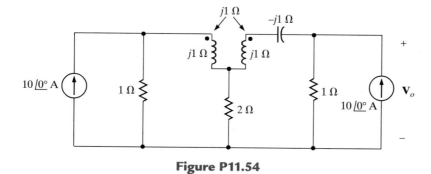

Figure P11.54

11.55 Determine the impedance seen by the source in the network shown in Fig. P11.55.

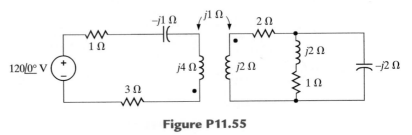

Figure P11.55

11.56 Determine the impedance seen by the source in the network shown in Fig. P11.56.

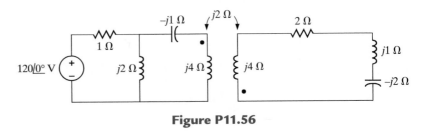

Figure P11.56

11.57 Determine the impedance seen by the source in the network shown in Fig. P11.57.

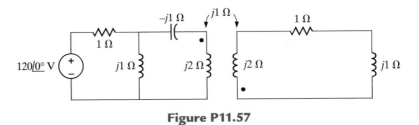

Figure P11.57

11.58 Determine the input impedance Z_{in} of the circuit in Fig. P11.58.

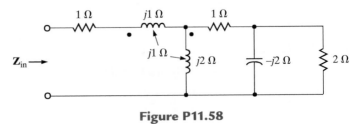

Figure P11.58

11.59 Determine the input impedance $\mathbf{Z}_{in}$ in the network in Fig. P11.59.

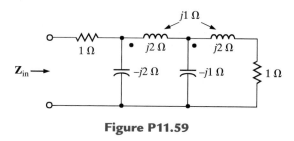

Figure P11.59

11.60 Given the network shown in Fig. P11.60, determine the value of the capacitor C that will cause the impedance to be purely resistive. $f = 60$ Hz.

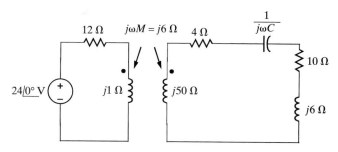

Figure P11.60

11.61 Analyze the network in Fig. P11.61 and determine if a value of X_C can be found such that the output voltage is equal to twice the input voltage.

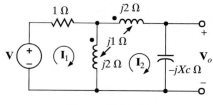

Figure P11.61

11.62 Determine I_o in the network in Fig. P11.62.

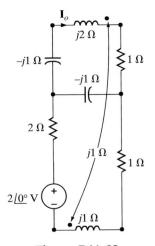

Figure P11.62

11.63 Determine the input impedance seen by the source in the network shown in Fig. P11.63.

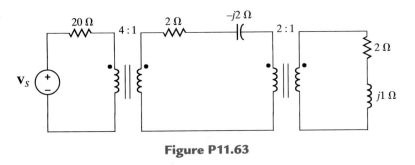

Figure P11.63

11.64 Determine the input impedance seen by the source in the network shown in Fig. P11.64.

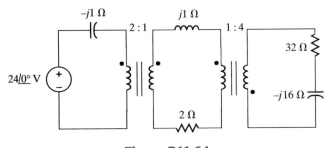

Figure P11.64

11.65 Find the current **I** in the network in Fig. P11.65.

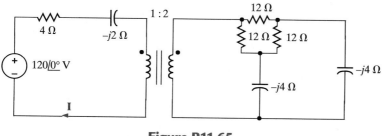

Figure P11.65

11.66 Find the current **I** in the network in Fig. P11.66.

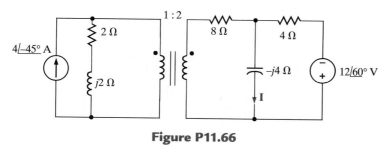

Figure P11.66

11.67 Find the voltage **V**$_o$ in the network in Fig. P11.67.

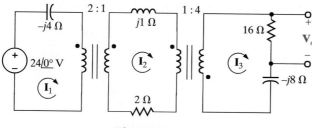

Figure P11.67

11.68 Find $\mathbf{V}_o$ in the circuit in Fig. P11.68.

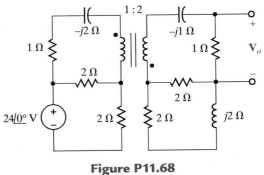

Figure P11.68

11.69 In the circuit in Fig. P11.69, if $\mathbf{I}_x = 6\ \underline{/45°}$ A, find $\mathbf{V}_o$.

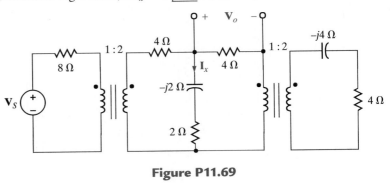

Figure P11.69

11.70 In the network in Fig. P11.70, if $\mathbf{I}_1 = 4\ \underline{/0°}$ A, find $\mathbf{V}_S$.

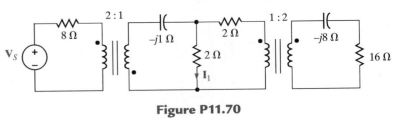

Figure P11.70

11.71 Find $\mathbf{V}_o$ in the network in Fig. P11.71.

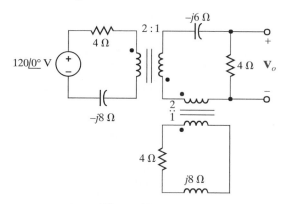

Figure P11.71

11.72 Determine the average power delivered to each resistor in the network shown in Fig. P11.72.

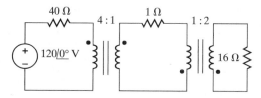

Figure P11.72

11.73 The diagram in Fig. P11.73 represents a power distribution system from the generating plant to a single residence. Power is generated at 20,000 V and then proceeds through several transformers before reaching the consumer. Find the required turns ratio for each transformer.

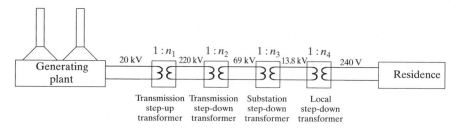

Figure P11.73

11.74 Find $\mathbf{V}_o$ in the circuit in Fig. P11.74.

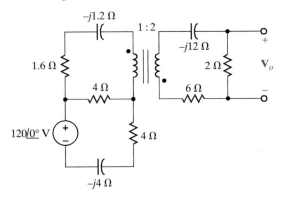

Figure P11.74

11.75 In the network shown in Fig. P11.75, determine the value of the load impedance for maximum power transfer.

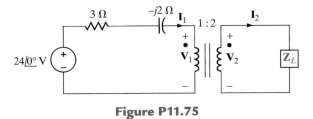

Figure P11.75

11.76 In the circuit shown in Fig. P11.76, if $\mathbf{V}_2 = 120\underline{/0°}$ V rms and the load $\mathbf{Z}_L$ absorbs 400 W at 0.9 pf lagging, determine the wattmeter reading.

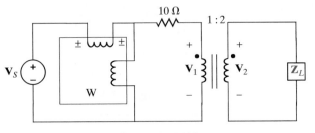

Figure P11.76

11.77 In the network shown in Fig. P11.77, $\mathbf{V}_L = 120\underline{/0°}$ V rms. Z_L absorbs 500 W at 0.85 pf lagging. Find $\mathbf{V}_S$.

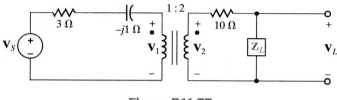

Figure P11.77

11.78 Find $\mathbf{V}_S$ in the network in Fig. P11.78.

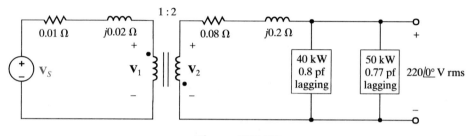

Figure P11.78

11.79 Find $\mathbf{V}_S$ in the network in Fig. P11.79.

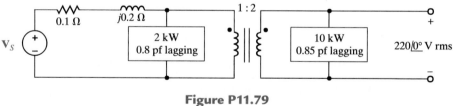

Figure P11.79

11.80 Determine the wattmeter reading in the network in Fig. P11.80.

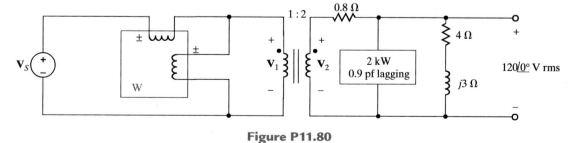

Figure P11.80

11.81 Determine the input impedance of the circuit shown in Fig. P11.81.

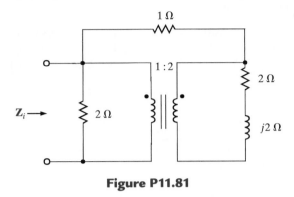

Figure P11.81

11.82 Given the circuit in Fig. P11.82, we wish to find the value of the load R_o for maximum power transfer and the value of the maximum power delivered to this load.

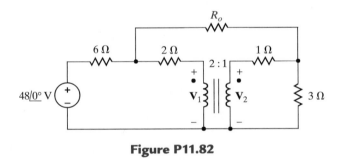

Figure P11.82

11.83 Find $\mathbf{V}_o$ in the network in Fig. P11.83.

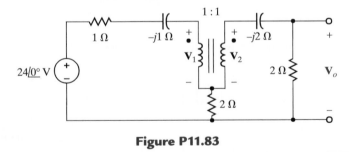

Figure P11.83

Variable-Frequency Network Performance

12

*I*n Chapter 8 we demonstrated that a network containing a capacitor and an inductor operated differently if the frequency was changed from the U.S. power grid frequency of 60 Hz to the aircraft frequency of 400 Hz. This phenomenon, while not surprising since the impedance of both these circuit elements is frequency dependent, indicates that if the frequency of the network sources is varied over some range, we can also expect the network to undergo variations in response to these changes in frequency.

Consider for a moment your stereo amplifier. The input signal contains sound waves of frequencies that range from soup to nuts; and yet, the amplifier must amplify each frequency component exactly the same amount in order to achieve perfect sound reproduction. Achieving perfect sound reproduction is a nontrivial task; and when you buy a very good amplifier, part of the price reflects the design necessary to achieve constant amplification over a wide range of frequencies.

In this chapter we will examine the performance of electrical networks when excited from variable-frequency sources. Such effects are important in the analysis and design of networks such as filters, tuners, and amplifiers which have wide application in communication and control systems. Terminology and techniques for frequency-response analysis are introduced including standard plots (Bode plots) for describing network performance graphically. In particular, Bode plot construction and interpretation are discussed in detail.

The concept of resonance is introduced in reference to frequency selectivity and tuning. The various parameters used to define selectivity such as bandwidth, cutoff frequency, and quality factor are defined and discussed. Network scaling for both magnitude and phase is also presented.

Networks with special filtering properties are examined. Specifically, low-pass, high-pass, band-pass, and band-elimination filters are discussed. Techniques for designing active filters (containing op-amps) are presented.

12.1 VARIABLE FREQUENCY-RESPONSE ANALYSIS

In previous chapters we investigated the response of RLC networks to sinusoidal inputs. In particular, we considered 60-Hz sinusoidal inputs. In this chapter we will allow the frequency of excitation to become a variable and evaluate network performance as a function of frequency. To begin, let us consider the effect of varying frequency on elements with which we are already quite familiar—the resistor, inductor, and capacitor. The frequency-domain impedance of the resistor shown in Fig. 12.1a is

$$\mathbf{Z}_R = R = R \, \underline{/0°}$$

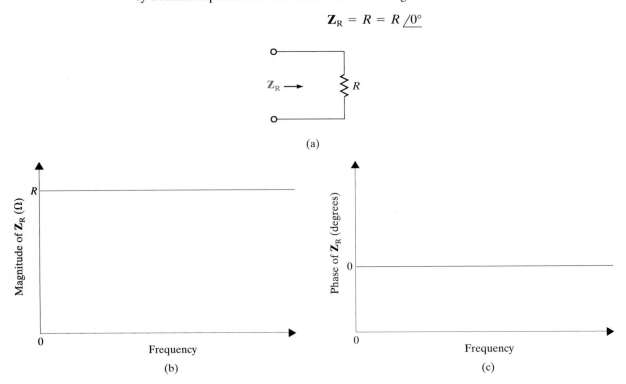

(a)

(b)
(c)

Figure 12.1 Frequency-dependent impedance of a resistor.

The magnitude and phase are constant and independent of frequency. Sketches of the magnitude and phase of $\mathbf{Z}_R$ are shown in Figs. 12.1b and c. Obviously, this is a very simple situation.

For the inductor in Fig. 12.2a, the frequency-domain impedance $\mathbf{Z}_L$ is

$$\mathbf{Z}_L = j\omega L = \omega L \, \underline{/90°}$$

The phase is constant at 90° but the magnitude of $\mathbf{Z}_L$ is directly proportional to frequency. Figures 12.2b and c show sketches of the magnitude and phase of $\mathbf{Z}_L$ versus frequency. Notice that at low frequencies the inductor's impedance is quite small. In fact, at dc, $\mathbf{Z}_L$ is zero and the inductor appears as a short circuit. Conversely, as frequency increases, the impedance also increases.

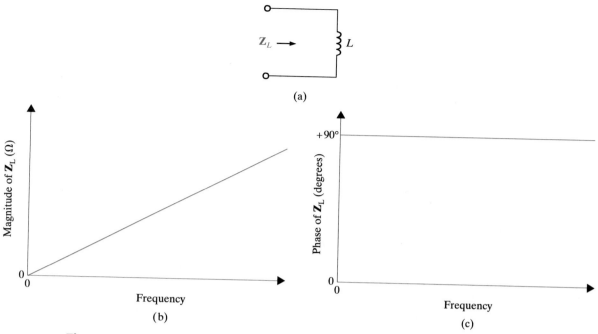

Figure 12.2 Frequency-dependent impedance of an inductor.

Next consider the capacitor of Fig. 12.3a. The impedance is

$$\mathbf{Z}_C = \frac{1}{j\omega C} = \frac{1}{\omega C} \underline{/-90°}$$

Once again the phase of the impedance is constant but now the magnitude is inversely proportional to frequency, as shown in Figs. 12.3b and c. Notice that the impedance approaches infinity, or an open circuit, as ω approaches zero and $\mathbf{Z}_C$ approaches zero as ω approaches infinity.

Now let us investigate a more complex circuit: the RLC series network in Fig. 12.4a. The equivalent impedance is

$$\mathbf{Z}_{eq} = R + j\omega L + \frac{1}{j\omega C}$$

or

$$\mathbf{Z}_{eq} = \frac{(j\omega)^2 LC + j\omega RC + 1}{j\omega C}$$

Sketches of the magnitude and phase of this function are shown in Figs. 12.4b and c.

Note that at very low frequencies, the capacitor appears as an open circuit and, therefore, the impedance is very large in this range. At high frequencies, the capacitor has very little effect and the impedance is dominated by the inductor, whose impedance keeps rising with frequency.

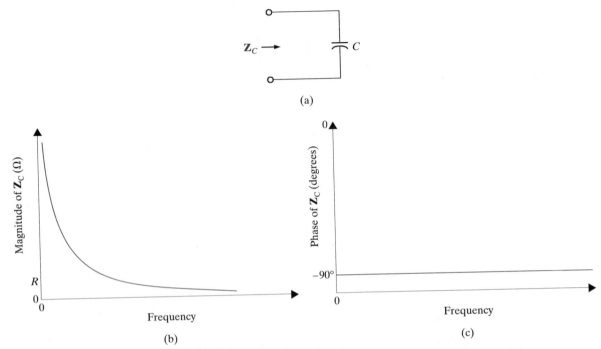

Figure 12.3 Frequency-dependent impedance of a capacitor.

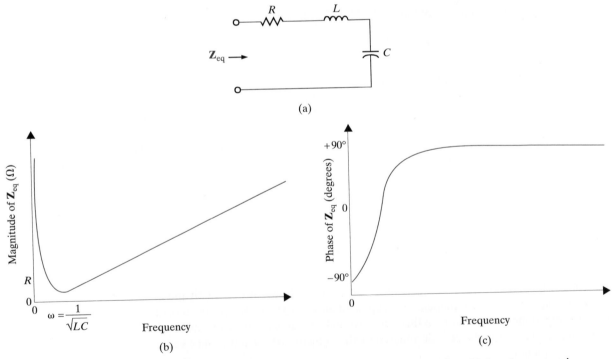

Figure 12.4 Frequency-dependent impedance of an *RLC* series network.

As the circuits become more complicated, the equations become more cumbersome. In an attempt to simplify them, let us make the substitution $j\omega = s$. (This substitution has a more important meaning, which we will describe in later chapters.) With this substitution, the expression for $\mathbf{Z}_{eq}$ becomes

$$\mathbf{Z}_{eq} = \frac{s^2 LC + sRC + 1}{sC}$$

If we review the four circuits we investigated thus far, we will find that in every case the impedance is the ratio of two polynomials in s and is of the general form

$$\mathbf{Z}(s) = \frac{N(s)}{D(s)} = \frac{a_m s^m + a_{m-1} s^{m-1} + \cdots + a_1 s + a_0}{b_n s^n + b_{n-1} s^{n-1} + \cdots + b_1 s + b_0} \qquad \textbf{12.1}$$

where $N(s)$ and $D(s)$ are polynomials of order m and n, respectively. An extremely important aspect of Eq. (12.1) is that it holds not only for impedances but also for all voltages, currents, admittances, and gains in the network. The only restriction is that the values of all circuit elements (resistors, capacitors, inductors, and dependent sources) must be real numbers.

Let us now demonstrate the manner in which the voltage across an element in a series RLC network varies with frequency.

EXAMPLE 12.1

Consider the network in Fig. 12.5a. We wish to determine the variation of the output voltage as a function of frequency over the range from 0 to 1 kHz.

SOLUTION   Using voltage division, the output can be expressed as

$$\mathbf{V}_o = \left(\frac{R}{R + j\omega L + \dfrac{1}{j\omega C}} \right) \mathbf{V}_S$$

or, equivalently,

$$\mathbf{V}_o = \left(\frac{j\omega CR}{(j\omega)^2 LC + j\omega CR + 1} \right) \mathbf{V}_S$$

Using the element values, the equation becomes

$$\mathbf{V}_o = \left(\frac{(j\omega)(37.95 \times 10^{-3})}{(j\omega)^2 (2.53 \times 10^{-4}) + j\omega(37.95 \times 10^{-3}) + 1} \right) 10 \underline{/0^\circ}$$

At this point we can simply substitute different values of frequency, over the range of interest, into the equation and determine the resultant magnitude and phase of the voltage. By using a large number of these points we can plot both the magnitude and phase of the output voltage as a function of frequency. This effective but tedious approach can be vastly simplified by using PSPICE.

The resulting magnitude and phase plots are shown in Figs. 12.5b and c. Let us perform a simple mental check to see if the results are plausible. At dc, we know that

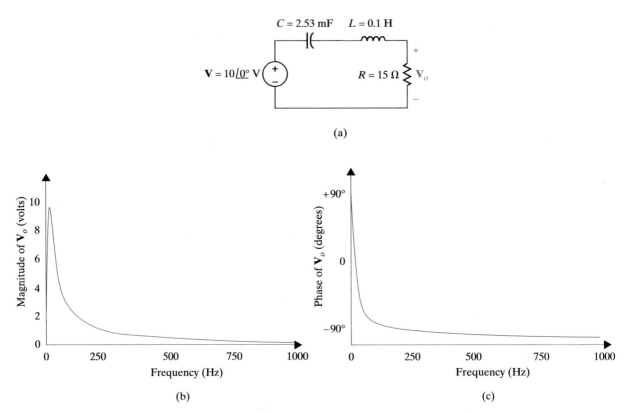

Figure 12.5 Frequency-response simulation.

the inductor acts as a short circuit and the capacitor as an open circuit. Since the capacitor impedance is infinite, all of the source voltage is dropped there and none appears at $\mathbf{V}_o$. This is consistent with the PSPICE plot of the magnitude plot of the magnitude at 0 Hz. If we now let the frequency approach infinity, the capacitor acts as a short circuit and the inductor as an open circuit. Now all of the source voltage is dropped across the inductor and again $\mathbf{V}_o$ is zero, consistent with fig. 12.5b. Thus, our results appear to be correct.

Figures 12.5b and c indicate that much of the variation in $\mathbf{V}_o$ occurs at the lower frequencies and is difficult to resolve. A standard technique employed by engineers is to plot the frequency on a log axis, which expands the low-frequency portion of the graphs, resulting in more meaningful plots. The resulting magnitude and phase plots are shown in Figs. 12.6a and b. Notice how the lower-frequency information is expanded and thus is easier to extract. Note also that the low-frequency end of the semilog plot is 1 Hz and not 0 Hz. ❏

We will illustrate in subsequent sections that the use of a semilog plot is a very useful tool in deriving frequency-response information.

As an introductory application of variable frequency-response analysis and characterization, let us consider a stereo amplifier. In particular we should consider first the fre-

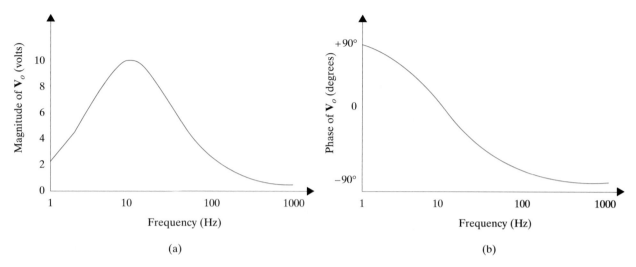

(a)   (b)

Figure 12.6 Magnitude and phase plots with frequency on a log axis.

quency range over which the amplifier must perform and then exactly what kind of performance we desire. The frequency range of the amplifier must exceed that of the human ear, which is roughly 50 Hz to 15,000 Hz. Accordingly, typical stereo amplifiers are designed to operate in the frequency range from 50 Hz to 20,000 Hz. Furthermore, we want to preserve the fidelity of the signal as it passes through the amplifier. Thus, the output signal should be an exact duplicate of the input signal times a gain factor. This requires that the gain be independent of frequency over the specified frequency range of 50 Hz to 20,000 Hz. An ideal sketch of this requirement for a gain of 1000 is shown in Fig. 12.7, where the midband region is defined as that portion of the plot where the gain is constant and is bounded by two points, which we will refer to as f_{LO} and f_{HI}. Notice once again that the frequency axis is a log axis and, thus, the frequency response is displayed on a semilog plot.

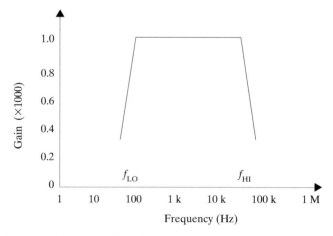

Figure 12.7 Amplifier frequency-response requirements.

A model for the amplifier described graphically in Fig. 12.7 is shown in Fig. 12.8a with the frequency-domain equivalent circuit in Fig. 12.8b.

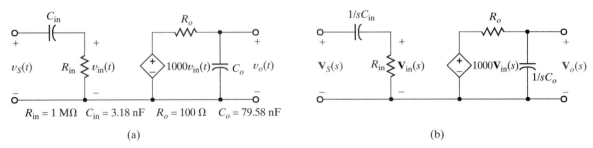

$R_{in} = 1\ M\Omega$ $C_{in} = 3.18\ nF$ $R_o = 100\ \Omega$ $C_o = 79.58\ nF$

(a)   (b)

Figure 12.8 Amplifier equivalent network.

If the input is a steady-state sinusoid, we can use frequency-domain analysis to find the gain

$$G_v(j\omega) = \frac{V_o(j\omega)}{V_s(j\omega)}$$

which with the substitution $s = j\omega$ can be expressed as

$$G_v(s) = \frac{V_o(s)}{V_s(s)}$$

Using voltage division, we find that the gain is

$$G_v(s) = \frac{V_o(s)}{V_s(s)} = \frac{V_{in}(s)}{V_s(s)} \frac{V_o(s)}{V_{in}(s)} = \left[\frac{R_{in}}{R_{in} + 1/sC_{in}}\right](1000)\left[\frac{1/sC_o}{R_o + 1/sC_o}\right]$$

or

$$G_v(s) = \left[\frac{sC_{in}R_{in}}{1 + sC_{in}R_{in}}\right](1000)\left[\frac{1}{1 + sC_oR_o}\right]$$

Using the element values in Fig. 12.8a,

$$G_v(s) = \left[\frac{s}{s + 100\pi}\right](1000)\left[\frac{40{,}000\pi}{s + 40{,}000\pi}\right]$$

where 100π and $40{,}000\pi$ are the radian equivalents of 50 Hz and 20,000 Hz, respectively. Since $s = j\omega$, the network function is indeed complex. An exact plot of $G_v(s)$ is shown in Fig. 12.9 superimposed over the sketch of Fig. 12.7. The exact plot exhibits smooth transitions at f_{LO} and f_{HI}; otherwise the plots match fairly well.

Let us examine our expression for $G_v(s)$ more closely with respect to the plot in Fig. 12.9. Assume that f is well within the midband frequency range; that is,

$$f_{LO} \ll f \ll f_{HI}$$

or

$$100\pi \ll |s| \ll 40{,}000\pi$$

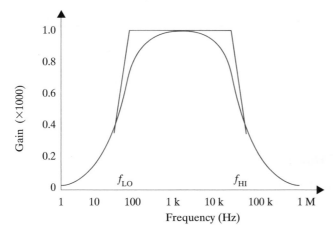

Figure 12.9 Exact and approximate amplifier gain versus frequency plots.

Under these conditions, the network function becomes

$$\mathbf{G}_v(S) \approx \left[\frac{s}{s}\right](1000)\left[\frac{1}{1 + 0}\right]$$

or

$$\mathbf{G}_v(S) = 1000$$

Thus, well within midband, the gain is constant. However, if the frequency of excitation decreases toward f_{LO}, then $|s|$ is comparable to 100π and

$$\mathbf{G}_v(s) \approx \left[\frac{s}{s + 100\pi}\right](1000)$$

Since $R_{\text{in}} C_{\text{in}} = 1/100\pi$, we see that C_{in} causes the rolloff in gain at low frequencies. Similarly, when the frequency approaches f_{HI}, the gain rolloff is due to C_o.

Through this amplifier example, we have introduced the concept of frequency-dependent networks and have demonstrated that frequency-dependent network performance is caused by the reactive elements in a network.

Network Functions

In the previous section, we introduced the term *voltage gain*, $\mathbf{G}_v(s)$. This term is actually only one of several network functions, designated generally as $\mathbf{H}(s)$, which define the ratio of response to input. Since the function describes a reaction due to an excitation at some other point in the circuit, network functions are also called *transfer functions*. Furthermore, transfer functions are not limited to voltage ratios. Since in electrical networks inputs and outputs can be either voltages or currents, there are four possible network functions, as listed in Table 12.1.

There are also *driving point functions*, which are impedances or admittances defined at a single pair of terminals. For example, the input impedance of a network is a driving point function.

Table 12.1 Network transfer functions

Input	Output	Transfer Function	Symbol
Voltage	Voltage	Voltage Gain	$G_v(s)$
Current	Voltage	Transimpedance	$Z(s)$
Current	Current	Current Gain	$G_i(s)$
Voltage	Current	Transadmittance	$Y(s)$

EXAMPLE 12.2

We wish to determine the transfer admittance $[I_2(s)/V_1(s)]$ and the voltage gain of the network shown in Fig. 12.10.

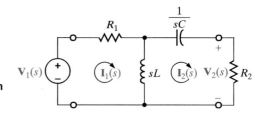

Figure 12.10 Circuit employed in Example 12.2.

SOLUTION The mesh equations for the network are

$$(R_1 + sL)I_1(s) - sLI_2(s) = V_1(s)$$

$$-sLI_1(s) + \left(R_2 + sL + \frac{1}{sC}\right)I_2(s) = 0$$

$$V_2(s) = I_2(s)R_2$$

Solving the equations for $I_2(s)$ yields

$$I_2(s) = \frac{sLV_1(s)}{(R_1 + sL)(R_2 + sL + 1/sC) - s^2 L^2}$$

Therefore, the transfer admittance $[I_2(s)/V_1(s)]$ is

$$\mathbf{Y}_T(s) = \frac{I_2(s)}{V_1(s)} = \frac{LCs^2}{(R_1 + R_2)LCs^2 + (L + R_1 R_2 C)s + R_1}$$

and the voltage gain is

$$\mathbf{G}_v(s) = \frac{V_2(s)}{V_1(s)} = \frac{LCR_2 s^2}{(R_1 + R_2)LCs^2 + (L + R_1 R_2 C)s + R_1} \qquad \square$$

Poles and Zeros

As we have indicated, the network function can be expressed as the ratio of the two polynomials in s. In addition, we note that since the values of our circuit elements, or controlled sources, are real numbers, the coefficients of the two polynomials will be real. Therefore, we will express a network function in the form

$$\mathbf{H}(s) = \frac{N(s)}{D(s)} = \frac{a_m s^m + a_{m-1} s^{m-1} + \cdots + a_1 s + a_0}{b_n s^n + b_{n-1} s^{n-1} + \cdots + b_1 s + b_0} \qquad \textbf{12.2}$$

where $N(s)$ is the numerator polynomial of degree m and $D(s)$ is the denominator polynomial of degree n. Equation (12.2) can also be written in the form

$$\mathbf{H}(s) = \frac{K_0(s - z_1)(s - z_2) \cdots (s - z_m)}{(s - p_1)(s - p_2) \cdots (s - p_n)} \qquad \textbf{12.3}$$

where K_0 is a constant, $z_1, \ldots, z_m$ are the roots of $N(s)$, and $p_1, \ldots, p_n$ are the roots of $D(s)$. Note that if $s = z_1$, or $z_2, \ldots, z_m$, then $\mathbf{H}(s)$ becomes zero and hence $z_1, \ldots, z_m$ are called zeros of the transfer function. Similarly, if $s = p_1$, or $p_2, \ldots, p_n$, then $\mathbf{H}(s)$ becomes infinite and, therefore, $p_1, \ldots, p_n$ are called poles of the function. The zeros or poles may actually be complex. However, if they are complex, they must occur in conjugate pairs since the coefficients of the polynomial are real. The representation of the network function specified in Eq. (12.3) is extremely important and is generally employed to represent any linear time-invariant system. The importance of this form stems from the fact that the dynamic properties of a system can be gleaned from an examination of the system poles.

EXTENSION EXERCISES

E12.1 Find the driving point impedance at $\mathbf{V}_S(s)$ in the amplifier shown in Fig. 12.8b.

ANSWER: $Z(s) = R_{in} + \dfrac{1}{sC_{in}} = \left[1 + \left(\dfrac{100\pi}{S} \right) \right] M\Omega$

E12.2 Find the pole and zero locations in hertz and the value of K_0 for the amplifier network in Fig. 12.8.

ANSWER: $z_1 = 0$ Hz (dc), $p_1 = -50$ Hz, $p_2 = -20{,}000$ Hz, $K_0 = (4 \times 10^7)\pi$.

12.2 SINUSOIDAL FREQUENCY ANALYSIS

Although there are specific cases in which a network operates at only one frequency (e.g., a power system network), in general we are interested in the behavior of a network as a function of frequency. In a sinusoidal steady-state analysis, the network function can be expressed as

$$\mathbf{H}(j\omega) = M(\omega)e^{j\phi(\omega)} \qquad \textbf{12.4}$$

where $M(\omega) = |\mathbf{H}(j\omega)|$ and $\phi(\omega)$ is the phase. A plot of these two functions, which are commonly called the *magnitude* and *phase characteristics*, displays the manner in which the response varies with the input frequency ω. We will now illustrate the manner in which to perform a frequency-domain analysis by simply evaluating the function at various frequencies within the range of interest.

Frequency Response Using a Bode Plot

If the network characteristics are plotted on a semilog scale (that is, a linear scale for the ordinate and a logarithmic scale for the abscissa), they are known as *Bode plots* (named after Hendrik W. Bode). This graph is a powerful tool in both the analysis and design of frequency-dependent systems and networks such as filters, tuners, and amplifiers. In using the graph, we plot $20 \log_{10} M(\omega)$ versus $\log_{10}(\omega)$ instead of $M(\omega)$ versus ω. The advantage of this technique is that rather than plotting the characteristic point by point, we can employ straight-line approximations to obtain the characteristic very efficiently. The ordinate for the magnitude plot is the decibel (dB). This unit was originally employed to measure the ratio of powers; that is,

$$\text{number of dB} = 10 \log_{10} \frac{P_2}{P_1} \tag{12.5}$$

If the powers are absorbed by two equal resistors, then

$$\text{number of dB} = 10 \log_{10} \frac{|\mathbf{V}_2|^2 / R}{|\mathbf{V}_1|^2 / R} = 10 \log_{10} \frac{|\mathbf{I}_2|^2 R}{|\mathbf{I}_1|^2 R} \tag{12.6}$$

$$= 20 \log_{10} \frac{|\mathbf{V}_2|}{|\mathbf{V}_1|} = 20 \log_{10} \frac{|\mathbf{I}_2|}{|\mathbf{I}_1|}$$

The term "dB" has become so popular that it now is used for voltage and current ratios, as illustrated in Eq. (12.6), without regard to the impedance employed in each case.

In the sinusoidal steady-state case, $\mathbf{H}(j\omega)$ in Eq. (12.3) can be expressed in general as

$$\mathbf{H}(j\omega) = \frac{K_0(j\omega)^{\pm N}(1 + j\omega\tau_1)\left[1 + 2\zeta_3(j\omega\tau_3) + (j\omega\tau_3)^2\right]\cdots}{(1 + j\omega\tau_a)\left[1 + 2\zeta_b(j\omega\tau_b) + (j\omega\tau_b)^2\right]\cdots} \tag{12.7}$$

Note that this equation contains the following typical factors:

1. A frequency-independent factor $K_0 > 0$
2. Poles or zeros at the origin of the form $j\omega$; that is, $(j\omega)^{+N}$ for zeros and $(j\omega)^{-N}$ for poles
3. Poles or zeros of the form $(1 + j\omega\tau)$
4. Quadratic poles or zeros of the form $1 + 2\zeta(j\omega\tau) + (j\omega\tau)^2$

Taking the logarithm of the magnitude of the function $\mathbf{H}(j\omega)$ in Eq. (12.7) yields

$$\begin{aligned}
20 \log_{10}|\mathbf{H}(j\omega)| &= 20 \log_{10} K_0 \pm 20N \log_{10}|j\omega| \\
&\quad + 20 \log_{10}|1 + j\omega\tau_1| \\
&\quad + 20 \log_{10}\left|1 + 2\zeta_3(j\omega\tau_3) + (j\omega\tau_3)^2\right| \\
&\quad + \cdots - 20 \log_{10}|1 + j\omega\tau_a| \\
&\quad - 20 \log_{10}\left|1 + 2\zeta_b(j\omega\tau_b) + (j\omega\tau_b)^2\right|\cdots
\end{aligned} \tag{12.8}$$

Note that we have used the fact that the log of the product of two or more terms is equal to the sum of the logs of the individual terms, the log of the quotient of two terms is equal to the difference of the logs of the individual terms, and $\log_{10} A^n = n \log_{10} A$.

The phase angle for $\mathbf{H}(j\omega)$ is

$$\underline{/\mathbf{H}(j\omega)} = 0 \pm N(90°) + \tan^{-1}\omega\tau_1$$

$$+ \tan^{-1}\left(\frac{2\zeta_3\,\omega\tau_3}{1 - \omega^2\,\tau_3^2}\right) + \cdots$$

$$- \tan^{-1}\omega\tau_a - \tan^{-1}\left(\frac{2\zeta_b\,\omega\tau_b}{1 - \omega^2\,\tau_b^2}\right)\cdots \qquad \textbf{12.9}$$

As Eqs. (12.8) and (12.9) indicate, we will simply plot each factor individually on a common graph and then sum them algebraically to obtain the total characteristic. Let us examine some of the individual terms and illustrate an efficient manner in which to plot them on the Bode diagram.

Constant Term.   The term $20\log_{10}K_0$ represents a constant magnitude with zero phase shift, as shown in Fig. 12.11a.

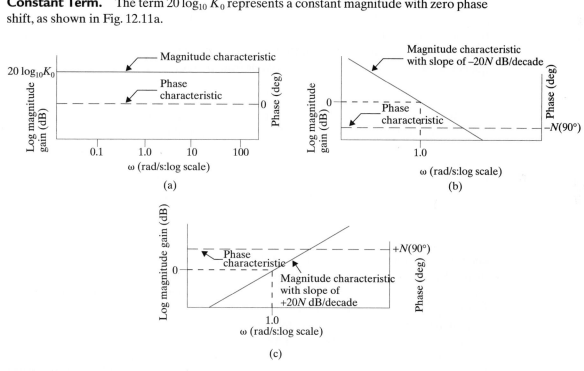

Figure 12.11 Magnitude and phase characteristics for a constant term and poles and zeros at the origin.

Poles or Zeros at the Origin.   Poles or zeros at the origin are of the form $(j\omega)^{\pm N}$, where $+$ is used for a zero and $-$ is used for a pole. The magnitude of this function is $\pm 20N\log_{10}\omega$, which is a straight line on semilog paper with a slope of $\pm 20N$ dB/decade; that is, the value will change by $20N$ each time the frequency is multiplied by 10, and the phase of this function is a constant $\pm N(90°)$. The magnitude and phase characteristics for poles and zeros at the origin are shown in Figs. 12.11b and c, respectively.

Simple Pole or Zero.   Linear approximations can be employed when a simple pole or zero of the form $(1 + j\omega\tau)$ is present in the network function. For $\omega\tau \ll 1$, $(1 + j\omega\tau) \approx 1$, and therefore, $20 \log_{10}|(1 + j\omega\tau)| = 20 \log_{10} 1 = 0$ dB. Similarly, if $\omega\tau \gg 1$, then $(1 + j\omega\tau) \approx j\omega\tau$, and hence $20 \log_{10}|(1 + j\omega\tau)| \approx 20 \log_{10} \omega\tau$. Therefore, for $\omega\tau \ll 1$ the response is 0 dB and for $\omega\tau \gg 1$ the response has a slope that is the same as that of a simple pole or zero at the origin. The intersection of these two asymptotes, one for $\omega\tau \ll 1$ and one for $\omega\tau \gg 1$, is the point where $\omega\tau = 1$ or $\omega = 1/\tau$, which is called the *break frequency*. At this break frequency, where $\omega = 1/\tau$, $20 \log_{10}|(1 + j1)| = 20 \log_{10}(2)^{1/2} = 3$ dB. Therefore, the actual curve deviates from the asymptotes by 3 dB at the break frequency. It can be shown that at one-half and twice the break frequency, the deviations are 1 dB. The phase angle associated with a simple pole or zero is $\phi = \tan^{-1} \omega\tau$, which is a simple arctangent curve. Therefore, the phase shift is 45° at the break frequency and 26.6° and 63.4° at one-half and twice the break frequency, respectively. The actual magnitude curve for a pole of this form is shown in Fig. 12.12a. For a zero the magnitude curve and the asymptote for $\omega\tau \gg 1$ have a positive slope, and the phase curve extends from 0° to $+90°$, as shown in Fig. 12.12b. If multiple poles or zeros of the form $(1 + j\omega\tau)^N$ are present, then the slope of the high-frequency asymptote is multiplied by N, the deviation between the actual curve and the asymptote at the break frequency is $3N$ dB, and the phase curve extends from 0 to $N(90°)$ and is $N(45°)$ at the break frequency.

Quadratic Poles or Zeros.   Quadratic poles or zeros are of the form $1 + 2\zeta(j\omega\tau) + (j\omega\tau)^2$. This term is a function not only of ω, but the dimensionless term ζ, which is called the *damping ratio*. If $\zeta > 1$ or $\zeta = 1$, the roots are real and unequal or real and equal, respectively, and these two cases have already been addressed. If $\zeta < 1$, the roots are complex conjugates, and it is this case that we will examine now. Following the preceding argument for a simple pole or zero, the log magnitude of the quadratic factor is 0 dB for $\omega\tau \ll 1$. For $\omega\tau \gg 1$,

$$20 \log_{10}|1 - (\omega\tau)^2 + 2j\zeta(\omega\tau)| \approx 20 \log_{10}|(\omega\tau)^2| = 40 \log_{10}|\omega\tau|$$

and therefore, for $\omega\tau \gg 1$, the slope of the log magnitude curve is +40 dB/decade for a quadratic zero and −40 dB/decade for a quadratic pole. Between the two extremes, $\omega\tau \ll 1$ and $\omega\tau \gg 1$, the behavior of the function is dependent on the damping ratio ζ. Figure 12.13a illustrates the manner in which the log magnitude curve for a quadratic *pole* changes as a function of the damping ratio. The phase shift for the quadratic factor is $\tan^{-1} 2\zeta\omega\tau/[1 - (\omega\tau)^2]$. The phase plot for quadratic *poles* is shown in Fig. 12.13b. Note that in this case the phase changes from 0° at frequencies for which $\omega\tau \ll 1$ to −180° at frequencies for which $\omega\tau \gg 1$. For quadratic zeros the magnitude and phase curves are inverted; that is, the log magnitude curve has a slope of +40 dB/decade for $\omega\tau \gg 1$, and the phase curve is 0° for $\omega\tau \ll 1$ and +180° for $\omega\tau \gg 1$.

EXAMPLE 12.3

We want to generate the magnitude and phase plots for the transfer function

$$\mathbf{G}_v(j\omega) = \frac{10(0.1j\omega + 1)}{(j\omega + 1)(0.02j\omega + 1)}$$

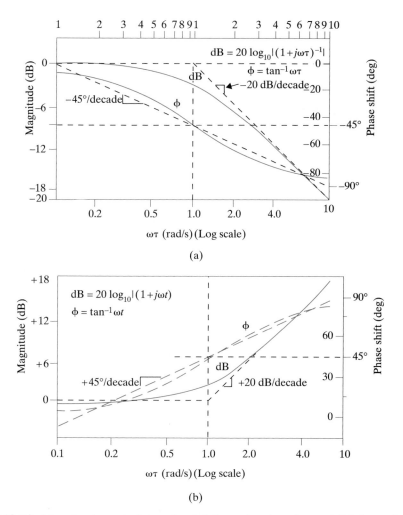

Figure 12.12 Magnitude and phase plot (a) for a simple pole, and (b) for a simple zero.

SOLUTION Note that this function is in standard form, since every term is of the form $(j\omega\tau + 1)$. In order to determine the composite magnitude and phase characteristics, we will plot the individual asymptotic terms and then add them as specified in Eqs. (12.8) and (12.9). Let us consider the magnitude plot first. Since $K_0 = 10$, $20 \log_{10} 10 = 20$ dB, which is a constant independent of frequency, as shown in Fig. 12.14a. The zero of the transfer function contributes a term of the form $+20 \log_{10} |1 + 0.1 j\omega|$, which is 0 dB for 0.1 $\omega \ll 1$, has a slope of $+20$ dB/decade for $0.1\omega \gg 1$, and has a break frequency at $\omega = 10$ rad/s. The poles have break frequencies at $\omega = 1$ and $\omega = 50$ rad/s. The pole with break frequency at $\omega = 1$ rad/s contributes a term of the form $-20 \log_{10} |1 + j\omega|$, which is 0 dB for $\omega \ll 1$ and has a slope of -20 dB/decade for $\omega \gg 1$. A similar argument can be made for the pole

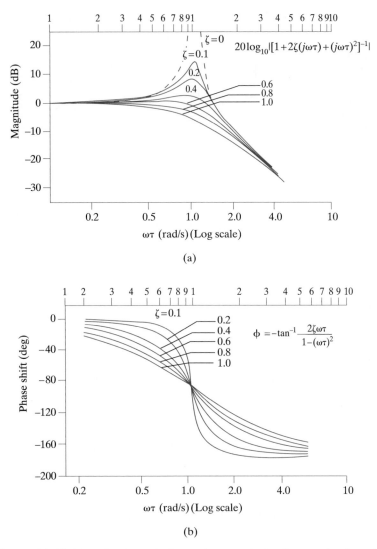

Figure 12.13 Magnitude and phase characteristics for quadratic poles.

that has a break frequency at $\omega = 50$ rad/s. These factors are all plotted individually in Fig. 12.14a.

Consider now the individual phase curves. The term K_0 is not a function of ω and does not contribute to the phase of the transfer function. The phase curve for the zero is $+\tan^{-1} 0.1\omega$, which is an arctangent curve that extends from $0°$ for $0.1\omega \ll 1$ to $+90°$ for $0.1\omega \gg 1$ and has a phase of $+45°$ at the break frequency. The phase curves for the two poles are $-\tan^{-1} \omega$ and $-\tan^{-1} 0.02\omega$. The term $-\tan^{-1} \omega$ is $0°$ for $\omega \ll 1, -90°$ for $\omega \gg 1$, and $-45°$ at the break frequency $\omega = 1$. The phase curve for the remaining pole is plotted in a similar fashion. All the individual phase curves are shown in Fig. 12.14a.

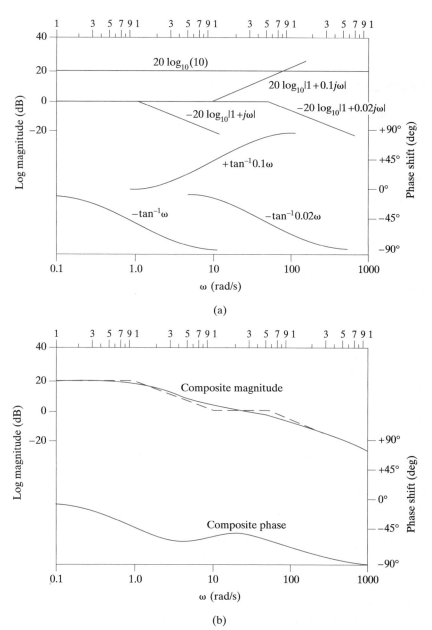

Figure 12.14 (a) Magnitude and phase components for the poles and zeros of the transfer function in Example 12.3; (b) Bode plot for the transfer function in Example 12.3.

As specified in Eqs. (12.8) and (12.9), the composite magnitude and phase of the transfer function are obtained simply by adding the individual terms. The composite curves are plotted in Fig. 12.14b. Note that the actual magnitude curve (solid line) differs from the straight-line approximation (dashed line) by 3 dB at the break frequencies and 1 dB at one-half and twice the break frequencies. ❏

EXAMPLE 12.4

Let us draw the Bode plot for the following transfer function:

$$\mathbf{G}_v(j\omega) = \frac{25(j\omega + 1)}{(j\omega)^2(0.1j\omega + 1)}$$

SOLUTION   Once again all the individual terms for both magnitude and phase are plotted in Fig. 12.15a. The straight line with a slope of -40 dB/decade is generated by the double pole at the origin. This line is a plot of $-40 \log_{10} \omega$ versus ω and therefore passes through 0 dB at $\omega = 1$ rad/s. The phase for the double pole is a constant $-180°$ for all frequencies. The remainder of the terms is plotted as illustrated in Example 12.3.

The composite plots are shown in Fig. 12.15b. Once again they are obtained simply by adding the individual terms in Fig. 12.15a. Note that for frequencies for which $\omega \ll 1$, the slope of the magnitude curve is -40 dB/decade. At $\omega = 1$ rad/s, which is the break frequency of the zero, the magnitude curve changes slope to -20 dB/decade. At $\omega = 10$ rad/s, which is the break frequency of the pole, the slope of the magnitude curve changes back to -40 dB/decade.

The composite phase curve starts at $-180°$ due to the double pole at the origin. Since the first break frequency encountered is a zero, the phase curve shifts toward $-90°$. However, before the composite phase reaches $-90°$, the pole with break frequency $\omega = 10$ rad/s begins to shift the composite curve back toward $-180°$. ❏

Example 12.4 illustrates the manner in which to plot directly terms of the form $K_0/(j\omega)^N$. For terms of this form, the initial slope of $-20N$ dB/decade will intersect the 0-dB axis at a frequency of $(K_0)^{1/N}$ rad/s; that is, $-20 \log_{10}|K_0/(j\omega)^N| = 0$ dB implies that $K_0/(\omega)^N = 1$, and therefore, $\omega = (K_0)^{1/N}$ rad/s. Note that the projected slope of the magnitude curve in Example 12.4 intersects the 0-dB axis at $\omega = (25)^{1/2} = 5$ rad/s.

Similarly, it can be shown that for terms of the form $K_0(j\omega)^N$, the initial slope of $+20N$ dB/decade will intersect the 0-dB axis at a frequency of $\omega = (1/K_0)^{1/N}$ rad/s; that is, $+20 \log_{10}|K_0/(j\omega)^N| = 0$ dB implies that $K_0/(\omega)^N = 1$, and therefore $\omega = (1/K_0)^{1/N}$ rad/s.

By applying the concepts we have just demonstrated, we can normally plot the log magnitude characteristic of a transfer function directly in one step.

EXTENSION EXERCISES

E12.3   Sketch the magnitude characteristic of the Bode plot, labeling all critical slopes and points for the function

$$\mathbf{G}(j\omega) = \frac{10^4(j\omega + 2)}{(j\omega + 10)(j\omega + 100)}$$

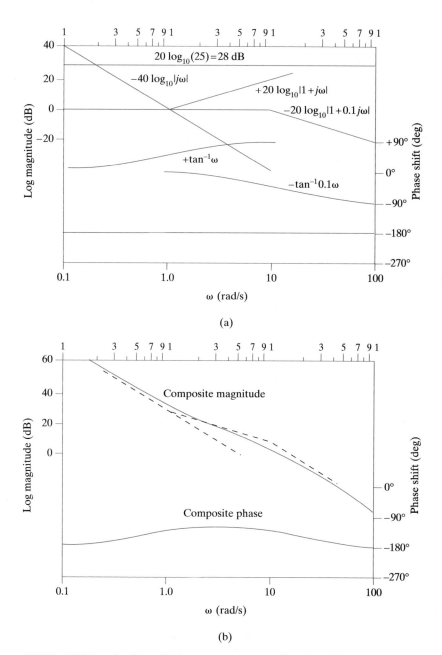

Figure 12.15 (a) Magnitude and phase components for the poles and zeros of the transfer function in Example 12.4; (b) Bode plot for the transfer function in Example 12.4.

ANSWER:

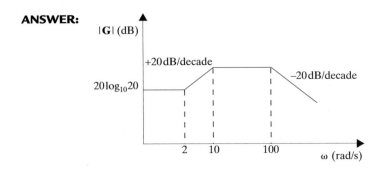

E12.4   Sketch the magnitude characteristic of the Bode plot, labeling all critical slopes and points for the function

$$\mathbf{G}(j\omega) = \frac{100(0.02j\omega + 1)}{(j\omega)^2}$$

ANSWER:

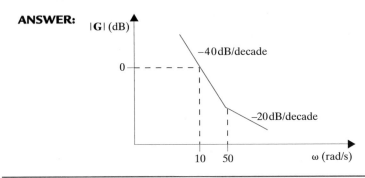

E12.5   Sketch the magnitude characteristic of the Bode plot, labeling all critical slopes and points for the function

$$\mathbf{G}(j\omega) = \frac{10j\omega}{(j\omega + 1)(j\omega + 10)}$$

ANSWER:

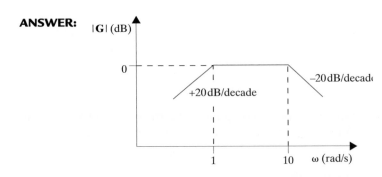

EXAMPLE 12.5

We wish to generate the Bode plot for the following transfer function:

$$\mathbf{G}_v(j\omega) = \frac{25j\omega}{(j\omega + 0.5)[(j\omega)^2 + 4j\omega + 100]}$$

SOLUTION   Expressing this function in standard form, we obtain

$$\mathbf{G}_v(j\omega) = \frac{0.5j\omega}{(2j\omega + 1)[(j\omega/10)^2 + j\omega/25 + 1]}$$

The Bode plot is shown in Fig. 12.16. The initial low-frequency slope due to the zero at the origin is +20 dB/decade, and this slope intersects the 0-dB line at $\omega = 1/K_0 = 2$ rad/s. At $\omega = 0.5$ rad/s the slope changes from +20 dB/decade to 0 dB/decade due to the presence of the pole with a break frequency at $\omega = 0.5$ rad/s. The quadratic term has a center frequency of $\omega = 10$ rad/s (i.e., $\tau = 1/10$). Since

$$2\zeta\tau = \frac{1}{25}$$

and

$$\tau = 0.1$$

then

$$\zeta = 0.2$$

Plotting the curve in Fig. 12.13a with a damping ratio of $\zeta = 0.2$ at the center frequency $\omega = 10$ rad/s completes the composite magnitude curve for the transfer function.

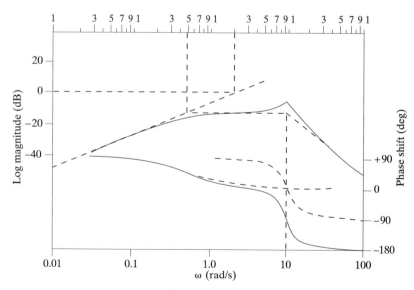

Figure 12.16 Bode plot for the transfer function in Example 12.5.

The initial low-frequency phase curve is +90°, due to the zero at the origin. This curve and the phase curve for the simple pole and the phase curve for the quadratic term, as defined in Fig. 12.13b, are combined to yield the composite phase curve.

❏

EXTENSION EXERCISE

E12.6   Given the following function $\mathbf{G}(j\omega)$, sketch the magnitude characteristic of the Bode plot, labeling all critical slopes and points.

$$\mathbf{G}(j\omega) = \frac{0.2(j\omega + 1)}{j\omega\big[(j\omega/12)^2 + j\omega/36 + 1\big]}$$

ANSWER:

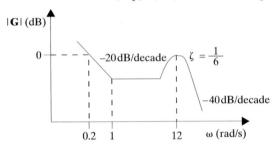

Deriving the Transfer Function from the Bode Plot

EXAMPLE 12.6

Given the asymptotic magnitude characteristic shown in Fig. 12.17, we wish to determine the transfer function $\mathbf{G}_v(j\omega)$.

SOLUTION   Since the initial slope is 0 dB/decade, and the level of the characteristic is 20 dB, the factor K_0 can be obtained from the expression

$$20 \text{ dB} = 20 \log_{10} K_0$$

and hence

$$K_0 = 10$$

The −20-dB/decade slope starting at $\omega = 0.1$ rad/s indicates that the first pole has a break frequency at $\omega = 0.1$ rad/s, and therefore one of the factors in the denominator is $(10j\omega + 1)$. The slope changes by +20 dB/decade at $\omega = 0.5$ rad/s, indicating that there is a zero present with a break frequency at $\omega = 0.5$ rad/s, and therefore the numerator has a factor of $(2j\omega + 1)$. Two additional poles are present with break frequencies at $\omega = 2$ rad/s and $\omega = 20$ rad/s. Therefore, the composite transfer function is

$$\mathbf{G}_v(j\omega) = \frac{10(2j\omega + 1)}{(10j\omega + 1)(0.5j\omega + 1)(0.05j\omega + 1)}$$

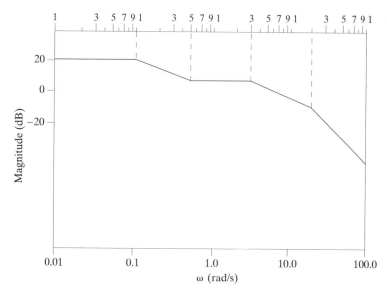

Figure 12.17 Straight-line magnitude plot employed in Example 12.6.

The reader should note carefully the ramifications of this example with regard to network design.   ❏

EXTENSION EXERCISE

E12.7   Determine the transfer function $\mathbf{G}(j\omega)$ if the straight-line magnitude characteristic approximation for this function is as shown in Fig. E12.7.

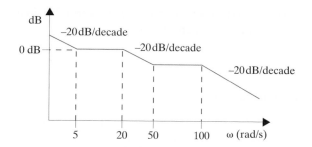

Figure E12.7

ANSWER: $\mathbf{G}(j\omega) = \dfrac{5\left(\dfrac{j\omega}{5} + 1\right)\left(\dfrac{j\omega}{50} + 1\right)}{j\omega\left(\dfrac{j\omega}{20} + 1\right)\left(\dfrac{j\omega}{100} + 1\right)}$

12.3 RESONANT CIRCUITS

Two circuits with extremely important frequency characteristics are shown in Fig. 12.18. The input impedance for the series RLC circuit is

$$\mathbf{Z}(j\omega) = R + j\omega L + \frac{1}{j\omega C} \qquad \text{12.10}$$

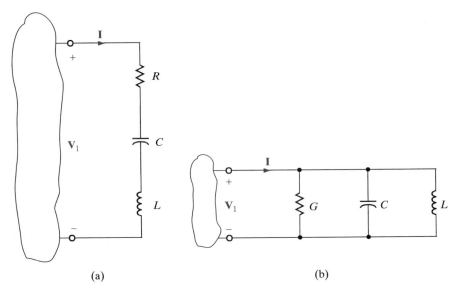

(a)   (b)

Figure 12.18 Series and parallel RLC circuits.

and the input admittance for the parallel RLC circuit is

$$\mathbf{Y}(j\omega) = G + j\omega C + \frac{1}{j\omega L} \qquad \text{12.11}$$

Note that these two equations have the same general form. The imaginary terms in both of the preceding equations will be zero if

$$\omega L = \frac{1}{\omega C}$$

The value of ω that satisfied this equation is

$$\omega_0 = \frac{1}{\sqrt{LC}} \qquad \text{12.12}$$

and at this value of ω the impedance of the series circuit becomes

$$\mathbf{Z}(j\omega_0) = R \qquad \text{12.13}$$

Recall that ω_0 is the undamped resonant frequency.

and the admittance of the parallel circuit is

$$\mathbf{Y}(j\omega_0) = G \qquad\qquad \textbf{12.14}$$

This frequency ω_0, at which the impedance of the series circuit or the admittance of the parallel circuit is purely real, is called the *resonant frequency*, and the circuits themselves, at this frequency, are said to be *in resonance*. Resonance is a very important consideration in engineering design. For example, engineers designing the attitude control system for the Saturn vehicles had to ensure that the control system frequency did not excite the body bending (resonant) frequencies of the vehicle. Excitation of the bending frequencies would cause oscillations that, if continued unchecked, would result in a buildup of stress until the vehicle would finally break apart.

At resonance the voltage and current are in phase and, therefore, the phase angle is zero and the power factor is unity. In the series case, at resonance the impedance is a minimum and, therefore, the current is maximum for a given voltage. Figure 12.19 illustrates the frequency response of both the series and parallel RLC circuits. Note that at low frequencies the impedance of the series circuit is dominated by the capacitive term and the admittance of the parallel circuit is dominated by the inductive term. At high frequencies the impedance of the series circuit is dominated by the inductive term, and the admittance of the parallel circuit is dominated by the capacitive term.

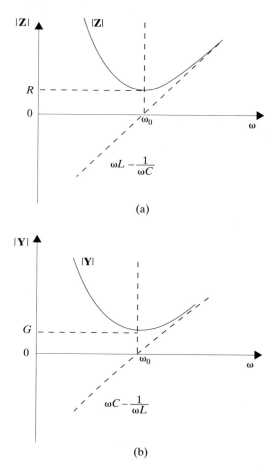

Figure 12.19 Frequency response of (a) a series, and (b) a parallel RLC circuit.

Resonance can be viewed from another perspective—that of the phasor diagram. Once again we will consider the series and parallel cases together in order to illustrate the similarities between them. In the series case the current is common to every element, and in the parallel case the voltage is a common variable. Therefore, the current in the series circuit and the voltage in the parallel circuit are employed as references. Phasor diagrams for both circuits are shown in Fig. 12.20 for the three frequency values $\omega < \omega_0, \omega = \omega_0$, and $\omega > \omega_0$.

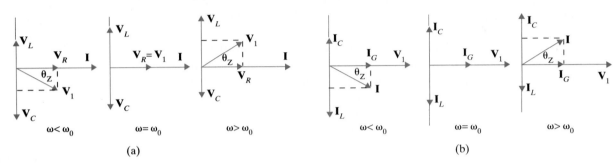

Figure 12.20 Phasor diagrams for (a) a series RLC circuit, and (b) a parallel GLC circuit.

In the series case when $\omega < \omega_0, \mathbf{V}_C > \mathbf{V}_L, \theta_Z$ is negative and the voltage $\mathbf{V}_1$ lags the current. If $\omega = \omega_0, \mathbf{V}_L = \mathbf{V}_C, \theta_Z$ is zero, and the voltage $\mathbf{V}_1$ is in phase with the current. If $\omega > \omega_0, \mathbf{V}_L > \mathbf{V}_C, \theta_Z$ is positive, and the voltage $\mathbf{V}_1$ leads the current. Similar statements can be made for the parallel case in Fig. 12.20b. Because of the close relationship between series and parallel resonance, as illustrated by the preceding material, we will concentrate most of our discussion on the series case in the following developments.

For the series circuit we define what is commonly called the *quality factor Q* as

> The quality factor is an important descriptor for resonant circuits.

$$Q = \frac{\omega_0 L}{R} = \frac{1}{\omega_0 CR} = \frac{1}{R}\sqrt{\frac{L}{C}} \qquad \textbf{12.15}$$

Q is a very important factor in resonant circuits, and its ramifications will be illustrated throughout the remainder of this section.

EXAMPLE 12.7

Consider the network shown in Fig. 12.21. Let us determine the resonant frequency, the voltage across each element at resonance, and the value of the quality factor.

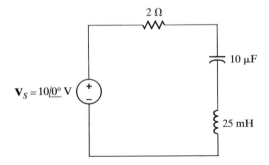

Figure 12.21 Series circuit.

SOLUTION   The resonant frequency is obtained from the expression

$$\omega_0 = \frac{1}{\sqrt{LC}}$$

$$= \frac{1}{\sqrt{(25)(10^{-3})(10)(10^{-6})}}$$

$$= 2000 \text{ rad/s}$$

At this resonant frequency

$$\mathbf{I} = \frac{\mathbf{V}}{\mathbf{Z}} = \frac{\mathbf{V}}{R} = 5 \underline{/0°} \text{ A}$$

Therefore,

$$\mathbf{V}_R = (5 \underline{/0°})(2) = 10 \underline{/0°} \text{ V}$$
$$\mathbf{V}_L = j\omega_0 L\mathbf{I} = 250 \underline{/90°} \text{ V}$$

$$\mathbf{V}_C = \frac{\mathbf{I}}{j\omega_0 C} = 250 \underline{/-90°} \text{ V}$$

Note the magnitude of the voltages across the inductor and capacitor with respect to the input voltage. Note also that these voltages are equal and are 180° out of phase with one another. Therefore, the phasor diagram for this condition is shown in Fig. 12.20a for $\omega = \omega_0$. The quality factor Q derived from Eq. (12.15) is

$$Q = \frac{\omega_0 L}{R} = \frac{(2)(10^3)(25)(10^{-3})}{2} = 25$$

It is interesting to note that the voltages across the inductor and capacitor can be written in terms of Q as

$$|\mathbf{V}_L| = \omega_0 L|\mathbf{I}| = \frac{\omega_0 L}{R} \mathbf{V}_S = Q\mathbf{V}_S$$

and

$$|\mathbf{V}_C| = \frac{|\mathbf{I}|}{\omega_0 C} = \frac{1}{\omega_0 CR} \mathbf{V}_S = Q\mathbf{V}_S$$

This analysis indicates that for a given current there is a resonant voltage rise across the inductor and capacitor that is equal to the product of Q and the applied voltage.

❏

EXAMPLE 12.8

In an undergraduate circuits laboratory, students are asked to construct an RLC network that will demonstrate resonance at $f = 1000$ Hz given a 0.02 H inductor that has a Q of 200. One student produced the circuit shown in Fig. 12.22, where the inductor's internal resistance is represented by R.

If the capacitor chosen to demonstrate resonance was an oil-impregnated paper capacitor rated at 300 V, let us determine the network parameters and the effect of this choice of capacitor.

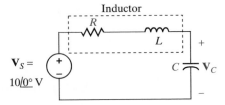

Figure 12.22 *RLC* series resonant network.

SOLUTION For resonance at 1000 Hz, the student found the required capacitor value using the expression

$$\omega_0 = 2\pi f_0 = \frac{1}{\sqrt{LC}}$$

which yields

$$C = 1.27 \ \mu F$$

The student selected an oil-impregnated paper capacitor rated at 300 V. The resistor value was found using the expression for Q

$$Q = \frac{\omega_0 L}{R} = 200$$

or

$$R = 1.59 \ \Omega$$

At resonance, the current would be

$$\mathbf{I} = \frac{\mathbf{V}_S}{R}$$

or

$$\mathbf{I} = 6.28 \ \underline{/0°} \ A$$

When constructed, the current was measured to be only

$$\mathbf{I} \sim 1 \ \underline{/0°} \ mA$$

This measurement clearly indicated that the impedance seen by the source was about 10 kΩ of resistance instead of 1.59 Ω—quite a drastic difference. Suspecting that the capacitor that was selected was the source of trouble, the student calculated what the capacitor voltage should be. If operated as designed, then at resonance,

$$\mathbf{v}_C = \frac{\mathbf{V}_S}{R}\left(\frac{1}{j\omega C}\right) = Q\mathbf{V}_S$$

or

$$\mathbf{V}_C = 2000 \ \underline{/-90°} \ V$$

which is more than six times the capacitor's rated voltage! This overvoltage had damaged the capacitor so that it did not function properly. When a new capacitor

was selected and the source voltage reduced by a factor of 10, the network performed properly as a high Q circuit.   ❑

EXTENSION EXERCISES

E12.8   Given the network in Fig. E12.8, find the value C that will place the circuit
in resonance at 1800 rad/s.

ANSWER: $C = 3.09 \ \mu\text{F}$.

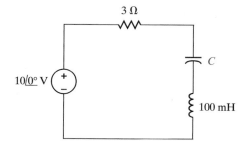

Figure E12.8

E12.9   Given the network in E12.8, determine the Q of the network and the
magnitude of the voltage across the capacitor.

ANSWER: $Q = 60, |\mathbf{V_C}| = 600$ V.

The impedance of the circuit in Fig. 12.18a is given by Eq. (12.10), which can be expressed as an admittance,

$$
\mathbf{Y}(j\omega) = \frac{1}{R[1 + j(1/R)(\omega L - 1/\omega C)]}
$$

$$
= \frac{1}{R[1 + j(\omega L/R - 1/\omega CR)]}
$$

$$
= \frac{1}{R[1 + jQ(\omega L/RQ - 1/\omega CRQ)]} \qquad \textbf{12.16}
$$

Using the fact that $Q = \omega_0 L/R = 1/\omega_0 CR$, Eq. (12.16) becomes

$$
\mathbf{Y}(j\omega) = \frac{1}{R[1 + jQ(\omega/\omega_0 - \omega_0/\omega)]} \qquad \textbf{12.17}
$$

Since $\mathbf{I} = \mathbf{Y}\mathbf{V}_1$ and the voltage across the resistor is $\mathbf{V}_R = \mathbf{I}R$, then

$$
\frac{\mathbf{V}_R}{\mathbf{V}_1} = \mathbf{G}_v(j\omega) = \frac{1}{1 + jQ(\omega/\omega_0 - \omega_0/\omega)} \qquad \textbf{12.18}
$$

and the magnitude and phase are

$$M(\omega) = \frac{1}{\left[1 + Q^2(\omega/\omega_0 - \omega_0/\omega)^2\right]^{1/2}}$$ **12.19**

and

$$\phi(\omega) = -\tan^{-1} Q\left(\frac{\omega}{\omega_0} - \frac{\omega_0}{\omega}\right)$$ **12.20**

The sketches for these functions are shown in Fig. 12.23. Note that the circuit has the form of a band-pass filter. The bandwidth as shown is the difference between the two half-power frequencies. Since power is proportional to the square of the magnitude, these two frequencies may be derived by setting the magnitude $M(\omega) = 1/\sqrt{2}$; that is,

$$\left| \frac{1}{1 + jQ(\omega/\omega_0 - \omega_0/\omega)} \right| = \frac{1}{\sqrt{2}}$$

Therefore,

$$Q\left(\frac{\omega}{\omega_0} - \frac{\omega_0}{\omega}\right) = \pm 1$$ **12.21**

Solving this equation, we obtain four frequencies,

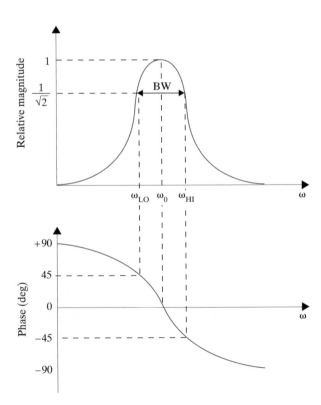

Figure 12.23 Magnitude and phase curves for Eq. (12.18).

$$\omega = \pm \frac{\omega_0}{2Q} \pm \omega_0 \sqrt{\left(\frac{1}{2Q}\right)^2 + 1} \qquad \textbf{12.22}$$

Taking only the positive values, we obtain

$$\omega_{LO} = \omega_0\left[-\frac{1}{2Q} + \sqrt{\left(\frac{1}{2Q}\right)^2 + 1}\right]$$

$$\omega_{HI} = \omega_0\left[\frac{1}{2Q} + \sqrt{\left(\frac{1}{2Q}\right)^2 + 1}\right] \qquad \textbf{12.23}$$

Half-power frequencies and their dependence on ω_0 and Q

Subtracting these two equations yields the bandwidth as shown in Fig. 12.22:

$$BW = \omega_{HI} - \omega_{LO} = \frac{\omega_0}{Q} \qquad \textbf{12.24}$$

The bandwidth is the difference between the half-power frequencies and a function of ω_0 and Q.

and multiplying the two equations yields

$$\omega_0^2 = \omega_{LO}\,\omega_{HI} \qquad \textbf{12.25}$$

which illustrates that the resonant frequency is the geometric mean of the two half-power frequencies. Recall that the half-power frequencies are the points at which the log-magnitude curve is down 3 dB from its maximum value. Therefore, the difference between the 3-dB frequencies, which is, of course, the bandwidth, is often called the 3-dB bandwidth.

EXTENSION EXERCISE

E12.10　For the network in Fig. E12.8, compute the two half-power frequencies and the bandwidth of the network.

ANSWER:　$\omega_{HI} = 1815$ rad/s, $\omega_{LO} = 1785$ rad/s, $BW = 30$ rad/s.

Equation (12.15) indicates the dependence of Q on R. A high-Q series circuit has a small value of R, and, as we will illustrate later, a high-Q parallel circuit has a relatively large value of R.

Equation (12.24) illustrates that the bandwidth is inversely proportional to Q. Therefore, the frequency selectivity of the circuit is determined by the value of Q. A high-Q circuit has a small bandwidth and, therefore, the circuit is very selective. The manner in which Q affects the frequency selectivity of the network is graphically illustrated in Fig. 12.24. Hence, if we pass a signal with a wide frequency range through a high-Q circuit, only the frequency components within the bandwidth of the network will not be attenuated; that is, the network acts like a band-pass filter.

Q has a more general meaning that we can explore via an energy analysis of the series circuit. Recall from Chapter 5 that an inductor stores energy in its magnetic field and a capacitor stores energy in its electric field. When a network is in resonance, there is a continuous exchange of energy between the magnetic field of the inductor and the electric field of the capacitor. During each half-cycle the energy stored in the inductor's magnetic field will vary from zero to a maximum value and back to zero again. The capacitor

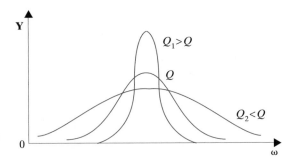

Figure 12.24 Network frequency response as a function of Q.

operates in a similar manner. The energy exchange takes place in the following way. During one quarter-cycle the capacitor absorbs energy as quickly as the inductor gives it up, and during the following one quarter-cycle the inductor absorbs energy as fast as it is released by the capacitor. Although the energy stored in each element is continuously varying, the total energy stored in the resonant circuit is constant and therefore not changing with time.

In order to define Q in terms of energy, suppose that the excitation is at the frequency ω_0 and that

$$i(t) = I_M \cos \omega_0 t$$

The energies stored in the inductor and capacitor are

$$w_L(t) = \tfrac{1}{2} L I_M^2 \cos^2 \omega_0 t$$

and

$$w_C(t) = \frac{1}{2} \frac{I_M^2}{\omega_0^2 C} \sin^2 \omega_0 t$$

The total energy stored is then

$$w_S(t) = w_L(t) + w_C(t)$$

$$= \tfrac{1}{2} I_M^2 \left(L \cos^2 \omega_0 t + \frac{1}{\omega_0^2 C} \sin^2 \omega_0 t \right)$$

However, since $\omega_0 = 1/\sqrt{LC}$, the preceding equation reduces to

$$w_S(t) = \tfrac{1}{2} L I_M^2 \equiv W_S \qquad\qquad 12.26$$

which is the maximum energy stored at resonance.

The energy dissipated per cycle can be derived by multiplying the average power absorbed by the resistor by the period of one cycle:

$$W_D = \left(\tfrac{1}{2} I_M^2 R \right) \frac{2\pi}{\omega_0} \qquad\qquad 12.27$$

Then

$$\frac{W_S}{W_D} = \frac{\frac{1}{2} L I_M^2}{\frac{1}{2} I_M^2 R (2\pi/\omega_0)}$$

$$= \frac{\omega_0 L}{R} \frac{1}{2\pi}$$

12.28

Using Eq. (12.16), we find that this equation becomes

$$\frac{W_S}{W_D} = \frac{Q}{2\pi}$$

or

$$Q = 2\pi \frac{W_S}{W_D}$$

12.29

where we recall that W_S is the maximum energy stored at resonance and W_D is the energy dissipated per cycle. The importance of this definition of Q stems from the fact that this expression is applicable to acoustic, electrical, and mechanical systems and therefore is generally considered to be the basic definition of Q.

EXAMPLE 12.9

Given a series circuit with $R = 2\ \Omega$, $L = 2$ mH, and $C = 5\ \mu$F, we wish to determine the resonant frequency, the quality factor, and the bandwidth for the circuit. Then we will determine the change in Q and the BW if R is changed from 2 to 0.2 Ω.

SOLUTION   Using Eq. (12.12), we have

$$\omega_0 = \frac{1}{\sqrt{LC}} = \frac{1}{[(2)(10^{-3})(5)(10^{-6})]^{1/2}}$$

$$= 10^4\ \text{rad/s}$$

and therefore, the resonant frequency is $10^4/2\pi = 1592$ Hz.

The quality factor is

$$Q = \frac{\omega_0 L}{R} = \frac{(10^4)(2)(10^{-3})}{2}$$

$$= 10$$

and the bandwidth is

$$\text{BW} = \frac{\omega_0}{Q} = \frac{10^4}{10}$$

$$= 10^3\ \text{rad/s}$$

If R is changed to $R = 0.2\ \Omega$, the new value of Q is 100, and therefore the new BW is 10^2 rad/s.   ❏

EXTENSION EXERCISES

E12.11 A series circuit is composed of $R = 2\,\Omega$, $L = 40\,$mH, and $C = 100\,\mu$F. Determine the bandwidth of this circuit about its resonant frequency.

ANSWER: BW $= 50\,$rad/s, $\omega_0 = 500\,$rad/s.

E12.12 A series RLC circuit has the following properties: $R = 4\,\Omega$, $\omega_0 = 4000\,$rad/s, and the BW $= 100\,$rad/s. Determine the values of L and C.

ANSWER: $L = 40\,$mH, $C = 1.56\,\mu$F.

EXAMPLE 12.10

We wish to determine the parameters R, L, and C so that the circuit shown in Fig. 12.25 operates as a band-pass filter with an ω_0 of 1000 rad/s and a bandwidth of 100 rad/s.

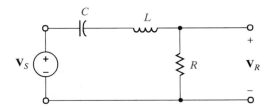

Figure 12.25 Series RLC circuit.

SOLUTION The voltage gain for the network is

$$\mathbf{G}_v(j\omega) = \frac{(R/L)j\omega}{(j\omega)^2 + (R/L)j\omega + 1/LC}$$

Hence,

$$\omega_0 = \frac{1}{\sqrt{LC}}$$

and since $\omega_0 = 10^3$,

$$\frac{1}{LC} = 10^6$$

The bandwidth is

$$\text{BW} = \frac{\omega_0}{Q}$$

Then

$$Q = \frac{\omega_0}{\text{BW}} = \frac{1000}{100}$$

$$= 10$$

However,

$$Q = \frac{\omega_0 L}{R}$$

Therefore,

$$\frac{1000L}{R} = 10$$

Note that we have two equations in the three unknown circuit parameters R, L, and C. Hence, if we select $C = 1\ \mu\text{F}$, then

$$L = \frac{1}{10^6 C} = 1\ \text{H}$$

and

$$\frac{1000(1)}{R} = 10$$

yields

$$R = 100\ \Omega$$

Therefore, the parameters $R = 100\ \Omega$, $L = 1\ \text{H}$, and $C = 1\ \mu\text{F}$ will produce the proper filter characteristics.   ❏

In Examples 12.7 and 12.8 we found that the voltage across the capacitor or inductor in the series resonant circuit could be quite high. In fact, it was equal to Q times the magnitude of the source voltage. With this in mind, let us reexamine this network as shown in Fig. 12.26. The output voltage for the network is

$$\mathbf{V}_o = \left(\frac{1/j\omega C}{R + j\omega L + 1/j\omega C} \right) \mathbf{V}_S$$

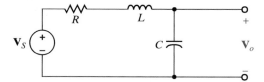

Figure 12.26 Series resonant circuit.

which can be written as

$$\mathbf{V}_o = \frac{\mathbf{V}_S}{1 - \omega^2 LC + j\omega CR}$$

The magnitude of this voltage can be expressed as

$$|\mathbf{V}_o| = \frac{|\mathbf{V}_S|}{\sqrt{(1 - \omega^2 LC)^2 + (\omega CR)^2}} \qquad \textbf{12.30}$$

In view of the previous discussion, we might assume that the maximum value of the output voltage would occur at the resonant frequency ω_0. Let's see if this assumption is correct. The frequency at which $|\mathbf{V}_o|$ is maximum is the nonzero value of ω, which satisfies the equation

$$\frac{d|\mathbf{V}_o|}{d\omega} = 0 \qquad \textbf{12.31}$$

If we perform the indicated operation and solve for the nonzero $\omega_{\max}$, we obtain

$$\omega_{\max} = \sqrt{\frac{1}{LC} - \frac{1}{2}\left(\frac{R}{L}\right)^2} \qquad \textbf{12.32}$$

By employing the relationships $\omega_0^2 = 1/LC$ and $Q = \omega_0 L/R$, the expression for $\omega_{\max}$ can be written as

$$\omega_{\max} = \sqrt{\omega_0^2 - \frac{1}{2}\left(\frac{\omega_0}{Q}\right)^2}$$

$$= \omega_0\sqrt{1 - \frac{1}{2Q^2}} \qquad \textbf{12.33}$$

Clearly, $\omega_{\max} \neq \omega_0$; however, ω_0 closely approximates $\omega_{\max}$ if the Q is high. In addition, if we substitute Eq. (12.33) into Eq.(12.30) and use the relationships $\omega_0^2 = 1/LC$ and $\omega_0^2 C^2 R^2 = 1/Q^2$, we find that

$$|\mathbf{V}_o|_{\max} = \frac{Q|\mathbf{V}_S|}{\sqrt{1 - 1/4Q^2}} \qquad \textbf{12.34}$$

Again, we see that $|\mathbf{V}_o|_{\max} \approx Q|\mathbf{V}_S|$ if the network has a high Q.

EXAMPLE 12.11

Given the network in Fig. 12.26, we wish to determine ω_0 and $\omega_{\max}$ for $R = 50\ \Omega$ and $R = 1\ \Omega$ if $L = 50$ mH and $C = 5\ \mu$F.

SOLUTION   The network parameters yield

$$\omega_0 = \frac{1}{\sqrt{LC}}$$

$$= \frac{1}{\sqrt{(5)(10^{-2})(5)(10^{-6})}}$$

$$= 2000 \text{ rad/s}$$

If $R = 50\ \Omega$, then

$$Q = \frac{\omega_0 L}{R}$$

$$= \frac{(2000)(0.05)}{50}$$

$$= 2$$

and

$$\omega_{max} = \omega_0 \sqrt{1 - \frac{1}{2Q^2}}$$

$$= 2000 \sqrt{1 - \frac{1}{8}}$$

$$= 1871 \text{ rad/s}$$

If $R = 1 \, \Omega$, then $Q = 100$ and $\omega_{max} = 2000$ rad/s.

A plot of $|\mathbf{V}_o|$ versus frequency for the network with $R = 50 \, \Omega$ and $R = 1 \, \Omega$ is shown in Figs. 12.27a and b, respectively. Note that when the Q of the network is small, the frequency response is not selective and $\omega_0 \neq \omega_{max}$. However, if the Q is large, the frequency response is very selective and $\omega_0 \simeq \omega_{max}$.

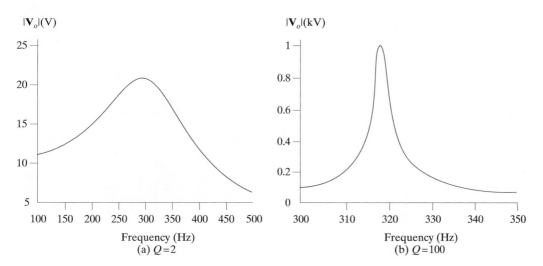

Figure 12.27 Frequency response plots for the network in Fig. 12.26 with (a) $R = 50 \, \Omega$ and (b) $R = 1 \, \Omega$.

EXAMPLE 12.12

On July 1, 1940, the third longest bridge in the nation, the Tacoma Narrows Bridge, was opened to traffic across Puget Sound in Washington. On November 7, 1940, the structure collapsed in what has become the most celebrated structural failure of this century. A photograph of the bridge, taken as it swayed back and forth just prior to breaking apart, is shown in Fig. 12.28. Explaining the disaster in quantitative terms is a feat for civil engineers and structures experts, and several theories have been presented. However, the one common denominator in each explanation is that wind blowing across the bridge caused the entire structure to resonate to such an extent that the bridge tore itself apart. One can theorize that the wind, fluctuating at a frequency near the natural frequency of the bridge (0.2 Hz), drove the structure into resonance. Thus, the bridge can be roughly modeled as a second-order system. Let us design an *RLC* resonance network to demonstrate the bridge's vertical movement and investigate the effect of the wind's frequency.

Figure 12.28 Tacoma Narrows Bridge on the verge of collapse. (Used with permission from Special Collection Division, University of Washington Libraries. Photo by Farguharson, Negative number 12.)

SOLUTION   The *RLC* network shown in Fig. 12.29 is a second-order system in which $v_{in}(t)$ is analogous to vertical deflection of the bridge's roadway (1 volt = 1 foot). The values of C, L, R_1, and R_2 can be derived from the data taken at the site and from scale models, as follows:

$$\text{vertical deflection at failure} \approx 4 \text{ feet}$$

$$\text{Wind speed at failure} \approx 42 \text{ mph}$$

$$\text{Resonant frequency} = f_0 \approx 0.2 \text{ Hz}$$

The output voltage can be expressed as

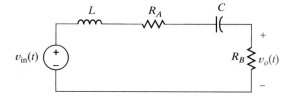

Figure 12.29 *RLC* resonance network for a simple Tacoma Narrows Bridge simulation.

$$\mathbf{V}_o(j\omega) = \frac{j\omega\left(\dfrac{R_B}{L}\right)\mathbf{V}_{\text{in}}(j\omega)}{-\omega^2 + j\omega\left(\dfrac{R_A + R_B}{L}\right) + \dfrac{1}{LC}}$$

from which we can easily extract the following expressions:

$$\omega_0 = \frac{1}{\sqrt{LC}} = 2\pi(0.2)\text{rad/s}$$

$$2\zeta\omega_o = \frac{R_A + R_B}{L}$$

and

$$\frac{\mathbf{V}_o(j\omega_0)}{\mathbf{V}_{\text{in}}(j\omega_0)} = \frac{R_B}{R_A + R_B} \approx \frac{4\text{ feet}}{42\text{ mph}}$$

Let us choose $R_B = 1\ \Omega$ and $R_A = 9.5\ \Omega$. Having no data for the damping ratio, ζ, we will select $L = 20$ H, which yields $\zeta = 0.209$ and $Q = 2.39$, which seem reasonable for such a large structure. Given the aforementioned choices, the required capacitor value is $C = 31.66$ mF. Using these circuit values, we now simulate the effect of 42 mph winds fluctuating at 0.05 Hz, 0.1 Hz, and 0.2 Hz using an ac analysis at the three frequencies of interest.

The results are shown in Fig. 12.30. Note that at 0.05 Hz the vertical deflection (1 ft/V) is only 0.44 feet, while at 0.1 Hz the bridge undulates about 1.07 feet. Finally, at the bridge's resonant frequency of 0.2 Hz, the bridge is oscillating 3.77 feet—catastrophic failure.

Clearly, we have used an extremely simplistic approach to modeling something as complicated as the Tacoma Narrows Bridge. However, we will revisit this event in Chapter 14 and examine it more closely with a more accurate model (K. Y. Billah and R. H. Scalan, "Resonance, Tacoma Narrows Bridge Failure, and Undergraduate Physics Textbooks," *American Journal of Physics* vol. 59, no. 2, pp. 118–124).   ❏

In our presentation of resonance thus far, we have focused most of our discussion on the series resonant circuit. We should recall, however, that the equations for the impedance of the series circuit and the admittance of the parallel circuit are similar. Therefore, the networks possess similar properties, as we will illustrate in the following examples.

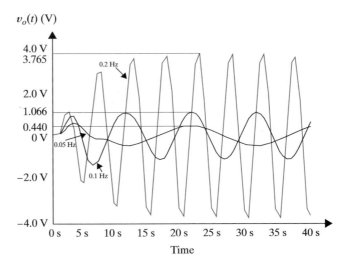

Figure 12.30 Simulated vertical deflection (1 volt = 1 foot) for the Tacoma Narrows Bridge for wind shift frequencies of 0.05, 0.1, and 0.2 Hz.

Consider the network shown in Fig. 12.31. The source current $\mathbf{I}_S$ can be expressed as

$$\mathbf{I}_S = \mathbf{I}_G + \mathbf{I}_C + \mathbf{I}_L$$

$$= \mathbf{V}_S G + j\omega C \mathbf{V}_S + \frac{\mathbf{V}_S}{j\omega L}$$

$$= \mathbf{V}_S \left[G + j\left(\omega C - \frac{1}{\omega L} \right) \right]$$

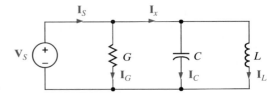

Figure 12.31 Parallel *RLC* circuit.

When the network is in resonance,

$$\mathbf{I}_S = G\mathbf{V}_S$$

that is, all the source current flows through the conductance G. Does this mean that there is no current in L or C? Definitely not! $\mathbf{I}_C$ and $\mathbf{I}_L$ are equal in magnitude but 180° out of phase with one another. Therefore, $\mathbf{I}_x$, as shown in Fig. 12.31, is zero. In addition, if $G = 0$, the source current is zero. What is actually taking place, however, is an energy exchange between the electric field of the capacitor and the magnetic field of the inductor. As one increases, the other decreases, and vice versa.

EXAMPLE 12.13

The network in Fig. 12.31 has the following parameters:

$$\mathbf{V}_S = 120 \underline{/0°}\ \text{V}, G = 0.01\ S, C = 600\ \mu\text{F, and } L = 120\ \text{mH}$$

If the source operates at the resonant frequency of the network, compute all the branch currents.

SOLUTION   The resonant frequency for the network is

$$\omega_0 = \frac{1}{\sqrt{LC}}$$

$$= \frac{1}{\sqrt{(120)(10^{-3})(600)(10^{-6})}}$$

$$= 117.85\ \text{rad/s}$$

At this frequency

$$\mathbf{Y}_C = j\omega_0 C = j7.07 \times 10^{-2}\ \text{S}$$

and

$$\mathbf{Y}_L = -j\left(\frac{1}{\omega_0 L}\right) = -j7.07 \times 10^{-2}\ \text{S}$$

The branch currents are then

$$\mathbf{I}_G = G\mathbf{V}_S = 1.2\ \underline{/0°}\ \text{A}$$
$$\mathbf{I}_C = \mathbf{Y}_C\mathbf{V}_S = 8.49\ \underline{/90°}\ \text{A}$$
$$\mathbf{I}_L = \mathbf{Y}_L\mathbf{V}_S = 8.49\ \underline{/-90°}\ \text{A}$$

and

$$\mathbf{I}_S = \mathbf{I}_G + \mathbf{I}_C + \mathbf{I}_L = \mathbf{I}_G = 1.2\ \underline{/0°}\ \text{A}$$

As the analysis indicates, the source supplies only the losses in the resistive element. In addition, the source voltage and current are in phase and, therefore, the power factor is unity.   ❑

EXAMPLE 12.14

Given the parallel RLC circuit in Fig. 12.32,

(a) Derive the expression for the resonant frequency, the half-power frequencies, the bandwidth, and the quality factor for the transfer characteristic $\mathbf{V}_{\text{out}}/\mathbf{I}_{\text{in}}$ in terms of the circuit parameters R, L, and C.

(b) Compute the quantities in part (a) if $R = 1\ \text{k}\Omega, L = 10\ \text{mH, and } C = 100\ \mu\text{F}$.

SOLUTION   (a) The output voltage can be written as

$$\mathbf{V}_{\text{out}} = \frac{\mathbf{I}_{\text{in}}}{\mathbf{Y}_T}$$

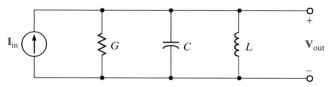

Figure 12.32 Circuit used in Example 12.14.

and, therefore, the magnitude of the transfer characteristic can be expressed as

$$\left|\frac{\mathbf{V}_{out}}{\mathbf{I}_{in}}\right| = \frac{1}{\sqrt{(1/R^2) + (\omega C - 1/\omega L)^2}}$$

The transfer characteristic is a maximum at the resonant frequency

$$\omega_0 = \frac{1}{\sqrt{LC}} \qquad\qquad \textbf{12.35}$$

and at this frequency

$$\left|\frac{\mathbf{V}_{out}}{\mathbf{I}_{in}}\right|_{max} = R \qquad\qquad \textbf{12.36}$$

As demonstrated earlier, at the half-power frequencies the magnitude is equal to $1/\sqrt{2}$ of its maximum value, and hence the half-power frequencies can be obtained from the expression

$$\frac{1}{\sqrt{(1/R^2) + (\omega C - 1/\omega L)^2}} = \frac{R}{\sqrt{2}}$$

Solving this equation and taking only the positive values of ω yields

$$\omega_{LO} = -\frac{1}{2RC} + \sqrt{\frac{1}{(2RC)^2} + \frac{1}{LC}} \qquad\qquad \textbf{12.37}$$

and

$$\omega_{HI} = \frac{1}{2RC} + \sqrt{\frac{1}{(2RC)^2} + \frac{1}{LC}} \qquad\qquad \textbf{12.38}$$

Subtracting these two half-power frequencies yields the bandwidth

$$BW = \omega_{HI} - \omega_{LO} \qquad\qquad \textbf{12.39}$$

$$= \frac{1}{RC}$$

Therefore, the quality factor is

$$Q = \frac{\omega_0}{BW}$$

$$= \frac{RC}{\sqrt{LC}} \qquad\qquad \textbf{12.40}$$

$$= R\sqrt{\frac{C}{L}}$$

Using Eqs. (12.35), (12.39), and (12.40), we can write Eqs. (12.37) and (12.38) as

$$\omega_{LO} = \omega_0\left[\frac{-1}{2Q} + \sqrt{\frac{1}{(2Q)^2} + 1}\right] \qquad \textbf{12.41}$$

$$\omega_{HI} = \omega_0\left[\frac{1}{2Q} + \sqrt{\frac{1}{(2Q)^2} + 1}\right] \qquad \textbf{12.42}$$

(b) Using the values given for the circuit components, we find that

$$\omega_0 = \frac{1}{\sqrt{(10^{-2})(10^{-4})}} = 10^3 \text{ rad/s}$$

The half-power frequencies are

$$\omega_{LO} = \frac{-1}{(2)(10^3)(10^{-4})} + \sqrt{\frac{1}{[(2)(10^{-1})]^2} + 10^6}$$

$$= 995 \text{ rad/s}$$

and

$$\omega_{HI} = 1005 \text{ rad/s}$$

Therefore, the bandwidth is

$$\text{BW} = \omega_{HI} - \omega_{LO} = 10 \text{ rad/s}$$

and

$$Q = 10^3 \sqrt{\frac{10^{-4}}{10^{-2}}}$$

$$= 100 \qquad \qquad \square$$

EXAMPLE 12.15

Two radio stations, WHEW and WHAT, broadcast in the same listening area: WHEW broadcasts at 100 MHz and WHAT at 98 MHz. A single-stage tuned amplifier, such as that shown in Fig. 12.33, can be used as a tuner to filter out one of the stations. However, single-stage tuned amplifiers have poor selectivity due to their wide bandwidths. In order to reduce the bandwidth (increase the quality factor) of single-stage tuned amplifiers, designers employ a technique called synchronous tuning. In this process, identical tuned amplifiers are cascaded. To demonstrate this phenomenon, let us generate a bode plot for the amplifier shown in Fig. 12.33 when it is tuned to WHEW (100 mhz), using one, two, three, and four stages of amplification.

SOLUTION   Using the circuit for a single-stage amplifier shown in Fig. 12.33, we can cascade the stages to form a four-stage synchronously tuned amplifier. If we now plot the frequency response over the range from 90 MHz to 110 MHz, which is easily done using PSPICE, we obtain the Bode plot shown in Fig. 12.34.

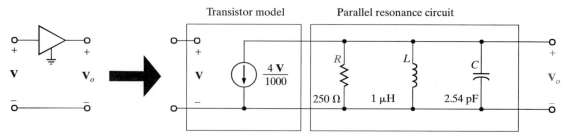

Figure 12.33 Single-stage tuned amplifier.

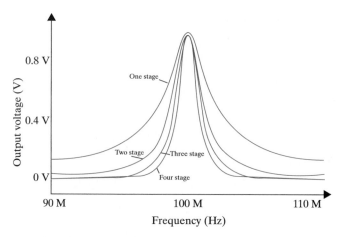

Figure 12.34 Bode plots for one-, two-, three-, and four-stage tuned amplifiers.

From the Bode plot in Fig. 12.34 we see that increasing the number of stages does indeed decrease the bandwidth without altering the center frequency. As a result, the quality factor and selectivity increase. Accordingly, as we add stages, the gain at 98 MHz (WHAT's frequency) decreases and that station is "tuned out." ❑

EXTENSION EXERCISES

E12.13   A parallel RLC circuit has the following parameters: $R = 2\ \text{k}\Omega$, $L = 20\ \text{mH}$, and $C = 150\ \mu\text{F}$. Determine the resonant frequency, the Q, and the bandwidth of the circuit.

ANSWER: $\omega_0 = 577\ \text{rad/s}$, $Q = 173$, and BW $= 3.33\ \text{rad/s}$.

E12.14   A parallel RLC circuit has the following parameters: $R = 6\ \text{k}\Omega$, BW $= 1000\ \text{rad/s}$, and $Q = 120$. Determine the values of L, C, and ω_0.

ANSWER: $L = 417.5\ \mu\text{H}$, $C = 0.167\ \mu\text{F}$, and $\omega_0 = 119{,}760\ \text{rad/s}$.

In general, the resistance of the winding of an inductor cannot be neglected, and hence a more practical parallel resonant circuit is the one shown in Fig. 12.35. The input admittance of this circuit is

$$\mathbf{Y}(j\omega) = j\omega C + \frac{1}{R + j\omega L}$$

$$= j\omega C + \frac{R - j\omega L}{R^2 + \omega^2 L^2}$$

$$= \frac{R}{R^2 + \omega^2 L^2} + j\left(\omega C - \frac{\omega L}{R^2 + \omega^2 L^2}\right)$$

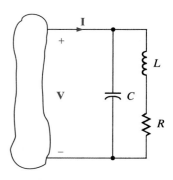

Figure 12.35 Practical parallel resonant circuit.

The resonant frequency at which the admittance is purely real is

$$\omega_r C - \frac{\omega_r L}{R^2 + \omega_r^2 L^2} = 0$$

$$\omega_r = \sqrt{\frac{1}{LC} - \frac{R^2}{L^2}}$$ **12.43**

EXAMPLE 12.16

Given the tank circuit in Fig. 12.36, let us determine ω_0 and ω_r for $R = 50\ \Omega$ and $R = 5\ \Omega$.

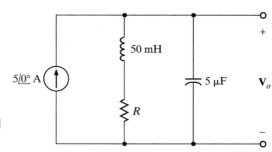

Figure 12.36 Tank circuit used in Example 12.16.

SOLUTION   Using the network parameter values, we obtain

$$\omega_0 = \frac{1}{\sqrt{LC}}$$

$$= \frac{1}{\sqrt{(0.05)(5)(10^{-6})}}$$

$$= 2000\ \text{rad/s}$$

$$= 318.3\ \text{Hz}$$

If $R = 50\ \Omega$, then

$$\omega_r = \sqrt{\frac{1}{LC} - \frac{R^2}{L^2}}$$

$$= \sqrt{\frac{1}{(0.05)(5)(10^{-6})} - \left(\frac{50}{0.05}\right)^2}$$

$$= 1732 \text{ rad/s}$$

$$= 275.7 \text{ Hz}$$

If $R = 5 \, \Omega$, then

$$\omega_r = \sqrt{\frac{1}{(0.05)(5)(10^{-6})} - \left(\frac{5}{0.05}\right)^2}$$

$$= 1997 \text{ rad/s}$$

$$= 317.9 \text{ Hz}$$

Note that as $R \rightarrow 0, \omega_r \rightarrow \omega_0$. This fact is also illustrated in the frequency-response curves in Figs. 12.37a and b, where we have plotted $|\mathbf{V}_o|$ versus frequency for $R = 50 \, \Omega$ and $R = 5 \, \Omega$, respectively.

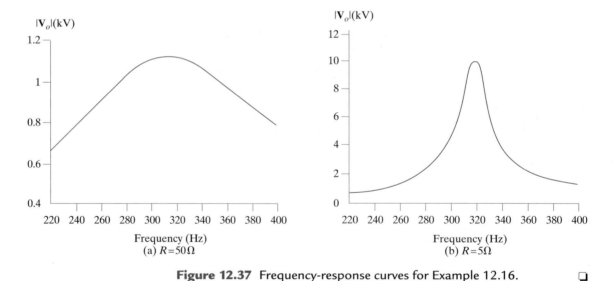

Figure 12.37 Frequency-response curves for Example 12.16. ❏

Let us now try to relate some of the things we have learned about resonance to the Bode plots we presented earlier. The admittance for the series resonant circuit is

$$\mathbf{Y}(j\omega) = \frac{1}{R + j\omega L + 1/j\omega C}$$

$$= \frac{j\omega C}{(j\omega)^2 LC + j\omega CR + 1} \qquad \textbf{12.44}$$

The standard form for the quadratic factor is

$$(j\omega\tau)^2 + 2\zeta\omega\tau j + 1$$

where $\tau = 1/\omega_0$, and hence in general the quadratic factor can be written as

$$\frac{(j\omega)^2}{\omega_0^2} + \frac{2\zeta\omega}{\omega_0}j + 1 \qquad \textbf{12.45}$$

If we now compare this form of the quadratic factor with the denominator of $\mathbf{Y}(j\omega)$, we find that

$$\omega_0^2 = \frac{1}{LC}$$

$$\frac{2\zeta}{\omega_0} = CR$$

and therefore,

$$\zeta = \frac{R}{2}\sqrt{\frac{C}{L}}$$

However, from Eq. (12.15),

$$Q = \frac{1}{R}\sqrt{\frac{L}{C}}$$

and hence,

$$Q = \frac{1}{2\zeta} \qquad \textbf{12.46}$$

To illustrate the significance of this equation, consider the Bode plot for the function $\mathbf{Y}(j\omega)$. The plot has an initial slope of +20 dB/decade due to the zero at the origin. If $\zeta > 1$, the poles represented by the quadratic factor in the denominator will simply roll off the frequency response, as illustrated in Fig. 12.13a, and at high frequencies the slope of the composite characteristic will be −20 dB/decade. Note from Eq. (12.46) that if $\zeta > 1$, the Q of the circuit is very small. However, if $0 < \zeta < 1$, the frequency response will peak as shown in Fig. 12.13a, and the sharpness of the peak will be controlled by ζ. If ζ is very small, the peak of the frequency response is very narrow, the Q of the network is very large, and the circuit is very selective in filtering the input signal. Equation (12.46) and Fig. 12.24 illustrate the connections among the frequency response, the Q, and the ζ of a network.

12.4 SCALING

Throughout this book we have employed a host of examples to illustrate the concepts being discussed. In many cases the actual values of the parameters were unrealistic in a practical sense, even though they may have simplified the presentation. In this section we illustrate how to *scale* the circuits to make them more realistic.

There are two ways to scale a circuit: *magnitude or impedance scaling* and *frequency scaling*. To magnitude scale a circuit, we simply multiply the impedance of each element by a scale factor K_M. Therefore, a resistor R becomes $K_M R$. Multiplying the impedance of an inductor $j\omega L$ by K_M yields a new inductor $K_M L$, and multiplying the impedance of a capacitor $1/j\omega C$ by K_M yields a new capacitor C/K_M. Therefore, in magnitude scaling,

Magnitude or impedance scaling

$$R' \to K_M R$$
$$L' \to K_M L$$
$$C' \to \frac{C}{K_M}$$

12.47

since

$$\omega_0' = \frac{1}{\sqrt{L'C'}} = \frac{1}{\sqrt{K_M LC/K_M}} = \omega_0$$

and Q' is

$$Q' = \frac{\omega_0 L'}{R'} = \frac{\omega_0 K_M L}{K_M R} = Q$$

The resonant frequency, the quality factor, and therefore the bandwidth are unaffected by magnitude scaling.

In frequency scaling the scale factor is denoted as K_F. The resistor is frequency independent and, therefore, unaffected by this scaling. The new inductor L', which has the same impedance at the scaled frequency ω_1', must satisfy the equation

$$j\omega_1 L = j\omega_1' L'$$

where $\omega_1' = k_F \omega_1$. Therefore,

$$j\omega_1 L = jK_F \omega_1 L'$$

Hence, the new inductor value is

$$L' = \frac{L}{K_F}$$

Using a similar argument, we find that

$$C' = \frac{C}{K_F}$$

Frequency scaling

Therefore, to frequency scale by a factor K_F,

$$R' \to R$$
$$L' \to \frac{L}{K_F}$$
$$C' \to \frac{C}{K_F}$$

12.48

Note that

$$\omega_0' = \frac{1}{\sqrt{(L/K_F)(C/K_F)}} = K_F\omega_0$$

and

$$Q' = \frac{K_F\omega_0 L}{RK_F} = Q$$

and therefore,

$$BW' = K_F(BW)$$

Hence, the resonant frequency and bandwidth of the circuit are affected by frequency scaling.

EXAMPLE 12.17

If the values of the circuit parameters in Fig 12.35 are $R = 2\,\Omega$, $L = 1\,H$, and $C = \frac{1}{2}\,F$, let us determine the values of the elements if the circuit is magnitude scaled by a factor $K_M = 10^2$ and frequency scaled by a factor $K_F = 10^2$.

SOLUTION   The magnitude scaling yields

$$R' = 2K_M = 200\,\Omega$$
$$L' = (1)K_M = 100\,H$$
$$C' = \frac{1}{2}\frac{1}{K_M} = \frac{1}{200}\,F$$

Applying frequency scaling to these values yields the final results:

$$R'' = 200\,\Omega$$
$$L'' = \frac{100}{K_F} = 100\,\mu H$$
$$C'' = \frac{1}{200}\frac{1}{K_F} = 0.005\,\mu F$$

EXTENSION EXERCISE

E12.15   An RLC network has the following parameter values: $R = 10\,\Omega$, $L = 1\,H$, and $C = 2\,F$. Determine the values of the circuit elements if the circuit is magnitude scaled by a factor of 100 and freqency scaled by a factor of 10,000.

ANSWER: $R = 1\,k\Omega$, $L = 10\,mH$, $C = 2\,\mu F$.

12.5 FILTER NETWORKS

Passive Filters

A filter network is generally designed to pass signals with a specific frequency range and reject or attenuate signals whose frequency spectrum is outside this passband. The most common filters are *low-pass* filters, which pass low frequencies and reject high frequencies; *high-pass* filters, which pass high frequencies and block low frequencies; *band-pass* filters, which pass some particular band of frequencies and reject all frequencies outside the range; and *band-rejection* filters, which are specifically designed to reject a particular band of frequencies and pass all other frequencies.

The ideal frequency characteristic for a low-pass filter is shown in Fig. 12.38a. Also shown is a typical or physically realizable characteristic. Ideally, we would like the low-pass filter to pass all frequencies to some frequency ω_0 and pass no frequency above that value; however, it is not possible to design such a filter with linear circuit elements. Hence, we must be content to employ filters that we can actually build in the laboratory, and these filters have frequency characteristics that are simply not ideal.

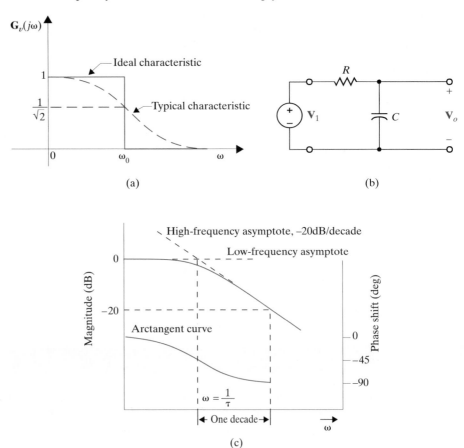

Figure 12.38 Low-pass filter circuit and its frequency characteristics.

A simple low-pass filter network is shown in Fig. 12.38b. The voltage gain for the network is

$$\mathbf{G}_v(j\omega) = \frac{1}{1 + j\omega RC} \qquad \textbf{12.49}$$

which can be written as

$$\mathbf{G}_v(j\omega) = \frac{1}{1 + j\omega\tau} \qquad \textbf{12.50}$$

where $\tau = RC$, the time constant. The amplitude characteristic is

$$M(\omega) = \frac{1}{\left[1 + (\omega\tau)^2\right]^{1/2}} \qquad \textbf{12.51}$$

and the phase characteristic is

$$\phi(\omega) = -\tan^{-1}\omega\tau \qquad \textbf{12.52}$$

Note that at the break frequency, $\omega = 1/\tau$ and the amplitude is

$$M\left(\omega = \frac{1}{\tau}\right) = \frac{1}{\sqrt{2}} \qquad \textbf{12.53}$$

The break frequency is also commonly called the *half-power frequency*. This name is derived from the fact that if the voltage or current is $1/\sqrt{2}$ of its maximum value, then the power, which is proportional to the square of the voltage or current, is one-half its maximum value.

The magnitude, in decibels, and phase curves for this simple low-pass circuit are shown in Fig. 12.38c. Note that the magnitude curve is flat for low frequencies and rolls off at high frequencies. The phase shifts from $0°$ at low frequencies to $-90°$ at high frequencies.

The ideal frequency characteristic for a high-pass filter is shown in Fig. 12.39a, together with a typical characteristic that we could achieve with linear circuit components. Ideally, the high-pass filter passes all frequencies above some frequency ω_0 and no frequencies below that value.

A simple high-pass filter network is shown in Fig. 12.39b. This is the same network as shown in Fig. 12.38b, except that the output voltage is taken across the resistor. The voltage gain for this network is

$$\mathbf{G}_v(j\omega) = \frac{j\omega\tau}{1 + j\omega\tau} \qquad \textbf{12.54}$$

where once again $\tau = RC$. The magnitude of this function is

$$M(\omega) = \frac{\omega\tau}{\left[1 + (\omega\tau)^2\right]^{1/2}} \qquad \textbf{12.55}$$

and the phase is

$$\phi(\omega) = \frac{\pi}{2} - \tan^{-1}\omega\tau \qquad \textbf{12.56}$$

The half-power frequency is $\omega = 1/\tau$, and the phase at this frequency is $45°$.

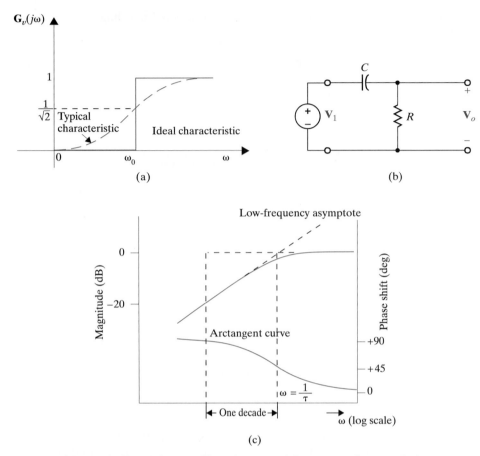

Figure 12.39 High-pass filter circuit and frequency characteristics.

The magnitude and phase curves for this high-pass filter are shown in Fig. 12.39c. At low frequencies the magnitude curve has a slope of +20 dB/decade due to the term $\omega\tau$ in the numerator of Eq. (12.55). Then at the break frequency the curve begins to flatten out. The phase curve is derived from Eq. (12.56).

Ideal and typical amplitude characteristics for simple band-pass and band-rejection filters are shown in Figs. 12.40a and b, respectively. Simple networks that are capable of realizing the typical characteristics of each filter are shown below the characteristics in Figs. 12.40c and d. ω_0 is the center frequency of the pass or rejection band and the frequency at which the maximum or minimum amplitude occurs. ω_{LO} and ω_{HI} are the lower and upper break frequencies or *cutoff frequencies*, where the amplitude is $1/\sqrt{2}$ of the maximum value. The width of the pass or rejection band is called *bandwidth*, and hence

$$BW = \omega_{HI} - \omega_{LO} \qquad \textbf{12.57}$$

To illustrate these points, let us consider the band-pass filter. The voltage transfer function is

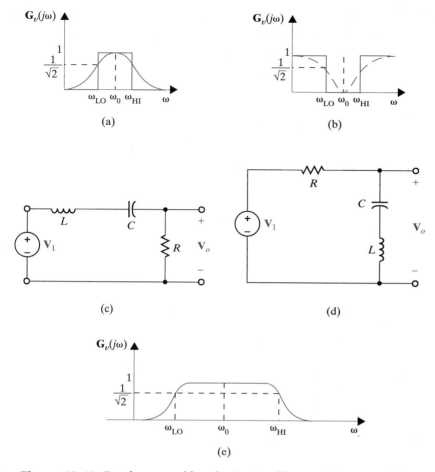

Figure 12.40 Band-pass and band-rejection filters and characteristics.

$$\mathbf{G}_v(j\omega) = \frac{R}{R + j(\omega L - 1/\omega C)}$$

and, therefore, the amplitude characteristic is

$$M(\omega) = \frac{RC\omega}{\sqrt{(RC\omega)^2 + (\omega^2 LC - 1)^2}}$$

At low frequencies

$$M(\omega) \approx \frac{RC\omega}{1} \approx 0$$

At high frequencies

$$M(\omega) \approx \frac{RC\omega}{\omega^2 LC} \approx \frac{R}{\omega L} \approx 0$$

In the midfrequency range $(RC\omega)^2 \gg (\omega^2 LC - 1)^2$, and thus $M(\omega) \approx 1$. Therefore, the frequency characteristic for this filter is shown in Fig. 12.40e. The center frequency is $\omega_0 = 1/\sqrt{LC}$. At the lower cutoff frequency

$$\omega^2 LC - 1 = -RC\omega$$

or

$$\omega^2 + \frac{R\omega}{L} - \omega_0^2 = 0$$

Solving this expression for ω_{LO}, we obtain

$$\omega_{LO} = \frac{-(R/L) + \sqrt{(R/L)^2 + 4\omega_0^2}}{2}$$

At the upper cutoff frequency

$$\omega^2 LC - 1 = +RC\omega$$

or

$$\omega^2 + \frac{R}{L}\omega - \omega_0^2 = 0$$

Solving this expression for ω_{HI}, we obtain

$$\omega_{HI} = \frac{+(R/L) + \sqrt{(R/L)^2 + 4\omega_0^2}}{2}$$

Therefore, the bandwidth of the filter is

$$BW = \omega_{HI} - \omega_{LO} = \frac{R}{L}$$

EXAMPLE 12.18

Consider the frequency-dependent network in Fig. 12.41. Given the following circuit parameter values: $L = 159\ \mu H$, $C = 159\ \mu F$, and $R = 10\ \Omega$, let us demonstrate that this one network can be used to produce a low-pass, high-pass, or band-pass filter.

SOLUTION The voltage gain $\mathbf{V}_R/\mathbf{V}_S$ is found by voltage division to be

$$\frac{\mathbf{V}_R}{\mathbf{V}_S} = \frac{R}{j\omega L + R + 1/(j\omega C)} = \frac{j\omega\left(\dfrac{R}{L}\right)}{(j\omega)^2 + j\omega\left(\dfrac{R}{L}\right) + \dfrac{1}{LC}}$$

$$= \frac{(62.8 \times 10^3)j\omega}{-\omega^2 + (62.8 \times 10^3)j\omega + 39.4 \times 10^6}$$

which is the transfer function for a band-pass filter. At resonance, $\omega^2 = 1/LC$, and hence

$$\frac{\mathbf{V}_R}{\mathbf{V}_S} = 1$$

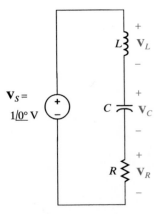

Figure 12.41 Circuit used in Example 12.18.

Now consider the gain $\mathbf{V}_L/\mathbf{V}_S$:

$$\frac{\mathbf{V}_L}{\mathbf{V}_S} = \frac{j\omega L}{j\omega L + R + 1/(j\omega C)} = \frac{-\omega^2}{(j\omega)^2 + j\omega\left(\dfrac{R}{L}\right) + \dfrac{1}{LC}}$$

$$= \frac{-\omega^2}{-\omega^2 + (62.8 \times 10^3)j\omega + 39.4 \times 10^6}$$

which is a second-order high-pass filter transfer function. Again, at resonance,

$$\frac{\mathbf{V}_L}{\mathbf{V}_S} = \frac{j\omega L}{R} = jQ = j0.1$$

Similarly, the gain $\mathbf{V}_C/\mathbf{V}_S$ is

$$\frac{\mathbf{V}_C}{\mathbf{V}_S} = \frac{1/(j\omega C)}{j\omega L + R + 1/(j\omega C)} = \frac{\dfrac{1}{LC}}{(j\omega)^2 + j\omega\left(\dfrac{R}{L}\right) + \dfrac{1}{LC}}$$

$$= \frac{39.4 \times 10^6}{-\omega^2 + (62.8 \times 10^3)j\omega + 39.4 \times 10^6}$$

which is a second-order low-pass filter transfer function. At resonant frequency,

$$\frac{\mathbf{V}_C}{\mathbf{V}_S} = \frac{1}{j\omega CR} = -jQ = -j0.1$$

Thus, one circuit produces three different filters depending on where the output is taken. This can be seen in the Bode plot for each of the three voltages in Fig. 12.42, where $\mathbf{V}_S$ is set to $1 \underline{/0°}$ V.

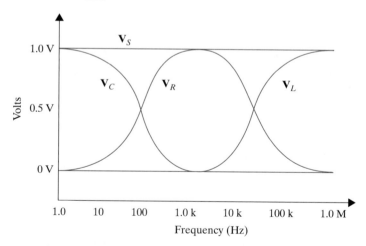

Figure 12.42 Bode plots for network in Fig. 12.41.

We know that Kirchoff's voltage law must be satisfied at all times. Note from the Bode plot that the $\mathbf{V}_R + \mathbf{V}_C + \mathbf{V}_L$ also equals $\mathbf{V}_S$ at all frequencies! Finally, let us demonstrate KVL by adding $\mathbf{V}_R, \mathbf{V}_L$, and $\mathbf{V}_C$.

$$\mathbf{V}_L + \mathbf{V}_R + \mathbf{V}_C = \frac{\left((j\omega)^2 + j\omega\left(\dfrac{R}{L}\right) + \dfrac{1}{\sqrt{LC}}\right)\mathbf{V}_S}{(j\omega)^2 + j\omega\left(\dfrac{R}{L}\right) + \dfrac{1}{\sqrt{LC}}} = \mathbf{V}_S$$

Thus, even though $\mathbf{V}_S$ is distributed between the resistor, capacitor, and inductor based on frequency, the sum of the three voltages completely reconstructs $\mathbf{V}_S$. ❑

EXAMPLE 12.19

A telephone transmission system suffers from 60-Hz interference caused by nearby power utility lines. Let us use the network in Fig. 12.43 to design a simple notch filter to eliminate the 60-Hz interference.

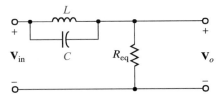

Figure 12.43 Circuit used in Example 12.19.

SOLUTION The resistor R_{eq} represents the equivalent resistance of the telephone system to the right of the LC combination. The LC parallel combination has an equivalent impedance of

$$\mathbf{Z} = (j\omega L)//(1/j\omega C) = \frac{(L/C)}{j\omega L + 1/(j\omega C)}$$

Now the voltage transfer function is

$$\frac{\mathbf{V}_o}{\mathbf{V}_{in}} = \frac{R_{eq}}{R_{eq} + \mathbf{Z}} = \frac{R_{eq}}{R_{eq} + \dfrac{(L/C)}{j\omega L + (1/j\omega C)}}$$

which can be written

$$\frac{\mathbf{V}_o}{\mathbf{V}_{in}} = \frac{(j\omega)^2 + \dfrac{1}{LC}}{(j\omega)^2 + \left(\dfrac{j\omega}{R_{eq}C}\right) + \dfrac{1}{LC}}$$

Notice that at resonance, the numerator and thus $\mathbf{V}_o$ go to zero. We want resonance to occur at 60 Hz. Thus,

$$\omega_0 = \frac{1}{\sqrt{LC}} = 2\pi(60) = 120\pi$$

If we select $C = 100\ \mu F$, then the required value for L is 70.3 mH—both are reasonable values. To demonstrate the effectiveness of the filter, let the input voltage consist of a 60-Hz sinusoid and a 1000-Hz sinusoid of the form

$$v_{in}(t) = 1 \sin[(2\pi)60t] + 0.2 \sin[(2\pi)1000t]$$

The input and output waveforms are both shown in Fig. 12.44. Note that the output voltage, as desired, contains none of the 60-Hz interference.

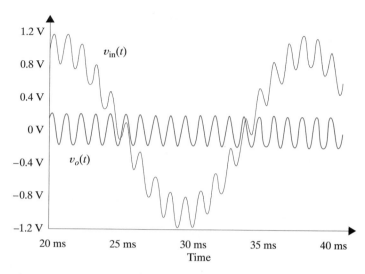

Figure 12.44 Transient analysis of the network in Fig. 12.43.

EXTENSION EXERCISES

E12.16   Given the filter network shown in Fig. E12.16, sketch the magnitude characteristic of the Bode plot for $\mathbf{G}_v(j\omega)$.

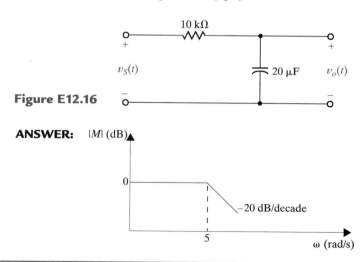

Figure E12.16

ANSWER:

E12.17   Given the filter network in Fig. E12.17, sketch the magnitude characteristic of the Bode plot for $\mathbf{G}_v(j\omega)$.

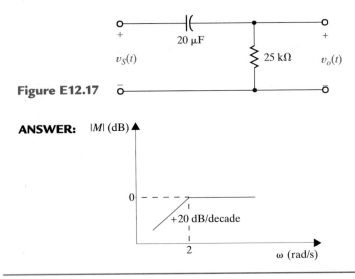

Figure E12.17

ANSWER:

E12.18   A band-pass filter network is shown in Fig. E12.18. Sketch the magnitude characteristic of the Bode plot for $\mathbf{G}_v(j\omega)$.

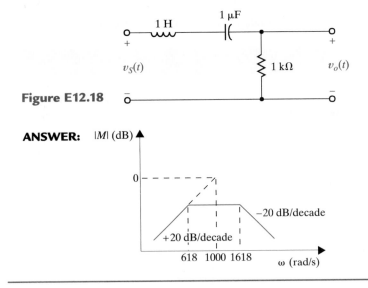

Figure E12.18

ANSWER:

Active Filters

In the preceding section we saw that the four major classes of filters (*low pass*, *high pass*, *band pass*, and *band rejection*) are realizable with simple, passive element circuits. However, passive filters do have some serious drawbacks. One obvious problem is the inability to generate a network with a gain > 1 since a purely passive network cannot add energy to a signal. Another serious drawback of passive filters is the need in many topologies for inductive elements. Inductors are generally expensive and are not usually available in precise values. In addition, inductors usually come in odd shapes (toroids, bobbins, E-cores, etc.) and are not easily handled by existing automated printed circuit board assembly machines. By applying operational amplifiers in linear feedback circuits, one can generate all of the primary filter types using only resistors, capacitors, and the op-amp integrated circuits themselves. The only exceptions to this are applications that involve very high frequency and/or high power.

The equivalent circuits for the operational amplifiers derived in Chapter 3 are valid in the sinusoidal steady-state case also, when we replace the attendant resistors with impedances. The equivalent circuits for the basic inverting and noninverting op-amp circuits are shown in Figs. 12.45a and b, respectively. Particular filter characteristics are obtained by judiciously selecting the impedances $\mathbf{Z}_1$ and $\mathbf{Z}_2$.

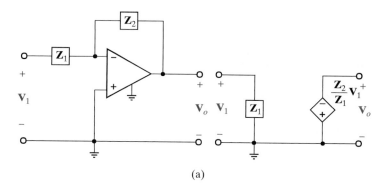

(a)

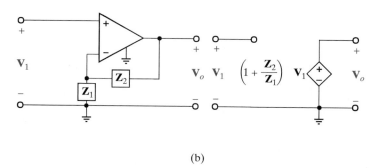

(b)

Figure 12.45 Equivalent circuits for the (a) inverting and (b) noninverting operational amplifier circuits.

EXAMPLE 12.20

Let us determine the filter characteristics of the network shown in Fig. 12.46.

SOLUTION   The impedances as illustrated in Fig. 12.45a are

$$\mathbf{Z}_1 = R_1$$

and

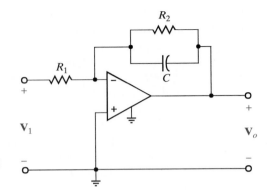

Figure 12.46 Operational amplifier filter circuit.

$$\mathbf{Z}_2 = \frac{R_2/j\omega C}{R_2 + 1/j\omega C} = \frac{R_2}{j\omega R_2 C + 1}$$

Therefore, the voltage gain of the network is

$$\mathbf{G}_v(j\omega) = \frac{\mathbf{V}_0(j\omega)}{\mathbf{V}_1(j\omega)} = \frac{-R_2/R_1}{j\omega R_2 C + 1}$$

Note that the transfer function is that of a low-pass filter. ❏

EXAMPLE 12.21

We will show that the amplitude characteristic for the filter network in Fig. 12.47a is as shown in Fig. 12.47b.

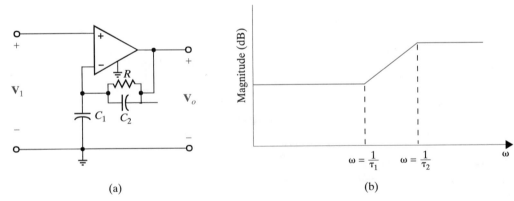

(a) (b)

Figure 12.47 Operational amplifier circuit and its amplitude characteristic.

SOLUTION Comparing this network with that in Fig. 12.45b, we see that

$$\mathbf{Z}_1 = \frac{1}{j\omega C_1}$$

and

$$\mathbf{Z}_2 = \frac{R}{j\omega RC_2 + 1}$$

Therefore, the voltage gain for the network as a function of frequency is

$$G_v(j\omega) = \frac{\mathbf{V}_0(j\omega)}{\mathbf{V}_1(j\omega)} = 1 + \frac{R/(j\omega RC_2 + 1)}{1/j\omega C_1}$$

$$= \frac{j\omega(RC_1 + RC_2) + 1}{j\omega RC_2 + 1}$$

$$= \frac{j\omega\tau_1 + 1}{j\omega\tau_2 + 1}$$

where $\tau_1 = R(C_1 + C_2)$ and $\tau_2 = RC_2$. Since $\tau_1 > \tau_2$, the amplitude characteristic is of the form shown in Fig. 12.47b. Note that the low frequencies have a gain of 1; however, the high frequencies are amplified. The exact amount of amplification is determined through selection of the circuit parameters.   ❏

EXAMPLE 12.22

A low-pass filter is shown in Fig. 12.48, together with the op-amp subcircuit. We wish to plot the frequency response of the filter over the range from 1 to 10,000 Hz.

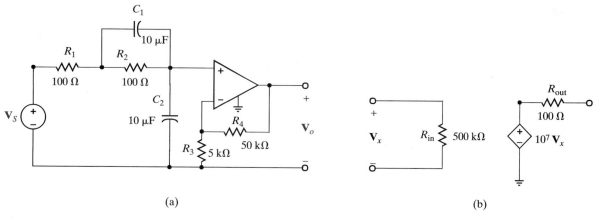

(a)   (b)

Figure 12.48 Circuit used in Example 12.22: (a) low-pass filter; (b) op-amp subcircuit.

SOLUTION   The frequency-response plot, which can be determined by any convenient method, such as a PSPICE simulation, is shown in Fig. 12.49.   ❏

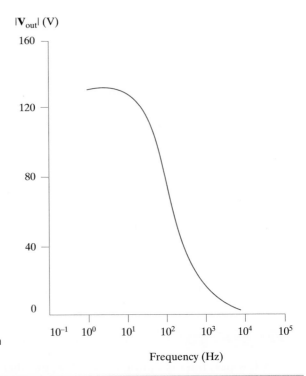

Figure 12.49 Frequency-response plot for the network in Example 12.22.

EXAMPLE 12.23

A high-pass filter network is shown in Fig. 12.50 together with the op-amp subcircuit. We wish to plot the frequency response of the filter over the range from 1 to 100 kHz.

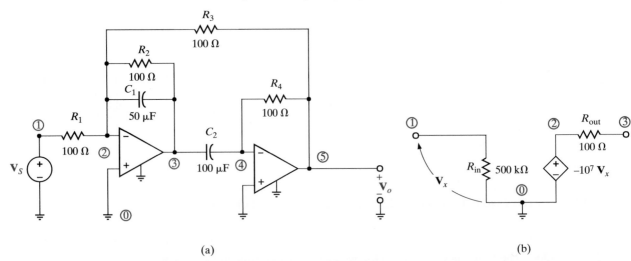

(a)   (b)

Figure 12.50 Circuits used in Example 12.23: (a) high-pass filter; (b) op-amp subcircuit.

SOLUTION   The frequency-response plot for this high-pass filter is shown in Fig. 12.51.   ❑

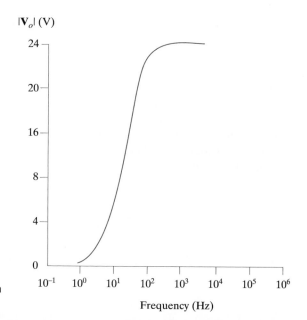

Figure 12.51 Frequency-response plot for the network in Example 12.23.

Frequency (Hz)

EXTENSION EXERCISE

E12.19   Given the filter network shown in Fig. E12.19, determine the transfer function $\mathbf{G}_v(j\omega)$, sketch the magnitude characteristic of the Bode plot for $\mathbf{G}_v(j\omega)$, and identify the filter characteristics of the network.

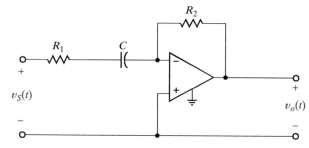

Figure E12.19

ANSWER: $\mathbf{G}_v(j\omega) = \dfrac{-j\omega C R_2}{1 + j\omega C R_1}$; this is a high-pass filter.

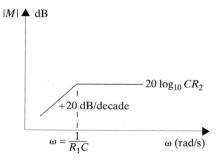

All the circuits considered so far in this section have been first-order filters. In other words, they all had no more than one pole and/or one zero. In many applications it is desired to generate a circuit with a frequency selectivity greater than that afforded by first-order circuits. The next logical step is to consider the class of second-order filters. For most active-filter applications, if an order greater than two is desired, one usually takes two or more active filter circuits and places them in series so that the total response is the desired higher-order response. This is done because first- and second-order filters are well understood and easily obtained with single op-amp circuits.

In general, second-order filters will have a transfer function with a denominator containing quadratic poles of the form $s^2 + As + B$. For high-pass and low-pass circuits, $B = \omega_c^2$ and $A = 2\zeta\omega_c$. For these circuits, ω_c is the cutoff frequency, and ζ is the damping ratio discussed earlier.

For band-pass circuits, $B = \omega_0^2$ and $A = \omega_0/Q$, where ω_0 is the center frequency and Q is the quality factor for the circuit. Notice that $Q = 1/2\zeta$. Q is a measure of the selectivity of these circuits. The bandwidth is ω_0/Q, as discussed previously.

The transfer function of the second-order low-pass active filter can generally be written as

$$H(s) = \frac{H_0\omega_c^2}{s^2 + 2\zeta\omega_c s + \omega_c^2} \qquad 12.58$$

where H_0 is the dc gain. A circuit that exhibits this transfer function is illustrated in Fig. 12.52 and has the following transfer function:

$$H(s) = \frac{V_o(s)}{V_{in(s)}} = \frac{-\left(\dfrac{R_3}{R_1}\right)\left(\dfrac{1}{R_3 R_2 C_1 C_2}\right)}{s^2 + s\left(\dfrac{1}{R_1 C_1} + \dfrac{1}{R_2 C_1} + \dfrac{1}{R_3 C_1}\right) + \dfrac{1}{R_3 R_2 C_1 C_2}} \qquad 12.59$$

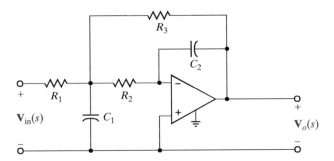

Figure 12.52 Second-order low-pass filter.

EXAMPLE 12.24

We wish to determine the damping ratio, cutoff frequency, and dc gain H_o for the network in Fig. 12.52 if $R_1 = R_2 = R_3 = 5\text{ k}\Omega$ and $C_1 = C_2 = 0.1\ \mu\text{F}$.

SOLUTION   Comparing Eqs. (12.58) and (12.59), we find that

$$\omega_c = \frac{1}{\sqrt{R_3 R_2 C_1 C_2}}$$

$$2\zeta\omega_c = \frac{1}{C_1}\left(\frac{1}{R_1} + \frac{1}{R_2} + \frac{1}{R_3}\right)$$

and therefore,

$$\zeta = \frac{1}{2}\sqrt{\frac{C_2}{C_1}}\left(\frac{1}{R_1} + \frac{1}{R_2} + \frac{1}{R_3}\right)\sqrt{R_2 R_3}$$

In addition, we note that

$$H_o = -\frac{R_3}{R_1}$$

Substituting the given parameter values into the preceding equation yields

$$\omega_o = 2000 \text{ rad/s} = 318 \text{ Hz}$$
$$\zeta = 1.5$$

and

$$H_o = -1 \qquad\qquad \square$$

EXAMPLE 12.25

We wish to vary the capacitors C_1 and C_2 in Example 12.24 to achieve damping ratios of 1.0, 0.75, 0.50, and 0.25 while maintaining ω_c constant at 2000 rad/s.

SOLUTION   As shown in the cutoff-frequency equation in Example 12.24, if ω_c is to remain constant at 2000 rad/s, the product of C_1 and C_2 must remain constant. Using the capacitance values in Example 12.24, we have

$$C_1 C_2 = (10)^{-4}$$

or

$$C_2 = \frac{(10)^{-14}}{C_1}$$

Substituting this expression into the equation for the damping ratio yields

$$\zeta = \frac{\sqrt{10^{-14}}}{\sqrt{C_1}\sqrt{C_1}}\left[\frac{1}{2}\left(\frac{1}{R_1} + \frac{1}{R_2} + \frac{1}{R_3}\right)\right]\sqrt{R_2 R_3}$$

$$= \frac{(0.15)(10^{-6})}{C_1}$$

or

$$C_1 = \frac{(0.15)(10^{-6})}{\zeta}$$

Therefore, for $\zeta = 1.0, 0.75, 0.50$, and 0.25, the corresponding values for C_1 are 0.15, 0.20, 0.30, and 0.6 μF, respectively. The values of C_2 that correspond to these values of C_1 are 67, 50, 33, and 17 nF, respectively. ❏

This example illustrates that we can adjust the network parameters to achieve a specified transient response while maintaining the cutoff frequency of the filter constant. In fact, as a general rule we design filters with specific characteristics through proper manipulation of the network parameters.

EXAMPLE 12.26

We will now demonstrate that the transient response of the circuits generated in Example 12.25 will exhibit increasing overshoot and ringing as ζ is decreased. We will apply a -1-V step function to the input of the network and employ the op-amp model with $R_i = \infty \ \Omega, R_o = 0 \ \Omega$, and $A = 10^5$.

SOLUTION The transient response for the four cases of damping, including the associated values for the capacitors, can be computed using any convenient method (e.g., PSPICE).

The results are shown in Fig. 12.53. The curves indicate that a $\zeta = 0.75$ might be a good design compromise between rapid step response and minimum overshoot.

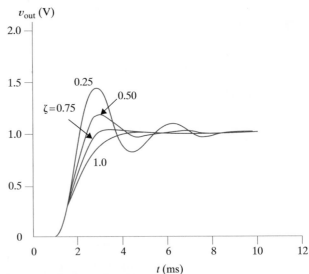

Figure 12.53 Transient analysis of Example 12.25.

EXTENSION EXERCISE

E12.20 Verify that Eq. (12.59) is the transfer function for the network in Fig. 12.52.

The general transfer function for the second-order band-pass filter is

$$\frac{V_o(s)}{V_s(s)} = \frac{sH_0}{s^2 + \dfrac{\omega_0}{Q}s + \omega_0^2}$$

12.60

As discussed earlier, ω_0 is the center frequency of the band-pass characteristic and Q is the quality factor. Recall that for low-pass filters, H_0 was the passband or dc gain. For a band-pass filter, the gain is maximum at the center frequency, ω_0. To find this maximum gain we substitute $s = j\omega_0$ in the preceding expression to obtain

$$\frac{V_o(j\omega_0)}{V_s(j\omega_0)} = \frac{j\omega_0 H_0}{-\omega_0^2 + j\omega_0(\omega_0/Q) + \omega_0^2}$$

12.61

$$= \frac{QH_0}{\omega_0}$$

In addition, the difference between the high and low half-power frequencies (i.e., $\omega_{HI} - \omega_{LO}$) is, of course, the bandwidth

$$\omega_{HI} - \omega_{LO} = BW = \frac{\omega_0}{Q}$$

12.62

Q is a measure of the selectivity of the band-pass filter, and as the equation indicates, as Q is increased, the bandwidth is decreased.

An op-amp implementation of a band-pass filter is shown in Fig. 12.54. The transfer function for this network is

$$\frac{V_o(s)}{V_s(s)} = \frac{-\left(\dfrac{1}{R_1 C_1}\right)s}{s^2 + \left(\dfrac{1}{R_2 C_1} + \dfrac{1}{R_2 C_2}\right)s + \dfrac{1 + R_1/R_3}{R_1 R_2 C_1 C_2}}$$

12.63

Comparing this expression to the more general expression for the band-pass filter yields the following definitions:

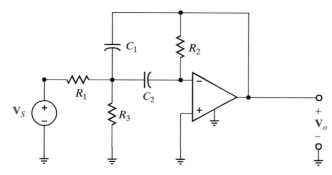

Figure 12.54 Second-order band-pass filter.

$$\omega_0 = \left(\frac{1 + R_1/R_3}{R_1 R_2 C_1 C_2}\right)^{1/2}$$

$$\frac{Q}{\omega_0} = \frac{R_2 C_1 C_2}{C_1 + C_2}$$

$$\frac{Q}{\omega_0}\omega_0 = \frac{R_2 C_1 C_2}{C_1 + C_2}\left(\frac{1 + R_1/R_3}{R_1 R_2 C_1 C_2}\right)^{1/2}$$

12.64

These expressions can be simplified to yield

$$Q = \frac{(1 + R_1/R_3)^{1/2}}{1 + C_1/C_2}\left(\frac{R_2 C_1}{R_1 C_2}\right)^{1/2}$$

12.65

$$BW = \frac{\omega_0}{Q} = \frac{1}{R_2}\left(\frac{1}{C_1} + \frac{1}{C_2}\right)$$

12.66

and

$$\left.\frac{\mathbf{V}_o}{\mathbf{V}_S}\right|_{\omega = \omega_0} = \frac{QH_0}{\omega_0} = -\frac{R_2}{R_1}\left(\frac{1}{1 + C_1/C_2}\right)$$

12.67

EXAMPLE 12.27

We wish to find a new expression for Eqs. (12.64) to (12.67) under the condition that $C_1 = C_2 = C$.

SOLUTION   Using the condition the equations reduce to

$$\omega_0 = \frac{1}{C}\sqrt{\frac{1 + R_1/R_3}{R_1 R_2}}$$

$$Q = \frac{1}{2}\sqrt{\frac{R_2}{R_1}}\sqrt{1 + \frac{R_1}{R_3}}$$

$$BW = \frac{2}{R_2 C}$$

and

$$\left.\frac{\mathbf{V}_o}{\mathbf{V}_S}\right|_{\omega = \omega_0} = -\frac{R_2}{2R_1}$$ ❑

EXAMPLE 12.28

Let us use the equations in Example 12.27 to design a band-pass filter of the form shown in Fig. 12.54 with a BW = 2000 rad/s, $(\mathbf{V}_o/\mathbf{V}_S)(\omega_0) = -5$, and $Q = 3$. Use $C = 0.1\ \mu\text{F}$, and determine the center frequency of the filter.

SOLUTION   Using the filter equations, we find that

$$BW = \frac{2}{R_2 C}$$

$$2000 = \frac{2}{R_2 (10)^{-7}}$$

$$R_2 = 10 \text{ k}\Omega$$

$$\frac{\mathbf{V}_o}{\mathbf{V}_S}(\omega_0) = -\frac{R_2}{2R_1}$$

$$-5 = -\frac{10,000}{2R_1}$$

$$R_1 = 1 \text{ k}\Omega$$

and

$$Q = \frac{1}{2}\sqrt{\frac{R_2}{R_1}}\sqrt{1 + \frac{R_1}{R_3}}$$

$$3 = \frac{1}{2}\sqrt{\frac{10,000}{1000}}\sqrt{1 + \frac{1000}{R_3}}$$

or

$$R_3 = 385 \ \Omega$$

Therefore, $R = 1 \text{ k}\Omega, R_2 = 10 \text{ k}\Omega, R_3 = 385 \ \Omega$, and $C = 0.1 \ \mu\text{F}$ completely define the band-pass filter shown in Fig. 12.54. The center frequency of the filter is

$$\omega_0 = \frac{1}{C}\sqrt{\frac{1 + R_1/R_3}{R_1 R_2}}$$

$$= \frac{1}{10^{-7}}\sqrt{\frac{1 + (1000/385)}{(1000)(10,000)}}$$

$$= 6000 \text{ rad/s} \qquad \qquad \square$$

EXAMPLE 12.29

We wish to obtain the Bode plot for the filter designed in Example 12.28. We will employ the op-amp model, in which $R_i = \infty, R_o = 0$, and $A = 10^5$, and plot over the frequency range from 600 to 60 kHz.

SOLUTION   The equivalent circuit for the filter is shown in Fig. 12.55a. The Bode plot is shown in Fig. 12.55b. As can be seen from the plot, the center frequency is 6 krad/s and BW = 2 krad/s.

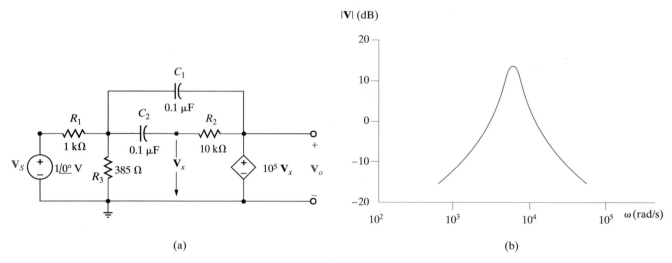

(a)

(b)

Figure 12.55 Figures employed in Example 12.29: (a) band-pass filter equivalent circuits; (b) Bode plot. ❏

EXTENSION EXERCISE

E12.21   Verify that Eq. (12.63) is the transfer function for the band-pass filter in Fig. 12.54.

It is instructive to examine briefly the frequency-response limitations of an op-amp. Thus far, our modeling of the op-amp has been limited to finite gain, input resistance, and output resistance. The astute reader may have noticed that this model has infinite frequency response (i.e., there are no poles or zeros in the forward gain, A). As developing engineers, we should be skeptical when an infinite response is associated with a real, physical system. Of course, real operational amplifiers do have a finite frequency response, and their forward gain is more realistically represented as

$$A(j\omega) = \frac{A_o[1 + j(f/f_{z_1})][1 + j(f/f_{z_2})]\cdots}{[1 + j(f/f_{p_1})][1 + j(f/f_{p_2})]}\cdots$$

where $1 + j(f/f_{p_1}) = 1 + j\omega\tau_{p_1}$, where $\omega = 2\pi f$ and $\tau_{p_1} = 1/2\tau f_{p_1}$.

There is, however, a large class of op-amps that is specifically designed to have a single-pole forward gain response, which ensures stable and reliable operation in a wide range of applications. The single-pole gain of the amplifier can be written as

$$A = \frac{A_o}{1 + j(f/f_p)}$$

Typical values for f_p are in the range from 5 to 30 Hz. These data prompt us to ask if such a low-frequency pole does not severely limit the usefulness of the amplifier. The following example will illustrate that the frequency response is actually limited by a factor known as the gain–bandwidth product, GBW, which for a single-pole op-amp is GBW $= A_o f_p$.

EXAMPLE 12.30

Consider the unity-gain op-amp buffer circuit shown in Fig. 12.56a. Let us show that the device acts like a low-pass filter.

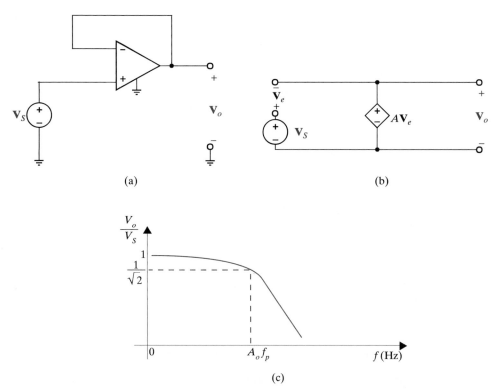

(a)

(b)

(c)

Figure 12.56 Figures used in Example 12.30: (a) buffer amplifier; (b) buffer amplifier equivalent circuit; (c) unity-gain buffer amplifier frequency response.

SOLUTION   The equivalent circuit for the amplifier is shown in Fig. 12.56b, where we have assumed that $R_i = \infty$, $R_o = 0$, and $\mathbf{A} = A_o/[1 + j(f/f_p)]$. KVL yields

$$\mathbf{V}_e = \mathbf{V}_S - \mathbf{AV}_e$$

or

$$(\mathbf{A} + 1)\mathbf{V}_e = \mathbf{V}_S$$

However,

$$\mathbf{V}_e = \mathbf{V}_S - \mathbf{V}_o$$

and therefore,

$$(\mathbf{A} + 1)\mathbf{V}_o = \mathbf{A}\mathbf{V}_S$$

and

$$\frac{\mathbf{V}_o}{\mathbf{V}_S} = \frac{\mathbf{A}}{1 + \mathbf{A}}$$

Substituting for **A** yields

$$\frac{\mathbf{V}_o}{\mathbf{V}_S} = \frac{A_o/[1 + j(f/f_p)]}{1 + A_o/[1 + j(f/f_p)]}$$

$$= \frac{A_o}{1 + j(f/f_p) + A_o}$$

Assuming that $A_o \gg 1$, the function reduces to

$$\frac{\mathbf{V}_o}{\mathbf{V}_S} \simeq \frac{A_o}{A_o + j(f/f_p)}$$

$$\simeq \frac{1}{1 + j(f/f_p A_o)}$$

Thus, at low frequencies $\mathbf{V}_o/\mathbf{V}_S = 1$ and at very high frequencies $(f \gg A_o f_p)$, the gain is $\mathbf{V}_o/\mathbf{V}_S \equiv 1/j(f/f_p A_o) = -j(A_o f_p/f)$. Therefore, the op-amp looks like a low-pass filter with a cutoff frequency equal to $f_c = A_o f_p$ (i.e., $\omega_c = 2\pi A_o f$), as shown in Fig. 12.56c. ❑

The equivalent circuit for a single-pole op-amp is shown in Fig. 12.57. The low-frequency pole is set at $f_p = 1/2\pi RC$. For example, if $R = 100$ kΩ and $C = 159$ nF, the op-amp would have a pole at 10 Hz. In the equivalent network the first dependent source models the dc gain. The second dependent source buffers the output, so that the pole frequency will not be affected by the output load impedance.

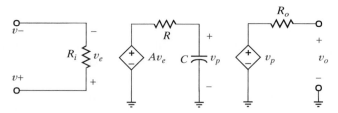

Figure 12.57 Single-pole op-amp equivalent circuit.

EXAMPLE 12.31

Given the high-pass filter network shown in Fig. E12.19 with $R_1 = 1$ kΩ, $R_2 = 10$ kΩ, and $C = 1.59$ nF, we wish to determine the cutoff frequency f_c and generate the Bode plot for both the simple and single-pole op-amp models. We will use $A = (2)\,10^5$, $R_i = \infty$, and $R_o = 0$ for the simple op-amp model. For the single-pole model we will use $f_p = 5$ Hz.

SOLUTION   The transfer function for the filter is

$$\frac{\mathbf{V}_o}{\mathbf{V}_S}(s) = \frac{R_2 C_S}{R_1 C_S + 1}$$

Therefore,

$$\omega_c = \frac{1}{R_1 C}$$

or

$$f_c = \frac{1}{2\pi R_1 C}$$

$$= \frac{1}{2\pi (10)^3 (1.59)(10)^{-9}}$$

$$= 100 \text{ kHz}$$

The equivalent circuit for the filter using the simple op-amp model is shown in Fig. 12.58a.

The equivalent circuit for the filter with the single-pole op-amp model is shown in Fig. 12.58b, where we have used $R_{P_1} = 100 \text{ k}\Omega$ and $C_{P_1} = 318.3 \text{ nF}$ to satisfy the condition $f_p = 5 \text{ Hz}$.

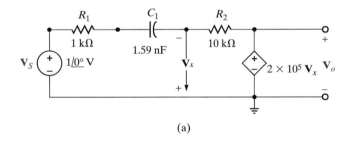

(a)

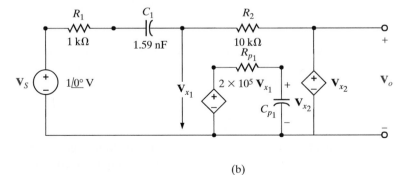

(b)

Figure 12.58 Circuits used in Example 12.31: (a) high-pass filter with simple op-amp model; (b) high-pass filter with single-pole op-amp model.

The Bode plots for the filter with both op-amp models are shown in Fig. 12.59. Note that the op-amp with a single internal pole rolls off the high-frequency characteristic, resulting in a composite frequency response similar to that of a band-pass filter.

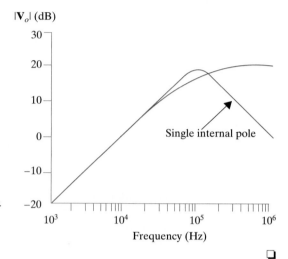

Figure 12.59 Bode plots for a high-pass filter using different op-amp models.

12.6 APPLICATIONS

EXAMPLE 12.32

The ac-dc converter in Fig. 12.60a is designed for use with a hand-held calculator. We will ignore the output load for now; it is addressed in Problem 12.74. Ideally, the circuit should convert a 120-V rms sinusoidal voltage to a 9-V dc output. In actuality the output is

$$v_o(t) = 9 + 5 \sin 377t$$

Let us use a low-pass filter to reduce the 60-Hz component of $v_o(t)$.

SOLUTION   The Thévenin equivalent circuit for the converter is shown in Fig. 12.60b. By placing a capacitor across the output terminals, as shown in Fig. 12.60c, we create a low-pass filter at the output. The transfer function of the filtered converter is

$$\frac{\mathbf{V}_{OF}}{\mathbf{V}_{Th}} = \frac{1}{1 + sR_{Th}C}$$

which has a pole at a frequency of $f = 1/2\pi R_{Th}C$. To obtain significant attenuation at 60 Hz, we will choose to place the pole at 6 Hz, yielding the equation

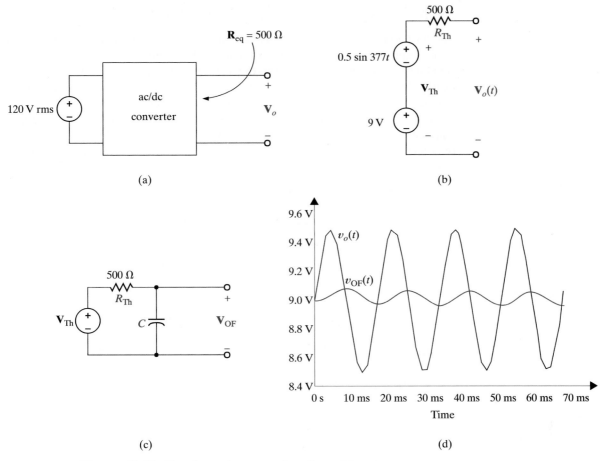

Figure 12.60 Circuits and output plots for ac/dc converter.

$$\frac{1}{2\pi R_{\text{Th}} C} = 6$$

or

$$C = 53.05 \ \mu\text{F}$$

A transient simulation of the converter is used to verify performance.

Figure 12.60d shows the output without filtering $v_o(t)$, and with filtering, $v_{\text{OF}}(t)$. The filter has successfully reduced the unwanted 60-Hz component by a factor of roughly six. ◻

EXAMPLE 12.33

The antenna of an FM radio picks up stations across the entire FM frequency range—approximately 87.5 MHz to 108 MHz. The radio's circuitry must have the capability to first reject all of the stations except the one that the listener wants to hear and then to boost the minute antenna signal. A tuned amplifier incorporating parallel resonance can perform both tasks simultaneously.

The network in Fig. 12.61a is a circuit model for a single-stage tuned transistor amplifier where the resistor, capacitor, and inductor are discrete elements. Let us find the transfer function $\mathbf{V}_o(s)/\mathbf{V}_A(s)$, where $\mathbf{V}_A(s)$ is the antenna voltage and the value of C for maximum gain at 91.1 MHz. Finally, we will simulate the results.

SOLUTION   Since $\mathbf{V}(s) = \mathbf{V}_A(s)$, the transfer function is

$$\frac{\mathbf{V}_o(s)}{\mathbf{V}_A(s)} = -\frac{4}{1000}\left[R//sL//\frac{1}{sC}\right]$$

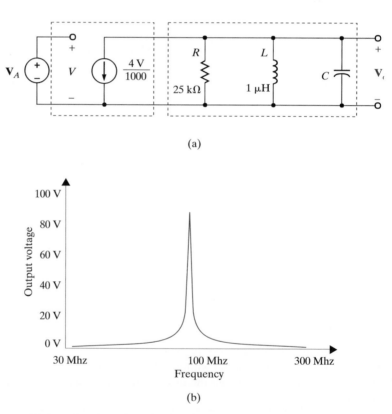

(a)

(b)

Figure 12.61 Circuit and Bode plot for parallel resonance tuned amplifier.

$$\frac{\mathbf{V}_o(s)}{\mathbf{V}_A(s)} = -\frac{4}{1000}\left[\frac{s/C}{s^2 + \dfrac{s}{RC} + \dfrac{1}{LC}}\right]$$

The parallel resonance network is actually a band-pass filter. Maximum gain occurs at the center frequency, f_0. This condition corresponds to a minimum value in the denominator. Isolating the denominator polynomial, $D(s)$, and letting $s = j\omega$, we have

$$D(j\omega) = \frac{R}{LC} - \omega^2 + \frac{j\omega}{C}$$

which has a minimum value when the real part goes to zero, or

$$\frac{1}{LC} - \omega_o^2 = 0$$

yielding a center frequency of

$$\omega_o = \frac{1}{\sqrt{LC}}$$

Thus, for a center frequency of 91.1 MHz, we have

$$2\pi\left(91.1 \times 10^6\right) = \frac{1}{\sqrt{LC}}$$

and the required capacitor value is

$$C = 3.05 \text{ pF}$$

The Bode plot for the tuned amplifier, as shown in Fig. 12.61b, confirms the design, since the center frequency is 91.1 MHz, as specified.   ❏

12.7 CIRCUIT DESIGN

Throughout this chapter we have presented a number of design examples. In this section we consider some additional ones that also have practical ramifications.

EXAMPLE 12.34

Compact disks (CDs) have become a very popular medium for recording and playing music. CDs store information in a digital manner; that is, the music is sampled at a very high rate, and the samples are recorded on the disc. The trick is to sample so quickly that the reproduction sounds continuous. The industry standard sampling rate is 44.1 kHz—one sample every 22.7 μs.

One interesting aspect regarding the analog-to-digital conversion that takes place inside the unit recording a CD is called the Nyquist criterion. This criterion states that in the analog conversion, any signal components at frequencies above half the sampling rate (22.05 kHz in this case) cannot be faithfully reproduced. Therefore, recording technicians filter these frequencies out before any sampling occurs, yielding higher fidelity to the listener.

Let us design a series of low-pass filters to perform this task.

SOLUTION Suppose, for example, that our specification for the filter is unity gain at dc and 20 dB of attenuation at 22.05 kHz. Let us consider first the simple RC filter in Fig. 12.62.

The transfer function is easily found to be

$$\mathbf{G}_{v1}(s) = \frac{\mathbf{V}_{o1}}{\mathbf{V}_{in}} = \frac{1}{1 + sRC}$$

Since a single-pole transfer function attenuates at 20 dB/decade, we should place the pole frequency one decade before the −20 dB point of 22.05 kHz.

Thus,

$$f_p = \frac{1}{2\pi RC} = 2.205 \text{ kHz}$$

If we arbitrarily choose $C = 1$ nF, the resulting value for R is 72.18 kΩ, which is reasonable. A Bode plot of the magnitude of $\mathbf{G}_{v1}(s)$ is shown in Fig. 12.63. All specifications are met but at the cost of severe attenuation in the audible frequency range. This is undesirable.

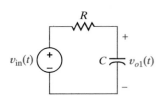

Figure 12.62 Single-pole low-pass filter.

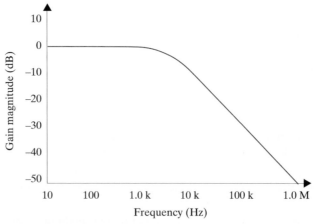

Figure 12.63 Bode plot for single-pole filter.

An improved filter is shown in Fig. 12.64. It is a two-stage low-pass filter with identical filter stages separated by a unity-gain buffer.

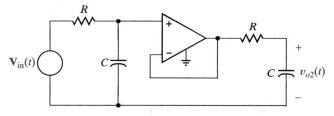

Figure 12.64 Two-stage buffered filter.

The presence of the op-amp permits us to consider the stages independently. Thus, the transfer function becomes

$$\mathbf{G}_{v2}(s) = \frac{\mathbf{V}_{o2}}{\mathbf{V}_{in}} = \frac{1}{[1 + sRC]^2}$$

To find the required pole frequencies, let us employ the equation for $\mathbf{G}_{v2}(s)$ at 22.05 kHz, since we know that the gain must be 0.1 (attenuated 20 dB) at that frequency. Using the substitution $s = j\omega$, we can express the magnitude of $\mathbf{G}_{v2}(s)$ as

$$|\mathbf{G}_{v2}| = \left\{ \frac{1}{1 + (22{,}050/f_p)^2} \right\} = 0.1$$

and the pole frequency is found to be 7.35 kHz. The corresponding resistor value is 21.65 kΩ. Bode plots for $\mathbf{G}_{v1}$ and $\mathbf{G}_{v2}$ are shown in Fig. 12.65. Note that the two-stage filter has a wider bandwidth, which improves the fidelity of the recording.

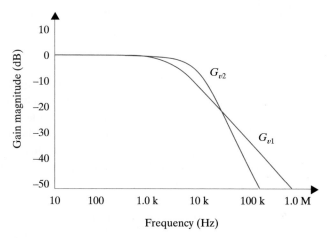

Figure 12.65 Bode plot for single- and two-stage filters.

Let us try one more improvement—expanding the two-stage filter to a four-stage filter. Again, the gain magnitude is 0.1 at 22.05 kHz and can be written

$$|\mathbf{G}_{v3}| = \left\{ \frac{1}{[1 + (22{,}050/f_p)^2]^2} \right\} = 0.1$$

The resulting pole frequencies are at 15 kHz and the required resistor value is 10.61 kΩ. Figure 12.66 shows all three Bode plots. Obviously, the four-stage filter, having the widest bandwidth, is the best option.

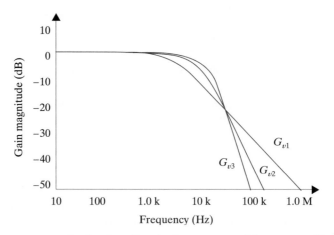

Figure 12.66 Bode plots for single-, two-, and four-stage filters.   ❏

<hr>

EXAMPLE 12.35

The circuit in Fig. 12.67a is called a notch filter. From a sketch of its Bode plot in Fig. 12.67b, we see that at the notch frequency, f_n, the transfer function gain is zero, while at frequencies above and below f_n the gain is unity. Let us design a notch filter to remove an annoying 60-Hz hum from the output voltage of a cassette tape player and generate its Bode plot.

SOLUTION   Figure 12.67c shows a block diagram for the filter implementation. The voltage $\mathbf{V}_{\text{tape}}$ contains both the desired music and the undesired hum. After filtering, the voltage $\mathbf{V}_{\text{amp}}$ will have no 60-Hz component as well as some attenuation at frequencies around 60 Hz. An equivalent circuit for the block diagram including a Thévenin equivalent for the tape deck and an equivalent resistance for the power amp is shown in Fig. 12.67d. Applying voltage division, the transfer function is found to be

$$\frac{\mathbf{V}_{\text{amp}}}{\mathbf{V}_{\text{tape}}} = \frac{R_{\text{amp}}}{R_{\text{amp}} + R_{\text{tape}} + \left(sL // \dfrac{1}{Cs} \right)}$$

After some manipulation, the transfer function can be written as

$$\frac{\mathbf{V}_{\text{amp}}}{\mathbf{V}_{\text{tape}}} = \frac{R_{\text{amp}}}{R_{\text{amp}} + R_{\text{tape}}} \left[\frac{s^2 LC + 1}{s^2 LC + s\left(\dfrac{L}{R_{\text{tape}} + R_{\text{amp}}} \right) + 1} \right]$$

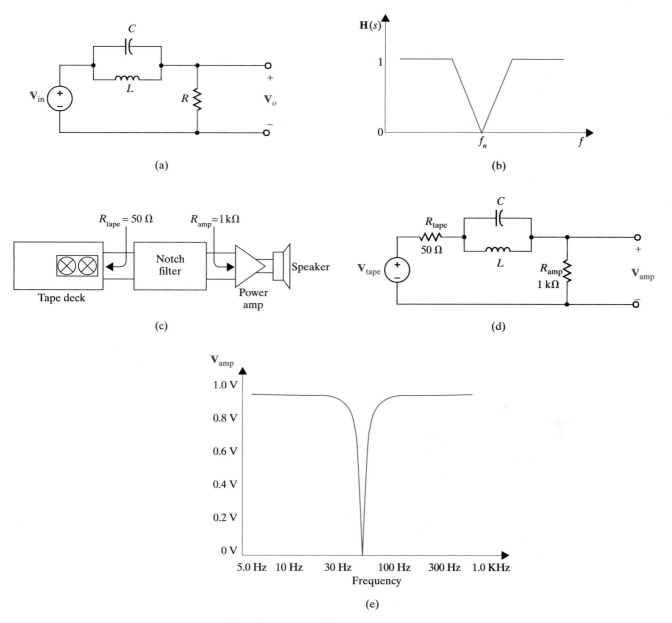

Figure 12.67 Circuits and Bode plots for 60-Hz notch filter.

We see that the transfer function contains two zeros and two poles. Letting $s = j\omega$, the zero frequencies, ω_z, are found to be at

$$\omega_z = \pm \frac{1}{\sqrt{LC}}$$

Obviously, we would like the zero frequencies to be at 60 Hz. If we arbitrarily choose $C = 10 \ \mu F$, then $L = 0.704 \ mH$.

The Bode plot, shown in Fig. 12.67e, confirms that there is indeed zero transmission at 60 Hz.   ❏

EXAMPLE 12.36

In spite of the popularity of CDs, many audiophiles still prefer LP albums. In the music recording industry, modifications are made to the frequency content of the music before pressing the vinyl to create these records. Specifically, low-frequency (bass) sounds are attenuated (deemphazised) and high frequencies are boosted (emphasized). When you play a record on your phonograph, the frequency content must be reconstructed to preserve fidelity. Thus, the RIAA (Recording Industry Association of America) created standard transfer functions for the recording and playback filters. The recording filter transfer function is

$$\mathbf{H}_R(s) = \frac{A_o(1 + s\tau_{z1})(1 + s\tau_{z2})}{1 + s\tau_p}$$

where the time constants are $\tau_{z1} = 75 \ \mu s$, $\tau_{z2} = 3180 \ \mu s$, and $\tau_p = 318 \ \mu s$. The constant A_o is chosen such that $\mathbf{H}_R$ has a magnitude of 1 at 1000 Hz. Of course, for perfect reconstruction, the playback filter, which is inside your stereo, must have the following transfer function:

$$\mathbf{H}_p(s) = \frac{1}{\mathbf{H}_R(s)}$$

Using the circuit in Fig. 12.68, where $\mathbf{V}_{Phono}$ is the output from the phonograph, let us design the playback filter.

SOLUTION   First, we will find the gain from $\mathbf{V}_{Phono}$ to $\mathbf{V}_1$. Since the input resistance into the op-amp is infinite, we can write

$$\frac{\mathbf{V}_1}{\mathbf{V}_{Phono}} = \frac{\left(R_2 // \dfrac{1}{sC_1}\right)}{\left(R_2 // \dfrac{1}{sC_1}\right) + R_1}$$

which, after some rearranging, can be expressed as

$$\frac{\mathbf{V}_1}{\mathbf{V}_{Phono}} = \left(\frac{R_2}{R_2 + R_1}\right)\frac{1}{sC_1(R_2//R_1)} + 1$$

The first op-amp has a gain of

$$\frac{\mathbf{V}_2}{\mathbf{V}_1} = 1 + \left(\frac{R_3 // \dfrac{1}{sC_3}}{\dfrac{1}{sC_2}}\right)$$

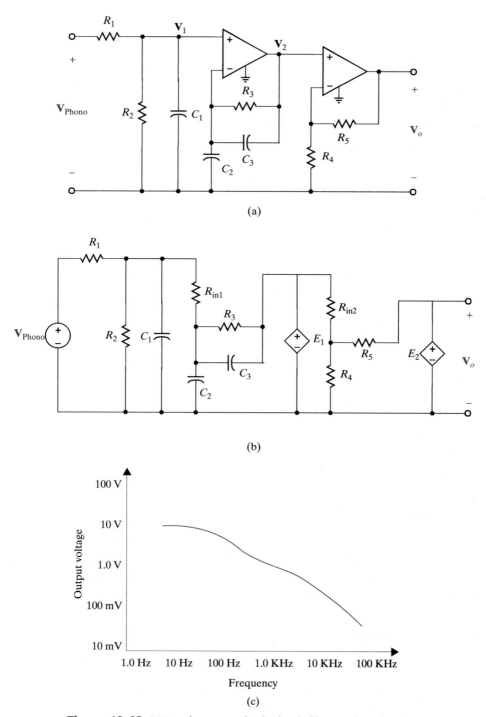

(a)

(b)

(c)

Figure 12.68 RIAA phonograph playback filter and Bode plot.

which can be expressed in the form

$$\frac{\mathbf{V}_2}{\mathbf{V}_1} = \left(\frac{1 + sR_3(C_2 + C_3)}{1 + sR_3C_3}\right)$$

Since the second op-amp is a simple noninverting gain stage, the overall transfer function is

$$\frac{\mathbf{V}_o}{\mathbf{V}_{Phono}} = \left(\frac{R_2}{R_2 + R_1}\right)\left(1 + \frac{R_5}{R_4}\right)\left(\frac{1}{sC_1(R_2//R_1) + 1}\right)\left(\frac{1 + sR_3(C_2 + C_3)}{1 + sR_3C_3}\right)$$

Based on the RIAA standards listed earlier, the following time constant relationship must hold.

$$R_3(C_2 + C_3) = 318 \ \mu s$$
$$R_3C_3 = 75 \ \mu s$$
$$(R_1//R_2)C_1 = 3180 \ \mu s$$

Also, the constant A_o is

$$\frac{1}{A_o} = \left(\frac{R_2}{R_2 + R_1}\right)\left(\frac{R_4 + R_5}{R_4}\right)$$

and must be chosen for a gain magnitude of 1 at $f = 1000$ Hz. Let us arbitrarily select $R_3 = 15 \ k\Omega$. This results in $C_3 = 16.2$ nF and $C_2 = 5$ nF. Next, if we let $C_1 = 50$ nF, the parallel combination $R_1//R_2$ will be 63.6 kΩ. For convenience sake, let $R_1 = R_2 = R_5 = 127.2 \ k\Omega$. To find R_4 we must know the magnitude of $\mathbf{H}_p$ at 1000 Hz. Letting $s = j2000\pi$, we find

$$|\mathbf{H}_p(j2000\pi)| = 1 = \left(\frac{R_2}{R_2 + R_1}\right)\left(1 + \frac{R_5}{R_4}\right)(0.1010)$$

Using the preceding values for R_1, R_2, and R_5,

$$R_5 = 18.80 \ R_4$$

or

$$R_4 = 6.77 \ k\Omega$$

Obviously, with eight external components, there are an infinite number of solutions to this application.

The equivalent circuit is shown in Fig. 12.68b.

The op-amp has $R_{in} = 10 \ M\Omega$, $R_o = 0 \ \Omega$, and $A = 10^5$. The resulting bode plot in Fig. 12.68c displays the attributes of a two-pole, single-zero system and has the required unity gain at 1000 Hz.

12.8 SUMMARY

- There are four types of network or transfer functions:
  1. **$\mathbf{Z}(j\omega)$**: the ratio of the output voltage to the input current
  2. **$\mathbf{Y}(j\omega)$**: the ratio of the output current to the input voltage
  3. **$\mathbf{G}_v(j\omega)$**: the ratio of the output voltage to the input voltage
  4. **$\mathbf{G}_i(j\omega)$**: the ratio of the output current to the input current

- Driving-point functions are impedances or admittances defined at a single pair of terminals, such as the input impedance of a network.
- When the network function is expressed in the form

$$\mathbf{H}(s) = \frac{N(s)}{D(s)}$$

the roots of $N(s)$ cause $\mathbf{H}(s)$ to become zero and are called zeros of the function, and the roots of $D(s)$ cause $\mathbf{H}(s)$ to become infinite and are called poles of the function.

- Bode plots are semilog plots of the magnitude and phase of a transfer function as a function of frequency. Straight-line approximations can be used to sketch quickly the magnitude characteristic. The error between the actual characteristic and the straight-line approximation can be calculated when necessary.
- The resonant frequency, given by the expression

$$\omega_0 = \frac{1}{\sqrt{LC}}$$

is the frequency at which the impedance of a series RLC circuit or the admittance of a parallel RLC circuit is purely real.

- The quality factor is a measure of the sharpness of the resonant peak. The higher the Q, the sharper the peak.
    For series RLC circuits, $Q = 1/R\sqrt{L/C}$. For parallel RLC circuits, $Q = R\sqrt{C/L}$.
- The half-power, cutoff, or break frequencies are the frequencies at which the magnitude characteristic of the Bode plot is $1/\sqrt{2}$ of its maximum value.
- The parameter values for passive circuit elements can be both magnitude and frequency scaled.
- The four common types of filters are low pass, high pass, band pass, and band rejection.
- The bandwidth of a band-pass or band-rejection filter is the difference in frequency between the half-power points; that is,

$$BW = \omega_{HI} - \omega_{LO}$$

For a series RLC circuit, $BW = R/L$. For a parallel RLC circuit, $BW = 1/RC$.

PROBLEMS

Section 12.1

12.1 Determine the driving point impedance at the input terminals of the network shown in Fig. P12.1 as a function of s.

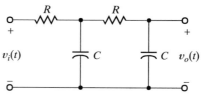

Figure P12.1

Combine impedances that are in series or parallel.

12.2 Determine the driving point impedance at the input terminals of the network shown in Fig. P12.2 as a function of s.

Similar to P12.1

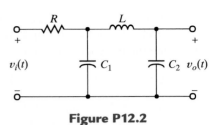

Figure P12.2

12.3 Find the transfer impedance $\mathbf{V}_o(s)/\mathbf{I}_S(s)$ for the network shown in Fig. P12.3.

Similar to Example 12.2

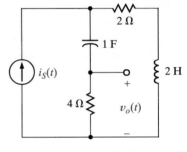

Figure P12.3

12.4 Determine the voltage transfer function $\mathbf{V}_o/\mathbf{V}_i$ as a function of s for the network shown in Fig. P12.4.

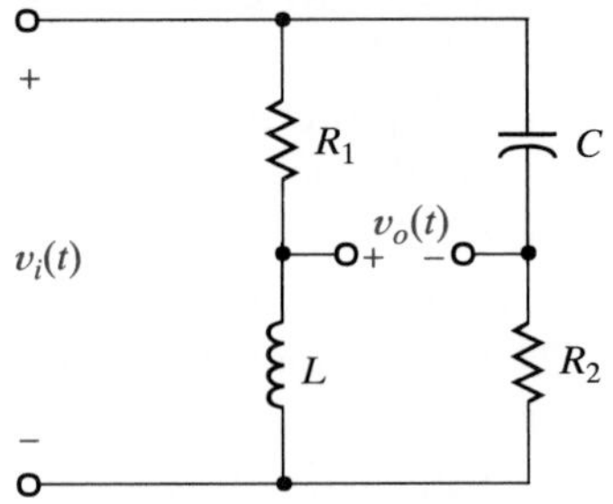

Similar to Example 12.2

Figure P12.4

Section 12.2

12.5 Draw the Bode plot for the network function

Follow the technique demonstrated in Example 12.3.

$$\mathbf{H}(j\omega) = \frac{j\omega 4 + 1}{j\omega 20 + 1}$$

12.6 Draw the Bode plot for the network function

Similar to P12.5

$$\mathbf{H}(j\omega) = \frac{j\omega 5 + 1}{j\omega 10 + 1}$$

12.7 Draw the Bode plot for the network function

$$\mathbf{H}(j\omega) = \frac{10(10j\omega + 1)}{(100j\omega + 1)(j\omega + 1)}$$

Similar to P12.5

12.8 Draw the Bode plot for the network function

$$\mathbf{H}(j\omega) = \frac{10j\omega + 1}{j\omega(0.1j\omega + 1)}$$

Similar to P12.5

12.9 Draw the Bode plot for the network function

$$\mathbf{H}(j\omega) = \frac{j\omega}{(j\omega + 1)(0.1j\omega + 1)}$$

Similar to P12.5

12.10 Draw the Bode plot for the network function

$$\mathbf{H}(j\omega) = \frac{16}{(j\omega)^2(j\omega 2 + 1)}$$

Follow the technique outlined in Example 12.4.

12.11 Sketch the magnitude characteristic of the Bode plot for the transfer function

$$\mathbf{H}(j\omega) = \frac{640(j\omega + 1)(0.01j\omega + 1)}{(j\omega)^2(j\omega + 10)}$$

Similar to P12.10

12.12 Sketch the magnitude characteristic of the Bode plot for the transfer function

$$\mathbf{G}(j\omega) = \frac{10j\omega}{(j\omega + 1)(j\omega + 10)^2}$$

Similar to P12.10

12.13 Sketch the magnitude characteristic of the Bode plot for the transfer function

$$\mathbf{G}(j\omega) = \frac{400(j\omega + 2)(j\omega + 50)}{-\omega^2(j\omega + 100)^2}$$

Similar to P12.10

12.14 Sketch the magnitude characteristic of the Bode plot for the transfer function

$$\mathbf{H}(j\omega) = \frac{10^5(5j\omega + 1)^2}{(j\omega)^2(j\omega + 10)(j\omega + 100)^2}$$

Similar to P12.10

12.15 Sketch the magnitude characteristic of the Bode plot for the transfer function

$$\mathbf{H}(j\omega) = \frac{10^2(j\omega)^2}{(j\omega + 1)(j\omega + 10)^2(j\omega + 50)}$$

Similar to P12.10

12.16 Sketch the magnitude characteristic of the Bode plot for the transfer function

Similar to P12.10

$$G(j\omega) = \frac{64(j\omega + 1)^2}{-j\omega^3(0.1j\omega + 1)}$$

12.17 Sketch the magnitude characteristic of the Bode plot for the transfer function

Similar to P12.10

$$G(j\omega) = \frac{-\omega^2}{(j\omega + 1)^3}$$

12.18 Sketch the magnitude characteristic of the Bode plot for the transfer function

Similar to P12.10

$$G(j\omega) = \frac{-\omega^2 10^4}{(j\omega + 1)^2(j\omega + 10)(j\omega + 100)^2}$$

12.19 Sketch the magnitude characteristic of the Bode plot for the transfer function

Follow the method shown in Example 12.5.

$$G(j\omega) = \frac{10(j\omega + 2)(j\omega + 100)}{j\omega(-\omega^2 + 4j\omega + 100)}$$

12.20 Draw the Bode plot for the network function

Similar to P12.19

$$H(j\omega) = \frac{72(j\omega + 2)}{j\omega[(j\omega)^2 + 2.4j\omega + 144]}$$

12.21 Sketch the magnitude characteristic of the Bode plot for the transfer function

Similar to P12.19

$$G(j\omega) = \frac{j\omega(j\omega + 100)}{(j\omega + 1)(-\omega^2 + 6j\omega + 400)}$$

12.22 Sketch the magnitude characteristic of the Bode plot for the transfer function

Similar to P12.19

$$G(j\omega) = \frac{10^4(j\omega + 1)(-\omega^2 + 6j\omega + 225)}{j\omega(j\omega + 50)^2(j\omega + 450)}$$

12.23 Determine $\mathbf{H}(j\omega)$ if the amplitude characteristic for $\mathbf{H}(j\omega)$ is shown in Fig. P12.23.

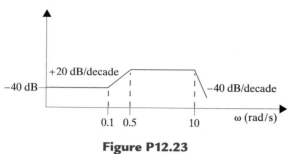

Figure P12.23

Follow the technique outlined in Example 12.6.

12.24 The magnitude characteristic of a band-elimination filter is shown in Fig. P12.24. Determine $\mathbf{H}(j\omega)$.

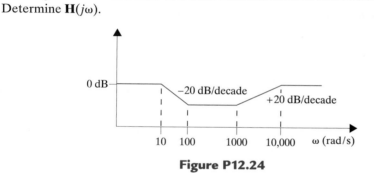

Figure P12.24

Similar to P12.23

12.25 Find $\mathbf{H}(j\omega)$ if its magnitude characteristic is shown in Fig. P12.25.

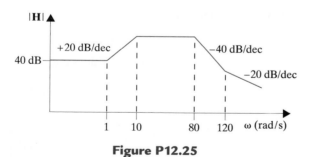

Figure P12.25

Similar to P12.23

12.26 Find $\mathbf{H}(j\omega)$ if its magnitude characteristic is shown in Fig. P12.26.

Similar to P12.23

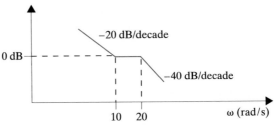

Figure P12.26

12.27 Find $\mathbf{H}(j\omega)$ if its amplitude characteristic is shown in Fig. P12.27.

Similar to P12.23

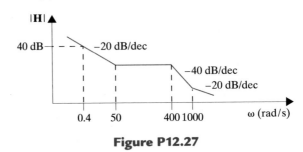

Figure P12.27

12.28 Determine $\mathbf{H}(j\omega)$ if its magnitude characteristic is shown in Fig. P12.28.

Similar to P12.23

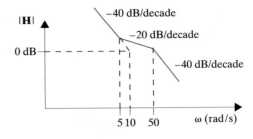

Figure P12.28

12.29 Find $\mathbf{H}(j\omega)$ if its amplitude characteristic is shown in P12.29.

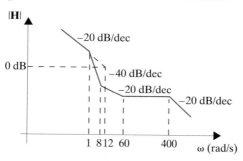

Figure P12.29

Similar to P12.23

12.30 Given the magnitude characteristic in Fig. P12.30, find $\mathbf{G}(j\omega)$.

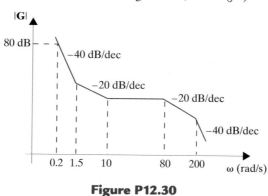

Figure P12.30

Similar to P12.23

12.31 Given the magnitude characteristic in Fig. P12.31, find $\mathbf{H}(j\omega)$.

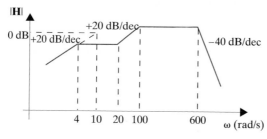

Figure P12.31

Similar to P12.23

12.32 Find $\mathbf{H}(j\omega)$ if its magnitude characteristic is shown in Fig. P12.32.

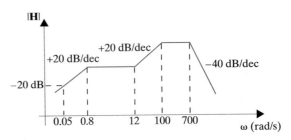

Figure P12.32

Similar to P12.23

Section 12.3

Use the equations that relate ω_0, Q, and BW to the circuit parameters.

12.33 A series RLC circuit resonates at 1000 rad/s. If $C = 20\ \mu\mathrm{F}$ and it is known that the impedance at resonance is 2.4 Ω, compute the value of L, the Q of the circuit, and the bandwidth.

Similar to P12.33

12.34 A series resonant circuit has a Q of 120 and a resonant frequency of 10,000 rad/s. Determine the half-power frequencies and the bandwidth of the circuit.

Example 12.13 is helpful in this case.

12.35 The series RLC circuit in Fig. P12.35 is driven by a variable-frequency source. If the resonant frequency of the network is selected as $\omega_0 = 1600$ rad/s, find the value of C. In addition, compute the current at resonance and at $\omega_0/4$ and 4 ω_0.

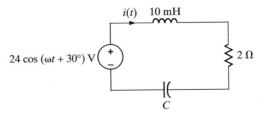

Figure P12.35

12.36 Given the series RLC circuit in Fig. P12.36, (a) derive the expression for the half-power frequencies, the resonant frequency, the bandwidth, and the quality factor for the transfer characteristic $\mathbf{I}/\mathbf{V}_{in}$ in terms of R, L, C. (b) Compute the quantities in part (a) if $R = 10\ \Omega, L = 50$ mH, and $C = 10\ \mu$F.

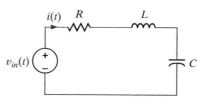

Figure P12.36

Similar to P12.33

12.37 Given the network in Fig. P12.37, find $\omega_0, Q, \omega_{max}$, and $|\mathbf{V}_o|_{max}$.

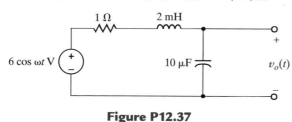

Figure P12.37

Follow the procedure outlined in Example 12.11.

12.38 Repeat Problem 12.37 if the value of R is changed to 0.1 Ω.

Similar to P12.37

12.39 In the network in Fig. P12.39, $\omega_0 = 1000$ rad/s and at that frequency the impedance seen by the source is 4 Ω. If $L = 10$ mH, determine the Q of the network and the bandwidth.

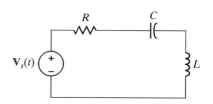

Figure P12.39

Similar to P12.33

12.40 A series RLC circuit is driven by a signal generator. The resonant frequency of the network is known to be 1600 rad/s, and at that frequency the impedance seen by the signal generator is 5 Ω. If $C = 20\ \mu$F, find L, Q, and the bandwidth.

Similar to P12.33

12.41 Given the series RLC circuit in Fig. P12.41, if $R = 20\,\Omega$, find the values of L and C such that the network will have a resonant frequency of 100 kHz and a bandwidth of 1 kHz.

Similar to P12.33

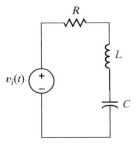

Figure P12.41

12.42 A variable-frequency voltage source drives the network in Fig. P12.42. Determine the resonant frequency, Q, BW, and the average power dissipated by the network at resonance.

Use the equation for ω_0, Q and W_D.

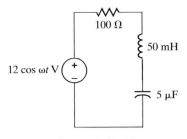

Figure P12.42

12.43 The network in Fig. P12.43 is driven by a variable-frequency source. If the magnitude of the current at resonance is 10A, $\omega_0 = 1000$ rad/s, and $L = 10$ mH, find C, Q, and the bandwidth of the circuit.

Similar to P 12.42

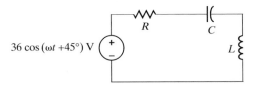

Figure P12.43

12.44 A parallel RLC resonant circuit with a resonant frequency of 20,000 rad/s has an admittance at resonance of 1 mS. If the capacitance of the network is 2 μF, find the values of R and L.

Equations (12.12) and (12.14) are useful in this case.

12.45 A parallel RLC resonant circuit has a resistance of 100 Ω. If it is known that the bandwidth is 80 rad/s and the lower half-power frequency is 800 rad/s, find the values of the parameters L and C.

The equations for parallel resonance are useful in this problem.

12.46 A parallel RLC circuit, which is driven by a variable-frequency 2-A current source, has the following values: $R = 1\ \text{k}\Omega$, $L = 400\ \text{mH}$, and $C = 10\ \mu\text{F}$. Find the bandwidth of the network, the half-power frequencies, and the voltage across the network at the half-power frequencies.

Similar to P12.45

12.47 A parallel RLC circuit, which is driven by a variable-frequency 10-A source, has the following parameters: $R = 1\ \text{k}\Omega$, $L = 0.5\ \text{mH}$, and $C = 20\ \mu\text{F}$. Find the resonant frequency, the Q, the average power dissipated at the resonant frequency, the BW, and the average power dissipated at the half-power frequencies.

Similar to P12.45

12.48 Consider the network in Fig. P12.48. If $R = 1\text{k}\Omega$, $L = 20\ \text{mH}$, $C = 50\ \mu\text{F}$, and $R_S = \infty$, determine the resonant frequency ω_0, the Q of the network, and the bandwidth of the network. What impact does an R_S of 10 kΩ have on the quantities determined?

Similar to P12.45

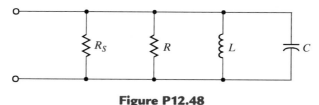

Figure P12.48

12.49 The source in the network in Fig. P12.49 is $i_S(t) = \cos 1000t + \cos 1500t$ A. $R = 200\ \Omega$ and $C = 500\ \mu\text{F}$. If $\omega_0 = 1000\ \text{rad/s}$, find L, Q, and the BW. Compute the output voltage $v_o(t)$ and discuss the magnitude of the output voltage at the two input frequencies.

Similar to P12.48

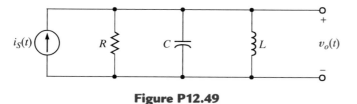

Figure P12.49

12.50 Determine the parameters of a parallel resonant circuit that has the following properties: $\omega_0 = 1\ \text{Mrad/s}$, BW $= 20\ \text{rad/s}$, and an impedance at resonance of 2000 Ω.

The parallel resonance equations are useful in this case.

Section 12.4

12.51 Determine the new parameters of the network shown in Fig. P12.51 if $\mathbf{Z}_{new} = 10^4 \mathbf{Z}_{old}$.

See Example 12.17.

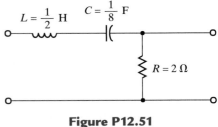

Figure P12.51

Similar to P12.51

12.52 Determine the new parameters of the network in Problem 12.51 if $\omega_{new} = 10^4 \omega_{old}$.

Section 12.5

12.53 Given the network in Fig. P12.53, sketch the magnitude characteristic of the transfer function

$$\mathbf{G}_v(j\omega) = \frac{\mathbf{V}_0}{\mathbf{V}_1}(j\omega)$$

Identify the type of filter.

See Example 12.18.

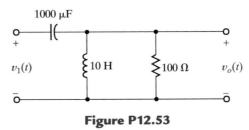

Figure P12.53

12.54 Given the network in Fig. P12.54, sketch the magnitude characteristic of the transfer function

$$\mathbf{G}_v(j\omega) = \frac{\mathbf{V}_0}{\mathbf{V}_1}(j\omega)$$

Identify the type of filter.

Similar to P12.53

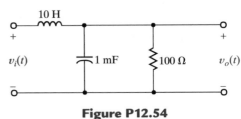

Figure P12.54

12.55 Determine what type of filter the network shown in Fig. P12.55 represents by determining the voltage transfer function.

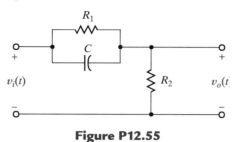

Figure P12.55

See Example 12.19.

12.56 Determine what type of filter the network shown in Fig. P12.56 represents by determining the voltage transfer function.

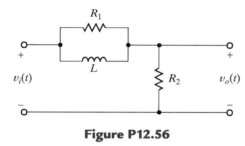

Figure P12.56

Similar to P12.55

12.57 Given the network in Fig. P12.57, and employing the voltage follower analyzed in Chapter 3, determine the voltage transfer function and its magnitude characteristic. What type of filter does the network represent?

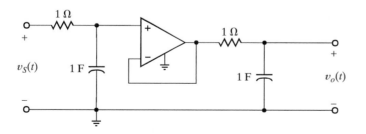

Figure P12.57

See Example 12.20.

12.58 Repeat Problem 12.53 for the network shown in Fig. P12.58.

See Example.12.21.

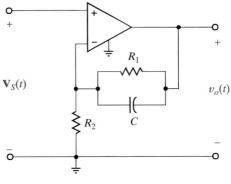

Figure P12.58

12.59 Design a first-order low-pass active filter of the form shown in Fig. 12.38 that has a dc gain of -20 and a cutoff frequency of 50 kHz. Assume that $R_1 = 1\ \text{k}\Omega$.

See Example 12.20.

12.60 For the low-pass active filter in Fig. P12.60, choose R_2 and C such that $H_0 = -7$ and $f_c = 10$ kHz.

Similar to Example 12.20

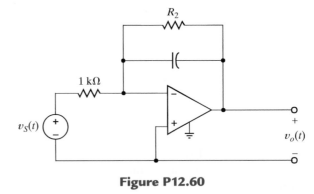

Figure P12.60

12.61 For the high-pass active filter in Fig. P12.61, choose C, R_1, and R_2 such that $H_0 = 5$ and $f_c = 3$ kHz.

Similar to Example 12.31

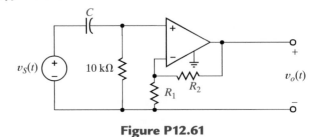

Figure P12.61

Advanced Problems

12.62 Find $\mathbf{H}(j\omega)$ for the magnitude characteristic shown in Fig. P12.62.

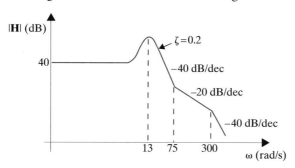

Figure P12.62

12.63 Find $\mathbf{G}(j\omega)$ if the amplitude characteristic for this function is shown in Fig. P12.63.

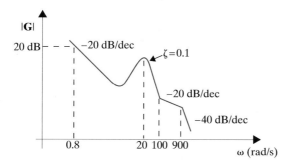

Figure P12.63

12.64 Determine ω_0 and Q for the network in Fig. P12.64.

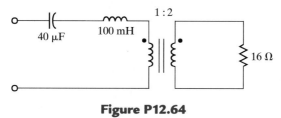

Figure P12.64

12.65 Find the value of n in the circuit in Fig. P12.65 to obtain a Q of 50.

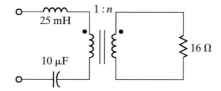

Figure P12.65

12.66 Determine the value of C in the network shown in Fig. P12.66 in order for the circuit to be in resonance.

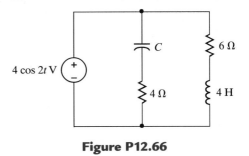

Figure P12.66

12.67 Determine the equation for the nonzero resonant frequency of the impedance shown in Fig. P12.67.

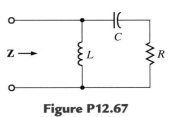

Figure P12.67

12.68 Given the lattice network shown in Fig. P.12.68, determine what type of filter this network represents by determining the voltage transfer function.

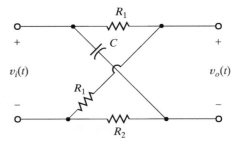

Figure P12.68

12.69 Sketch the frequency response of the active filter shown in Fig. P12.69. For the op-amp model use $R_{in} = 500 \text{ k}\Omega$, $R_{out} = 100\Omega$, and $A = 10^7$. What type of filter is this?

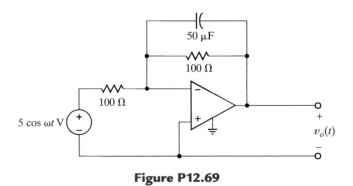

Figure P12.69

12.70 If two first-order low-pass filters are placed in series, the result is a second-order filter. Find H_0, ζ, and ω_c for the network in Fig. P12.70.

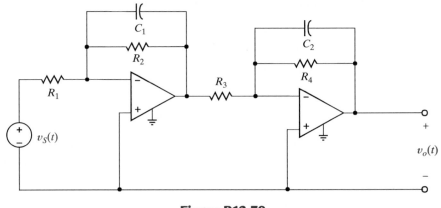

Figure P12.70

12.71 Given the second-order low-pass filter in Fig. 12.70, design a filter that has $H_0 = 100$ and $f_c = 5 \text{ kHz}$. Set $R_1 = R_3 = 1 \text{ k}\Omega$, and let $R_2 = R_4$ and $C_1 = C_2$. Use an op-amp model with $R_i = \infty$, $R_o = 0$, and $A = (2)10^5$.

12.72 The second-order low-pass filter shown in Fig. P12.72 has the transfer function

$$\frac{\mathbf{V}_o}{\mathbf{V}_1}(s) = \frac{\dfrac{-R_3}{R_1}\left(\dfrac{1}{R_2 R_3 C_1 C_2}\right)}{s^2 + \dfrac{s}{C_1}\left(\dfrac{1}{R_1} + \dfrac{1}{R_2} + \dfrac{1}{R_3}\right) + \dfrac{1}{R_2 R_3 C_1 C_2}}$$

Design a filter with $H_0 = -10$ and $f_c = 5\text{ kHz}$, assuming that $C_1 = C_2 = 10\text{ nF}$ and $R_1 = 1\text{ k}\Omega$.

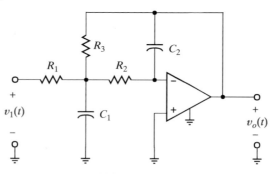

Figure P12.72

12.73 Given the circuit in Figure 12.54, design a second-order bandpass filter with a center frequency gain of -5, $\omega_0 = 50\text{ krad/s}$, and a BW $= 10\text{ krad/s}$. Let $C_1 = C_2 = C$ and $R_1 = 1\text{ k}\Omega$. What is the Q of this filter? Sketch the Bode plot for the filter. Use the ideal op-amp model.

12.74 Refer to the ac/dc converter low-pass filter application of Example 12.32. If we put the converter to use powering a calculator, the load current can be modeled by a resistor as shown in Fig. P12.74. The load resistor will affect both the magnitude of the dc component of V_{OF} and the pole frequency. Plot both the pole frequency and the ratio of the 60-Hz component of the output voltage to the dc component of V_{OF} versus R_L for $100\ \Omega \le R_L \le 100\text{ k}\Omega$. Comment on the advisable limitations on R_L if **(a)** the dc component of V_{OF} is to remain within 20% of its 9-V ideal value; **(b)** the 60-Hz component of V_{OF} remains less than 15% of the dc component.

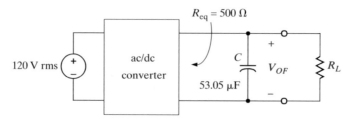

Figure P12.74

12.75 Referring to Example 12.35, design a notch filter for the tape deck for use in Europe, where power utilities generate at 50 Hz.

The Laplace Transform

13

W e will now introduce the Laplace transform. This is an extremely important technique in that for a given set of initial conditions, it will yield the total response of the circuit consisting of both the natural and forced responses in one operation.

Our use of the Laplace transform to solve circuit problems is analogous to that of using phasors in sinusoidal steady-state analysis. Using the Laplace transform, we transform the circuit problem from the time domain to the complex frequency domain, solve the problem using algebra in the complex frequency domain, and then convert the solution in the complex frequency domain back to the time domain. Therefore, as we shall see, the Laplace transform is an integral transform that converts a set of linear simultaneous integrodifferential equations to a set of simultaneous algebraic equations.

Our approach is to define the Laplace transform, derive some of the transform pairs, consider some of the important properties of the transform, illustrate the inverse transform operation, introduce the convolution integral, and finally apply the transform to circuit analysis.

13.1 DEFINITION

The Laplace transform of a function $f(t)$ is defined by the equation

$$\mathcal{L}[f(t)] = \mathbf{F}(s) = \int_0^\infty f(t)e^{-st}\, dt \qquad \textbf{13.1}$$

where s is the complex frequency

$$s = \sigma + j\omega \qquad \textbf{13.2}$$

and the function $f(t)$ is assumed to possess the property that

$$f(t) = 0 \quad \text{for} \quad t < 0$$

Note that the Laplace transform is unilateral ($0 \leq t < \infty$), in contrast to the Fourier transform (see Chapter 15), which is bilateral ($-\infty < t < \infty$). In our analysis of circuits using the Laplace transform, we will focus our attention on the time interval $t \geq 0$. It is important to note that it is the initial conditions that account for the operation of the circuit prior to $t = 0$, and therefore our analyses will describe the circuit operation for $t \geq 0$.

In order for a function $f(t)$ to possess a Laplace transform, it must satisfy the condition

$$\int_0^\infty e^{-\sigma t}\,|f(t)|\, dt < \infty \qquad \textbf{13.3}$$

for some real value of σ. Because of the convergence factor $e^{-\sigma t}$, there are a number of important functions that have Laplace transforms, even though Fourier transforms for these functions do not exist. All of the inputs we will apply to circuits possess Laplace transforms. Functions that do not have Laplace transforms $\left(\text{e.g., } e^{t^2}\right)$ are of no interest to us in circuit analysis.

The inverse Laplace transform, which is analogous to the inverse Fourier transform, is defined by the relationship

$$\mathcal{L}^{-1}[\mathbf{F}(s)] = f(t) = \frac{1}{2\pi j}\int_{\sigma_1 - j\infty}^{\sigma_1 + j\infty} \mathbf{F}(s)e^{st}\, ds \qquad \textbf{13.4}$$

where σ_1 is real and $\sigma_1 > \sigma$ in Eq. (13.3). The evaluation of this integral is based on complex variable theory, and therefore we will circumvent its use by developing and using a set of Laplace transform pairs.

13.2 TWO IMPORTANT SINGULARITY FUNCTIONS

There are two singularity functions that are very important in circuit analysis: (1) the unit step function, $u(t)$, discussed in Chapter 6, and (2) the unit impulse or delta function, $\delta(t)$.

They are called *singularity functions* because they are either not finite or they do not possess finite derivatives everywhere. They are mathematical models for signals that we employ in circuit analysis.

The *unit step function* $u(t)$ shown in Fig. 13.1a was defined in section 6.4 as

$$u(t) = \begin{cases} 0 & t < 0 \\ 1 & t > 0 \end{cases}$$

Recall that the physical analogy of this function, as illustrated earlier, corresponds to closing a switch at $t = 0$ and connecting a voltage source of 1 V or a current source of 1 A to a given circuit. The following example illustrates the calculation of the Laplace transform for unit step functions.

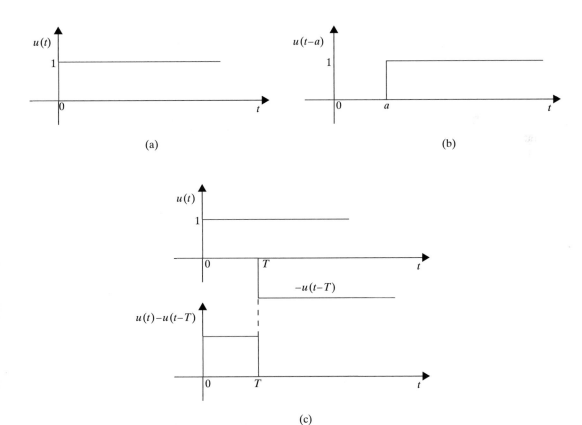

Figure 13.1 Representations of the unit step function.

EXAMPLE 13.1

Let us determine the Laplace transform for the waveforms in Fig. 13.1.

SOLUTION The Laplace transform for the unit step function in Fig. 13.1a is

$$\mathbf{F}(s) = \int_0^\infty u(t)e^{-st}\,dt$$

$$= \int_0^\infty 1e^{-st}\,dt$$

$$= -\frac{1}{s}e^{-st}\bigg|_0^\infty$$

$$= \frac{1}{s} \quad \sigma > 0$$

Therefore,

$$\mathcal{L}[u(t)] = \mathbf{F}(s) = \frac{1}{s}$$

The Laplace transform of the time-shifted unit step function shown in Fig. 13.1b is

$$\mathbf{F}(s) = \int_0^\infty u(t - a)e^{-st}\,dt$$

Note that

$$u(t - a) = \begin{cases} 1 & a < t < \infty \\ 0 & t < a \end{cases}$$

Therefore,

$$\mathbf{F}(s) = \int_a^\infty e^{-st}\,dt$$

$$= \frac{e^{-as}}{s}, \sigma > 0$$

Finally, the Laplace transform of the pulse shown in Fig. 13.1c is

$$\mathbf{F}(s) = \int_0^\infty [u(t) - u(t - T)]e^{-st}\,dt$$

$$= \frac{1 - e^{-Ts}}{s}, \sigma > 0 \qquad \square$$

The unit impulse function can be represented in the limit by the rectangular pulse shown in Fig. 13.2a as $a \rightarrow 0$. The function is defined by the following:

$$\delta(t - t_0) = 0 \qquad t \neq t_0$$

$$\int_{t_0-\varepsilon}^{t_0+\varepsilon} \delta(t - t_0)\,dt = 1 \qquad \varepsilon > 0$$

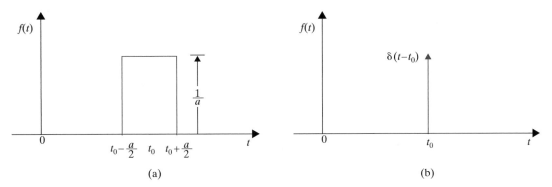

Figure 13.2 Representations of the unit impulse.

The unit impulse is zero except at $t = t_0$, where it is undefined, but it has unit area (sometimes referred to as *strength*). We represent the unit impulse function on a plot as shown in Fig. 13.2b.

An important property of the unit impulse function is what is often called the *sampling property*, which is exhibited by the following integral:

$$\int_{t_1}^{t_2} f(t)\delta(t - t_0)\, dt = \begin{cases} f(t_0) & t_1 < t_0 < t_2 \\ 0 & t_0 < t_1, t_0 > t_2 \end{cases}$$

for a finite t_0 and any $f(t)$ continuous at t_0. Note that the unit impulse function simply samples the value of $f(t)$ at $t = t_0$.

D13.1   Evaluate the integral

$$\int_0^{2\pi} \cos t\, \delta(t - \pi)\, dt$$

ANSWER: -1

EXAMPLE 13.2

Let us determine the Laplace transform of an impulse function.

SOLUTION   The Laplace transform of the impulse function is

$$\mathbf{F}(s) = \int_0^\infty \delta(t - t_0)e^{-st}\, dt$$

Using the sampling property of the delta function, we obtain

$$\mathcal{L}[\delta(t - t_0)] = e^{-t_0 s}$$

In the limit as $t_0 \to 0$, $e^{-t_0 s} \to 1$, and therefore

$$\mathcal{L}[\delta(t)] = \mathbf{F}(s) = 1 \qquad \square$$

D13.2   Evaluate the integral

$$\int_0^{2\pi} e^{-t}\cos t\, \delta(t - 10\pi)\, dt$$

ANSWER: 0

13.3 TRANSFORM PAIRS

We will now illustrate the development of a number of basic transform pairs that are very useful in circuit analysis.

EXAMPLE 13.3

Let us find the Laplace transform of $f(t) = t$.

SOLUTION The Laplace transform of the function $f(t) = t$ is

$$\mathbf{F}(s) = \int_0^\infty t e^{-st}\, dt$$

Integrating the function by parts, we let

$$u = t \quad \text{and} \quad dv = e^{-st}\, dt$$

Then

$$du = dt \quad \text{and} \quad v = \int e^{-st}\, dt = -\frac{1}{s} e^{-st}$$

Therefore,

$$\mathbf{F}(s) = \left. \frac{-t}{s} e^{-st} \right|_0^\infty + \int_0^\infty \frac{e^{-st}}{s}\, dt$$

$$= \frac{1}{s^2}, \sigma > 0$$

$t \leftrightarrow \dfrac{1}{s^2}$

EXAMPLE 13.4

Let us determine the Laplace transform of the cosine function.

SOLUTION The Laplace transform for the cosine function is

$$\mathbf{F}(s) = \int_0^\infty \cos \omega t\, e^{-st}\, dt$$

$$= \int_0^\infty \frac{e^{-j\omega t} + e^{-j\omega t}}{2} e^{-st}\, dt$$

$$= \int_0^\infty \frac{e^{-(s-j\omega)t} + e^{-(s+j\omega)t}}{2}\, dt$$

$$= \frac{1}{2} \left(\frac{1}{s - j\omega} + \frac{1}{s + j\omega} \right) \qquad \sigma > 0$$

$$= \frac{s}{s^2 + \omega^2}$$

$\cos \omega t \leftrightarrow \dfrac{s}{s^2 + \omega^2}$

A short table of useful Laplace transform pairs is shown in Table 13.1.

Once the transform pairs are known, we can easily move back and forth between the time domain and the complex frequency domain without having to use Eqs. (13.1) and (13.4).

Table 13.1 Short Table of Laplace Transform Pairs

$f(t)$	$F(s)$
$\delta(t)$	1
$u(t)$	$\dfrac{1}{s}$
e^{-at}	$\dfrac{1}{s + a}$
t	$\dfrac{1}{s^2}$
$\dfrac{t^n}{n!}$	$\dfrac{1}{s^{n+1}}$
te^{-at}	$\dfrac{1}{(s + a)^2}$
$\dfrac{t^n e^{-at}}{n!}$	$\dfrac{1}{(s + a)^{n+1}}$
$\sin bt$	$\dfrac{b}{s^2 + b^2}$
$\cos bt$	$\dfrac{s}{s^2 + b^2}$
$e^{-at} \sin bt$	$\dfrac{b}{(s + a)^2 + b^2}$
$e^{-at} \cos bt$	$\dfrac{s + a}{(s + a)^2 + b^2}$

EXTENSION EXERCISES

E13.1   If $f(t) = e^{-at}$, show that $\mathbf{F}(s) = 1/(s + a)$.

E13.2   If $f(t) = \sin \omega t$, show that $\mathbf{F}(s) = \omega/(s^2 + \omega^2)$.

13.4 PROPERTIES OF THE TRANSFORM

We now present a number of important properties of the Laplace transform and illustrate their usefulness by examples.

THEOREM 1

$$\mathcal{L}[Af(t)] = A\mathbf{F}(s) \qquad\qquad \textbf{13.5}$$

The transform is defined as

$$\mathcal{L}[Af(t)] = \int_0^\infty Af(t)e^{-st}\, dt$$

Since A is not a function of time,

$$\mathcal{L}[Af(t)] = A\int_0^\infty f(t)e^{-st}\, dt$$
$$= A\mathbf{F}(s) \qquad \square$$

These theorems describe linear properties of the transform.

THEOREM 2

$$\mathcal{L}[f_1(t) \pm f_2(t)] = \mathbf{F}_1(s) \pm \mathbf{F}_2(s) \qquad \textbf{13.6}$$

By definition,

$$\mathcal{L}[f_1(t) \pm f_2(t)] = \int_0^\infty [f_1(t) \pm f_2(t)]e^{-st}\, dt$$
$$= \int_0^\infty f_1(t)e^{-st}\, dt \pm \int_0^\infty f_2(t)e^{-st}\, dt$$
$$= \mathbf{F}_1(s) \pm \mathbf{F}_2(s) \qquad \square$$

Time-scaling theorem

Time-scaling theorem

THEOREM 3

$$\mathcal{L}[f(at)] = \frac{1}{a}\mathbf{F}\left(\frac{s}{a}\right) \qquad a > 0 \qquad \textbf{13.7}$$

The Laplace tranform of $f(at)$ is

$$\mathcal{L}[f(at)] = \int_0^\infty f(at)e^{-st}\, dt$$

Now let $\lambda = at$ and $d\lambda = a\, dt$. Then

$$\mathcal{L}[f(at)] = \int_0^\infty f(\lambda)e^{-(\lambda/a)s}\, \frac{d\lambda}{a}$$
$$= \frac{1}{a}\int_0^\infty f(\lambda)e^{-(s/a)\lambda}\, d\lambda$$
$$= \frac{1}{a}\mathbf{F}\left(\frac{s}{a}\right) \qquad a > 0 \qquad \square$$

EXAMPLE 13.5

We wish to find the Laplace transform of $f(t) = \cos \omega(t/2)$.

SOLUTION We have found that $\mathcal{L}[\cos \omega t] = s/(s^2 + \omega^2)$. Therefore, using Theorem 3 yields

$$\cos \omega \left(\frac{t}{2} \right) = \frac{4s}{(2s)^2 + \omega^2}$$

$$= \frac{s}{s^2 + \omega^2/4}$$

which should be fairly obvious, since $\mathcal{L}[\cos \alpha t] = s/(s^2 + \alpha^2)$, where in this case $\alpha = \omega/2$.   ❑

THEOREM 4

$$\mathcal{L}[f(t - t_0)u(t - t_0)] = e^{-t_0 s}\mathbf{F}(s) \quad t_0 \geq 0 \qquad \textbf{13.8}$$

Time-shifting theorem
Time-shifting theorem

This theorem, commonly known as the shifting theorem, is illustrated as follows:

$$\mathcal{L}[f(t - t_0)u(t - t_0)] = \int_0^\infty f(t - t_0)u(t - t_0)e^{-st}\,dt$$

$$= \int_{t_0}^\infty f(t - t_0)e^{-st}\,dt$$

If we now let $\lambda = t - t_0$ and $d\lambda = dt$, then

$$\mathcal{L}[f(t - t_0)u(t - t_0)] = \int_0^\infty f(\lambda)e^{-s(\lambda + t_0)}\,d\lambda$$

$$= e^{-t_0 s}\int_0^\infty f(\lambda)e^{-s\lambda}\,d\lambda$$

$$= e^{-t_0 s}\mathbf{F}(s) \quad t_0 \geq 0 \qquad ❑$$

EXAMPLE 13.6

Let us determine the Laplace transform of $f(t) = (t - 1)u(t - 1)$.

SOLUTION   If $f(t) = (t - 1)u(t - 1)$, then by employing Theorem 4,

$$\mathcal{L}[(t - 1)u(t - 1)] = e^{-s}\mathcal{L}[t] = \frac{e^{-s}}{s^2} \qquad ❑$$

THEOREM 5

$$\mathcal{L}[f(t)u(t - t_0)] = e^{-t_0 s}\mathcal{L}[f(t + t_0)] \qquad \textbf{13.9}$$

By definition,

Note the difference between theorems 4 and 5.

$$\mathcal{L}[f(t)u(t - t_0)] = \int_0^\infty f(t)u(t - t_0)e^{-st}\,dt$$

Now let $\lambda = t - t_0$ and $d\lambda = dt$, and therefore,

$$\int_0^\infty f(t)u(t - t_0)e^{-st}\,dt = \int_{-t_0}^\infty f(\lambda + t_0)u(\lambda)e^{-s(\lambda + t_0)}\,d\lambda$$

$$= \int_0^\infty f(\lambda + t_0)e^{-s(\lambda + t_0)}\,d\lambda$$

$$= e^{-t_0 s}\,\mathcal{L}[f(t + t_0)] \qquad \square$$

EXAMPLE 13.7

Let us find the Laplace transform of $f(t) = tu(t - 1)$.

SOLUTION   If $f(t) = tu(t - 1)$, then $\mathcal{L}[f(t)]$ can be found from Theorem 5, as

$$\mathcal{L}[tu(t - 1)] = e^{-s}\,\mathcal{L}[t + 1]$$

$$= e^{-s}\left(\frac{1}{s^2} + \frac{1}{s}\right) \qquad \square$$

THEOREM 6

$$\mathcal{L}[e^{-at}f(t)] = \mathbf{F}(s + a) \qquad \qquad \textbf{13.10}$$

Frequency-shifting or modulation theorem

By definition,

$$\mathcal{L}[e^{-at}f(t)] = \int_0^\infty e^{-at}f(t)e^{-st}\,dt$$

$$= \int_0^\infty f(t)e^{-(s+a)t}\,dt$$

$$= \mathbf{F}(s + a) \qquad \square$$

EXAMPLE 13.8

Use the Laplace transform of $\cos \omega t$ to find the Laplace transform of $e^{-\alpha t} \cos \omega t$.

SOLUTION   Since the Laplace transform of $\cos \omega t$ is known to be

$$\mathcal{L}[\cos \omega t] = \frac{s}{s^2 + \omega^2}$$

then

$$\mathcal{L}[e^{-at} \cos \omega t] = \frac{s + a}{(s + a)^2 + \omega^2} \qquad \square$$

THEOREM 7

$$\mathcal{L}[tf(t)] = \frac{-d\mathbf{F}(s)}{ds} \qquad \qquad \textbf{13.11}$$

Theorem for differentiation in
the frequency domain

By definition,

$$\mathbf{F}(s) = \int_0^\infty f(t)e^{-st}\, dt$$

Then

$$\frac{d\mathbf{F}(s)}{ds} = \int_0^\infty f(t)\left(-te^{-st}\right) dt$$

$$= -\int_0^\infty tf(t)e^{-st}\, dt$$

$$= -\mathcal{L}[tf(t)] \qquad \qquad \square$$

EXAMPLE 13.9

Let us demonstrate the use of Theorem 7.

SOLUTION   To begin, we know that

$$\mathcal{L}[u(t)] = \frac{1}{s}$$

Then

$$\mathcal{L}[tu(t)] = \frac{-d}{ds}\left(\frac{1}{s}\right) = \frac{1}{s^2}$$

Continuing, we note that

$$\mathcal{L}[t^2 u(t)] = \frac{d}{ds}\left(\frac{1}{s^2}\right) = \frac{2}{s^3}$$

and

$$\mathcal{L}[t^3 u(t)] = \frac{-d}{ds}\left(\frac{2}{s^3}\right) = \frac{3!}{s^4}$$

and in general,

$$\mathcal{L}[t^n u(t)] = \frac{n!}{s^{n+1}} \qquad \qquad \square$$

THEOREM 8

$$\mathcal{L}\left[\frac{f(t)}{t}\right] = \int_S^\infty \mathbf{F}(\lambda)d\lambda \qquad \textbf{13.12}$$

Theorem for integration in the frequency domain

By definition,

$$\int_0^\infty f(t)e^{-\lambda t}\, dt = \mathbf{F}(\lambda)$$

Therefore,

$$\int_S^\infty \int_S^\infty f(t)e^{-\lambda t}\, dt\, d\lambda = \int_S^\infty \mathbf{F}(\lambda)d\lambda$$

Since $f(t)$ is Laplace transformable, we can change the order of integration so that

$$\int_0^\infty f(t)\int_S^\infty e^{-\lambda t}\, d\lambda\, dt = \int_S^\infty \mathbf{F}(\lambda)d\lambda$$

and hence,

$$\int_0^\infty \frac{f(t)}{t}e^{-st}\, dt = \int_S^\infty \mathbf{F}(\lambda)d\lambda$$

Therefore,

$$\mathcal{L}\left[\frac{f(t)}{t}\right] = \int_S^\infty \mathbf{F}(\lambda)d\lambda \qquad \square$$

EXAMPLE 13.10

Let us demonstrate Theorem 8.

SOLUTION   If $f(t) = te^{-at}$, then

$$\mathbf{F}(\lambda) = \frac{1}{(\lambda + a)^2}$$

Therefore,

$$\int_S^\infty \mathbf{F}(\lambda)d\lambda = \int_S^\infty \frac{1}{(\lambda + a)^2}\, d\lambda = \frac{-1}{\lambda + a}\Big|_S^\infty = \frac{1}{s + a}$$

Hence,

$$f_1(t) = \frac{f(t)}{t} = \frac{te^{-at}}{t} = e^{-at} \quad \text{and} \quad \mathbf{F}_1(s) = \frac{1}{s + a} \qquad \square$$

The remaining two theorems in this section are very useful in the solution of integrodifferential equations, and we will employ them later in this chapter in analyzing circuits described by these equations.

THEOREM 9

$$\mathcal{L}\left[\frac{d^n f(t)}{dt^n}\right] = s^n \, \mathbf{F}(s) - s^{n-1} f_{(0)} - s^{n-2} f'_{(0)} \cdots - s^0 f^{n-1}(0) \qquad \textbf{13.13}$$

where

Theorem for differentiation in the time domain

$$f^{(k)}(0) = \left.\frac{d^k f(t)}{dt^k}\right|_{t=0}$$

Let us begin by examining $\mathcal{L}[df(t)/dt]$. By definition,

$$\mathcal{L}\left[\frac{df(t)}{dt}\right] = \int_0^\infty \frac{df(t)}{dt} e^{-st} \, dt$$

Using integration by parts gives us

$$u = e^{-st} \qquad\qquad dv = \frac{df(t)}{dt} \, dt = df(t)$$

$$du = -se^{-st} \, dt \qquad v = f(t)$$

Hence,

$$\mathcal{L}\left[\frac{df(t)}{dt}\right] = f(t)e^{-st}\Big|_0^\infty + s\int_0^\infty f(t)e^{-st} \, dt$$

If we assume that the Laplace transform of the function $f(t)$ exists so that

$$\lim_{t\to\infty} e^{-st} f(t) = 0$$

then

$$\mathcal{L}\left[\frac{df(t)}{dt}\right] = -f(0) + s\mathbf{F}(s)$$

Using this result, we can write

$$\mathcal{L}\left[\frac{d^2 f(t)}{dt^2}\right] = s\mathcal{L}[f'(t)] - f'(0)$$

$$= s[s\mathbf{F}(s) - f(0)] - f'(0)$$

$$= s^2\mathbf{F}(s) - sf(0) - f'(0)$$

By continuing in this manner we can demostrate the original statement.   ❑

THEOREM 10

$$\mathcal{L}\left[\int_0^t f(\lambda)d\lambda\right] = \frac{1}{s}\mathbf{F}(s) \qquad \textbf{13.14}$$

We begin with the expression

Theorem for integration in the time domain

$$\mathcal{L}\left[\int_0^t f(\lambda)d\lambda\right] = \int_0^\infty \int_0^t f(\lambda)d\lambda e^{-st} \, dt$$

Integrating by parts yields

$$u = \int_0^t f(\lambda)d\lambda \qquad dv = e^{-st} \, dt$$

$$du = f(t) \, dt \qquad v = \frac{-1}{s} e^{-st}$$

Therefore,

$$\mathcal{L}\left[\int_0^t f(\lambda)d\lambda \right] = \frac{-e^{-st}}{s} \int_0^t f(\lambda)d\lambda \Big|_0^\infty + \frac{1}{s} \int_0^\infty f(t)e^{-st} \, dt$$

$$= \frac{1}{s} \mathbf{F}(s) \qquad\qquad\qquad \square$$

We employ this theorem later in the chapter when we examine integrodifferential equations. The theorems we have presented are listed in Table 13.2 for quick reference.

Table 13.2 Some Useful Properties of the Laplace Transform

$f(t)$	$F(s)$
$Af(t)$	$A\mathbf{F}(s)$
$f_1(t) \pm f_2(t)$	$\mathbf{F}_1(s) \pm \mathbf{F}_2(s)$
$f(at)$	$\dfrac{1}{a}\mathbf{F}\left(\dfrac{s}{a}\right), a > 0$
$f(t - t_0)u(t - t_0), t_0 \geq 0$	$e^{-t_0 s}\mathbf{F}(s)$
$f(t)u(t - t_0)$	$e^{-t_0 s}\mathcal{L}[f(t - t_0)]$
$e^{-at}f(t)$	$\mathbf{F}(s + a)$
$\dfrac{d^n f(t)}{dt^n}$	$s^n\mathbf{F}(s) - s^{n-1}f(0) - s^{n-2}f'(0)\cdots s^0 f^{n-1}(0)$
$tf(t)$	$-\dfrac{d\mathbf{F}(s)}{ds}$
$\dfrac{f(t)}{t}$	$\displaystyle\int_s^\infty \mathbf{F}(\lambda)d\lambda$
$\displaystyle\int_0^t f(\lambda)d\lambda$	$\dfrac{1}{s}\mathbf{F}(s)$
$\displaystyle\int_0^t f_1(\lambda)f_2(t - \lambda)d\lambda$	$\mathbf{F}_1(s)\mathbf{F}_2(s)$

EXAMPLE 13.11

If $f(t) = te^{-at}u(t - 1)$, find $\mathbf{F}(s)$.

SOLUTION   Using Theorem 5,

$$\mathbf{F}(s) = e^{-s}\mathcal{L}\left[(t + 1)e^{-a(t+1)}\right]$$
$$\mathbf{F}(s) = e^{-s}\mathcal{L}\left[e^{-a}te^{-at} + e^{-a}e^{-at}\right]$$
$$= e^{-s}\left(\frac{e^{-a}}{(s + a)^2} + \frac{e^{-a}}{s + a}\right)$$ ❏

EXAMPLE 13.12

If $f(t) = e^{-at}\sin \omega t\, u(t - 1)$, find $\mathbf{F}(s)$.

SOLUTION

$$\mathbf{F}(s) = \mathcal{L}\left[e^{-at}\sin \omega t\, u(t - 1)\right]$$

Then, using Theorem 5,

$$\mathbf{F}(s) = e^{-s}\mathcal{L}\left[e^{-a(t+1)}\sin \omega(t + 1)\right]$$
$$= e^{-(s+a)}\mathcal{L}\left[e^{-at}\sin (\omega t + \omega)\right]$$
$$= e^{-(s+a)}\mathcal{L}\left[e^{-at}\sin \omega t \cos \omega + e^{-at}\cos \omega t \sin \omega\right]$$
$$= e^{-(s+a)}\left[\frac{\omega \cos \omega}{(s + a)^2 + \omega^2} + \frac{(s + a)\sin \omega}{(s + a)^2 + \omega^2}\right]$$ ❏

EXAMPLE 13.13

If $f(t) = t\cos(\omega t)u(t - 1)$, find $\mathbf{F}(s)$.

SOLUTION

$$\mathcal{L}\left[t\cos \omega t\, u(t - 1)\right] = e^{-s}\mathcal{L}\left[(t + 1)\cos(\omega(t + 1))\right]$$
$$= e^{-s}\mathcal{L}\left[t\cos(\omega t)\cos \omega - t\sin(\omega t)\sin \omega + \cos \omega t\cos \omega - \sin \omega t\sin \omega\right]$$

Since

$$\mathcal{L}\left[\cos \omega t\cos \omega\right] = \frac{s\cos \omega}{s^2 + \omega^2}$$

$$\mathcal{L}\left[\sin \omega t\sin \omega\right] = \frac{\omega \sin \omega}{s^2 + \omega^2}$$

$$\mathcal{L}\left[t\cos \omega t\cos \omega\right] = \frac{-d}{ds}\frac{s\cos \omega}{s^2 + \omega^2} = \frac{(s^2 - \omega^2)\cos \omega}{(s^2 + \omega^2)^2}$$

and

$$\mathcal{L}[t \sin \omega t \sin \omega] = \frac{-d}{ds} \frac{\omega \sin \omega}{s^2 + \omega^2} = \frac{\omega \sin \omega(2s)}{\left(s^2 + \omega^2\right)^2}$$

therefore,

$$\mathcal{L}[t \cos(\omega t)u(t-1)] = e^{-s}\left[\frac{(s^2 - \omega^2)\cos\omega - 2\omega s \sin\omega}{\left(s^2 + \omega^2\right)^2} \right.$$
$$\left. + \frac{s\cos\omega - \omega\sin\omega}{s^2 + \omega^2} \right]$$

❏

EXAMPLE 13.14

Let us employ the Laplace transform to solve the equation

$$\frac{dy(t)}{dt} + 2y(t) + \int_0^t y(\lambda)e^{-2(t-\lambda)}\,d\lambda = 10u(t) \qquad y(0) = 0$$

SOLUTION   Applying the transform, we obtain

$$s\mathbf{Y}(s) + 2\mathbf{Y}(s) + \frac{\mathbf{Y}(s)}{s+2} = \frac{10}{s}$$

$$\mathbf{Y}(s)\left(s + 2 + \frac{1}{s+2}\right) = \frac{10}{s}$$

$$\mathbf{Y}(s) = \frac{10(s+2)}{s(s^2 + 4s + 5)}$$

❏

This is the solution of the linear constant-coefficient integrodifferential equation in the s-domain. However, we want the solution $y(t)$ in the time domain. $y(t)$ is obtained by performing the inverse transform, which is the topic of the next section, and the solution $y(t)$ is derived in Example 13.16.

EXTENSION EXERCISES

E13.3   Find $\mathbf{F}(s)$ if $f(t) = \frac{1}{2}(t - 4e^{-2t})$.

ANSWER: $\mathbf{F}(s) = \dfrac{1}{2s^2} - \dfrac{2}{s+2}$.

E13.4   If $f(t) = te^{-(t-1)}u(t-1) - e^{-(t-1)}u(t-1)$, determine $\mathbf{F}(s)$ using the shifting theorem.

ANSWER: $\mathbf{F}(s) = \dfrac{e^{-s}}{(s+1)^2}$.

E13.5   Find $\mathbf{F}(s)$ if $f(t) = e^{-4t}(t - e^{-t})$. Use Theorem 2.

ANSWER: $\mathbf{F}(s) = \dfrac{1}{(s + 4)^2} - \dfrac{1}{s + 5}$.

E13.6   Find the Laplace transform of the function $te^{-4x}/(a^2 + 4)$.

ANSWER: $\mathbf{F}(s) = \dfrac{e^{-4x}}{s^2(a^2 + 4)}$.

E13.7   If $f(t) = \cos \omega\, tu(t - 1)$, find $\mathbf{F}(s)$ using Theorem 5.

ANSWER: $\mathbf{F}(s) = e^{-s}\left(\dfrac{s \cos \omega}{s^2 + \omega^2} - \dfrac{\omega \sin \omega}{s^2 + \omega^2}\right)$.

E13.8   Use Theorem 7 to demonstrate that $\mathcal{L}[te^{-at}] = 1/(s + a)^2$.

13.5 PERFORMING THE INVERSE TRANSFORM

As we begin our discussion of this topic, let us outline the procedure we will use in applying the Laplace transform to circuit analysis. First, we will transform the problem from the time domain to the complex frequency domain (that is, s-domain). Next we will solve the circuit equations algebraically in the complex frequency domain. Finally, we will transform the solution from the s-domain back to the time domain. It is this latter operation that we discuss now.

The algebraic solution of the circuit equations in the complex frequency domain results in a rational function of s of the form

$$\mathbf{F}(s) = \frac{\mathbf{P}(s)}{\mathbf{Q}(s)} = \frac{a_m s^m + a_{m-1} s^{m-1} + \cdots + a_1 s + a_0}{b_n s^n + b_{n-1} s^{n-1} + \cdots + b_1 s + b_0} \qquad \textbf{13.15}$$

The roots of the polynomial $\mathbf{P}(s)$ (i.e., $-z_1, -z_2 \cdots -z_m$) are called the *zeros* of the function $\mathbf{F}(s)$ because at these values of s, $\mathbf{F}(s) = 0$. Similarly, the roots of the polynomial $\mathbf{Q}(s)$ (i.e., $-p_1, -p_2 \cdots -p_n$) are called *poles* of $\mathbf{F}(s)$, since at these values of s, $\mathbf{F}(s)$ becomes infinite.

If $\mathbf{F}(s)$ is a proper rational function of s, then $n > m$. However, if this is not the case, we simply divide $\mathbf{P}(s)$ by $\mathbf{Q}(s)$ to obtain a quotient and a remainder; that is,

$$\frac{\mathbf{P}(s)}{\mathbf{Q}(s)} = C_{m-n} s^{m-n} + \cdots + C_2 s^2 + C_1 s + C_0 + \frac{\mathbf{P}_1(s)}{\mathbf{Q}(s)} \qquad \textbf{13.16}$$

Now $\mathbf{P}_1(s)/\mathbf{Q}(s)$ is a proper rational function of s. Let us examine the possible forms of the roots of $\mathbf{Q}(s)$.

DRILL

D13.3   Find the partial fraction form of the function

$$\mathbf{F}(s) = \frac{s^2 + 4s + 5}{s^2 + 3s + 2}$$

ANSWER:

$$\mathbf{F}(s) = 1 + \frac{K_1}{s + 1} + \frac{K_2}{s + 2}$$

DRILL

D13.4   Find the partial fraction form of the function

$$\mathbf{F}(s) = \frac{s + 5}{s^3 + 4s^2 + 4s}$$

ANSWER: $\mathbf{F}(s) =$

$$\frac{K_1}{s} + \frac{K_{21}}{s + 2} + \frac{K_{22}}{(s + 2)^2}$$

1. If the roots are simple, $\mathbf{P}_1(s)/\mathbf{Q}(s)$ can be expressed in partial fraction form as

$$\frac{\mathbf{P}_1(s)}{\mathbf{Q}(s)} = \frac{K_1}{s + p_1} + \frac{K_2}{s + p_2} + \cdots + \frac{K_n}{s + p_n} \qquad \textbf{13.17}$$

2. If $\mathbf{Q}(s)$ has simple complex roots, they will appear in complex-conjugate pairs and the partial fraction expansion of $\mathbf{P}_1(s)/\mathbf{Q}(s)$ for each pair of complex-conjugate roots will be of the form

$$\frac{\mathbf{P}_1(s)}{\mathbf{Q}_1(s)(s + \alpha - j\beta)(s + \alpha + j\beta)} = \frac{K_1}{s + \alpha - j\beta} + \frac{K_1^*}{s + \alpha + j\beta} + \cdots \qquad \textbf{13.18}$$

where $\mathbf{Q}(s) = \mathbf{Q}_1(s)(s + \alpha - j\beta)(s + \alpha + j\beta)$ and K_1^* in the complex conjugate of K_1.

3. If $\mathbf{Q}(s)$ has a root of multiplicity r, the partial fraction expansion for each such root will be of the form

$$\frac{\mathbf{P}_1(s)}{\mathbf{Q}_1(s)(s + p_1)^r} = \frac{K_{11}}{(s + p_1)} + \frac{K_{12}}{(s + p_1)^2} + \cdots + \frac{K_{1r}}{(s + p_1)^r} + \cdots \qquad \textbf{13.19}$$

The importance of these partial fraction expansions stems from the fact that once the function $\mathbf{F}(s)$ is expressed in this form, the individual inverse Laplace transforms can be obtained from known and tabulated transform pairs. The sum of these inverse Laplace transforms then yields the desired time function, $f(t) = \mathcal{L}^{-1}[\mathbf{F}(s)]$.

Simple Poles

Let us assume that all the poles of $\mathbf{F}(s)$ are simple, so that the partial fraction expansion of $\mathbf{F}(s)$ is of the form

$$\mathbf{F}(s) = \frac{\mathbf{P}(s)}{\mathbf{Q}(s)} = \frac{K_1}{s + p_1} + \frac{K_2}{s + p_2} + \cdots + \frac{K_n}{s + p_n} \qquad \textbf{13.20}$$

Then the constant K_i can be computed by multiplying both sides of this equation by $(s + p_i)$ and evaluating the equation at $s = p_i$; that is,

$$\left. \frac{(s + p_i)\mathbf{P}(s)}{\mathbf{Q}(s)} \right|_{s=-P_i} = 0 + \cdots + 0 + K_i + 0 + \cdots + 0 \qquad i = 1, 2, \ldots, n \qquad \textbf{13.21}$$

Once all of the K_i terms are known, the time function $f(t) = \mathcal{L}^{-1}[\mathbf{F}(s)]$ can be obtained using the Laplace transform pair.

$$\mathcal{L}^{-1}\left[\frac{1}{s + a}\right] = e^{-at} \qquad \textbf{13.22}$$

EXAMPLE 13.15

Given that

$$\mathbf{F}(s) = \frac{12(s + 1)(s + 3)}{s(s + 2)(s + 4)(s + 5)}$$

let us find the function $f(t) = \mathcal{L}^{-1}[\mathbf{F}(s)]$.

SOLUTION   Expressing $\mathbf{F}(s)$ in a partial fraction expansion, we obtain

$$\frac{12(s + 1)(s + 3)}{s(s + 2)(s + 4)(s + 5)} = \frac{K_0}{s} + \frac{K_1}{s + 2} + \frac{K_2}{s + 4} + \frac{K_3}{s + 5}$$

To determine K_0, we multiply both sides of the equation by s to obtain the equation

$$\frac{12(s + 1)(s + 3)}{s(s + 2)(s + 4)(s + 5)} = K_0 + \frac{K_1 s}{s + 2} + \frac{K_2 s}{s + 4} + \frac{K_3 s}{s + 5}$$

Evaluating the equation at $s = 0$ yields

$$\frac{(12)(1)(3)}{(2)(4)(5)} = K_0 + 0 + 0 + 0$$

or

$$K_0 = \frac{36}{40}$$

Similarly,

$$(s + 2)\mathbf{F}(s)\Big|_{S=-2} = \frac{12(s + 1)(s + 3)}{s(s + 4)(s + 5)}\Big|_{S=-2} = K_1$$

or

$$K_1 = 1$$

Using the same approach, we find that $K_2 = \frac{36}{8}$ and $K_3 = -\frac{32}{5}$. Hence $\mathbf{F}(s)$ can be written as

$$\mathbf{F}(s) = \frac{36/40}{s} + \frac{1}{s + 2} + \frac{36/8}{s + 4} - \frac{32/5}{s + 5}$$

Then $f(t) = \mathcal{L}^{-1}\big[\mathbf{F}(s)\big]$ is

$$f(t) = \left(\frac{36}{40} + 1e^{-2t} + \frac{36}{8}e^{-4t} - \frac{32}{5}e^{-5t}\right)u(t) \qquad \square$$

EXTENSION EXERCISES

E13.9   Find $f(t)$ if $\mathbf{F}(s) = 10(s + 6)/(s + 1)(s + 3)$.

ANSWER: $f(t) = (25e^{-t} - 15e^{-3t})u(t)$.

E13.10   If $\mathbf{F}(s) = 12(s + 2)/s(s + 1)$, find $f(t)$.

ANSWER: $f(t) = (24 - 12e^{-t})u(t)$.

Complex-Conjugate Poles

Let us assume that $\mathbf{F}(s)$ has one pair of complex-conjugate poles. The partial fraction expansion of $\mathbf{F}(s)$ can then be written as

$$\mathbf{F}(s) = \frac{\mathbf{P}_1(s)}{\mathbf{Q}_1(s)(s + \alpha - j\beta)(s + \alpha + j\beta)} = \frac{K_1}{s + \alpha - j\beta} + \frac{K_1^*}{s + \alpha + j\beta} + \cdots \qquad \textbf{13.23}$$

The constant K_1 can then be determined using the procedure employed for simple poles; that is,

$$(s + \alpha - j\beta)\mathbf{F}(s)\Big|_{s=-\alpha+j\beta} = K_1 \qquad \textbf{13.24}$$

In this case K_1 is in general a complex number that can be expressed as $|K_1| \underline{/\theta}$. Then $K_1^* = |K_1| \underline{/-\theta}$. Hence, the partial fraction expansion can be expressed in the form

$$\mathbf{F}(s) = \frac{|K_1| \underline{/\theta}}{s + \alpha - j\beta} + \frac{|K_1| \underline{/-\theta}}{s + \alpha + j\beta} + \cdots \qquad \textbf{13.25}$$

$$= \frac{|K_1|e^{j\theta}}{s + \alpha - j\beta} + \frac{|K_1|e^{-j\theta}}{s + \alpha + j\beta} + \cdots$$

Recall that

$$\cos x = \frac{e^{jx} + e^{-jx}}{2}$$

The corresponding time function is then of the form

$$f(t) = \mathcal{L}^{-1}[\mathbf{F}(s)] = |K_1|e^{j\theta}e^{-(\alpha - j\beta)t} + |K_1|e^{-j\theta}e^{-(\alpha + j\beta)t} + \cdots$$

$$= |K_1|e^{-\alpha t}[e^{j(\beta t + \theta)} + e^{-j(\beta t + \theta)}] + \cdots \qquad \textbf{13.26}$$

$$= 2|K_1|e^{-\alpha t}\cos(\beta t + \theta) + \cdots$$

EXAMPLE 13.16

Let us determine the time function $y(t)$ for the function

$$\mathbf{Y}(s) = \frac{10(s + 2)}{s(s^2 + 4s + 5)}$$

SOLUTION   Expressing the function in a partial fraction expansion, we obtain

$$\frac{10(s + 2)}{s(s + 2 - j1)(s + 2 + j1)} = \frac{K_0}{s} + \frac{K_1}{s + 2 - j1} + \frac{K_1^*}{s + 2 - j1}$$

$$\frac{10(s + 2)}{s^2 + 4s + 5}\Big|_{s=0} = K_0$$

$$4 = K_0$$

In a similar manner,

$$\frac{10(s + 2)}{s(s + 2 + j1)}\Big|_{s=-2+j1} = K_1$$

$$2.236 \underline{/-153.43°} = K_1$$

Therefore,

$$2.236 \underline{/153.43°} = K_1^*$$

The partial fraction expansion of $\mathbf{Y}(s)$ is then

$$\mathbf{Y}(s) = \frac{4}{s} + \frac{2.236\ \underline{/-153.43°}}{s + 2 - j1} + \frac{2.236\ \underline{/153.43°}}{s + 2 + j1}$$

and therefore,

$$y(t) = [4 + 4.472e^{-2t}\cos(t - 153.43°)]u(t) \qquad \square$$

EXTENSION EXERCISE

E13.11   Determine $f(t)$ if $\mathbf{F}(s) = s/(s^2 + 4s + 8)$.

ANSWER: $f(t) = 1.41e^{-2t}\cos(2t + 45°)u(t)$.

Multiple Poles

Let us suppose that $\mathbf{F}(s)$ has a pole of multiplicity r. Then $\mathbf{F}(s)$ can be written in a partial fraction expansion of the form

$$\mathbf{F}(s) = \frac{\mathbf{P}_1(s)}{\mathbf{Q}_1(s)(s + p_1)^r}$$

$$= \frac{K_{11}}{s + p_1} + \frac{K_{12}}{(s + p_1)^2} + \cdots + \frac{K_{1r}}{(s + p_1)^r} + \cdots \qquad \mathbf{13.27}$$

Employing the approach for a simple pole, we can evaluate K_{1r} as

$$(s + p_1)^r \mathbf{F}(s)\Big|_{s=-p_1} = K_{1r} \qquad \mathbf{13.28}$$

In order to evaluate K_{1r-1} we multiply $\mathbf{F}(s)$ by $(s + p_1)^r$ as we did to determine K_{1r}; however, prior to evaluating the equation at $s = p_1$, we take the derivative with respect to s. The proof that this will yield K_{1r-1} can be obtained by multiplying both sides of Eq. (13.27) by $(s + p_1)^r$ and then taking the derivative with respect to s. Now when we evaluate the equation at $s = -p_1$, the only term remaining on the right side of the equation is K_{1r-1}, and therefore,

$$\frac{d}{ds}\left[(s + p_1)^r \mathbf{F}(s)\right]\Big|_{s=-p_1} = K_{1r-1} \qquad \mathbf{13.29}$$

K_{1r-2} can be computed in a similar fashion, and in that case the equation is

$$\frac{d^2}{ds^2}\left[(s + p_1)^r \mathbf{F}(s)\right]\Big|_{s=-p_1} = (2!)K_{1r-2} \qquad \mathbf{13.30}$$

The general expression for this case is

$$K_{1j} = \frac{1}{(r - j)!}\frac{d^{r-j}}{ds^{r-j}}\left[(s + p_1)^r \mathbf{F}(s)\right]\Big|_{s=-p_1} \qquad \mathbf{13.31}$$

Let us illustrate this procedure with an example.

EXAMPLE 13.17

Given the following function $\mathbf{F}(s)$, let us determine the corresponding time function $f(t) = \mathcal{L}^{-1}[\mathbf{F}(s)]$.

$$\mathbf{F}(s) = \frac{10(s + 3)}{(s + 1)^3(s + 2)}$$

SOLUTION Expressing $\mathbf{F}(s)$ as a partial fraction expansion, we obtain

$$\mathbf{F}(s) = \frac{10(s + 3)}{(s + 1)^3(s + 2)} = \frac{K_{11}}{s + 1} + \frac{K_{12}}{(s + 1)^2} + \frac{K_{13}}{(s + 1)^3} + \frac{K_2}{s + 2}$$

Then

$$(s + 1)^3 \mathbf{F}(s)\bigg|_{s=-1} = K_{13}$$

$$20 = K_{13}$$

K_{12} is now determined by the equation

$$\frac{d}{ds}\big[(s + 1)^3 \mathbf{F}(s)\big]\bigg|_{s=-1} = K_{12}$$

$$\frac{-10}{(s + 2)^2}\bigg|_{s=-1} = -10 = K_{12}$$

In a similar fashion K_{11} is computed from the equation

$$\frac{d^2}{ds^2}\big[(s + 1)^3 \mathbf{F}(s)\big]\bigg|_{s=-1} = 2K_{11}$$

$$\frac{20}{(s + 2)^3}\bigg|_{s=-1} = 20 = 2K_{11}$$

Therefore,

$$10 = K_{11}$$

In addition,

$$(s + 2)\mathbf{F}(s)\bigg|_{s=-2} = K_2$$

$$-10 = K_2$$

Hence, $\mathbf{F}(s)$ can be expressed as

$$\mathbf{F}(s) = \frac{10}{s + 1} - \frac{10}{(s + 1)^2} + \frac{20}{(s + 1)^3} - \frac{10}{s + 2}$$

Now we employ the transform pair

$$\mathcal{L}^{-1}\left[\frac{1}{(s + a)^{n+1}}\right] = \frac{t^n}{n!} e^{-at}$$

and hence,

$$f(t) = \big(10e^{-t} - 10te^{-t} + 10t^2 e^{-t} - 10e^{-2t}\big)u(t) \qquad \square$$

E13.12   Determine $f(t)$ if $\mathbf{F}(s) = s/(s+1)^2$.

ANSWER: $f(t) = (e^{-t} - te^{-t})u(t)$.

E13.13   If $\mathbf{F}(s) = (s+2)/s^2(s+1)$, find $f(t)$.

ANSWER: $f(t) = (-1 + 2t + e^{-t})u(t)$.

13.6 CONVOLUTION INTEGRAL

Convolution is a very important concept and has wide application in circuit and systems analysis. We will first illustrate the connection that exists between the convolution integral and the Laplace transform. We then indicate the manner in which the convolution integral is applied in circuit analysis.

THEOREM 11

If

$$f(t) = \int_0^t f_1(t-\lambda)f_2(\lambda)\,d\lambda = \int_0^t f_1(\lambda)f_2(t-\lambda)\,d\lambda \qquad \textbf{13.32}$$

and

$$\mathcal{L}[f(t)] = \mathbf{F}(s), \mathcal{L}[f_1(t)] = \mathbf{F}_1(s) \quad and \quad \mathcal{L}[f_2(t)] = \mathbf{F}_2(s)$$

then

$$\mathbf{F}(s) = \mathbf{F}_1(s)\mathbf{F}_2(s) \qquad \textbf{13.33}$$

Our demonstration begins with the definition

$$\mathcal{L}[f(t)] = \int_0^\infty \left[\int_0^t f_1(t-\lambda)f_2(\lambda)\,d\lambda\right]e^{-st}\,dt$$

We now force the function into the proper format by introducing into the integral within the brackets the unit step function $u(t-\lambda)$. We can do this because

$$u(t-\lambda) = \begin{cases} 1 & \text{for } \lambda < t \\ 0 & \text{for } \lambda > t \end{cases} \qquad \textbf{13.34}$$

The first condition in Eq. (13.34) ensures that the insertion of the unit step function has no impact within the limits of integration. The second condition in Eq. (13.34) allows us to change the upper limit of integration from t to ∞. Therefore,

$$\mathcal{L}[f(t)] = \int_0^\infty \left[\int_0^\infty f_1(t - \lambda)u(t - \lambda)f_2(\lambda)\,d\lambda \right] e^{-st}\,dt$$

which can be written as

$$\mathcal{L}[f(t)] = \int_0^\infty f_2(\lambda) \left[\int_0^\infty f_1(t - \lambda)u(t - \lambda)e^{-st}\,dt \right] d\lambda$$

Note that the integral within the brackets is the shifting theorem illustrated in Example 13.6. Hence, the equation can be written as

$$\mathcal{L}[f(t)] = \int_0^\infty f_2(\lambda)\mathbf{F}_1(s)e^{-s\lambda}\,d\lambda$$

$$= \mathbf{F}_1(s) \int_0^\infty f_2(\lambda)e^{-s\lambda}\,d\lambda$$

$$= \mathbf{F}_1(s)\mathbf{F}_2(s)$$

Note that convolution in the time domain corresponds to multiplication in the frequency domain.

Let us now illustrate the use of Theorem 11 in the evaluation of an inverse Laplace transform.   ❑

EXAMPLE 13.18

The transfer function for a network is given by the expression

$$\mathbf{H}(s) = \frac{\mathbf{V}_0(s)}{\mathbf{V}_s(s)} = \frac{10}{s + 5}$$

The input is a unit step function $\mathbf{V}_s(s) = \dfrac{1}{s}$. Let us use convolution to determine the output voltage $v_0(t)$.

SOLUTION   Since $\mathbf{H}(s) = \dfrac{10}{(s + 5)}$, $h(t) = 10e^{-5t}$ and therefore

$$v_o(t) = \int_0^t 10u(\lambda)e^{-5(t-\lambda)}\,d\lambda$$

$$= 10e^{-5t} \int_0^\infty e^{5\lambda}\,d\lambda$$

$$= \frac{10e^{-5t}}{5}[e^{5t} - 1]$$

$$= 2[1 - e^{-5t}]u(t)$$

For comparison, let us determine $v_0(t)$ from $\mathbf{H}(s)$ and $\mathbf{V}_s(s)$ using the partial fraction expansion method. $\mathbf{V}_0(s)$ can be written as

$$\mathbf{V}_0(s) = \mathbf{H}(s)\mathbf{V}_s(s)$$

$$= \frac{10}{s(s + 5)} = \frac{K_0}{s} + \frac{K_1}{s + 5}$$

Evaluating the constants, we obtain $K_0 = 2$ and $K_1 = -2$. Therefore,

$$\mathbf{V}_0(s) = \frac{2}{s} - \frac{2}{s+5}$$

and hence

$$v_0(t) = 2\left[1 - e^{-5t}\right]u(t)$$ ❏

Although we can employ convolution to derive an inverse Laplace transform, the example, although quite simple, illustrates that this is a very poor approach. If the function $F(s)$ is very complicated, the mathematics can become unwieldy. Convolution is, however, a very powerful and useful tool. To understand its usefulness, let us interpret the operation of the convolution integral by examining it from a graphical standpoint.

Suppose that the functions $f_1(\lambda)$ and $f_2(\lambda)$ are as shown in Fig. 13.3. If $f_2(\lambda) = e^{-\alpha\lambda}u(\lambda)$, the time-shifted function

$$f_2(\lambda - t) = e^{-\alpha(\lambda - t)}u(\lambda - t)$$

is as shown in Fig. 13.4a. If we now change the sign of the argument so that the function is

$$f_2(t - \lambda) = e^{-\alpha(t - \lambda)}u(t - \lambda)$$

then the function is folded or reflected about the point $\lambda = t$, as shown in Fig. 13.4b. The integrand of the convolution integral, $f_1(\lambda)f_2(t - \lambda)$, will be nonzero only when the two functions overlap. The overlap, or product of the two functions, for several values of t is shown shaded in Fig. 13.5. The convolution integral for various values of t is equal to the shaded area shown in Fig. 13.5. A plot of these areas as t varies is shown in Fig. 13.6. If we had shifted and folded the function $f_1(\lambda)$, multipied it by $f_2(\lambda)$, and integrated, we would again obtain the curve in Fig. 13.6.

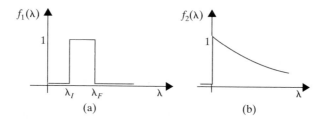

Figure 13.3 Two functions $f_1(\lambda)$ and $f_2(\lambda)$.

(a)   (b)

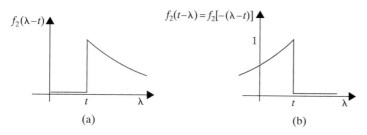

(a)   (b)

Figure 13.4 Shifting and folding operation on $f_2(\lambda)$.

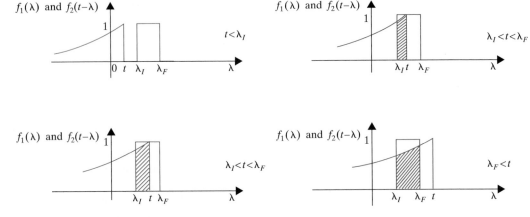

Figure 13.5 Product of the functions $f_1(\lambda)$ and $f_2(t - \lambda)$ shown shaded in the figures for different values of t.

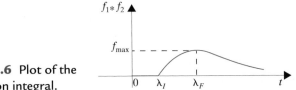

Figure 13.6 Plot of the convolution integral.

This graphical explanation of convolution has hopefully provided some additional insight. It also serves, however, to illustrate that *convolution is a very important simulation tool*. For example, if we know the impulse response of a network, we can use convolution to determine the network's response to an input that may be available only as an experimental curve obtained in the laboratory. Thus, convolution permits us to obtain the network response to inputs that cannot be written as analytical functions, but can be simulated on a digital computer.

13.7 INITIAL-VALUE AND FINAL-VALUE THEOREMS

Suppose that we wish to determine the initial or final value of a circuit response in the time domain from the Laplace transform of the function in the s-domain without performing the inverse transform. If we determine the function $f(t) = \mathcal{L}^{-1}[\mathbf{F}(s)]$, we can find the initial value by evaluating $f(t)$ as $t \to 0$ and the final value by evaluating $f(t)$ as $t \to \infty$. It would be very convenient, however, if we could simply determine the initial and final values from $\mathbf{F}(s)$ without having to perform the inverse transform. The initial- and final-value theorems allow us to do just that.

The *initial-value theorem* states that

$$\lim_{t \to 0} f(t) = \lim_{s \to \infty} s\mathbf{F}(s) \tag{13.35}$$

provided that $f(t)$ and its first derivative are transformable.

The proof of this theorem employs the Laplace transform of the function $df(t)/dt$.

$$\int_0^\infty \frac{df(t)}{dt} e^{-st} \, dt = s\mathbf{F}(s) - f(0)$$

Taking the limit of both sides as $s \to \infty$, we find that

$$\lim_{S \to \infty} \int_0^\infty \frac{df(t)}{dt} e^{-st} \, dt = \lim_{S \to \infty}\left[s\mathbf{F}(s) - f(0)\right]$$

and since

$$\int_0^\infty \frac{df(t)}{dt} \lim_{S \to \infty} e^{-st} \, dt = 0$$

then

$$f(0) = \lim_{S \to \infty} s\mathbf{F}(s)$$

which is, of course,

$$\lim_{t \to 0} f(t) = \lim_{S \to \infty} s\mathbf{F}(s)$$

The *final-value theorem* states that

$$\lim_{t \to \infty} f(t) = \lim_{S \to 0} s\mathbf{F}(s) \qquad\qquad \textbf{13.36}$$

provided that $f(t)$ and its first derivative are transformable and that $f(\infty)$ exists. This latter requirement means that the poles of $\mathbf{F}(s)$ must have negative real parts with the exception that there can be a simple pole at $s = 0$.

The proof of this theorem also involves the Laplace transform of the function $df(t)/dt$.

$$\int_0^\infty \frac{df(t)}{dt} e^{-st} \, dt = s\mathbf{F}(s) - f(0)$$

Taking the limit of both sides as $s \to 0$ gives us

$$\lim_{S \to 0} \int_0^\infty \frac{df(t)}{dt} e^{-st} \, dt = \lim_{S \to 0}\left[s\mathbf{F}(s) - f(0)\right]$$

Therefore,

$$\int_0^\infty \frac{df(t)}{dt} \, dt = \lim_{S \to 0}\left[s\mathbf{F}(s) - f(0)\right]$$

and

$$f(\infty) - f(0) = \lim_{S \to 0} s\mathbf{F}(s) - f(0)$$

and hence,

$$f(\infty) = \lim_{t \to \infty} f(t) = \lim_{S \to 0} s\mathbf{F}(s)$$

EXAMPLE 13.19

Let us determine the initial and final values for the function

$$\mathbf{F}(s) = \frac{10(s+1)}{s(s^2 + 2s + 2)}$$

and corresponding time function

$$f(t) = 5 + 5\sqrt{2}\, e^{-t} \cos(t - 135°)$$

SOLUTION   Applying the initial-value theorem, we have

$$f(0) = \lim_{s \to \infty} s\mathbf{F}(s)$$

$$= \lim_{s \to \infty} \frac{10(s+1)}{s^2 + 2s + 2}$$

$$= 0$$

The poles of $\mathbf{F}(s)$ are $s = 0$ and $s = -1 \pm j1$, so the final-value theorem is applicable. Thus,

$$f(\infty) = \lim_{s \to 0} s\mathbf{F}(s)$$

$$= \lim_{s \to 0} \frac{10(s+1)}{s^2 + 2s + 2}$$

$$= 5$$

Note that these values could be obtained directly from the time function $f(t)$.   ❏

EXTENSION EXERCISE

E13.14   Find the initial and final values of the function $f(t)$ if $\mathbf{F}(s) = \mathcal{L}[f(t)]$ is given by the expression

$$\mathbf{F}(s) = \frac{(s+1)^2}{s(s+2)(s^2 + 2s + 2)}$$

ANSWER: $f(0) = 0$ and $f(\infty) = \frac{1}{4}$.

13.8 APPLICATIONS

As a prelude to the following chapter in which we will employ the power and versatility of the Laplace transform in a wide variety of circuit analysis problems, we will now demonstrate how the techniques outlined in this chapter can be used in the solution of a circuit problem via the differential equation that describes the network.

EXAMPLE 13.20

Consider the network shown in Fig. 13.7a. Assume the network is in steady state prior to $t = 0$. Let us find the current $i(t)$ for $t > 0$.

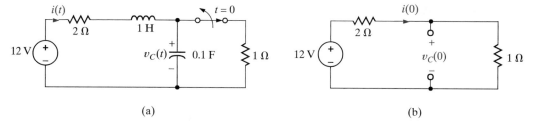

(a)   (b)

Figure 13.7 Circuits used in Example 13.20.

SOLUTION   In steady state prior to $t = 0$, the network is as shown in Fig. 13.7b, since the inductor acts like a short circuit to dc and the capacitor acts like an open circuit to dc. From Fig. 13.7b we note that $i(0) = 4$ A and $v_C(0) = 4$ V. For $t > 0$, the KVL equation for the network is

$$12\,u(t) = 2i(t) + 1\frac{di(t)}{dt} + \frac{1}{0.1}\int_0^t i(x)\,dx + v_C(0)$$

Using the results of Example 13.1 and Theorems 9 and 10, the transformed expression becomes

$$\frac{12}{s} = 2\mathbf{I}(s) + s\mathbf{I}(s) - i(0) + \frac{10}{s}\mathbf{I}(s) + \frac{v_C(0)}{s}$$

Using the initial conditions, the equation becomes

$$\frac{12}{s} = \mathbf{I}(s)\left(2 + s + \frac{10}{s}\right) - 4 + \frac{4}{s}$$

or

$$\mathbf{I}(s) = \frac{4(s + 2)}{s^2 + 2s + 10} = \frac{4(s + 2)}{(s + 1 - j3)(s + 1 + j3)}$$

and then

$$K_1 = \frac{4(s + 2)}{s + 1 + j3}\bigg|_{s=-1+j3}$$
$$= 2.11\ \underline{/-18.4^\circ}$$

Therefore,

$$i(t) = 2(2.11)e^{-t}\cos(3t - 18.4^\circ)\ \text{A}$$

Note that this expression satisfies the initial condition $i(0) = 4$ A.   ❑

E13.15   Assuming the network in Fig. E13.15 is in steady state prior to $t = 0$, find $i(t)$ for $t > 0$.

ANSWER: $i(t) = 3 - e^{-2t}$ A.

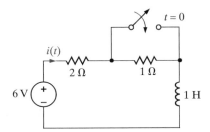

Figure E13.15

13.9 SUMMARY

- In applying the Laplace transform, we convert an integrodifferential equation in the time domain to an algebraic equation, which includes initial conditions, in the s-domain. We solve for the unknowns in the s-domain and convert the results back to the time domain.

- The Laplace transform is defined by the expression

$$\mathcal{L}[f(t)] = \mathbf{F}(s) = \int_0^\infty f(t)e^{-st}\, dt$$

- Laplace tranform pairs, as listed in Table 13.1, can be used to convert back and forth between the time and frequency domains.

- The Laplace transform theorems, as listed in Table 13.2, are useful in performing the Laplace tranform and its inverse.

- The partial fraction expansion of a function in the s-domain permits the use of the transform pairs in Table 13.1 and the theorems in Table 13.2 to convert the function to the time domain.

- The convolution of two functions in the time domain corresponds to a simple multiplication of the two functions in the s-domain.

- The initial and final values of a time-domain function can be obtained from its Laplace transform in the frequency domain.

PROBLEMS

Section 13.2

Use the sampling property.

13.1   Find the Laplace transform of the function $f(t) = te^{-a(t-1)}\, \delta(t - 1)$.

Similar to P15.1.

13.2   Find the Laplace transform of the function $f(t) = te^{-at} \sin(\omega t)\, \delta(t - 4)$.

Section 13.3

13.3 If $f(t) = te^{-at}\cos\omega t$, show that $\mathbf{F}(s) = s + a/(s + a)^2 + \omega^2$.

Similar to Example 13.4

Section 13.4

13.4 Use Theorem 5 to find $\mathcal{L}[f(t)]$ if $f(t) = e^{-at}u(t - 1)$.

13.5 Use Theorem 6 to find $\mathcal{L}[f(t)]$ if $f(t) = te^{-at}u(t - 1)$.

13.6 If $f(t) = te^{-at}u(t - 2)$, find $\mathbf{F}(s)$.

Similar to Example 13.11

13.7 Use the shifting theorem to determine $\mathcal{L}[f(t)]$, where
$f(t) = [e^{-(t-2)} - e^{-2(t-2)}]u(t - 2)$.

Use Theorem 4.

13.8 Use the shifting theorem to determine $\mathcal{L}[f(t)]$, where
$f(t) = [t - 1 + e^{-(t-1)}]u(t - 1)$.

Use Theorem 4.

13.9 If $f(t) = e^{-at}\cos\omega\, tu(t - 1)$, find $\mathbf{F}(s)$.

Similar to Example 13.12

13.10 If $f(t) = t\sin(\omega t)u(t - 1)$, find $\mathbf{F}(s)$.

Similar to Example 13.13

13.11 If $f(t) = te^{-(t-a)}u(t - a) - e^{-t}u(t - a)$, find $\mathbf{F}(s)$.

Use Theorems 4 and 5.

13.12 If $f(t) = te^{-t}\cos(\omega t)(a^2 + 1)$, find $\mathbf{F}(s)$.

Use Theorems 6 and 7.

Section 13.5

13.13 Given the following functions $\mathbf{F}(s)$, find $f(t)$.

(a) $\mathbf{F}(s) = \dfrac{4}{(s + 1)(s + 2)}$

See Example 13.15.

(b) $\mathbf{F}(s) = \dfrac{9s}{(s + 1)(s + 4)}$

13.14 Given the following functions $\mathbf{F}(s)$, find $f(t)$.

(a) $\mathbf{F}(s) = \dfrac{s + 10}{(s + 4)(s + 6)}$

Similar to P13.13

(b) $\mathbf{F}(s) = \dfrac{24}{(s + 2)(s + 6)}$

13.15 Given the following functions $F(s)$, find $f(t)$.

Similar to P13.13

(a) $F(s) = \dfrac{s + 6}{s(s + 2)(s + 3)}$

(b) $F(s) = \dfrac{s^2 + s + 1)}{s(s + 1)(s + 2)}$

13.16 Given the following functions $F(s)$, find $f(t)$.

Similar to P13.13

(a) $F(s) = \dfrac{s^2 + 5s + 4}{(s + 2)(s + 4)(s + 6)}$

(b) $F(s) = \dfrac{(s + 3)(s + 6)}{s(s^2 + 8s + 12)}$

13.17 Given the following functions $F(s)$, find $f(t)$.

See Example 13.16.

(a) $F(s) = \dfrac{4}{s^2 + 2s + 2}$

(b) $F(s) = \dfrac{2s + 4}{s^2 + 4s + 5}$

13.18 Given the following functions $F(s)$, find the inverse Laplace transform of each function.

Similar to P13.17

(a) $F(s) = \dfrac{10(s + 1)}{s^2 + 2s + 2}$

(b) $F(s) = \dfrac{s + 1}{s(s^2 + 4s + 5)}$

13.19 Given the following functions $F(s)$, find $f(t)$.

Similar to P13.17

(a) $F(s) = \dfrac{s(s + 6)}{(s + 3)(s^2 + 6s + 18)}$

(b) $F(s) = \dfrac{(s + 4)(s + 8)}{s(s^2 + 4s + 8)}$

13.20 Given the following functions $F(s)$, find $f(t)$.

Similar to P13.17

(a) $F(s) = \dfrac{6s + 12}{(s^2 + 4s + 5)(s^2 + 4s + 8)}$

(b) $F(s) = \dfrac{s(s + 2)}{s^2 + 2s + 2}$

13.21 Given the following functions $\mathbf{F}(s)$, find $f(t)$.

(a) $\mathbf{F}(s) = \dfrac{(s + 1)(s + 3)}{(s + 2)(s^2 + 2s + 2)}$ Similar to P13.17

(b) $\mathbf{F}(s) = \dfrac{(s + 2)^2}{s^2 + 4s + 5}$

13.22 Given the following functions $\mathbf{F}(s)$, find $f(t)$.

(a) $\mathbf{F}(s) = \dfrac{s^2 + 4s + 8}{(s + 1)(s + 4)}$ Similar to P13.13.

(b) $\mathbf{F}(s) = \dfrac{s + 4}{s^2}$ See Example 13.17.

13.23 Given the following functions $\mathbf{F}(s)$, find the inverse Laplace transform of each function.

(a) $\mathbf{F}(s) = \dfrac{s + 6}{s^2(s + 2)}$ Similar to P13.22(b)

(b) $\mathbf{F}(s) = \dfrac{s + 3}{(s + 1)^2(s + 3)}$

13.24 Given the following functions $\mathbf{F}(s)$, find $f(t)$.

(a) $\mathbf{F}(s) = \dfrac{s + 8}{s^2(s + 4)}$ Similar to P13.22(b)

(b) $\mathbf{F}(s) = \dfrac{1}{s^2(s + 1)^2}$

13.25 Given the following functions $\mathbf{F}(s)$, find $f(t)$.

(a) $\mathbf{F}(s) = \dfrac{s + 3}{(s + 2)^2}$ Similar to P13.22(b)

(b) $\mathbf{F}(s) = \dfrac{s + 6}{s(s + 2)^2}$

13.26 Given the following functions $\mathbf{F}(s)$, find $f(t)$.

(a) $\mathbf{F}(s) = \dfrac{s^2}{(s + 1)^2(s + 2)}$ Similar to P13.22(b)

(b) $\mathbf{F}(s) = \dfrac{s^2 + 10s + 24}{s(s + 4)^3(s + 5)}$

Similar to P13.22(b)

13.27 Find $f(t)$ if $F(s)$ is given by the expression

$$F(s) = \frac{s(s + 1)}{(s + 2)^3(s + 3)}$$

See P13.17 and P13.22(b).

13.28 Find $f(t)$ if $F(s)$ is given by the expression

$$F(s) = \frac{12(s + 2)}{s^2(s + 1)(s^2 + 2s + 2)}$$

13.29 Find the inverse Laplace transform of following functions.

(a) $F(s) = \dfrac{e^{-s}}{s + 1}$

Use Theorem 5.

(b) $F(s) = \dfrac{1 - e^{-2s}}{s}$

(c) $F(s) = \dfrac{1 - e^{-s}}{s + 2}$

13.30 Find the inverse Laplace transform of following functions.

(a) $F(s) = \dfrac{(s + 2)e^{-s}}{s(s + 2)}$

See P13.13 and P13.29.

(b) $F(s) = \dfrac{e^{-10s}}{(s + 2)(s + 3)}$

13.31 Find the inverse Laplace transform $f(t)$ if $F(s)$ is

Similar to P13.30

$$F(s) = \frac{se^{-s}}{(s + 1)(s + 2)}$$

13.32 Find $f(t)$ if $F(s)$ is given by the following functions:

(a) $F(s) = \dfrac{(s^2 + 2s + 1)e^{-2s}}{s(s + 1)(s + 2)}$

Similar to P13.30

(b) $F(s) = \dfrac{(s + 1)e^{-4s}}{s^2(s + 2)}$

13.33 Find $f(t)$ if $F(s)$ is given by the following functions:

(a) $F(s) = \dfrac{2(s + 3)e^{-s}}{(s + 2)(s + 4)}$

Similar to P13.30

(b) $F(s) = \dfrac{10(s + 2)e^{-2s}}{(s + 3)(s + 4)}$

13.34 Solve the following differential equations using Laplace transforms.

(a) $\dfrac{dx(t)}{dt} + 3x(t) = e^{-2t}$, $x(0) = 1$ See Examples 13.14 and 13.15.

(b) $\dfrac{dx(t)}{dt} + 3x(t) = 2u(t)$, $x(0) = 2$

13.35 Determine $y(t)$ in the following equation if all initial conditions are zero.

$$\frac{d^3 y(t)}{dt^3} + \frac{4d^2 y(t)}{dt^2} + \frac{3dy(t)}{dt} = 10e^{-2t}$$

Similar to P13.34

13.36 Use Laplace transforms to find $y(t)$ if

$$\frac{dy(t)}{dt} + 3y(t) + 2\int_0^t y(x)\, dx = u(t), \quad y(0) = 0, \quad t > 0$$

Similar to P13.34

13.37 Solve the following integrodifferential equation using Laplace transforms.

$$\frac{dy(t)}{dt} + 4y(t) + \int_0^t 4y(\lambda)d\lambda = 1 - e^{-3t}, \quad y(0) = 0, \quad t > 0$$

See Examples 13.14 and 13.15.

13.38 Solve the following differential equations using Laplace transforms.

(a) $\dfrac{d^2 y(t)}{dt^2} + \dfrac{2dy(t)}{dt} + y(t) = e^{-2t}, \quad y(0) = y'(0) = 0$

Similar to P13.37

(b) $\dfrac{d^2 y(t)}{dt^2} + \dfrac{4dy(t)}{dt} + 4y(t) = u(t), \quad y(0) = 0, \quad y'(0) = 1$

Section 13.6

13.39 Find $f(t)$ using convolution if $F(s)$ is

$$F(s) = \frac{1}{(s + 1)(s + 2)}$$

Separate $F(s)$ into two parts. Convert each to the time domain. Perform the convolution.

13.40 Use convolution to find $f(t)$ if

$$F(s) = \frac{1}{(s + 1)(s + 2)^2}$$

Similar to P13.39

Section 13.7

13.41 Determine the initial and final values of $f(t)$ if $\mathbf{F}(s)$ is given by the expressions

Use the initial-value and final-value theorems.

(a) $\mathbf{F}(s) = \dfrac{2(s + 2)}{s(s + 3)}$

(b) $\mathbf{F}(s) = \dfrac{(s^2 + 2s + 4)}{(s + 1)(s + 2)(s + 3)}$

(c) $\mathbf{F}(s) = \dfrac{2s^2}{(s + 3)(s^2 + 2s + 2)}$

13.42 Find the initial and final values of the time function $f(t)$ if $\mathbf{F}(s)$ is given as

(a) $\mathbf{F}(s) = \dfrac{10(s + 2)}{(s + 1)(s + 4)}$

Similar to P13.41

(b) $\mathbf{F}(s) = \dfrac{s^2 + 2s + 2}{(s + 6)(s^3 + 4s^2 + 8s + 4)}$

(c) $\mathbf{F}(s) = \dfrac{2s}{s^2 + 2s + 3}$

13.43 Find the final values of the time function $f(t)$ given that

(a) $\mathbf{F}(s) = \dfrac{10(s + 1)}{(s + 2)(s + 3)}$

Similar to P13.41

(b) $\mathbf{F}(s) = \dfrac{10}{s^2 + 4s + 4}$

Section 13.8

13.44 In the network in Fig. P13.44, the switch opens at $t = 0$. Use Laplace transforms to find $i(t)$ for $t > 0$.

See Example 13.20.

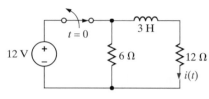

Figure P13.44

13.45 In the circuit in Fig. P13.45, the switch moves from position 1 to position 2 at $t = 0$. Use Laplace transforms to find $v(t)$ for $t > 0$.

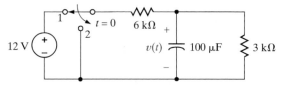

Similar to P13.44

Figure P13.45

Advanced Problems

13.46 If $\mathbf{F}(s) = (s + 2)^2/(s + 1)(s + 3)$, find $f(t)$.

13.47 If $\mathbf{F}(s) = (s + 1)^2(s + 3)^2/(s + 2)(s + 4)$, find $f(t)$.

13.48 If $f(t) = d/dt(e^{-5t} \cos 2t)$, find $\mathbf{F}(s)$.

13.49 If $f(t) = d/dt(te^{-5t} \sin 5t)$, find $\mathbf{F}(s)$.

13.50 The switch in the circuit in Fig. P13.50 opens at $t = 0$. Find $i(t)$ for $t > 0$ using Laplace transforms.

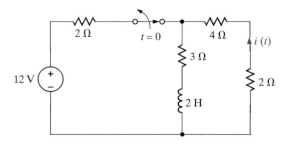

Figure P13.50

13.51 In the network in Fig. P13.51, the switch opens at $t = 0$. Use Laplace transforms to find $v_0(t)$ for $t > 0$.

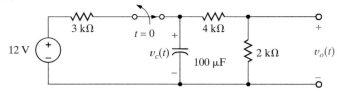

Figure P13.51

13.52 In the network in Fig. P13.52, the switch closes at $t = 0$. Use Laplace transforms to find $v_C(t)$ for $t > 0$.

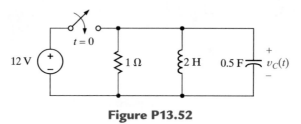

Figure P13.52

13.53 In the network in Fig. P13.53, the switch opens at $t = 0$. Use Laplace transforms to find $i_L(t)$ for $t > 0$.

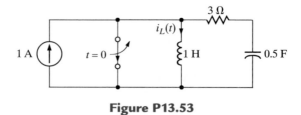

Figure P13.53

Application of the Laplace Transform to Circuit Analysis

14

We have demonstrated in Chapter 13 how the Laplace transform can be used to solve linear constant-coefficient differential equations. Since all of the linear networks with which we are dealing can be described by linear constant-coefficient differential equations, use of the Laplace transform for circuit analysis would appear to be a feasible approach. The terminal characteristics for each circuit element can be described in the *s*-domain by transforming the appropriate time domain equations. Kirchhoff's laws when applied to a circuit produce a set of integrodifferential equations in terms of the terminal characteristics of the network elements, which when transformed yield a set of algebraic equations in the *s*-domain. Therefore, a complex frequency domain analysis, in which the passive network elements are represented by their transform impedance or admittance and sources (independent or dependent) are represented in terms of their transform variables, can be performed. Such an analysis in the *s*-domain is algebraic, and all of the techniques derived in dc analysis will apply. Therefore, the analysis is similar to the dc analysis of resistive networks, and all of the network analysis techniques and network theorems that were applied in dc analysis are valid in the *s*-domain (e.g., node analysis, loop analysis, superposition, source transformation, Thévenin's theorem, Norton's theorem, and combinations of impedance or admittance). Therefore, our approach will be to transform each of the circuit elements, draw an *s*-domain equivalent circuit, and then, using the transformed network, solve the circuit equations algebraically in the *s*-domain. Finally, a table of transform pairs can be employed to obtain a complete (transient plus steady-state) solution. Our approach in this chapter is to employ the circuit element models and demonstrate, using a variety of examples, many of the concepts and techniques that have been presented earlier.

14.1 LAPLACE CIRCUIT SOLUTIONS

To introduce the utility of the Laplace transform in circuit analysis, let us consider the RL series circuit shown in Fig. 14.1. In particular, let us find the current, $i(t)$.

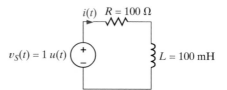

Figure 14.1 *RL* series network.

Using Kirchhoff's voltage law, we can write the time domain differential equation,

$$v_S(t) = L\left(\frac{di(t)}{dt}\right) + Ri(t)$$

The complementary differential equation is

$$L\left(\frac{di(t)}{dt}\right) + Ri(t) = 0 \qquad\qquad \textbf{14.1}$$

and has the solution

$$i_C(t) = K_C e^{-\alpha t}$$

Substituting $i_C(t)$ into the complementary equation yields the relationship

$$R - \alpha L = 0$$

or

$$\alpha = \frac{R}{L} = 1000$$

The particular solution is of the same form as the forcing function, $v_S(t)$.

$$i_p(t) = K_p$$

Substituting $i_p(t)$ into the original differential equation yields the expression

$$1 = RK_p$$

or

$$K_p = 1/R = 1/100$$

The final solution is the sum of $i_p(t)$ and $i_C(t)$,

$$i(t) = K_p + K_C e^{-\alpha t} = \frac{1}{100} + K_C e^{-1000t}$$

To find K_C, we must use the value of the current at some particular instant of time. For $t < 0$, the unit step function is zero and so is the current. At $t = 0$, the unit step goes to one; however, the inductor forces the current to instantaneously remain at zero. Therefore, at $t = 0$, we can write

$$i(0) = 0 = K_p + K_C$$

or

$$K_C = -K_p = -\frac{1}{100}$$

Thus, the current is

$$i(t) = 10(1 - e^{-1000t})u(t) \text{ mA}$$

Let us now try a different approach to the same problem. Making use of Table 13.2, let us take the Laplace transform of both sides of Eq. (14.1).

$$\mathcal{L}[v_S(t)] = \mathbf{V}_S(s) = L[s\mathbf{I}(s)] + R\mathbf{I}(s)$$

Now the circuit is represented not by a time domain differential equation, but rather by an algebraic expression in the s-domain. Solving for $\mathbf{I}(s)$, we can write

$$\mathbf{I}(s) = \frac{\mathbf{V}_S(s)}{sL + R} = \frac{1}{s[sL + R]}$$

We find $i(t)$ using the inverse Laplace transform. First, let us express $\mathbf{I}(s)$ as a sum of partial products

$$\mathbf{I}(s) = \frac{1/L}{s\left[s + \dfrac{R}{L}\right]} = \frac{1}{sR} - \frac{1}{R\left[s + \dfrac{R}{L}\right]}$$

The inverse transform is simply

$$i(t) = \frac{1}{R}\left(1 - e^{-Rt/L}\right)$$

Given the circuit element values in Fig. 14.1, the current is

$$i(t) = 10(1 - e^{-1000t})u(t) \text{ mA}$$

which is exactly the same as that obtained using the differential equation approach. Note carefully that the solution using the Laplace transform approach yields the entire solution in one step.

We have shown that the Laplace transform can be used to transform a differential equation into an algebraic equation. Since the voltage–current relationships for resistors, capacitors, and inductors involve only constants, derivatives, and integrals, we can represent and solve any circuit in the s-domain.

14.2 CIRCUIT ELEMENT MODELS

The Laplace transform technique employed earlier implies that the terminal characteristics of circuit elements can be expressed as algebraic expressions in the s-domain. Let us examine these characteristics for the resistor, capacitor, and inductor.

DRILL

D14.1 Find $v(t)$ for $t > 0$ in the following network:

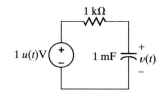

ANSWER:
$v(t) = 1 - e^{-t}$ V.

The voltage–current relationship for a resistor in the time domain using the passive sign convention is

$$v(t) = Ri(t) \qquad \textbf{14.2}$$

Using the Laplace transform, this relationship in the s-domain is

$$\mathbf{V}(s) = R\mathbf{I}(s) \qquad \textbf{14.3}$$

Therefore, the time domain and complex frequency domain representations of this element are as shown in Fig. 14.2a.

The time domain relationships for a capacitor using the passive sign convention are

$$v(t) = \frac{1}{C}\int_0^t i(x)\,dx + v(0) \qquad \textbf{14.4}$$

$$i(t) = C\frac{dv(t)}{dt} \qquad \textbf{14.5}$$

The s-domain equations for the capacitor are then

$$\mathbf{V}(s) = \frac{\mathbf{I}(s)}{sC} + \frac{v(0)}{s} \qquad \textbf{14.6}$$

$$\mathbf{I}(s) = sC\mathbf{V}(s) - Cv(0) \qquad \textbf{14.7}$$

and hence the s-domain representation of this element is as shown in Fig. 14.2b.

For the inductor, the voltage–current relationships using the passive sign convention are

$$v(t) = L\frac{di(t)}{dt} \qquad \textbf{14.8}$$

$$i(t) = \frac{1}{L}\int_0^t v(x)\,dx + i(0) \qquad \textbf{14.9}$$

The relationships in the s-domain are then

$$\mathbf{V}(s) = sL\mathbf{I}(s) - Li(0) \qquad \textbf{14.10}$$

$$\mathbf{I}(s) = \frac{\mathbf{V}(s)}{sL} + \frac{i(0)}{s} \qquad \textbf{14.11}$$

The s-domain representation of this element is shown in Fig. 14.2c.

Using the passive sign convention, we find that the voltage–current relationships for the coupled inductors shown in Fig. 14.2d are

$$v_1(t) = L_1\frac{di_1(t)}{dt} + M\frac{di_2(t)}{dt} \qquad \textbf{14.12}$$

$$v_2(t) = L_2\frac{di_2(t)}{dt} + M\frac{di_1(t)}{dt}$$

The relationships in the s-domain are then

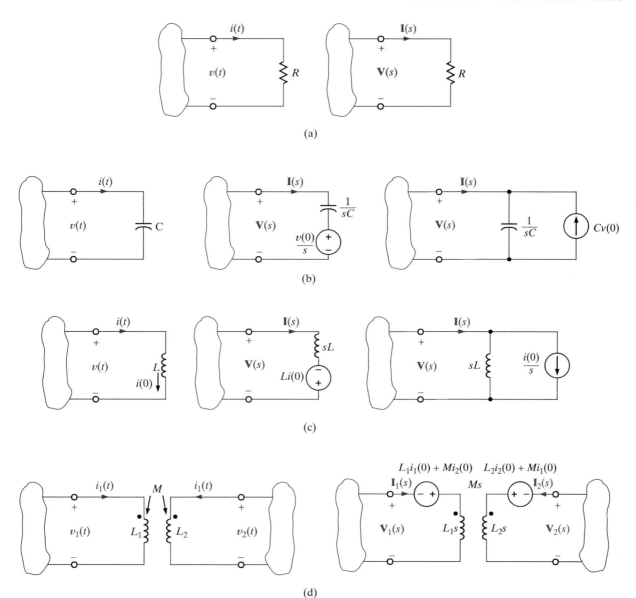

Figure 14.2 Time domain and s-domain representations of circuit elements.

$$\mathbf{V}_1(s) = L_1 s\mathbf{I}_1(s) - L_1 i_1(0) + Ms\mathbf{I}_2(s) - Mi_2(0)$$
$$\mathbf{V}_2(s) = L_2 s\mathbf{I}_2(s) - L_2 i_2(0) + Ms\mathbf{I}_1(s) - Mi_1(0)$$

14.13

Independent and dependent voltage and current sources can also be represented by their transforms; that is,

$$\mathbf{V}_1(s) = \mathcal{L}[v_1(t)]$$
$$\mathbf{I}_2(s) = \mathcal{L}[i_2(t)]$$

14.14

and if $v_1(t) = Ai_2(t)$, which represents a current-controlled voltage source, then

$$\mathbf{V}_1(s) = A\mathbf{I}_2(s)$$

14.15

The reader should note carefully the direction of the current sources and the polarity of the voltage sources in the transformed network which result from the initial conditions. If the polarity of the initial voltage or direction of the initial current is reversed, the sources in the transformed circuit that results from the initial condition are also reversed.

DRILL

D14.2   The s-domain model for the network in (a) for $t > 0$ is shown in (b). Find the s-domain expression for the voltage across the inductor.

(a)

(b)

ANSWER: $\mathbf{V}(s) = \dfrac{-1}{s + 1}$.

14.3 ANALYSIS TECHNIQUES

Now that we have the s-domain representation for the circuit elements, we are in a position to analyze networks using a transformed circuit.

EXAMPLE 14.1

Given the network in Fig. 14.3a, let us draw the s-domain equivalent circuit and find the output voltage in both the s and time domains.

SOLUTION   The s-domain network is shown in Fig. 14.3b. We can write the output voltage as

$$\mathbf{V}_o(s) = \left[R // \frac{1}{sC} \right] \mathbf{I}_s(s)$$

or

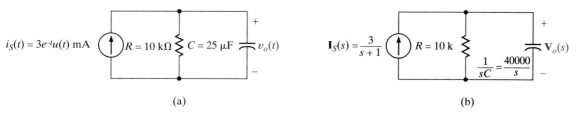

Figure 14.3 Time domain and s-domain representations of an RC parallel network.

$$\mathbf{V}_o(s) = \left[\frac{1/C}{s + (1/RC)}\right]\mathbf{I}_s(s)$$

Given the element values, $\mathbf{V}_o(s)$ becomes

$$\mathbf{V}_o(s) = \left(\frac{40{,}000}{s + 4}\right)\left(\frac{0.003}{s + 1}\right) = \frac{120}{(s + 4)(s + 1)}$$

Expanding $\mathbf{V}_o(s)$ into partial fractions yields

$$\mathbf{V}_o(s) = \frac{120}{(s + 4)(s + 1)} = \frac{40}{s + 1} - \frac{40}{s + 4}$$

Performing the inverse Laplace transform yields the time domain representation

$$v_o(t) = 40\left[e^{-t} - e^{-4t}\right]u(t) \text{ V} \qquad \square$$

Now that we have demonstrated the use of the Laplace transform in the solution of a simple circuit, let us consider the more general case. Note that in Fig. 14.2 we have shown two models for the capacitor and inductor when initial conditions are present. Let us now consider an example in which we will illustrate the use of these models in deriving both the node and loop equations for the circuit.

EXAMPLE 14.2

Given the circuits in Figs. 14.4a and b, we wish to write the mesh equations in the s-domain for the network in Fig. 14.4a and the node equations in the s-domain for the network in Fig. 14.4b.

SOLUTION   The transformed circuit for the network in Fig. 14.4a is shown in Fig. 14.4c. The mesh equations for this network are

$$\left(R_1 + \frac{1}{sC_1} + \frac{1}{sC_2} + sL_1\right)\mathbf{I}_1(s) - \left(\frac{1}{sC_2} + sL_1\right)\mathbf{I}_2(s)$$

$$= \mathbf{V}_A(s) - \frac{v_1(0)}{s} + \frac{v_2(0)}{s} - L_1 i_1(0)$$

$$-\left(\frac{1}{sC_2} + sL_1\right)\mathbf{I}_1(s) + \left(\frac{1}{sC_2} + sL_1 + sL_2 + R_2\right)\mathbf{I}_2(s)$$

$$= L_1 i_1(0) - \frac{v_2(0)}{s} - L_2 i_2(0) + \mathbf{V}_B(s)$$

Note that the equations employ the same convention as that used in dc analysis.

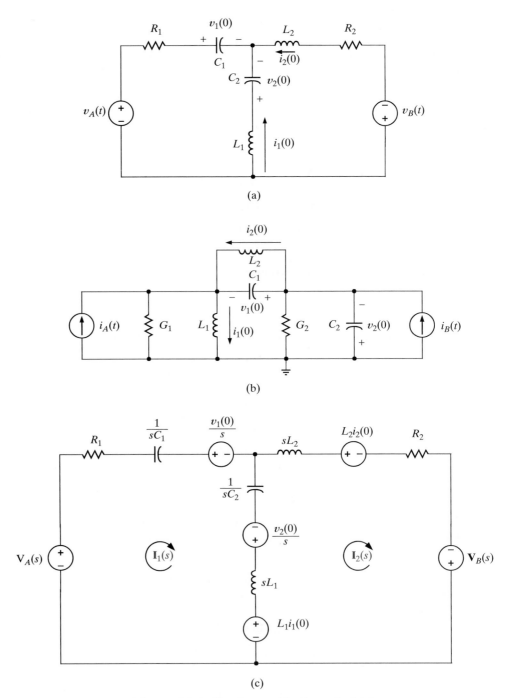

Figure 14.4 Circuits used in Example 14.2.

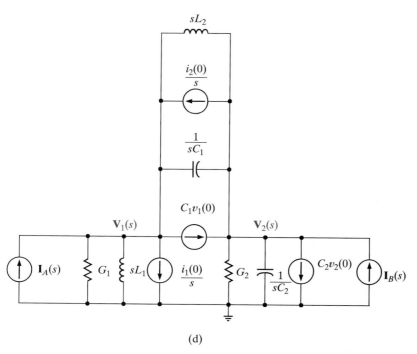

(d)

Figure 14.4 (*continued*)

The transformed circuit for the network in Fig. 14.4b is shown in Fig. 14.4d. The node equations for this network are

$$\left(G_1 + \frac{1}{sL_1} + sC_1 + \frac{1}{sL_2}\right)\mathbf{V}_1(s) - \left(\frac{1}{sL_2} + sC_1\right)\mathbf{V}_2(s)$$

$$= \mathbf{I}_A(s) - \frac{i_1(0)}{s} + \frac{i_2(0)}{s} - C_1 v_1(0)$$

$$-\left(\frac{1}{sL_2} + sC_1\right)\mathbf{V}_1(s) + \left(\frac{1}{sL_2} + sC_1 + G_2 + sC_2\right)\mathbf{V}_2(s)$$

$$= C_1 v_1(0) - \frac{i_2(0)}{s} - C_2 v_2(0) + \mathbf{I}_B(s) \qquad \square$$

Example 14.2 attempts to illustrate the manner in which to employ the two *s*-domain representations of the inductor and capacitor circuit elements when initial conditions are present. In the following examples we illustrate the use of a number of analysis techniques in obtaining the complete response of a transformed network. The circuits analyzed have been specifically chosen to demonstrate the application of the Laplace transform to circuits with a variety of passive and active elements.

EXAMPLE 14.3

Consider the network shown in Fig. 14.5a. Let us determine the node voltages $v_1(t)$ and $v_2(t)$ using nodal analysis and superposition.

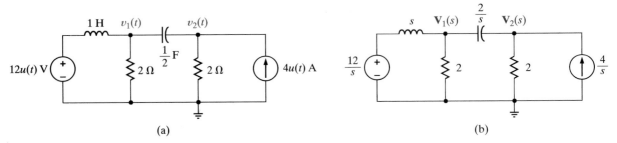

(a)   (b)

Figure 14.5 Circuits used in Example 14.3.

SOLUTION   The transformed network is shown in Fig. 14.5b. The node equations for the network are

Summing the currents leaving a node. Outbound currents have a positive sign.

$$\mathbf{V}_1(s)\left(\frac{1}{s} + \frac{1}{2} + \frac{s}{2}\right) - \mathbf{V}_2(s)\left(\frac{s}{2}\right) = \frac{12}{s^2}$$

$$-\mathbf{V}_1(s)\left(\frac{s}{2}\right) + \mathbf{V}_2(s)\left(\frac{s}{2} + \frac{1}{2}\right) = \frac{4}{s}$$

The equations in matrix form are

$$\begin{bmatrix} \dfrac{s^2 + s + 2}{2s} & \dfrac{-s}{2} \\[3mm] \dfrac{-s}{2} & \dfrac{s + 1}{2} \end{bmatrix} \begin{bmatrix} \mathbf{V}_1(s) \\ \mathbf{V}_2(s) \end{bmatrix} = \begin{bmatrix} \dfrac{12}{s^2} \\[2mm] \dfrac{4}{s} \end{bmatrix}$$

Solving for the node voltages, we obtain

$$\begin{bmatrix} \mathbf{V}_1(s) \\ \mathbf{V}_2(s) \end{bmatrix} = \begin{bmatrix} \dfrac{s^2 + s + 2}{2s} & \dfrac{-s}{2} \\[3mm] \dfrac{-s}{2} & \dfrac{s + 1}{2} \end{bmatrix}^{-1} \begin{bmatrix} \dfrac{12}{s^2} \\[2mm] \dfrac{4}{s} \end{bmatrix}$$

$$= \frac{4s}{2s^2 + 3s + 2} \begin{bmatrix} \dfrac{s + 1}{2} & \dfrac{s}{2} \\[3mm] \dfrac{s}{2} & \dfrac{s^2 + s + 2}{2s} \end{bmatrix} \begin{bmatrix} \dfrac{12}{s^2} \\[2mm] \dfrac{4}{s} \end{bmatrix}$$

$$= \begin{bmatrix} \dfrac{4(s^2 + 3s + 3)}{s(s^2 + 3s/2 + 1)} \\[4mm] \dfrac{4(s^2 + 4s + 2)}{s(s^2 + 3s/2 + 1)} \end{bmatrix}$$

Let us now determine $v_1(t) = \mathcal{L}^{-1}[\mathbf{V}_1(s)]$ and $v_2(t) = \mathcal{L}^{-1}[\mathbf{V}_2(s)]$. $\mathbf{V}_1(s)$ can be written as

$$\mathbf{V}_1(s) = \frac{4(s^2 + 3s + 3)}{s(s^2 + \frac{3}{2}s + 1)} = \frac{4(s^2 + 3s + 3)}{s[s + \frac{3}{4} + j(\sqrt{7}/4)][s + \frac{3}{4} - j(\sqrt{7}/4)]}$$

Therefore,

$$\frac{4(s^2 + 3s + 3)}{s[s + \frac{3}{4} - j(\sqrt{7}/4)][s + \frac{3}{4} + j(\sqrt{7}/4)]}$$

$$= \frac{K_0}{s} + \frac{K_1}{s + \frac{3}{4} - j(\sqrt{7}/4)} + \frac{K_1^*}{s + \frac{3}{4} + j(\sqrt{7}/4)}$$

Evaluating the constants K_0 and K_1, we obtain

$$\frac{4(s^2 + 3s + 3)}{s^2 + \frac{3}{2}s + 1}\bigg|_{s=0} = K_0$$

$$12 = K_0$$

and

$$\frac{4(s^2 + 3s + 3)}{s[s + \frac{3}{4} + j(\sqrt{7}/4)]}\bigg|_{s=-3/4+j(\sqrt{7}/4)} = K_1$$

$$4\,\underline{/180°} = K_1$$

Therefore,

$$v_1(t) = \left[12 + 8e^{-(3/4)t}\cos\left(\frac{\sqrt{7}}{4}t + 180°\right)\right]u(t)\text{ V}$$

In a similar manner,

$$\mathbf{V}_2(s) = \frac{4(s^2 + 4s + 2)}{s(s^2 + \frac{3}{2}s + 1)} = \frac{K_0}{s} + \frac{K_1}{s + \frac{3}{4} - j(\sqrt{7}/4)} + \frac{K_1^*}{s + \frac{3}{4} + j(\sqrt{7}/4)}$$

Evaluating the constants K_0 and K_1, we obtain

$$\frac{4(s^2 + 4s + 2)}{s^2 + \frac{3}{2}s + 1}\bigg|_{s=0} = K_0$$

$$8 = K_0$$

and

$$\frac{4(s^2 + 4s + 2)}{s[s + \frac{3}{4} + j(\sqrt{7}/4)]}\bigg|_{s=-3/4+j(\sqrt{7}/4)} = K_1$$

$$5.66\,\underline{/-110.7°} = K_1$$

Therefore,

$$v_2(t) = \left[8 + 11.32e^{-(3/4)t}\cos\left(\frac{\sqrt{7}}{4}t - 110.7°\right)\right]u(t)\text{ V}$$

It is interesting to note that as $t \to \infty$, the sources in the circuit appear to be dc sources. In the dc case the inductor acts like a short circuit and the capacitor acts like an open circuit. Under these conditions the voltage $v_1(\infty) = 12$ V, and the voltage $v_2(\infty) = 8$ V. It is reassuring to realize that the equations for $v_1(t)$ and $v_2(t)$ yield these results also. ❏

EXAMPLE 14.4

Consider the network shown in Fig. 14.6a. We wish to determine the output voltage $v_o(t)$.

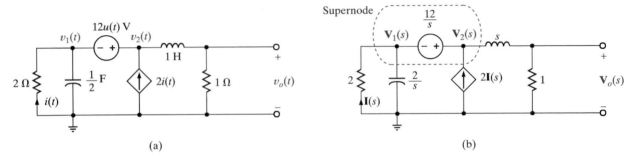

Figure 14.6 Circuits used in Example 14.4.

SOLUTION As we begin to attack the problem we note two things. First, because the source $12u(t)$ is connected between $v_1(t)$ and $v_2(t)$, we have a supernode. Second, if $v_2(t)$ is known, $v_o(t)$ can be easily obtained by voltage division. Hence, we will use nodal analysis in conjunction with voltage division to obtain a solution.

The transformed network is shown in Fig. 14.6b. KCL for the supernode is

Summing the currents leaving the supernode

$$\frac{\mathbf{V}_1(s)}{2} + \mathbf{V}_1(s)\frac{s}{2} - 2\mathbf{I}(s) + \frac{\mathbf{V}_2(s)}{s+1} = 0$$

However,

$$\mathbf{I}(s) = -\frac{\mathbf{V}_1(s)}{2}$$

and

$$\mathbf{V}_1(s) = \mathbf{V}_2(s) - \frac{12}{s}$$

Substituting the last two equations into the first equation yields

$$\left[\mathbf{V}_2(s) - \frac{12}{s}\right]\frac{s+3}{2} + \frac{\mathbf{V}_2(s)}{s+1} = 0$$

or

$$\mathbf{V}_2(s) = \frac{12(s + 1)(s + 3)}{s(s^2 + 4s + 5)}$$

Employing a voltage divider, we obtain

$$\mathbf{V}_o(s) = \mathbf{V}_2(s) \frac{1}{s + 1}$$

$$= \frac{12(s + 3)}{s(s^2 + 4s + 5)}$$

$$= \frac{12(s + 3)}{s(s + 2 - j1)(s + 2 + j1)}$$

which can be written as

$$\frac{12(s + 3)}{s(s + 2 - j1)(s + 2 + j1)} = \frac{K_0}{s} + \frac{K_1}{s + 2 - j1} + \frac{K_1^*}{s + 2 + j1}$$

Evaluating the constants, we obtain

$$\left.\frac{12(s + 3)}{s^2 + 4s + 5}\right|_{S=0} = K_0$$

$$\frac{36}{5} = K_0$$

and

$$\left.\frac{12(s + 3)}{s(s + 2 + j1)}\right|_{S=-2+j1} = K_1$$

$$3.79 \,\underline{/161.57°} = K_1$$

Therefore,

$$v_o(t) = \left[7.2 + 7.58e^{-2t} \cos(t + 161.57°)\right]u(t) \text{ V} \qquad \square$$

EXAMPLE 14.5

Let us examine the network in Fig. 14.7a. We wish to determine the output voltage $v_o(t)$.

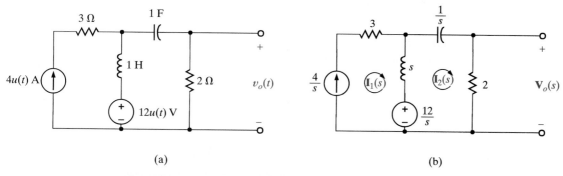

(a)   (b)

Figure 14.7 Circuits used in Example 14.5.

SOLUTION   We will solve this problem using mesh equations.
The transformed network is shown in Fig. 14.7b. KVL for the right-hand loop is

$$\frac{12}{s} - \left[\mathbf{I}_2(s) - \mathbf{I}_1(s) \right]s - \frac{\mathbf{I}_2(s)}{s} - 2\mathbf{I}_2(s) = 0$$

However, $\mathbf{I}_1(s) = 4/s$, and hence

$$\mathbf{I}_2(s) = \frac{4(s + 3)}{(s + 1)^2}$$

Therefore,

$$\mathbf{V}_o(s) = \frac{8(s + 3)}{(s + 1)^2}$$

$\mathbf{V}_o(s)$ can be written as

$$\mathbf{V}_o(s) = \frac{8(s + 3)}{(s + 1)^2} = \frac{K_{11}}{(s + 1)^2} + \frac{K_{12}}{s + 1}$$

Evaluating the constants, we obtain

$$8(s + 3)|_{s=-1} = K_{11}$$
$$16 = K_{11}$$

and

$$\frac{d}{ds}\left[8(s + 3) \right]\Big|_{s=-1} = K_{12}$$
$$8 = K_{12}$$

Therefore,

$$v_o(t) = \left(16te^{-t} + 8e^{-t} \right)u(t) \text{ V} \qquad \square$$

EXAMPLE 14.6

Let us determine the voltage $v_o(t)$ in the network in Fig. 14.8a using Thévenin's theorem.

SOLUTION   The transformed network is shown in Fig. 14.8b. The open-circuit voltage can be computed from the network in Fig. 14.8c. Note that

$$\mathbf{V}_{oc}(s) + \mathbf{V}_x'(s) = \frac{\mathbf{V}_x'(s)}{2}$$

Therefore,

$$\mathbf{V}_x'(s) = -2\mathbf{V}_{oc}(s)$$

KCL at the supernode including the two independent sources is

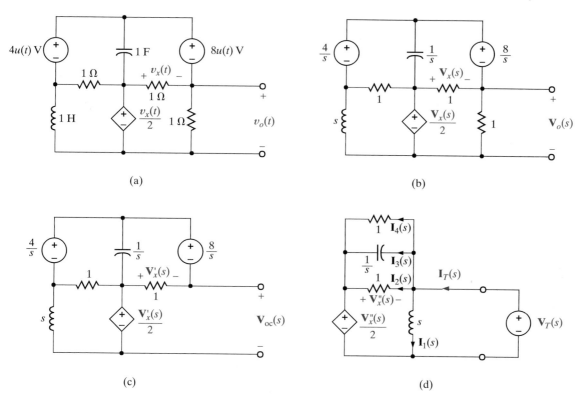

Figure 14.8 Circuits used in Example 14.6.

$$\frac{\mathbf{V}_{oc}(s) + (4/s)}{s} + \frac{\mathbf{V}_{oc}(s) + (4/s) - \frac{1}{2}(-2\mathbf{V}_{oc}(s))}{1}$$

$$+ \frac{\mathbf{V}_{oc}(s) + (8/s) - \frac{1}{2}(-2\mathbf{V}_{oc}(s))}{1/s} + \frac{\mathbf{V}_{oc}(s) - \frac{1}{2}(-2\mathbf{V}_{oc}(s))}{1} = 0$$

Summing the currents leaving the supernode

Solving this equation for $\mathbf{V}_{oc}(s)$ yields

$$\mathbf{V}_{oc}(s)\left(\frac{1}{2} + 2 + 2s + 2\right) = -\left(\frac{4}{s^2} + \frac{4}{s} + 8\right)$$

or

$$\mathbf{V}_{oc}(s) = \frac{-(8s^2 + 4s + 4)}{s(2s^2 + 4s + 1)}$$

In order to determine the Thévenin equivalent impedance we will zero the independent sources and apply a test source $\mathbf{V}_T(s)$ at the open-circuit terminals, as shown in Fig. 14.8d. KVL illustrates that

$$\mathbf{V}_T(s) + \mathbf{V}_x''(s) - \frac{\mathbf{V}_x''(s)}{2} = 0$$

so that

$$-\frac{\mathbf{V}''_x(s)}{2} = \mathbf{V}_T(s)$$

Then

D14.3 $\mathbf{Z}_{\text{Th}}(s)$ can also be found from $\mathbf{V}_{\text{oc}}(s)/\mathbf{I}_{\text{sc}}(s)$. Find $\mathbf{I}_{\text{sc}}(s)$ in this example using loop analysis.

ANSWER:

$$\mathbf{I}_{\text{sc}}(s) = -\frac{8s^2 + 4s + 4}{s^2}.$$

$$\mathbf{I}_T(s) = \mathbf{I}_1(s) + \mathbf{I}_2(s) + \mathbf{I}_3(s) + \mathbf{I}_4(s)$$

$$= \frac{\mathbf{V}_T(s)}{s} + \frac{\mathbf{V}_T(s) - (-\mathbf{V}_T(s))}{1} + \frac{\mathbf{V}_T(s) - (-\mathbf{V}_T(s))}{1/s} + \frac{\mathbf{V}_T(s) - (-\mathbf{V}_T(s))}{1}$$

$$= \mathbf{V}_T(s)\left(\frac{1}{s} + 2 + 2s + 2\right)$$

and hence,

$$\mathbf{Z}_{\text{Th}}(s) = \frac{\mathbf{V}_T(s)}{\mathbf{I}_T(s)} = \frac{s}{2s^2 + 4s + 1}$$

Then

$$\mathbf{V}_o(s) = \frac{\mathbf{V}_{\text{oc}}(s)}{\mathbf{Z}_{\text{Th}}(s) + 1} \quad (1)$$

$$= \frac{-8(8s^2 + 4s + 4)}{s(2s^2 + 5s + 1)}$$

This function can be expressed in the form

$$\frac{-2(2s^2 + 1s + 1)}{s(s^2 + \frac{5}{2}s + \frac{1}{2})} = \frac{K_0}{s} + \frac{K_1}{s + 0.22} + \frac{K_2}{s + 2.28}$$

Solving for K_0, K_1, K_2 in the manner illustrated earlier yields

$$K_0 = -4$$

$$K_1 = 2.9$$

and

$$K_2 = -2.9$$

Therefore,

$$v_o(t) = \left[-4 + 2.9e^{-0.22t} - 2.9e^{-2.28t}\right]u(t) \text{ V}$$

EXTENSION EXERCISES

E14.1   Find $i_o(t)$ in the network in Fig. E14.1 using node equations.

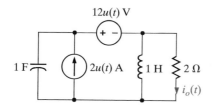

Figure E14.1

ANSWER: $i_o(t) = 6.35e^{-t/4} \cos\left[\left(\sqrt{15}/4\right)t - 156.72°\right]u(t)$ A.

E14.2   Find $v_o(t)$ in the network in Fig. E14.2 using node equations.

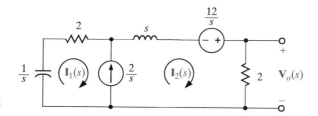

Figure E14.2

ANSWER: $v_o(t) = \left(4 - 8.93e^{-3.73t} + 4.93e^{-0.27t}\right)u(t)$ V.

E14.3   Solve Extension Exercises E14.2 using Thévenin's theorem.

We will now illustrate the use of the Laplace transform in the transient analysis of circuits. We will analyze networks such as those considered in Chapters 6 and 7. Our approach will first be to determine the initial conditions for the capacitors and inductors in the network, and then we will employ the element models that were specified at the beginning of this chapter together with the circuit analysis techniques to obtain a solution. The following examples will demonstrate the approach.

EXAMPLE 14.7

Let us determine the output voltage of the network shown in Fig. 14.9a.

SOLUTION   At $t = 0$ the initial voltage across the capacitor is 1 V, and the initial current drawn through the inductor is 1 A. The circuit for $t > 0$ is shown in Fig. 14.9b with the initial conditions. The transformed network is shown in Fig. 14.9c.

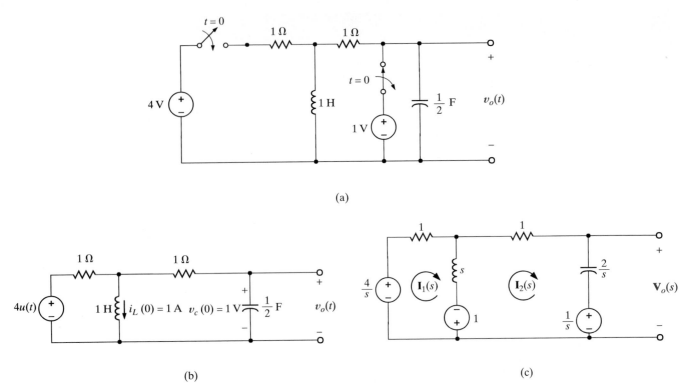

Figure 14.9 Circuits employed in Example 14.7.

The mesh equations for the transformed network are

$$(s + 1)\mathbf{I}_1(s) - s\mathbf{I}_2(s) = \frac{4}{s} + 1$$

$$-s\mathbf{I}_1(s) + \left(s + \frac{2}{s} + 1\right)\mathbf{I}_2(s) = \frac{-1}{s} - 1$$

which can be written in matrix form as

$$\begin{bmatrix} s + 1 & -s \\ -s & \dfrac{s^2 + s + 2}{s} \end{bmatrix} \begin{bmatrix} \mathbf{I}_1(s) \\ \mathbf{I}_2(s) \end{bmatrix} = \begin{bmatrix} \dfrac{s + 4}{s} \\ \dfrac{-(s + 1)}{s} \end{bmatrix}$$

Solving for the currents, we obtain

$$
\begin{bmatrix} \mathbf{I}_1(s) \\ \mathbf{I}_2(s) \end{bmatrix} = \begin{bmatrix} s+1 & -s \\ -s & \dfrac{s^2+s+2}{s} \end{bmatrix}^{-1} \begin{bmatrix} \dfrac{s+4}{s} \\ \dfrac{-(s+1)}{s} \end{bmatrix}
$$

$$
= \frac{s}{2s^2+3s+2} \begin{bmatrix} \dfrac{s^2+s+2}{s} & s \\ s & s+1 \end{bmatrix} \begin{bmatrix} \dfrac{s+4}{s} \\ \dfrac{-(s+1)}{s} \end{bmatrix}
$$

$$
= \begin{bmatrix} \dfrac{4s^2+6s+8}{s(2s^2+3s+2)} \\ \dfrac{2s-1}{2s^2+3s+2} \end{bmatrix}
$$

The output voltage is then

$$
\mathbf{V}_o(s) = \frac{2}{s}\mathbf{I}_2(s) + \frac{1}{s}
$$

$$
= \frac{2}{s}\left(\frac{2s-1}{2s^2+3s+2}\right) + \frac{1}{s}
$$

$$
= \frac{s+\frac{7}{2}}{s^2+\frac{3}{2}s+1}
$$

This function can be written in a partial fraction expansion as

$$
\frac{s+\frac{7}{2}}{s^2+\frac{3}{2}s+1} = \frac{K_1}{s+\frac{3}{4}-j(\sqrt{7}/4)} + \frac{K_1^*}{s+\frac{3}{4}+j(\sqrt{7}/4)}
$$

Evaluating the constants, we obtain

$$
\left.\frac{s+\frac{7}{2}}{s+\frac{3}{4}+j(\sqrt{7}/4)}\right|_{s=-(3/4)+j(\sqrt{7}/4)} = K_1
$$

$$
2.14\ \underline{/-76.5°} = K_1
$$

Therefore,

$$
v_o(t) = \left[4.29e^{-(3/4)t}\cos\left(\frac{\sqrt{7}}{4}t-76.5°\right)\right]u(t)\ \text{V} \qquad \square
$$

EXAMPLE 14.8

We wish to find the output voltage $v_o(t)$ for $t > 0$ in the network in Fig. 14.10a.

SOLUTION   In steady state prior to $t = 0$, the current $i_1(0)$ is

$$
i_1(0) = \frac{12}{2+4} = 2\ \text{A}
$$

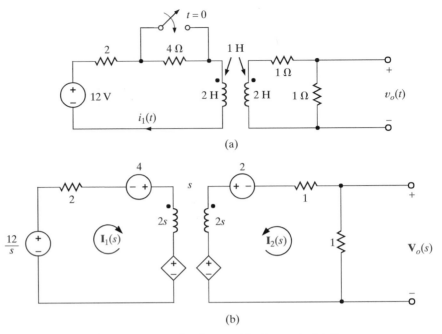

Figure 14.10 Circuits used in Example 14.8.

The dependent sources result from magnetic coupling. See Eq. (14.13).

Hence,

$$L_1 i_1(0) = 4$$
$$M i_1(0) = 2$$

Using these data, we obtain the transformed network as shown in Fig. 14.10b. The mesh equations for this network are

$$(2s + 2)\mathbf{I}_1(s) + s\mathbf{I}_2(s) = \frac{12}{s} + 4$$

$$s\mathbf{I}_1(s) + (2s + 2)\mathbf{I}_2(s) = 2$$

Solving these equations for the mesh currents, we obtain

$$\begin{bmatrix} \mathbf{I}_1(s) \\ \mathbf{I}_2(s) \end{bmatrix} = \frac{1}{3s^2 + 8s + 4} \begin{bmatrix} 2s + 2 & -s \\ -s & 2s + 2 \end{bmatrix} \begin{bmatrix} 4s + 12 \\ s \\ 2 \end{bmatrix}$$

which yields

$$\mathbf{I}_2(s) = \frac{-8}{3s^2 + 8s + 4}$$

and

$$\mathbf{V}_o(s) = -1\mathbf{I}_2(s) = \frac{\frac{8}{3}}{s^2 + \frac{8}{3}s + \frac{4}{3}}$$

This equation can be written as

$$\frac{\frac{8}{3}}{s^2 + \frac{8}{3}s + \frac{4}{3}} = \frac{K_1}{s + 2} + \frac{K_2}{s + \frac{2}{3}}$$

Solving for the constants yields $K_1 = -2$ and $K_2 = 2$. Therefore,

$$v_o(t) = 2\left(e^{-(2/3)t} - e^{-2t}\right)u(t) \text{ V}$$ ❏

EXTENSION EXERCISES

E14.4 Solve Extension Exercise E6.2 using Laplace transforms.

ANSWER: $i_1(t) = 1e^{-9t}$ A.

E14.5 Solve Extension Exercise E6.4 using Laplace transforms.

ANSWER: $v_o(t) = 6 - \frac{10}{3}e^{-2t}$ V.

E14.6 Find $v_o(t)$ for $t > 0$ in the network in Fig. E14.6.

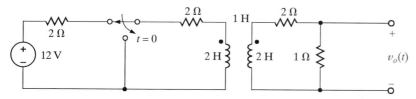

Figure E14.6

ANSWER: $v_o(t) = 1.13\left(e^{-2.55t} - e^{-0.78t}\right)u(t)$ V.

EXAMPLE 14.9

In this example we analyze the circuit in Example 6.8 using Laplace transforms. The circuit and the input are redrawn in Fig. 14.11. We wish to find the expression for the output voltage $v_o(t)$ for $t > 0$.

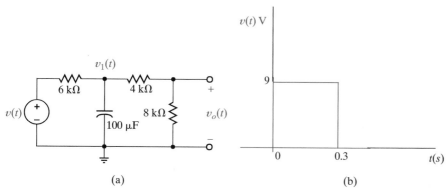

Figure 14.11 Network (a) and input (b) used in Example 14.9.

SOLUTION KCL can be used to compute the node voltage $V_1(s)$ and then a simple voltage divider can be employed to find $V_o(s)$. KCL yields

$$\frac{V_1(s) - V(s)}{6000} + \frac{V_1(s)}{1/10^{-4}s} + \frac{V_1(s)}{12,000} = 0$$

Solving this equation for $V_1(s)$, we obtain

$$V_1(s) = \frac{V(s)}{6000(10^{-4}s + 2.5 \times 10^{-4})}$$

Note that

$$V_o(s) = \frac{8000}{4000 + 8000} V_1(s)$$

Therefore,

$$V_o(s) = \frac{2V(s)}{1.8s + 4.5}$$

$$= \frac{1}{0.9} \frac{V(s)}{s + 2.5}$$

The expression for $V(s) = \mathcal{L}[v(t)]$, where $v(t)$ is as shown in Fig. 14.11, is

$$V(s) = 9\left(\frac{1}{s} - \frac{e^{-0.3s}}{s}\right)$$

Substituting this equation into the expression for $V_o(s)$ yields

$$V_o(s) = \left(\frac{1}{0.9}\right) \frac{1}{s + 2.5} (9) \frac{1 - e^{-0.3s}}{s}$$

$$= \frac{10(1 - e^{-0.3s})}{s(s + 2.5)}$$

Since

$$\frac{10}{s(s + 2.5)} = \frac{4}{s} - \frac{4}{s + 2.5}$$

then

$$V_o(s) = \left(\frac{4}{s} - \frac{4}{s + 2.5}\right)(1 - e^{-0.3s})$$

and therefore

$$v_o(t) = (4 - 4e^{-2.5t})u(t) - (4 - 4e^{-2.5(t-0.3)})u(t - 0.3) \text{ V}$$

This expression is of course identical to that obtained in Example 6.8. ❑

EXTENSION EXERCISES

E14.2 Find the expression for the voltage $v_o(t)$ for $t > 0$ in the network in Fig. E14.7a if the input is as shown in Fig. E14.6.

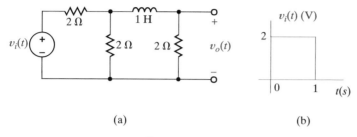

(a)   (b)

Figure E14.7

ANSWER: $v_o(t) = \frac{2}{3}\left[\left(1 - e^{-3t}\right)u(t) - \left(1 - e^{-3(t-1)}\right)u(t - 1)\right]$ V.

14.4 TRANSFER FUNCTION

In Chapter 12 we introduced the concept of network or transfer function. It is essentially nothing more than the ratio of some output variable to some input variable. If both variables are voltages, the transfer function is a voltage gain. If both variables are currents, the transfer function is a current gain. If one variable is a voltage and the other is a current, the transfer function becomes a transfer admittance or impedance.

In deriving a transfer function, all initial conditions are set equal to zero. In addition, if the output is generated by more than one input source in a network, superposition can be employed in conjunction with the transfer function for each source.

To present this concept in a more formal manner, let us assume that the input/output relationship for a linear circuit is

$$b_n \frac{d^n y_o(t)}{dt^n} + b_{n-1} \frac{d^{n-1} y_o(t)}{dt^{n-1}} + \cdots + b_1 \frac{dy_o(t)}{dt} + b_0 y_o(t)$$
$$= a_m \frac{d^m x_i(t)}{dt^m} + a_{m-1} \frac{d^{m-1} x_i(t)}{dt^{m-1}} + \cdots + a_1 \frac{dx_i(t)}{dt} + a_0 x_i(t)$$

If all the initial conditions are zero, the transform of the equation is

$$\left(b_n s^n + b_{n-1} s^{n-1} + \cdots + b_1 s + b_0\right)\mathbf{Y}_o(s)$$
$$= \left(a_m s^m + a_{m-1} s^{m-1} + \cdots + a_1 s + a_0\right)\mathbf{X}_i(s)$$

or

$$\frac{\mathbf{Y}_o(s)}{\mathbf{X}_i(s)} = \frac{a_m s^m + a_{m-1} s^{m-1} + \cdots + a_1 s + a_0}{b_n s^n + b_{n-1} s^{n-1} + \cdots + b_1 s + b_0}$$

This ratio of $\mathbf{Y}_o(s)$ to $\mathbf{X}_i(s)$ is called the *transfer* or *network function*, which we denote as $\mathbf{H}(s)$; that is,

$$\frac{\mathbf{Y}_o(s)}{\mathbf{X}_i(s)} = \mathbf{H}(s)$$

or

$$\mathbf{Y}_o(s) = \mathbf{H}(s)\mathbf{X}_i(s) \qquad \text{14.16}$$

This equation states that the output response $\mathbf{Y}_o(s)$ is equal to the network function multiplied by the input $\mathbf{X}_i(s)$. Note that if $x_i(t) = \delta(t)$ and therefore $\mathbf{X}_i(s) = 1$, the impulse response is equal to the inverse Laplace transform of the network function. This is an extremely important concept because it illustrates that if we know the impulse response of a network, we can find the response due to some other forcing function using Eq. (14.16).

At this point it is informative to review briefly the natural response of both first-order and second-order networks. We have demonstrated in Chapter 6 that if only a single storage element is present, the natural response of a network to an initial condition is always of the form

$$x(t) = X_0 e^{-t/\tau}$$

where $x(t)$ can be either $v(t)$ or $i(t)$, X_0 is the initial value $x(t)$, and τ is the time constant of the network.

As illustrated in Chapter 7, the natural response of a second-order network is controlled by the roots of the *characteristic equation*, which is of the form

$$s^2 + 2\zeta\omega_0 s + \omega_0^2 = 0$$

where ζ is the *damping ratio* and ω_0 is the *undamped natural frequency*. These two key factors, ζ and ω_0, control the response, and there are basically three cases of interest.

Case 1, $\zeta > 1$: Overdamped Network.   The roots of the characteristic equation are $s_1, s_2 = -\zeta\omega_0 \pm \omega_0\sqrt{\zeta^2 - 1}$ and, therefore, the network response is of the form

$$x(t) = K_1 e^{-(\zeta\omega_0 + \omega_0\sqrt{\zeta^2-1})t} + K_2 e^{-(\zeta\omega_0 - \omega_0\sqrt{\zeta^2-1})t}$$

Case 2, $\zeta < 1$: Underdamped Network.   The roots of the characteristic equation are $s_1, s_2 = -\zeta\omega_0 \pm j\omega_0\sqrt{1 - \zeta^2}$ and, therefore, the network response is of the form

$$x(t) = K e^{-\zeta\omega_0 t} \cos(\omega_0\sqrt{1 - \zeta^2}\, t + \phi)$$

Case 3, $\zeta = 1$: Critically Damped Network.   The roots of the characteristic equation are $s_1, s_2 = -\omega_0$ and, hence, the response is of the form

$$x(t) = K_1 t e^{-\omega_0 t} + K_2 e^{-\omega_0 t}$$

It is very important that the reader note that the characteristic equation is the denominator of the transfer function $\mathbf{H}(s)$, and the roots of this equation, which are the poles of the network, determine the form of the network's natural response.

A convenient method for displaying the network's poles and zeros in graphical form is the use of a pole-zero plot. A pole-zero plot of a function can be accomplished using what is commonly called the *complex* or *s-plane*. In the complex plane the abscissa is σ and the ordinate is $j\omega$. Zeros are represented by 0's and poles are represented by $\times$'s. Although we are concerned only with the finite poles and zeros specified by the network or response function, we should point out that a rational function must have the same number of poles and zeros. Therefore, if $n > m$, there are $n - m$ zeros at the point at infinity, and if $n < m$, there are $m - n$ poles at the point at infinity. A systems engineer can tell a lot about the operation of a network or system by simply examining its pole-zero plot.

In order to correlate the natural response of a network to an initial condition with the network's pole locations, we have illustrated in Fig. 14.12 the correspondence for all three cases: overdamped, underdamped, and critically damped. Note that if the network poles are real and unequal, the response is slow and, therefore, $x(t)$ takes a long time to reach zero. If the network poles are complex conjugates, the response is fast; however, it overshoots and is eventually damped out. The dividing line between the overdamped and underdamped cases is the critically damped case in which the roots are real and equal. In this case the transient response dies out as quickly as possible, with no overshoot.

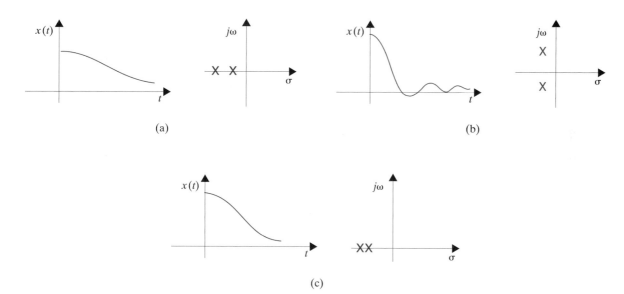

Figure 14.12 Natural response of a second-order network together with network pole locations for the three cases: (a) overdamped; (b) underdamped; (c) critically damped.

EXAMPLE 14.10

If the impulse response of a network is $h(t) = e^{-t}$, let us determine the response $v_o(t)$ to an input $v_i(t) = 10e^{-2t}u(t)$ V.

SOLUTION The transformed variables are

$$\mathbf{H}(s) = \frac{1}{s + 1}$$

$$\mathbf{V}_i(s) = \frac{10}{s + 2}$$

Therefore,

$$\mathbf{V}_o(s) = \mathbf{H}(s)\mathbf{V}_i(s)$$

$$= \frac{10}{(s + 1)(s + 2)}$$

and hence,

$$v_o(t) = 10(e^{-t} - e^{-2t})u(t) \text{ V}$$

The importance of the transfer function stems from the fact that it provides the systems engineer with a great deal of knowledge about the system's operation, since its dynamic properties are governed by the system poles.

EXAMPLE 14.11

Let us derive the transfer function $\mathbf{V}_o(s)/\mathbf{V}_i(s)$ for the network in Fig. 14.13.

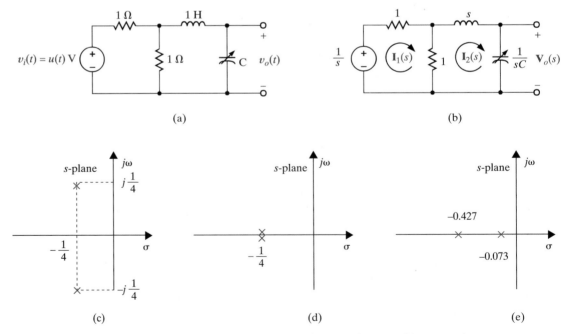

Figure 14.13 Networks and pole zero plots used in Example 14.11.

SOLUTION   Our output variable is the voltage across a variable capacitor and the input voltage is a unit step. The transformed network is shown in Fig. 14.13b. The mesh equations for the network are

$$2I_1(s) - I_2(s) = V_i(s)$$

$$-I_1(s) + \left(s + \frac{1}{sC} + 1\right)I_2(s) = 0$$

and the output equation is

$$V_o(s) = \frac{1}{sC}I_2(s)$$

From these equations we find that the transfer function is

$$\frac{V_o(s)}{V_i(s)} = \frac{1/2C}{s^2 + \frac{1}{2}s + 1/C}$$

Since the transfer function is dependent on the value of the capacitor, let us examine the transfer function and the output response for three values of the capacitor.

(a) $C = 8\,\text{F}$

$$\frac{V_o(s)}{V_i(s)} = \frac{\frac{1}{16}}{\left(s^2 + \frac{1}{2}s + \frac{1}{8}\right)}$$

$$= \frac{\frac{1}{16}}{\left(s + \frac{1}{4} - j\frac{1}{4}\right)\left(s + \frac{1}{4} + j\frac{1}{4}\right)}$$

The output response is

$$V_o(s) = \frac{\frac{1}{16}}{s\left(s + \frac{1}{4} - j\frac{1}{4}\right)\left(s + \frac{1}{4} + j\frac{1}{4}\right)}$$

As illustrated in Chapter 7, the poles of the transfer function, which are the roots of the characteristic equation, are complex conjugates, as shown in Fig. 14.13c; therefore, the output response will be *underdamped*. The output response as a function of time is

$$v_o(t) = \left[\frac{1}{2} + \frac{1}{\sqrt{2}}e^{-t/4}\cos\left(\frac{t}{4} + 135°\right)\right]u(t)\,\text{V}$$

Note that for large values of time the transient oscillations, represented by the second term in the response, become negligible and the output settles out to a value of $\frac{1}{2}$ V. This can also be seen directly from the circuit since for large values of time the input looks like a dc source, the inductor acts like a short circuit, the capacitor acts like an open circuit, and the resistors form a voltage divider.

(b) $C = 16\,\text{F}$

$$\frac{V_o(s)}{V_i(s)} = \frac{\frac{1}{32}}{s^2 + \frac{1}{2}s + \frac{1}{16}}$$

$$= \frac{\frac{1}{32}}{\left(s + \frac{1}{4}\right)^2}$$

The output response is

$$\mathbf{V}_o(s) = \frac{\frac{1}{32}}{s\left(s + \frac{1}{4}\right)^2}$$

Since the poles of the transfer function are real and equal as shown in Fig. 14.13d, the output response will be *critically damped*. $v_o(t) = \mathcal{L}^{-1}[\mathbf{V}_o(s)]$ is

$$v_o(t) = \left[\frac{1}{2} - \left(\frac{t}{8} + \frac{1}{2}\right)e^{-t/4}\right]u(t)\ \text{V}$$

(c) $C = 32\ \text{F}$

$$\frac{\mathbf{V}_o(s)}{\mathbf{V}_i(s)} = \frac{\frac{1}{64}}{s^2 + \frac{1}{2}s + \frac{1}{32}}$$

$$= \frac{\frac{1}{64}}{(s + 0.427)(s + 0.073)}$$

The output response is

$$\mathbf{V}_o(s) = \frac{\frac{1}{64}}{s(s + 0.427)(s + 0.073)}$$

The poles of the transfer function are real and unequal, as shown in Fig. 14.13e and, therefore, the output response will be *overdamped*. The response as a function of time is

$$v_o(t) = \left(0.5 + 0.103e^{-0.427t} - 0.603e^{-0.073t}\right)u(t)\ \text{V}$$

Although the values selected for the network parameters are not very practical, the reader is reminded that both magnitude and frequency scaling, as outlined in Chapter 12, can be applied here also.   ❏

EXAMPLE 14.12

For the network in Fig. 14.14a let us compute (a) the transfer function, (b) the type of damping exhibited by the network, and (c) the unit step response.

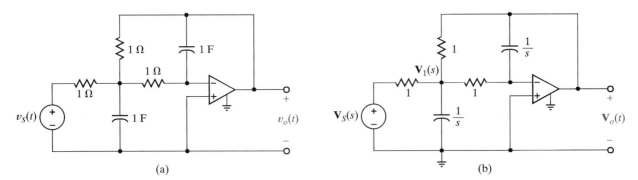

Figure 14.14 Circuits used in Example 14.12.

SOLUTION   Recall that the voltage across the op-amp input terminals is zero and therefore KCL at the node labeled $\mathbf{V}_1(s)$ in Fig. 14.14b yields the following equation:

$$\frac{\mathbf{V}_S(s) - \mathbf{V}_1(s)}{1} = s\mathbf{V}_1(s) + \frac{\mathbf{V}_1(s) - \mathbf{V}_o(s)}{1} + \frac{\mathbf{V}_1(s)}{1}$$

Since the current into the negative input terminal of the op-amp is zero, KCL requires that

$$s\mathbf{V}_o(s) = -\frac{\mathbf{V}_1(s)}{1}$$

Combining the two equations yields the transfer function

$$\frac{\mathbf{V}_o(s)}{\mathbf{V}_S(s)} = \frac{-1}{s^2 + 3s + 1}$$

which can be expressed in the form

$$\frac{\mathbf{V}_o(s)}{\mathbf{V}_S(s)} = \frac{-1}{(s + 2.62)(s + 0.38)}$$

Since the roots are real and unequal, the step response of the network will be overdamped. The step response is

$$\mathbf{V}_o(s) = \frac{-1}{s(s + 2.62)(s + 0.38)}$$

$$= \frac{-1}{s} + \frac{-0.17}{s + 2.62} + \frac{1.17}{s + 0.38}$$

Therefore,

$$v_o(t) = \left(-1 - 0.17e^{-2.62t} + 1.17e^{-0.38t}\right)u(t) \text{ V} \qquad \square$$

EXTENSION EXERCISES

E14.8   If the unit impulse response of a network is known to be $\frac{10}{9}\left(e^{-t} - e^{-10t}\right)$, determine the unit step response

ANSWER: $x(t) = \left(1 - \frac{10}{9}e^{-t} + \frac{1}{9}e^{-10t}\right)u(t).$

E14.9   The transfer function for a network is

$$\mathbf{H}(s) = \frac{s + 10}{s^2 + 4s + 8}$$

Determine the pole-zero plot of $\mathbf{H}(s)$, the type of dampling exhibited by the network, and the unit step response of the network.

ANSWER: The network is underdamped;
$x(t) = \left[\frac{10}{8} + 1.46e^{-2t}\cos(2t - 210.96°)\right]u(t).$

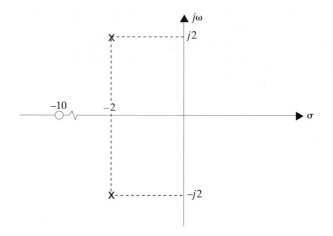

Recall from our previous discussion that if a second-order network is underdamped, the characteristic equation of the network is of the form

$$s^2 + 2\zeta\omega_0 s + \omega_0^2 = 0$$

and the roots of this equation, which are the network poles, are of the form

$$s_1, s_2 = -\zeta\omega_0 \pm j\omega_0 \sqrt{1 - \zeta^2}$$

The roots s_1 and s_2, when plotted in the s-plane, generally appear as shown in Fig. 14.15, where

$$\zeta = \text{damping ratio}$$
$$\omega_0 = \text{undamped natural frequency}$$

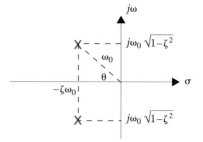

Figure 14.15 Pole locations for second-order underdamped network.

and as shown in Fig. 14.15,

$$\zeta = \cos\theta$$

It is important for the reader to note that the damping ratio and the undamped natural frequency are exactly the same quantities as those employed in Chapter 12 when determining a network's frequency response. We find that it is these same quantities that govern the network's transient response.

EXAMPLE 14.13

Let us examine the effect of pole position in the s-plane on the transient response of the second-order RLC series network shown in Fig. 14.16.

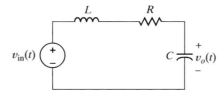

Figure 14.16 RLC series network.

SOLUTION   The voltage gain transfer function is

$$\mathbf{G}_v(s) = \frac{\dfrac{1}{LC}}{s^2 + s\left(\dfrac{R}{L}\right) + \dfrac{1}{LC}} = \frac{\omega_0^2}{s^2 + 2\zeta\omega_0 s + \omega_0^2}$$

For this analysis we will let $\omega_0 = 2000$ rad/s for $\zeta = 0.25, 0.50, 0.75$, and 1.0. From the preceding equation we see that

$$LC = \frac{1}{\omega_0^2} = 2.5 \times 10^{-7}$$

and

$$R = 2\zeta\sqrt{\frac{L}{C}}$$

If we arbitrarily let $L = 10$ mH, then $C = 25$ µF. Also, for $\zeta = 0.25, 0.50, 0.75$, and $1.0, R = 10\ \Omega, 20\ \Omega, 30\ \Omega$, and $40\ \Omega$, respectively. Over the range of ζ values, the network ranges from underdamped to critically damped. Since poles are complex for underdamped systems, the real and imaginary components and the magnitude of the poles of $\mathbf{G}_v(s)$ are given in Table 14.1 for the ζ values listed previously.

Table 14.1 Pole locations for $\zeta = 0.25$ to 1.0

Damping Ratio	Real	Imaginary	Magnitude
1.00	2000.0	0.0	2000.0
0.75	1500.0	1322.9	2000.0
0.50	1000.0	1732.1	2000.0
0.25	500.0	1936.5	2000.0

Fig. 14.17 shows the pole-zero diagrams for each value of ζ. Notice first that all the poles lie on a circle; thus, the pole magnitudes are constant, consistent with

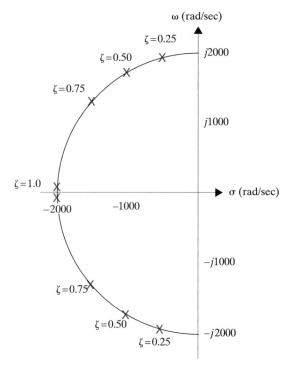

Figure 14.17 Pole-zero diagrams for $\zeta = 0.25$ to 1.0.

Table 14.1. Second, as ζ decreases, the real part of the pole decreases while the imaginary part increases. In fact, when ζ goes to zero, the poles become imaginary.

A PSPICE simulation of a unit step transient excitation for all four values of R is shown in Fig. 14.18. we see that as ζ decreases, the overshoot in the output voltage increases. furthermore, when the network is critically damped ($\zeta = 1$) there is no overshoot at all. In most applications excessive overshoot is not desired. To correct this, the damping ratio, ζ, should be increased, which for this circuit would require an increase in the resistor value. ❏

EXAMPLE 14.14

Let us revisit the Tacoma Narrows Bridge disaster examined in Example 12.12. A photograph of the bridge as it collapsed is shown in Fig. 14.19.

In Chapter 12, we assumed that the bridge's demise was brought on by winds oscillating back and forth at a frequency near that of the bridge (0.2 Hz). We found that we could create an RLC circuit, shown in Fig. 12.29, that resonates at 0.2 Hz and has an output voltage consistent with the vertical deflection of the bridge. This kind of forced resonance never happened at Tacoma Narrows. The real culprit was not so much wind fluctuations, but the bridge itself. This is thoroughly explained in the paper "Resonance, Tacoma Narrows Bridge Failure, and Undergraduate Physics Textbooks," by K. Y. Billah and R. H. Scalan published in the *American Journal of*

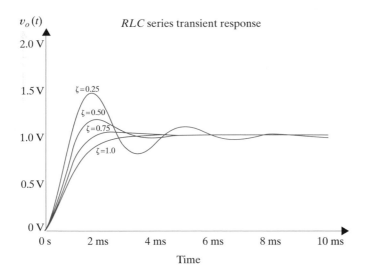

Figure 14.18 PSPICE transient response output for $\zeta = 0.25$ to 1.0.

Figure 14.19 Tacoma Narrows Bridge as it collapsed on November 7, 1940. (Used with Permission from Special Collection Division, University of Washington Libraries. Photo by Farguharson, negative number 12.)

Physics, vol. 59, no. 2 p. 118–124, in which the authors determined that changes in wind speed affected the coefficients of the second order differential equation that models the resonant behavior. In particular, the damping ratio, ζ, was dependent on the wind speed and is roughly given as

$$\zeta = 0.00460 - 0.00013U \qquad \textbf{14.17}$$

where U is the wind speed in mph. Note, as shown in Fig. 14.20, that ζ becomes negative at wind speeds in excess of 35 mph—a point we will demonstrate later. Furthermore, Billah and Scalan report that the bridge resonated in a twisting mode, which can be easily seen in Fig. 12.28 and is described by the differential equation

$$\frac{d^2\theta(t)}{dt^2} + 2\zeta\omega_0\frac{d\theta(t)}{dt} + \omega_0^2\,\theta(t) = 0$$

or

$$\ddot{\theta} + 2\zeta\omega_0\dot{\theta} + \omega_0^2\theta = 0 \qquad \textbf{14.18}$$

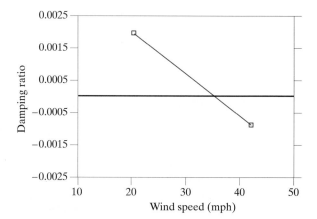

Figure 14.20 Damping ratio versus wind speed for second-order twisting model of the Tacoma Narrows Bridge.

where $\theta(t)$ is the angle of twist in degrees and wind speed is implicit in ζ through Eq. (14.17). Billah and Scalan list the following data obtained either by direct observation at the bridge site or through scale model experiments afterward.

$$\text{Wind speed at failure} \approx 42 \text{ mph}$$
$$\text{Twist at failure} \approx \pm 12°$$
$$\text{Time to failure} \approx 45 \text{ minutes}$$

We will start the twisting oscillations using an initial condition on $\theta(0)$ and see if the bridge oscillations decrease or increase over time. Let us now design a network that will simulate the true Tacoma Narrows disaster.

SOLUTION   First, we solve for $\ddot{\theta}(t)$ in Eq. (14.18)

$$\ddot{\theta} = -2\zeta\omega_0\dot{\theta} - \omega_0^2\,\theta \qquad \textbf{14.19}$$
$$\ddot{\theta} = -2(2\pi)(0.2)(0.0046 - 0.00013U)\dot{\theta} - \left[2(2\pi)(0.2)\right]^2\theta$$

or

$$\ddot{\theta} = -(0.01156 - 0.00033U)\dot{\theta} - 1.579\theta$$

We now wish to model this equation to produce a voltage proportional to $\ddot{\theta}(t)$. We can accomplish this using the op-amp integrator circuit shown in Fig. 14.21.

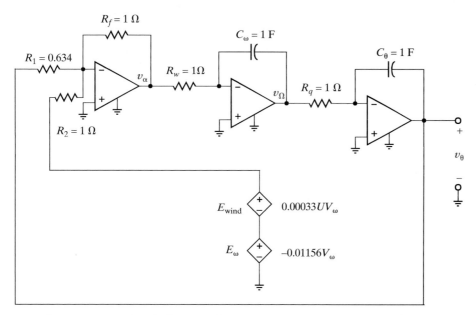

Figure 14.21 Circuit diagram for Tacoma Narrows Bridge simulations.

The circuit's operation can perhaps be best understood by first assigning the voltage v_α to be proportional to $\ddot{\theta}(t)$, where 1 V represents 1 deg/s². Thus, the output of the first integrator, v_ω must be

$$v_\omega = -\frac{1}{R_\omega C_\omega} \int v_\alpha \, dt$$

or, since $R_\omega = 1\ \Omega$ and $C_\omega = 1$ F,

$$v_\omega = -\int v_\alpha \, dt$$

So v_ω is proportional to $-\ddot{\theta}(t)$ and 1 V equals -1 deg/s. Similarly, the output of the second integrator must be

$$v_\theta = -\int v_\omega \, dt$$

where $v_\theta(t)$ is proportional to $\theta(t)$ and 1 V equals 1 degree. The outputs of the integrators are then fed back as inputs to the summing op-amp. Note that the dependent sources, E_ω and E_{wind}, re-create the coefficient on $\dot{\theta}(t)$ in Eq. (14.17); that is,

$$2\zeta\omega_0 = (2)(0.2)(2\pi)\zeta = 0.01156 - 0.00033U$$

To simulate various wind speeds, we need only change the gain factor of E_{wind}. Finally, we can solve the circuit for $v_\alpha(t)$.

$$v_\alpha(t) = -\left(\frac{R_f}{R_2}\right)(E_\omega - E_{\text{wind}}) - \left(\frac{R_f}{R_1}\right)v_\theta$$

which matches Eq. (14.19) if

$$\frac{R_f}{R_1} = \omega_0^2 = \left[2\pi(0.2)\right]^2 = 1.579$$

and

$$\frac{R_f}{R_2}\left[E_\omega - E_{\text{wind}}\right] = 2\zeta\omega_o$$

or

$$\frac{R_f}{R_2} = 1$$

Thus, if $R_f = R_2 = 1$ and $R_1 = 0.634$, the circuit will simulate the bridge's twisting motion. We will start the twisting oscillations using an initial condition $\theta(0)$ and see if the bridge oscillations decrease or increase over time.

The first simulation is for a wind speed of 20 mph and one degree of twist. The corresponding output voltage is shown in Fig. 14.22. Notice that the bridge twists at a frequency of 0.2 Hz and the oscillations decrease exponentially, indicating a non-destructive situation.

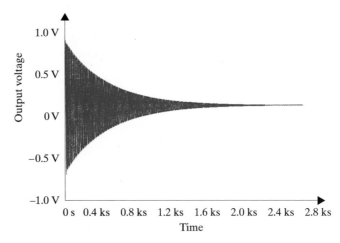

Figure 14.22 Tacoma Narrows Bridge simulation at 20 mph wind speed and one degree twist initial condition.

Fig. 14.23 shows the output for 35-mph winds and an initial twist of one degree. Notice that the oscillations neither increase nor decrease. This indicates that the damping ratio is zero.

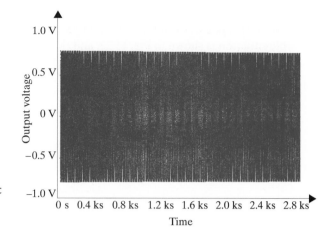

Figure 14.23 Tacoma Narrows Bridge simulation at 35-mph winds and one degree of initial twist.

Finally, the simulation at a wind speed of 42 mph and 1 degree initial twist is shown in Fig. 14.24. The twisting becomes worse and worse until after 45 minutes, the bridge is twisting $\pm$ 12.5 degrees, which matches values reported by Billah and Scalan for collapse.

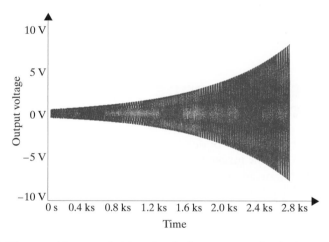

Figure 14.24 Tacoma Narrows Bridge simulation at 42 mph wind speed and one degree of initial twist.

The dependency of the damping ratio on wind speed can also be demonstrated by investigating how the system poles change with the wind. The characteristic equation for the system is

$$s^2 + 2\zeta\omega_0 s + \omega_0^2 = 0$$

or

$$s^2 + (0.01156 - 0.00033U)s + 1.579 = 0$$

The roots of the characteristic equation yield the pole locations. Fig. 14.25 shows the system poles at wind speeds of 20, 35, and 42 mph. Note that at 20 mph, the stable situation is shown in Fig. 14.22, and the poles are in the left-half of the s-plane. At 35 mph ($\zeta = 0$) the poles are on the $j\omega$ axis and the system is oscillatory, as shown in Fig. 14.23. Finally, at 42 mph, we see that the poles are in the right half of the s-plane and, from Fig. 14.24 we know this is an unstable system. This relationship between pole location and transient response is true for all systems—right-half plane poles result in unstable systems.

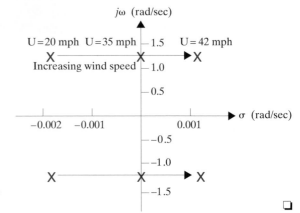

Figure 14.25 Pole-zero plot for Tacoma Narrows Bridge second-order model at wind speeds of 20, 35, and 42 mph.

14.5 POLE-ZERO PLOT/BODE PLOT CONNECTION

In Chapter 12 we introduced the Bode plot as an analysis tool for sinusoidal frequency-response studies. Let us now investigate the relationship between the s-plane pole-zero plot and the Bode plot. As an example, consider the transfer function of the RLC high-pass filter shown in Fig. 14.26.

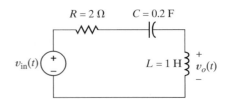

Figure 14.26 RLC high-pass filter.

The transfer function is

$$\mathbf{G}_v(s) = \frac{sL}{sL + R + \dfrac{1}{sC}} = \frac{s^2}{s^2 + s\left(\dfrac{R}{L}\right) + \dfrac{1}{LC}}$$

Using the element values, the transfer function becomes

$$\mathbf{G}_v(s) = \frac{s^2}{s^2 + 2s + 5} = \frac{s^2}{(s + 1 + j2)(s + 1 - j2)}$$

We see the transfer function has two zeros at the origin ($s = 0$) and two complex-conjugate poles $s = -1 \pm j2$. The standard pole-zero plot for this function is shown in Fig. 14.27a. A three-dimensional s-plane plot of the magnitude of $\mathbf{G}_v(s)$ is shown in Fig. 14.27b. Note carefully that when $s = 0$, $\mathbf{G}_v(s) = 0$ and when $s = -1 \pm j2$, the function is infinite.

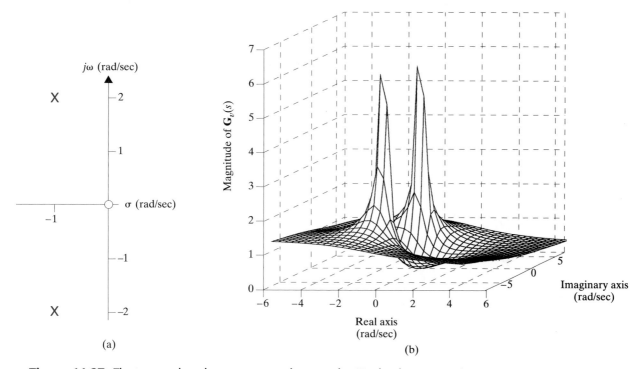

Figure 14.27 Figures used to demonstrate pole-zero plot/Bode plot connection.

Recall that the Bode plot of a transfer function's magnitude is in reality a plot of the magnitude of the gain versus frequency. The frequency domain, where $s = j\omega$, corresponds to the $j\omega$-axis in the s-plane obtained by setting σ, the real part of s, to zero. Thus, the frequency domain corresponds directly to that part of the s-domain where $\sigma = 0$, as illustrated in the three-dimensional plot in Fig. 14.27c.

Let us develop the Bode plot by first rotating Fig. 14.27c such that the real axis is perpendicular to the page as shown in Fig. 14.27d. Note that the transfer function maximum occurs at $\omega = \sqrt{5} = 2.24$ rad/s, which is the magnitude of the complex pole frequencies. In addition, the symmetry of the pole around the real axis becomes readily apparent. As a result of this symmetry, we can restrict our analysis to positive values of $j\omega$, with no loss of information. This plot for $\omega \geq 0$ is shown in Fig. 14.27e where frequency is plotted in Hz rather than rad/s. Finally, converting the transfer function magnitude to dB and using a log axis for frequency, we produce the Bode plot in Fig. 14.27f.

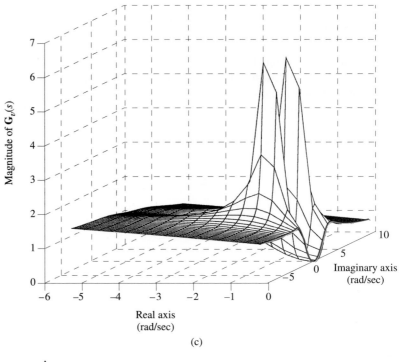

(c)

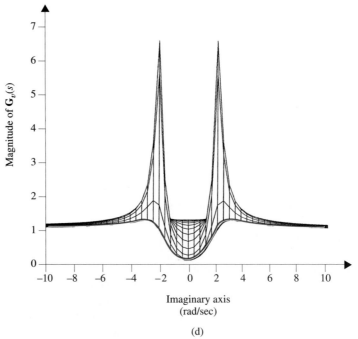

(d)

Figure 14.27 (*continued*)

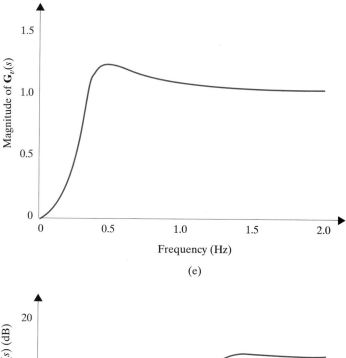

(e)

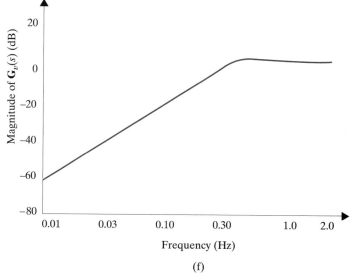

(f)

Figure 14.27 (*continued*)

14.6 STEADY-STATE RESPONSE

In Section 14.3 we have demonstrated, using a variety of examples, the power of the Laplace transform technique in determining the complete response of a network. This complete response is composed of transient terms, which disappear as $t \rightarrow \infty$, and steady-state terms, which are present at all times. Let us now examine a method by which to determine the steady-state response of a network directly. Recall from previous examples that the network response can be written as

$$\mathbf{Y}(s) = \mathbf{H}(s)\mathbf{X}(s) \tag{14.20}$$

where $\mathbf{Y}(s)$ is the output or response, $\mathbf{X}(s)$ is the input or forcing function, and $\mathbf{H}(s)$ is the network function or transfer function defined in Section 12.1. The transient portion of the response $\mathbf{Y}(s)$ results from the poles of $\mathbf{H}(s)$, and the steady-state portion of the response results from the poles of the input or forcing function.

As a direct parallel to the sinusoidal response of a network as outlined in Section 8.2, we assume that the forcing function is of the form

$$x(t) = X_M e^{j\omega_0 t} \tag{14.21}$$

which by Euler's identity can be written as

$$x(t) = X_M \cos \omega_0 t + j X_M \sin \omega_0 t \tag{14.22}$$

The Laplace transform of Eq. (14.21) is

$$\mathbf{X}(s) = \frac{X_M}{s - j\omega_0} \tag{14.23}$$

and therefore,

$$\mathbf{Y}(s) = \mathbf{H}(s)\left(\frac{X_M}{s - j\omega_0}\right) \tag{14.24}$$

At this point we tacitly assume that $\mathbf{H}(s)$ does not have any poles of the form $(s - j\omega_k)$. If, however, this is the case, we simply encounter difficulty in defining the steady-state response.

Performing a partial fraction expansion of Eq. (14.24) yields

$$\mathbf{Y}(s) = \frac{X_M \mathbf{H}(j\omega_0)}{s - j\omega_0} + \text{terms that occur due to the poles of } \mathbf{H}(s) \tag{14.25}$$

The first term to the right of the equal sign can be expressed as

$$\mathbf{Y}(s) = \frac{X_M |\mathbf{H}(j\omega_0)| e^{j\phi(j\omega_0)}}{s - j\omega_0} + \cdots \tag{14.26}$$

since $\mathbf{H}(j\omega_0)$ is a complex quantity with a magnitude and phase that are a function of $j\omega_0$.

Performing the inverse transform of Eq. (14.26), we obtain

$$y(t) = X_M |\mathbf{H}(j\omega_0)| e^{j\omega_0 t} e^{j\phi(j\omega_0)} + \cdots$$
$$= X_M |\mathbf{H}(j\omega_0)| e^{j(\omega_0 t + \phi(j\omega_0))} + \cdots \tag{14.27}$$

and hence the steady-state response is

$$y_{ss}(t) = X_M |\mathbf{H}(j\omega_0)| e^{j(\omega_0 t + \phi(j\omega_0))} \tag{14.28}$$

The transient terms disappear in steady-state.

Since the actual forcing function is $X_M \cos \omega_0(t)$, which is the real part of $X_M e^{j\omega_0 t}$, the steady-state response is the real part of Eq. (14.28).

$$y_{ss}(t) = X_M |\mathbf{H}(j\omega_0)| \cos[\omega_0 t + \phi(j\omega_0)] \tag{14.29}$$

In general, the forcing function may have a phase angle θ. In this case, θ is simply added to $\phi(j\omega_0)$ so that the resultant phase of the response is $\phi(j\omega_0) + \theta$.

EXAMPLE 14.15

For the circuit shown in Fig. 14.28a, we wish to determine the steady-state voltage $v_{oss}(t)$ for $t > 0$ if the initial conditions are zero.

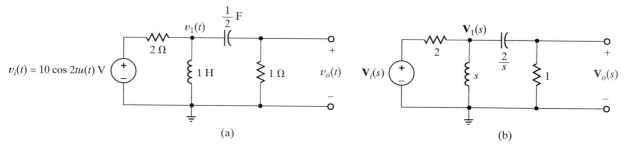

Figure 14.28 Circuit used in Example 14.15.

SOLUTION   As illustrated earlier, this problem could be solved using a variety of techniques, such as node equations, mesh equations, source transformation, and Thévenin's theorem. We will employ node equations to obtain the solution. The transformed network using the impedance values for the parameters is shown in Fig. 14.28b. The node equations for this network are

$$\left(\frac{1}{2} + \frac{1}{s} + \frac{s}{2}\right)\mathbf{V}_1(s) - \left(\frac{s}{2}\right)\mathbf{V}_o(s) = \frac{1}{2}\mathbf{V}_i(s)$$

$$-\left(\frac{s}{2}\right)\mathbf{V}_1(s) + \left(\frac{s}{2} + 1\right)\mathbf{V}_o(s) = 0$$

Solving these equations for $\mathbf{V}_o(s)$, we obtain

$$\mathbf{V}_o(s) = \frac{s^2}{3s^2 + 4s + 4}\mathbf{V}_i(s)$$

Note that this equation is in the form of Eq. (14.20), where $\mathbf{H}(s)$ is

$$\mathbf{H}(s) = \frac{s^2}{3s^2 + 4s + 4}$$

Since the forcing function is $10 \cos 2t\, u(t)$, then $V_M = 10$ and $\omega_0 = 2$. Hence,

$$\mathbf{H}(j2) = \frac{(j2)^2}{3(j2)^2 + 4(j2) + 4}$$

$$= 0.354\ \underline{/45°}$$

Therefore,

$$|\mathbf{H}(j2)| = 0.354$$

$$\phi(j2) = 45°$$

and, hence, the steady-state response is

$$v_{oss}(t) = V_M|\mathbf{H}(j2)| \cos[2t + \phi(j2)]$$
$$= 3.54 \cos(2t + 45°)$$

The complete (transient plus steady-state) response can be obtained from the expression

$$\mathbf{V}_o(s) = \frac{s^2}{3s^2 + 4s + 4} \mathbf{V}_i(s)$$

$$= \frac{s^2}{3s^2 + 4s + 4}\left(\frac{10s}{s^2 + 4}\right)$$

$$= \frac{10s^3}{(s^2 + 4)(3s^2 + 4s + 4)}$$

Determining the inverse Laplace transform of this function using the techniques of Chapter 13, we obtain

$$v_o(t) = 3.54 \cos(2t + 45°) + 1.44e^{-(2/3)t} \cos\left(\frac{2\sqrt{2}}{3}t - 55°\right) \text{V}$$

Note that as $t \to \infty$ the second term approaches zero, and thus the steady-state response is

$$v_{oss}(t) = 3.54 \cos(2t + 45°) \text{V}$$

which can easily be checked using a phasor analysis.   ❑

It is interesting to note that we can use a pole-zero plot to determine the magnitude and phase of a function $\mathbf{F}(s)$ at some specific frequency ω. The following example illustrates the graphical technique.

EXAMPLE 14.16

Given a network that has the voltage transfer function

$$\mathbf{G}_v(s) = \frac{20s(s + 1)}{(s + 3)(s^2 + 2s + 2)}$$

Let us draw the pole-zero plot for this function and graphically find the magnitude and phase at $s = j4 \ r/s$.

SOLUTION   The transfer function can be written in the form

$$\mathbf{G}_v(s) = \frac{20s(s + 1)}{(s + 3)(s + 2 + j2)(s + 2 - j2)}$$

which has the pole-zero plot shown in Fig. 14.29a. Let us examine the term $s + 3$. At the desired frequency $s = j4$, this term becomes $3 + j4$, which has a magnitude of

$$C = |3 + j4| = \sqrt{3^2 + 4^2} = 5$$

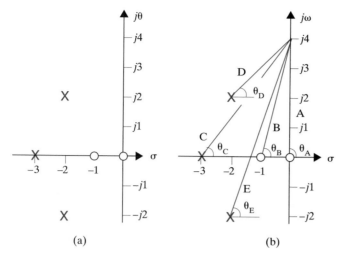

Figure 14.29 Pole-zero plots for Example 14.16.

and a phase angle of

$$\theta_C = \underline{/3 + j4} = \tan^{-1}\left(\frac{4}{3}\right) = 53.1°$$

Thus the term $(s + 3)$ evaluated at $s = j4$ is

$$(s + 3)|_{s=j4} = C\ \underline{/\theta_C} = \underline{/53.1°}$$

This term is represented by the appropriate vector shown in Fig. 14.29b. Using this technique to represent the remaining terms, we find that

$$\mathbf{G}_v(j4) = \frac{20(A\ \underline{/\theta_A})(B\ \underline{/\theta_B})}{(C\ \underline{/\theta_C})(D\ \underline{/\theta_D})(E\ \underline{/\theta_E})}$$

where each vector in the equation is shown in Fig. 14.29b. Using the value for each vector yields

$$\begin{aligned}
\mathbf{G}_v(j4) &= \frac{20(4\ \underline{/90°})(4.12\ \underline{/75.96°})}{(5\ \underline{/53.13°})(6.32\ \underline{/71.57°})(2.83\ \underline{/45°})} \\
&= 3.7\ \underline{/-3.7°}
\end{aligned}$$

and hence the value of $\mathbf{G}_v(s)$ at $s = j4$ is $3.69\ \underline{/-3.7°}$. ❑

EXTENSION EXERCISES

E14.10 Determine the steady-state voltage $v_{oss}(t)$ in the network in Fig. E14.10 for $t > 0$ if the initial conditions in the network are zero.

ANSWER: $v_{oss}(t) = 3.95 \cos(2t - 99.46°)$ V.

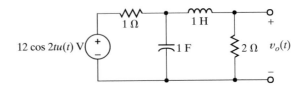

Figure E14.10

14.7 APPLICATIONS

EXAMPLE 14.17

As mentioned in Section 12.8, the Recording Industry Association of America (RIAA) uses standardized recording and playback filters to improve the quality of phonographic disk recordings. This process is demonstrated in Fig. 14.30. During a recording session, the voice or music signal is passed through the recording filter, which deemphasizes the bass content. This filtered signal is then recorded into the vinyl. Upon playback, the phonograph needle assembly senses the recorded message and reproduces the filtered signal, which proceeds to the playback filter. The purpose of the playback filter is to emphasize the bass content and reconstruct the original voice/music signal. Next, the reconstructed signal can be amplified and sent on to the speakers.

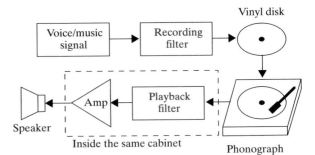

Figure 14.30 Block diagram for phonograph disk recording and playback.

Let us examine the pole-zero diagrams for the record and playback filters.

SOLUTION   The transfer function for the recording filter is

$$\mathbf{G}_{vR}(s) = \frac{K(1 + s\tau_{z1})(1 + s\tau_{z2})}{1 + s\tau_p}$$

where the time constants are $\tau_{z1} = 75$ μs, $\tau_{z2} = 3180$ μs, and $\tau_p = 318$ μs and K is a constant chosen such that $\mathbf{G}_{vR}(s)$ has a magnitude of 1 at 1000 Hz. The resulting pole and zero frequencies in radians/second are

$$\omega_{z1} = 1/\tau_{z1} = 13.33 \text{ kr/s}$$
$$\omega_{z2} = 1/\tau_{z2} = 314.46 \text{ r/s}$$
$$\omega_p = 1/\tau_p = 3.14 \text{ kr/s}$$

Fig. 14.31a shows the pole-zero diagram for the recording filter.

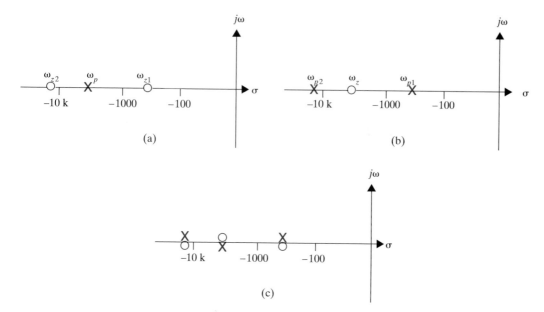

Figure 14.31 Pole-zero diagrams for RIAA phonographic filters.

The playback filter transfer function is the reciprocal of the record transfer function.

$$\mathbf{G}_{vP}(s) = \frac{1}{\mathbf{G}_{vR}(s)} = \frac{A_o(1 + s\tau_z)}{(1 + s\tau_{p1})(1 + s\tau_{p2})}$$

where the time constants are now $\tau_{p1} = 75$ µs, $\tau_{p2} = 3180$ µs, $\tau_z = 318$ µs and A_o is $1/K$. Pole and zero frequencies, in radians/second, are

$$\omega_{p1} = 1/\tau_{z1} = 13.33 \text{ krad/s}$$

$$\omega_{p2} = 1/\tau_{z2} = 314.46 \text{ krad/s}$$

$$\omega_z = 1/\tau_p = 3.14 \text{ krad/s}$$

which yields the pole-zero diagram in Fig. 14.31b. The voice/music signal eventually passes through both filters before proceeding to the amplifier. In the s-domain, this is equivalent to multiplying $\mathbf{V}_s(s)$ by both $\mathbf{G}_{vR}(s)$, and $\mathbf{G}_{vP}(s)$. In the pole-zero diagram, we simply superimpose the pole-zero diagrams of the two filters, as shown in Fig. 14.31c. Notice that at each pole frequency there is a zero and vice versa. The pole-zero pairs cancel one another, yielding a pole-zero diagram that contains no poles and no zeros. This effect can be seen mathematically by multiplying the two transfer functions, $\mathbf{G}_{vR}(s)\mathbf{G}_{vP}(s)$, which yields a product independent of s. Thus, the original voice/music signal is reconstructed and fidelity is preserved.   ❑

14.8 CIRCUIT DESIGN

EXAMPLE 14.18

In a large computer network, two computers are transferring digital data on a single wire at a rate of 1000 bits/s. The voltage waveform, v_{data}, in Fig. 14.32 shows a possible sequence of bits alternating between "high" and "low" values. Also present in the environment is a source of 100 kHz (628 krad/s) noise, which is corrupting the data.

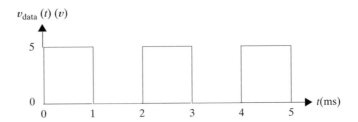

Figure 14.32 1000 bit/s digital data waveform.

It is necessary to filter out the high-frequency noise without destroying the data waveform. Let us place the second-order low-pass active filter of Fig. 14.33 in the data path so that the data and noise signals will pass through it.

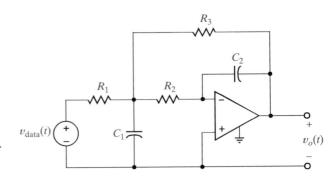

Figure 14.33 Second-order low-pass filter.

SOLUTION   The filter's transfer function is found to be

$$G_v(s) = \frac{V_o(s)}{V_{data}(s)} = \frac{-\left(\dfrac{R_3}{R_1}\right)\left(\dfrac{1}{R_2 R_3 C_1 C_2}\right)}{s^2 + s\left(\dfrac{1}{R_1 C_1} + \dfrac{1}{R_2 C_1} + \dfrac{1}{R_3 C_1}\right) + \dfrac{1}{R_2 R_3 C_1 C_2}}$$

To simplify our work, let $R_1 = R_2 = R_3 = R$. From our work in Chapter 12, we know that the characteristic equation of a second-order system can be expressed

$$s^2 + 2s\zeta\omega_o + \omega_o^2 = 0$$

Comparing the two preceding equations, we find that

$$\omega_o = \frac{1}{R\sqrt{C_1 C_2}}$$

$$2\zeta\omega_o = \frac{3}{RC_1}$$

and therefore,

$$\zeta = \frac{3}{2}\sqrt{\frac{C_2}{C_1}}$$

The poles of the filter are at

$$s_1, s_2 = -\zeta\omega_o \pm \omega_o \sqrt{\zeta^2 - 1}$$

To eliminate the 100-kHz noise, at least one pole should be well below 100 kHz, as shown in the Bode plot sketched in Fig. 14.34. By placing a pole well below 100 kHz, the gain of the filter will be quite small at 100 kHz, effectively filtering the noise.

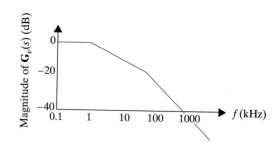

Figure 14.34 Bode plot sketch for a second-order low-pass filter.

If we arbitrarily choose an overdamped system with $\omega_0 = 25$ krad/s and $\zeta = 2$, the resulting filter is overdamped with poles at $s_1 = -6.7$ krad/s and $s_2 = -93.3$ krad/s. The pole-zero diagram for the filter is shown in Fig. 14.35.

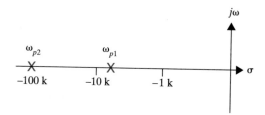

Figure 14.35 Pole-zero diagram for low-pass filter.

If we let $R = 40$ kΩ, then we may write

$$\omega_o = 25,000 = \frac{1}{40,000\sqrt{C_1 C_2}}$$

or

$$C_1 C_2 = 10^{-18}$$

Also,

$$\zeta = 2 = \frac{3}{2}\sqrt{\frac{C_2}{C_1}}$$

which can be expressed

$$\frac{C_2}{C_1} = \frac{16}{9}$$

Solving for C_1 and C_2 yields

$$C_1 = 0.75 \text{ nF}$$
$$C_2 = 1.33 \text{ nF}$$

The circuit used to simulate the filter is shown in Fig. 14.36. The sinusoidal source has a frequency of 100 kHz and is used to represent the noise source.

Plots for the input to the filter and the output voltage for 2 ms are shown in Fig. 14.37. Notice that output indeed contains much less of the 100-kHz noise. Also, the fast rise and fall times of the data signal are slower in the output voltage. Despite this slower response, the output voltage is fast enough to keep pace with the 1000-bits/s transfer rate.

Let us now increase the data transfer rate from 1000 to 25,000 bits/s, as shown in Fig. 14.38. The total input and output signals are plotted in Fig. 14.39 for 200 μs. Now the output cannot keep pace with the input and the data information is lost. Let us investigative why this occurs. We know that the filter is second order with poles at s_1 and s_2. If we represent the data input as a 5-V step function, the output voltage is

$$\mathbf{V}_o(s) = \mathbf{G}_v(s)\left(\frac{5}{s}\right) = \frac{K}{(s + s_1)(s + s_2)}\left(\frac{5}{s}\right)$$

where K is a constant. Since the filter is overdamped, s_1 and s_2 are real and positive. A partial fraction expansion of $\mathbf{V}_o(s)$ is of the form

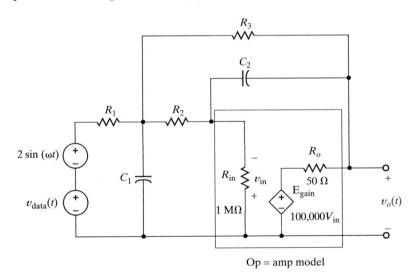

Op = amp model

Figure 14.36 Circuit for second-order filter.

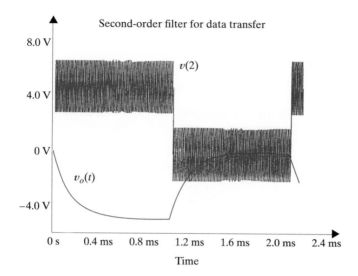

Figure 14.37 Simulation outputs for node 2 and $v_o(t)$.

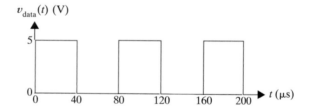

Figure 14.38 25,000-bit/s digital data waveform.

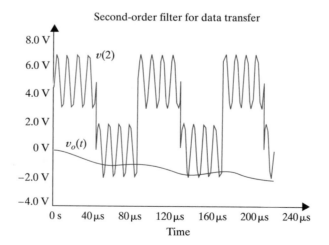

Figure 14.39 Simulation output for node 2 and $v_o(t)$ with 25,000-bit/s data transfer rate.

$$\mathbf{V}_o(s) = \frac{K_1}{s} + \frac{K_2}{(s + s_1)} + \frac{K_3}{(s + s_2)}$$

yielding the time domain expression

$$v_o(t) = \left[K_1 + K_2 e^{-s_1 t} + K_3 e^{-s_2 t} \right] u(t)$$

where K_1, K_2, and K_3 are real constants. The exponential time constants are the reciprocals of the pole frequencies.

$$\tau_1 = \frac{1}{s_1} = \frac{1}{6.7K} = 149 \ \mu s$$

$$\tau_2 = \frac{1}{s_2} = \frac{1}{93.3K} = 10.7 \ \mu s$$

Since exponentials reach steady state in roughly 5τ, the exponential associated with τ_2 affects the output for about 50 μs and the τ_1 exponential will reach steady state after about 750 μs. From Fig. 14.38 we see that at a 25,000-bits/s data transfer rate, each bit (a "high" or "low" voltage value) occupies a 40-μs time span. Therefore, the exponential associated with s_1, and thus $v_o(t)$, is still far from its steady-state condition when the next bit is transmitted. In short, s_1 is too small.

Let us remedy this situation by increasing the pole frequencies and changing to a critically damped system, $\zeta = 1$. If we select $\omega_o = 125$ krad/s, the poles will be at $s_1 = s_2 = -125$ krad/s or 19.9 kHz—both below the 100-kHz noise we wish to filter out. Figure 14.40 shows the new pole positions moved to the left of their earlier positions, which we expect will result in a quicker response to the v_{data} pulse train.

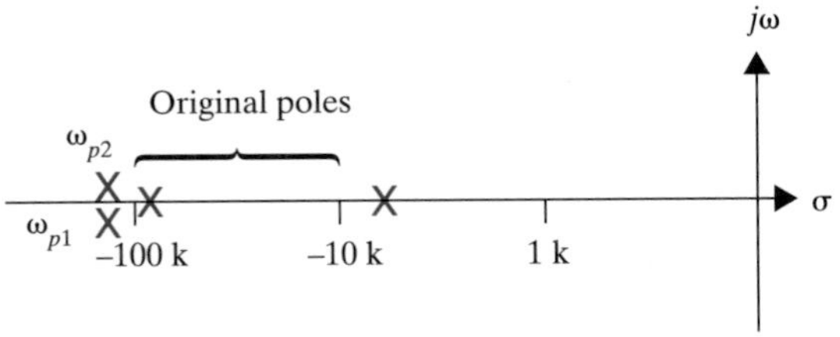

Figure 14.40 Pole-zero diagram for both original and critically damped systems.

Now the expressions for ω_o and ζ are

$$\omega_o = 125,000 = \frac{1}{40,000 \sqrt{C_1 C_2}}$$

or

$$C_1 C_2 = 4 \times 10^{-20}$$

Also,

$$\zeta = 1 = \frac{3}{2} \sqrt{\frac{C_2}{C_1}}$$

which can be expressed

$$\frac{C_2}{C_1} = \frac{4}{9}$$

Solving for C_1 and C_2 yields

$$C_1 = 300 \text{ pF}$$
$$C_2 = 133.3 \text{ pF}$$

A simulation using these new capacitor values produces the input-output data shown in Fig. 14.41. Now the output voltage just reaches the "high" and "low" levels just before v_{data} makes its next transition and the 100-kHz noise is still much reduced.

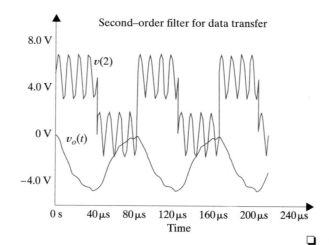

Figure 14.41 Simulation outputs for node 2 and $v_o(t)$ for the critically damped system.

14.9 SUMMARY

- The use of s-domain models for circuit elements permits us to describe them with algebraic, rather than differential, equations.

- All the dc analysis techniques, including the network theorems, are applicable in the s-domain. Once the s-domain solution is obtained, the inverse transform is used to obtain a time domain solution.

- The dc properties of the storage elements, L and C, can be used to obtain initial and final conditions. The initial conditions are required as a part of the s-domain model, and final conditions are often useful in verifying a solution.

- The Laplace transform solution for the network response is composed of transient terms, which disappear as $t \rightarrow \infty$, and steady-state terms, which are present at all times.

- The roots of the network's characteristic equation (i.e., the poles) determine the type of network response. A plot of these roots in the left half of the s-plane provides an immediate indication of the network's behavior. The relationship between the pole-zero plot and the Bode plot provides further insight.

- The transfer (network) function for a network is expressed as

$$\mathbf{H}(s) = \frac{\mathbf{Y}(s)}{\mathbf{X}(s)}$$

where $\mathbf{Y}(s)$ is the network response and $\mathbf{X}(s)$ is the input forcing function. If the transfer function is known, the output response is simply given by the product $\mathbf{H}(s)\mathbf{X}(s)$. If the input is an impulse function so that $\mathbf{X}(s) = 1$, the impulse response is equal to the inverse Laplace transform of the network function.

- The network response can be expressed as

$$\mathbf{Y}(s) = \mathbf{H}(s)\mathbf{X}(s)$$

- The transient portion of the response $\mathbf{Y}(s)$ results from the poles of $\mathbf{H}(s)$, and the steady-state portion of the response results from the poles of the forcing function $\mathbf{X}(s)$.

PROBLEMS

Section 14.3

14.1   Given the network in Fig. P14.1, determine the value of the output voltage as $t \to \infty$.

See Examples 14.1, 14.3, and 14.5.

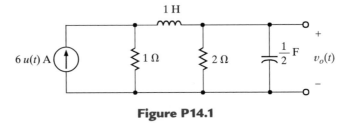

6 $u(t)$ A   1 H   1 Ω   2 Ω $\frac{1}{2}$ F $v_o(t)$

Figure P14.1

14.2   For the network shown in Fig. P14.2, determine the value of the output voltage as $t \to \infty$.

Similar to P14.1

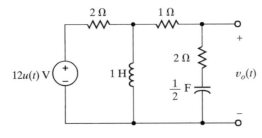

12$u(t)$ V   2 Ω   1 Ω   2 Ω   1 H $\frac{1}{2}$ F $v_o(t)$

Figure P14.2

14.3 For the network shown in Fig. P14.3, determine the output voltage $v_o(t)$ as $t \to \infty$.

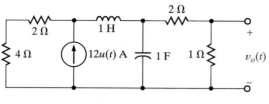

Figure P14.3

Similar to P14.1

14.4 Given the network shown in Fig. P14.4, determine the value of the output voltage $v_o(t)$ as $t \to \infty$.

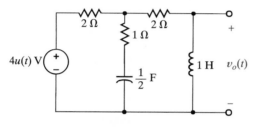

Figure P14.4

Similar to P14.1

14.5 Use Laplace transforms to find $v(t)$ for $t > 0$ in the network shown in Fig. P14.5. Assume zero initial conditions.

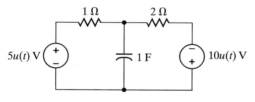

Figure P14.5

Similar to Example 14.3

14.6 Use Laplace transforms and node analysis to find $i_1(t)$ for $t > 0$ in the network shown in Fig. P14.6. Assume zero initial conditions.

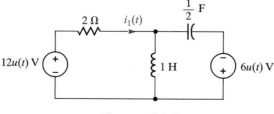

Figure P14.6

Similar to P14.5

14.7   For the network in Fig. P14.7, find $v(t)$ for $t > 0$.

Similar to P14.5

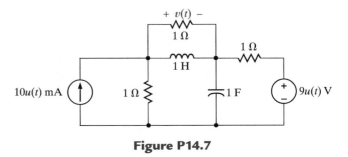

Figure P14.7

14.8   For the network shown in Fig. P14.8, find $v_o(t), t > 0$, using node equations.

Similar to P14.5

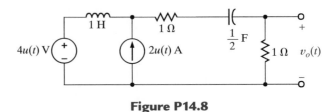

Figure P14.8

14.9   Find $v_o(t), t > 0$, in the network in Fig. P14.9 using node equations.

Similar to Example 14.5

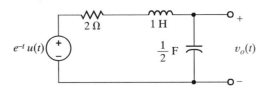

Figure P14.9

14.10   Find $v_o(t), t > 0$, in the network shown in Fig. P14.10 using node equations.

Similar to P14.9

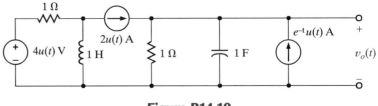

Figure P14.10

14.11 Use mesh equations to find $v_o(t), t > 0,$ in the network in Fig. P14.11.

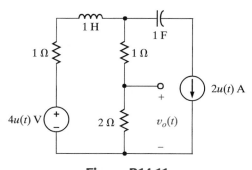

Figure P14.11

Similar to P14.5

14.12 For the network shown in Fig. P14.12, find $v_o(t), t > 0,$ using loop equations.

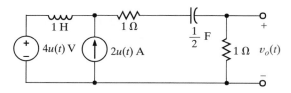

Figure P14.12

Similar to Example 14.5

14.13 For the network shown in Fig. P14.13, find $v_o(t), t > 0,$ using mesh equations.

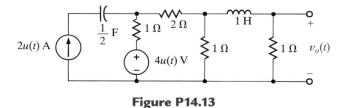

Figure P14.13

Similar to Example 14.5

14.14 Given the network in Fig. P14.14, find $i_o(t), t > 0,$ using mesh equations.

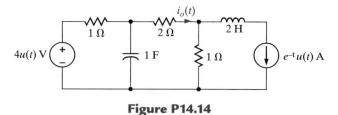

Figure P14.14

Similar to Example 14.5

14.15 Use loop equations to find $i_o(t), t > 0,$
in the network shown in Fig. P14.15.

Similar to Example 14.5

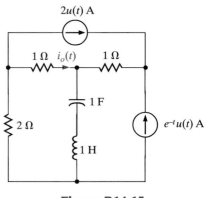

Figure P14.15

14.16 Use Source Transformation to find $v_o(t), t > 0$ in the circuit in Fig. P14.16.

Convert the voltage source to a
current source.

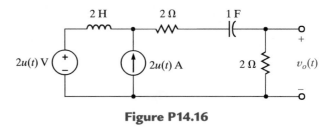

Figure P14.16

Remove the load. Find $V_{oc}(s)$
and $Z_{Th}(s)$. Form a Thévenin
equivalent circuit to find $V_o(s)$.
Perform the inverse transform.

14.17 Use Thévenin's theorem to solve Problem 14.16.

14.18 Use Thévenin's theorem to find $v_o(t), t > 0$, in the network in Fig. P14.18.

Similar to P14.17

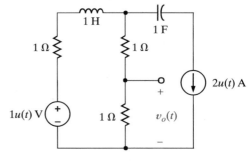

Figure P14.18

14.19 Use Thévenin's theorem to determine $i_o(t), t > 0$ in the circuit shown in Fig. P14.19.

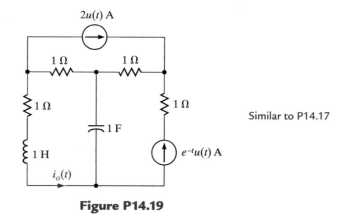

Similar to P14.17

Figure P14.19

14.20 Use Thévenin's theorem to find $v_o(t), t > 0$ in the network in Fig. P14.20.

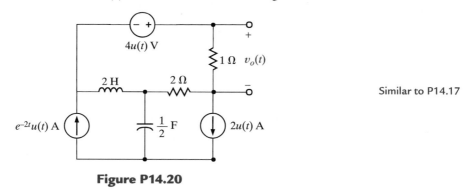

Similar to P14.17

Figure P14.20

14.21 Find $v_o(t)$, for $t > 0$ in the network in Fig. P14.21 using Thévenin's theorem.

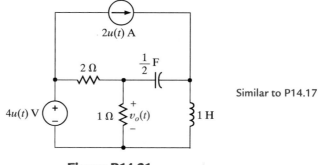

Similar to P14.17

Figure P14.21

14.22 For the network shown in Fig. P14.22, find $v_o(t), t > 0$, using node equations.

See Example 14.4.

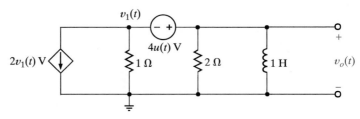

Figure P14.22

14.23 Use loop equations to find $v_o(t), t > 0$, in the network shown in Fig. P14.23.

Similar to P14.22

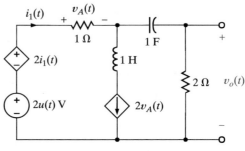

Figure P14.23

14.24 Use Thévenin's theorem to find $v_o(t)$, $t > 0$, in the network shown in Fig. P14.24.

Similar to P14.22

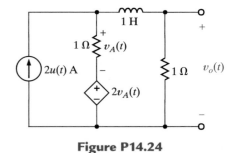

Figure P14.24

14.25 Find $v_o(t), t > 0$, in the network shown in Fig. P14.25 using Laplace transforms. Assume that the circuit has reached steady state at $t = 0-$.

See Example 14.8.

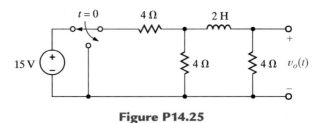

Figure P14.25

14.26 Find $i_o(t), t > 0$, in the network shown in Fig. P14.26.

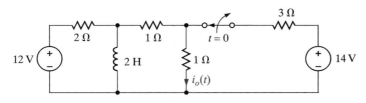

Figure P14.26

Similar to P14.25

14.27 Find $v_o(t), t > 0$, in the circuit shown in Fig. P14.27.

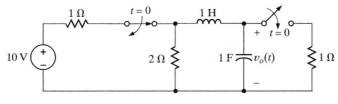

Figure P14.27

Similar to P14.25

14.28 Find $i_o(t), t > 0$, in the network shown in Fig. P14.28.

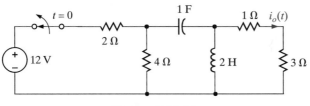

Figure P14.28

Similar to P14.25.

14.29 Find $i_o(t), t > 0$, in the network shown in Fig. P14.29.

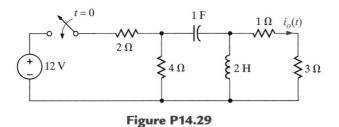

Figure P14.29

Similar to P14.25

14.30 Find $i_o(t), t > 0$, in the network in Fig. P14.30.

Similar to P14.25

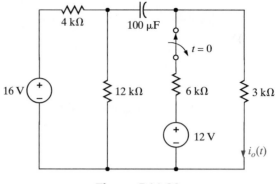

Figure P14.30

14.31 Find $v_o(t)$ for $t > 0$, in the network in Fig. P14.31.

Similar to P14.25

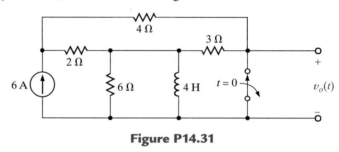

Figure P14.31

14.32 Find $v_o(t)$ for $t > 0$, in the network in Fig. P14.32.

See Example 14.8.

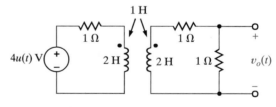

Figure P14.32

14.33 Find $i_o(t)$, for $t > 0$, in the network in Fig. P14.33.

Transform the primary side of the transformer to the secondary and solve for $i_o(t)$.

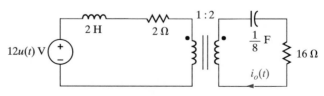

Figure P14.33

14.34 Find $v_o(t)$, for $t > 0$, in the network in Fig. P14.34.

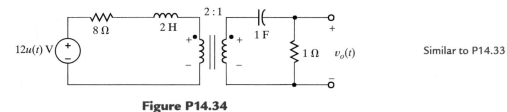

Figure P14.34

Similar to P14.33

14.35 Find $v_o(t)$ for $t > 0$, in the network in Fig. P14.35.

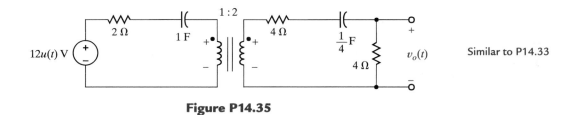

Figure P14.35

Similar to P14.33

14.36 Find $v_o(t)$ for $t > 0$ in the network in Fig. P14.36.

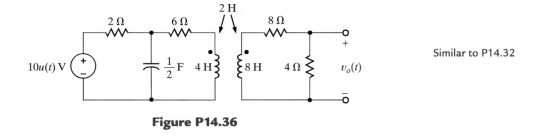

Figure P14.36

Similar to P14.32

14.37 Determine the initial and final values of the voltage $v_o(t)$ in the network in Fig. P14.37.

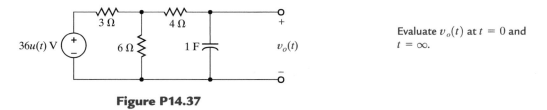

Figure P14.37

Evaluate $v_o(t)$ at $t = 0$ and $t = \infty$.

14.38 Determine the initial and final values of the voltage $v_o(t)$ in the network in Fig. P14.38.

Similar to P14.37

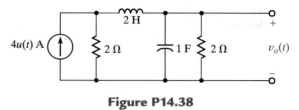

Figure P14.38

14.39 Find the initial and final values of the voltage $v_o(t)$ in the network in Fig. P14.39.

Similar to P14.37

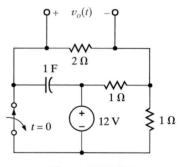

Figure P14.39

14.40 Find the initial and final values of the current $i_o(t)$ in the network in Fig. P14.40.

Similar to P14.37

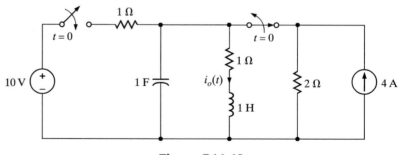

Figure P14.40

14.41 Determine the output voltage $v_o(t)$ in the network in Fig. P14.41a if the input is given by the source in Fig. P14.41b.

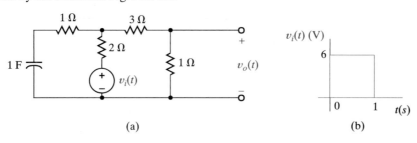

(a)   (b)

Figure P14.41

See Example 14.9.

14.42 Find the output voltage $v_o(t), t > 0$, in the network in Fig. P14.42a if the input is represented by the waveform shown in Fig. P14.42b.

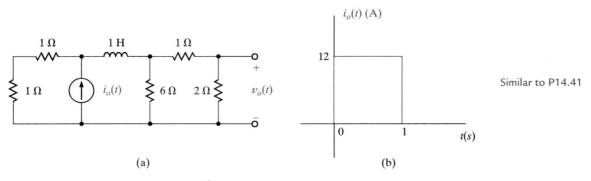

(a)   (b)

Figure P14.42

Similar to P14.41

Section 14.4

14.43 Find the transfer function $\mathbf{V}_o(s)/\mathbf{V}_i(s)$ for the network shown in Fig. P14.43.

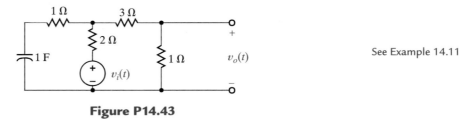

Figure P14.43

See Example 14.11

14.44 Determine the transfer function $\mathbf{I}_o(s)/\mathbf{V}_i(s)$ for the network shown in Fig. P14.44.

Similar to P14.43

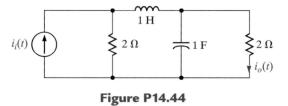

Figure P14.44

14.45 Find the transfer function for the network in Fig. P14.45.

See Example 14.12.

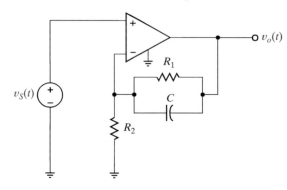

Figure P14.45

14.46 Find the transfer function for the network shown in Fig. P14.46.

Similar to P14.45

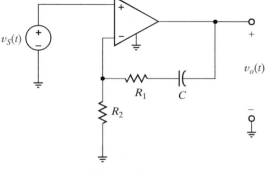

Figure P14.46

14.47 Find the transfer function for the network shown in Fig. P14.47.

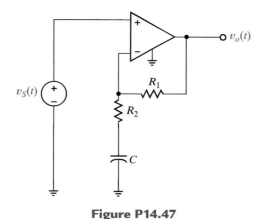

Figure P14.47

Similar to P14.45

14.48 The voltage response of the network to a unit step input is

$$\mathbf{V}_o(s) = \frac{2(s+1)}{s(s^2+10s+25)}$$

Is the response overdamped?

Compute the damping ratio.

14.49 The transfer function of the network is given by the expression

$$\mathbf{G}(s) = \frac{100s}{s^2+13s+40}$$

Determine the damping ratio, the undamped natural frequency, and the type of response that will be exhibited by the network.

Compare the denominator to the characteristic equation in standard form.

14.50 The current response of a network to a unit step input is

$$\mathbf{I}_o(s) = \frac{10(s+2)}{s^2(s^2+12s+30)}$$

Is the response underdamped?

Similar to P14.48

14.51 The voltage response of a network to a unit step input is

$$\mathbf{V}_o(s) = \frac{10}{s(s^2+8s+18)}$$

Is the response critically damped?

Similar to P14.48

Section 14.6

14.52 Find the steady-state response $v_o(t)$ for the network in Fig. P14.52.

See Example 14.15.

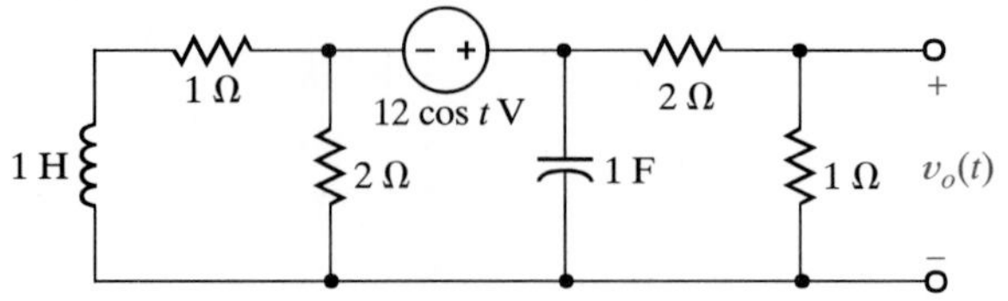

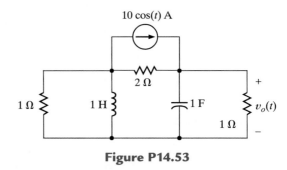

Figure P14.52

14.53 Find the steady-state response $v_o(t)$ for the circuit shown in Fig. P14.53.

Similar to P14.52

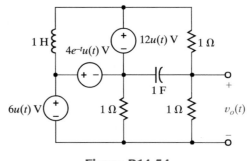

Figure P14.53

Advanced Problems

14.54 Find $v_o(t), t > 0$, in the network in Fig. P14.54 using nodal analysis.

Figure P14.54

14.55 Find $v_o(t), t > 0$, in the network in Fig. P14.55.

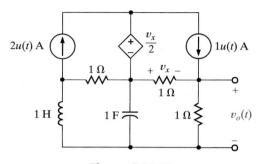

Figure P14.55

14.56 Find $v_o(t)$, for $t > 0$, in the network in Fig. P14.56.

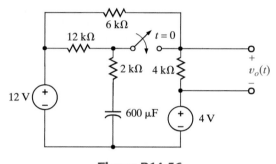

Figure P14.56

14.57 Find $v_o(t)$, for $t > 0$, in the network in Fig. P14.57.

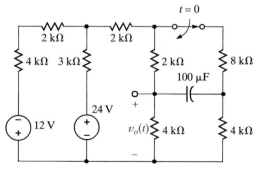

Figure P14.57

14.58 Determine the output voltage, $v_o(t)$, in the circuit in Fig. P14.58a if the input is given by the source described in Fig. P14.58b.

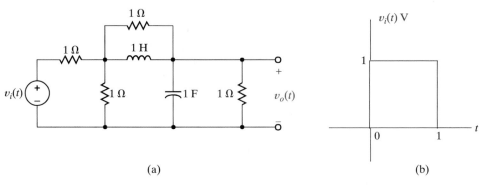

(a)   (b)

Figure P14.58

14.59 Determine the transfer function for the network shown in Fig. P14.59. If a step function is applied to the network, what type of damping will the network exhibit?

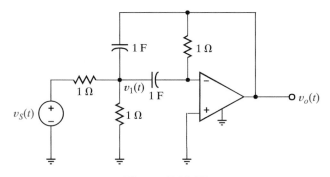

Figure P14.59

14.60 Find the transfer function for the network in Fig. P14.60 If a step function is applied to the network, will the response be overdamped, underdamped, or critically damped?

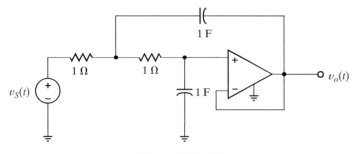

Figure P14.60

14.61 Determine the steady-state response $v_o(t)$ for the network in Fig. P14.61.

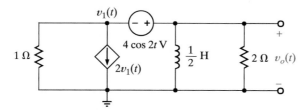

Figure P14.61

14.62 Find the steady-state response $i_o(t)$ for the network shown in Fig. P14.62.

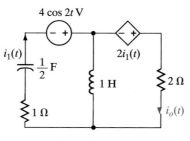

Figure P14.62

14.63 Determine the steady-state response $i_o(t)$ for the network in Fig. P14.63.

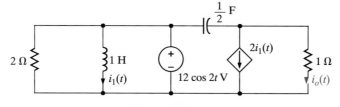

Figure P14.63

14.64 Find the steady-state response $v_o(t)$, for $t > 0$, in the network in Fig. P14.64.

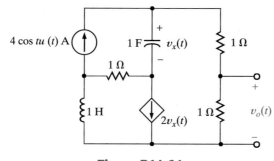

Figure P14.64

14.65 Find the steady-state response $v_o(t)$, for $t > 0$, in the network in Fig. P14.65.

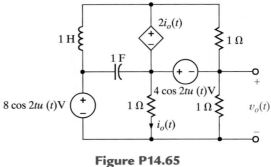

Figure P14.65

14.66 For the network in Fig. P14.66, choose the value of C for critical damping.

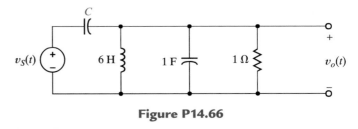

Figure P14.66

14.67 For the filter in Fig. P14.67, choose the values of C_1 and C_2 to place poles at $s = -2$ and $s = -5$ rad/s.

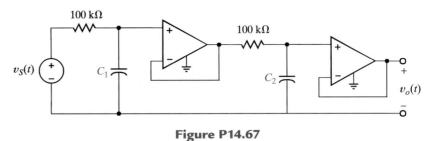

Figure P14.67

Fourier Analysis Techniques

15

In this chapter we examine two very important topics: the Fourier series and the Fourier transform. These two techniques vastly expand our circuit analysis capabilities because they provide a means of effectively dealing with nonsinusoidal periodic signals and aperiodic signals. Using the Fourier series, we show that we can determine the steady-state response of a network to a nonsinusoidal periodic input. The Fourier transform will allow us to analyze circuits with aperiodic inputs by transforming the problem to the frequency domain, solving it algebraically, and then transforming back to the time domain in a manner similar to that used with Laplace transforms.

15.1 FOURIER SERIES

A periodic function is one that satisfies the relationship

$$f(t) = f(t + nT_0), \quad n = \pm 1, \pm 2, \pm 3, \ldots$$

for every value of t where T_0 is the period. As we have shown in previous chapters, the sinusoidal function is a very important periodic function. However, there are many other periodic functions that have wide applications. For example, laboratory signal generators produce the pulse-train and square-wave signals shown in Figs. 15.1a and b, respectively, which are used for testing circuits. The oscilloscope is another laboratory instrument and the sweep of its electron beam across the face of the cathode ray tube is controlled by a triangular signal of the form shown in Fig. 15.1c.

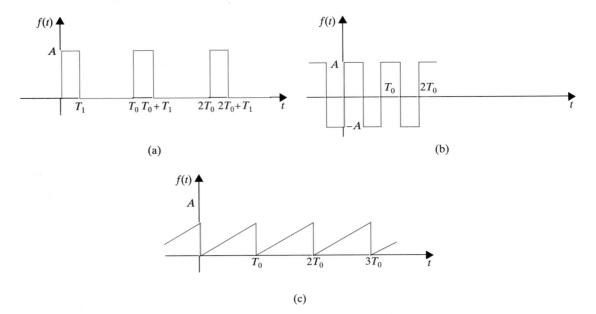

(a)

(b)

(c)

Figure 15.1 Some useful periodic signals.

The techniques we will explore are based on the work of Jean Baptiste Joseph Fourier. Although our analyses will be confined to electric circuits, it is important to point out that the techniques are applicable to a wide range of engineering problems. In fact, it was Fourier's work in heat flow that led to the techniques that will be presented here.

In his work, Fourier demonstrated that a periodic function $f(t)$ could be expressed as a sum of sinusoidal functions. Therefore, given this fact and the fact that if a periodic function is expressed as a sum of linearly independent functions, each function in the sum must be periodic with the same period, and the function $f(t)$ can be expressed in the form

$$f(t) = a_0 + \sum_{n=1}^{\infty} D_n \cos(n\omega_0 t + \theta_n) \qquad \textbf{15.1}$$

where $\omega_0 = 2\pi/T_0$ and a_0 is the average value of the waveform. An examination of this expression illustrates that all sinusoidal waveforms that are periodic with period T_0 have been included. For example, for $n = 1$, one cycle covers T_0 seconds and $D_1 \cos(\omega_0 t + \theta_1)$ is called the *fundamental*. For $n = 2$, two cycles fall within T_0 seconds and the term $D_2 \cos(2\omega_0 t + \theta_2)$ is called the *second harmonic*. In general, for $n = k, k$ cycles fall within T_0 seconds and $D_k \cos(k\omega_0 t + \theta_k)$ is the *kth harmonic term*.

Since the function $\cos(n\omega_0 t + \theta_k)$ can be written in exponential form using Euler's identity or as a sum of cosine and sine terms of the form $\cos n\omega_0 t$ and $\sin n\omega_0 t$ as demonstrated in Chapter 8, the series in Eq. (15.1) can be written as

$$f(t) = a_0 + \sum_{\substack{n=-\infty \\ n\neq 0}}^{\infty} \mathbf{c}_n e^{jn\omega_0 t} = \sum_{n=-\infty}^{\infty} \mathbf{c}_n e^{jn\omega_0 t} \qquad \textbf{15.2}$$

Using the real-part relationship employed as a transformation between the time domain and the frequency domain, we can express $f(t)$ as

$$f(t) = a_0 + \sum_{n=1}^{\infty} \text{Re}\left[\left(D_n \underline{/\theta_n}\right)e^{jn\omega_0 t}\right] \qquad \textbf{15.3}$$

$$= a_0 + \sum_{n=1}^{\infty} \text{Re}\left(2\mathbf{c}_n e^{jn\omega_0 t}\right) \qquad \textbf{15.4}$$

$$= a_0 + \sum_{n=1}^{\infty} \text{Re}\left[\left(a_n - jb_n\right)e^{jn\omega_0 t}\right] \qquad \textbf{15.5}$$

$$= a_0 + \sum_{n=1}^{\infty} a_n \cos n\omega_0 t + b_n \sin n\omega_0 t \qquad \textbf{15.6}$$

These equations allow us to write the Fourier series in a number of equivalent forms. Note that the *phasor* for the *n*th harmonic is

$$D_n \underline{/\theta_n} = 2\mathbf{c}_n = a_n - jb_n \qquad \textbf{15.7}$$

The approach we will take will be to represent a nonsinusoidal periodic input by a sum of complex exponential functions, which because of Euler's identity is equivalent to a sum of sines and cosines. We will then use (1) the superposition property of linear systems and (2) our knowledge that the steady-state response of a time-invariant linear system to a sinusoidal input of frequency ω_0 is a sinusoidal function of the same frequency to determine the response of such a system.

In order to illustrate the manner in which a nonsinusoidal periodic signal can be represented by a Fourier series, consider the periodic function shown in Fig. 15.2a. In Figs. 15.2b–d we can see the impact of using a specific number of terms in the series to represent the original function. Note that the series more closely represents the original function as we employ more and more terms.

Exponential Fourier Series

Any physically realizable periodic signal may be represented over the interval $t_1 < t < t_1 + T_0$ by the *exponential Fourier series*

$$f(t) = \sum_{n=-\infty}^{\infty} \mathbf{c}_n e^{jn\omega_0 t} \qquad \textbf{15.8}$$

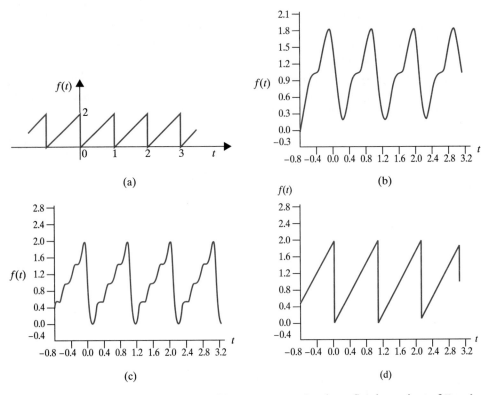

Figure 15.2 Periodic function (a) and its representation by a fixed number of Fourier series terms; (b) 2 terms; (c) 4 terms; (d) 100 terms.

where the $\mathbf{c}_n$ are the complex (phasor) Fourier coefficients. These coefficients are derived as follows. Multiplying both sides of Eq. (15.8) by $e^{-jk\omega_0 t}$ and integrating over the interval t_1 to $t_1 + T_0$, we obtain

$$\int_{t_1}^{t_1+T_0} f(t)e^{-jk\omega_0 t}\, dt = \int_{t_1}^{t_1+T_0}\left(\sum_{n=-\infty}^{\infty} \mathbf{c}_n e^{jn\omega_0 t}\right)e^{-jk\omega_0 t}\, dt$$

$$= \mathbf{c}_k T_0$$

since

$$\int_{t_1}^{t_1+T_0} e^{j(n-k)\omega_0 t}\, dt = \begin{cases} 0 & \text{for } n \neq k \\ T_0 & \text{for } n = k \end{cases}$$

Therefore, the Fourier coefficients are defined by the equation

$$\mathbf{c}_n = \frac{1}{T_0}\int_{t_1}^{t_1+T_0} f(t)e^{-jn\omega_0 t}\, dt \qquad\qquad \textbf{15.9}$$

The following example illustrates the manner in which we can represent a periodic signal by an exponential Fourier series.

EXAMPLE 15.1

We wish to determine the exponential Fourier series for the periodic voltage waveform shown in Fig. 15.3.

SOLUTION   The Fourier coefficients are determined using Eq. (15.9) by integrating over one complete period of the waveform.

$$
\mathbf{c}_n = \frac{1}{T} \int_{-T/2}^{T/2} f(t) e^{-jn\omega_0 t} \, dt
$$

$$
= \frac{1}{T} \int_{-T/2}^{-T/4} -V e^{-jn\omega_0 t} \, dt + \int_{-T/4}^{T/4} V e^{-jn\omega_0 t} \, dt + \int_{T/4}^{T/2} -V e^{-jn\omega_0 t} \, dt
$$

$$
= \frac{V}{jn\omega_0 T} \left[+e^{-jn\omega_0 t}\Big|_{-T/2}^{-T/4} - e^{-jn\omega_0 t}\Big|_{-T/4}^{T/4} + e^{-jn\omega_0 t}\Big|_{T/4}^{T/2} \right]
$$

$$
= \frac{\mathbf{V}}{jn\omega_0 T} \left(2e^{jn\pi/2} - 2e^{-jn\pi/2} + e^{jn\pi} - e^{+jn\pi} \right)
$$

$$
= \frac{V}{n\omega_0 T} \left[4 \sin \frac{n\pi}{2} - 2 \sin (n\pi) \right]
$$

$$
= 0 \quad \text{for } n \text{ even}
$$

$$
= \frac{2V}{n\pi} \sin \frac{n\pi}{2} \quad \text{for } n \text{ odd}
$$

$\mathbf{c}_0$ corresponds to the average value of the waveform. This term can be evaluated using the original equation for $\mathbf{c}_n$. Therefore,

$$
c_0 = \frac{1}{T} \int_{-\frac{T}{2}}^{\frac{T}{2}} v(t) \, dt
$$

$$
= \frac{1}{T} \left[\int_{-\frac{T}{2}}^{-\frac{T}{4}} -V \, dt + \int_{-\frac{T}{4}}^{\frac{T}{4}} V \, dt + \int_{\frac{T}{4}}^{\frac{T}{2}} -V \, dt \right]
$$

$$
= \frac{1}{T} \left[-\frac{VT}{4} + \frac{VT}{2} - \frac{VT}{4} \right] = 0
$$

Therefore,

$$
v(t) = \sum_{\substack{n=-\infty \\ n \neq 0 \\ n \text{ odd}}}^{\infty} \frac{2V}{n\pi} \sin \frac{n\pi}{2} e^{jn\omega_0 t}
$$

This equation can be written as

$$
v(t) = \sum_{\substack{n=1 \\ n \text{ odd}}}^{\infty} \frac{2V}{n\pi} \sin \frac{n\pi}{2} e^{jn\omega_0 t} + \sum_{\substack{n=-1 \\ n \text{ odd}}}^{-\infty} \frac{2V}{n\pi} \sin \frac{n\pi}{2} e^{jn\omega_0 t}
$$

$$
= \sum_{\substack{n=1 \\ n \text{ odd}}}^{\infty} \left(\frac{2V}{n\pi} \sin \frac{n\pi}{2} \right) e^{jn\omega_0 t} + \left(\frac{2V}{n\pi} \sin \frac{n\pi}{2} \right)^{*} e^{-jn\omega_0 t}
$$

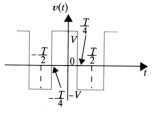

Figure 15.3 Periodic voltage waveform.

Since a number plus its complex conjugate is equal to two times the real part of the number, $v(t)$ can be written as

$$v(t) = \sum_{\substack{n=1 \\ n \text{ odd}}}^{\infty} 2 \operatorname{Re}\left(\frac{2V}{n\pi} \sin \frac{n\pi}{2} e^{jn\omega_0 t}\right)$$

or

$$v(t) = \sum_{\substack{n=1 \\ n \text{ odd}}}^{\infty} \frac{4V}{n\pi} \sin \frac{n\pi}{2} \cos n\omega_0 t$$

Note that this same result could have been obtained by integrating over the interval $-T/4$ to $3T/4$.   ❏

EXTENSION EXERCISES

E15.1   Find the Fourier coefficients for the waveform in Fig. E15.1.

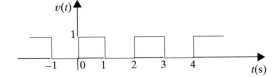

Figure E15.1

ANSWER: $c_n = \dfrac{1 - e^{-jn\pi}}{j2\pi n}$, $c_0 = \dfrac{1}{2}$.

E15.2   Find the Fourier coefficients for the waveform in Fig. E15.2.

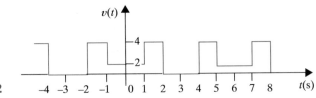

Figure E15.2

ANSWER: $c_n = \dfrac{2}{n\pi}\left(2 \sin \dfrac{2\pi n}{3} - \sin \dfrac{n\pi}{3}\right)$, $c_0 = 2$.

Trigonometric Fourier Series

Let us now examine another form of the Fourier series. Since

$$2c_n = a_n - jb_n \qquad \textbf{15.10}$$

we will examine this quantity $2c_n$ and separate it into its real and imaginary parts. Using Eq. (15.9), we find that

$$2\mathbf{c}_n = \frac{2}{T_0} \int_{t_1}^{t_1+T_0} f(t) e^{-jn\omega_0 t}\, dt \qquad\qquad \textbf{15.11}$$

Using Euler's identity, we can write this equation in the form

$$2\mathbf{c}_n = \frac{2}{T_0} = \int_{t_1}^{t_1+T_0} f(t)\big(\cos n\omega_0 t - j \sin n\omega_0 t\big)\, dt$$

$$= \frac{2}{T_0} \int_{t_1}^{t_1+T_0} f(t)\cos n\omega_0 t\, dt - j\frac{2}{T_0} \int_{t_1}^{t_1+T_0} f(t)\sin n\omega_0 t\, dt$$

From Eq. (15.10) we note then that

$$a_n = \frac{2}{T_0} \int_{t_1}^{t_1+T_0} f(t) \cos n\omega_0 t\, dt \qquad\qquad \textbf{15.12}$$

$$b_n = \frac{2}{T_0} \int_{t_1}^{t_1+T_0} f(t) \sin n\omega_0 t\, dt \qquad\qquad \textbf{15.13}$$

These are the coefficients of the Fourier series described by Eq. (15.3), which we call the *trigonometric Fourier series*. These equations are derived directly in most textbooks using the orthogonality properties of the cosine and sine functions. Note that we can now evaluate $\mathbf{c}_n, a_n, b_n$, and since

$$2\mathbf{c}_n = D_n \underline{/\theta_n} \qquad\qquad \textbf{15.14}$$

we can derive the coefficients for the *cosine Fourier series* described by Eq. (15.1). This form of the Fourier series is particularly useful because it allows us to represent each harmonic of the function as a phasor.

From Eq. (15.9) we note that $\mathbf{c}_0$, which is written as a_0, is

$$a_0 = \frac{1}{T} \int_{t_1}^{t_1+T_0} f(t)\, dt \qquad\qquad \textbf{15.15}$$

This is the average value of the signal $f(t)$ and can often be evaluated directly from the waveform.

Symmetry and the Trigonometric Fourier Series

If a signal exhibits certain symmetrical properties, we can take advantage of these properties to simplify the calculations of the Fourier coefficients. There are three types of symmetry: (1) even-function symmetry, (2) odd-function symmetry, and (3) half-wave symmetry.

Even-Function Symmetry.   A function is said to be even if

$$f(t) = f(-t) \qquad\qquad \textbf{15.16}$$

An even function is symmetrical about the vertical axis and a notable example is the function $\cos n\omega_0 t$. Note that the waveform in Fig. 15.3 also exhibits even-function symmetry. Let us now determine the expressions for the Fourier coefficients if the function satisfies Eq. (15.16).

If we let $t_1 = -T_0/2$ in Eq. (15.15), we obtain

$$a_0 = \frac{1}{T_0} \int_{-T_0/2}^{T_0/2} f(t) \, dt$$

which can be written as

$$a_0 = \frac{1}{T_0} \int_{-T_0/2}^{0} f(t) \, dt + \frac{1}{T_0} \int_{0}^{T_0/2} f(t) \, dt$$

If we now change the variable on the first integral (i.e., let $t = -x$), then $f(-x) = f(x)$, $dt = -dx$, and the range of integration is from $x = T_0/2$ to 0. Therefore, the preceding equation becomes

$$\begin{aligned}
a_0 &= \frac{1}{T_0} \int_{T_0/2}^{0} f(x)(-dx) + \frac{1}{T_0} \int_{0}^{T_0/2} f(t) \, dt \\
&= \frac{1}{T_0} \int_{0}^{T_0/2} f(x) \, dx + \frac{1}{T_0} \int_{0}^{T_0/2} f(t) \, dt \\
&= \frac{2}{T_0} \int_{0}^{T_0/2} f(t) \, dt
\end{aligned}$$

15.17

The other Fourier coefficients are derived in a similar manner. The a_n coefficient can be written

$$a_n = \frac{2}{T_0} \int_{-T_0/2}^{0} f(t) \cos n\omega_0 t \, dt + \frac{2}{T_0} \int_{0}^{T_0/2} f(t) \cos n\omega_0 t \, dt$$

Employing the change of variable that led to Eq. (15.17), we can express the preceding equation as

$$\begin{aligned}
a_n &= \frac{2}{T_0} \int_{T_0/2}^{0} f(x) \cos (-n\omega_0 x)(-dx) + \frac{2}{T_0} \int_{0}^{T_0/2} f(t) \cos n\omega_0 t \, dt \\
&= \frac{2}{T_0} \int_{0}^{T_0/2} f(x) \cos n\omega_0 x \, dx + \frac{2}{T_0} \int_{0}^{T_0/2} f(t) \cos n\omega_0 t \, dt
\end{aligned}$$

$$a_n = \frac{4}{T_0} \int_{0}^{T_0/2} f(t) \cos n\omega_0 t \, dt$$

15.18

Once again, following the preceding development, we can write the equation for the b_n coefficient as

$$b_n = \frac{2}{T_0} \int_{-T_0/2}^{0} f(t) \sin n\omega_0 t \, dt + \int_{0}^{T_0/2} f(t) \sin n\omega_0 t \, dt$$

The variable change employed previously yields

$$\begin{aligned}
b_n &= \frac{2}{T_0} \int_{T_0/2}^{0} f(x) \sin(-n\omega_0 x)(-dx) + \frac{2}{T_0} \int_{0}^{T_0/2} f(t) \sin n\omega_0 t \, dt \\
&= \frac{-2}{T_0} \int_{0}^{T_0/2} f(x) \sin n\omega_0 x \, dx + \frac{2}{T_0} \int_{0}^{T_0/2} f(t) \sin n\omega_0 t \, dt
\end{aligned}$$

$$b_n = 0 \qquad\qquad\qquad\qquad\text{15.19}$$

The preceding analysis indicates that the Fourier series for an even periodic function consists only of a constant term and cosine terms. Therefore, if $f(t)$ is even, $b_n = 0$ and from Eqs. (15.10) and (15.14), $\mathbf{c}_n$ are real and θ_n are multiples of $180°$.

Odd-Function Symmetry. A function is said to be odd if

$$f(t) = -f(-t) \qquad\qquad\qquad\qquad\text{15.20}$$

An example of an odd function is $\sin n\omega_0 t$. Another example is the waveform in Fig. 15.4a. Following the mathematical development that led to Eqs. (15.17) to (15.19), we can show that for an odd function the Fourier coefficients are

$$a_0 = 0 \qquad\qquad\qquad\qquad\text{15.21}$$
$$a_n = 0 \qquad \text{for all } n > 0 \qquad\qquad\text{15.22}$$
$$b_n = \frac{4}{T_0} \int_0^{T_0/2} f(t) \sin n\omega_0 t \, dt \qquad\qquad\text{15.23}$$

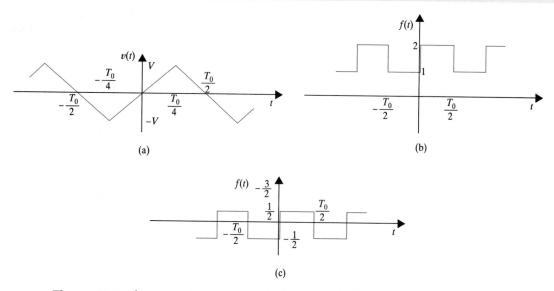

(a)

(b)

(c)

Figure 15.4 Three waveforms; (a) and (c) possess half-wave symmetry.

Therefore, if $f(t)$ is odd, $a_n = 0$ and, from Eqs. (15.10) and (15.14), $\mathbf{c}_n$ are pure imaginary and θ_n are odd multiples of $90°$.

Half-Wave Symmetry. A function is said to possess *half-wave symmetry* if

$$f(t) = -f\left(t - \frac{T_0}{2}\right) \qquad\qquad\qquad\qquad\text{15.24}$$

Basically, this equation states that each half-cycle is an inverted version of the adjacent half-cycle; that is, if the waveform from $-T_0/2$ to 0 is inverted, it is identical to the waveform from 0 to $T_0/2$. The waveforms shown in Figs. 15.4a and c possess half-wave symmetry.

Once again we can derive the expressions for the Fourier coefficients in this case by repeating the mathematical development that led to the equations for even-function symmetry using the change of variable $t = x + T_0/2$ and Eq. (15.24). The results of this development are the following equations:

$$a_0 = 0 \qquad\qquad\qquad\qquad\qquad\qquad\qquad\qquad \textbf{15.25}$$

$$a_n = b_n = 0 \qquad\qquad\quad \text{for } n \text{ even} \qquad\qquad \textbf{15.26}$$

$$a_n = \frac{4}{T_0} \int_0^{T_0/2} f(t) \cos n\omega_0 t \, dt \quad \text{for } n \text{ odd} \qquad\qquad \textbf{15.27}$$

$$b_n = \frac{4}{T_0} \int_0^{T_0/2} f(t) \sin n\omega_0 t \, dt \quad \text{for } n \text{ odd} \qquad\qquad \textbf{15.28}$$

The following equations are often useful in the evaluation of the trigonometric Fourier series coefficients:

$$\int \sin ax \, dx = -\frac{1}{a} \cos ax$$

$$\int \cos ax \, dx = \frac{1}{a} \sin ax$$

$$\int x \sin ax \, dx = \frac{1}{a^2} \sin ax - \frac{1}{a} x \cos ax$$

$$\int x \cos ax \, dx = \frac{1}{a^2} \cos ax + \frac{1}{a} x \sin ax \qquad\qquad \textbf{15.29}$$

EXAMPLE 15.2

We wish to find the trigonometric Fourier series for the periodic signal in Fig. 15.3.

SOLUTION The waveform exhibits even-function symmetry and therefore

$$a_0 = 0$$
$$b_n = 0 \quad \text{for all } n$$

The waveform exhibits half-wave symmetry and therefore

$$a_n = 0 \quad \text{for } n \text{ even}$$

Hence,

$$
\begin{aligned}
a_n &= \frac{4}{T_0} \int_0^{T/2} f(t) \cos n\omega_0 t \, dt \quad \text{for } n \text{ odd} \\
&= \frac{4}{T} \left(\int_0^{T/4} V \cos n\omega_0 t \, dt - \int_{T/4}^{T/2} V \cos n\omega_0 t \, dt \right) \\
&= \frac{4V}{n\omega_0 T} \left(\sin n\omega_0 t \Big|_0^{T/4} - \sin n\omega_0 t \Big|_{T/4}^{T/2} \right)
\end{aligned}
$$

$$= \frac{4V}{n\omega_0 T} \left(\sin \frac{n\pi}{2} - \sin n\pi + \sin \frac{n\pi}{2} \right)$$

$$= \frac{8V}{n2\pi} \sin \frac{n\pi}{2} \qquad \text{for } n \text{ odd}$$

$$= \frac{4V}{n\pi} \sin \frac{n\pi}{2} \qquad \text{for } n \text{ odd}$$

The reader should compare this result with that obtained in Example 15.1.   ❏

EXAMPLE 15.3

Let us determine the trigonometric Fourier series expansion for the waveform shown in Fig. 15.4a.

SOLUTION   The function not only exhibits odd-function symmetry, but it possesses half-wave symmetry as well. Therefore, it is necessary to determine only the coefficients b_n for n odd. Note that

$$v(t) = \begin{cases} \dfrac{4Vt}{T_0} & 0 \leq t \leq T_0/4 \\[2mm] 2V - \dfrac{4Vt}{T_0} & T_0/4 < t \leq T_0/2 \end{cases}$$

The b_n coefficients are then

$$b_n = \frac{4}{T_0} \int_0^{T_0/4} \frac{4Vt}{T_0} \sin n\omega_0 t \, dt + \frac{4}{T_0} \int_{T_0/4}^{T_0/2} \left(2V - \frac{4Vt}{T_0} \right) \sin n\omega_0 t \, dt$$

The evaluation of these integrals is tedious but straightforward and yields

$$b_n = \frac{8V}{n^2 \pi^2} \sin \frac{n\pi}{2} \qquad \text{for } n \text{ odd}$$

Hence, the Fourier series expansion is

$$v(t) = \sum_{\substack{n=1 \\ n \text{ odd}}}^{\infty} \frac{8V}{n^2 \pi^2} \sin n\omega_0 t \qquad\qquad ❏$$

EXAMPLE 15.4

We wish to find the trigonometric Fourier series expansion of the waveform in Fig. 15.4b.

SOLUTION   Note that this waveform has an average value of $\frac{3}{2}$. Therefore, instead of determining the Fourier series expansion of $f(t)$, we will determine the Fourier series for $f(t) - 3/2$, which is the waveform shown in Fig. 15.4c. The latter waveform possesses half-wave symmetry. The function is also odd and therefore

$$b_n = \frac{4}{T_0} \int_0^{T_0/2} \tfrac{1}{2} \sin n\omega_0 t \, dt$$

$$= \frac{2}{T_0} \left(\frac{-1}{n\omega_0} \cos n\omega_0 t \Big|_0^{T_0/2} \right)$$

$$= \frac{-2}{n\omega_0 T_0} (\cos n\pi - 1)$$

$$= \frac{2}{n\pi} \, n \, \text{odd}$$

Therefore, the Fourier series expansion for $f(t) - \frac{3}{2}$ is

$$f(t) - \tfrac{3}{2} = \sum_{\substack{n=1 \\ n \text{ odd}}}^{\infty} \frac{2}{n\pi} \sin n\omega_0 t$$

or

$$f(t) = \tfrac{3}{2} + \sum_{\substack{n=1 \\ n \text{ odd}}}^{\infty} \frac{2}{n\pi} \sin n\omega_0 t \qquad \square$$

EXTENSION EXERCISES

E15.3   Determine the type of symmetry exhibited by the waveform in Figs. E15.2 and E15.3.

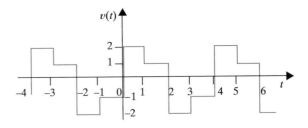

Figure E15.3

ANSWER: Figure E15.2, even symmetry; Fig. E15.3, half-wave symmetry.

E15.4   Find the trigonometric Fourier series for the voltage waveform in Fig. E15.2.

ANSWER: $v(t) = 2 + \sum_{n=1}^{\infty} \frac{4}{n\pi} \left(2 \sin \frac{2\pi n}{3} - \sin \frac{n\pi}{3} \right) \cos \frac{n\pi}{3} t \ \text{V}.$

E15.5   Find the trigonometric Fourier series for the voltage waveform in Fig. E15.3.

ANSWER: $v(t) = \displaystyle\sum_{\substack{n=1 \\ n \text{ odd}}}^{\infty} \frac{2}{n\pi} \sin \frac{n\pi}{2} \cos \frac{n\pi}{2} t$

$$+ \frac{2}{n\pi}(2 - \cos n\pi) \sin \frac{n\pi}{2} t \text{ V.}$$

We have tried to indicate the advantage of recognizing symmetry in a waveform. However, it is interesting for us to note that any signal $f(t)$ can be resolved into two components: an even component $f_e(t)$ and an odd component $f_o(t)$. In other words,

$$f(t) = f_e(t) + f_o(t) \tag{15.30}$$

From this equation we can write

$$f(-t) = f_e(-t) + f_o(-t)$$

However, for an even function $f_e(-t) = f_e(t)$ and for an odd function $f_o(-t) = -f_o(t)$. Hence,

$$f(-t) = f_e(t) - f_o(t) \tag{15.31}$$

Solving Eqs. (15.30) and (15.31) for the even and odd components of the signal yields

$$f_e(t) = \tfrac{1}{2}[f(t) + f(-t)] \tag{15.32}$$
$$f_o(t) = \tfrac{1}{2}[f(t) - f(-t)] \tag{15.33}$$

Let us now indicate how these equations can be used in determining the Fourier coefficients of a periodic waveform.

EXAMPLE 15.5

Let us determine the trigonometric Fourier series coefficients for the waveform in Fig. 15.5a via Eqs. (15.26) and (15.27).

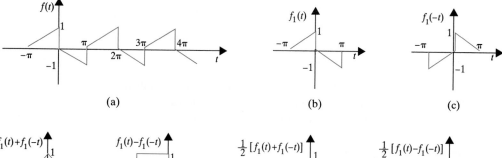

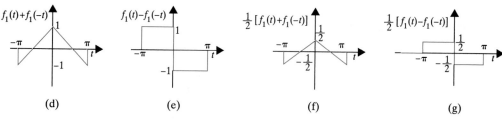

Figure 15.5 Figures used in Example 15.5.

SOLUTION   One period of the function is shown in Fig. 15.5b. Figures 15.5b–g are used to compute the even and odd functions $f_e(t)$ and $f_o(t)$. Note that

$$f_e(t) = \frac{1}{2} - \frac{t}{\pi} \qquad 0 \le t \le \pi$$

and

$$f_o(t) = -\frac{1}{2} \qquad 0 \le t \le \pi$$

From the original waveform we note that $\omega_0 = 2\pi/2\pi = 1$ and the average value of $f(t) = 0$ and hence $a_0 = 0$. For $f_e(t)$,

$$a_n = \frac{4}{2\pi} \int_0^\pi \left(-\frac{t}{\pi} + \frac{1}{2} \right) \cos nt \, dt$$

$$= \frac{-2}{\pi^2} \int_0^\pi t \cos nt \, dt + \frac{1}{\pi} \int_0^\pi \cos nt \, dt$$

$$= \frac{-2}{\pi^2} \left(\frac{1}{n^2} \cos nt + \frac{t}{n} \sin nt \right)_0^\pi + \frac{1}{\pi} \left(\frac{1}{n} \sin nt \right)_0^\pi$$

$$= \begin{cases} 0 & \text{for } n \text{ even} \\ \dfrac{4}{\pi^2 n^2} & \text{for } n \text{ odd} \end{cases}$$

For $f_o(t)$

$$b_n = \frac{4}{2\pi} \int_0^\pi \frac{-1}{2} \sin nt \, dt$$

$$= \frac{-1}{\pi} \int_0^\pi \sin nt \, dt$$

$$= \frac{1}{\pi n} (\cos nt) \Big|_0^\pi$$

$$= \begin{cases} 0 & \text{for } n \text{ even} \\ \dfrac{-2}{\pi n} & \text{for } n \text{ odd} \end{cases}$$

❑

Although the use of the technique employed in the example does not necessarily provide a feasible alternative for calculating the trigonometric Fourier coefficients, it does provide additional insight when examining waveforms for symmetry.

Time Shifting

Let us now examine the effect of time shifting a periodic waveform $f(t)$ defined by the equation

$$f(t) = \sum_{n=-\infty}^{\infty} c_n e^{jn\omega_0 t}$$

Note that

$$f(t - t_0) = \sum_{n=-\infty}^{\infty} \mathbf{c}_n e^{jn\omega_0(t-t_0)}$$

$$f(t - t_0) = \sum_{n=-\infty}^{\infty} \left(\mathbf{c}_n e^{-jn\omega_0 t_0}\right) e^{jn\omega_0 t} \qquad \textbf{15.34}$$

Since $e^{-jn\omega_0 t_0}$ corresponds to a phase shift, the Fourier coefficients of the time-shifted function are the Fourier coefficients of the original function with the angle shifted by an amount directly proportional to frequency. Therefore, time shift in the time domain corresponds to phase shift in the frequency domain.

EXAMPLE 15.6

Let us time delay the waveform in Fig. 15.3 by a quarter period and compute the Fourier series.

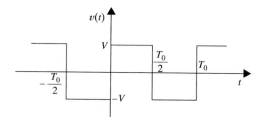

Figure 15.6 Waveform in Fig. 15.3 time shifted by $T_0/4$.

SOLUTION   The waveform in Fig. 15.3 time delayed by $T_0/4$ is shown in Fig. 15.6. Since the time delay is $T_0/4$,

$$n\omega_0 t_d = n \frac{2\pi}{T_0} \frac{T_0}{4} = n \frac{\pi}{2} = n\,90°$$

Therefore, using Eq. (15.34) and the results of Example 15.1, the Fourier coefficients for the time-shifted waveform are

$$\mathbf{c}_n = \frac{2V}{n\pi} \sin \frac{n\pi}{2} \underline{/-n\,90°} \qquad n \text{ odd}$$

and therefore,

$$v(t) = \sum_{\substack{n=1 \\ n \text{ odd}}}^{\infty} \frac{4V}{n\pi} \sin \frac{n\pi}{2} \cos\left(n\omega_0 t - n\,90°\right)$$

If we compute the Fourier coefficients for the time-shifted waveform in Fig. 15.6, we obtain

$$c_n = \frac{1}{T_0} \int_{-T_0/2}^{T_0/2} f(t)e^{-jn\omega_0 t}\, dt$$

$$= \frac{1}{T_0} \int_{-T_0/2}^{0} -V e^{-jn\omega_0 t}\, dt + \frac{1}{T_0} \int_{0}^{T_0/2} V e^{-jn\omega_0 t}\, dt$$

$$= \frac{2V}{jn\pi} \quad \text{for } n \text{ odd}$$

Therefore,

$$\mathbf{c}_n = \frac{2V}{n\pi}\; \underline{/-90^\circ} \qquad n \text{ odd}$$

Since n is odd, we can show that this expression is equivalent to the one obtained earlier. ❏

In general, we can compute the phase shift in degrees using the expression

$$\text{phase shift(deg)} = \omega_0 t_d = (360^\circ)\frac{t_d}{T_0} \qquad \textbf{15.35}$$

so that a time shift of one-quarter period corresponds to a 90° phase shift.

As another interesting facet of the time shift, consider a function $f_1(t)$ that is nonzero in the interval $0 \le t \le T_0/2$ and is zero in the interval $T_0/2 < t \le T_0$. For purposes of illustration, let us assume that $f_1(t)$ is the triangular waveform shown in Fig. 15.7a. $f_1(t - T_0/2)$ is then shown in Fig. 15.7b. Then the function $f(t)$ defined as

$$f(t) = f_1(t) - f_1\!\left(t - \frac{T_0}{2}\right) \qquad \textbf{15.36}$$

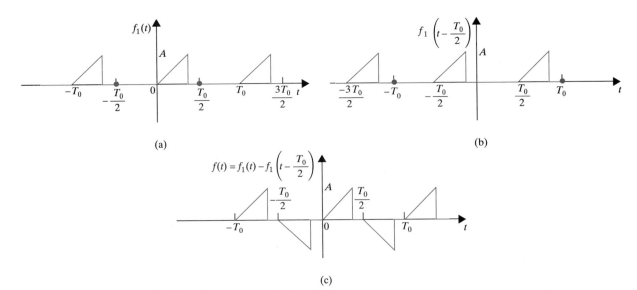

Figure 15.7 Waveforms that illustrate the generation of half-wave symmetry.

is shown in Fig. 15.7c. Note that $f(t)$ has half-wave symmetry. In addition, note that if

$$f_1(t) = \sum_{n=-\infty}^{\infty} \mathbf{c}_n e^{-jn\omega_0 t}$$

then

$$f(t) = f_1(t) - f_1\left(t - \frac{T_0}{2}\right) = \sum_{n=-\infty}^{\infty} \mathbf{c}_n\left(1 - e^{-jn\pi}\right)e^{jn\omega_0 t}$$

$$= \begin{cases} \sum_{n=-\infty}^{\infty} 2\mathbf{c}_n e^{jn\omega_0 t} & n \text{ odd} \\ 0 & n \text{ even} \end{cases} \qquad \textbf{15.37}$$

Therefore, we see that any function with half-wave symmetry can be expressed in the form of Eq. (15.36), where the Fourier series is defined by Eq. (15.37), and $\mathbf{c}_n$ is the Fourier coefficient for $f_1(t)$.

EXTENSION EXERCISE

E15.6   If the waveform in Fig. E15.1 is time delayed 1 s, we obtain the waveform in Fig. E15.6. Compute the exponential Fourier coefficients for the waveform in Fig. E15.6 and show that they differ from the coefficients for the waveform in Fig. E15.1 by an angle $n(180°)$.

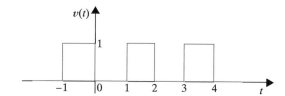

Figure E15.6

ANSWER: $\mathbf{c}_0 = \dfrac{1}{2}, \mathbf{c}_n = -\left(\dfrac{1 - e^{-jn\pi}}{j2\pi n}\right).$

Waveform Generation

The magnitude of the harmonics in a Fourier series is independent of the time scale for a given wave shape. Therefore, the equations for a variety of waveforms can be given in tabular form without expressing a specific time scale. Table 15.1 is a set of commonly occurring periodic waves where the advantage of symmetry has been used to simplify the coefficients. These waveforms can be used to generate other waveforms. The level of a wave can be adjusted by changing the average value component; the time can be shifted by adjusting the angle of the harmonics; and two waveforms can be added to produce a

Table 15.1 Fourier Series for Some Common Waveforms

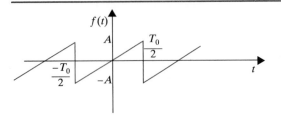

$$f(t) = \sum_{n=1}^{\infty} (-1)^{n+1} \frac{2A}{n\pi} \sin n\omega_0 t$$

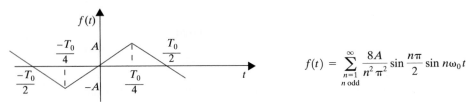

$$f(t) = \sum_{\substack{n=1 \\ n \text{ odd}}}^{\infty} \frac{8A}{n^2 \pi^2} \sin \frac{n\pi}{2} \sin n\omega_0 t$$

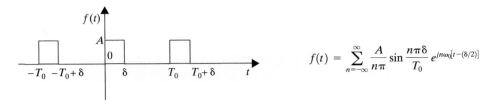

$$f(t) = \sum_{n=-\infty}^{\infty} \frac{A}{n\pi} \sin \frac{n\pi\delta}{T_0} e^{jn\omega_0[t-(\delta/2)]}$$

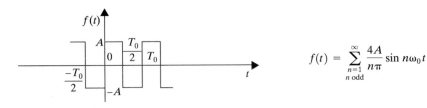

$$f(t) = \sum_{\substack{n=1 \\ n \text{ odd}}}^{\infty} \frac{4A}{n\pi} \sin n\omega_0 t$$

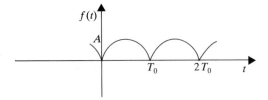

$$f(t) = \frac{2A}{\pi} + \sum_{n=1}^{\infty} \frac{4A}{\pi(1 - 4n^2)} \cos n\omega_0 t$$

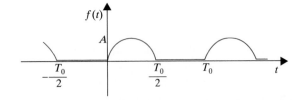

$$f(t) = \frac{A}{\pi} + \frac{A}{2} \sin \omega_0 t + \sum_{\substack{n=2 \\ n \text{ even}}}^{\infty} \frac{2A}{\pi(1 - n^2)} \cos n\omega_0 t$$

Table 15.1 (continued)

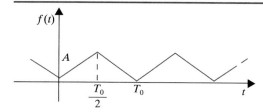

$$f(t) = \frac{A}{2} + \sum_{\substack{n=-\infty \\ n \neq 0 \\ n \text{ odd}}}^{\infty} \frac{-2A}{n^2 \pi^2} e^{jn\omega_0 t}$$

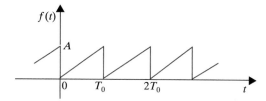

$$f(t) = \frac{A}{2} + \sum_{n=1}^{\infty} \frac{-A}{n\pi} \sin n\omega_0 t$$

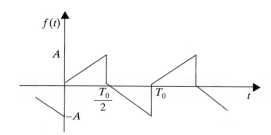

$$f(t) = \sum_{n=1}^{\infty} \frac{-4A}{\pi^2 n^2} \cos n\omega_0 t + \frac{2A}{\pi n} \sin n\omega_0 t$$

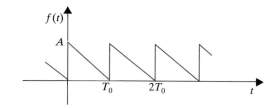

$$f(t) = \frac{A}{2} + \sum_{n=1}^{\infty} \frac{A}{\pi n} \sin n\omega_0 t$$

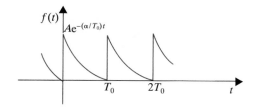

$$f(t) = \sum_{n=-\infty}^{\infty} \frac{A(1 - e^{-\alpha})}{\alpha + j2\pi n} e^{jn\omega_0 t}$$

third waveform; for example, the waveforms in Figs. 15.8a and b can be added to produce the waveform in Fig. 15.8c.

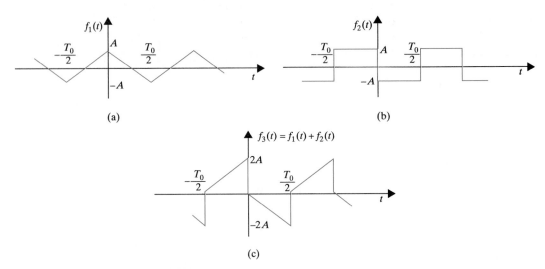

Figure 15.8 Example of waveform generation.

EXTENSION EXERCISE

E15.7 Two periodic waveforms are shown in Fig. E15.7. Compute the exponential Fourier series for each waveform, and then add the results to obtain the Fourier series for the waveform in Fig. E15.2.

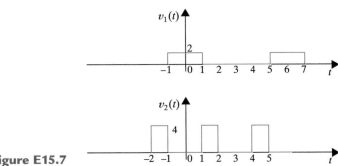

Figure E15.7

ANSWER: $v_1(t) = \dfrac{2}{3} + \displaystyle\sum_{\substack{n=-\infty \\ n\neq 0}}^{\infty} \dfrac{2}{n\pi} \sin \dfrac{n\pi}{3} e^{jn\omega_0 t},$

$v_2(t) = \dfrac{4}{3} + \displaystyle\sum_{n=-\infty}^{\infty} -\dfrac{4}{n\pi}\left(\sin \dfrac{n\pi}{3} - \sin \dfrac{2n\pi}{3} \right) e^{jn\omega_0 t}.$

Frequency Spectrum

The *frequency spectrum* of the function $f(t)$ expressed as a Fourier series consists of a plot of the amplitude of the harmonics versus frequency, which we call the *amplitude spectrum*, and a plot of the phase of the harmonics versus frequency, which we call the *phase spectrum*. Since the frequency components are discrete, the spectra are called *line spectra*. Such spectra illustrate the frequency content of the signal. Plots of the amplitude and phase spectra are based on Eqs. (15.1), (15.3), and (15.7) and represent the amplitude and phase of the signal at specific frequencies.

EXAMPLE 15.7

The Fourier series for the triangular-type waveform shown in Fig. 15.8c with $A = 5$ is given by the equation

$$v(t) = \sum_{\substack{n=1 \\ n \text{ odd}}}^{\infty} \left(\frac{20}{n\pi} \sin n\omega_0 t - \frac{40}{n^2 \pi^2} \cos n\omega_0 t \right)$$

We wish to plot the first four terms of the amplitude and phase spectra for this signal.

SOLUTION   Since $D_n = \underline{/\theta_n} = a_n - jb_n$, the first four terms for this signal are

$$D_1 \underline{/\theta_1} = -\frac{40}{\pi^2} - j\frac{20}{\pi} = 7.5 \underline{/-122°}$$

$$D_3 \underline{/\theta_3} = -\frac{40}{9\pi^2} - j\frac{20}{3\pi} = 2.2 \underline{/-102°}$$

$$D_5 \underline{/\theta_5} = -\frac{40}{25\pi^2} - j\frac{20}{5\pi} = 1.3 \underline{/-97°}$$

$$D_7 \underline{/\theta_7} = -\frac{40}{49\pi^2} - j\frac{20}{7\pi} = 0.91 \underline{/-95°}$$

Therefore, the plots of the amplitude and phase versus ω are as shown in Fig. 15.9.   ❏

EXTENSION EXERCISE

E15.8   Determine the trigonometric Fourier series for the voltage waveform in Fig. E15.8 and plot the first four terms of the amplitude and phase spectra for this signal.

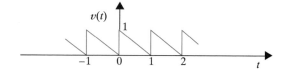

Figure E15.8

ANSWER: $a_0 = \frac{1}{2}$, $D_1 = -j(1/\pi)$, $D_2 = -j(1/2\pi)$, $D_3 = -j(1/3\pi)$, $D_4 = -j(1/4\pi)$.

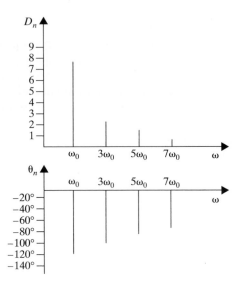

Figure 15.9 Amplitude and phase spectra.

Steady-State Network Response

If a periodic signal is applied to a network, the steady-state voltage or current response at some point in the circuit can be found in the following manner. First, we represent the periodic forcing function by a Fourier series. If the input forcing function for a network is a voltage, the input can be expressed in the form

$$v(t) = v_0 + v_1(t) + v_2(t) + \cdots$$

and therefore represented in the time domain as shown in Fig. 15.10. Each source has its own amplitude and frequency. Next we determine the response due to each component of the input Fourier series; that is, we use phasor analysis in the frequency domain to determine the network response due to each source. The network response due to each source in the frequency domain is then transformed to the time domain. Finally, we add the time domain solutions due to each source using the principle of superposition to obtain the Fourier series for the total *steady-state* network response.

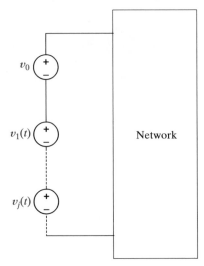

Figure 15.10 Network with a periodic voltage forcing function.

EXAMPLE 15.8

We wish to determine the steady-state voltage $v_o(t)$ in Fig. 15.11 if the input voltage $v(t)$ is given by the expression

$$v(t) = \sum_{\substack{n=1 \\ n \text{ odd}}}^{\infty} \left(\frac{20}{n\pi} \sin 2nt - \frac{40}{n^2 \pi^2} \cos 2nt \right) \text{V}$$

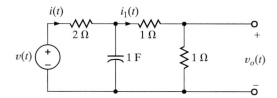

Figure 15.11 *RC* circuit employed in Example 15.8.

SOLUTION   Note that this source has no constant term, and therefore its dc value is zero. The amplitude and phase for the first four terms of this signal are given in Example 15.7, and therefore the signal $v(t)$ can be written as

$$v(t) = 7.5 \cos(2t - 122°) + 2.2 \cos(6t - 102°) + 1.3 \cos(10t - 97°)$$
$$+ \, 0.91 \cos(14t - 95°) + \cdots$$

From the network we find that

$$\mathbf{I} = \frac{\mathbf{V}}{2 + \dfrac{2/j\omega}{2 + 1/j\omega}} = \frac{\mathbf{V}(1 + 2j\omega)}{4 + 4j\omega}$$

$$\mathbf{I}_1 = \frac{\mathbf{I}(1/j\omega)}{2 + 1/j\omega} = \frac{\mathbf{I}}{1 + 2j\omega}$$

$$\mathbf{V}_o = (1)\mathbf{I}_1 = 1 \cdot \frac{\mathbf{V}(1 + 2j\omega)}{4 + 4j\omega} \cdot \frac{1}{1 + 2j\omega} = \frac{\mathbf{V}}{4 + 4j\omega}$$

Therefore, since $\omega_0 = 2$,

$$\mathbf{V}_o(n) = \frac{\mathbf{V}(n)}{4 + j8n}$$

The individual components of the output due to the components of the input source are then

$$\mathbf{V}_o(\omega_0) = \frac{7.5 \,\underline{/-122°}}{4 + j8} = 0.84 \,\underline{/-185.4°} \text{ V}$$

$$\mathbf{V}_o(3\omega_0) = \frac{2.2 \,\underline{/-102°}}{4 + j24} = 0.09 \,\underline{/-182.5°} \text{ V}$$

$$\mathbf{V}_o(5\omega_0) = \frac{1.3 \,\underline{/-97°}}{4 + j40} = 0.03 \,\underline{/-181.3°} \text{ V}$$

$$\mathbf{V}_o(7\omega_0) = \frac{0.91 \,\underline{/-95°}}{4 + j56} = 0.016 \,\underline{/-181°} \text{ V}$$

Hence, the steady-state output voltage $v_o(t)$ can be written as

$$v_o(t) = 0.84 \cos(2t - 185.4°) + 0.09 \cos(6t - 182.5°)$$
$$+ 0.03 \cos(10t - 181.3°) + 0.016 \cos(14t - 181°) + \cdots \qquad \square$$

EXTENSION EXERCISE

E15.9   Determine the expression for the steady-state current $i(t)$ in Fig. E15.9 if the input voltage $v_S(t)$ is given by the expression

$$v_S(t) = \frac{20}{\pi} + \sum_{n=1}^{\infty} \frac{-40}{\pi(4n^2 - 1)} \cos 2nt \text{ V}$$

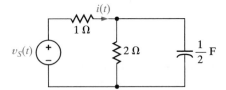

Figure E15.9

ANSWER: $i(t) = 2.12 + \displaystyle\sum_{n=1}^{\infty} \frac{-40}{\pi(4n^2 - 1)} \frac{1}{A_n} \cos\left(2nt - \theta_n\right)$ A.

Average Power

We have shown that when a linear network is forced with a nonsinusoidal periodic signal, voltages and currents throughout the network are of the form

$$v(t) = V_{dc} + \sum_{n=1}^{\infty} V_n \cos(n\omega_0 t - \theta_{v_n})$$

and

$$i(t) = I_{dc} + \sum_{n=1}^{\infty} I_n \cos(n\omega_0 t - \theta_{i_n})$$

If we employ the passive sign convention and assume that the voltage across an element and the current through it are given by the preceding equations, then from Eq. (9.6),

$$P = \frac{1}{T} \int_{t_0}^{t_0+T} p(t) \, dt$$
$$= \frac{1}{T} \int_{t_0}^{t_0+T} v(t)i(t) \, dt$$

Note that the integrand involves the product of two infinite series. However, the determination of the average power is actually easier than it appears. First, note that the product $V_{dc}I_{dc}$ when integrated over a period and divided by the period is simply $V_{dc}I_{dc}$.

Second, the product of V_{dc} and any harmonic of the current or I_{dc} and any harmonic of the voltage when integrated over a period yields zero. Third, the product of any two *different* harmonics of the voltage and the current when integrated over a period yields zero. Finally, nonzero terms result only from the products of voltage and current at the *same* frequency. Hence, using the mathematical development that follows Eq. (9.6), we find that

$$P = V_{dc} I_{dc} + \sum_{n=1}^{\infty} \frac{V_n I_n}{2} \cos \left(\theta_{v_n} - \theta_{i_n} \right) \qquad\qquad \textbf{15.38}$$

EXAMPLE 15.9

In the network in Fig. 15.12, $v(t) = 42 + 16 \cos(377t + 30°) + 12 \cos(754t - 20°)$ V. We wish to compute the current $i(t)$ and determine the average power absorbed by the network.

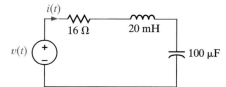

Figure 15.12 Network used in Example 15.9.

SOLUTION   The capacitor acts as an open circuit to dc, and therefore $I_{dc} = 0$. At $\omega = 377$ rad/s.

$$\frac{1}{j\omega C} = \frac{1}{j(377)(100)(10)^{-6}} = -j26.53 \ \Omega$$

$$j\omega L = j(377)(20)10^{-3} = j7.54 \ \Omega$$

Hence,

$$\mathbf{I}_{377} = \frac{16 \ \underline{/30°}}{16 + j7.54 - j26.53} = 0.64 \ \underline{/79.88°} \ \text{A}$$

At $\omega = 377$ rad/s,

$$\frac{1}{j\omega C} = \frac{1}{j(754)(100)(10)^{-6}} = -j13.26 \ \Omega$$

$$j\omega L = j(754)(20)10^{-3} = j15.08 \ \Omega$$

Hence,

$$\mathbf{I}_{754} = \frac{12 \ \underline{/20°}}{16 + j15.08 - j13.26} = 0.75 \ \underline{/-26.49°} \ \text{A}$$

Therefore, the current $i(t)$ is

$$i(t) = 0.64 \cos(377t + 79.88°) + 0.75 \cos(754t + 26.49°) \ \text{A}$$

and the average power absorbed by the network is

$$P = (42)(0) + \frac{(16)(0.64)}{2} \cos{(30° - 79.88°)}$$

$$+ \frac{(12)(0.75)}{2} \cos{(-20° + 26.49°)}$$

$$= 7.77 \text{ W}$$ ❏

EXTENSION EXERCISE

E15.10 At the terminals of a two-port network, the voltage $v(t)$ and the current $i(t)$ are given by the following expression:

$$v(t) = 64 + 36 \cos{(377t + 60°)} - 24 \cos{(754t + 102°)} \text{ V}$$
$$i(t) = 1.8 \cos{(377t + 45°)} + 1.2 \cos{(754t + 100°)} \text{ A}$$

Find the average power absorbed by the network.

ANSWER: $P_{\text{Ave}} = 16.91 \text{ W}$.

15.2 FOURIER TRANSFORM

The preceding sections of this chapter have illustrated that the exponential Fourier series can be used to represent a periodic signal for all time. We will now consider a technique for representing an aperiodic signal for all values of time.

Suppose that an aperiodic signal $f(t)$ is as shown in Fig. 15.13a. We now construct a new signal $f_p(t)$ that is identical to $f(t)$ in the interval $-T/2$ to $T/2$ but is *periodic* with period T, as shown in Fig. 15.13b. Since $f_p(t)$ is periodic, it can be represented in the interval $-\infty$ to ∞ by an exponential Fourier series;

$$f_p(t) = \sum_{n=-\infty}^{\infty} \mathbf{c}_n e^{jn\omega_0 t}$$ **15.39**

where

$$\mathbf{c}_n = \frac{1}{T} \int_{-T/2}^{T/2} f_p(t) e^{-jn\omega_0 t} \, dt$$ **15.40**

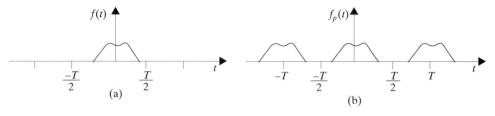

Figure 15.13 Aperiodic and periodic signals.

and

$$\omega_0 = \frac{2\pi}{T} \qquad \textbf{15.41}$$

At this point we note that if we take the limit of the function $f_p(t)$ as $T \to \infty$, the periodic signal in Fig. 15.13b approaches the aperiodic signal in Fig. 15.13a; that is, the repetitious signals centered at $-T$ and $+T$ in Fig. 15.13b are moved to infinity.

The line spectrum for the periodic signal exists at harmonic frequencies $(n\omega_0)$ and the incremental spacing between the harmonics is

$$\Delta\omega = (n + 1)\omega_0 - n\omega_0 = \omega_0 = \frac{2\pi}{T} \qquad \textbf{15.42}$$

As $T \to \infty$ the lines in the frequency spectrum for $f_p(t)$ come closer and closer together, $\Delta\omega$ approaches the differential $d\omega$ and $n\omega_0$ can take on any value of ω. Under these conditions the line spectrum becomes a continuous spectrum. Since as $T \to \infty, \mathbf{c}_n \to 0$ in Eq. (15.40), we will examine the product $\mathbf{c}_n T$, where

$$\mathbf{c}_n T = \int_{-T/2}^{T/2} f_p(t) e^{-jn\omega_0 t}\, dt$$

In the limit as $T \to \infty$,

$$\lim_{T \to \infty} (\mathbf{c}_n T) = \lim_{T \to \infty} \int_{-T/2}^{T/2} f_p(t) e^{-jn\omega_0 t}\, dt$$

which in view of the previous discussion can be written as

$$\lim_{T \to \infty} (\mathbf{c}_n T) = \int_{-\infty}^{\infty} f(t) e^{-j\omega t}\, dt$$

This integral is the Fourier transform of $f(t)$, which we will denote as $\mathbf{F}(\omega)$, and hence

$$\mathbf{F}(\omega) = \int_{-\infty}^{\infty} f(t) e^{-j\omega t}\, dt \qquad \textbf{15.43}$$

Similarly, $f_p(t)$ can be expressed as

$$f_p(t) = \sum_{n=-\infty}^{\infty} \mathbf{c}_n e^{jn\omega_0 t}$$

$$= \sum_{n=-\infty}^{\infty} (\mathbf{c}_n T) e^{jn\omega_0 t} \frac{1}{T}$$

$$= \sum_{n=-\infty}^{\infty} (\mathbf{c}_n T) e^{jn\omega_0 t} \frac{\Delta\omega}{2\pi}$$

which in the limit as $T \to \infty$ becomes

$$f(t) = \frac{1}{2\pi} \int_{-\infty}^{\infty} \mathbf{F}(\omega)e^{j\omega t}\, d\omega \qquad \textbf{15.44}$$

Equations (15.43) and (15.44) constitute what is called the *Fourier transform pair*. Since $\mathbf{F}(\omega)$ is the Fourier transform of $f(t)$ and $f(t)$ is the inverse Fourier transform of $\mathbf{F}(\omega)$, they are normally expressed in the form

$$\mathbf{F}(\omega) = \mathcal{F}[f(t)] = \int_{-\infty}^{\infty} f(t)e^{-j\omega t}\, dt \qquad \textbf{15.45}$$

$$f(t) = \mathcal{F}^{-1}[\mathbf{F}(\omega)] = \frac{1}{2\pi} \int_{-\infty}^{\infty} \mathbf{F}(\omega)e^{j\omega t}\, d\omega \qquad \textbf{15.46}$$

Some Important Transform Pairs

There are a number of important Fourier transform pairs, and in the following material we will derive a number of them and then list some of the more common ones in tabular form.

EXAMPLE 15.10

We wish to derive the Fourier transform for the voltage pulse shown in Fig. 15.14a.

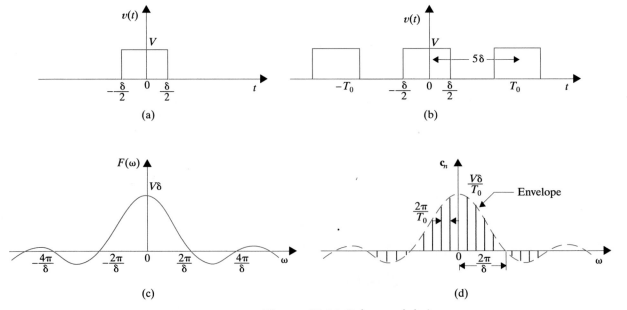

Figure 15.14 Pulses and their spectra.

SOLUTION   Using Eq. (15.45), the Fourier transform is

$$\mathbf{F}(\omega) = \int_{-\delta/2}^{\delta/2} V e^{-j\omega t}\, dt$$

$$= \frac{V}{-j\omega} e^{-j\omega t} \Big|_{-\delta/2}^{\delta/2}$$

$$= V \frac{e^{-j\omega\delta/2} - e^{+j\omega\delta/2}}{-j\omega}$$

$$= V\delta \frac{\sin(\omega\delta/2)}{\omega\delta/2}$$

Therefore, the Fourier transform for the function

$$f(t) = \begin{cases} 0 & -\infty < t \le -\dfrac{\delta}{2} \\[2mm] V & -\dfrac{\delta}{2} < t \le \dfrac{\delta}{2} \\[2mm] 0 & \dfrac{\delta}{2} < t < \infty \end{cases}$$

is

$$\mathbf{F}(\omega) = V\delta \frac{\sin(\omega\delta/2)}{\omega\delta/2}$$

A plot of this function is shown in Fig. 15.14c. Let us explore this example even further. Consider now the pulse train shown in Fig. 15.14b. Using the techniques that have been demonstrated earlier, we can show that the Fourier coefficients for this waveform are

$$\mathbf{c}_n = \frac{V\delta}{T_0} \frac{\sin(n\omega_0 \delta/2)}{n\omega_0 \delta/2}$$

The line spectrum for $T_0 = 5\delta$ is shown in Fig. 15.14d.

What these equations and figures in this example indicate are that as $T_0 \to \infty$ and the periodic function becomes aperiodic, the lines in the discrete spectrum become denser and the amplitude gets smaller, and the amplitude spectrum changes from a line spectrum to a continuous spectrum. Note that the envelope for the discrete spectrum has the same shape as the continuous spectrum. Since the Fourier series represents the amplitude and phase of the signal at specific frequencies, the Fourier transform also specifies the frequency content of a signal.   ❏

EXAMPLE 15.11

Find the Fourier transform for the unit impulse function $\delta(t)$.

SOLUTION   The Fourier transform of the unit impulse function $\delta(t - a)$ is

$$\mathbf{F}(\omega) = \int_{-\infty}^{\infty} \delta(t - a) e^{-j\omega t}\, dt$$

$$\delta(t) \leftrightarrow 1$$

$$\delta(t - a) \leftrightarrow e^{-j\omega a}$$

Using the sampling property of the unit impulse, we find that

$$\mathbf{F}(\omega) = e^{-j\omega a}$$

and if $a = 0$, then

$$\mathbf{F}(\omega) = 1$$

Note then that the $\mathbf{F}(\omega)$ for $f(t) = \delta(t)$ is *constant for all frequencies*. This is an important property, as we shall see later. ❏

EXAMPLE 15.12

We wish to determine the Fourier transform of the function $f(t) = e^{j\omega_0 t}$.

SOLUTION In this case note that if $\mathbf{F}(\omega) = 2\pi\delta(\omega - \omega_0)$, then

$$f(t) = \frac{1}{2\pi} \int_{-\infty}^{\infty} 2\pi\delta(\omega - \omega_0)e^{j\omega t}\, d\omega$$

$$= e^{j\omega_0 t}$$

$$e^{j\omega_0 t} \leftrightarrow 2\pi\delta(\omega - \omega_0)$$

Therefore, $f(t) = e^{j\omega_0 t}$ and $\mathbf{F}(\omega) = 2\pi\delta(\omega - \omega_0)$ represent a Fourier transform pair. ❏

EXAMPLE 15.13

Let us determine the Fourier transform of the function $f(t) = \cos \omega_0 t$.

SOLUTION The Fourier transform of the function $f(t) = \cos \omega_0 t$ is

$$\mathbf{F}(\omega) = \int_{-\infty}^{\infty} \cos \omega_0 t\, e^{-j\omega t}\, dt$$

$$\cos \omega_0 t \leftrightarrow \pi\delta(\omega - \omega_0) \\ +\pi\delta(\omega + \omega_0)$$

$$= \frac{1}{2} \int_{-\infty}^{\infty} e^{j\omega_0 t} e^{-j\omega t}\, dt + \frac{1}{2} \int_{-\infty}^{\infty} e^{-j\omega_0 t} e^{-j\omega t}\, dt$$

Using the results of Example 15.12, we obtain

$$\mathbf{F}(\omega) = \pi\delta(\omega - \omega_0) + \pi\delta(\omega + \omega_0)$$ ❏

EXAMPLE 15.14

Let us determine the Fourier transform of the function $f(t) = e^{-at}u(t)$.

SOLUTION   The Fourier transform of this function is

$$\mathbf{F}(\omega) = \int_{-\infty}^{\infty} e^{-at}u(t)e^{-j\omega t}\,dt$$

$$= \int_{0}^{\infty} e^{-at}e^{-j\omega t}\,dt$$

$$= \frac{-1}{a + j\omega}\,e^{-at}e^{-j\omega t}\Big|_{0}^{\infty}$$

$$= \frac{1}{a + j\omega}$$

for $a > 0$ since $e^{-at} \to 0$ as $t \to \infty$.   ❑

A number of useful Fourier transform pairs are shown in Table 15.2.

Table 15.2   Fourier Transform Pairs

$f(t)$	$\mathbf{F}(\omega)$
$\delta(t - a)$	$e^{-j\omega a}$
A	$2\pi A\delta(\omega)$
$e^{j\omega_0 t}$	$2\pi\delta(\omega - \omega_0)$
$\cos \omega_0 t$	$\pi\delta(\omega - \omega_0) + \pi\delta(\omega - \omega_0)$
$\sin \omega_0 t$	$j\pi\delta(\omega - \omega_0) + j\pi\delta(\omega - \omega_0)$
$e^{-at}u(t), a > 0$	$\dfrac{1}{a + j\omega}$
$e^{-a|t|}, a > 0$	$\dfrac{2a}{a^2 + \omega^2}$
$e^{-at}\cos \omega_0 t u(t), a > 0$	$\dfrac{j\omega + a}{(j\omega + a)^2 + \omega_0^2}$
$e^{-at}\sin \omega_0 t u(t), a > 0$	$\dfrac{\omega_0}{(j\omega + a)^2 + \omega_0^2}$

EXTENSION EXERCISE

E15.11   If $f(t) = \sin \omega_0 t$, find $\mathbf{F}(\omega)$.

ANSWER: $\mathbf{F}(\omega) = \pi j\big[\delta(\omega + \omega_0) - \delta(\omega - \omega_0)\big]$.

Some Properties of the Fourier Transform

Let us examine now some of the properties of the Fourier transform defined by the equation

$$\mathbf{F}(\omega) = \int_{-\infty}^{\infty} f(t)e^{-j\omega t}\, dt$$

Using Euler's identity, we can write this function as

$$\mathbf{F}(\omega) = A(\omega) + jB(\omega) \qquad\qquad \textbf{15.47}$$

where

$$A(\omega) = \int_{-\infty}^{\infty} f(t)\cos \omega t\, dt \qquad\qquad \textbf{15.48}$$

$$B(\omega) = -\int_{-\infty}^{\infty} f(t)\sin \omega t\, dt \qquad\qquad \textbf{15.49}$$

From Eq. (15.47) we note that

$$\mathbf{F}(\omega) = |\mathbf{F}(\omega)|e^{j\theta(\omega)} \qquad\qquad \textbf{15.50}$$

where

$$|\mathbf{F}(\omega)| = \sqrt{A^2(\omega) + B^2(\omega)} \qquad\qquad \textbf{15.51}$$

$$\theta(\omega) = \tan^{-1}\frac{B(\omega)}{A(\omega)} \qquad\qquad \textbf{15.52}$$

Given the preceding definitions, we can show that

$$A(\omega) = A(-\omega)$$
$$B(\omega) = -B(-\omega) \qquad\qquad \textbf{15.53}$$
$$\mathbf{F}(-\omega) = \mathbf{F}^*(\omega)$$

Equations (15.53) simply state that $A(\omega)$ is an even function of ω, $B(\omega)$ is an odd function of ω, and $\mathbf{F}(\omega)$ evaluated at $-\omega$ is the complex conjugate of $\mathbf{F}(\omega)$. Therefore, $|\mathbf{F}(\omega)|$ is an even function of ω and $\theta(\omega)$ is an odd function of ω. Some additional properties of the Fourier transform are now presented in rapid succession.

Linearity. If $\mathcal{F}[f_1(t)] = \mathbf{F}_1(\omega)$ and $\mathcal{F}[f_2(t)] = \mathbf{F}_2(\omega)$, then

$$\mathcal{F}[af_1(t) + bf_2(t)] = \int_{-\infty}^{\infty}[af_1(t) + bf_2(t)]e^{-j\omega t}\, dt$$

$$= \int_{-\infty}^{\infty} af_1(t)e^{-j\omega t}\, dt + \int_{-\infty}^{\infty} bf_2(t)e^{-j\omega t}\, dt$$

$$= a\mathbf{F}_1(\omega) + b\mathbf{F}_2(\omega) \qquad\qquad \textbf{15.54}$$

Time Scaling. If $\mathcal{F}[f(t)] = \mathbf{F}(\omega)$, then $a < 0$,

$$\mathcal{F}[f(at)] = \int_{-\infty}^{\infty} f(at)e^{-j\omega t}\, dt$$

If we now let $x = at$, then

$$\mathcal{F}[f(at)] = \int_{-\infty}^{\infty} f(x)e^{-j\omega x/a} \frac{dx}{a}$$

$$= \frac{1}{a} \mathbf{F}\left(\frac{\omega}{a}\right) \qquad\qquad \textbf{15.55}$$

and if $a < 0$, then following the previous development,

$$\mathcal{F}[f(at)] = \int_{+\infty}^{-\infty} f(x)e^{-j\omega x/a} \frac{dx}{a}$$

$$= -\frac{1}{a} \mathbf{F}\left(\frac{\omega}{a}\right) \qquad\qquad \textbf{15.56}$$

Time Shifting.   If $\mathcal{F}[f(t)] = \mathbf{F}(\omega)$, then

$$\mathcal{F}[f(t - t_0)] = \int_{-\infty}^{\infty} f(t - t_0)e^{-j\omega t}\, dt$$

If we now let $x = t - t_0$, then

$$\mathcal{F}[f(t - t_0)] = \int_{-\infty}^{\infty} f(x)e^{-j\omega(x+t_0)}\, dx$$

$$= e^{-j\omega t_0}\mathbf{F}(\omega) \qquad\qquad \textbf{15.57}$$

Modulation.   If $\mathcal{F}[f(t)] = \mathbf{F}(\omega)$, then

$$\mathcal{F}[e^{j\omega_0 t} f(t)] = \int_{-\infty}^{\infty} f(t)e^{j\omega_0 t}\, e^{-j\omega t}\, dt$$

$$= \int_{-\infty}^{\infty} f(t)e^{-j(\omega - \omega_0)t}\, dt$$

$$= \mathbf{F}(\omega - \omega_0) \qquad\qquad \textbf{15.58}$$

Differentiation.   If $f(t)$ is a function that is continuous in any finite interval, is absolutely integrable, and its first derivative is piecewise continuous and absolutely integrable, then

$$\mathcal{F}[f'(t)] = \int_{-\infty}^{\infty} f'(t)e^{-j\omega t}\, dt$$

Using integration by parts, we obtain

$$\mathcal{F}[f'(t)] = f(t)e^{-j\omega t}\Big|_{-\infty}^{\infty} + \int_{-\infty}^{\infty} j\omega f(t)e^{-j\omega t}\, dt$$

Since $f(t)$ is absolutely integrable,

$$\lim_{t \to -\infty} |f(t)| = \lim_{t \to +\infty} |f(t)| = 0$$

and therefore,

$$\mathcal{F}[f'(t)] = j\omega\mathbf{F}(\omega) \qquad\qquad\qquad \mathbf{15.59}$$

Repeated application of this technique under the conditions listed previously yields

$$\mathcal{F}[f^{(n)}(t)] = (j\omega)^n\,\mathbf{F}(\omega) \qquad\qquad\qquad \mathbf{15.60}$$

It is important to note that Eq. (15.59) does not guarantee the existence of $\mathcal{F}[f'(t)]$. It simply states that *if* the transform exists, it is given by Eq. (15.59).

Time Convolution.   If $\mathcal{F}[f_1(t)] = \mathbf{F}_1(\omega)$ and $\mathcal{F}[f_2(t)] = \mathbf{F}_2(\omega)$, then

$$\mathcal{F}\left[\int_{-\infty}^{\infty} f_1(x)f_2(t-x)\,dx\right] = \int_{t=-\infty}^{\infty}\int_{x=-\infty}^{\infty} f_1(x)f_2(t-x)\,dx\,e^{-j\omega t}\,dt$$

$$= \int_{x=-\infty}^{\infty} f_1(x)\int_{t=-\infty}^{\infty} f_2(t-x)e^{-j\omega t}\,dt\,dx$$

If we now let $u = t - x$, then

$$\mathcal{F}\left[\int_{-\infty}^{\infty} f_1(x)f_2(t-x)\,dx\right] = \int_{x=-\infty}^{\infty} f_1(x)\int_{u=-\infty}^{\infty} f_2(u)e^{-j\omega(u+x)}\,du\,dx$$

$$= \int_{x=-\infty}^{\infty} f_1(x)e^{-j\omega x}\int_{u=-\infty}^{\infty} f_2(u)e^{-j\omega u}\,du\,dx$$

$$= \mathbf{F}_1(\omega)\mathbf{F}_2(\omega) \qquad\qquad\qquad \mathbf{15.61}$$

We should note very carefully the time convolution property of the Fourier transform. With reference to Fig. 15.15, this property states that if $\mathbf{V}_i(\omega) = \mathcal{F}[v_1(t)]$, $\mathbf{H}(\omega) = \mathcal{F}[h(t)]$, and $\mathbf{V}_o(\omega) = \mathcal{F}[v_o(t)]$, then

$$\mathbf{V}_o(\omega) = \mathbf{H}(\omega)\mathbf{V}_i(\omega) \qquad\qquad\qquad \mathbf{15.62}$$

where $\mathbf{V}_i(\omega)$ represents the input signal, $\mathbf{H}(\omega)$ is the network transfer function, and $\mathbf{V}_o(\omega)$ represents the output signal. Equation (15.62) tacitly assumes that the initial conditions of the network are zero.

Table 15.3 provides a short list of some of the Fourier transform properties.

Figure 15.15
Representation of the time
convolution property.

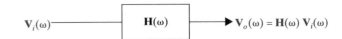

$\mathbf{V}_i(\omega)$ ——— [$\mathbf{H}(\omega)$] ——▶ $\mathbf{V}_o(\omega) = \mathbf{H}(\omega)\,\mathbf{V}_i(\omega)$

EXAMPLE 15.15

Let us determine the Fourier transform of the function $f(t - t_0) = e^{-(t-t_0)}u(t - t_0)$.

SOLUTION   By definition

$$\mathcal{F}[f(t - t_0)] = \int_{-\infty}^{\infty} f(t - t_0)e^{-j\omega t}\,dt$$

which can be written using the time-shifting property as

Table 15.3 Properties of the Fourier Transform

$f(t)$	$\mathbf{F}(\omega)$
$Af(t)$	$A\mathbf{F}(\omega)$
$f_1(t) \pm f_2(t)$	$\mathbf{F}_1(\omega) \pm \mathbf{F}_2(\omega)$
$f(at)$	$\dfrac{1}{a}\mathbf{F}\!\left(\dfrac{\omega}{a}\right),\ a > 0$
$f(t - t_0)$	$e^{-j\omega t_0}\mathbf{F}(\omega)$
$e^{j\omega_0 t}f(t)$	$\mathbf{F}(\omega - \omega_0)$
$\dfrac{d^n f(t)}{dt^n}$	$(j\omega)^n\mathbf{F}(\omega)$
$t^n f(t)$	$(j)^n\dfrac{d^n\mathbf{F}(\omega)}{d\omega^n}$
$f(t)\cos\omega_0 t$	$\tfrac{1}{2}\big[\mathbf{F}(\omega - \omega_0) + \mathbf{F}(\omega + \omega_0)\big]$
$\displaystyle\int_{-\infty}^{\infty} f_1(x)f_2(t - x)\,dx$	$\mathbf{F}_1(\omega)\mathbf{F}_2(\omega)$
$f_1(t)f_2(t)$	$\dfrac{1}{2\pi}\displaystyle\int_{-\infty}^{\infty}\mathbf{F}_1(x)\mathbf{F}_2(\omega - x)\,dx$

$$\mathcal{F}[f(t - t_0)] = e^{-j\omega t_0}\int_{-\infty}^{\infty} f(t)e^{-j\omega t}\,dt$$

$$= e^{-j\omega t_0}\int_{-\infty}^{\infty} e^{-t}u(t)e^{-j\omega t}\,dt$$

which from the results of Example 15.14 is

$$\mathcal{F}[f(t - t_0)] = \frac{e^{-j\omega t_0}}{1 + j\omega} \qquad \square$$

EXAMPLE 15.16

Let us determine the Fourier transform of the function

$$f(t) = \frac{d}{dt}\big[e^{-at}u(t)\big] = -ae^{-at}u(t) + e^{-at}\delta(t)$$

SOLUTION $\displaystyle \mathcal{F}[f(t)] = \int_{-\infty}^{\infty}\big[-ae^{-at}u(t) + e^{-at}\delta(t)\big]e^{-j\omega t}\,dt$

$$= -\int_{-\infty}^{\infty} ae^{-at}u(t)e^{-j\omega t}\,dt + \int_{-\infty}^{\infty} e^{-at}\delta(t)e^{-j\omega t}\,dt$$

$$= \frac{-a}{a + j\omega} + 1$$

$$= \frac{j\omega}{a + j\omega}$$

The transform could also be evaluated using the differentiation property; that is, if

$$\mathcal{F}\left[e^{-at} u(t)\right] = \frac{1}{a + j\omega}$$

then

$$\mathcal{F}\left[\frac{d}{dt}\left(e^{-at} u(t)\right)\right] = j\omega\left(\frac{1}{a + j\omega}\right) = \frac{j\omega}{a + j\omega} \qquad \square$$

EXTENSION EXERCISES

E15.12 If $\mathcal{F}\left[e^{-t} u(t)\right] = 1/(1 + j\omega)$, use the time-scaling property of the Fourier transform to find $\mathcal{F}\left[e^{-at} u(t)\right], a > 0$.

ANSWER: $\dfrac{1}{a + j\omega}$.

E15.13 Use the property $\mathcal{F}\left[t^n f(t)\right] = j^n\left[d^n \mathbf{F}(\omega)/d\omega^n\right]$ to determine the Fourier transform of $te^{-at} u(t), a > 0$.

ANSWER: $\dfrac{1}{(a + j\omega)^2}$.

E15.14 Determine the output $v_o(t)$ in Fig. E15.14 if the signal $v_i(t) = e^{-t} u(t)$, the network impulse response $h(t) = e^{-2t} u(t)$, and all initial conditions are zero.

Figure E15.14

ANSWER: $v_o(t) = \left(e^{-t} - e^{-2t}\right)u(t)$ V.

Parseval's Theorem

A mathematical statement of Parseval's theorem is

$$\int_{-\infty}^{\infty} f^2(t)\, dt = \frac{1}{2\pi} \int_{-\infty}^{\infty} |\mathbf{F}(\omega)|^2\, d\omega \qquad \textbf{17.63}$$

This relationship can be easily derived as follows:

$$\int_{-\infty}^{\infty} f^2(t)\, dt = \int_{-\infty}^{\infty} f(t) \frac{1}{2\pi} \int_{-\infty}^{\infty} \mathbf{F}(\omega) e^{j\omega t}\, d\omega\, dt$$

$$= \frac{1}{2\pi} \int_{-\infty}^{\infty} \mathbf{F}(\omega) \int_{-\infty}^{\infty} f(t) e^{-j(-\omega)t}\, dt\, d\omega$$

$$= \int_{-\infty}^{\infty} \frac{1}{2\pi} \mathbf{F}(\omega)\mathbf{F}(-\omega)\, d\omega$$

$$= \int_{-\infty}^{\infty} \frac{1}{2\pi} \mathbf{F}(\omega)\mathbf{F}^*(\omega)\, d\omega$$

$$= \int_{-\infty}^{\infty} \frac{1}{2\pi} |\mathbf{F}(\omega)|^2 d\omega$$

The importance of Parseval's theorem can be seen if we imagine that $f(t)$ represents the current in a 1-Ω resistor. Since $f^2(t)$ is power and the integral of power over time is energy, Eq. (15.63) shows that we can compute this 1-Ω energy or normalized energy in either the time domain or the frequency domain.

Applications

Let us now apply to circuit problems some of the things we have learned about the Fourier transform.

EXAMPLE 15.17

Using the transform technique, we wish to determine $v_o(t)$ in Fig. 15.16 if (a) $v_i(t) = 5e^{-2t}u(t)$ V and (b) $v_i(t) = 5\cos 2t$ V.

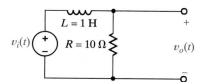

Figure 15.16 Simple *RL* circuit.

SOLUTION   (a) In this case since $v_i(t) = 5e^{-2t}u(t)$ V, then

$$\mathbf{V}_i(\omega) = \frac{5}{2 + j\omega}\ \text{V}$$

$\mathbf{H}(\omega)$ for the network is

$$\mathbf{H}(\omega) = \frac{R}{R + j\omega L}$$

$$= \frac{10}{10 + j\omega}$$

From Eq. (15.62),

$$\mathbf{V}_o(\omega) = \mathbf{H}(\omega)\mathbf{V}_i(\omega)$$

$$= \frac{50}{(2 + j\omega)(10 + j\omega)}$$

$$= \frac{50}{8}\left(\frac{1}{2 + j\omega} - \frac{1}{10 + j\omega}\right)\ \text{V}$$

Hence, from Table 15.3, we see that

$$v_o(t) = 6.25\big[e^{-2t}u(t) - e^{-10t}u(t)\big]\ \text{V}$$

(b) In this case, since $v_i(t) = 5\cos 2t$,

$$\mathbf{V}_i(\omega) = 5\pi\delta(\omega - 2) + 5\pi\delta(\omega + 2)\ \text{V}$$

The output voltage in the frequency domain is then

$$\mathbf{V}_o(\omega) = \frac{50\pi\big[\delta(\omega - 2) + \delta(\omega + 2)\big]}{(10 + j\omega)}$$

Using the inverse Fourier transform gives us

$$v_o(t) = \mathcal{F}^{-1}\big[\mathbf{V}_o(\omega)\big] = \frac{1}{2\pi}\int_{-\infty}^{\infty} 50\pi\,\frac{\delta(\omega - 2) + \delta(\omega + 2)}{10 + j\omega}\,e^{j\omega t}\,d\omega$$

Employing the sampling property of the unit impulse function, we obtain

$$v_o(t) = 25\left(\frac{e^{j2t}}{10 + j2} + \frac{e^{-j2t}}{10 - j2}\right)$$

$$= 25\left(\frac{e^{j2t}}{10.2e^{j11.31°}} + \frac{e^{-j2t}}{10.2e^{-j1.31°}}\right)$$

$$= 4.90\cos(2t - 11.31°)\ \text{V}$$

This result can be easily checked using phasor analysis. ❑

EXAMPLE 15.18

Consider the network shown in Fig. 15.17a. This network represents a simple low-pass filter as shown in Chapter 12. We wish to illustrate the impact of this network on the input signal by examining the frequency characteristics of the output signal and the relationship between the 1-Ω or normalized energy at the input and output of the network.

SOLUTION   The network transfer function is

$$\mathbf{H}(\omega) = \frac{1/RC}{1/RC + j\omega} = \frac{5}{5 + j\omega} = \frac{1}{1 + 0.2j\omega}$$

The Fourier transform of the input signal is

$$\mathbf{V}_i(\omega) = \frac{20}{20 + j\omega} = \frac{1}{1 + 0.05j\omega}$$

Then, using Eq. (15.62), the Fourier transform of the output is

$$\mathbf{V}_o(\omega) = \frac{1}{(1 + 0.2j\omega)(1 + 0.05j\omega)}$$

Using the techniques of Chapter 12, we note that the straight-line log-magnitude plot (frequency characteristic) for these functions is shown in Figs. 15.17b–d. Note that

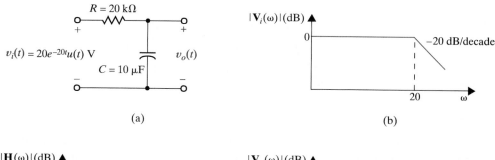

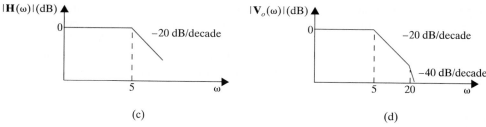

Figure 15.17 Low-pass filter, its frequency characteristic, and its and output spectra.

the low-pass filter passes the low frequencies of the input signal but attenuates the high frequencies.

The normalized energy at the filter input is

$$W_i = \int_0^\infty \left(20e^{-20t}\right)^2 dt$$

$$= \frac{400}{-40} e^{-40t} \Big|_0^\infty$$

$$= 10 \text{ J}$$

The normalized energy at the filter output can be computed using Parseval's theorem. Since

$$\mathbf{V}_o(\omega) = \frac{100}{(5 + j\omega)(20 + j\omega)}$$

and

$$\left|\mathbf{V}_o(\omega)\right|^2 = \frac{10^4}{\left(\omega^2 + 25\right)\left(\omega^2 + 400\right)}$$

$\left|\mathbf{V}_o(\omega)\right|^2$ is an even function, and therefore

$$W_o = 2\left(\frac{1}{2\pi}\right) \int_0^\infty \frac{10^4 \, d\omega}{\left(\omega^2 + 25\right)\left(\omega^2 + 400\right)}$$

However, we can use the fact that

$$\frac{10^4}{\left(\omega^2 + 25\right)\left(\omega^2 + 400\right)} = \frac{10^4/375}{\omega^2 + 25} - \frac{10^4/375}{\omega^2 + 400}$$

Then

$$W_o = \frac{1}{\pi}\left(\int_0^\infty \frac{10^4/375}{\omega^2 + 25}\,d\omega - \int_0^\infty \frac{10^4/375}{\omega^2 + 400}\,d\omega\right)$$

$$= \frac{10^4}{375}\left(\frac{1}{\pi}\right)\left[\frac{1}{5}\left(\frac{\pi}{2}\right) - \frac{1}{20}\left(\frac{\pi}{2}\right)\right]$$

$$= 2.0\text{ J} \qquad\qquad \square$$

Example 15.18 illustrates the effect that $\mathbf{H}(\omega)$ has on the frequency spectrum of the input signal. In general, $\mathbf{H}(\omega)$ can be selected to shape that spectrum in some prescribed manner. As an illustration of this effect, consider the *ideal* frequency spectrums shown in Fig. 15.18. Figure 15.18a shows an ideal input magnitude spectrum $|\mathbf{V}_i(\omega)|$. $|\mathbf{H}(\omega)|$ and the output magnitude spectrum $|\mathbf{V}_o(\omega)|$, which are related by Eq. (15.62), are shown in Figs. 15.18b–e for an *ideal* low-pass, high-pass, band-pass, and band-elimination filter, respectively.

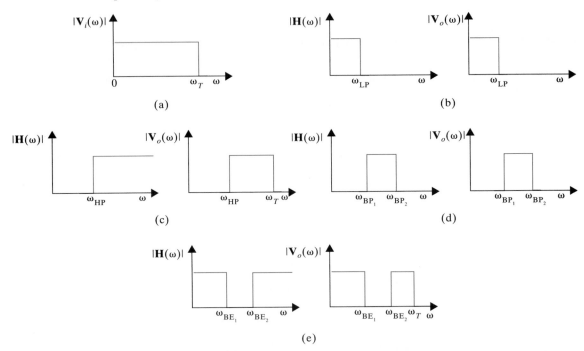

Figure 15.18 Frequency spectra for the input and output of ideal low-pass, high-pass, band-pass, and band-elimination filters.

EXTENSION EXERCISES

E15.15 Compute the total 1-Ω energy content of the signal $v_i(t) = e^{-2t}u(t)$ **V** using both the time domain and frequenmcy domain approaches.

ANSWER: $W_T = 0.25$ J.

E15.16   Compute the 1-Ω energy content of the signal $v_1(t) = e^{-2t}u(t)$ **V** in the frequency from 0 to 1 rad/s.

ANSWER: $W = 0.07$ J.

We note that by using Parseval's theorem we can compute the total energy content of a signal using either a time domain or frequency domain approach. However, the frequency domain is more flexible in that it permits us to determine the energy content of a signal within some specified frequency band.

15.3 APPLICATIONS

EXAMPLE 15.19

In AM (amplitude modulation) radio, there are two very important waveforms, the signal, $s(t)$, and the carrier. All of the information we desire to transmit, voice, music, and so on, is contained in the signal waveform, which is in essence transported by the carrier. Therefore, the Fourier transform of $s(t)$ contains frequencies from about 50 Hz to 20,000 Hz. The carrier, $c(t)$, is a sinusoid oscillating at a frequency much greater than those in $s(t)$. For example, the FCC (Federal Communications Commission) rules and regulations have allocated the frequency range 540 kHz to 1.7 MHz for AM radio station carrier frequencies. Even the lowest possible carrier frequency allocation of 540 kHz is much greater than the audio frequencies in $s(t)$. In fact, when a station broadcasts its call letters and frequency, they are telling you the carrier's frequency, which the FCC assigned to that station!

In simple cases, the signal, $s(t)$, is modified to produce a voltage of the form,

$$v(t) = [A + s(t)]\cos(\omega_c t)$$

where A is a constant and ω_c is the carrier frequency in rad/s. This voltage, $v(t)$, with the signal "coded" within, is sent to the antenna and is broadcast to the public whose radios "pick up" a faint replica of the waveform $v(t)$.

Let us plot the magnitude of the Fourier transform of both $s(t)$ and $v(t)$ given that $s(t)$ is

$$s(t) = \cos(2\pi f_a t)$$

where f_a is 1000 Hz, the carrier frequency is 900 kHz, and the constant A is unity.

SOLUTION   The Fourier transform of $s(t)$ is

$$\mathbf{S}(\omega) = \mathcal{F}[\cos(\omega_a t)] = \pi\delta(\omega - \omega_a) + \pi\delta(\omega + \omega_a)$$

and is shown in Fig. 15.19.

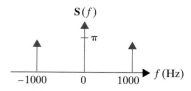

Figure 15.19 Fourier transform magnitude for $s(t)$ versus frequency.

The voltage $v(t)$ can be expressed in the form

$$v(t) = [1 + s(t)]\cos(\omega_c t) = \cos(\omega_c t) + s(t)\cos(\omega_c t)$$

The Fourier transform for the carrier is

$$\mathcal{F}[\cos(\omega_c t)] = \pi\delta(\omega - \omega_c) + \pi\delta(\omega + \omega_c)$$

The term $s(t)\cos(\omega_c t)$ can be written as

$$s(t)\cos(\omega_c t) = s(t)\left\{\frac{e^{j\omega_c t} + e^{-j\omega_c t}}{2}\right\}$$

Using the property of modulation as given in Eq. 15.58, we can express the Fourier transform of $s(t)\cos(\omega_c t)$ as

$$\mathcal{F}[s(t)\cos(\omega_c t)] = \frac{1}{2}\{\mathbf{S}(\omega - \omega_c) + \mathbf{S}(\omega + \omega_c)\}$$

Employing $\mathbf{S}(\omega)$, we find

$$\mathcal{F}[s(t)\cos(\omega_c t)] = \mathcal{F}[\cos(\omega_a t)\cos(\omega_c t)]$$

$$= \frac{\pi}{2}\{\delta(\omega - \omega_a - \omega_c) + \delta(\omega + \omega_a - \omega_c)$$

$$+ \delta(\omega - \omega_a + \omega_c) + \delta(\omega + \omega_a + \omega_c)\}$$

Finally, the Fourier transform of $v(t)$ is

$$\mathbf{V}(\omega) = \frac{\pi}{2}\{2\delta(\omega - \omega_c) + 2\delta(\omega + \omega_c) + \delta(\omega - \omega_a - \omega_c)$$

$$+ \delta(\omega + \omega_a - \omega_c) + \delta(\omega - \omega_a + \omega_c) + \delta(\omega + \omega_a + \omega_c)\}$$

which is shown in Fig. 15.20. Notice that $\mathbf{S}(\omega)$ is centered about the carrier frequency. This is the effect of modulation.

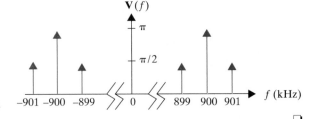

Figure 15.20 Fourier transform of the transmitted waveform $v(t)$ versus frequency.

15.4 CIRCUIT DESIGN

EXAMPLE 15.20

Two nearby AM stations are broadcasting at carrier frequencies

$$f_1 = 900 \text{ kHz}$$

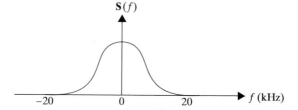

Figure 15.21 Sketch of an arbitrary AM Fourier transform.

and

$$f_2 = 960 \text{ kHz}$$

respectively. To simplify the analysis, we will assume that the information signals, $s_1(t)$ and $s_2(t)$, are identical. It follows that the Fourier transforms $\mathbf{S}_1(\omega)$ and $\mathbf{S}_2(\omega)$ are also identical, and a sketch of what they might look like is shown in Fig. 15.21.

The broadcast waveforms are

$$v_1(t) = \left[1 + s_1(t)\right] \cos(\omega_1 t)$$

and

$$v_2(t) = \left[1 + s_2(t)\right] \cos(\omega_2 t)$$

An antenna in the vicinity will "pick up" both broadcasts. Assuming $v_1(t)$ and $v_2(t)$ are of equal strength at the antenna, the voltage received is

$$v_r(t) = K\left[v_1(t) + v_2(t)\right]$$

where K is a constant much less than 1. (Typical antenna voltages are in the μV-to-mV range). A sketch of the Fourier transform of $v_r(t)$ is shown in Fig. 15.22.

Before passing $v_r(t)$ on to amplifying and decoding circuitry, we must first employ a tuner to select a particular station. Let us design an *RLC* band-pass filter that contains a variable capacitor to serve as our tuner. Such a circuit is shown in Fig. 15.23.

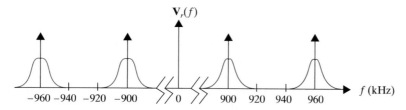

Figure 15.22 Fourier transform of the antenna waveform, $v_r(t)$.

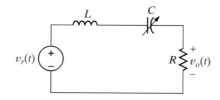

Figure 15.23 *RLC* band-pass filter-tuner circuit.

SOLUTION   The transfer function is easily found to be

$$\mathbf{G}_v(s) = \frac{\mathbf{V}_o(s)}{\mathbf{V}_r(s)} = \frac{s\left(\dfrac{R}{L}\right)}{s^2 + s\left(\dfrac{R}{L}\right) + \dfrac{1}{LC}}$$

As shown in Chapter 12, the center frequency and bandwidth can be expressed in hertz as

$$f_o = \frac{1}{2\pi\sqrt{LC}}$$

and

$$\mathrm{BW} = \frac{1}{2\pi}\frac{R}{L}$$

Since the two carrier frequencies are separated by only 60 kHz, the filter bandwidth should be less than 60 kHz. Let us arbitrarily choose a bandwidth of 10 kHz and $R = 10\ \Omega$. Based upon this selection, our design involves determining the resulting L and C values. From the expression for bandwidth,

$$L = \frac{1}{2\pi}\frac{R}{\mathrm{BW}}$$

or

$$L = 159.2\ \mu\mathrm{H}$$

Placing the center frequency at 900 kHz, we find C to be

$$C = \frac{1}{L[2\pi f_o]^2}$$

or

$$C = 196.4\ \mathrm{pF}$$

To tune to 960 kHz we need only change C to 172.6 pF and the bandwidth is unchanged. A Bode plot of the magnitude of $\mathbf{G}_v(s)$ tuned to 960 kHz is shown in Fig. 15.24. Notice that the broadcast at 900 kHz, while attenuated, is not completely eliminated. If this is a problem, we can either narrow the bandwidth through R and/or L or design a more complex tuner filter.   ❏

15.5 SUMMARY

- A periodic function, its representation using a Fourier series, and some of the useful properties of a Fourier series are outlined here.

A periodic function

$$f(t) = f(t + nT_0), \qquad n = 1, 2, 3, \dots \text{ and } T_0 \text{ is the period}$$

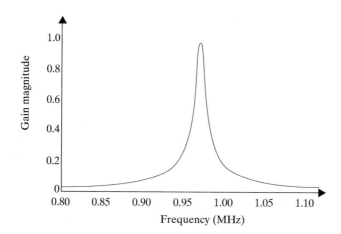

Figure 15.24 Bode plot for *RLC* tuner circuit of Figure 15.23.

Exponential Fourier series of a periodic function

$$f(t) = \sum_{n=-\infty}^{\infty} c_n e^{jn\omega_0 t}, \qquad c_n = \frac{1}{T_0} \int_{t_1}^{t_1+T_0} f(t) e^{-jn\omega_0 t} \, dt$$

Trigonometric Fourier series of a periodic function

$$f(t) = a_o + \sum_{n=1}^{\infty} \left(a_n \cos n\omega_0 t + b_n \sin n\omega_0 t \right)$$

$$a_n = \frac{2}{T_0} \int_{t_1}^{t_1+T_0} f(t) \cos n\omega_0 t \, dt, \qquad b_n = \frac{2}{T_0} \int_{t_1}^{t_1+T_0} f(t) \sin n\omega_0 t \, dt \qquad \text{and}$$

$$a_o = \frac{1}{T_0} \int_{t_1}^{t_1+T_0} f(t) \, dt$$

Even symmetry of a periodic function

$$f(t) = f(-t)$$

$$a_n = \frac{4}{T_0} \int_0^{T_0/2} f(t) \cos n\omega_0 t \, dt, \qquad b_n = 0, \qquad a_0 = \frac{2}{T_0} \int_0^{T_0/2} f(t) \, dt$$

Odd symmetry of a periodic function

$$f(t) = -f(-t)$$

$$a_n = 0, \qquad b_n = \frac{4}{T_0} \int_0^{T_0/2} f(t) \sin n\omega_0 t \, dt, \qquad \text{and} \qquad a_0 = 0$$

Half-wave symmetry of a periodic function

$$f(t) = -f(t - T_0/2)$$

$$a_n = b_n = 0, \qquad \text{for } n \text{ even}, \qquad a_n = \frac{4}{T_0} \int_0^{T_0/2} f(t) \cos n\omega_0 t \, dt \qquad \text{for } n \text{ odd}$$

$$b_n = \frac{4}{T_0} \int_0^{T_0/2} f(t) \sin n\omega_0 t \, dt \qquad \text{for } n \text{ odd} \qquad \text{and } a_0 = 0$$

Time shifting of a periodic function

$$f(t - t_0) = \sum_{n=-\infty}^{\infty} \left(c_n e^{-jn\omega_0 t_0} \right) e^{jn\omega_0 t}$$

Frequency spectrum of a periodic function   A Fourier series contains discrete frequency components, called line spectra.

Steady-state response of a periodic function input   The periodic function input is expressed as a Fourier series, and phasor analysis is used to determine the response of each component of the series. Each component is transformed to the time domain and superposition is used to determine the total output.

- The Fourier transform, its features and properties, as well as its use in circuit analysis is outlined here.

Fourier transform for an aperiodic function

$$F(\omega) = \int_{-\infty}^{\infty} f(t) e^{-j\omega t} \qquad \text{and} \qquad f(t) = \frac{1}{2\pi} \int_{-\infty}^{\infty} F(\omega) e^{j\omega t} \, d\omega$$

Fourier transform pairs and properties   The Fourier transform pairs in Table 15.2 and the properties in Table 15.3 can be used together to transform time domain functions to the frequency domain and vice versa.

Parseval's theorem for determining the energy content of a signal

$$\int_{-\infty}^{\infty} f^2(t) \, dt = \frac{1}{2\pi} \int_{-\infty}^{\infty} |\mathbf{F}(\omega)|^2 \, d\omega$$

Network response to an aperiodic input   An aperiodic input $x(t)$ can be transformed to the frequency domain as $\mathbf{X}(\omega)$. Then using the network transfer function $\mathbf{H}(\omega)$, the output can be computed as $\mathbf{Y}(\omega) = \mathbf{H}(\omega)\mathbf{X}(\omega)$. $y(t)$ can be obtained transforming $\mathbf{Y}(\omega)$ to the time domain.

PROBLEMS

15.1   Find the exponential Fourier series for the signal shown in Fig. P15.1.

Follow the technique outlined in Example 15.1.

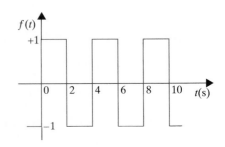

Figure P15.1

15.2 Find the exponential Fourier series for the periodic pulse train shown in Fig. P15.2.

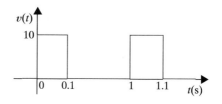

Figure P15.2

Similar to P15.1

15.3 Find the exponential Fourier series for the periodic signal shown in Fig. P15.3.

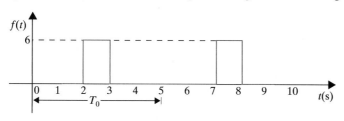

Figure P15.3

Similar to P 15.1

15.4 Compute the exponential Fourier series for the waveform that is the sum of the two waveforms in Fig. P15.4 by computing the exponential Fourier series of the two waveforms and adding them.

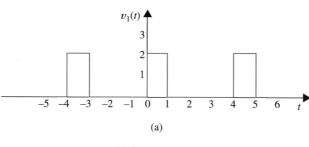

(a)

Similar to P15.1

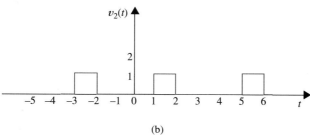

(b)

Figure P15.4

15.5   Find the exponential Fourier series for the signal shown in Fig. P15.5.

Similar to P15.1

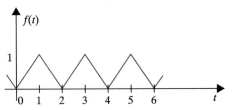

Figure P15.5

15.6   Given the waveform in Fig. P15.6, determine the type of symmetry that exists if the origin is selected at: (a) l_1 and (b) l_2.

Equations (15.16), (15.20), and (15.24) are useful here.

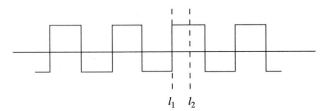

Figure P15.6

15.7   What type of symmetry is exhibited by the two waveforms in Fig. P15.7.

Similar to P15.6

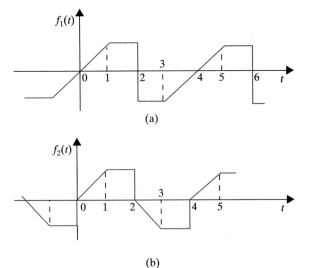

Figure P15.7

15.8   Given the waveform in Fig. P15.8 show that

$$f(t) = \frac{A}{2} + \sum_{n=1}^{\infty} \frac{-A}{n\pi} \sin \frac{2\pi n}{T_0} t$$

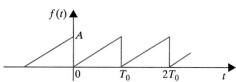

Figure P15.8

Similar to Example 15.4

15.9   Derive the trigonometric Fourier series for the waveform shown in Fig. P15.9.

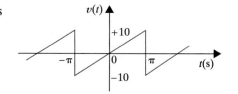

Figure P15.9

Similar to Example 15.3

15.10   Find the trigonometric Fourier series coefficients for the waveform in Fig. P15.10.

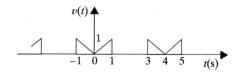

Figure P15.10

Similar to Example 15.3 with different symmetry.

15.11   Find the trigonometric Fourier series coefficients for the waveform in Fig. P15.11.

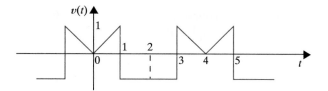

Figure P15.11

Similar to P15.10

15.12   Find the trigonometric Fourier series coefficients for the waveform in Fig. P15.12.

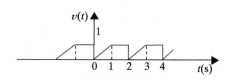

Figure P15.12

Similar to Example 15.5

15.13 Find the trigonometric Fourier series coefficients for the waveform in Fig. P15.13.

Similar to P15.12

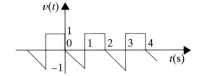

Figure P15.13

15.14 Find the trigonometric Fourier series coefficients for the waveform in Fig. P15.14.

Similar to P15.12

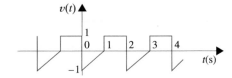

Figure P15.14

15.15 Find the trigonometric Fourier series coefficients for the waveform in Fig. P15.15.

Similar to P15.12

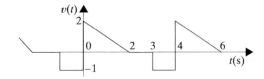

Figure P15.15

15.16 Find the trigonometric Fourier series for the waveform shown in Fig. P15.16.

Similar to P15.12

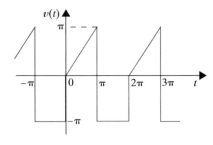

Figure P15.16

Similar to P15.9

15.17 Find the trigonometric Fourier series coefficients for the waveform in Fig. P15.7a.

15.18 Derive the trigonometric Fourier series for the function $v(t) = A|\sin t|$ as shown in Fig. P15.18.

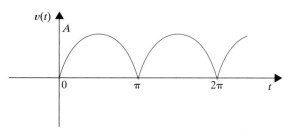

Figure P15.18

Similar to P15.10

15.19 Derive the trigonometric Fourier series for the waveform shown in Fig. P15.19.

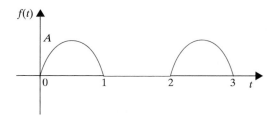

Figure P15.19

Similar to P15.12

15.20 Find the trigonometric Fourier series for the waveform shown in Fig. P15.20.

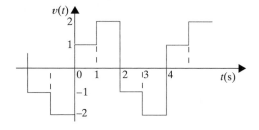

Figure P15.20

Similar to P15.8

15.21 Derive the trigonometric Fourier series for the function shown in Fig. P15.21.

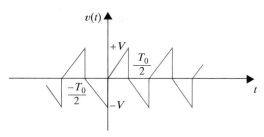

Figure P15.21

Similar to P15.12

15.22 Find the trigonometric Fourier series for the periodic waveform shown in Fig. P15.22.

Similar to P15.12

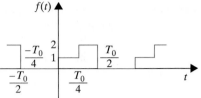

Figure P15.22

15.23 The discrete line spectrum for a periodic function $f(t)$ is shown in Fig. P15.23. Determine the expression for $f(t)$.

Reverse the technique in Example 15.7.

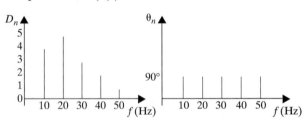

Figure P15.23

15.24 The amplitude and phase spectra for a periodic function $v(t)$ that has only a small number of terms is shown in Fig. P15.24. Determine the expression for $v(t)$ if $T_0 = 0.1$ s.

Similar to P15.23

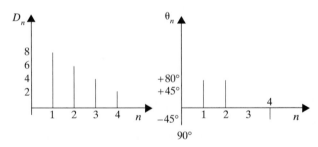

Figure P15.24

15.25 Plot the first four terms of the amplitude and phase spectra for the signal

Similar to Example 15.7

$$f(t) = \sum_{\substack{n=1 \\ n \text{ odd}}}^{\infty} \frac{-2}{n\pi} \sin \frac{n\pi}{2} \cos n\omega_0 t + \frac{6}{n\pi} \sin n\omega_0 t$$

15.26 Determine the steady-state response of the current $i_o(t)$ in the circuit shown in Fig. P15.26 if the input voltage is described by the waveform shown in Problem 15.9.

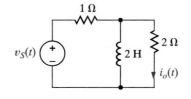

Similar to Example 15.8

Figure P15.26

15.27 Find the steady-state current $i(t)$ in the network in Fig. P15.27 if the input signal is shown in Fig. P15.29.

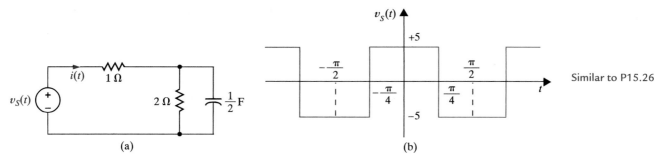

Similar to P15.26

(a)   (b)

Figure P15.27

15.28 If the input voltage in Problem 17.28 is

$$v_S(t) = 1 - \frac{2}{\pi} \sum_{n=1}^{\infty} \frac{1}{n} \sin 0.2\, \pi nt \text{ V}$$

Similar to P15.26

find the expression for the steady-state current $i_o(t)$.

15.29 Determine the first three terms of the steady-state voltage $v_o(t)$ in Fig. P15.29 if the input voltage is a periodic signal of the form

$$v(t) = \frac{1}{2} + \sum_{n=1}^{\infty} \frac{1}{n\pi} (\cos n\pi - 1) \sin nt \text{ V}$$

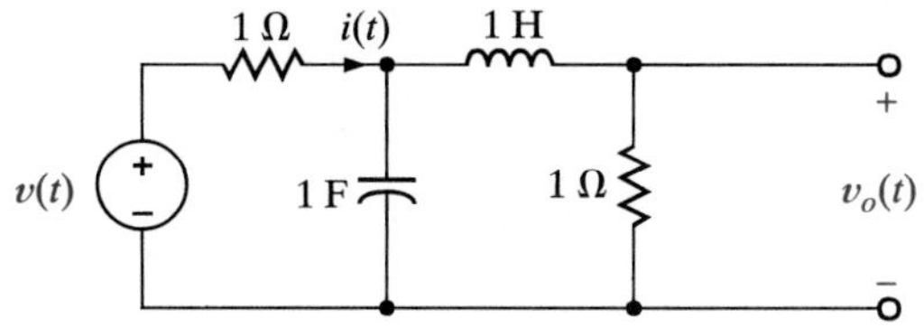

Similar to P15.26

Figure P15.29

15.30 The current $i_S(t)$ shown in Fig. P15.30a is applied to the circuit shown in Fig. P15.30b. Determine the expression for the steady-state current $i_o(t)$ using the first four harmonics.

Similar to P15.26

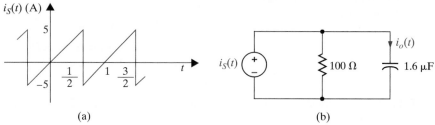

(a)   (b)

Figure P15.30

15.31 Determine the steady-state voltage $v_o(t)$ in the network in Fig. P15.31 if the input current is given in Fig. P15.31.

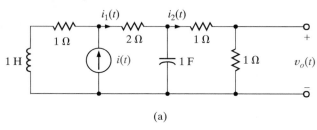

(a)

Similar to P15.26

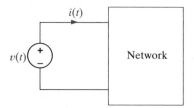

(b)

Figure P15.31

15.32 Find the average power absorbed by the network in Fig. P15.32 if

$$v(t) = 12 + 6\cos(377t - 10°) + 4\cos(754t - 60°) \text{ V}$$
$$i(t) = 0.2 + 0.4\cos(377t - 150°)$$
$$-0.2\cos(754t - 80°) + 0.1\cos(1131t - 60°) \text{ A}$$

Similar to Example 15.9

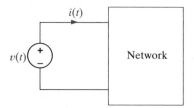

Figure P15.32

15.33 Find the average power absorbed by the network in Fig. P15.33 if
$v(t) = 60 + 36\cos(377t + 45°) + 24\cos(754t - 60°)$ V.

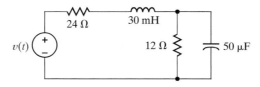

Figure P15.33

Similar to P15.32

15.34 Find the average power absorbed by the 12-Ω resistor in the network in Fig.
P15.33 if $v(t) = 50 + 25\cos(377t - 45°) + 12.5\cos(754t + 45°)$ V.

Similar to P15.32

Section 15.2

15.35 Determine the Fourier transform of the waveform shown in Fig. P15.35.

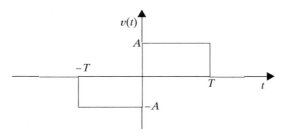

Figure P15.35

Use the technique in Example
15.10.

15.36 Derive the Fourier transform for the following functions:
(a) $f(t) = e^{-2t}\cos 4tu(t)$.
(b) $f(t) = e^{-2t}\sin 4tu(t)$.

Use Tables 15.1 and 15.3.

15.37 Show that

$$\mathcal{F}[f_1(t)f_2(t)] = \frac{1}{2\pi}\int_{-\infty}^{\infty} \mathbf{F}_1(x)\mathbf{F}_2(\omega - x)\, dx$$

Use time convolution.

15.38 Find the Fourier transform of the function $f(t) = e^{-a|t|}$.

Use symmetry.

15.39 Find the Fourier transform of the function $f(t) = 12e^{-2|t|}\cos 4t$.

Use symmetry.

Similar to Example 15.17

15.40 Determine the output signal $v_o(t)$ of a network with input signal $v_i(t) = 3e^{-t}u(t)$ and network impulse response $h(t) = e^{-2t}u(t)$. Assume that all initial conditions are zero.

Similar to P15.40

15.41 The input signal to a network is $v_i(t) = e^{-3t}u(t)$. The transfer function of the network is $\mathbf{H}(j\omega) = 1/(j\omega + 4)$. Find the output of the network $v_o(t)$ if the initial conditions are zero.

Similar to P15.40

15.42 Use the transform technique to find $v_o(t)$ in the network in Fig. P15.31a if (a) $i(t) = 4(e^{-t} - e^{-2t})u(t)$ A and (b) $i(t) = 12 \cos 4t$ A.

15.43 The input signal for the network in Fig. P15.43 is $v_i(t) = 10e^{-5t}u(t)$ V. Determine the total 1-Ω energy content of the output $v_o(t)$.

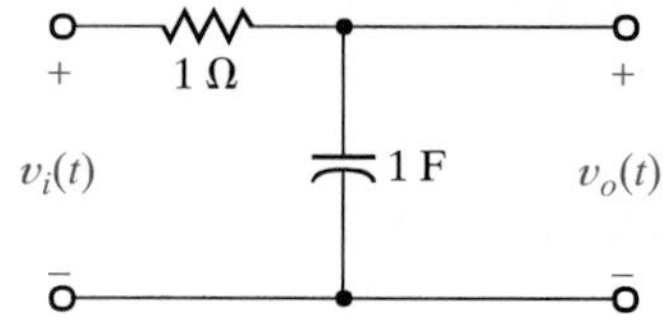

Similar to Example 15.18

Figure P15.43

15.44 Use the Fourier transform to find $i(t)$ in the network in Fig. P15.44 if $v_i(t) = 2e^{-t}u(t)$.

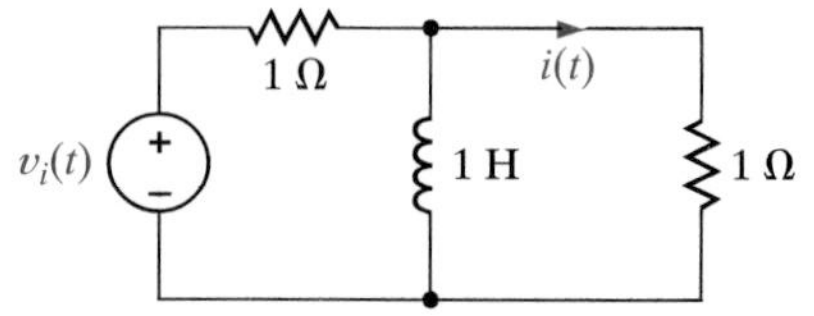

Similar to P15.40

Figure P15.44

15.45 Use the transform technique to find $v_o(t)$ in the network in Fig. P15.45 if (a) $v_1(t) = 4e^{-t}u(t)$ V and (b) $v_1(t) = 4(e^{-2t} + 2e^{-4t})u(t)$ V.

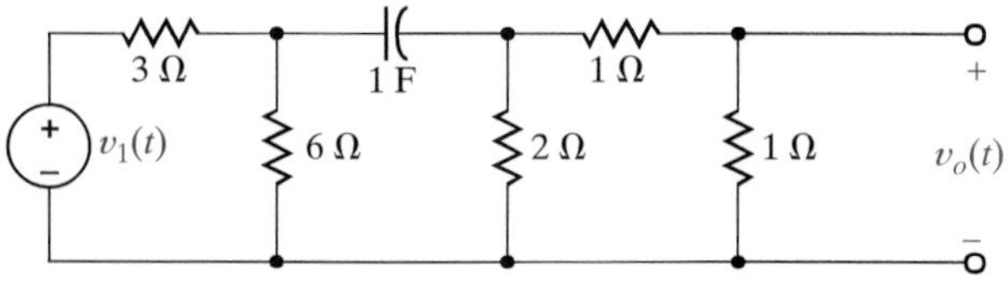

Similar to P15.40

Figure P15.45

Similar to P15.43

15.46 Compute the 1-Ω energy content of the signal $v_o(t)$ in Fig. P15.43 in the frequency range from $\omega = 2$ to $\omega = 4$ rad/s.

Similar to P15.43

15.47 Determine the 1-Ω energy content of the signal $v_o(t)$ in Fig. P15.43 in the frequency range from 0 to 1 rad/s.

15.48 Determine the voltage $v_o(t)$ in the circuit shown in Fig. 15.48, using the Fourier transform if
$i_i(t) = 2e^{-4t}u(t)$ A.

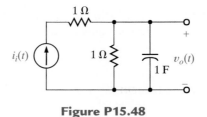

Figure P15.48

Similar to P15.40

15.49 Compare the 1-Ω energy at both the input and output of the network in Problem 15.48 for the given input forcing function.

Similar to P15.43

Advanced Problems

15.50 Determine the voltage $v_o(t)$ in the circuit shown in Fig P15.50a if the input signal is shown in Fig. P15.59.

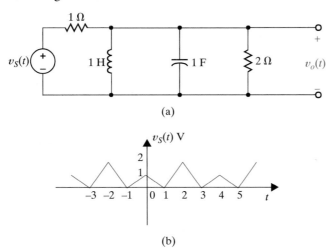

(a)

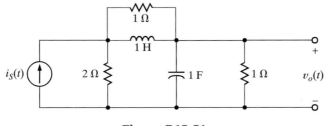

(b)

Figure P15.50

15.51 Determine $v_o(t)$ in the circuit shown in Fig. P15.51 using the Fourier transform   if the input signal is $i_S(t) = (e^{-2t} + \cos t)u(t)$ A.

Figure P15.51

15.52 Determine $v_o(t)$ in the circuit shown in Fig. P15.52 using the Fourier transform if the input signal is $v_S(t) = \cos(7 \times 10^7 t) + \cos(1 \times 10^5 t)\,\text{V}$.

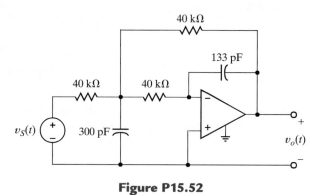

Figure P15.52

Circuit Equations Via Network Topology

We now generalize the methods of nodal and loop analyses that we have presented. The vehicle that we will employ to accomplish this is *topology*. For our purposes topology refers to the properties that relate to the geometry of the circuit. These properties remain unchanged even if the circuit is bent into any other shape provided that no parts are cut and no new connections are made.

A.1 BASIC DEFINITIONS

To provide a basis for the concepts to be discussed next, consider the network shown in Fig. A.1. Suppose that we are required to find all the currents and voltages in this circuit. This circuit seems to be much more complicated than those we considered earlier. It is more complicated because it is what is called a *nonplanar* circuit. A *planar* circuit is one that can be drawn on a plane surface with no crossovers; that is, no branch passes over any other branch. And, of course, a nonplanar circuit is one that is not planar.

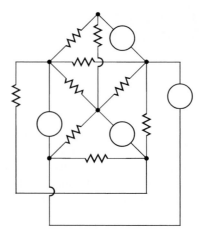

Figure A.1 Example of a nonplanar circuit.

Let us now present some topological concepts to aid us in our discussion. Since the geometrical properties of a circuit are independent of the circuit elements that are contained in each branch, we will simply represent each network branch by a line segment. Such a drawing of the circuit is called a *graph*.

EXAMPLE A.1

Find the graph for the circuit shown in Fig. A.2a.

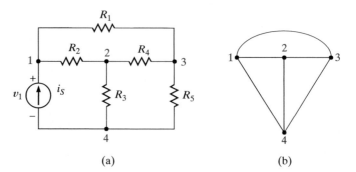

Figure A.2 Circuit and its associated graph.

SOLUTION   The graph associated with the circuit shown in Fig. A.2a is given in Fig. A.2b.   ❏

As can be seen from the example, the graph consists of a number of interconnected nodes. If a path exists from any node to every other node, the graph is said to be *connected*.

A *tree* of the graph is defined as a set of branches that connects every node to every other node via some path without forming a closed path. In general, there exists a number of different trees for any given circuit.

EXAMPLE A.2

Find three possible trees for the graph shown in Fig. A.2b.

SOLUTION   Three possible trees for the graph shown in Fig. A.2b are given in Fig. A.3.

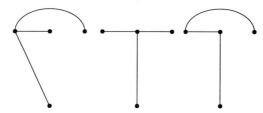

Figure A.3 Trees for the graph in Fig. A.2b.

If a graph for a network is known and a particular tree is specified, those branches of the graph that are not part of the tree form what is called the *cotree*. The cotree consists of what we call *links*.   ❏

EXAMPLE A.3

Find the links for the cotrees corresponding to the trees in Fig. A.3.

SOLUTION   The links that belong to the cotrees that correspond to the trees shown in Fig. A.3 are shown dashed in Fig. A.4.

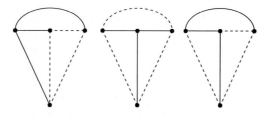

Figure A.4 Cotrees consisting of links for the trees shown in Fig. A.3.

Note that it is impossible to specify categorically a particular branch of a graph as a link, since it may be a link for one choice of tree and not for another.

Finally, we wish to define what is called a *cut set*. A cut set is simply a minimum set of branches that, when cut, will divide the graph into two separate parts. Hence it is impossible to go from a node in one part of the graph to a node in another part without passing through a branch of the cut set. ❑

EXAMPLE A.4

Find two cut sets for the graph shown in Fig. A.2b.

SOLUTION   Two cut sets for the graph shown in Fig. A.2b are shown in Fig. A.5. Note that in one case node 4 is separated from nodes 1, 2, and 3; in the other case nodes 1 and 2 are separated from nodes 3 and 4.

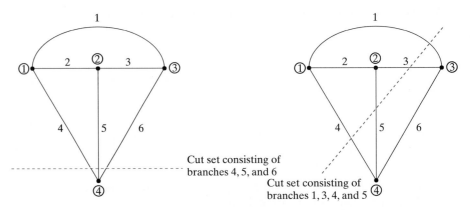

Figure A.5 Two cut sets for the graph in Fig. A.2b. ❑

Suppose that we have a graph that represents a circuit, and let us define the following parameters:

$$L = \text{number of links}$$
$$B = \text{number of branches in the graph}$$
$$N = \text{number of nodes in the graph}$$

Now suppose that we remove all the branches from the graph so that only nodes remain. In order to construct a tree, we begin by placing one branch between two nodes. We continue adding branches to the tree without forming any loops. Each time we add an additional branch we add one more node. Note that the addition of every successive branch connects one node, except for the first branch, which connects two nodes. Hence, there is one more node than there are branches on the tree. Therefore, in general, a tree consists of $N - 1$ branches. Since the entire graph contains B branches, the number of links is given by the expression

$$L = B - (N - 1) = B - N + 1 \qquad \textbf{A.1}$$

EXAMPLE A.5

Find a tree for the network shown in Fig. A.1.

SOLUTION   A specific tree for the network shown in Fig. A.1 is illustrated in Fig. A.6. Note that the graph of the network satisfies the relationship in Eq. A.1.

$$L = B - N + 1$$
$$8 = 13 - 6 + 1$$

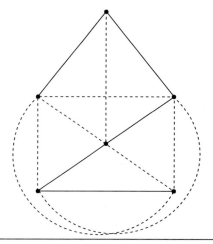

Figure A.6 Tree and its corresponding links for the network in Fig. A.1.

❑

EXTENSION EXERCISES

EA.1   Given the graph for a network in Fig. EA.1a, state whether the tree–cotree combinations in Fig. EA.1b to *f* form a valid set of tree branches and links.

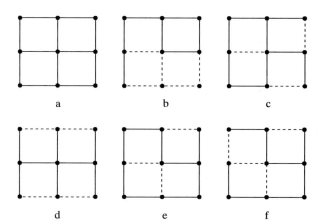

Figure EA.1

ANSWER: (b) No, (c) yes, (d) yes, (e) no, (f) yes.

EA.2   Compute the proper number of links for the graphs in Fig. EA.1a.

ANSWER: $L = 4$.

A.2   GENERAL NODAL ANALYSIS EQUATIONS

We have shown that if we had a *linearly independent* set of $N - 1$ nodal equations for an N-node network, we could determine the necessary node voltages. With this in mind, let us redraw the graphs in Fig. A.5 with assumed directions of currents as shown in Fig. A.7. KCL for either part of the circuit divided by a cut set is called a *cut-set equation*. Note that each cut-set equation is nothing more than a sum of KCL equations written for all nodes in either part of the circuit. For example, the cut-set equation for Fig. A.7a is

$$i_4 - i_5 + i_6 = 0$$

which is KCL at node 4, where $i_k(t)$ represents the kth branch. Similarly, the cut-set equation for Fig. A.7b is

$$i_4 - i_5 + i_3 + i_1 = 0$$

which is the sum of the KCL equations at nodes 3 and 4, or equivalently at nodes 1 and 2:

$$\text{For node 3: } \quad i_1 + i_3 - i_6 = 0$$
$$\text{For node 4: } \quad i_4 - i_5 + i_6 = 0$$

Adding these two equations yields

$$i_4 - i_5 + i_3 + i_1 = 0$$

Because of this relationship between KCL equations and the cut-set equations, we can determine the independent KCL equations for a network by determining the independent

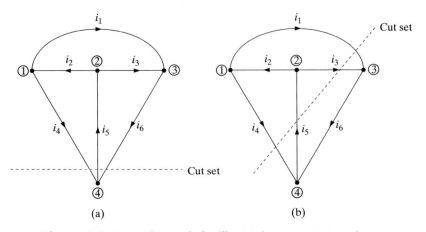

Figure A.7 Example graph for illustrating cut-set equations.

cut-set equations. Furthermore, we can prove [1] that the number of independent cut-set equations (and therefore KCL equations) is exactly $N - 1$.

The cut-set equations are derived from a tree that, as we have shown, contains $N - 1$ branches. If any branch of the tree is cut, the tree will be divided into two parts. Each tree branch that is cut, together with the links that connect the two separate parts of the graph, form a cut set. For example, the cut sets for a specific tree for the graph shown in Fig. A.2 are shown in Fig. A.8. Note that cut set 1 includes tree branch 5 and links 4 and 6. Cut set 2 includes tree branch 2 and links 4, 3, and 6. Cut set 3 includes tree branch 1 and links 3 and 6. Cut sets containing only a single tree branch are called *fundamental cut sets*, and there are $N - 1$ fundamental sets because there are $N - 1$ tree branches.

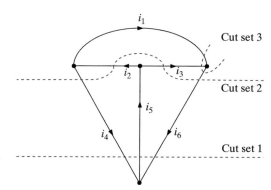

Figure A.8 Illustration of fundamental cut sets.

Let us now demonstrate the use of topology in determining the nodal equations for a network.

EXAMPLE A.6

Write the node equations for the network in Fig. A.2a.

SOLUTION   The circuit in Fig. A.2a is redrawn in Fig. A.9a with assumed directions for currents. A corresponding tree with fundamental cut sets is shown in Fig. A.9b. The cut-set equations for cut sets 1, 2, and 3 are

$$i_1 + i_2 - i_S = 0 \qquad\qquad \textbf{A.2}$$

$$i_1 + i_4 - i_5 = 0 \qquad\qquad \textbf{A.3}$$

$$i_S - i_3 - i_5 = 0 \qquad\qquad \textbf{A.4}$$

Subtracting the second equation from the first and adding the third equation yields

$$i_2 - i_4 - i_3 = 0 \qquad\qquad \textbf{A.5}$$

Any three of these four equations are linearly independent. Selecting equations (A.2), (A.5), and (A.3) and writing the branch currents in terms of the node voltages yields

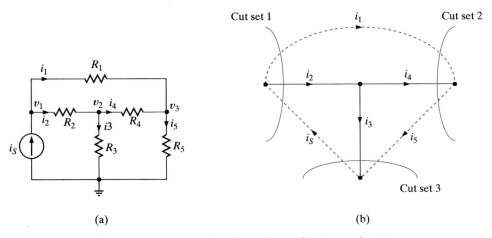

Figure A.9 Network and graph used in Example A.6.

$$\frac{v_1 - v_3}{R_1} + \frac{v_1 - v_2}{R_2} - i_S = 0$$

$$\frac{v_1 - v_2}{R_2} - \frac{v_2}{R_3} + \frac{(v_2 - v_3)}{R_4} = 0$$

$$\frac{v_1 - v_3}{R_1} + \frac{v_2 - v_3}{R_4} - \frac{v_3}{R_5} = 0$$

or

$$v_1\left(\frac{1}{R_1} + \frac{1}{R_2}\right) - v_2\left(\frac{1}{R_2}\right) - v_3\left(\frac{1}{R_1}\right) = i_S$$

$$-v_1\left(\frac{1}{R_2}\right) + v_2\left(\frac{1}{R_2} + \frac{1}{R_3} + \frac{1}{R_4}\right) - v_3\left(\frac{1}{R_4}\right) = 0$$

$$-v_1\left(\frac{1}{R_1}\right) - v_2\left(\frac{1}{R_4}\right) + v_3\left(\frac{1}{R_1} + \frac{1}{R_4} + \frac{1}{R_5}\right) = 0$$

Note that the three equations selected were chosen because they can be recognized as the nodal equations for the network. ❏

EXAMPLE A.7

We wish to derive the nodal equations for the network in Fig. 3.6. A graph for this network is shown in Fig. A.10.

SOLUTION   The cut-set equations for this graph are

$$i_1 - i_A + i_2 - i_3 = 0$$

$$i_3 - i_2 + i_4 + i_B = 0$$

$$i_3 + i_5 + i_B = 0$$

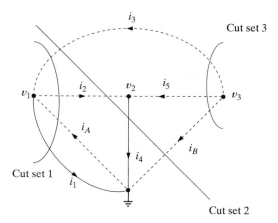

Figure A.10 Graph and cut sets for the network shown in Fig. 3.6.

Subtracting the third equation from the second equation and using the resultant equation with the first and third equations yields

$$i_1 + i_2 - i_3 = i_A$$
$$-i_2 + i_4 - i_5 = 0$$
$$i_3 + i_5 = -i_B$$

which are identical to the equations derived earlier.   ❏

Finally, if there are voltage sources present in the network, then, in general, the number of independent KCL equations is $N - 1 - $ the number of voltage sources.

When using node equations to solve circuits containing voltage sources, the following rules govern the formulation of the equations required:

1. Choose one node as the reference and assign a voltage variable to all other nodes. Let N be the total number of nodes.

2. Form a tree including all voltage sources (dependent and independent). Note that there are $N - 1$ tree branches and an equal number of fundamental cut sets. Let M be the number of voltage sources. Hence, there are M fundamental cut sets, each of which contains a voltage source.

3. For each voltage source, write the corresponding constraint equation. There are M constraint equations.

4. For each fundamental cut set containing no voltage source, write the corresponding cut-set equation. There are $N - 1 - M$ linear independent nodal equations.

EXTENSION EXERCISE

EA.3   Given the network in Fig. EA.3a, use the specified graph for the network in Fig. EA.3b to write a proper set of nodal equations for the network.

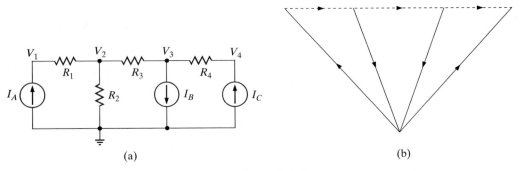

(a)   (b)

Figure EA.3

ANSWER:

$$I_A = \frac{V_1 - V_2}{R_1}, \frac{V_1 - V_2}{R_1} = \frac{V_2}{R_2} + \frac{V_2 - V_3}{R_3},$$

$$\frac{V_2 - V_3}{R_3} = \frac{V_3 - V_4}{R_4} + I_B, \frac{V_3 - V_4}{R_4} = -I_C.$$

A.3 GENERAL LOOP ANALYSIS EQUATIONS

Consider once again the graph shown in Fig. A.8. Imagine that we begin only with the tree branches 1, 2, and 5. If we now add one link at a time to the tree, we create a new loop with each link. For example, adding link 3 creates the loop consisting of that link and tree branches 1 and 2. The other links added one at a time create similar loops. Since we have shown that the number of links is equal to $B - N + 1$, we will construct this same number of loops, each one of which contains only one link. The loops constructed in this manner are called the *fundamental loops*. If KVL equations are written for these fundamental loops, these equations will be independent, since each one contains a link voltage that is not present in any other equation. Hence there are at least $B - N + 1$ independent KVL equations, and in fact, we can prove [1] that there are exactly $B - N + 1$ independent KVL equations.

If there are current sources present in the network, then, in general, the number of independent KVL equations is $B - N + 1 -$ the number of current sources.

When using loop equations to solve circuits containing current sources, the following rules govern the formulation of the equations required:

1. Form a tree excluding all current sources (dependent and independent). Let P be the total number of current sources.

2. Assign each fundamental loop a loop current variable. Let B represent the total number of branches. Note that there are $B - (N - 1)$ links and an equal number of fundamental loops. Hence, there are P fundamental loops, each of which contains a current source.

3. For each fundamental loop containing a current source, write the corresponding constraint equation that relates the loop current to the value of the current source. There are P such constraint equations.

4. For each fundamental loop containing no current source, write the corresponding loop equation. There are $B - (N - 1) - P$ linearly independent loop equations.

Thus far, our analysis has been very general. In the special case of planar networks, windowpanes can be employed to write the necessary KVL equations. Figure A.11, which is a redrawn version of Fig. A.9, illustrates this approach. If all the mesh currents are assumed to be in the same direction and no controlled sources are present, the KVL equations for the windowpanes can be written by inspection, as demonstrated earlier.

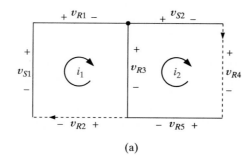

Figure A.11 Graph illustrating windowpanes for circuit in Fig. A.9.

EXAMPLE A.8

Using the topology methods we have just described, we wish to derive the loop equations for the network in Fig. 3.17. A graph of this network is shown in Fig. A.12a. Branch voltages have been selected and the links together with the appropriate tree branches form the fundamental loops.

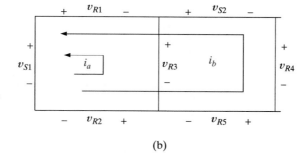

(a)

(b)

Figure A.12 Graph for the network in Fig. 3.17 illustrating fundamental loops.

SOLUTION   The fundamental loop current variables are i_1 and i_2 and the fundamental loop equations are

$$-v_{S1} + v_{R_1} + v_{R_3} + v_{R_2} = 0$$
$$-v_{R_3} + v_{S2} + v_{R_4} + v_{R_5} = 0$$

Using the relationship $v = iR$, we obtain

$$i_1 R_1 + (i_1 - i_2)R_3 + i_1 R_2 = v_{S1}$$
$$-(i_1 - i_2)R_3 + i_2 R_4 + i_2 R_5 = -v_{S2}$$

or

$$i_1(R_1 + R_2 + R_3) - i_2 R_3 = v_{S1}$$
$$-i_1 R_3 + i_2(R_3 + R_4 + R_5) = -v_{S2}$$

which are the same as the equations obtained earlier.

Suppose, however, that we had selected tree branches and corresponding links for the network as shown in Fig. A.12b. The fundamental loop currents are i_a and i_b, and the fundamental loop equations in this case are

$$v_{S1} - v_{R_1} - v_{R_3} - v_{R_2} = 0$$
$$v_{S1} - v_{R_1} - v_{S2} - v_{R_4} - v_{R_5} - v_{R_2} = 0$$

Since $v_{R_1} = (i_a + i_b)R_1$, and so on, the preceding equations can be written in terms of the loop currents as

$$(R_1 + R_2 + R_3)i_a + (R_1 + R_2)i_b = v_{S1}$$
$$(R_1 + R_2)i_a + (R_1 + R_2 + R_4 + R_5)i_b = v_{S1} - v_{S2}$$

These two linearly independent equations will also yield all the currents and voltages in the network. ❏

EXTENSION EXERCISE

EA.4   Given the network in Fig. EA.4a and the graph for the network in Fig. EA.4b, write the proper set of loop equations that will yield all the currents in the network.

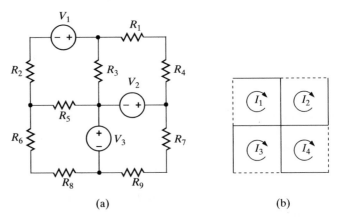

(a)   (b)

Figure EA.4

ANSWER: $I_1(R_2 + R_3 + R_5) - I_2(R_3) - I_3(R_5) - I_4(0) = V_1,$
$-I_1(R_3) + I_2(R_1 + R_3 + R_4) - I_3(0) - I_4(0) = -V_2,$
$-I_1(R_5) - I_2(0) + I_3(R_5 + R_6 + R_8) - I_4(0) = -V_3,$
$-I_1(0) - I_2(0) - I_3(0) + I_4(R_7 + R_9) = V_2 + V_3.$

A.4 SUMMARY

- The standard definitions used in our presentation of network topology (e.g., graph, tree, link, and cut set) are defined.
- The use of cut-set equations in developing the proper node equations for a network is demonstrated.
- The use of links in obtaining the proper set of loop equations for a circuit is explained.

PROBLEMS

A.1 For the graphs in Fig. PA.1, determine the number of tree branches and the number of links.

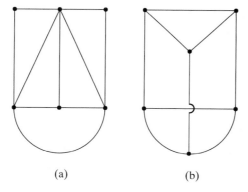

(a)   (b)

Figure PA.1

A.2 For the graphs shown in Fig. PA.2, with a specified tree, find an independent set of KCL equations using cut sets.

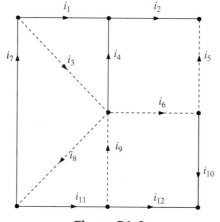

Figure PA.2

A.3 For the graphs shown in Fig. PA.3, with a specified tree, find an independent set of KCL equations using cut sets.

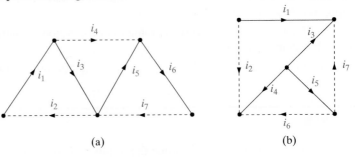

(a) (b)

Figure PA.3

A.4 For the graphs shown in Fig. PA.4, with a specified tree, find an independent set of KVL equations using fundamental loops.

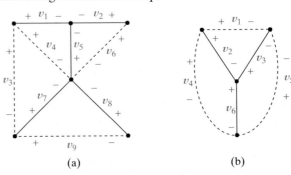

(a) (b)

Figure PA.4

A.5 For the graphs shown in Fig. PA.5 with a specified tree, find an independent set of KVL equations using fundamental loops.

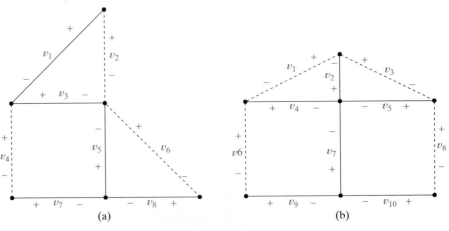

(a) (b)

Figure PA.5

A.6   The tree for the network in Fig. PA.6a is given in Fig. PA.6b. Write the KCL equations in terms of the unknown node voltages.

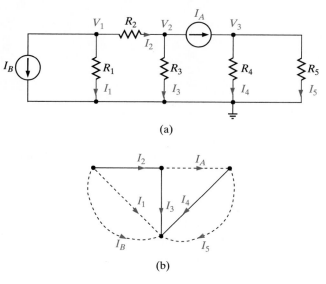

(a)

(b)

Figure PA.6

A.7   Given the tree in Fig. PA.7b for the network in Fig. PA.7a, write the loop equations necessary to solve for all the unknown voltages.

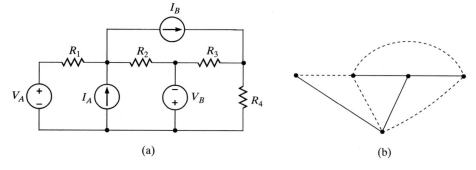

(a)   (b)

Figure PA.7

Complex Numbers

Complex numbers are typically represented in three forms: exponential, polar, or rectangular. In the exponential form a complex number **A** is written as

$$\mathbf{A} = ze^{j\theta} \tag{B.1}$$

The real quantity z is known as the amplitude or magnitude, the real quantity θ is called the *angle* as shown in Fig. B.1 and j is the imaginary operator $j = \sqrt{-1}$. θ, which is the angle between the real axis and **A**, may be expressed in either radians or degrees.

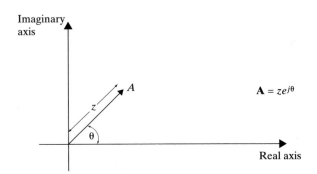

Figure B.1 The exponential form of a complex number.

The polar form of a complex number **A**, which is symbolically equivalent to the exponential form, is written as

$$\mathbf{A} = z \underline{/\theta} \tag{B.2}$$

and the rectangular representation of a complex number is written as

$$\mathbf{A} = x + jy \tag{B.3}$$

where x is the real part of **A** and y is the imaginary part of **A**.

The connection between the various representations of **A** can be seen via Euler's identity, which is

$$e^{j\theta} = \cos\theta + j\sin\theta \tag{B.4}$$

Fig. B.2 illustrates that this function in rectangular form is a complex number with a unit amplitude.

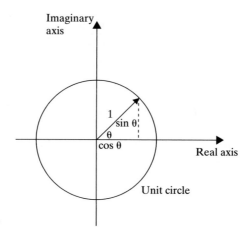

Figure B.2 A graphical interpretation of Euler's identity.

Using this identity, the complex number **A** can be written as

$$\mathbf{A} = ze^{j\theta} = z\cos\theta + jz\sin\theta \qquad \text{B.5}$$

which, as shown in Fig. B.3, can be written as

$$\mathbf{A} = x + jy$$

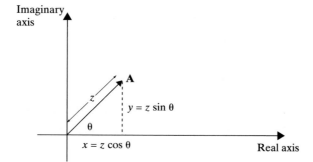

Figure B.3. The relationship between the exponential and rectangular representation of a complex number.

Equating the real and imaginary parts of these two equations yields

$$x = z\cos\theta \qquad \text{B.6}$$
$$y = z\sin\theta$$

From these equations we obtain

$$x^2 + y^2 = z^2\cos^2\theta + z^2\sin^2\theta = z^2 \qquad \text{B.7}$$

Therefore,

$$z = \sqrt{x^2 + y^2} \tag{B.8}$$

Additionally,

$$\frac{z \sin \theta}{z \cos \theta} = \tan \theta = \frac{y}{x}$$

and hence,

$$\theta = \tan^{-1} \frac{y}{x} \tag{B.9}$$

The interrelationships among the three representations of a complex number are as follows.

Exponential	Polar	Rectangular
$ze^{j\theta}$	$z\,\underline{/\theta}$	$x + jy$
$\theta = \tan^{-1} y/x$	$\theta = \tan^{-1} y/x$	$x = z \cos \theta$
$z = \sqrt{x^2 + y^2}$	$z = \sqrt{x^2 + y^2}$	$y = z \sin \theta$

We will now show that the operations of addition, subtraction, multiplication, and division apply to complex numbers in the same manner that they apply to real numbers.

The *sum* of two complex numbers $\mathbf{A} = x_1 + jy_1$ and $\mathbf{B} = x_2 + jy_2$ is

$$\mathbf{A} + \mathbf{B} = x_1 + jy_1 + x_2 + jy_2 \tag{B.10}$$
$$= (x_1 + x_2) + j(y_1 + y_2)$$

That is, we simply add the individual real parts, and we add the individual imaginary parts to obtain the components of the resultant complex number.

EXAMPLE B.1

Addition and subtraction of complex numbers are most easily performed when the numbers are in rectangular form.

Suppose we wish to calculate the sum $\mathbf{A} + \mathbf{B}$ if $\mathbf{A} = 5\,\underline{/36.9°}$ and $\mathbf{B} = 5\,\underline{/53.1°}$.

SOLUTION   We must first convert from polar to rectangular form.

$$\mathbf{A} = 5\,\underline{/36.9°} = 4 + j3$$
$$\mathbf{B} = 5\,\underline{/53.1°} = 3 + j4$$

Therefore,

$$\mathbf{A} + \mathbf{B} = 4 + j3 + 3 + j4 = 7 + j7$$
$$= 9.9\,\underline{/45°} \qquad\qquad\Box$$

The *difference* of two complex numbers $\mathbf{A} = x_1 + jy_1$ and $\mathbf{B} = x_2 + jy_2$ is

$$\mathbf{A} - \mathbf{B} = (x_1 + jy_1) - (x_2 + jy_2) \tag{B.11}$$
$$= (x_1 - x_2) + j(y_1 - y_2)$$

That is, we simply subtract the individual real parts and we subtract the individual imaginary parts to obtain the components of the resultant complex number.

EXAMPLE B.2

Let us calculate the difference $\mathbf{A} - \mathbf{B}$ if $\mathbf{A} = 5\,\underline{/36.9°}$ and $\mathbf{B} = 5\,\underline{/53.1°}$.

SOLUTION   Converting both numbers from polar to rectangular form,

$$\mathbf{A} = 5\,\underline{/36.9°} = 4 + j3$$
$$\mathbf{B} = 5\,\underline{/53.1°} = 3 + j4$$

Then

$$\mathbf{A} - \mathbf{B} = (4 + j3) - (3 + j4) = 1 - j1 = \sqrt{2}\,\underline{/-45°} \qquad \square$$

The *product* of two complex numbers $\mathbf{A} = z_1\,\underline{/\theta_1} = x_1 + jy$ and $\mathbf{B} = z_2\,\underline{/\theta_2} = x_2 + jy_2$ is

$$\mathbf{AB} = \left(z_1\,e^{j\theta_1}\right)\!\left(z_2\,e^{j(\theta_2)}\right) = z_1\,z_2\,\underline{/\theta_1 + \theta_2} \qquad \textbf{B.12}$$

EXAMPLE B.3

Given $\mathbf{A} = 5\,\underline{/36.9°}$ and $\mathbf{B} = 5\,\underline{/53.1°}$, we wish to calculate the product in both polar and rectangular forms.

Multiplication and division of complex numbers are most easily performed when the numbers are in exponential or polar form.

SOLUTION

$$\mathbf{AB} = \left(5\,\underline{/36.9°}\right)\!\left(5\,\underline{/53.1°}\right) = 25\,\underline{/90°}$$
$$= (4 + j3)(3 + j4)$$
$$= 12 + j16 + j9 + j^2\,12$$
$$= 25j$$
$$= 25\,\underline{/90°} \qquad \square$$

EXAMPLE B.4

Given $\mathbf{A} = 2 + j2$ and $\mathbf{B} = 3 + j4$, we wish to calculate the product $\mathbf{AB}$.

SOLUTION

$$\mathbf{A} = 2 + j2 = 2.828\,\underline{/45°}$$
$$\mathbf{B} = 3 + j4 = 5\,\underline{/53°}$$

and

$$\mathbf{AB} = \left(2.828\,\underline{/45°}\right)\!\left(5\,\underline{/53°}\right) = 14.14\,\underline{/98°}$$

The *quotient* of two complex numbers $\mathbf{A} = z_1 \, \underline{/\theta_1} = x_1 + jy_1$ and $\mathbf{B} = z_2 \, \underline{/\theta_2} = x_2 + jy_2$ is

$$\frac{\mathbf{A}}{\mathbf{B}} = \frac{z_1 e^{j\theta_1}}{z_2 e^{j\theta_2}} = \frac{z_1}{z_2} e^{j(\theta_1 - \theta_2)} = \frac{z_1}{z_2} \, \underline{/\theta_1 - \theta_2} \qquad \textbf{B.13} \quad \square$$

EXAMPLE B.5

Given $\mathbf{A} = 10 \, \underline{/30°}$ and $\mathbf{B} = 5 \, \underline{/53.1°}$, we wish to determine the quotient $\mathbf{A/B}$ in both polar and rectangular forms.

SOLUTION

$$\frac{\mathbf{A}}{\mathbf{B}} = \frac{10 \, \underline{/30°}}{5 \, \underline{/53.1°}}$$

$$= 2 \, \underline{/-23.1°}$$

$$= 1.84 - j0.79 \qquad \square$$

EXAMPLE B.6

Given $\mathbf{A} = 3 + j4$ and $\mathbf{B} = 1 + j2$, we wish calculate the quotient $\mathbf{A/B}$.

SOLUTION

$$\mathbf{A} = 3 + j4 = 5 \, \underline{/53°}$$
$$\mathbf{B} = 1 + j2 = 2.236 \, \underline{/63°}$$

and

$$\frac{\mathbf{A}}{\mathbf{B}} = \frac{5 \, \underline{/53°}}{2.236 \, \underline{/63°}} = 2.236 \, \underline{/-10°} \qquad \square$$

EXAMPLE B.7

If $\mathbf{A} = 3 + j4$, let us compute $1/\mathbf{A}$.

SOLUTION

$$\mathbf{A} = 3 + j4 = 5 \, \underline{/53°}$$

and

$$1/\mathbf{A} = \frac{1 \, \underline{/0°}}{5 \, \underline{/53°}} = 0.2 \, \underline{/-53°}$$

or

$$\frac{1}{\mathbf{A}} = \frac{1}{3 + j4} = \frac{3 - j4}{(3 + j4)(3 - j4)} = \frac{3 - j4}{25} = 0.12 - j0.16 \qquad \square$$

Most of the modern scientific calculators have built-in functions for complex number manipulation. It is highly recommended that one be obtained. They are fast and accurate and an excellent time-saving device.

PROBLEMS

B.1   Find $\mathbf{A} + \mathbf{B}$ if

(1) $\mathbf{A} = 8 + j2,$ $\mathbf{B} = 5 - j4$

(2) $\mathbf{A} = 10\ \underline{/45°},$ $\mathbf{B} = 9\ \underline{/30°}$ See Example B.1.

(3) $\mathbf{A} = 6\ \underline{/60°},$ $\mathbf{B} = -5 + j3$

(4) $\mathbf{A} = 5\ \underline{/-30°},$ $\mathbf{B} = -4 - j3$

B.2   Find $\mathbf{A} - \mathbf{B}$ if

(1) $\mathbf{A} = 5 - j2,$ $\mathbf{B} = 8 + j3$

(2) $\mathbf{A} = 3\ \underline{/75°},$ $\mathbf{B} = \sqrt{2}\ \underline{/45°}$ See Example B.2.

(3) $\mathbf{A} = 8\ \underline{/45°},$ $\mathbf{B} = -4 - j2$

(4) $\mathbf{A} = 9 - j6,$ $\mathbf{B} = 4\ \underline{/-30°}$

B.3   Find $\mathbf{AB}$ if

(1) $\mathbf{A} = 6 - j5,$ $\mathbf{B} = 6 + j5$

(2) $\mathbf{A} = 5\ \underline{/30°},$ $\mathbf{B} = 4\ \underline{/-15°}$ See Examples B.3 and B.4.

(3) $\mathbf{A} = 9\ \underline{/45°},$ $\mathbf{B} = -5 - j6$

(4) $\mathbf{A} = 7\ \underline{/-40°},$ $\mathbf{B} = 6 - j2$

B.4   Find $\mathbf{A/B}$ is

(1) $\mathbf{A} = 6 - j7,$ $\mathbf{B} = 3 + j3$

(2) $\mathbf{A} = 5\ \underline{/60°},$ $\mathbf{B} = 2\ \underline{/40°}$ See Examples B.5 and B.6.

(3) $\mathbf{A} = 9\ \underline{/45°},$ $\mathbf{B} = 2 + j4$

(4) $\mathbf{A} = -4 - j3,$ $\mathbf{B} = 6\ \underline{/15°}$

B.5   Find $1/\mathbf{A}$ if

(1) $\mathbf{A} = 5 - j4$ See Example B.7.

(2) $\mathbf{A} = 8\ \underline{/-30°}$

Two–Port Networks

We say that the linear network in Fig. C.1a has a single *port*; that is, a single pair of terminals. The pair of terminals *A-B* that constitutes this port could represent a single element (e.g., *R*, *L*, or *C*), or it could be some interconnection of these elements. The linear network in Fig. C.1b is called a two-port. As a general rule the terminals *A-B* represent the input port and the terminals *C-D* represent the output port.

We study two-ports and the parameters that describe them for a number of reasons. For example, most circuits or systems have at least two ports. We may put an input signal into one port and obtain an output signal from another. The parameters of the two-port completely describe its behavior in terms of the voltage and current at each port. Thus, knowing the parameters of a two-port network permits us to describe its operation when it is connected into a larger network. Two-port networks are also important in modeling electronic devices and system components. For example, in electronics, two-port networks are employed to model such things as transistors and op-amps. Other examples of electrical components modeled by two-ports are transformers and transmission lines.

In general, we describe the two-port as a network consisting of *R*, *L*, and *C* elements, transformers, op-amps, dependent sources, but no independent sources. The network has an input port and an output port, and as is the case with an op-amp, one terminal may be common to both ports.

Four popular types of two-port parameters are examined: admittance, impedance, hybrid, and transmission. We will demonstrate the usefulness of the admittance and impedance parameters, describe the remaining parameters, and provide a table that relates one set a parameters to another.

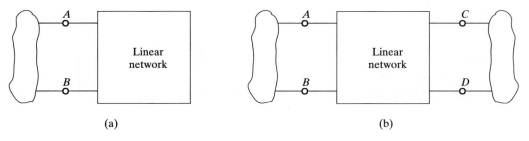

(a) (b)

Figure C.1 (a) Single-port network; (b) two-port network.

C.1 ADMITTANCE PARAMETERS

In the two-port network shown in Fig. C.2, it is customary to label the voltages and currents as shown; that is, the upper terminals are positive with respect to the lower terminals and the currents are into the two-port at the upper terminals and, because KCL must be satisfied at each port, the current is out of the two-port at the lower terminals. Since the network is linear and contains no independent sources, the principle of superposition can be applied to determine the current $\mathbf{I}_1$, which can be written as the sum of two components, one due to $\mathbf{V}_1$ and one due to $\mathbf{V}_2$. Using this principle, we can write

$$\mathbf{I}_1 = \mathbf{y}_{11}\,\mathbf{V}_1 + \mathbf{y}_{12}\,\mathbf{V}_2$$

where $\mathbf{y}_{11}$ and $\mathbf{y}_{12}$ are essentially constants of proportionality with units of siemens. In a similar manner $\mathbf{I}_2$ can be written as

$$\mathbf{I}_2 = \mathbf{y}_{21}\,\mathbf{V}_1 + \mathbf{y}_{22}\,\mathbf{V}_2$$

Therefore, the two equations that describe the two-port are

$$\mathbf{I}_1 = \mathbf{y}_{11}\,\mathbf{V}_1 + \mathbf{y}_{12}\,\mathbf{V}_2 \qquad\qquad \textbf{C.1}$$
$$\mathbf{I}_2 = \mathbf{y}_{21}\,\mathbf{V}_1 + \mathbf{y}_{22}\,\mathbf{V}_2$$

or in matrix form,

$$\begin{bmatrix} \mathbf{I}_1 \\ \mathbf{I}_2 \end{bmatrix} = \begin{bmatrix} \mathbf{y}_{11} & \mathbf{y}_{12} \\ \mathbf{y}_{21} & \mathbf{y}_{22} \end{bmatrix} \begin{bmatrix} \mathbf{V}_1 \\ \mathbf{V}_2 \end{bmatrix}$$

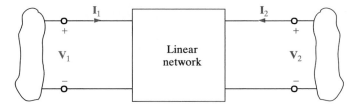

Figure C.2 Generalized two-port network.

Note that subscript 1 refers to the input port and subscript 2 refers to the output port, and the equations describe what we will call the *Y parameters* for a network. If these parameters y_{11}, y_{12}, y_{21}, and y_{22}, are known, the input/output operation of the two-port is completely defined.

From Eq. (C.1) we can determine the *Y* parameters in the following manner. Note from the equations that y_{11} is equal to I_1 divided by V_1 with the output short circuited (i.e., $V_2 = 0$).

$$y_{11} = \frac{I_1}{V_1}\bigg|_{V_2=0} \qquad \text{C.2}$$

Since y_{11} is an admittance at the input measured in siemens with the output short circuited, it is called the *short-circuit input admittance*. The equations indicate that the other *Y* parameters can be determined in a similar manner:

$$y_{12} = \frac{I_1}{V_2}\bigg|_{V_1=0}$$

$$y_{21} = \frac{I_2}{V_1}\bigg|_{V_2=0} \qquad \text{C.3}$$

$$y_{22} = \frac{I_2}{V_2}\bigg|_{V_1=0}$$

y_{12} and y_{21} are called the *short-circuit transfer admittances*, and y_{22} is called the *short-circuit output admittance*. As a group the *Y* parameters are referred to as the *short-circuit admittance parameters*. Note that by applying the preceding definitions, these parameters could be determined experimentally for a two-port whose actual configuration is unknown.

EXAMPLE C.1

We wish to determine the *Y* parameters for the two-port shown in Fig. C.3a. Once these parameters are known, we will determine the current in a 4-Ω load, which is connected to the output port when a 2-A current source is applied at the input port.

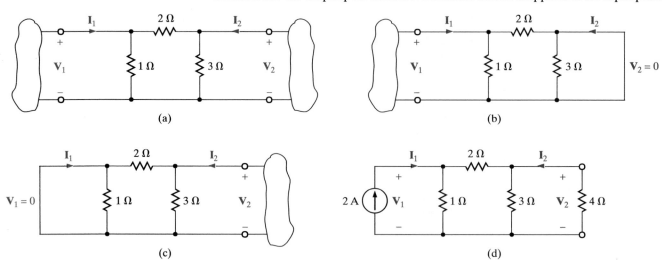

Figure C.3 Networks employed in Example C.1.

SOLUTION   From Fig. C.3b, we note that

$$\mathbf{I}_1 = \mathbf{V}_1(\tfrac{1}{1} + \tfrac{1}{2})$$

Therefore,

$$\mathbf{y}_{11} = \tfrac{3}{2}\,\mathrm{S}$$

As shown in Fig. C.3c,

$$\mathbf{I}_1 = -\frac{\mathbf{V}_2}{2}$$

and hence,

$$\mathbf{y}_{12} = -\tfrac{1}{2}\,\mathrm{S}$$

Also, $\mathbf{y}_{21}$ is computed from Fig. C.3b using the equation

$$\mathbf{I}_2 = -\frac{\mathbf{V}_1}{2}$$

and therefore,

$$\mathbf{y}_{21} = -\tfrac{1}{2}\,\mathrm{S}$$

Finally, $\mathbf{y}_{22}$ can be derived from Fig. C.3c using

$$\mathbf{I}_2 = \mathbf{V}_2(\tfrac{1}{3} + \tfrac{1}{2})$$

and

$$\mathbf{y}_{22} = \tfrac{5}{6}\,\mathrm{S}$$

Therefore, the equations that describe the two-port itself are

$$\mathbf{I}_1 = \tfrac{3}{2}\mathbf{V}_1 - \tfrac{1}{2}\mathbf{V}_2$$
$$\mathbf{I}_2 = -\tfrac{1}{2}\mathbf{V}_1 + \tfrac{5}{6}\mathbf{V}_2$$

These equations can now be employed to determine the operation of the two-port for some given set of terminal conditions. The terminal conditions we will examine are shown in Fig. C.3d. From this figure we note that

$$\mathbf{I}_1 = 2\ \mathrm{A} \quad \text{and} \quad \mathbf{V}_2 = -4\mathbf{I}_2$$

Combining these with the preceding two-port equations yields

$$2 = \tfrac{3}{2}\mathbf{V}_1 - \tfrac{1}{2}\mathbf{V}_2$$
$$0 = -\tfrac{1}{2}\mathbf{V}_1 + \tfrac{13}{12}\mathbf{V}_2$$

or in matrix form,

$$\begin{bmatrix} \tfrac{3}{2} & -\tfrac{1}{2} \\ -\tfrac{1}{2} & \tfrac{13}{12} \end{bmatrix} \begin{bmatrix} \mathbf{V}_1 \\ \mathbf{V}_2 \end{bmatrix} = \begin{bmatrix} 2 \\ 0 \end{bmatrix}$$

Note carefully that these equations are simply the nodal equations for the network in Fig. C.3d. Solving the equations, we obtain $\mathbf{V}_2 = \tfrac{8}{11}\,\mathrm{V}$ and therefore $\mathbf{I}_2 = -\tfrac{2}{11}\,\mathrm{A}$.

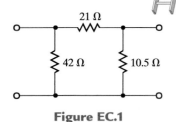

Figure EC.1

EC.1 Find the Y parameters for the two-port network shown in Fig. EC.1.

ANSWER: $y_{11} = \frac{1}{14}$ S, $y_{12} = y_{21} = -\frac{1}{21}$ S, $y_{22} = \frac{1}{7}$ S.

EC.2 If a 10-A source is connected to the input of the two-port in Fig. EC.1, find the current in a 5-Ω resistor connected to the output port.

ANSWER: $I_2 = -4.29$ A.

C.2 IMPEDANCE PARAMETERS

Once again if we assume that the two-port is a linear network that contains no independent sources, then by means of superposition we can write the input and output voltages as the sum of two components, one due to I_1 and one due to I_2:

$$V_1 = z_{11} I_1 + z_{12} I_2$$
$$V_2 = z_{21} I_1 + z_{22} I_2$$

$\qquad$**C.4**

These equations, which describe the two-port, can also be written in matrix form as

$$\begin{bmatrix} V_1 \\ V_2 \end{bmatrix} = \begin{bmatrix} z_{11} & z_{12} \\ z_{21} & z_{22} \end{bmatrix} \begin{bmatrix} I_1 \\ I_2 \end{bmatrix}$$

$\qquad$**C.5**

Like the Y parameters, these Z parameters can be derived as follows:

$$z_{11} = \left. \frac{V_1}{I_1} \right|_{I_2=0}$$
$$z_{12} = \left. \frac{V_1}{I_2} \right|_{I_1=0}$$
$$z_{21} = \left. \frac{V_2}{I_1} \right|_{I_2=0}$$
$$z_{22} = \left. \frac{V_2}{I_2} \right|_{I_1=0}$$

$\qquad$**C.6**

In the preceding equations, setting I_1 or $I_2 = 0$ is equivalent to open circuiting the input or output port. Therefore, the Z parameters are called the *open-circuit impedance parameters*. z_{11} is called the *open-circuit input impedance*, z_{22} is called the *open-circuit output impedance*, and z_{12} and z_{21} are termed *open-circuit transfer impedances*.

EXAMPLE C.2

We wish to find the Z parameters for the network in Fig. C.4a. Once the parameters are known, we will use them to find the current in a 4-Ω resistor that is connected

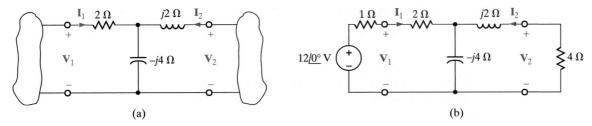

Figure C.4 Circuits employed in Example C.2.

to the output terminals when a $12 \underline{/0°}$-V source with an internal impedance of $1 + j0 \; \Omega$ is connected to the input.

SOLUTION   From Fig. C.4a we note that

$$\mathbf{z}_{11} = 2 - j4 \; \Omega$$

$$\mathbf{z}_{12} = -j4 \; \Omega$$

$$\mathbf{z}_{21} = -j4 \; \Omega$$

$$\mathbf{z}_{22} = -j4 + j2 = -j2 \; \Omega$$

The equations for the two-port are, therefore,

$$\mathbf{V}_1 = (2 - j4)\mathbf{I}_1 - j4\mathbf{I}_2$$
$$\mathbf{V}_2 = -j4\mathbf{I}_1 - j2\mathbf{I}_2$$

The terminal conditions for the network shown in Fig. C.4b are

$$\mathbf{V}_1 = 12 \underline{/0°} - (1)\mathbf{I}_1$$
$$\mathbf{V}_2 = -4\mathbf{I}_2$$

Combining these with the two-port equations yields

$$12 \underline{/0°} = (3 - j4)\mathbf{I}_1 - j4\mathbf{I}_2$$
$$0 = -j4\mathbf{I}_1 + (4 - j2)\mathbf{I}_2$$

It is interesting to note that these equations are the mesh equations for the network. If we solve the equations for $\mathbf{I}_2$, we obtain $\mathbf{I}_2 = 1.61 \underline{/137.73°}$ A, which is the current in the 4-Ω load.

EXTENSION EXERCISES

EC.3   Find the Z parameters for the network in Fig. EC.3. Then compute the current in a 4-Ω load if a $24 \underline{/0°}$-V source is connected at the input port.

ANSWER: $\mathbf{I}_2 = -0.73 \underline{/0°}$ A.

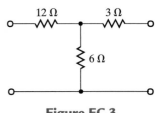

Figure EC.3

C.3 HYBRID AND TRANSMISSION PARAMETERS

Under the assumptions used to develop the Y and Z parameters, we can obtain what are commonly called the *hybrid parameters*. In the pair of equations that define these parameters, $\mathbf{V}_1$ and $\mathbf{I}_2$ are the independent variables. Therefore, the two-port equations in terms of the hybrid parameters are

$$\mathbf{V}_1 = \mathbf{h}_{11}\mathbf{I}_1 + \mathbf{h}_{12}\mathbf{V}_2 \tag{C.7}$$
$$\mathbf{I}_2 = \mathbf{h}_{21}\mathbf{I}_1 + \mathbf{h}_{22}\mathbf{V}_2$$

or in matrix form,

$$\begin{bmatrix} \mathbf{V}_1 \\ \mathbf{I}_2 \end{bmatrix} = \begin{bmatrix} \mathbf{h}_{11} & \mathbf{h}_{12} \\ \mathbf{h}_{21} & \mathbf{h}_{22} \end{bmatrix} \begin{bmatrix} \mathbf{I}_1 \\ \mathbf{V}_2 \end{bmatrix} \tag{C.8}$$

These parameters are especially important in transistor circuit analysis.

The *transmission parameters* are defined by the equations

$$\mathbf{V}_1 = \mathbf{A}\mathbf{V}_2 - \mathbf{B}\mathbf{I}_2$$
$$\mathbf{I}_1 = \mathbf{C}\mathbf{V}_2 - \mathbf{D}\mathbf{I}_2 \tag{C.9}$$

or in matrix form,

$$\begin{bmatrix} \mathbf{V}_1 \\ \mathbf{I}_1 \end{bmatrix} = \begin{bmatrix} \mathbf{A} & \mathbf{B} \\ \mathbf{C} & \mathbf{D} \end{bmatrix} \begin{bmatrix} \mathbf{V}_2 \\ -\mathbf{I}_2 \end{bmatrix} \tag{C.10}$$

These parameters are very useful in the analysis of circuits connected in cascade.

C.4 PARAMETER CONVERSION

If all the two-port parameters for a network exist, it is possible to relate one set of parameters to another since the parameters interrelate the variables $\mathbf{V}_1, \mathbf{I}_1, \mathbf{V}_2,$ and $\mathbf{I}_2$.

Table C.1 lists all the conversion formulas that relate one set of two-port parameters to another. Note that $\Delta_Z, \Delta_Y, \Delta_H,$ and Δ_T refer to the determinants of the matrices for the $Z, Y,$ hybrid and $ABCD$ parameters, respectively. Therefore, given one set of parameters for a network, we can use Table C.1 to find others.

EXTENSION EXERCISE

EC.4   Determine the Y parameters for a two-port if the Z parameters are

$$\mathbf{Z} = \begin{bmatrix} 18 & 6 \\ 6 & 9 \end{bmatrix}$$

ANSWER: $\mathbf{y}_{11} = \frac{1}{14}\,\text{S}, \; \mathbf{y}_{12} = \mathbf{y}_{21} = -\frac{1}{21}\,\text{S}, \; \mathbf{y}_{22} = \frac{1}{7}\,\text{S}.$

Table C.1 Two-Port Parameter Conversion Formulas

$$
\begin{bmatrix} z_{11} & z_{12} \\ z_{21} & z_{22} \end{bmatrix}
\qquad
\begin{bmatrix} \dfrac{y_{22}}{\Delta_Y} & \dfrac{-y_{12}}{\Delta_Y} \\[2mm] \dfrac{-y_{21}}{\Delta_Y} & \dfrac{y_{11}}{\Delta_Y} \end{bmatrix}
\qquad
\begin{bmatrix} \dfrac{A}{C} & \dfrac{\Delta_T}{C} \\[2mm] \dfrac{1}{C} & \dfrac{D}{C} \end{bmatrix}
\qquad
\begin{bmatrix} \dfrac{\Delta_H}{h_{22}} & \dfrac{h_{12}}{h_{22}} \\[2mm] \dfrac{-h_{21}}{h_{22}} & \dfrac{1}{h_{22}} \end{bmatrix}
$$

$$
\begin{bmatrix} \dfrac{z_{22}}{\Delta_Z} & \dfrac{-z_{12}}{\Delta_Z} \\[2mm] \dfrac{-z_{21}}{\Delta_Z} & \dfrac{z_{11}}{\Delta_Z} \end{bmatrix}
\qquad
\begin{bmatrix} y_{11} & y_{12} \\ y_{21} & y_{22} \end{bmatrix}
\qquad
\begin{bmatrix} \dfrac{D}{B} & \dfrac{-\Delta_T}{B} \\[2mm] -\dfrac{1}{B} & \dfrac{A}{B} \end{bmatrix}
\qquad
\begin{bmatrix} \dfrac{1}{h_{11}} & \dfrac{-h_{12}}{h_{11}} \\[2mm] \dfrac{h_{21}}{h_{11}} & \dfrac{\Delta_H}{h_{11}} \end{bmatrix}
$$

$$
\begin{bmatrix} \dfrac{z_{11}}{z_{21}} & \dfrac{\Delta_Z}{z_{21}} \\[2mm] \dfrac{1}{z_{21}} & \dfrac{z_{22}}{z_{21}} \end{bmatrix}
\qquad
\begin{bmatrix} \dfrac{-y_{22}}{y_{21}} & \dfrac{-1}{y_{21}} \\[2mm] \dfrac{-\Delta_Y}{y_{21}} & \dfrac{-y_{11}}{y_{21}} \end{bmatrix}
\qquad
\begin{bmatrix} A & B \\ C & D \end{bmatrix}
\qquad
\begin{bmatrix} \dfrac{-\Delta_H}{h_{21}} & \dfrac{-h_{11}}{h_{21}} \\[2mm] \dfrac{-h_{22}}{h_{21}} & \dfrac{-1}{h_{21}} \end{bmatrix}
$$

$$
\begin{bmatrix} \dfrac{\Delta_Z}{z_{22}} & \dfrac{z_{12}}{z_{22}} \\[2mm] \dfrac{-z_{21}}{z_{22}} & \dfrac{1}{z_{22}} \end{bmatrix}
\qquad
\begin{bmatrix} \dfrac{1}{y_{11}} & \dfrac{-y_{12}}{y_{11}} \\[2mm] \dfrac{y_{21}}{y_{11}} & \dfrac{\Delta_Y}{y_{11}} \end{bmatrix}
\qquad
\begin{bmatrix} \dfrac{B}{D} & \dfrac{\Delta_T}{D} \\[2mm] \dfrac{1}{D} & \dfrac{C}{D} \end{bmatrix}
\qquad
\begin{bmatrix} h_{11} & h_{12} \\ h_{21} & h_{22} \end{bmatrix}
$$

C.5 SUMMARY

- Four of the most common parameters used to describe a two-port network are the admittance, impedance, hybrid, and transmission parameters.
- If all the two-port parameters for a network exist, a set of conversion formulas can be used to relate one of two-port parameters to another.

PROBLEMS

C.1 Given the two networks in Fig. PC.1, find the Y parameters for the circuit in (a) and the Z parameters for the circuit in (b).

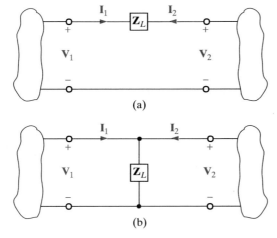

(a)

(b)

Figure PC.1

C.2   Find the Y parameters for the two-port
network shown in Fig. PC.2.

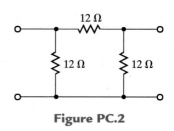

Figure PC.2

C.3   Find the Y parameters for the two-port
network shown in Fig. PC.3.

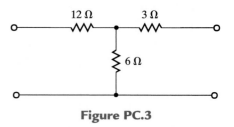

Figure PC.3

C.4   If a 12-A source is connected at the input port of the network shown in Fig.
PC.2, find the current in a 4-Ω load resistor.

C.5   Find the Y parameters for the two-port network shown in Fig. PC.5.

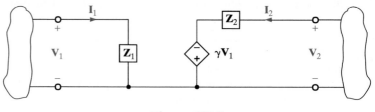

Figure PC.5

C.6   Determine the Y parameters for the network shown in Fig. PC.6.

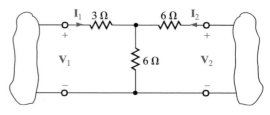

Figure PC.6

C.7   Find the Z parameters for the two-port in Fig. PC.7.

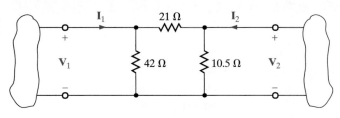

Figure PC.7

C.8   Find the Z parameters of the two-port in Fig. PC.8.

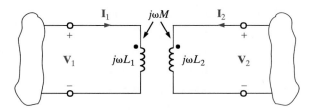

Figure PC.8

C.9   Following are the hybrid parameters for a network.

$$\begin{bmatrix} \mathbf{h}_{11} & \mathbf{h}_{12} \\ \mathbf{h}_{21} & \mathbf{h}_{22} \end{bmatrix} = \begin{bmatrix} \frac{11}{5} & \frac{2}{5} \\ -\frac{2}{5} & \frac{1}{5} \end{bmatrix}$$

Determine the Y parameters for the network.

C.10   If the Y parameters for a network are known to be

$$\begin{bmatrix} \mathbf{y}_{11} & \mathbf{y}_{12} \\ \mathbf{y}_{21} & \mathbf{y}_{22} \end{bmatrix} = \begin{bmatrix} \frac{5}{11} & -\frac{2}{11} \\ -\frac{2}{11} & \frac{3}{11} \end{bmatrix}$$

find the Z parameters.

Answers to Selected Problems

1.1	$I = 60$ A	**2.85**	$V_0 = 2$ V
1.5	$\Delta Q = 1.4$ C	**2.89**	$I_S = 7$ mA
1.8	$\Delta W = 432$ kJ	**2.93**	$V_S = 36$ V
1.11	$P_a = 4.5$ kW, $P_b = 1.25$ W, $P_c = 52.08$ mW	**2.96**	$V_0 = 12$ V
1.14	$P_1 = 12$ W, $P_2 = 8$ W	**2.99**	$G_A = V_0/V_S = -32.4$ V/V
1.17	(a) $P_1 = 8$ W absorbed, $P_2 = 16$ W absorbed,	**2.102**	$P_{10K} = 10$ mW
	$P_{12v} = 24$ W delivered	**2.106**	$R_3 = 70\ \Omega$
	(b) $P_1 = 16$ W absorbed, $P_{8v} = 8$ W absorbed,	**3.1**	$I_0 = 2.78$ mA
	$P_{24v} = 24$ W delivered	**3.3**	$V_2 = 22$ V
1.20	$i(t) = 2$ A	**3.7**	$I_0 = 0.6$ mA
1.24	$I_x = 2$ A	**3.11**	$V_0 = -20$ V
1.28	$W = 0.62$ J	**3.14**	$V_0 = 0.27$ V
2.1	$I = 0.75$ A, $P = 6.75$ W	**3.18**	$I_0 = 1.67$ mA
2.3	$V = 24$ V, $P = 48$ W	**3.22**	$V_0 = 5$ V
2.7	$P_{\text{BATT}} = 72$ W	**3.26**	$I_0 = -0.25$ mA
2.11	$R_x = 5$ kΩ	**3.30**	$V_0 = 20$ V
2.15	$I_1 = 8$ mA	**3.34**	$V_0 = 4$ V
2.19	$I_0 = 6$ mA	**3.37**	$V_0 = 3.43$ V
2.23	$I_x = 2$ mA	**3.41**	$I_0 = 2.67$ mA
2.27	$V_x = 3$ V	**3.45**	$I_0 = 5.2$ mA
2.31	$V_{\text{ac}} = 3$ V	**3.49**	$V_0 = 6$ V
2.35	$V_x = 5$ V	**3.53**	$\dfrac{v_0}{i_s} = -1$
2.38	$I_0 = 90$ mA		
2.42	$R_{AB} = 9$ kΩ	**3.57**	$R_1 = 2$ kΩ
2.45	$R_{AB} = 3$ kΩ	**3.61**	$V_0 = -0.43$ V
2.48	$I_1 = -0.5$ mA, $V_0 = -4.5$ V	**3.64**	$V_0 = 3.56$ V
2.52	$I_0 = -1.33$ mA	**3.68**	$V_0 = 1.6$ V
2.55	$V_s = 7$ V	**3.72**	$V_0 = -0.88$ V
2.59	$I_s = 6$ mA	**3.77**	$I_0 = 4.36$ mA
2.63	$V_0 = 2$ V	**3.80**	$V_0 = -11.2$ V
2.67	$R_{\text{SH}} = 0.385\ \Omega$	**3.83**	$R_A = 10\ \Omega$
2.71	$V_{\text{MAX}} = 1.05$ V		$R_B = 100\ \Omega$
2.74	$R_{AB} = 3$ kΩ	**4.1**	$V_0 = 0.75$ V
2.78	$V_0 = 2$ V	**4.4**	$I_0 = 1.2$ mA
2.81	$I_0 = -3$ mA	**4.7**	$V_0 = 7.5$ V

4.11 $V_0 = -3\text{ V}$

4.15 $V_0 = 8\text{ V}$

4.18 $I_0 = -0.3\text{ mA}$

4.22 $V_0 = 8\text{ V}$

4.26 $I_0 = 0.39\text{ mA}$

4.30 $I_0 = 5\text{ mA}$

4.34 $V_0 = 2.25\text{ V}$

4.37 $V_0 = 2\text{ V}$

4.41 $I_0 = 2\text{ mA}$

4.45 $R_L = 1.2\text{ k}\Omega$

4.48 $P_L = 2.53\text{ mW}$

4.52 $V_0 = 1.33\text{ V}$

4.56 $I_0 = 0.75\text{ mA}$

4.60 $I_0 = -0.2\text{ mA}$

4.64 $I_0 = 0.5\text{ mA}$

4.67 $V_0 = 0.75\text{ V}$

4.70 $I_0 = -0.6\text{ mA}$

4.74 $V_0 = 258\text{ mV}$

4.77 $P_L = 8.1\text{ mW}$

4.80 $I = 4\text{ mA}$

5.1 $Q = 72\ \mu\text{C}$

5.3 $V = 40\text{ V}$

5.6 $V(t_2) = 40\text{ V}$

5.8 (a) $v(t) = 2.65\sin 377t\text{ V}$
     (b) $p(t) = 13.3\sin 754t\text{ mW}$

5.13 $i(t) = 0.6\text{ A}$ $0 < t < 2\text{ s}$
        $= -2.4\text{ A}$ $2 < t < 3\text{ s}$
        $= 0.6\text{ A}$ $3 < t < 5\text{ s}$
        $= 0$ $5 < t$

5.17 $i(t) = 377\cos 2000\pi t\text{ mA}$

5.21 $v_{ind} = -1\text{ V}$

5.25 (a) $i(t) = 31.83\sin 377t\text{ A}$
     (b) $p(t) = 1910\sin 754t\text{ W}$

5.29 $v(1) = 91.97\ \mu\text{V}, w(1) = 0.5\ \mu\text{J}$

5.33 $v(t) = -192\text{ V}$ $0 < t < 2\text{ ms}$
        $= 192\text{ V}$ $2 < t < 5\text{ ms}$
        $= 0\text{ V}$ $5 < t < 9\text{ ms}$
        $= -192\text{ V}$ $9 < t < 11\text{ ms}$
        $= 192\text{ V}$ $11 < t < 12\text{ ms}$
        $= 0\text{ V}$ $12 < t$

5.38 $v(t) = -3.02\cos 2000\pi t\text{ mV}$

5.42 $V_0 = 12\text{ V}$

5.45 $C_{eq} = 2\ \mu\text{F}$

5.48 $L_{eq} = 32\text{ mH}$

5.51 $C = 1.25\ \mu\text{F}$

5.54 $C_T = 2\ \mu\text{F}$

5.59 $L_{AB} = 5\text{ mH}$

5.63 $L_T = 3\text{ mH}$

5.67 $v_0(t) = -21.22\sin 377t\text{ V}$

6.1 $v_0(t) = 4.36(1 - e^{-1.83t})\text{ V}$ $t > 0$
      $v_0(t) = 0\text{ V}$ $t \le 0$

6.4 $v_0(t) = 2.67(1 + e^{-0.96t})\text{ V}$ $t > 0$
      $v_0(t) = 5.33\text{ V}$ $t \le 0$

6.7 $i_L(t) = 1.33e^{-9t}\text{ A}$ $t > 0$
      $i_L(t) = 1.33\text{ A}$ $t \le 0$

6.10 $i_0(t) = 1\text{ mA}$ for all time

6.13 $i_0(t) = 2e^{-4t}\text{ A}$ $t > 0$
        $i_0(t) = -2\text{ A}$ $t \le 0$

6.16 $i_0(t) = 5.33e^{-3.33t}\text{ A}$ $t > 0$
        $i_0(t) = 5.33\text{ A}$ $t \le 0$

6.19 $v_0(t) = -9.6e^{-9.6t}\text{ V}$ $t > 0$
        $v_0(t) = 0\text{ V}$ $t \le 0$

6.22 $v_0(t) = 9e^{-6.67t}\text{ V}$ $t > 0$
        $v_0(t) = 9\text{ V}$ $t \le 0$

6.25 $i_0(t) = e^{-t/0.6}\text{ mA}$ $t > 0$
        $i_0(t) = 2\text{ mA}$ $t \le 0$

6.28 $i_0(t)3 - 0.33e^{-t/0.6}\text{ mA}$ $t > 0$
        $i_0(t) = 2\text{ mA}$ $t \le 0$

6.31 $i_0(t) = 5/6\text{ mA}$ for all time

6.34 $v_0(t) - 6e^{-4t}\text{ V}$ $t > 0$
        $v_0(t) = 2\text{ V}$ $t \le 0$

6.37 $i_0(t) = -2.4e^{-4.8t}\text{ A}$ $t > 0$
        $i_0(t) = 0\text{ A}$ $t \le 0$

6.40 $v_0(t) = 3 + (9/11)e^{-t/0.55}\text{ V}$ $t > 0$
        $v_0(t) = 6\text{ V}$ $t \le 0$

6.43 $v_0(t) = 0.14e^{-50t}\text{ V}$ $t > 0$

6.46 $v_0(t) = 5e^{-10t}\text{ V}$ $0 < t < 50\text{ ms}$
        $v_0(t) = 1.97e^{-10(t-0.05)}\text{ V}$ $t \ge 50\text{ ms}$

6.49 $t_{discharge} = 2.76\text{ s}$

6.53 $i_0(t) = -1.33e^{-1.61t}$ mA $t > 0$

6.56 $v_0(t) = -3.6e^{-8t}$ V $t > 0$

6.59 $v_0(t) = 6e^{-6.67t}$ V $t > 0$

6.62 $v_0(t) = 18 - 9e^{-6t}$ V $t > 0$

6.65 $v_0(t) = 16/3 + (4/51)e^{-(27/34)t}$ V $t > 0$

6.68 $i_0(t) = 1/9 + (1/18)e^{-t/0.066}$ mA $t > 0$

6.71 $v_0(t) = 4 - 10e^{-2t}$ V $t > 0$

7.1 (a) $s^2 + 6s + 4 = 0$

(b) 5.24 rad/s and 0.76 rad/s

(c) $i_0(t) = k_1e^{-5.24t} + k_2e^{-0.76t}$

7.4 (a) $s^2 + 8s + 10 = 0$

(b) 6.45 rad/s and 1.55 rad/s

(c) $v_0(t) = k_1e^{-6.45t} + k_2e^{-1.55t}$

7.7 $C = 0.5$ F

7.10 $v(t) = 5e^{-5t} - 225te^{-5t}$ V

7.14 $v(t) = 12 - 29.25e^{-8873t} - 17.25e^{-1127t}$ V

7.17 $i(t) = -(4/7)e^{-t} + (32/7)e^{-8t}$ A

7.19 $v(t) = -5e^{-t} + e^{-5t} + 8$ V

7.22 Choose $R = 4$ kΩ and then $L = 100$ H, $C = 25$ pF

7.25 $v_0(t) = 2e^{-5\times10^5 t} \sin(10^6 t)$ V

7.28 $v_0(t) = 1.2 \times 10^7 te^{-2\times10^6 t}$ V

7.31 $v_0(t) = 19.2\left[e^{-10^6 t} - e^{-3\times10^6 t}\right]$ V

8.1 $f = 60$ Hz, $T = 16.67$ ms

8.3 $v(t)$ leads $i(t)$ by 140°

8.6 (a) $\mathbf{I} = 26.53\ \underline{/-45°}$ A

(b) $\mathbf{I} = 13.26\ \underline{/+90°}$ A

8.10 $\mathbf{Z} = 1 + j2$ Ω

8.13 $\mathbf{Z} = 2$ Ω

8.16 $i(t) = 0.36 \cos(377t - 43.3°)$ A

8.20 $i(t) = 0.99 \cos(1000t - 54.35°)$ A

8.27 $C = 1.41$ mF

8.31 $L = 704$ μH

8.35 $\mathbf{V} = 0.89\ \underline{/56.57°}$ V

8.39 $\mathbf{I}_0 = 6.20\ \underline{/21.34°}$ A

8.43 $\mathbf{V}_0 = 10.71\ \underline{/-153.43°}$ V

8.47 $\mathbf{V}_0 = 17.46\ \underline{/86.97°}$ V

8.51 $\mathbf{I}_0 = 2.12\ \underline{/-45°}$ A

8.55 $\mathbf{I}_0 = 2.69\ \underline{/116.57°}$ A

8.59 $\mathbf{V}_0 = 7.34\ \underline{/34.6°}$ V

8.63 $\mathbf{V}_2 = 35.78\ \underline{/-71.57°}$ V

8.67 $\mathbf{V}_0 = 10.46\ \underline{/-88.36°}$ V

8.71 $\mathbf{V}_0 = 5.41\ \underline{/4.57°}$ V

8.75 $\mathbf{I}_0 = 2\ \underline{/-36.87°}$ A

8.79 $\mathbf{V}_0 = 3.58\ \underline{/153.43°}$ A

8.83 $\mathbf{V}_0 = 1.30\ \underline{/12.53°}$ V

8.87 $\mathbf{Z} = 2.83\ \underline{/16.93°}$ Ω

8.90 $\mathbf{C} = 265$ μF

8.94 $v_0(t) = 7.07 \cos(800\pi t + 45°)$ V

8.97 $\mathbf{V}_0 = 8.00\ \underline{/-53.14°}$ V

9.1 $P = 410.52$ W

9.3 $P = 150$ W

9.7 $P_I = 8.66$ W

$P_V = 41.35$ W

9.10 $P_R = 4.5$ W

9.13 $P_{4\Omega} = 10.40$ W

9.17 $P_R = 2.5$ W

9.20 $\mathbf{Z}_L = 1 - j1$ Ω

9.24 $\mathbf{Z}_L = 1$ Ω

9.28 $\mathbf{Z}_L = 1.79\ \underline{/153.43°}$ Ω

9.32 V rms = 1.87 V rms

9.36 V rms = 3.98 V rms

9.39 V rms = 0.71 V rms

9.43 pf = 0.83

9.46 $P_{\text{LINE}} = 1.6$ kW

9.50 $\mathbf{V}_s = 516.31\ \underline{/-8.85°}$ V rms

9.53 $\mathbf{V}_s = 279.24\ \underline{/3.48°}$ V rms

9.57 $C = 1.92$ mF

9.61 $P_m = 239.3$ W

9.65 $I = 18$ A

9.69 $P_{2\Omega} = 3.6$ W

9.72 $P_{\text{OUT}} = 4.5$ mW

9.76 $P_{\text{OUT}} = 18$ W

9.80 $\mathbf{Z}_L = 1.5$ Ω

$P_{\text{OUT}} = 13.5$ W

9.84 $P_{\text{METER}} = 20.8$ W

10.4 $\mathbf{V}_{ab} = 415.7\ \underline{/120°}$ V rms

$\mathbf{V}_{bc} = 415.7\ \underline{/0°}$ V rms

$\mathbf{V}_{ca} = 415.7\ \underline{/-120°}$ V rms

10.7 $\mathbf{I}_{aA} = 3.16 \underline{/-18.43°}$ A rms
 $\mathbf{I}_{bB} = 3.16 \underline{/-138.43°}$ A rms
 $\mathbf{I}_{cC} = 3.16 \underline{/101.57°}$ A rms

10.11 $\mathbf{V}_{an} = 121.54 \underline{/80.19°}$ V rms
 $\mathbf{V}_{bn} = 121.54 \underline{/-39.81°}$ V rms
 $\mathbf{V}_{cn} = 121.54 \underline{/-159.81°}$ V rms

10.15 $\mathbf{V}_{an} = 120$ V rms

10.19 $\mathbf{I}_{sc} = 56.6$ A rms

10.23 $\mathbf{Z}_y = 38.60 \underline{/33.25°}$ Ω

10.27 $\mathbf{I}_{aA} = 16.84 \underline{/9.39°}$ A rms
 $\mathbf{I}_{bB} = 16.84 \underline{/-110.61°}$ A rms
 $\mathbf{I}_{cC} = 16.84 \underline{/129.39°}$ A rms

10.31 $\mathbf{V}_{AB} = 32.30 \underline{/26.57°}$ V rms

10.34 $\mathbf{I}_{AB} = 4.94 \underline{/42.89°}$ A rms
 $\mathbf{I}_{BC} = 4.94 \underline{/-77.11°}$ A rms
 $\mathbf{I}_{CA} = 4.94 \underline{/162.89°}$ A rms

10.38 $\mathbf{Z}_\Delta = 9.54 \underline{/30.32°}$ Ω

10.42 $\mathbf{Z}_Y = 12 \underline{/75°}$ Ω

10.46 $I_{aA} = 82.58$ A rms

10.50 pf = 0.91 lagging

10.54 $P_1 = 6.68$ kW, $P_2 = 2.73$ kW

10.58 pf = 0.97 lagging

10.62 $C = 15$ μF

10.66 $\mathbf{Z} = 3 - j1$ Ω

10.68 $\mathbf{V}_{AB} = 138.56 \underline{/24.68°}$ V rms

10.72 $\mathbf{V}_{ab} = 236.10 \underline{/71.57°}$ V rms

10.75 $\mathbf{S}_3 = 56.83 \underline{/0°}$ kVA

10.79 $\mathbf{V}_{ab} = 216.02$ V rms
 pf = 0.92 lagging

11.1 (a) $v_a(t) = -L_1 \dfrac{di_1(t)}{dt} - M \dfrac{di_2(t)}{dt}$

  $v_b(t) = -L_2 \dfrac{di_2(t)}{dt} - M \dfrac{di_1(t)}{dt}$

 (b) $v_c(t) = -v_a(t)$
  $v_d(t) = -v_b(t)$

11.4 $\mathbf{V}_0 = 2.98 \underline{/26.57°}$ V

11.8 $\mathbf{V}_0 = 1.25 \underline{/180°}$ V

11.11 $\mathbf{V}_0/\mathbf{V}_1 = 0.20 \underline{/23.96°}$

11.15 $\mathbf{V}_0 = 1.36 \underline{/-85.4°}$ V

11.19 $\mathbf{V}_0 = 4.48 \underline{/-26.57°}$ V

11.23 $\mathbf{V}_0 = 0.97 \underline{/13.73°}$ V

11.27 $\mathbf{I}_2 = 2.71 \underline{/-49.40°}$ A

11.30 $M = 12.73$ mH

11.34 $W(0.002) = 3.71$ μJ

11.38 $\mathbf{I}_1 = 3.16 \underline{/-41.56°}$ A, $\mathbf{V}_1 = 4.47 \underline{/3.44°}$ V
 $\mathbf{I}_2 = 1.58 \underline{/138.44°}$ A, $\mathbf{V}_2 = 8.94 \underline{/3.44°}$ V

11.42 $\mathbf{V}_0 = 7.43 \underline{/-38.20°}$ V

11.46 $\mathbf{I}_1 = 4.02 \underline{/26.57°}$ A

11.48 $\mathbf{I} = 1.72 \underline{/54.77°}$ A

11.52 $\mathbf{V}_0 = 3.92 \underline{/11.31°}$ V

11.56 $\mathbf{Z}_{\text{SOURCE}} = 1.56 \underline{/42.27°}$ Ω

11.59 $\mathbf{Z}_{\text{IN}} = 1.94 \underline{/-33.69°}$ Ω

11.63 $\mathbf{Z}_S = 180 + j33$ Ω

11.67 $\mathbf{V}_0 = 15.78 \underline{/189.46°}$ V

11.71 $\mathbf{V}_0 = 28.28 \underline{/-135°}$ V

11.75 $\mathbf{Z}_L = 12 + j8$ Ω

11.79 $\mathbf{V}_S = 135.73 \underline{/6.27°}$ V rms

11.83 $\mathbf{V}_0 = 11.31 \underline{/45°}$ V

12.1 $\mathbf{Z}(s) = \dfrac{s^2(RC)^2 + 3sCR + 1}{sC[2 + sCR]}$

12.5 Zero at 0.25 rad/s, pole at 0.05 rad/s

12.9 Zero at dc, poles at 1 and 10 rad/s

12.13 Zeros: 2 at 50 rad/s; poles: 2 at dc and 2 at 100 rad/s

 Gain factor is 12 dB

12.17 Zeros: 2 at dc; poles: 3 at 1 rad/s

12.20 Zeros: 1 at 2 rad/s, simple pole: 1 at dc, complex pole: $\omega_0 = 12$ rad/s and $\zeta = 0.1$

12.24 $\mathbf{H}(j\omega) = \dfrac{(0.01j\omega + 1)(10^{-3}j\omega + 1)}{(0.1j\omega + 1)(10^{-4}j\omega + 1)}$

12.28 $\mathbf{H}(j\omega) = \dfrac{100(0.2j\omega + 1)}{(j\omega)^2(0.02j\omega + 1)}$

12.32 $\mathbf{H}(j\omega) =$

 $\dfrac{2(j\omega)\big((j\omega/12) + 1\big)}{(1.25j\omega + 1)\big((j\omega/100) + 1\big)\big((j\omega/700) + 1\big)^2}$

12.37 $\omega_0 = 7071$ rad/s, $Q = 14.14$, $\omega_{\text{MAX}} = 7062$ rad/s, $|V_0|_{\text{MAX}} = 84.89$ V

12.40 $L = 19.5$ mH, BW $= 256$ rad/s, $Q = 6.25$

12.44 $L = 1.25$ mH, $R = 1$ kΩ

12.48 $R_S = \infty$: $\omega_0 = 1$ krad/s, BW $= 20$ rad/s, $Q = 50$
$R_S = 10$ kΩ: $\omega_0 = 1$ krad/s, BW $= 22$ rad/s,
$Q = 45.5$

12.52 $R_{NEW} = 2\ \Omega$, $L_{NEW} = 50\ \mu$H, $C_{NEW} = 12.5\ \mu$F

12.56 Network is a low-pass filter

12.60 $R_2 = 7$ kΩ, $C = 2.27$ nF

12.64 $\omega_0 = 500$ rad/s, $Q = 12.5$

12.69 A low-pass filter

12.72 $R_3 = 10$ kΩ, $R_2 = 1013\ \Omega$

12.75 $C = 100\ \mu$F, $L = 101$ mH

13.1 $F(s) = e^{-s}$

13.4 $F(s) = e^{-(sta)}/s + 1$

13.7 $F(s) = e^{-2s}\left(\dfrac{1}{s+1} - \dfrac{1}{s+2}\right)$

13.10 $F(s) = e^{-s}$

13.13 (a) $f(t) = (4e^{-t} - 4e^{-2t})u(t)$
(b) $f(t) = (12e^{-4t} - 3e^{-t})u(t)$

13.16 (a) $f(t) = (1/4)\left[5e^{-6t} - e^{-2t}\right]u(t)$
(b) $f(t) = (\frac{1}{2})(3 - e^{-2t})u(t)$

13.19 (a) $f(t) = (2e^{-3t}\cos(3t) - e^{-3t})u(t)$
(b) $f(t) = \left[4 + \sqrt{10}\, e^{-2t}\cos(2t - 161.6°)\right]u(t)$

13.22 (a) $f(t) = \left[\delta(t) + 5/3e^{-t} - 8/3e^{-4t}\right]u(t)$
(b) $f(t) = (4t + 1)u(t)$

13.25 (a) $f(t) = (te^{-2t} + e^{-2t})u(t)$
(b) $f(t) = (3/2 - 2te^{-2t} - 3/2e^{-2t})u(t)$

13.28 $f(t) = \left[30 + 12t - 12e^{-t}\right.$
$\left. + \sqrt{2}\, e^{-t}\cos(t - 45°)\right]u(t)$

13.31 $f(t) = \left[2e^{-(t-1)} - e^{-(t-1)}\right]u(t - 1)$

13.34 (a) $x(t) = e^{-2t}u(t)$
(b) $x(t) = 2/3 + (4/3)e^{-3t}u(t)$

13.37 $y(t) = \left[3te^{-2t} - 3e^{-2t} + 3e^{-3t}\right]u(t)$

13.40 $f(t) = \left[e^{-t} - e^{-2t} - te^{-2t}\right]u(t)$

13.43 (a) $f(\infty) = 0$
(b) $f(\infty) = 0$

13.46 $f(t) = \{\delta(t) + \frac{1}{2}[e^{-t} - e^{-3t}]\}u(t)$

13.47 $f(t) = \left[\dfrac{d^2\delta(t)}{dt^2} + 2\dfrac{d\delta(t)}{dt}\right.$
$\left. + 2\delta(t) + (9/2)e^{-4t} - (1/2)e^{-2t}\right]u(t)$

13.49 $F(s) = \dfrac{10(s + 5)}{\left[(s + 5)^2 + 5^2\right]^2}$

13.52 $v_c(t) = 24\left[e^{-t/2}\cos\left(\sqrt{3/4}t\right)\right.$
$\left. - (1/3)e^{-t/2}\sin\left(\sqrt{3/4}t\right)\right]u(t)$

14.2 As $t \to \infty$, $v_0(t) \to 0$

14.4 As $t \to \infty$, $v_0(t) \to 0$

14.6 $i_1(t) = \left[6 + 6.42e^{-t/2}\cos\left(\dfrac{\sqrt{7}}{4}t + 62.2°\right)\right]u(t)$

14.8 $v_0(t) = \left[2\sqrt{2}\, e^{-t}\cos(t - 45°)\right]u(t)$

14.11 $v_0(t) = (1 - 5e^{-4t})u(t)$

14.14 $i_0(t) = \left[1 - (2/3)e^{-(4/3)t}\right]u(t)$

14.17 $v_0(t) = (4.83e^{-0.29t} - 0.83e^{-1.71t})u(t)$

14.20 $v_0(t) = \left[8/3 + 4e^{-2t} - (17/3)e^{-(3/2)t}\right]u(t)$

14.23 $v_0(t) = (4/3)e^{-t/3}u(t)$

14.26 $i_0(t) = -e^{-t/2}u(t)$

14.29 $i_0(t) = \left[1.61e^{-0.82t} - 0.61e^{-0.31t}\right]u(t)$

14.32 $v_0(t) = 1.15(e^{-0.42t} - e^{-1.58t})u(t)$

14.35 $v_0(t) = 6e^{-t/2}u(t)$

14.38 $v_0(0) = 0$ and $v_0(\infty) = 4$ V

14.41 $v_0(t) = (1 + (4/7)e^{-3t/7})u(t)$
$- (1 + (4/7)e^{-3/7(t-1)})u(t - 1)$

14.44 $I_0/I_1 = \dfrac{2}{2s^2 + 5s + 4}$

14.47 $v_0/v_s = \dfrac{R_1 + R_2 + 1/s\,C}{R_2 + 1/s\,C}$

14.50 Roots of $I_0(s)$ are 0, and $-3 \pm \sqrt{6}$. Response is overdamped.

14.53 $v_0(t) = 7.12e^{-t/2}\cos\left[\left(\dfrac{\sqrt{7}}{2}\right)t + 20.7°\right]u(t)$

14.56 $v_0(t) = (4 + 2e^{-5t/12})u(t)$

14.59 $v_0/v_s = \dfrac{-s}{(s + 1)^2 + 1}$

14.62 $i_0(t) = 4/(\sqrt{2})\cos(2t - 45°)$ A

14.65 $v_0(t) = 5.23\cos(2t + 97.8°)$ V

15.1 $f(t) = \displaystyle\sum_{n=-\infty}^{\infty} \dfrac{2}{jn\pi}e^{jn\omega_0 t}$ $n \neq 0$ and n odd

15.4 $C_0 = \dfrac{3}{4}$ $C_n = \dfrac{1}{jn\pi}\left[1 - \dfrac{e^{\frac{-jn\pi}{2}}}{2} - \dfrac{e^{-jn\pi}}{2}\right]$

15.7 (a) Odd symmetry

(b) Half-wave symmetry

15.10 $v(t) = \dfrac{1}{4} - \sum_{n=1}^{\infty} \dfrac{4}{n^2 \pi^2}$

$\left[\cos\left(\dfrac{n\pi}{2}\right) - 1 \right] + \dfrac{2}{n\pi} \sin\left(\dfrac{n\pi}{2}\right)$

15.13 $a_n = \dfrac{1}{n^2 \pi^2} \left[1 - \cos(n\pi)\right]$

$b_n = \dfrac{1}{n\pi} \left[2\cos(n\pi) - 1\right]$

$a_0 = 1/4$

15.16 $v(t) = \dfrac{-\pi}{4} + \sum_{n=1}^{\infty} \left[\dfrac{\cos(n\pi) - 1}{\pi n^2} \right] \cos(nt)$

$+ \dfrac{1 - 2\cos(n\pi)}{\pi} \sin(nt)$

15.19 $f(t) = \dfrac{1}{\pi} + \dfrac{1}{2} \sin(\pi t)$

$+ \sum_{\substack{n=2 \\ n \text{ even}}}^{\infty} \dfrac{1 - \cos(n+1)\pi}{\pi(1 - n^2)} \cos(n\pi t)$

15.22 $f(t) = \dfrac{3}{4} - \dfrac{1}{\pi} \sum_{n=1}^{\infty} \dfrac{1}{n} \sin\left(\dfrac{n\pi}{2}\right) \cos(n\omega_0 t)$

$+ \dfrac{1}{n} \left[2\cos(n\pi) - \cos\left(\dfrac{n\pi}{2}\right) - 1\right] \sin(n\omega_0 t)$

15.24 $f(t) = 4 + 8\cos(\omega_0 t + 80°)$

$+ 6\cos(2\omega_0 t + 45°)$

$+ 4\cos(3\omega_0 t) + 2\cos(4\omega_0 t - 45°)$

where $\omega_0 = 2\pi f_0 = 20\pi$ rad/s

15.26 $i_0(t) = \sum_{n=1}^{\infty} (-1)^{n+1} \left(\dfrac{20}{n\pi}\right) \left| \dfrac{jn}{1 + 3jn} \right|$

$\cos(nt - 90° - \theta_n)$ A

where $\theta_n = \tan^{-1}\left(\dfrac{jn}{1 + 3jn}\right)$

15.28 $i_0(t) = \sum_{n=1}^{\infty} \dfrac{-2}{n\pi |A_n|} \cos(0.2n\pi t - 90° - \theta_n)$ A

where $A_n = \dfrac{2 + 1.2jn\pi}{0.4jn\pi}$

and $\theta_n = \tan^{-1}\left(\dfrac{0.4jn\pi}{2 + 1.2jn\pi}\right)$

15.31 $v_0(t) = 1.2 + 0.28\cos(2\pi t + 21.7°)$

$+ 0.07\cos(4\pi t + 11.25°)$

$+ 0.03\cos(6\pi t + 7.55°)$

$+ 0.02\cos(8\pi t + 5.68°)$ V

15.34 $P = 29.6$ W

15.36 (a) $F(\omega) = \dfrac{j\omega + 2}{(j\omega + 2)^2 + 16}$

(b) $F(\omega) = \dfrac{4}{(j\omega + 2)^2 + 16}$

15.39 $F(\omega) = \dfrac{24}{4 + (\omega - 4)^2} + \dfrac{24}{4 + (\omega + 4)^2}$

15.42 (a) $v_0(t) = 2\left[\dfrac{2}{3}e^{-t} - 2e^{-2t} + \dfrac{4}{3}e^{-5t/2}\right]u(t)$ V

(b) $v_0(t) = 1.27\cos(4t - 58°)$ V

15.45 (a) $v_0(t) = (2/9)\left[3e^{-t} - e^{-t/3}\right]u(t)$ V

(b) $v_0(t) = (4/9)\left[\dfrac{63}{55}e^{-t/3} - \dfrac{3}{5}e^{-2t}\right.$

$\left. - \dfrac{6}{11}e^{-4t}\right]u(t)$ V

15.48 $v_0(t) = (2/3)\left[e^{-t} - e^{-4t}\right]u(t)$ V

15.51 $v_0(t) = (2/3)\left[e^{-t} - e^{-2t} + \cos(t - 45°)\right]u(t)$ V

Index

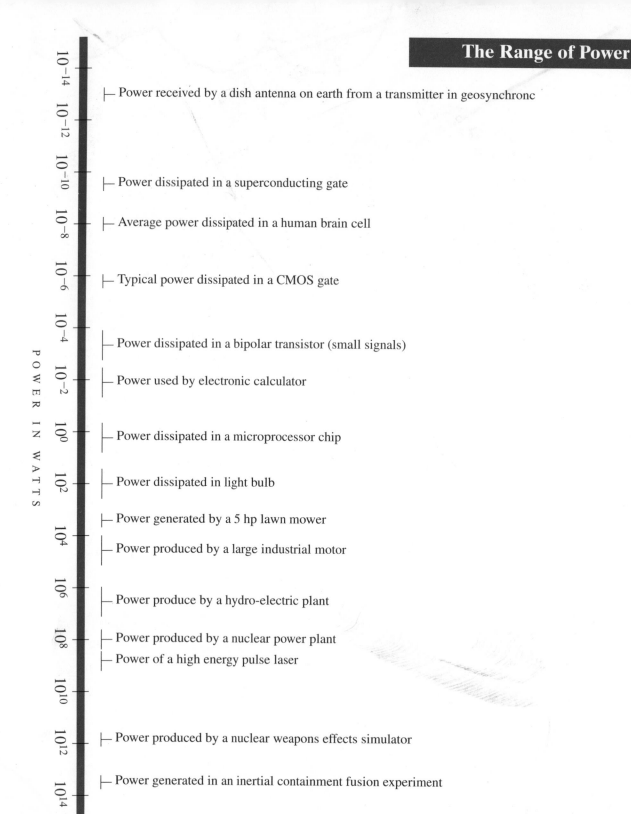

The Range of Power

POWER IN WATTS

10^{-14}
10^{-12} — Power received by a dish antenna on earth from a transmitter in geosynchronc

10^{-10} — Power dissipated in a superconducting gate

10^{-8} — Average power dissipated in a human brain cell

10^{-6} — Typical power dissipated in a CMOS gate

10^{-4} — Power dissipated in a bipolar transistor (small signals)

10^{-2} — Power used by electronic calculator

10^{0} — Power dissipated in a microprocessor chip

10^{2} — Power dissipated in light bulb

— Power generated by a 5 hp lawn mower

10^{4} — Power produced by a large industrial motor

10^{6} — Power produce by a hydro-electric plant

10^{8} — Power produced by a nuclear power plant
— Power of a high energy pulse laser

10^{10}

10^{12} — Power produced by a nuclear weapons effects simulator

10^{14} — Power generated in an inertial containment fusion experiment

OSCAR HALECKI

FROM FLORENCE TO BREST

(1439-1596)

SECOND EDITION

ARCHON BOOKS
1968

First published 1958
Second Edition 1968
© 1968, Oscar Halecki

SBN: 208 00702 4
Library of Congress Catalog Card Number: 68–26103
Printed in the United States of America